Modern Physical Metallurgy and Mater<del>ials Engineering</del>

About the authors

Professor R. E. Smallman

After gaining his PhD in 1953, Professor Smallman spent five years at the Atomic Energy Research Establishment at Harwell, before returning to the University of Birmingham where he became Professor of Physical Metallurgy in 1964 and Feeney Professor and Head of the Department of Physical Metallurgy and Science of Materials in 1969. He subsequently became Head of the amalgamated Department of Metallurgy and Materials (1981), Dean of the Faculty of Science and Engineering, and the first Dean of the newly-created Engineering Faculty in 1985. For five years he was Vice-Principal of the University (1987–92).

He has held visiting professorship appointments at the University of Stanford, Berkeley, Pennsylvania (USA), New South Wales (Australia), Hong Kong and Cape Town and has received Honorary Doctorates from the University of Novi Sad (Yugoslavia) and the University of Wales. His research work has been recognized by the award of the Sir George Beilby Gold Medal of the Royal Institute of Chemistry and Institute of Metals (1969), the Rosenhain Medal of the Institute of Metals for contributions to Physical Metallurgy (1972) and the Platinum Medal, the premier medal of the Institute of Materials (1989).

He was elected a Fellow of the Royal Society (1986), a Fellow of the Royal Academy of Engineering (1990) and appointed a Commander of the British Empire (CBE) in 1992. A former Council Member of the Science and Engineering Research Council, he has been Vice President of the Institute of Materials and President of the Federated European Materials Societies. Since retirement he has been academic consultant for a number of institutions both in the UK and overseas.

R. J. Bishop

After working in laboratories of the automobile, forging, tube-drawing and razor blade industries (1944–59), Ray Bishop became a Principal Scientist of the British Coal Utilization Research Association (1959–68), studying superheater-tube corrosion and mechanisms of ash deposition on behalf of boiler manufacturers and the Central Electricity Generating Board. He specialized in combustor simulation of conditions within pulverized-fuel-fired power station boilers and fluidized-bed combustion systems. He then became a Senior Lecturer in Materials Science at the Polytechnic (now University), Wolverhampton, acting at various times as leader of C&G, HNC, TEC and CNAA honours Degree courses and supervising doctoral researches. For seven years he was Open University Tutor for materials science and processing in the West Midlands. In 1986 he joined the School of Metallurgy and Materials, University of Birmingham as a part-time Lecturer and was involved in administration of the Federation of European Materials Societies (FEMS). In 1995 and 1997 he gave lecture courses in materials science at the Naval Postgraduate School, Monterey, California. Currently he is an Honorary Lecturer at the University of Birmingham.

Modern Physical Metallurgy and Materials Engineering

Science, process, applications

Sixth Edition

R. E. Smallman, *CBE, DSc, FRS, FREng, FIM*

R. J. Bishop, *PhD, CEng, MIM*

OXFORD AUCKLAND BOSTON JOHANNESBURG MELBOURNE NEW DELHI

Butterworth-Heinemann
Linacre House, Jordan Hill, Oxford OX2 8DP
225 Wildwood Avenue, Woburn, MA 01801-2041
A division of Reed Educational and Professional Publishing Ltd

-℞ A member of the Reed Elsevier plc group

First published 1962
Second edition 1963
Reprinted 1965, 1968
Third edition 1970
Reprinted 1976 (twice), 1980, 1983
Fourth edition 1985
Reprinted 1990, 1992
Fifth edition 1995
Sixth edition 1999

Ref
TN
690
.S56
1999

British Library Cataloguing in Publication Data
A catalogue record for this book is available from the British Library

Library of Congress Cataloguing in Publication Data
A catalogue record for this book is available from the Library of Congress

ISBN 0 7506 4564 4

Composition by Scribe Design, Gillingham, Kent, UK
Typeset by Laser Words, Madras, India
Printed and bound in Great Britain by Bath Press, Avon

Contents

Preface

It is less than five years since the last edition of *Modern Physical Metallurgy* was enlarged to include the related subject of Materials Science and Engineering, appearing under the title *Metals and Materials: Science, Processes, Applications*. In its revised approach, it covered a wider range of metals and alloys and included ceramics and glasses, polymers and composites, modern alloys and surface engineering. Each of these additional subject areas was treated on an individual basis as well as against unifying background theories of structure, kinetics and phase transformations, defects and materials characterization.

In the relatively short period of time since that previous edition, there have been notable advances in the materials science and engineering of biomaterials and sports equipment. Two new chapters have now been devoted to these topics. The subject of biomaterials concerns the science and application of materials that must function effectively and reliably whilst in contact with living tissue; these vital materials feature increasingly in modern surgery, medicine and dentistry. Materials developed for sports equipment must take into account the demands peculiar to each sport. In the process of writing these additional chapters, we became increasingly conscious that engineering aspects of the book were coming more and more into prominence. A new form of title was deemed appropriate. Finally, we decided to combine the phrase 'physical metallurgy', which expresses a sense of continuity with earlier editions, directly with 'materials engineering' in the book's title.

Overall, as in the previous edition, the book aims to present the science of materials in a relatively concise form and to lead naturally into an explanation of the ways in which various important materials are processed and applied. We have sought to provide a useful survey of key materials and their interrelations, emphasizing, wherever possible, the underlying scientific and engineering principles. Throughout we have indicated the manner in which powerful tools of characterization, such as optical and electron microscopy, X-ray diffraction, etc. are used to elucidate the vital relations between the structure of a material and its mechanical, physical and/or chemical properties. Control of the microstructure/property relation recurs as a vital theme during the actual processing of metals, ceramics and polymers; production procedures for ostensibly dissimilar materials frequently share common principles.

We have continued to try and make the subject area accessible to a wide range of readers. Sufficient background and theory is provided to assist students in answering questions over a large part of a typical Degree course in materials science and engineering. Some sections provide a background or point of entry for research studies at postgraduate level. For the more general reader, the book should serve as a useful introduction or occasional reference on the myriad ways in which materials are utilized. We hope that we have succeeded in conveying the excitement of the atmosphere in which a life-altering range of new materials is being conceived and developed.

R. E. Smallman
R. J. Bishop

Chapter 1

The structure and bonding of atoms

1.1 The realm of materials science

In everyday life we encounter a remarkable range of engineering materials: metals, plastics and ceramics are some of the generic terms that we use to describe them. The size of the artefact may be extremely small, as in the silicon microchip, or large, as in the welded steel plate construction of a suspension bridge. We acknowledge that these diverse materials are quite literally the stuff of our civilization and have a determining effect upon its character, just as cast iron did during the Industrial Revolution. The ways in which we use, or misuse, materials will obviously also influence its future. We should recognize that the pressing and interrelated global problems of energy utilization and environmental control each has a substantial and inescapable 'materials dimension'.

The engineer is primarily concerned with the function of the component or structure, frequently with its capacity to transmit working stresses without risk of failure. The secondary task, the actual choice of a suitable material, requires that the materials scientist should provide the necessary design data, synthesize and develop new materials, analyse failures and ultimately produce material with the desired shape, form and properties at acceptable cost. This essential collaboration between practitioners of the two disciplines is sometimes expressed in the phrase 'Materials Science and Engineering (MSE)'. So far as the main classes of available materials are concerned, it is initially useful to refer to the type of diagram shown in Figure 1.1. The principal sectors represent metals, ceramics and polymers. All these materials can now be produced in non-crystalline forms, hence a glassy 'core' is shown in the diagram. Combining two or more materials of very different properties, a centuries-old device, produces important composite materials: carbon-fibre-reinforced polymers (CFRP) and metal-matrix composites (MMC) are modern examples.

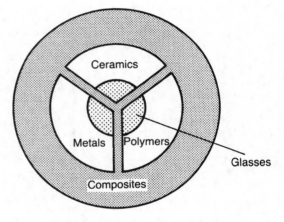

Figure 1.1 *The principal classes of materials (after Rice, 1983).*

Adjectives describing the macroscopic behaviour of materials naturally feature prominently in any language. We write and speak of materials being hard, strong, brittle, malleable, magnetic, wear-resistant, etc. Despite their apparent simplicity, such terms have depths of complexity when subjected to scientific scrutiny, particularly when attempts are made to relate a given property to the internal structure of a material. In practice, the search for bridges of understanding between macroscopic and microscopic behaviour is a central and recurrent theme of materials science. Thus Sorby's metallurgical studies of the structure/property relations for commercial irons and steel in the late nineteenth century are often regarded as the beginning of modern materials science. In more recent times, the enhancement of analytical techniques for characterizing structures in fine detail has led to the development and acceptance of polymers and ceramics as trustworthy engineering materials.

Having outlined the place of materials science in our highly material-dependent civilization, it is now appropriate to consider the smallest structural entity in materials and its associated electronic states.

1.2 The free atom

1.2.1 The four electron quantum numbers

Rutherford conceived the atom to be a positively-charged nucleus, which carried the greater part of the mass of the atom, with electrons clustering around it. He suggested that the electrons were revolving round the nucleus in circular orbits so that the centrifugal force of the revolving electrons was just equal to the electrostatic attraction between the positively-charged nucleus and the negatively-charged electrons. In order to avoid the difficulty that revolving electrons should, according to the classical laws of electrodynamics, emit energy continuously in the form of electromagnetic radiation, Bohr, in 1913, was forced to conclude that, of all the possible orbits, only certain orbits were in fact permissible. These discrete orbits were assumed to have the remarkable property that when an electron was in one of these orbits, no radiation could take place. The set of stable orbits was characterized by the criterion that the angular momenta of the electrons in the orbits were given by the expression $n\mathbf{h}/2\pi$, where $\mathbf{h}$ is Planck's constant and n could only have integral values ($n = 1, 2, 3$, etc.). In this way, Bohr was able to give a satisfactory explanation of the line spectrum of the hydrogen atom and to lay the foundation of modern atomic theory.

In later developments of the atomic theory, by de Broglie, Schrödinger and Heisenberg, it was realized that the classical laws of particle dynamics could not be applied to fundamental particles. In classical dynamics it is a prerequisite that the position and momentum of a particle are known exactly: in atomic dynamics, if either the position or the momentum of a fundamental particle is known exactly, then the other quantity cannot be determined. In fact, an uncertainty must exist in our knowledge of the position and momentum of a small particle, and the product of the degree of uncertainty for each quantity is related to the value of Planck's constant ($\mathbf{h} = 6.6256 \times 10^{-34}$ J s). In the macroscopic world, this fundamental uncertainty is too small to be measurable, but when treating the motion of electrons revolving round an atomic nucleus, application of Heisenberg's Uncertainty Principle is essential.

The consequence of the Uncertainty Principle is that we can no longer think of an electron as moving in a fixed orbit around the nucleus but must consider the motion of the electron in terms of a wave function. This function specifies only the probability of finding one electron having a particular energy in the space surrounding the nucleus. The situation is further complicated by the fact that the electron behaves not only as if it were revolving round the nucleus but also as if it were spinning about its own axis. Consequently, instead of specifying the motion of an electron in an atom by a single integer n, as required by the Bohr theory, it is now necessary to specify the electron state using four numbers. These numbers, known as electron quantum numbers, are n, l, m and s, where n is the principal quantum number, l is the orbital (azimuthal) quantum number, m is the magnetic quantum number and s is the spin quantum number. Another basic premise of the modern quantum theory of the atom is the Pauli Exclusion Principle. This states that no two electrons in the same atom can have the same numerical values for their set of four quantum numbers.

If we are to understand the way in which the Periodic Table of the chemical elements is built up in terms of the electronic structure of the atoms, we must now consider the significance of the four quantum numbers and the limitations placed upon the numerical values that they can assume. The most important quantum number is the principal quantum number since it is mainly responsible for determining the energy of the electron. The principal quantum number can have integral values beginning with $n = 1$, which is the state of lowest energy, and electrons having this value are the most stable, the stability decreasing as n increases. Electrons having a principal quantum number n can take up integral values of the orbital quantum number l between 0 and $(n - 1)$. Thus if $n = 1$, l can only have the value 0, while for $n = 2$, $l = 0$ or 1, and for $n = 3$, $l = 0$, 1 or 2. The orbital quantum number is associated with the angular momentum of the revolving electron, and determines what would be regarded in non-quantum mechanical terms as the shape of the orbit. For a given value of n, the electron having the lowest value of l will have the lowest energy, and the higher the value of l, the greater will be the energy.

The remaining two quantum numbers m and s are concerned, respectively, with the orientation of the electron's orbit round the nucleus, and with the orientation of the direction of spin of the electron. For a given value of l, an electron may have integral values of the inner quantum number m from $+l$ through 0 to $-l$. Thus for $l = 2$, m can take on the values $+2$, $+1$, 0, -1 and -2. The energies of electrons having the same values of n and l but different values of m are the same, provided there is no magnetic field present. When a magnetic field is applied, the energies of electrons having different m values will be altered slightly, as is shown by the splitting of spectral lines in the Zeeman effect. The spin quantum number s may, for an electron having the same values of n, l and m, take one of two values, that is, $+\frac{1}{2}$ or $-\frac{1}{2}$. The fact that these are non-integral values need not concern us for the present purpose. We need only remember that two electrons in an atom can have the same values for the three quantum numbers n, l and m, and that these two electrons will have their spins oriented in opposite directions. Only in a magnetic field will the

Table 1.1 Allocation of states in the first three quantum shells

Shell	n	l	m	s	Number of states	Maximum number of electrons in shell
1st K	1	0	0	$\pm 1/2$	Two 1s-states	2
		0	0	$\pm 1/2$	Two 2s-states	
2nd L	2	1	+1	$\pm 1/2$		8
			0	$\pm 1/2$	Six 2p-states	
			−1	$\pm 1/2$		
		0	0	$\pm 1/2$	Two 3s-states	
3rd M		1	+1	$\pm 1/2$		
			0	$\pm 1/2$	Six 3p-states	
			−1	$\pm 1/2$		
	3					18
		2	+2	$\pm 1/2$		
			+1	$\pm 1/2$		
			0	$\pm 1/2$	Ten 3d-states	
			−1	$\pm 1/2$		
			−2	$\pm 1/2$		

energies of the two electrons of opposite spin be different.

1.2.2 Nomenclature for the electronic states

Before discussing the way in which the periodic classification of the elements can be built up in terms of the electronic structure of the atoms, it is necessary to outline the system of nomenclature which enables us to describe the states of the electrons in an atom. Since the energy of an electron is mainly determined by the values of the principal and orbital quantum numbers, it is only necessary to consider these in our nomenclature. The principal quantum number is simply expressed by giving that number, but the orbital quantum number is denoted by a letter. These letters, which derive from the early days of spectroscopy, are s, p, d and f, which signify that the orbital quantum numbers l are 0, 1, 2 and 3, respectively.[1]

When the principal quantum number $n = 1$, l must be equal to zero, and an electron in this state would be designated by the symbol 1s. Such a state can only have a single value of the inner quantum number $m = 0$, but can have values of $+\frac{1}{2}$ or $-\frac{1}{2}$ for the spin quantum number s. It follows, therefore, that there are only two electrons in any one atom which can be in a 1s-state, and that these electrons will spin in opposite directions. Thus when $n = 1$, only s-states

can exist and these can be occupied by only two electrons. Once the two 1s-states have been filled, the next lowest energy state must have $n = 2$. Here l may take the value 0 or 1, and therefore electrons can be in either a 2s-or a 2p-state. The energy of an electron in the 2s-state is lower than in a 2p-state, and hence the 2s-states will be filled first. Once more there are only two electrons in the 2s-state, and indeed this is always true of s-states, irrespective of the value of the principal quantum number. The electrons in the p-state can have values of $m = +1$, 0, −1, and electrons having each of these values for m can have two values of the spin quantum number, leading therefore to the possibility of six electrons being in any one p-state. These relationships are shown more clearly in Table 1.1.

No further electrons can be added to the state for $n = 2$ after two 2s- and six 2p-state are filled, and the next electron must go into the state for which $n = 3$, which is at a higher energy. Here the possibility arises for l to have the values 0, 1 and 2 and hence, besides s- and p-states, d-states for which $l = 2$ can now occur. When $l = 2$, m may have the values $+2, +1, 0, −1, −2$ and each may be occupied by two electrons of opposite spin, leading to a total of ten d-states. Finally, when $n = 4$, l will have the possible values from 0 to 4, and when $l = 4$ the reader may verify that there are fourteen 4f-states.

Table 1.1 shows that the maximum number of electrons in a given shell is $2n^2$. It is accepted practice to retain an earlier spectroscopic notation and to label the states for which $n = 1, 2, 3, 4, 5, 6$ as K-, L-, M- N-, O- and P-shells, respectively.

[1]The letters, s, p, d and f arose from a classification of spectral lines into four groups, termed sharp, principal, diffuse and fundamental in the days before the present quantum theory was developed.

1.3 The Periodic Table

The Periodic Table provides an invaluable classification of all chemical elements, an element being a collection of atoms of one type. A typical version is shown in Table 1.2. Of the 107 elements which appear, about 90 occur in nature; the remainder are produced in nuclear reactors or particle accelerators. The atomic number (Z) of each element is stated, together with its chemical symbol, and can be regarded as either the number of protons in the nucleus or the number of orbiting electrons in the atom. The elements are naturally classified into periods (horizontal rows), depending upon which electron shell is being filled, and groups (vertical columns). Elements in any one group have the electrons in their outermost shell in the same configuration, and, as a direct result, have similar chemical properties.

The building principle (*Aufbauprinzip*) for the Table is based essentially upon two rules. First, the Pauli Exclusion Principle (Section 1.2.1) must be obeyed. Second, in compliance with Hund's rule of maximum multiplicity, the ground state should always develop maximum spin. This effect is demonstrated diagrammatically in Figure 1.2. Suppose that we supply three electrons to the three 'empty' $2p$-orbitals. They will build up a pattern of parallel spins (a) rather than paired spins (b). A fourth electron will cause pairing (c). Occasionally, irregularities occur in the 'filling' sequence for energy states because electrons always enter the lowest available energy state. Thus, $4s$-states, being at a lower energy level, fill before the $3d$-states.

We will now examine the general process by which the Periodic Table is built up, electron by electron, in closer detail. The progressive filling of energy states can be followed in Table 1.3. The first period commences with the simple hydrogen atom which has a single proton in the nucleus and a single orbiting electron ($Z = 1$). The atom is therefore electrically neutral and for the lowest energy condition, the electron will be in the $1s$-state. In helium, the next element, the nucleus charge is increased by one proton and an additional electron maintains neutrality ($Z = 2$). These two electrons fill the $1s$-state and will necessarily have opposite spins. The nucleus of helium contains two neutrons as well as two protons, hence

its mass is four times greater than that of hydrogen. The next atom, lithium, has a nuclear charge of three ($Z = 3$) and, because the first shell is full, an electron must enter the $2s$-state which has a somewhat higher energy. The electron in the $2s$-state, usually referred to as the valency electron, is 'shielded' by the inner electrons from the attracting nucleus and is therefore less strongly bonded. As a result, it is relatively easy to separate this valency electron. The 'electron core' which remains contains two tightly-bound electrons and, because it carries a single net positive charge, is referred to as a monovalent cation. The overall process by which electron(s) are lost or gained is known as ionization.

The development of the first short period from lithium ($Z = 3$) to neon ($Z = 10$) can be conveniently followed by referring to Table 1.3. So far, the sets of states corresponding to two principal quantum numbers ($n = 1$, $n = 2$) have been filled and the electrons in these states are said to have formed closed shells. It is a consequence of quantum mechanics that, once a shell is filled, the energy of that shell falls to a very low value and the resulting electronic configuration is very stable. Thus, helium, neon, argon and krypton are associated with closed shells and, being inherently stable and chemically unreactive, are known collectively as the inert gases.

The second short period, from sodium ($Z = 11$) to argon ($Z = 18$), commences with the occupation of the $3s$-orbital and ends when the $3p$-orbitals are full (Table 1.3). The long period which follows extends from potassium ($Z = 19$) to krypton ($Z = 36$), and, as mentioned previously, has the unusual feature of the $4s$-state filling before the $3d$-state. Thus, potassium has a similarity to sodium and lithium in that the electron of highest energy is in an s-state; as a consequence, they have very similar chemical reactivities, forming the group known as the alkali-metal elements. After calcium ($Z = 20$), filling of the $3d$-state begins.

The $4s$-state is filled in calcium ($Z = 20$) and the filling of the $3d$-state becomes energetically favourable to give scandium ($Z = 21$). This belated filling of the five $3d$-orbitals from scandium to its completion in copper ($Z = 29$) embraces the first series of transition elements. One member of this series, chromium ($Z = 24$), obviously behaves in an unusual manner. Applying Hund's rule, we can reason

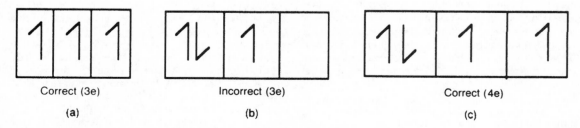

Correct (3e)   Incorrect (3e)   Correct (4e)

(a)   (b)   (c)

Figure 1.2 *Application of Hund's multiplicity rule to the electron-filling of energy states.*

Table 1.2 The Periodic Table of the elements (from Puddephatt and Monaghan, 1986; by permission of Oxford University Press)

← New IUPAC notation
← Previous IUPAC form

1	2	3	4	5	6	7	8	9	10	11	12	13	14	15	16	17	18
IA	IIA	IIIA	IVA	VA	VIA	VIIA	VIII	VIII	VIII	IB	IIB	IIIB	IVB	VB	VIB	VIIB	O
$_1$H 1.008																	$_2$He 4.003
$_3$Li 6.941	$_4$Be 9.012											$_5$B 10.81	$_6$C 12.01	$_7$N 14.01	$_8$O 16.00	$_9$F 19.00	$_{10}$Ne 20.18
$_{11}$Na 22.99	$_{12}$Mg 24.31											$_{13}$Al 26.98	$_{14}$Si 28.09	$_{15}$P 30.97	$_{16}$S 32.45	$_{17}$Cl 35.45	$_{18}$A 39.95
$_{19}$K 39.10	$_{20}$Ca 40.08	$_{21}$Sc 44.96	$_{22}$Ti 47.90	$_{23}$V 50.94	$_{24}$Cr 52.00	$_{25}$Mn 54.94	$_{26}$Fe 55.85	$_{27}$Co 58.93	$_{28}$Ni 58.71	$_{29}$Cu 63.55	$_{30}$Zn 65.37	$_{31}$Ga 69.72	$_{32}$Ge 72.92	$_{33}$Ge 74.92	$_{34}$Se 78.96	$_{35}$Br 79.90	$_{36}$Kr 83.80
$_{37}$Rb 85.47	$_{38}$Sr 87.62	$_{39}$Y 88.91	$_{40}$Zr 91.22	$_{41}$Nb 92.91	$_{42}$Mo 95.94	$_{43}$Tc 98.91	$_{44}$Ru 101.1	$_{45}$Rh 102.9	$_{46}$Pd 106.4	$_{47}$Ag 107.9	$_{48}$Cd 112.4	$_{49}$In 114.8	$_{50}$Sn 118.7	$_{51}$Sb 121.8	$_{52}$Te 127.6	$_{53}$I 126.9	$_{54}$Xe 131.3
$_{55}$Cs 132.9	$_{56}$Ba 137.3	$_{57}$La 138.9	$_{72}$Hf 178.5	$_{73}$Ta 180.9	$_{74}$W 183.9	$_{75}$Re 186.2	$_{76}$Os 190.2	$_{77}$Ir 192.2	$_{78}$Pt 195.1	$_{79}$Au 197.0	$_{80}$Hg 200.6	$_{81}$Tl 204.4	$_{82}$Pb 207.2	$_{83}$Bi 209.0	$_{84}$Po (210)	$_{85}$At (210)	$_{86}$Rn (222)
$_{87}$Fr (223)	$_{88}$Ra (226.0)	$_{89}$Ac (227)	$_{104}$Unq	$_{105}$Unp	$_{106}$Unh	$_{107}$Uns											

← s-block →   ← d-block →   ← p-block →

Lanthanides	$_{57}$La 138.9	$_{58}$Ce 140.1	$_{59}$Pr 140.9	$_{60}$Nd 144.2	$_{61}$Pm (147)	$_{62}$Sm 150.4	$_{63}$Eu 152.0	$_{64}$Gd 157.3	$_{65}$Tb 158.9	$_{66}$Dy 162.5	$_{67}$Ho 164.9	$_{68}$Er 167.3	$_{69}$Tm 168.9	$_{70}$Yb 173.0	$_{71}$Lu 175.0
Actinides	$_{89}$Ac (227)	$_{90}$Th 232.0	$_{91}$Pa 231.0	$_{92}$U 238.0	$_{93}$Np 237.0	$_{94}$Pu (242)	$_{95}$Am (243)	$_{96}$Cm (248)	$_{97}$Bk (247)	$_{98}$Cf (251)	$_{99}$Es (254)	$_{100}$Fm (253)	$_{101}$Md (256)	$_{102}$No (254)	$_{103}$Lr (257)

f-block

Table 1.3 Electron quantum numbers (Hume-Rothery, Smallman and Haworth, 1988)

Element and atomic number — *Principal and secondary quantum numbers*

Element	$n=1$	$n=2$		$n=3$			$n=4$			
	$l=0$	0	1	0	1	2	0	1	2	3
1 H	1									
2 He	2									
3 Li	2	1								
4 Be	2	2								
5 B	2	2	1							
6 C	2	2	2							
7 N	2	2	3							
8 O	2	2	4							
9 F	2	2	5							
10 Ne	2	2	6							
11 Na	2	2	6	1						
12 Mg	2	2	6	2						
13 Al	2	2	6	2	1					
14 Si	2	2	6	2	2					
15 P	2	2	6	2	3					
16 S	2	2	6	2	4					
17 Cl	2	2	6	2	5					
18 A	2	2	6	2	6					
19 K	2	2	6	2	6		1			
20 Ca	2	2	6	2	6		2			
21 Sc	2	2	6	2	6	1	2			
22 Ti	2	2	6	2	6	2	2			
23 V	2	2	6	2	6	3	2			
24 Cr	2	2	6	2	6	5	1			
25 Mn	2	2	6	2	6	5	2			
26 Fe	2	2	6	2	6	6	2			
27 Co	2	2	6	2	6	7	2			
28 Ni	2	2	6	2	6	8	2			
29 Cu	2	2	6	2	6	10	1			
30 Zn	2	2	6	2	6	10	2			
31 Ga	2	2	6	2	6	10	2	1		
32 Ge	2	2	6	2	6	10	2	2		
33 As	2	2	6	2	6	10	2	3		
34 Se	2	2	6	2	6	10	2	4		
35 Br	2	2	6	2	6	10	2	5		
36 Kr	2	2	6	2	6	10	2	6		

Element	$n=1$	2	3	4				5			6
	$l=-$	$-$	$-$	0	1	2	3	0	1	2	0
37 Rb	2	8	18	2	6			1			
38 Sr	2	8	18	2	6			2			
39 Y	2	8	18	2	6	1		2			
40 Zr	2	8	18	2	6	2		2			
41 Nb	2	8	18	2	6	4		1			
42 Mo	2	8	18	2	6	5		1			
43 Tc	2	8	18	2	6	5		2			
44 Ru	2	8	18	2	6	7		1			
45 Rh	2	8	18	2	6	8		1			
46 Pd	2	8	18	2	6	10		—			
47 Ag	2	8	18	2	6	10		1			
48 Cd	2	8	18	2	6	10		2			
49 In	2	8	18	2	6	10		2	1		
50 Sn	2	8	18	2	6	10		2	2		
51 Sb	2	8	18	2	6	10		2	3		
52 Te	2	8	18	2	6	10		2	4		
53 I	2	8	18	2	6	10		2	5		
54 Xe	2	8	18	2	6	10		2	6		
55 Cs	2	8	18	2	6	10		2	6		1
56 Ba	2	8	18	2	6	10		2	6		2
57 La	2	8	18	2	6	10		2	6	1	2
58 Ce	2	8	18	2	6	10	2	2	6		2
59 Pr	2	8	18	2	6	10	3	2	6		2
60 Nd	2	8	18	2	6	10	4	2	6		2
61 Pm	2	8	18	2	6	10	5	2	6		2
62 Sm	2	8	18	2	6	10	6	2	6		2
63 Eu	2	8	18	2	6	10	7	2	6		2
64 Gd	2	8	18	2	6	10	7	2	6	1	2
65 Tb	2	8	18	2	6	10	9	2	6		2
66 Dy	2	8	18	2	6	10	10	2	6		2
67 Ho	2	8	18	2	6	10	11	2	6		2
68 Er	2	8	18	2	6	10	12	2	6		2
69 Tm	2	8	18	2	6	10	13	2	6		2
70 Yb	2	8	18	2	6	10	14	2	6		2
71 Lu	2	8	18	2	6	10	14	2	6	1	2
72 Hf	2	8	18	2	6	10	14	2	6	2	2

Element	$n=1$	2	3	4	5				6			7
	$l=-$	$-$	$-$	$-$	0	1	2	3	0	1	2	0
73 Ta	2	8	18	32	2	6	3		2			
74 W	2	8	18	32	2	6	4		2			
75 Re	2	8	18	32	2	6	5		2			
76 Os	2	8	18	32	2	6	6		2			
77 Ir	2	8	18	32	2	6	7		2			
78 Pt	2	8	18	32	2	6	9		1			
79 Au	2	8	18	32	2	6	10		1			
80 Hg	2	8	18	32	2	6	10		2			
81 Tl	2	8	18	32	2	6	10		2	1		
82 Pb	2	8	18	32	2	6	10		2	2		
83 Bi	2	8	18	32	2	6	10		2	3		
84 Po	2	8	18	32	2	6	10		2	4		
85 At	2	8	18	32	2	6	10		2	5		
86 Rn	2	8	18	32	2	6	10		2	6		
87 Fr	2	8	18	32	2	6	10		2	6		1
88 Ra	2	8	18	32	2	6	10		2	6		2
89 Ac	2	8	18	32	2	6	10		2	6	1	2
90 Th	2	18	8	32	2	6	10		2	6	2	2
91 Pa	2	18	8	32	2	6	10	2	2	6	1	2
92 U	2	18	8	32	2	6	10	3	2	6	1	2
93 Np	2	18	8	32	2	6	10	4	2	6	1	2
94 Pu	2	18	8	32	2	6	10	5	2	6	1	2

The exact electronic configurations of the later elements are not always certain but the most probable arrangements of the outer electrons are:

95 Am	$(5f)^7(7s)^2$
96 Cm	$(5f)^7(6d)^1(7s)^2$
97 Bk	$(5f)^8(6d)^1(7s)^2$
98 Cf	$(5f)^{10}(7s)^2$
99 Es	$(5f)^{11}(7s)^2$
100 Fm	$(5f)^{12}(7s)^2$
101 Md	$(5f)^{13}(7s)^2$
102 No	$(5f)^{14}(7s)^2$
103 Lw	$(5f)^{14}(6d)^1(7s)^2$
104 —	$(5f)^{14}(6d)^2(7s)^2$

that maximization of parallel spin is achieved by locating six electrons, of like spin, so that five fill the $3d$-states and one enters the $4s$-state. This mode of fully occupying the $3d$-states reduces the energy of the electrons in this shell considerably. Again, in copper ($Z = 29$), the last member of this transition series, complete filling of all $3d$-orbitals also produces a significant reduction in energy. It follows from these explanations that the $3d$- and $4s$-levels of energy are very close together. After copper, the energy states fill in a straightforward manner and the first long period finishes with krypton ($Z = 36$). It will be noted that lanthanides ($Z = 57$ to 71) and actinides ($Z = 89$ to 103), because of their state-filling sequences, have been separated from the main body of Table 1.2. Having demonstrated the manner in which quantum rules are applied to the construction of the Periodic Table for the first 36 elements, we can now examine some general aspects of the classification.

When one considers the small step difference of one electron between adjacent elements in the Periodic Table, it is not really surprising to find that the distinction between metallic and non-metallic elements is imprecise. In fact there is an intermediate range of elements, the metalloids, which share the properties of both metals and non-metals. However, we can regard the elements which can readily lose an electron, by ionization or bond formation, as strongly metallic in character (e.g. alkali metals). Conversely, elements which have a strong tendency to acquire an electron and thereby form a stable configuration of two or eight electrons in the outermost shell are non-metallic (e.g. the halogens fluorine, chlorine, bromine, iodine). Thus electropositive metallic elements and the electronegative non-metallic elements lie on the left- and right-hand sides of the Periodic Table, respectively. As will be seen later, these and other aspects of the behaviour of the outermost (valence) electrons have a profound and determining effect upon bonding and therefore upon electrical, magnetic and optical properties.

Prior to the realization that the frequently observed periodicities of chemical behaviour could be expressed in terms of electronic configurations, emphasis was placed upon 'atomic weight'. This quantity, which is now referred to as relative atomic mass, increases steadily throughout the Periodic Table as protons and neutrons are added to the nuclei. Atomic mass[1] determines physical properties such as density, specific heat capacity and ability to absorb electromagnetic radiation: it is therefore very relevant to engineering practice. For instance, many ceramics are based upon the light elements aluminium, silicon and oxygen and consequently have a low density, i.e. <3000 kg m^{-3}.

1.4 Interatomic bonding in materials

Matter can exist in three states and as atoms change directly from either the gaseous state (desublimation) or the liquid state (solidification) to the usually denser solid state, the atoms form aggregates in three-dimensional space. Bonding forces develop as atoms are brought into proximity to each other. Sometimes these forces are spatially-directed. The nature of the bonding forces has a direct effect upon the type of solid structure which develops and therefore upon the physical properties of the material. Melting point provides a useful indication of the amount of thermal energy needed to sever these interatomic (or interionic) bonds. Thus, some solids melt at relatively low temperatures (m.p. of tin = 232°C) whereas many ceramics melt at extremely high temperatures (m.p. of alumina exceeds 2000°C). It is immediately apparent that bond strength has far-reaching implications in all fields of engineering.

Customarily we identify four principal types of bonding in materials, namely, metallic bonding, ionic bonding, covalent bonding and the comparatively much weaker van der Waals bonding. However, in many solid materials it is possible for bonding to be mixed, or even intermediate, in character. We will first consider the general chemical features of each type of bonding; in Chapter 2 we will examine the resultant disposition of the assembled atoms (ions) in three-dimensional space.

As we have seen, the elements with the most pronounced metallic characteristics are grouped on the left-hand side of the Periodic Table (Table 1.2). In general, they have a few valence electrons, outside the outermost closed shell, which are relatively easy to detach. In a metal, each 'free' valency electron is shared among all atoms, rather than associated with an individual atom, and forms part of the so-called 'electron gas' which circulates at random among the regular array of positively-charged electron cores, or cations (Figure 1.3a). Application of an electric potential gradient will cause the 'gas' to drift though the structure with little hindrance, thus explaining the outstanding electrical conductivity of the metallic state. The metallic bond derives from the attraction between the cations and the free electrons and, as would be expected, repulsive components of force develop when cations are brought into close proximity. However, the bonding forces in metallic structures are spatially non-directed and we can readily simulate the packing and space-filling characteristics of the atoms with modelling systems based on equal-sized spheres (polystyrene balls, even soap bubbles). Other properties such as ductility, thermal conductivity and the transmittance of electromagnetic radiation are also directly influenced by the non-directionality and high electron mobility of the metallic bond.

The ionic bond develops when electron(s) are transferred from atoms of active metallic elements to atoms of active non-metallic elements, thereby enabling each

---

[1] Atomic mass is now expressed relative to the datum value for carbon (12.01). Thus, a copper atom has 63.55/12.01 or 5.29 times more mass than a carbon atom.

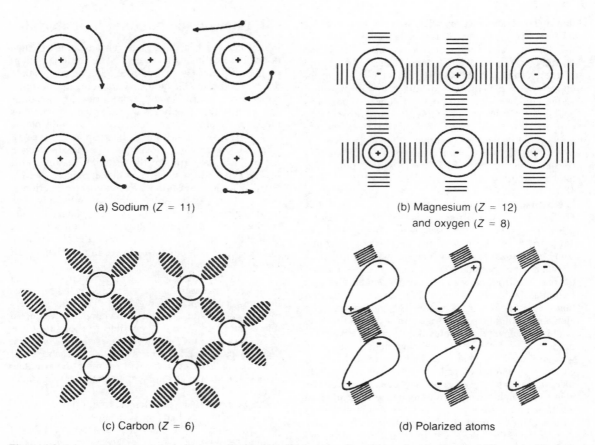

(a) Sodium ($Z = 11$)

(b) Magnesium ($Z = 12$) and oxygen ($Z = 8$)

(c) Carbon ($Z = 6$)

(d) Polarized atoms

Figure 1.3 *Schematic representation of (a) metallic bonding, (b) ionic bonding, (c) covalent bonding and (d) van der Waals bonding.*

of the resultant ions to attain a stable closed shell. For example, the ionic structure of magnesia (MgO), a ceramic oxide, forms when each magnesium atom ($Z = 12$) loses two electrons from its L-shell ($n = 2$) and these electrons are acquired by an oxygen atom ($Z = 8$), producing a stable octet configuration in its L-shell (Table 1.3). Overall, the ionic charges balance and the structure is electrically neutral (Figure 1.3b). Anions are usually larger than cations. Ionic bonding is omnidirectional, essentially electrostatic in character and can be extremely strong; for instance, magnesia is a very useful refractory oxide (m.p. $= 2930°C$). At low to moderate temperatures, such structures are electrical insulators but, typically, become conductive at high temperatures when thermal agitation of the ions increases their mobility.

Sharing of valence electrons is the key feature of the third type of strong primary bonding. Covalent bonds form when valence electrons of opposite spin from adjacent atoms are able to pair within overlapping spatially-directed orbitals, thereby enabling each atom to attain a stable electronic configuration (Figure 1.3c).

Being oriented in three-dimensional space, these localized bonds are unlike metallic and ionic bonds. Furthermore, the electrons participating in the bonds are tightly bound so that covalent solids, in general, have low electrical conductivity and act as insulators, sometimes as semiconductors (e.g. silicon). Carbon in the form of diamond is an interesting prototype for covalent bonding. Its high hardness, low coefficient of thermal expansion and very high melting point (3300°C) bear witness to the inherent strength of the covalent bond. First, using the (8 – N) Rule, in which N is the Group Number[1] in the Periodic Table, we deduce that carbon ($Z = 6$) is tetravalent; that is, four bond-forming electrons are available from the L-shell ($n = 2$). In accordance with Hund's Rule (Figure 1.2), one of the two electrons in the 2s-state is promoted to a higher 2p-state to give a maximum spin condition, producing an overall configuration of $1s^2\ 2s^1\ 2p^3$ in the carbon atom. The outermost second shell accordingly

---

[1] According to previous IUPAC notation: see top of Table 1.2.

has four valency electrons of like spin available for pairing. Thus each carbon atom can establish electron-sharing orbitals with four neighbours. For a given atom, these four bonds are of equal strength and are set at equal angles (109.5°) to each other and therefore exhibit tetrahedral symmetry. (The structural consequences of this important feature will be discussed in Chapter 2.)

This process by which s-orbitals and p-orbitals combine to form projecting hybrid sp-orbitals is known as hybridization. It is observed in elements other than carbon. For instance, trivalent boron ($Z = 5$) forms three co-planar sp^2-orbitals. In general, a large degree of overlap of sp-orbitals and/or a high electron density within the overlap 'cloud' will lead to an increase in the strength of the covalent bond. As indicated earlier, it is possible for a material to possess more than one type of bonding. For example, in calcium silicate (Ca_2SiO_4), calcium cations Ca^{2+} are ionically bonded to tetrahedral SiO_4^{4-} clusters in which each silicon atom is covalently-bonded to four oxygen neighbours.

The final type of bonding is attributed to the van der Waals forces which develop when adjacent atoms, or groups of atoms, act as electric dipoles. Suppose that two atoms which differ greatly in size combine to form a molecule as a result of covalent bonding. The resultant electron 'cloud' for the whole molecule can be pictured as pear-shaped and will have an asymmetrical distribution of electron charge. An electric dipole has formed and it follows that weak directed forces of electrostatic attraction can exist in an aggregate of such molecules (Figure 1.3d). There are no 'free' electrons hence electrical conduction is not favoured. Although secondary bonding by van der Waals forces is weak in comparison to the three forms of primary bonding, it has practical significance. For instance, in the technologically-important mineral talc, which is hydrated magnesium silicate $Mg_3Si_4O_{10}(OH)_2$, the parallel covalently-bonded layers of atoms are attracted to each other by van der Waals forces. These layers can easily be slid past each other, giving the mineral its characteristically slippery feel. In thermoplastic polymers, van der Waals forces of attraction exist between the extended covalently-bonded hydrocarbon chains; a combination of heat and applied shear stress will overcome these forces and cause the molecular chains to glide past each other. To quote a more general case, molecules of water vapour in the atmosphere each have an electric dipole and will accordingly tend to be adsorbed if they strike solid surfaces possessing attractive van der Waals forces (e.g. silica gel).

1.5 Bonding and energy levels

If one imagines atoms being brought together uniformly to form, for example, a metallic structure, then when the distance between neighbouring atoms approaches the interatomic value the outer electrons are no longer localized around individual atoms. Once

the outer electrons can no longer be considered to be attached to individual atoms but have become free to move throughout the metal then, because of the Pauli Exclusion Principle, these electrons cannot retain the same set of quantum numbers that they had when they were part of the atoms. As a consequence, the free electrons can no longer have more than two electrons of opposite spin with a particular energy. The energies of the free electrons are distributed over a range which increases as the atoms are brought together to form the metal. If the atoms when brought together are to form a stable metallic structure, it is necessary that the mean energy of the free electrons shall be lower than the energy of the electron level in the free atom from which they are derived. Figure 1.4 shows the broadening of an atomic electron level as the atoms are brought together, and also the attendant lowering of energy of the electrons. It is the extent of the lowering in mean energy of the outer electrons that governs the stability of a metal. The equilibrium spacing between the atoms in a metal is that for which any further decrease in the atomic spacing would lead to an increase in the repulsive interaction of the positive ions as they are forced into closer contact with each other, which would be greater than the attendant decrease in mean electron energy.

In a metallic structure, the free electrons must, therefore, be thought of as occupying a series of discrete energy levels at very close intervals. Each atomic level which splits into a band contains the same number of energy levels as the number N of atoms in the piece of metal. As previously stated, only two electrons of opposite spin can occupy any one level, so that a band can contain a maximum of $2N$ electrons. Clearly, in the lowest energy state of the metal all the lower energy levels are occupied.

The energy gap between successive levels is not constant but decreases as the energy of the levels increases. This is usually expressed in terms of the density of electronic states $N(E)$ as a function of the energy E. The quantity $N(E)dE$ gives the number of

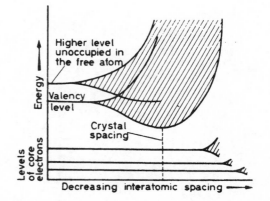

Figure 1.4 *Broadening of atomic energy levels in a metal.*

energy levels in a small energy interval dE, and for free electrons is a parabolic function of the energy, as shown in Figure 1.5.

Because only two electrons can occupy each level, the energy of an electron occupying a low-energy level cannot be increased unless it is given sufficient energy to allow it to jump to an empty level at the top of the band. The energy[1] width of these bands is commonly about 5 or 6 eV and, therefore, considerable energy would have to be put into the metal to excite a low-lying electron. Such energies do not occur at normal temperatures, and only those electrons with energies close to that of the top of the band (known as the Fermi level and surface) can be excited, and therefore only a small number of the free electrons in a metal can take part in thermal processes. The energy of the Fermi level E_F depends on the number of electrons N per unit volume V, and is given by $(\mathbf{h}^2/8m)(3N/\pi V)^{2/3}$.

The electron in a metallic band must be thought of as moving continuously through the structure with an energy depending on which level of the band it occupies. In quantum mechanical terms, this motion of the electron can be considered in terms of a wave with a wavelength which is determined by the energy of the electron according to de Broglie's relationship $\lambda = \mathbf{h}/m\mathrm{v}$, where $\mathbf{h}$ is Planck's constant and m and v are, respectively, the mass and velocity of the moving electron. The greater the energy of the electron, the higher will be its momentum $m\mathrm{v}$, and hence the smaller will be the wavelength of the wave function in terms of which its motion can be described. Because the movement of an electron has this wave-like aspect, moving electrons can give rise, like optical waves, to diffraction effects. This property of electrons is used in electron microscopy (Chapter 5).

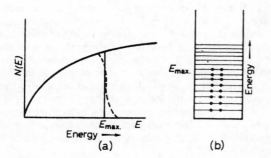

Figure 1.5 (a) Density of energy levels plotted against energy; (b) filling of energy levels by electrons at absolute zero. At ordinary temperatures some of the electrons are thermally excited to higher levels than that corresponding to E_{max} as shown by the broken curve in (a).

[1] An electron volt is the kinetic energy an electron acquires in falling freely through a potential difference of 1 volt (1 eV = 1.602×10^{-19} J; 1 eV per particle = $23\,050 \times 4.186$ J per mol of particles).

Further reading

Cottrell, A. H. (1975). *Introduction to Metallurgy*. Edward Arnold, London.

Huheey, J. E. (1983). *Inorganic Chemistry*, 3rd edn. Harper and Row, New York.

Hume-Rothery, W., Smallman, R. E. and Haworth, C. W. (1975). *The Structure of Metals and Alloys*, 5th edn (1988 reprint). Institute of Materials, London.

Puddephatt, R. J. and Monaghan, P. K. (1986). *The Periodic Table of the Elements*. Clarendon Press, Oxford.

van Vlack, L. H. (1985). *Elements of Materials Science*, 5th edn. Addison-Wesley, Reading, MA.

Chapter 2

Atomic arrangements in materials

2.1 The concept of ordering

When attempting to classify a material it is useful to decide whether it is crystalline (conventional metals and alloys), non-crystalline (glasses) or a mixture of these two types of structure. The critical distinction between the crystalline and non-crystalline states of matter can be made by applying the concept of ordering. Figure 2.1a shows a symmetrical two-dimensional arrangement of two different types of atom. A basic feature of this aggregate is the nesting of a small atom within the triangular group of three much larger atoms. This geometrical condition is called short-range ordering. Furthermore, these triangular groups are regularly arranged relative to each other so that if the aggregate were to be extended, we could confidently predict the locations of any added atoms. In effect, we are taking advantage of the long-range ordering characteristic of this array. The array of Figure 2.1a exhibits both short- and long-range

(a)   (b)

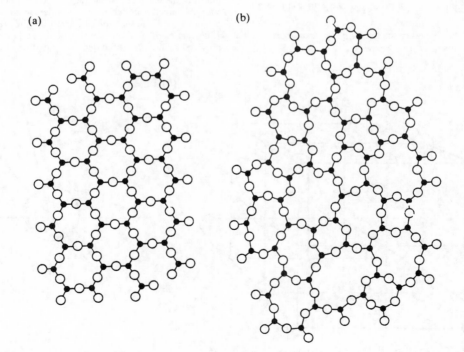

Figure 2.1 *Atomic ordering in (a) crystals and (b) glasses of the same composition (from Kingery, Bowen and Uhlmann, 1976; by permission of Wiley-Interscience).*

ordering and is typical of a single crystal. In the other array of Figure 2.1b, short-range order is discernible but long-range order is clearly absent. This second type of atomic arrangement is typical of the glassy state.[1]

It is possible for certain substances to exist in either crystalline or glassy forms (e.g. silica). From Figure 2.1 we can deduce that, for such a substance, the glassy state will have the lower bulk density. Furthermore, in comparing the two degrees of ordering of Figures 2.1a and 2.1b, one can appreciate why the structures of comparatively highly-ordered crystalline substances, such as chemical compounds, minerals and metals, have tended to be more amenable to scientific investigation than glasses.

2.2 Crystal lattices and structures

We can rationalize the geometry of the simple representation of a crystal structure shown in Figure 2.1a by adding a two-dimensional frame of reference, or space lattice, with line intersections at atom centres. Extending this process to three dimensions, we can construct a similar imaginary space lattice in which triple intersections of three families of parallel equidistant lines mark the positions of atoms (Figure 2.2a). In this simple case, three reference axes (x, y, z) are oriented at 90° to each other and atoms are 'shrunk', for convenience. The orthogonal lattice of Figure 2.2a defines eight unit cells, each having a shared atom at every corner. It follows from our recognition of the inherent order of the lattice that we can express the

[1]The terms glassy, non-crystalline, vitreous and amorphous are synonymous.

geometrical characteristics of the whole crystal, containing millions of atoms, in terms of the size, shape and atomic arrangement of the unit cell, the ultimate repeat unit of structure.[2]

We can assign the lengths of the three cell parameters (a, b, c) to the reference axes, using an internationally-accepted notation (Figure 2.2b). Thus, for the simple cubic case portrayed in Figure 2.2a, $x = y = z = 90°$; $a = b = c$. Economizing in symbols, we only need to quote a single cell parameter (a) for the cubic unit cell. By systematically changing the angles (α, β, γ) between the reference axes, and the cell parameters (a, b, c), and by four skewing operations, we derive the seven crystal systems (Figure 2.3). Any crystal, whether natural or synthetic, belongs to one or other of these systems. From the premise that each point of a space lattice should have identical surroundings, Bravais demonstrated that the maximum possible number of space lattices (and therefore unit cells) is 14. It is accordingly necessary to augment the seven primitive (P) cells shown in Figure 2.3 with seven more non-primitive cells which have additional face-centring, body-centring or end-centring lattice points. Thus the highly-symmetrical cubic system has three possible lattices: primitive (P), body-centred (I; from the German word *innenzentrierte*) and face-centred (F). We will encounter the latter two again in Section 2.5.1. True primitive space lattices, in which

[2]The notion that the striking external appearance of crystals indicates the existence of internal structural units with similar characteristics of shape and orientation was proposed by the French mineralogist Hauy in 1784. Some 130 years elapsed before actual experimental proof was provided by the new technique of X-ray diffraction analysis.

(a)

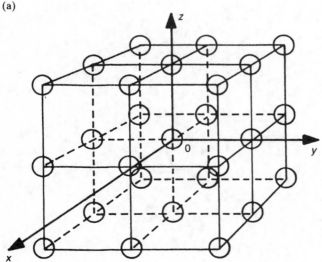

(b)

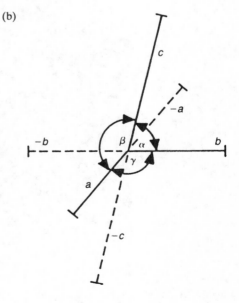

Figure 2.2 *Principles of lattice construction.*

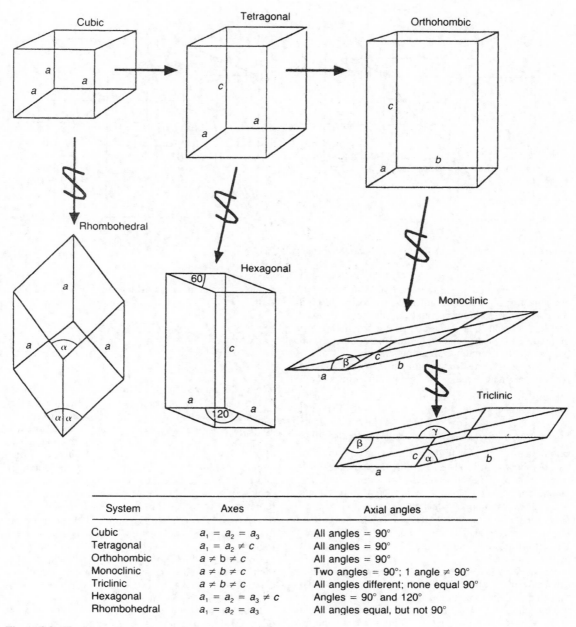

Figure 2.3 *The seven systems of crystal symmetry (S = skew operation).*

System	Axes	Axial angles
Cubic	$a_1 = a_2 = a_3$	All angles = 90°
Tetragonal	$a_1 = a_2 \neq c$	All angles = 90°
Orthohombic	$a \neq b \neq c$	All angles = 90°
Monoclinic	$a \neq b \neq c$	Two angles = 90°; 1 angle ≠ 90°
Triclinic	$a \neq b \neq c$	All angles different; none equal 90°
Hexagonal	$a_1 = a_2 = a_3 \neq c$	Angles = 90° and 120°
Rhombohedral	$a_1 = a_2 = a_3$	All angles equal, but not 90°

each lattice point has identical surroundings, can sometimes embody awkward angles. In such cases it is common practice to use a simpler orthogonal non-primitive lattice which will accommodate the atoms of the actual crystal structure.[1]

---

[1]Lattices are imaginary and limited in number; crystal structures are real and virtually unlimited in their variety.

2.3 Crystal directions and planes

In a structurally-disordered material, such as fully-annealed silica glass, the value of a physical property is independent of the direction of measurement; the material is said to be isotropic. Conversely, in many single crystals, it is often observed that a structurally-sensitive property, such as electrical conductivity, is strongly direction-dependent because of variations in

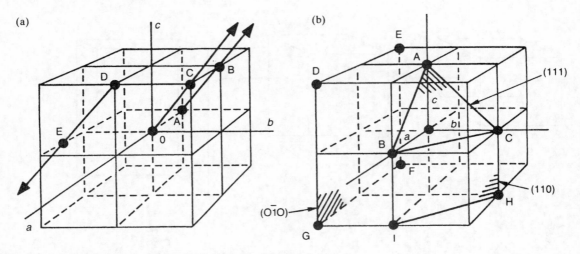

Figure 2.4 *Indexing of (a) directions and (b) planes in cubic crystals.*

the periodicity and packing of atoms. Such crystals are anisotropic. We therefore need a precise method for specifying a direction, and equivalent directions, within a crystal. The general method for defining a given direction is to construct a line through the origin parallel to the required direction and then to determine the coordinates of a point on this line in terms of cell parameters (a, b, c). Hence, in Figure 2.4a, the direction $\overrightarrow{AB}$ is obtained by noting the translatory movements needed to progress from the origin O to point C, i.e. $a = 1$, $b = 1$, $c = 1$. These coordinate values are enclosed in square brackets to give the direction indices [1 1 1]. In similar fashion, the direction $\overrightarrow{DE}$ can be shown to be $[\overline{1/2}\,\overline{1}\,\overline{1}]$ with the bar sign indicating use of a negative axis. Directions which are crystallographically equivalent in a given crystal are represented by angular brackets. Thus, $\langle 1\,0\,0 \rangle$ represents all cube edge directions and comprises [1 0 0], [0 1 0], [0 0 1], [$\overline{1}$ 0 0], [0 $\overline{1}$ 0] and [0 0 $\overline{1}$] directions. Directions are often represented in non-specific terms as [uvw] and $\langle uvw \rangle$.

Physical events and transformations within crystals often take place on certain families of parallel equidistant planes. The orientation of these planes in three-dimensional space is of prime concern; their size and shape is of lesser consequence. (Similar ideas apply to the corresponding external facets of a single crystal.) In the Miller system for indexing planes, the intercepts of a representative plane upon the three axes (x, y, z) are noted.[1] Intercepts are expressed relatively in terms of a, b, c. Planes parallel to an axis are said to intercept at infinity. Reciprocals of the three intercepts are taken and the indices enclosed by round brackets. Hence, in

Figure 2.4b, the procedural steps for indexing the plane ABC are:

	a	b	c
Intercepts	1	1	1
Reciprocals	$\frac{1}{1}$	$\frac{1}{1}$	$\frac{1}{1}$
Miller indices		(1 1 1)	

The Miller indices for the planes DEFG and BCHI are ($0\,\overline{1}\,0$) and (1 1 0), respectively. Often it is necessary to ignore individual planar orientations and to specify all planes of a given crystallographic type, such as the planes parallel to the six faces of a cube. These planes constitute a crystal form and have the same atomic configurations; they are said to be equivalent and can be represented by a single group of indices enclosed in curly brackets, or braces. Thus, {1 0 0} represents a form of six planar orientations, i.e. (1 0 0), (0 1 0), (0 0 1), ($\overline{1}$ 0 0), ($0\,\overline{1}\,0$) and (0 0 $\overline{1}$). Returning to the (1 1 1) plane ABC of Figure 2.4b, it is instructive to derive the other seven equivalent planes, centring on the origin O, which comprise {1 1 1}. It will then be seen why materials belonging to the cubic system often crystallize in an octahedral form in which octahedral {1 1 1} planes are prominent.

It should be borne in mind that the general purpose of the Miller procedure is to define the orientation of a family of parallel equidistant planes; the selection of a convenient representative plane is a means to this end. For this reason, it is permissible to shift the origin provided that the relative disposition of a, b and c is maintained. Miller indices are commonly written in the symbolic form (hkl). Rationalization of indices, either to reduce them to smaller numbers with the same ratio or to eliminate fractions, is unnecessary. This often-recommended step discards information; after all, there is a real difference between the two families of planes (1 0 0) and (2 0 0).

[1]For mathematical reasons, it is advisable to carry out all indexing operations (translations for directions, intercepts for planes) in the strict sequence a, b, c.

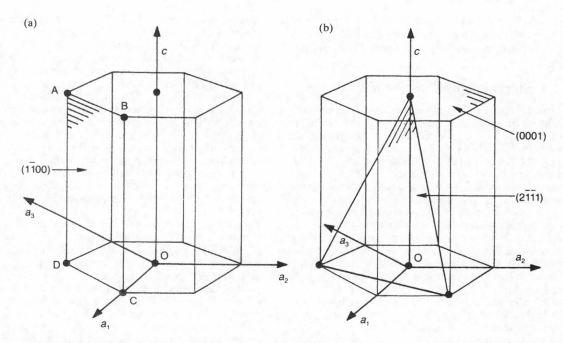

Figure 2.5 *Prismatic, basal and pyramidal planes in hexagonal structures.*

As mentioned previously, it is sometimes convenient to choose a non-primitive cell. The hexagonal structure cell is an important illustrative example. For reasons which will be explained, it is also appropriate to use a four-axis Miller-Bravais notation $(hkil)$ for hexagonal crystals, instead of the three-axis Miller notation (hkl). In this alternative method, three axes (a_1, a_2, a_3) are arranged at 120° to each other in a basal plane and the fourth axis (c) is perpendicular to this plane (Figure 2.5a). Hexagonal structures are often compared in terms of the axial ratio c/a. The indices are determined by taking intercepts upon the axes in strict sequence. Thus the procedural steps for the plane ABCD, which is one of the six prismatic planes bounding the complete cell, are:

	a_1	a_2	a_3	c
Intercepts	1	-1	∞	∞
Reciprocals	$\frac{1}{1}$	$-\frac{1}{1}$	0	0
Miller-Bravais indices		$(1\bar{1}00)$		

Comparison of these digits with those from other prismatic planes such as $(10\bar{1}0)$, $(01\bar{1}0)$ and $(1\bar{1}00)$ immediately reveals a similarity; that is, they are crystallographically equivalent and belong to the $\{10\bar{1}0\}$ form. The three-axis Miller method lacks this advantageous feature when applied to hexagonal structures. For geometrical reasons, it is essential to ensure that the plane indices comply with the condition $(h+k) = -i$. In addition to the prismatic planes, basal planes of (0001) type and pyramidal planes of the $(11\bar{2}1)$

type are also important features of hexagonal structures (Figure 2.5b).

The Miller-Bravais system also accommodates directions, producing indices of the form $[uvtw]$. The first three translations in the basal plane must be carefully adjusted so that the geometrical condition $u + v = -t$ applies. This adjustment can be facilitated by sub-dividing the basal planes into triangles (Figure 2.6). As before, equivalence is immediately

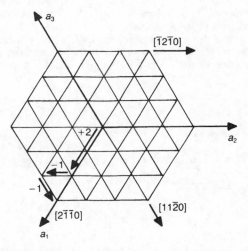

Figure 2.6 *Typical Miller-Bravais directions in (0001) basal plane of hexagonal crystal.*

revealed; for instance, the close-packed directions in the basal plane have the indices $[2\bar{1}\bar{1}0]$, $[11\bar{2}0]$, $[\bar{1}2\bar{1}0]$, etc. and can be represented by $\langle 21\bar{1}0 \rangle$.

2.4 Stereographic projection

Projective geometry makes it possible to represent the relative orientation of crystal planes and directions in three-dimensional space in a more convenient two-dimensional form. The standard stereographic projection is frequently used in the analysis of crystal behaviour; X-ray diffraction analyses usually provide the experimental data. Typical applications of the method are the interpretation of strain markings on crystal surfaces, portrayal of symmetrical relationships, determination of the axial orientations in a single crystal and the plotting of property values for anisotropic single crystals. (The basic method can also be adapted to produce a pole figure diagram which can show preferred orientation effects in polycrystalline aggregates.)

A very small crystal of cubic symmetry is assumed to be located at the centre of a reference sphere, as shown in Figure 2.7a, so that the orientation of a crystal plane, such as the (111) plane marked, may be represented on the surface of the sphere by the point of intersection, or pole, of its normal P. The angle ϕ between the two poles (001) and (111), shown in Figure 2.7b, can then be measured in degrees along the arc of the great circle between the poles P and P'. To represent all the planes in a crystal in this three-dimensional way is rather cumbersome; in the stereographic projection, the array of poles which represents the various planes in the crystal is projected from the reference sphere onto the equatorial plane. The pattern of poles projected on the equatorial, or primitive, plane then represents the stereographic projection of the crystal. As shown in Figure 2.7c, poles in the northern half of the reference sphere are projected onto the equatorial plane by joining the pole P to the south pole S, while those in the southern half of the reference sphere, such as Q, are projected in the same way in the direction of the north pole N. Figure 2.8a shows the

stereographic projection of some simple cubic planes, $\{100\}$, $\{110\}$ and $\{111\}$, from which it can be seen that those crystallographic planes which have poles in the southern half of the reference sphere are represented by circles in the stereogram, while those which have poles in the northern half are represented by dots.

As shown in Figure 2.7b, the angle between two poles on the reference sphere is the number of degrees separating them on the great circle passing through them. The angle between P and P' can be determined by means of a hemispherical transparent cap graduated and marked with meridian circles and latitude circles, as in geographical work. With a stereographic representation of poles, the equivalent operation can be performed in the plane of the primitive circle by using a transparent planar net, known as a Wulff net. This net is graduated in intervals of 2°, with meridians in the projection extending from top to bottom and latitude lines from side to side.[1] Thus, to measure the angular distance between any two poles in the stereogram, the net is rotated about the centre until the two poles lie upon the same meridian, which then corresponds to one of the great circles of the reference sphere. The angle between the two poles is then measured as the difference in latitude along the meridian. Some useful crystallographic rules may be summarized:

1. The Weiss Zone Law: the plane (hkl) is a member of the zone $[uvw]$ if $hu + kv + lw = 0$. A set of planes which all contain a common direction $[uvw]$ is known as a zone; $[uvw]$ is the zone axis (rather like the spine of an open book relative to the flat leaves). For example, the three planes $(1\bar{1}0)$, $(0\bar{1}1)$ and $(\bar{1}01)$ form a zone about the $[111]$ direction (Figure 2.8a). The pole of each plane containing $[uvw]$ must lie at 90° to $[uvw]$; therefore these three poles all lie in the same plane and upon the same great circle trace. The latter is known as the zone circle or zone trace. A plane trace is to a plane as a zone circle is to a zone. Uniquely, in the cubic

---

[1] A less-used alternative to the Wulff net is the polar net, in which the N–S axis of the reference sphere is perpendicular to the equatorial plane of projection.

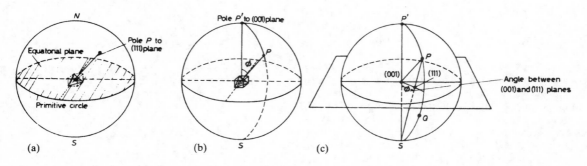

Figure 2.7 *Principles of stereographic projection, illustrating (a) the pole P to a (1 1 1) plane, (b) the angle between two poles, P, P' and (c) stereographic projection of P and P' poles to the (1 1 1) and (0 0 1) planes, respectively.*

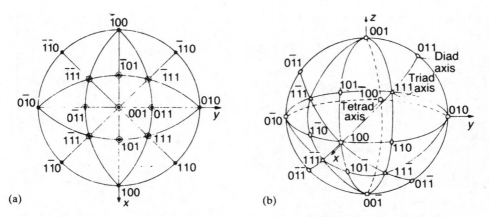

Figure 2.8 *Projections of planes in cubic crystals: (a) standard (0 0 1) stereographic projection and (b) spherical projection.*

system alone, zone circles and plane traces with the same indices lie on top of one another.

2. If a zone contains $(h_1k_1l_1)$ and $(h_2k_2l_2)$ it also contains any linear combination of them, e.g. $m(h_1k_1l_1) + n(h_2k_2l_2)$. For example, the zone $[1\,1\,1]$ contains $(1\,\bar{1}\,0)$ and $(0\,1\,\bar{1})$ and it must therefore contain $(1\,\bar{1}\,0) + (0\,1\,\bar{1}) = (1\,0\,\bar{1})$, $(1\,\bar{1}\,0) + 2(0\,1\,\bar{1}) = (1\,1\,\bar{2})$, etc. The same is true for different directions in a zone, provided that the crystal is cubic.

3. The Law of Vector Addition: the direction $[u_1v_1w_1] + [u_2v_2w_2]$ lies between $[u_1v_1w_1]$ and $[u_2v_2w_2]$.

4. The angle between two directions is given by:

$$\cos\theta = \frac{u_1u_2 + v_1v_2 + w_1w_2}{\sqrt{[(u_1^2 + v_1^2 + w_1^2)(u_2^2 + v_2^2 + w_2^2)]}}$$

where $u_1v_1w_1$ and $u_2v_2w_2$ are the indices for the two directions. Provided that the crystal system is cubic, the angles between planes may be found by substituting the symbols h, k, l and for u, v, w in this expression.

When constructing the standard stereogram of any crystal it is advantageous to examine the symmetry elements of that structure. As an illustration, consider a cubic crystal, since this has the highest symmetry of any crystal class. Close scrutiny shows that the cube has thirteen axes of symmetry; these axes comprise three fourfold (tetrad) axes, four threefold (triad) axes and six twofold (diad) axes, as indicated in Figure 2.9a. (This diagram shows the standard square, triangular and lens-shaped symbols for the three types of symmetry axis.) An n-fold axis of symmetry operates in such a way that after rotation through an angle $2\pi/n$, the crystal comes into an identical or self-coincident position in space. Thus, a tetrad axis passes through the centre of each face of the cube parallel to one of the edges, and a rotation of 90° in either direction about one of these axes turns the cube into a new

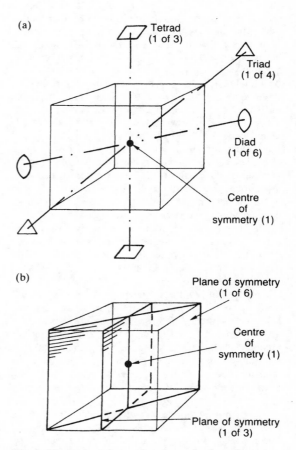

Figure 2.9 *Some elements of symmetry for the cubic system; total number of elements = 23.*

position which is crystallographically indistinguishable from the old position. Similarly, the cube diagonals form a set of four threefold axes, and each of the lines passing through the centre of opposite edges form a set of six twofold symmetry axes. Some tetrad, triad and diad axes are marked on the spherical projection of a cubic crystal shown in Figure 2.8b. The cube also has nine planes of symmetry (Figure 2.9b) and one centre of symmetry, giving, together with the axes, a total of 23 elements of symmetry.

In the stereographic projection of Figure 2.8a, planes of symmetry divide the stereogram into 24 equivalent spherical triangles, commonly called unit triangles, which correspond to the 48 (24 on the top and 24 on the bottom) seen in the spherical projection. The two-, three- and fourfold symmetry about the $\{110\}$, $\{111\}$ and $\{100\}$ poles, respectively, is apparent. It is frequently possible to analyse a problem in terms of a single unit triangle. Finally, reference to a stereogram (Figure 2.8a) confirms rule (2) which states that the indices of any plane can be found merely by adding simple multiples of other planes which lie in the same zone. For example, the (011) plane lies between the (001) and (010) planes and clearly $011 = 001 + 010$. Owing to the action of the symmetry elements, it can be reasoned that there must be a total of 12 $\{011\}$ planes because of the respective three- and fourfold symmetry about the $\{111\}$ and $\{100\}$ axes. As a further example, it is clear that the (112) plane lies between the (001) plane and (111) plane since $112 = 001 + 111$ and that the $\{112\}$ form must contain 24 planes, i.e. a icositetrahedron. The plane (123), which is an example of the most general crystal plane in the cubic system because its *hkl* indices are all different, lies between (112) and (011) planes; the 48 planes of the $\{123\}$ form make up a hexak-isoctahedron.

The tetrahedral form, a direct derivative of the cubic form, is often encountered in materials science (Figure 2.10a). Its symmetry elements comprise four triad axes, three diad axes and six 'mirror' planes, as shown in the stereogram of Figure 2.10b.

Concepts of symmetry, when developed systematically, provide invaluable help in modern structural analysis. As already implied, there are three basic elements, or operations, of symmetry. These operations involve translation (movement along parameters *a*, *b*, *c*), rotation (about axes to give diads, triads, etc.) and reflection (across 'mirror' planes). Commencing with an atom (or group of atoms) at either a lattice point or at a small group of lattice points, a certain combination of symmetry operations will ultimately lead to the three-dimensional development of any type of crystal structure. The procedure provides a unique identifying code for a structure and makes it possible to locate it among 32 point groups and 230 space groups of symmetry. This classification obviously embraces the seven crystal systems. Although many metallic structures can be defined relatively simply in terms of space lattice and one or more lattice constants, complex structures require the key of symmetry theory.

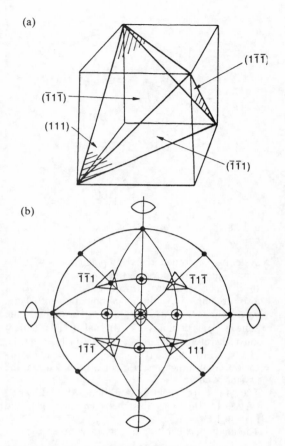

Figure 2.10 *Symmetry of the tetrahedral form.*

2.5 Selected crystal structures

2.5.1 Pure metals

We now examine the crystal structures of various elements (metallic and non-metallic) and compounds, using examples to illustrate important structure-building principles and structure/property relations.[1] Most elements in the Periodic Table are metallic in character; accordingly, we commence with them.

Metal ions are relatively small, with diameters in the order of 0.25 nm. A millimetre cube of metal therefore contains about 10^{20} atoms. The like ions in pure solid metal are packed together in a highly regular manner and, in the majority of metals, are packed so that ions collectively occupy the minimum volume. Metals are normally crystalline and for all of them, irrespective of whether the packing of ions is close or open, it

[1] Where possible, compound structures of engineering importance have been selected as illustrative examples. Prototype structures, such as NaCl, ZnS, CaF_2, etc., which appear in standard treatments elsewhere, are indicated as appropriate.

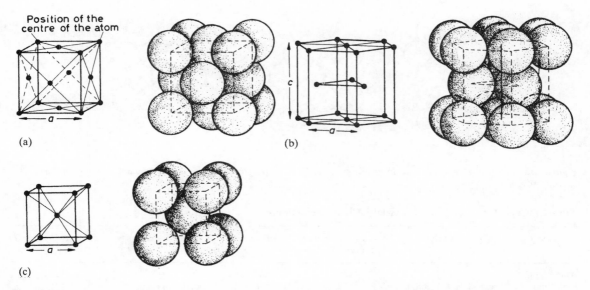

Figure 2.11 *Arrangement of atoms in (a) face-centred cubic structure, (b) close-packed hexagonal structure, and (c) body-centred cubic structure.*

is possible to define and express atomic arrangements in terms of structure cells (Section 2.2). Furthermore, because of the non-directional nature of the metallic bond, it is also possible to simulate these arrangements by simple 'hard-sphere' modelling.

There are two ways of packing spheres of equal size together so that they occupy the minimum volume. The structure cells of the resulting arrangements, face-centred cubic (fcc) and close-packed hexagonal (cph), are shown in Figures 2.11a and 2.11b. The other structure cell (Figure 2.11c) has a body-centred cubic (bcc) arrangement; although more 'open' and not based on close-packing, it is nevertheless adopted by many metals.

In order to specify the structure of a particular metal completely, it is necessary to give not only the type of crystal structure adopted by the metal but also the dimensions of the structure cell. In cubic structure cells it is only necessary to give the length of an edge a, whereas in a hexagonal cell the two parameters a and c must be given, as indicated in Figures 2.11a–c. If a hexagonal structure is ideally close-packed, the ratio c/a must be 1.633. In hexagonal metal structures, the axial ratio c/a is never exactly 1.633. These structures are, therefore, never quite ideally closed-packed, e.g. c/a (Zn) = 1.856, c/a(Ti) = 1.587. As the axial ratio approaches unity, the properties of cph metals begin to show similarities to fcc metals.

A knowledge of cell parameters permits the atomic radius r of the metal atoms to be calculated on the assumption that they are spherical and that they are in closest possible contact. The reader should verify that in the fcc structure $r = (a\sqrt{2})/4$ and in the bcc structure $r = (a\sqrt{3})/4$, where a is the cell parameter.

The coordination number (CN), an important concept in crystal analysis, is defined as the number of nearest equidistant neighbouring atoms around any atom in the crystal structure. Thus, in the bcc structure shown in Figure 2.11c the atom at the centre of the cube in surrounded by eight equidistant atoms, i.e. CN = 8. It is perhaps not so readily seen from Figure 2.11a that the coordination number for the fcc structure is 12. Perhaps the easiest method of visualizing this is to place two fcc cells side by side, and then count the neighbours of the common face-centring atom. In the cph structure with ideal packing ($c/a = 1.633$) the coordination number is again 12, as can be seen by once more considering two cells, one stacked on top of the other, and choosing the centre atom of the common basal plane. This $(0\,0\,0\,1)$ basal plane has the densest packing of atoms and has the same atomic arrangement as the closest-packed plane in the fcc structure.[1]

The cph and fcc structures represent two effective methods of packing spheres closely; the difference between them arises from the different way in which the close-packed planes are stacked. Figure 2.12a shows an arrangement of atoms in **A**-sites of a close-packed plane. When a second plane of close-packed atoms is laid down, its first atom may be placed in either a **B**-site or a **C**-site, which are entirely equivalent. However, once the first atom is placed in one of these two types of site, all other atoms in the second

---

[1] The Miller indices for the closest-packed (octahedral) planes of the fcc structure are {1 1 1}; these planes are best revealed by balancing a ball-and-stick model of the fcc cell on one corner.

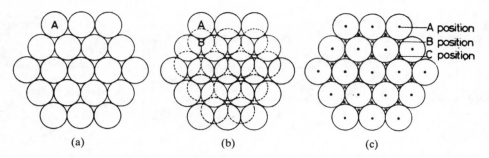

Figure 2.12 *(a) Arrangements of atoms in a close-packed plane, (b) registry of two close-packed planes, and (c) the stacking of successive planes.*

Table 2.1 Crystal structures of some metals at room temperature

Element	Crystal structure	Closest interatomic distance (nm)	Element	Crystal structure	Closest interatomic distance (nm)
Aluminium	fcc	0.286	Platinum	fcc	0.277
Beryllium	cph ($c/a = 1.568$)	0.223	Potassium	bcc	0.461
Cadmium	cph ($c/a = 1.886$)	0.298	Rhodium	fcc	0.269
Chromium	bcc	0.250	Rubidium	bcc	0.494
Cobalt	cph ($c/a = 1.623$)	0.250	Silver	fcc	0.289
Copper	fcc	0.255	Sodium	bcc	0.372
Gold	fcc	0.288	Tantalum	bcc	0.286
Iron	bcc	0.248	Thorium	fcc	0.360
Lead	fcc	0.350	Titanium	cph ($c/a = 1.587$)	0.299
Lithium	bcc	0.331	Tungsten	bcc	0.274
Magnesium	cph ($c/a = 1.623$)	0.320	Uranium	orthorhombic	0.275
Molybdenum	bcc	0.275	Vanadium	bcc	0.262
Nickel	fcc	0.249	Zinc	cph ($c/a = 1.856$)	0.266
Niobium	bcc	0.286	Zirconium	cph ($c/a = 1.592$)	0.318

plane must be in similar sites. (This is because neighbouring **B**- and **C** sites are too close together for both to be occupied in the same layer.) At this stage there is no difference between the cph and fcc structure; the difference arises only when the third layer is put in position. In building up the third layer, assuming that sites of type **B** have been used to construct the second layer, as shown in Figure 2.12b, either **A**-sites or **C**-sites may be selected. If **A**-sites are chosen, then the atoms in the third layer will be directly above those in the first layer, and the structure will be cph, whereas if **C**-sites are chosen this will not be the case and the structure will be fcc. Thus a cph structure consists of layers of close-packed atoms stacked in the sequence of **ABABAB** or, of course, equally well, **ACACAC**. An fcc structure has the stacking sequence **ABCABCABC** so that the atoms in the fourth layer lie directly above those in the bottom layer. The density of packing within structures is sometimes expressed as an atomic packing fraction (APF) which is the fraction of the cell volume occupied by atoms. The APF value for a bcc cell is 0.68; it rises to 0.74 for the more closely packed fcc and cph cells.

Table 2.1 gives the crystal structures adopted by some typical metals, the majority of which are either

fcc or bcc. As indicated previously, an atom does not have precise dimensions; however, it is convenient to express atomic diameters as the closest distance of approach between atom centres. Table 2.1 lists structures that are stable at room temperature; at other temperatures, some metals undergo transition and the atoms rearrange to form a different crystal structure, each structure being stable over a definite interval of temperature. This phenomenon is known as allotropy. The best-known commercially-exploitable example is that of iron, which is bcc at temperatures below 910°C, fcc in the temperature range 910–1400°C and bcc at temperatures between 1400°C and the melting point (1535°C). Other common examples include titanium and zirconium which change from cph to bcc at temperatures of 882°C and 815°C, respectively, tin, which changes from cubic (grey) to tetragonal (white) at 13.2°C, and the metals uranium and plutonium. Plutonium is particularly complex in that it has six different allotropes between room temperature and its melting point of 640°C.

These transitions between allotropes are usually reversible and, because they necessitate rearrangement of atoms, are accompanied by volume changes and either the evolution or absorption of thermal energy.

The transition can be abrupt but is often sluggish. Fortunately, tetragonal tin can persist in a metastable state at temperatures below the nominal transition temperature. However, the eventual transition to the friable low-density cubic form can be very sudden.[1]

Using the concept of a unit cell, together with data on the atomic mass of constituent atoms, it is possible to derive a theoretical value for the density of a pure single crystal. The parameter a for the bcc cell of pure iron at room temperature is 0.286 64 nm. Hence the volume of the unit cell is 0.023 55 nm^3. Contrary to first impressions, the bcc cell contains two atoms, i.e. $(8 \times \frac{1}{8}$ atom$) + 1$ atom. Using the Avogadro constant N_A,[2] we can calculate the mass of these two atoms as $2(55.85/N_A)$ or 185.46×10^{-24} kg, where 55.85 is the relative atomic mass of iron. The theoretical density (mass/volume) is thus 7875 kg m^{-3}. The reason for the slight discrepancy between this value and the experimentally-determined value of 7870 kg m^{-3} will become evident when we discuss crystal imperfections in Chapter 4.

2.5.2 Diamond and graphite

It is remarkable that a single element, carbon, can exist in two such different crystalline forms as diamond and graphite. Diamond is transparent and one of the

hardest materials known, finding wide use, notably as an abrasive and cutting medium. Graphite finds general use as a solid lubricant and writing medium (pencil 'lead'). It is now often classed as a highly refractory ceramic because of its strength at high temperatures and excellent resistance to thermal shock.

We can now progress from the earlier representation of the diamond structure (Figure 1.3c) to a more realistic version. Although the structure consists of two interpenetrating fcc sub-structures, in which one sub-structure is slightly displaced along the body diagonal of the other, it is sufficient for our purpose to concentrate on a representative structure cell (Figure 2.13a). Each carbon atom is covalently bonded to four equidistant neighbours in regular tetrahedral[3] coordination (CN = 4). For instance, the atom marked X occupies a 'hole', or interstice, at the centre of the group formed by atoms marked 1, 2, 3 and 4. There are eight equivalent tetrahedral sites of the X-type, arranged four-square within the fcc cell; however, in the case of diamond, only half of these sites are occupied. Their disposition, which also forms a tetrahedron, maximizes the intervening distances between the four atoms. If the fcc structure of diamond depended solely upon packing efficiency, the coordination number would be 12; actually CN = 4, because only four covalent bonds can form. Silicon ($Z = 14$), germanium ($Z = 32$) and grey tin ($Z = 50$) are fellow-members of Group IV in the Periodic Table and are therefore also tetravalent. Their crystal structures are identical in character, but obviously not in dimensions, to the diamond structure of Figure 2.13a.

---

[1]Historical examples of 'tin plague' abound (e.g. buttons, coins, organ pipes, statues).

[2]The Avogadro constant N_A is $0.602\,217 \times 10^{-24}$ mol^{-1}. The mole is a basic SI unit. It does not refer to mass and has been likened to terms such as dozen, score, gross, etc. By definition, it is the amount of substance which contains as many elementary units as there are atoms in 0.012 kg of carbon-12. The elementary unit must be specified and may be an atom, a molecule, an ion, an electron, a photon, etc. or a group of such entities.

[3]The stability and strength of a tetrahedral form holds a perennial appeal for military engineers: spiked iron caltrops deterred attackers in the Middle Ages and concrete tetrahedra acted as obstacles on fortified Normandy beaches in World War II.

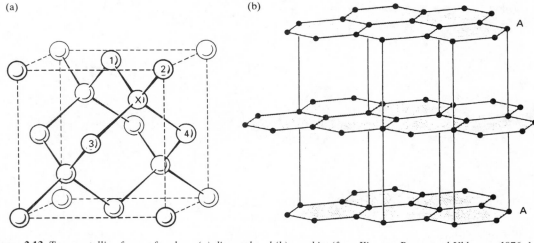

(a)   (b)

Figure 2.13 *Two crystalline forms of carbon: (a) diamond and (b) graphite (from Kingery, Bowen and Uhlmann, 1976; by permission of Wiley-Interscience).*

Graphite is less dense and more stable than diamond. In direct contrast to the cross-braced structure of diamond, graphite has a highly anisotropic layer structure (Figure 2.13b). Adjacent layers in the **ABABAB** sequence are staggered; the structure is not cph. A less stable rhombohedral **ABCABC** sequence has been observed in natural graphite. Charcoal, soot and lampblack have been termed 'amorphous carbon'; actually they are microcrystalline forms of graphite. Covalent-bonded carbon atoms, 0.1415 nm apart, are arranged in layers of hexagonal symmetry. These layers are approximately 0.335 nm apart. This distance is relatively large and the interlayer forces are therefore weak. Layers can be readily sheared past each other, thus explaining the lubricity of graphitic carbon. (An alternative solid lubricant, molybdenum disulphide, MoS_2, has a similar layered structure.).

The ratio of property values parallel to the *a*-axis and the *c*-axis is known as the anisotropy ratio. (For cubic crystals, the ratio is unity.) Special synthesis techniques can produce near-ideal graphite[1] with an anisotropy ratio of thermal conductivity of 200.

2.5.3 Coordination in ionic crystals

We have seen in the case of diamond how the joining of four carbon atoms outlines a tetrahedron which is smaller than the structure cell (Figure 2.13a). Before examining some selected ionic compounds, it is necessary to develop this aspect of coordination more fully. This approach to structure-building concerns packing and is essentially a geometrical exercise. It is subordinate to the more dominant demands of covalent bonding.

In the first of a set of conditional rules, assembled by Pauling, the relative radii of cation (*r*) and anion (*R*) are compared. When electrons are stripped from the outer valence shell during ionization, the remaining

[1]Applications range from rocket nozzles to bowl linings for tobacco pipes.

electrons are more strongly attracted to the nucleus; consequently, cations are usually smaller than anions. **Rule 1** states that the coordination of anions around a reference cation is determined by the geometry necessary for the cation to remain in contact with each anion. For instance, in Figure 2.14a, a radius ratio r/R of 0.155 signifies touching contact when three anions are grouped about a cation. This critical value is readily derived by geometry. If the r/R ratio for threefold coordination is less than 0.155 then the cation 'rattles' in the central interstice, or 'hole', and the arrangement is unstable. As r/R exceeds 0.155 then structural distortion begins to develop.

In the next case, that of fourfold coordination, the 'touching' ratio has a value of 0.225 and joining of the anion centres defines a tetrahedron (Figure 2.14b). For example, silicon and oxygen ions have radii of 0.039 nm and 0.132 nm, respectively, hence $r/R = 0.296$. This value is slightly greater than the critical value of 0.225 and it follows that tetrahedral coordination gives a stable configuration; indeed, the complex anion SiO_4^{4-} is the key structural feature of silica, silicates and silica glasses. The quadruple negative charge is due to the four unsatisfied oxygen bonds which project from the group.

In a feature common to many structures, the tendency for anions to distance themselves from each other as much as possible is balanced by their attraction towards the central cation. Each of the four oxygen anions is only linked by one of its two bonds to the silicon cation, giving an effective silicon/oxygen ratio of 1:2 and thus confirming the stoichiometric chemical formula for silica, SiO_2. Finally, as shown in Figure 2.14c, the next coordination polyhedron is an octahedron for which $r/R = 0.414$. It follows that each degree of coordination is associated with a nominal range of r/R values, as shown in Table 2.2. Caution is necessary in applying these ideas of geometrical packing because (1) range limits are approximative, (2) ionic radii are very dependent upon CN, (3) ions can be non-spherical in anisotropic crystals and

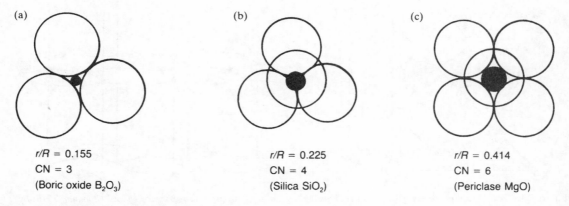

(a)

$r/R = 0.155$
CN = 3
(Boric oxide B_2O_3)

(b)

$r/R = 0.225$
CN = 4
(Silica SiO_2)

(c)

$r/R = 0.414$
CN = 6
(Periclase MgO)

Figure 2.14 *Nesting of cations within anionic groups.*

Table 2.2 Relation between radius ratio and coordination

r/R	Maximum coordination number (CN)	Form of coordination
<0.155	2	Linear
0.155–0.225	3	Equilateral triangle
0.225–0.414	4	Regular tetrahedron
0.414–0.732	6	Regular octahedron
0.732–1.0	8	Cube
1.00	12	Cuboctahedron

(4) considerations of covalent or metallic bonding can be overriding. The other four Pauling rules are as follows:

Rule II. In a stable coordinated structure the total valency of the anion equals the summated bond strengths of the valency bonds which extend to this anion from all neighbouring cations. Bond strength is defined as the valency of an ion divided by the actual number of bonds; thus, for Si^{4+} in tetrahedral coordination it is $\frac{4}{4} = 1$. This valuable rule, which expresses the tendency of each ion to achieve localized neutrality by surrounding itself with ions of opposite charge, is useful in deciding the arrangement of cations around an anion. For instance, the important ceramic barium titanate ($BaTiO_3$) has Ba^{2+} and Ti^{4+} cations bonded to a common O^{2-} anion. Given that the coordination numbers of O^{2-} polyhedra centred on Ba^{2+} and Ti^{4+} are 12 and 6, respectively, we calculate the corresponding strengths of the Ba–O and Ti–O bonds as $\frac{2}{12} = \frac{1}{6}$ and $\frac{4}{6} = \frac{2}{3}$. The valency of the shared anion is 2, which is numerically equal to $(4 \times \frac{1}{6}) + (2 \times \frac{2}{3})$. Accordingly, coordination of the common oxygen anion with four barium cations and two titanium cations is a viable possibility.

Rule III. An ionic structure tends to have maximum stability when its coordination polyhedra share corners; edge- and face-sharing give less stability. Any arrangement which brings the mutually-repelling central cations closer together tends to destabilize the structure. Cations of high valency (charge) and low CN (poor 'shielding' by surrounding anions) aggravate the destabilizing tendency.

Rule IV. In crystals containing different types of cation, cations of high valency and low CN tend to limit the sharing of polyhedra elements; for instance, such cations favour corner-sharing rather than edge-sharing.

Rule V. If several alternative forms of coordination are possible, one form usually applies throughout the structure. In this way, ions of a given type are more likely to have identical surroundings.

In conclusion, it is emphasized that the Pauling rules are only applicable to structures in which ionic bonding predominates. Conversely, any structure which fails to comply with the rules is extremely unlikely to be ionic.

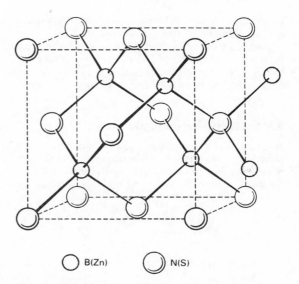

B(Zn)   N(S)

Figure 2.15 *Zinc blende (α-ZnS) structure, prototype for cubic boron nitride (BN) (from Kingery, Bowen and Uhlmann, 1976; by permission of Wiley-Interscience).*

The structure of the mineral zinc blende (α-ZnS) shown in Figure 2.15 is often quoted as a prototype for other structures. In accord with the radius ratio $r/R = 0.074/0.184 = 0.4$, tetrahedral coordination is a feature of its structure. Coordination tetrahedra share only corners (vertices). Thus one species of ion occupies four of the eight tetrahedral sites within the cell. These sites have been mentioned previously in connection with diamond (Section 2.5.2); in that case, the directional demands of the covalent bonds between like carbon atoms determined their location. In zinc sulphide, the position of unlike ions is determined by geometrical packing. Replacement of the Zn^{2+} and S^{2-} ions in the prototype cell with boron and nitrogen atoms produces the structure cell of cubic boron nitride (BN). This compound is extremely hard and refractory and, because of the adjacency of boron ($Z = 5$) and nitrogen ($Z = 7$) to carbon ($Z = 6$) in the Periodic Table, is more akin in character to diamond than to zinc sulphide. Its angular crystals serve as an excellent grinding abrasive for hardened steel. The precursor for cubic boron nitride is the more common and readily-prepared form, hexagonal boron nitride.[1]

This hexagonal form is obtained by replacing the carbon atoms in the layered graphite structure (Figure 2.13b) alternately with boron and nitrogen atoms and also slightly altering the stacking registry of the layer planes. It feels slippery like graphite and

[1] The process for converting hexagonal BN to cubic BN (*Borazon*) involves very high temperature and pressure and was developed by Dr R. H. Wentorf at the General Electric Company, USA (1957).

is sometimes called 'white graphite'. Unlike graphite, it is an insulator, having no free electrons.

Another abrasive medium, silicon carbide (SiC), can be represented in one of its several crystalline forms by the zinc blende structure. Silicon and carbon are tetravalent and the coordination is tetrahedral, as would be expected.

2.5.4 AB-type compounds

An earlier diagram (Figure 1.3b) schematically portrayed the ionic bonding within magnesium oxide (periclase). We can now develop a more realistic model of its structure and also apply the ideas of coordination.

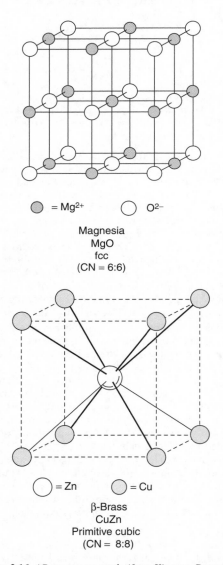

= Mg^{2+} ◯ O^{2-}

Magnesia
MgO
fcc
(CN = 6:6)

◯ = Zn   = Cu

β-Brass
CuZn
Primitive cubic
(CN = 8:8)

Figure 2.16 *AB-type compounds (from Kingery, Bowen and Uhlmann, 1976; by permission of Wiley-Interscience).*

Generically, MgO is a sodium chloride-type structure (Figure 2.16a), with Mg^{2+} cations and O^{2-} anions occupying two interpenetrating[1] fcc sub-lattices. Many oxides and halides have this type of structure (e.g. CaO, SrO, BaO, VO, CdO, MnO, FeO, CoO, NiO; NaCl, NaBr, NaI, NaF, KCl, etc.). The ratio of ionic radii $r/R = 0.065/0.140 = 0.46$ and, as indicated by Table 2.2, each Mg^{2+} cation is octahedrally coordinated with six larger O^{2-} anions, and vice versa (CN = 6:6). Octahedra of a given type share edges. The 'molecular' formula MgO indicates that there is an exact stoichiometric balance between the numbers of cations and anions; more specifically, the unit cell depicted contains $(8 \times \frac{1}{8}) + (6 \times \frac{1}{2}) = 4$ cations and $(12 \times \frac{1}{4}) + 1 = 4$ anions.

The second example of an AB-type compound is the hard intermetallic compound CuZn (β-brass) shown in Figure 2.16b. It has a caesium chloride-type structure in which two simple cubic sub-lattices interpenetrate. Copper ($Z = 29$) and zinc ($Z = 30$) have similar atomic radii. Each copper atom is in eightfold coordination with zinc atoms; thus CN = 8:8. The coordination cubes share faces. Each unit cell contains $(8 \times \frac{1}{8}) = 1$ corner atom and 1 central atom; hence the formula CuZn. In other words, this compound contains 50 at.% copper and 50 at.% zinc.

2.5.5 Silica

Compounds of the AB$_2$-type (stoichiometric ratio 1:2) form a very large group comprising many different types of structure. We will concentrate upon β-cristobalite, which, as Table 2.3 shows, is the high-temperature modification of one of the three principal forms in which silica (SiO$_2$) exists. Silica is a refractory ceramic which is widely used in the steel and glass industries. Silica bricks are prepared by kiln-firing quartz of low impurity content at a temperature of 1450°C, thereby converting at least 98.5% of it into a mixture of the more 'open', less dense forms, tridymite and cristobalite. The term 'conversion' is equivalent to that of allotropic transformation in metallic materials and refers to a transformation which is reconstructive in character, involving the breaking and re-establishment of inter-atomic bonds. These solid-state changes are generally rather sluggish and, as a consequence, crystal structures frequently persist in a metastable condition at temperatures outside the nominal ranges of stability given in Table 2.3. Transformations from one modification to another only involve displacement of bonds and reorientation of bond directions; they are known as inversions. As these changes are comparatively limited in range, they are usually quite rapid and reversible. However, the associated volume change can be substantial. For example, the $\alpha \rightarrow \beta$ transition in cristobalite at a

[1]Sub-lattices can be discerned by concentrating on each array of like atoms (ions) in turn.

Table 2.3 Principal crystalline forms of silica

Form	Range of stability (°C)	Modifications	Density (kg m^{-3})
Cristobalite	1470–1723 (m.p.)	β—(cubic)	2210
		α—(tetragonal)	2330
Tridymite	870–1470	γ—(?)	—
		β—(hexagonal)	2300
		α—(orthorhombic)	2270
Quartz	<870	β—(hexagonal)	2600
		α—(trigonal)	2650

temperature of 270°C is accompanied by a volume increase of 3% which is capable of disrupting the structure of a silica brick or shape. In order to avoid this type of thermal stress cracking, it is necessary to either heat or cool silica structures very slowly at temperatures below 700°C (e.g. at 20°Ch^{-1}). Above this temperature level, the structure is resilient and, as a general rule, it is recommended that silica refractory be kept above a temperature of 700°C during its entire working life. Overall, the structural behaviour of silica during kiln-firing and subsequent service is a complicated subject,[1] particularly as the presence of other substances can either catalyse or hinder transformations.

Substances which promote structural change in ceramics are known as mineralizers (e.g. calcium oxide (CaO)). The opposite effect can be produced by associated substances in the microstructure; for instance, an encasing envelope of glassy material can inhibit the cooling inversion of a small volume of β-cristobalite by opposing the associated contraction. The pronounced metastability of cristobalite and tridymite at relatively low temperatures is usually attributed to impurity atoms which, by their presence in the interstices, buttress these 'open' structures and inhibit conversions. However, irrespective of these complications, corner-sharing SiO_4^{4-} tetrahedra, with their short-range order, are a common feature of all these crystalline modifications of silica; the essential difference between modifications is therefore one of long-range ordering. We will use the example of the β-cristobalite structure to expand the idea of these versatile tetrahedral building units. (Later we will see that they also act as building units in the very large family of silicates.)

In the essentially ionic structure of β-cristobalite (Figure 2.17) small Si^{4+} cations are located in a cubic arrangement which is identical to that of diamond. The much larger O^{2-} anions form SiO_4^{4-} tetrahedra around each of the four occupied tetrahedral sites in such a way that each Si^{4+} lies equidistant between two anions.

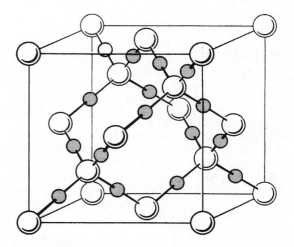

Figure 2.17 *Structure of β-cristobalite (from Kingery, Bowen and Uhlmann, 1976; by permission of Wiley-Interscience).*

The structure thus forms a regular network of corner-sharing tetrahedra. The coordination of anions around a cation is clearly fourfold; coordination around each anion can be derived by applying Pauling's Rule III. Thus, CN = 4:2 neatly summarizes the coordination in β-cristobalite. Oxygen anions obviously occupy much more volume than cations and consequently their grouping in space determines the essential character of the structure. In other words, the radius ratio is relatively small. As the anion and cation become progressively more similar in size in some of the other AB_2-type compounds, the paired coordination numbers take values of 6:3 and then 8:4. These paired values relate to structure groups for which rutile (TiO_2) and fluorite (CaF_2), respectively, are commonly quoted as prototypes. AB_2-type compounds have their alloy counterparts and later, in Chapter 3, we will examine in some detail a unique and important family of alloys (e.g. $MgCu_2$, $MgNi_2$, $MgZn_2$, etc.). In these so-called Laves phases, two dissimilar types of atoms pack so closely that the usual coordination maximum of 12, which is associated with equal-sized atoms, is actually exceeded.

---

[1] The fact that cristobalite forms at a kiln-firing temperature which is below 1470°C illustrates the complexity of the structural behaviour of commercial-quality silica.

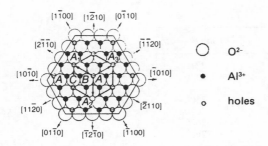

Figure 2.18 *Structure of α-alumina (corundum) viewed perpendicular to (0 0 0 1) basal plane (from Hume-Rothery, Smallman and Haworth, 1988).*

2.5.6 Alumina

Alumina exists in two forms: α-Al_2O_3 and γ-Al_2O_3. The former, often referred to by its mineral name corundum, serves as a prototype for other ionic oxides, such as α-Fe_2O_3 (haematite), Cr_2O_3, V_2O_3, Ti_2O_3, etc. The structure of α-Al_2O_3 (Figure 2.18) can be visualized as layers of close-packed O^{2-} anions with an **ABABAB**... sequence in which two-thirds of the octahedral holes or interstices are filled symmetrically with smaller Al^{3+} cations. Coordination is accordingly 6:4. This partial filling gives the requisite stoichiometric ratio of ions. The structure is not truly cph because all the octahedral sites are not filled.

α-A_2O_3 is the form of greatest engineering interest. The other term, γ-Al_2O_3, refers collectively to a number of variants which have O^{2-} anions in an fcc arrangement. As before, Al^{3+} cations fill two-thirds of the octahedral holes to give a structure which is conveniently regarded as a 'defect' spinel structure with a deficit, or shortage, of Al^{3+} cations; spinels will be described in Section 2.5.7. γ-Al_2O_3 has very useful adsorptive and catalytic properties and is sometimes referred to as 'activated alumina', illustrating yet again the way in which structural differences within the same compound can produce very different properties.

2.5.7 Complex oxides

The ABO_3-type compounds, for which the mineral perovskite ($CaTiO_3$) is usually quoted as prototype, form an interesting and extremely versatile family. Barium titanium oxide[1] ($BaTiO_3$) has been studied extensively, leading to the development of important synthetic compounds, notably the new generation of ceramic superconductors.[2] It is polymorphic,

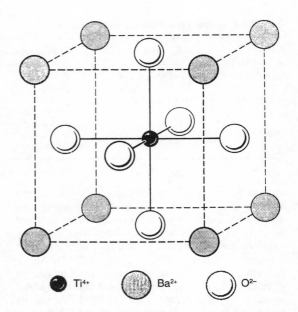

Figure 2.19 *Unit cell of cubic $BaTiO_3$ ($CN = 6:12$) (from Kingery, Bowen and Uhlmann, 1976; by permission of Wiley-Interscience).*

exhibiting at least four temperature-dependent transitions. The cubic form, which is stable at temperatures below 120°C, is shown in Figure 2.19. The large barium cations are located in the 'holes', or interstices, between the regularly stacked titanium-centred oxygen octahedra. Each barium cation is at the centre of a polyhedron formed by twelve oxygen anions. (Coordination in this structure was discussed in terms of Pauling's Rule II in Section 2.5.3).

Above the ferroelectric Curie point (120°C), the cubic unit cell of $BaTiO_3$ becomes tetragonal as Ti^{4+} cations and O^{2-} anions move in opposite directions parallel to an axis of symmetry. This slight displacement of approximately 0.005 nm is accompanied by a change in axial ratio (c/a) from unity to 1.04. The new structure develops a dipole of electric charge as it becomes less symmetrical; it also exhibits marked ferroelectric characteristics. The electrical and magnetic properties of perovskite-type structures will be explored in Chapter 6.

Inorganic compounds with structures similar to that of the hard mineral known as spinel, $MgAl_2O_4$, form an extraordinarily versatile range of materials (e.g. watch bearings, refractories). Numerous alternative combinations of ions are possible. Normal versions of these mixed oxides are usually represented by the general formula AB_2O_4; however, other combinations of the two dissimilar cations, A and B, are also

---

[1] The structure does not contain discrete $TiO_3{}^{2-}$ anionic groups; hence, strictly speaking, it is incorrect to imply that the compound is an inorganic salt by referring to it as barium 'titanate'.

[2] K. A. Muller and J. G. Bednorz, IBM Zurich Research Laboratory, based their researches upon perovskite-type structures. In 1986 they produced a complex

super-conducting oxide of lanthanum, barium and copper which had the unprecedentedly-high critical temperature of 35 K.

possible. Terms such as II-III spinels, II-IV spinels and I-VI spinels have been adopted to indicate the valencies of the first two elements in the formula; respective examples being $Mg^{2+}Al_2^{3+}O_4^{2-}$, $Mg_2^{2+}Ge^{4+}O_4^{2-}$ and $Ag_2^{1+}Mo^{6+}O_4^{2-}$. In each spinel formula, the total cationic charge balances the negative charge of the oxygen anions. (Analogous series of compounds are formed when the divalent oxygen anions are completely replaced by elements from the same group of the Periodic Table, i.e. sulphur, selenium and tellurium.)

The principle of substitution is a useful device for explaining the various forms of spinel structure.

Thus, in the case of II-III spinels, the Mg^{2+} cations of the reference spinel structure $MgAl_2O_4$ can be replaced by Fe^{2+}, Zn^{2+}, Ni^{2+} and Mn^{2+} and virtually any trivalent cation can replace Al^{3+} ions (e.g. Fe^{3+}, Cr^{3+}, Mn^{3+}, Ti^{3+}, V^{3+}, rare earth ions, etc.). The scope for extreme diversity is immediately apparent. The cubic unit cell, or true repeat unit, of the II-III prototype $MgAl_2O_4$ comprises eight fcc sub-cells and, overall, contains 32 oxygen anions in almost perfect fcc arrangement. The charge-compensating cations are distributed among the tetrahedral (CN = 4) and octahedral (CN = 6) interstices of these anions. (Each individual fcc sub-cell has eight tetrahedral sites within it, as explained for diamond, and 12 octahedral 'holes' located midway along each of the cube edges.) One eighth of the 64 tetrahedral 'holes' of the large unit cell are occupied by Mg^{2+} cations and one half of the 32 octahedral 'holes' are occupied by Al^{3+} cations. A similar distribution of divalent and trivalent cations occurs in other normal II-III spinels e.g. $MgCr_2O_4$, $ZnCr_2Se_4$. Most spinels are of the II-III type.

Ferrospinels ('ferrites'), such as $NiFe_2O_4$ and $CoFe_2O_4$, form an 'inverse' type of spinel structure in which the allocation of cations to tetrahedral and octahedral sites tends to change over, producing significant and useful changes in physical characteristics (e.g. magnetic and electrical properties). The generic formula for 'inverse' spinels takes the form $B(AB)O_4$, with the parentheses indicating the occupancy of octahedral sites by both types of cation. In this 'inverse' arrangement, B cations rather than A cations occupy tetrahedral sites. In the case of the two ferrospinels named, 'inverse' structures develop during slow cooling from sintering heat-treatment. In the first spinel, which we can now write as $Fe^{3+}(Ni^{2+}Fe^{3+})O_4$, half of the Fe^{3+} cations are in tetrahedral sites. The remainder, together with all Ni^{2+} cations, enter octahedral sites. Typically, these compounds respond to the conditions of heat-treatment: rapid cooling after sintering will affect the distribution of cations and produce a structure intermediate to the limiting normal and inverse forms. The partitioning among cation sites is often quantified in terms of the degree of inversion (λ) which states the fraction of B cations occupying tetrahedral sites. Hence, for normal and inverse spinels respectively, $\lambda = 0$ and $\lambda = 0.5$. Intermediate values of λ

between these limits are possible. Magnetite, the navigational aid of early mariners, is an inverse spinel and has the formula $Fe^{3+}(Fe^{2+}Fe^{3+})O_4$ and $\lambda = 0.5$. $Fe^{3+}(Mg^{2+}Fe^{3+})O_4$ is known to have a λ value of 0.45. Its structure is therefore not wholly inverse, but this formula notation does convey structural information. Other, more empirical, notations are sometimes used; for instance, this particular spinel is sometimes represented by the formulae $MgFe_2O_4$ and $MgO.Fe_2O_3$.

2.5.8 Silicates

Silicate minerals are the predominant minerals in the earth's crust, silicon and oxygen being the most abundant chemical elements. They exhibit a remarkable diversity of properties. Early attempts to classify them in terms of bulk chemical analysis and concepts of acidity/basicity failed to provide an effective and convincing frame of reference. An emphasis upon stoichiometry led to the practice of representing silicates by formulae stating the thermodynamic components. Thus two silicates which are encountered in refractories science, forsterite and mullite, are sometimes represented by the 'molecular' formulae $2MgO.SiO_2$ and $3Al_2O_3.2SiO_2$. (A further step, often adopted in phase diagram studies, is to codify them as M_2S and A_3S_2, respectively.) However, as will be shown, the summated counterparts of the above formulae, namely Mg_2SiO_4 and $Al_6Si_2O_{13}$, provide some indication of ionic grouping and silicate type. In keeping with this emphasis upon structure, the characterization of ceramics usually centres upon techniques such as X-ray diffraction analysis, with chemical analyses making a complementary, albeit essential, contribution.

The SiO_4 tetrahedron previously described in the discussion of silica (Section 2.5.5) provides a highly effective key to the classification of the numerous silicate materials, natural and synthetic. From each of the four corner anions projects a bond which is satisfied by either (1) an adjacent cation, such as Mg^{2+}, Fe^{2+}, Fe^{3+}, Ca^{2+} etc., or (2) by the formation of 'oxygen bridges' between vertices of tetrahedra. In the latter case an increased degree of cornersharing leads from structures in which isolated tetrahedra exist to those in which tetrahedra are arranged in pairs, chains, sheets or frameworks (Table 2.4). Let us briefly consider some examples of this structural method of classifying silicates.

In the nesosilicates, isolated SiO_4^{4-} tetrahedra are studded in a regular manner throughout the structure. Zircon (zirconium silicate) has the formula $ZrSiO_4$ which displays the characteristic silicon/oxygen ratio (1:4) of a nesosilicate. (It is used for the refractory kiln furniture which supports ceramic ware during the firing process.) The large family of nesosilicate minerals known as olivines has a generic formula $(Mg, Fe)_2SiO_4$, which indicates that the negatively-charged tetrahedra are balanced electrically by either

Table 2.4 Classification of silicate structures

Type of silicate		$(Si^{4+} + Al^{3+}) : O^{2-a}$	Arrangement of tetrahedra[b]	Examples
Mineralogical name	*Chemical name*			
Nesosilicate	'Orthosilicate'	1:4	Isolated	Zircon, olivines, garnets
Sorosilicate	'Pyrosilicate'	2:7	Pairing	Thortveitite
		1:3, 4:11	Linear chains	Amphiboles, pyroxenes
Inosilicate	'Metasilicate'	3:9, 6:18, etc.	Rings	Beryl
Phyllosilicate		2:5	Flat sheets	Micas, kaolin, talc
Tectosilicate		1:2	Framework	Feldspars, zeolites, ultramarines

[a]Only includes Al cations within tetrahedra.

[b]△ represents a tetrahedron.

Mg^{2+} or Fe^{2+} cations. This substitution, or replacement, among the available cation sites of the structure forms a solid solution.[1] This means that the composition of an olivine can lie anywhere between the compositions of the two end-members, forsterite (Mg_2SiO_4) and fayalite (Fe_2SiO_4). The difference in high-temperature performance of these two varieties of olivine is striking; white forsterite (m.p. 1890°C) is a useful refractory whereas brown/black fayalite (m.p. 1200°C), which sometimes forms by interaction between certain refractory materials and a molten furnace charge, is weakening and undesirable. Substitution commonly occurs in non-metallic compounds (e.g. spinels). Variations in its form and extent can be considerable and it is often found that samples can vary according to source, method of manufacture, etc. Substitution involving ions of different valency is found

[1]This important mixing effect also occurs in many metallic alloys; an older term, 'mixed crystal' (from the German word *Mischkristall*), is arguably more appropriate.

in the dense nesosilicates known as garnets. In their representational formula, $A_3{}^{II}B_2{}^{III}(SiO_4)_3$, the divalent cation A can be Ca^{2+}, Mg^{2+}, Mn^{2+} or Fe^{2+} and the trivalent cation B can be Al^{3+}, Cr^{3+}, Fe^{3+}, or Ti^{3+}. (Garnet is extremely hard and is used as an abrasive.)

Certain asbestos minerals are important examples of inosilicates. Their unique fibrous character, or asbestiform habit, can be related to the structural disposition of $SiO_4{}^{4-}$ tetrahedra. These impure forms of magnesium silicate are remarkable for their low thermal conductivity and thermal stability. However, all forms of asbestos break down into simpler components when heated in the temperature range 600–1000°C. The principal source materials are:

Amosite (brown asbestos)	$(Fe_2{}^{2+}Mg)_7(Si_4O_{11})_2(OH)_4$
Crocidolite (blue asbestos)	$Na_2Fe_2{}^{3+}(Fe^{2+}Mg)_3(Si_4O_{11})_2(OH)_4$
Chrysotile (white asbestos)	$Mg_3Si_2O_5(OH)_4$

These chemical formulae are idealized. Amosite and crocidolite belong to the amphibole group of minerals in which SiO_4^{4-} tetrahedra are arranged in double-strand linear chains (Table 2.4). The term (Si_4O_{11}) represents the repeat unit in the chain which is four tetrahedra wide. Being hydrous minerals, hydroxyl ions $(OH)^-$ are interspersed among the tetrahedra. Bands of cations separate the chains and, in a rather general sense, we can understand why these structures cleave to expose characteristic thread-like fracture surfaces. Each thread is a bundle of solid fibrils or filaments, 20–200 nm in breadth. The length/diameter ratio varies but is typically 100:1. Amphibole fibres are used for high-temperature insulation and have useful acid resistance; however, they are brittle and inflexible ('harsh') and are therefore difficult to spin into yarn and weave. In marked contrast, chrysotile fibres are strong and flexible and have been used specifically for woven asbestos articles, for friction surfaces and for asbestos/cement composites. Chrysotile belongs to the serpentine class of minerals in which SiO_4^{4-} tetrahedra are arranged in sheets or layers. It therefore appears paradoxical for it to have a fibrous fracture. High-resolution electron microscopy solved the problem by showing that chrysotile fibrils, sectioned transversely, were hollow tubes in which the structural layers were curved and arranged either concentrically or as scrolls parallel to the major axis of the tubular fibril.

Since the 1970s considerable attention has been paid to the biological hazards associated with the manufacture, processing and use of asbestos-containing materials. It has proved to be a complicated and highly emotive subject. Minute fibrils of asbestos are readily airborne and can cause respiratory diseases (asbestosis) and cancer. Crocidolite dust is particularly dangerous. Permissible atmospheric concentrations and safe handling procedures have been prescribed. Encapsulation and/or coating of fibres is recommended. Alternative materials are being sought but it is difficult to match the unique properties of asbestos. For instance, glassy 'wool' fibres have been produced on a commercial scale by rapidly solidifying molten rock but they do not have the thermal stability, strength and flexibility of asbestos. Asbestos continues to be widely used by the transportation and building industries. Asbestos textiles serve in protective clothing, furnace curtains, pipe wrapping, ablative nose cones for rockets, and conveyors for molten glass. Asbestos is used in friction components,[1] gaskets, gland packings, joints, pump seals, etc. In composite asbestos cloth/phenolic resin form, it is used for bearings, bushes, liners and aero-engine heat shields. Cement reinforced with asbestos fibres is used for roofing, cladding and for pressure pipes which distribute potable water.

The white mineral kaolinite is an important example of the many complex silicates which have a layered structure, i.e. Si:O = 2:5. As indicated previously, in the discussion of spinels, atomic grouping(s) within the structural formula can indicate actual structural groups. Thus, kaolinite is represented by $Al_2Si_2O_5(OH)_4$ rather than by $Al_2O_3.2SiO_2.2H_2O$, an older notation which uses 'waters of crystallization' and disregards the significant role of hydroxyl OH^- ions. Sometimes the formula is written as $[Al_2Si_2O_5(OH)_4]_2$ in order to give a truer picture of the repeat cell. Kaolinite is the commonest clay mineral and its small crystals form the major constituent of kaolin (china-clay), the rock that is a primary raw material of the ceramics industry. (It is also used for filling and coating paper.) Clays are the sedimentary products of the weathering of rocks and when one considers the possible variety of geological origins, the opportunities for the acquisition of impurity elements and the scope for ionic replacement it is not surprising to find that the compositions and structures of clay minerals show considerable variations. To quote one practical instance, only certain clays, the so-called fireclays, are suitable for manufacture into refractory firebricks for furnace construction.

Structurally, kaolinite provides a useful insight into the arrangement of ions in layered silicates. Essentially the structure consists of flat layers, several ions thick. Figure 2.20 shows, in section, adjacent vertically-stacked layers of kaolinite, each layer having five sub-layers or sheets. The lower side of each layer consists of SiO_4^{4-} tetrahedra arranged hexagonally in a planar net. Three of the four vertices of these tetrahedra are joined by 'oxygen bridges' and lie in the lowermost face; the remaining vertices all point upwards. The central Si^{4+} cations of the tetrahedra form the second sub-layer. The upward-pointing vertices, together with OH^- ions, form the close-packed third sub-layer. Al^{3+} cations occupy some of the octahedral 'holes' (CN = 6) between this third layer and a fifth close-packed layer of OH^- ions. The coordination of each

---

[1]Dust from asbestos friction components, such as brake linings, pads and clutches of cars, can contain 1–2% of asbestos fibres and should be removed by vacuum or damp cloth rather than by blasts of compressed air.

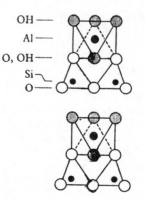

Figure 2.20 *Schematic representation of two layers of kaolinite structure (from Evans, 1966, by permission of Cambridge University Press).*

aluminium cation with two oxygen ions and four hydroxyl ions forms an octahedron, i.e. $AlO_2(OH)_4$. Thus, in each layer, a sheet of SiO_4^{4-} tetrahedra lies parallel to a sheet of $AlO_2(OH)_4$ octahedra, with the two sheets sharing common O^{2-} anions. Strong ionic and covalent bonding exists within each layer and each layer is electrically neutral. However, the uneven distribution of ionic charge across the five sub-layers has a polarizing effect, causing opposed changes to develop on the two faces of the layer. The weak van der Waals bonding between layers is thus explicable. This asymmetry of ionic structure also unbalances the bonding forces and encourages cleavage within the layer itself. In general terms, one can understand the softness, easy cleavage and mouldability (after moistening) of this mineral. The ionic radii of oxygen and hydroxyl ions are virtually identical. The much smaller Al^{3+} cations are shown located outside the SiO_4^{4-} tetrahedra. However, the radii ratio for aluminium and oxygen ions is very close to the geometrical boundary value of 0.414 and it is possible in other aluminosilicates for Al^{3+} cations to replace Si^{4+} cations at the centres of oxygen tetrahedra. In such structures, ions of different valency enter the structure in order to counterbalance the local decreases in positive charge. To summarize, the coordination of aluminium in layered aluminosilicates can be either four- or sixfold.

Many variations in layer structure are possible in silicates. Thus, talc (French chalk), $Mg_3Si_4O_{10}(OH)_2$, has similar physical characteristics to kaolinite and finds use as a solid lubricant. In talc, each layer consists of alternating Mg^{2+} and OH^- ions interspersed between the inwardly-pointing vertices of two sheets of SiO_4^{4-} tetrahedra. This tetrahedral-tetrahedral layering thus contrasts with the tetrahedral-octahedral layering of kaolinite crystals.

Finally, in our brief survey of silicates, we come to the framework structures in which the SiO_4^{4-} tetrahedra share all four corners and form an extended and regular three-dimensional network. Feldspars, which are major constituents in igneous rocks, are fairly compact but other framework silicates, such as the zeolites and ultramarine, have unusually 'open' structures with tunnels and/or polyhedral cavities. Natural and synthetic zeolites form a large and versatile family of compounds. As in other framework silicates, many of the central sites of the oxygen tetrahedra are occupied by Al^{3+} cations. The negatively charged framework of $(Si, Al)O_4$ tetrahedra is balanced by associated cations; being cross-braced in three dimensions, the structure is rigid and stable. The overall $(Al^{3+} + Si^{4+})$:O^{2-} ratio is always 1:2 for zeolites. In their formulae, (H_2O) appears as a separate term, indicating that these water molecules are loosely bound. In fact, they can be readily removed by heating without affecting the structure and can also be re-absorbed. Alternatively, dehydrated zeolites can be used to absorb gases, such as carbon dioxide (CO_2) and ammonia (NH_3). Zeolites are well-known for their ion-exchange capacity[1] but synthetic resins now compete in this application. Ion exchange can be accompanied by appreciable absorption so that the number of cations entering the zeolitic structure can actually exceed the number of cations being replaced.

Dehydrated zeolites have a large surface/mass ratio, like many other catalysts, and are used to promote reactions in the petrochemical industry. Zeolites can also serve as 'molecular sieves'. By controlling the size of the connecting tunnel system within the structure, it is possible to separate molecules of different size from a flowing gaseous mixture.

2.6 Inorganic glasses

2.6.1 Network structures in glasses

Having examined a selection of important crystalline structures, we now turn to the less-ordered glassy structures. Boric oxide (B_2O_3; m.p. 460°C) is one of the relatively limited number of oxides that can exist in either a crystalline or a glassy state. Figure 2.1, which was used earlier to illustrate the concept of ordering (Section 2.1), portrays in a schematic manner the two structural forms of boric oxide. In this figure, each planar triangular group (CN = 3) represents three oxygen anions arranged around a much smaller B^{3+} cation. Collectively, the triangles form a random network in three dimensions. Similar modelling can be applied to silica (m.p. 1725°C), the most important and common glass-forming oxide. In silica glass, SiO_4^{4-} tetrahedra form a three-dimensional network with oxygen 'bridges' joining vertices. Like boric oxide glass, the 'open' structure contains many 'holes' of irregular shape. The equivalent of metallic alloying is achieved by basing a glass upon a combination of two glass-formers, silica and boric oxide. The resulting network consists of triangular and tetrahedral anionic groups and, as might be anticipated, is less cohesive and rigid than a pure SiO_2 network. B_2O_3 therefore has a fluxing action. By acting as a network-former, it also has less effect upon thermal expansivity than conventional fluxes, such as Na_2O and K_2O, which break up the network. The expansion characteristics can thus be adjusted by control of the B_2O_3/Na_2O ratio.

Apart from chemical composition, the main variable controlling glass formation from oxides is the rate of cooling from the molten or fused state. Slow cooling provides ample time for complete ordering of atoms and groups of atoms. Rapid cooling restricts this physical process and therefore favours glass formation.[2] The

---

[1] In the *Permutite* water-softening system, calcium ions in 'hard' water exchange with sodium ions of a zeolite (e.g. thomsonite, $NaCa_2(Al_5Si_5O_{20})$). Spent zeolite is readily regenerated by contact with brine (NaCl) solution.

[2] The two states of aggregation may be likened to a stack of carefully arranged bricks (crystal) and a disordered heap of bricks (glass).

American Society for Testing and Materials (ASTM) defines glass as an inorganic product of fusion that has cooled to a rigid condition without crystallizing. The cooling rate can be influenced by a 'mass effect' with the chances of glass formation increasing as the size of particle or cross-section decreases. Accordingly, a more precise definition of a glass-former might also specify a minimum mass, say 20 mg, and free-cooling of the melt. As a consequence of their irregular and aperiodic network structures, glasses share certain distinctive characteristics. They are isotropic and have properties that change gradually with changing temperature. Bond strengths vary from region to region within the network so that the application of stress at an elevated temperature causes viscous deformation or flow. This remarkable ability to change shape without fracture is used to maximum advantage in the spinning, drawing, rolling, pressing and blowing operations of the glass industry (e.g. production of filaments, tubes, sheets, shapes and containers). Glasses do not cleave, because there are no crystallographic planes, and fracture to produce new surfaces that are smooth and shell-like (conchoidal). It is usually impossible to represent a glass by a stoichiometric formula. Being essentially metastable, the structure of a glass can change with the passage of time. Raising the temperature increases ionic mobility and hastens this process, being sometimes capable of inducing the nucleation and growth of crystalline regions within the glassy matrix. Controlled devitrification of special glasses produces the heat- and fracture-resistant materials known as glass-ceramics (Section 10.4.4). Finally, glasses lack a definite melting point. This feature is apparent when specific volume ($m^3\ kg^{-1}$), or a volume-related property, is plotted against temperature for the crystalline and glassy forms of a given substance (Figure 2.21). On cooling, the melt viscosity rapidly increases. Simultaneously, the specific volume decreases as a result of normal thermal contraction and contraction due to structural (configurational) rearrangement within the liquid. After supercooling below the crystalline melting point, a curved inflexion over a temperature range of roughly 50°C marks the decrease and eventual cessation of structural rearrangement. The final portion of the curve, of lesser slope, represents normal thermal contraction of the rigid glass structure. The fictive (imagined) temperature T_f shown in Figure 2.21 serves as an index of transition; however, it increases in value as the cooling rate is increased. Being disordered, a glass has a lower density than its corresponding crystalline form.

2.6.2 Classification of constituent oxides

After considering the relation of oxides to glass structure, Zachariasen categorized oxides as (1) network-formers, (2) intermediates and (3) network-modifiers. Oxides other than boric oxide and silica have the ability to form network structures. They are listed in Table 2.5.

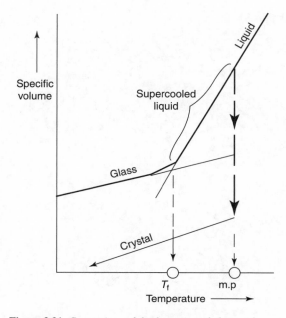

Figure 2.21 *Comparison of the formation of glass and crystals from a melt.*

Table 2.5 Classification of oxides in accordance with their ability to form glasses (after Tooley)

Network-formers	Intermediates	Network-modifiers
B_2O_3	Al_2O_3	MgO
SiO_2	Sb_2O_3	Li_2O
GeO_2	ZrO_2	BaO
P_2O_5	TiO_2	CaO
V_2O_5	PbO	Na_2O
As_2O_3	BeO	SrO
	ZnO	K_2O

This particular method of classification primarily concerns the glass-forming ability of an oxide; thus oxides classed as network-modifiers have little or no tendency to form network structures. Modifiers can have very important practical effects. For instance, the alkali-metal oxides of sodium and potassium are used to modify the glasses based on silica which account for 90% of commercial glass production. Sodium carbonate (Na_2CO_3) and calcium carbonate ($CaCO_3$) are added to the furnace charge of silica sand and cullet (recycled glass) and dissociate to provide the modifying oxides, releasing carbon dioxide. Eventually, after melting, fining and degassing operations in which the temperature can ultimately reach 1500–1600°C, the melt is cooled to the working temperature of 1000°C. Sodium ions become trapped in the network and reduce the number of 'bridges' between tetrahedra, as shown schematically in Figure 2.22a. These Na^+ cations influence 'hole' size and it has been proposed

(a)

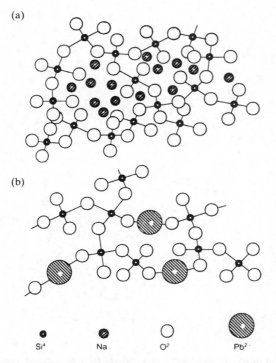

(b)

●	◍	○	◍
Si⁴	Na	O²	Pb²

Figure 2.22 *Schematic representation of action of modifiers in silica glass. (a) Na₂O breaking-up network; (b) PbO entering network.*

that they may cluster rather than distribute themselves randomly throughout the network. However, although acting as a flux, sodium oxide by itself renders the glass water-soluble. This problem is solved by adding a stabilising modifier, CaO, to the melt, a device known to the glassmakers of antiquity.[1] Ca^{2+} ions from dissociated calcium carbonate also enter the 'holes' of the network; however, for each nonbridging O^{2-} anion generated, there will be half as many Ca^{2+} ions as Na^+ ions (Figure 2.22a).

There are certain limits to the amounts of the various agents that can be added. As a general rule, the glass network becomes unstable and tends to crystallize if the addition of modifier or intermediate increases the numerical ratio of oxygen to silicon ions above a value of 2.5. Sometimes the tolerance of the network for an added oxide can be extremely high. For instance, up to 90% of the intermediate, lead oxide (PbO), can be added to silica glass. Pb^{2+} cations enter the network (Figure 2.22b). Glass formulations are discussed further in Sections 10.5 and 10.6.

---

[1]Extant 2000-year-old Roman vases are remarkable for their beauty and craftsmanship; the Portland vase, recently restored by the British Museum, London, and tentatively valued at £30 million, is a world-famous example.

2.7 Polymeric structures

2.7.1 Thermoplastics

Having examined the key role of silicon in crystalline silicates and glasses, we now turn to another tetravalent element, carbon, and examine its central contribution to the organic structures known as polymers, or, in a more general commercial sense, 'plastics'. These structures are based upon long-chain molecules and can be broadly classified in behavioural terms as thermoplastics, elastomers and thermosets. In order to illustrate some general principles of 'molecular engineering', we will first consider polyethylene (PE), a linear thermoplastic which can be readily shaped by a combination of heat and pressure. Its basic repeat unit of structure (mer) is derived from the ethene, or ethylene,[2] molecule C_2H_4 and has a relative mer mass M_{mon} of 28, i.e. $(12 \times 2) + (1 \times 4)$. This monomer has two free bonds and is said to be unsaturated and bifunctional; these mers can link up endwise to form a long-chain molecule $(C_2H_4)_n$, where n is the degree of polymerization or number of repeat units per chain. Thus the relative mass[3] of a chain molecule is $M = nM_{mon}$. The resultant chain has a strong spine of covalently-bonded carbon atoms that are arranged in a three-dimensional zigzag form because of their tetrahedral bonding. Polyethylene in bulk can be visualized as a tangled mass of very large numbers of individual chain molecules. Each molecule may contain thousands of mers, typically 10^3 to 10^5. The carbon atoms act rather like 'universal joints' and allow it to flex and twist.

The mass and shape of these linear molecules have a profound effect upon the physical, mechanical and chemical properties of the bulk polymer. As the length of molecules increases, the melting points, strength, viscosity and chemical insolubility also tend to increase. For the idealized and rare case of a

---

[2]The double bond of ethene

$$
\begin{array}{ccc}
H & & H \\
| & & | \\
C & = & C \\
| & & | \\
H & & H \\
\end{array}
$$

is essential for polymerization; it is opened up by heat, light, pressure and/or catalysts to form a reactive bifunctional monomer

$$
\begin{array}{ccc}
& H & H \\
& | & | \\
- & C - C & - \\
& | & | \\
& H & H \\
\end{array}
$$

[3]It is common practice to use relative molecular masses or 'molecular weights'. Strictly speaking, one should use molar masses; that is, amounts of substance containing as many elementary entities (molecules, mers), as there are atoms in 0.012 kg of the carbon isotope $_6C^{12}$.

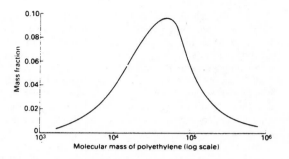

Figure 2.23 *The molecular mass distribution of a polyethylene, determined using GPC (from Mills, 1986; by permission of Edward Arnold).*

monodisperse polymer, all the chain molecules are of equal length and M is constant. However, in practice, polymers are usually polydisperse with a statistical distribution of chain lengths (Figure 2.23). The average molecular mass $\overline{M}$ and the 'spread' in values between short and long chains are important quantitative indicators of behaviour during processing.

A polymer sample may be regarded as a collection of fractions, or sub-ranges, of molecular size, with each fraction having a certain mid-value of molecular mass. Let us suppose that the ith fraction contains N_i molecules and that the mid-value of the fraction is M_i. Hence the total number of molecules for all fractions of the sample is $\sum N_i$. In calculating a single numerical value, the average molecular mass $\overline{M}$, which will characterize the distribution of chain sizes, it is necessary to distinguish between number-average fractions and mass-average fractions of molecules in the sample. Thus, in calculating the number-average molecular mass $\overline{M}_N$ of the sample, let the number fraction be α_i. Then $\alpha_i = N_i / \sum N_i$ and:

$$\overline{M}_N = \sum \alpha_i M_i = \sum \left(N_i \Big/ \sum N_i \right) M_i$$
$$= \sum N_i M_i \Big/ \sum N_i$$

$\overline{M}_N$ is very sensitive to the presence of low-mass molecules; accordingly, it is likely to correlate with any property that is sensitive to the presence of short-length molecules (e.g. tensile strength).

In similar fashion, the mass-average molecular mass $\overline{M}_W$ can be calculated from mass fractions ϕ_i. Since $\phi_i = m_i / \sum m_i$, it follows that:

$$\overline{M}_W = \sum \phi_i M_i = \sum \left(m_i \Big/ \sum m_i \right) M_i$$
$$= \sum m_i M_i \Big/ \sum m_i$$

Using the Avogadro constant N_A, we can relate mass and number fractions as follows:

$$m_i / M_i = N_i / N_A \text{ hence } m_i = N_i M_i / N_A$$

Substituting for m_i, the previous expression for mass-average molecular mass becomes:

$$\overline{M}_W = \sum N_i M_i^2 \Big/ \sum N_i M_i$$

The full molecular mass distribution, showing its 'spread', any skewness, as well as the two average values $\overline{M}_W$ and $\overline{M}_N$, can be determined by gel permeation chromatography (GPC). This indirect method is calibrated with data obtained from direct physical measurements on solutions of polymers (e.g. osmometry, light scattering, etc.). For the routine control of production processes, faster and less precise methods, such as melt flow index (MFI) measurement, are used to gauge the average molecular mass.

$\overline{M}_W$ is particularly sensitive to the long-chain molecules and therefore likely to relate to properties which are strongly influenced by their presence (e.g. viscosity). $\overline{M}_W$ always exceeds $\overline{M}_N$. (In a hypothetical monodisperse system, $\overline{M}_W = \overline{M}_N$.) This inequality occurs because a given mass of polymer at one end of the distribution contains many short molecules whereas, at the other end, the same mass need only contain a few molecules. $\overline{M}_W$ is generally more informative than $\overline{M}_N$ so far as bulk properties are concerned. The ratio $\overline{M}_W / \overline{M}_N$ is known as the polydispersivity index; an increase in its value indicates an increase in the 'spread', or dispersion, of the molecular mass distribution (MMD) in a polydisperse. In a relatively simple polymer, this ratio can be as low as 1.5 or 2 but, as a result of complex polymerization processes, it can rise to 50, indicating a very broad distribution of molecular size.

The development of engineering polymers usually aims at maximizing molecular mass. For a particular polymer, there is a threshold value for the average degree of polymerization ($\overline{n}$) beyond which properties such as strength and toughness develop in a potentially useful manner. (Either $\overline{M}_W$ or $\overline{M}_N$ can be used to calculate $\overline{n}$.) It is apparent that very short molecules of low mass can slip past each other fairly easily, to the detriment of mechanical strength and thermal stability. On the other hand, entanglement of chains becomes more prevalent as chains lengthen. However, improvement in properties eventually becomes marginal and the inevitable increase in viscosity can make processing very difficult. Thus, for many practical polymers, $\overline{n}$ values lie in the range 200–2000, roughly corresponding to molecular masses of 20 000 to 200 000.

In certain polymer systems it is possible to adjust the conditions of polymerization (e.g. pressure, temperature, catalyst type) and encourage side reactions at sites along the spine of each discrete chain molecule. The resultant branches can be short and/or long, even multiple. Polyethylene provides an important commercial example of this versatility. The original low-density form (LDPE) has a high degree of branching, with about 15–30 short and long branches per thousand carbon atoms, and a density less then 940 kg m^{-3}.

The use of different catalysts permitted lower polymerization pressures and led to the development of a high-density form (HDPE) with just a few short branches and a density greater than 940 kg m^{-3}. Being more linear and closely-packed than LDPE, HDPE is stronger, more rigid and has a melting point (135°C) which is 25°C higher.

Weak forces exist between adjacent chain molecules in polyethylene. Heating, followed by the application of pressure, causes the molecules to straighten and slide past each other easily in a viscous manner. Molecular mobility is the outstanding feature of thermoplastics and they are well suited to melt-extrusion and injection-moulding. These processes tend to align the chain molecules parallel to the direction of shear, producing a pronounced preferred orientation (anisotropy) in the final product. If the polymer is branched, rather than simply linear,

Table 2.6 Repeat units of typical thermoplastics

mer = ☐, Methyl radical = CH$_3$, Benzene ring ⬡ = C$_6$H$_6$,   Acetate radical (A$_c$) = $-$ O $-$ C $-$ CH$_3$
‖
O

branches on adjacent chains will hook on to each other and reduce their relative mobility. This effect underlines the fundamental importance of molecular shape.

In the important vinyl family of thermoplastics, one of the four hydrogen atoms in the C_2H_4 monomer of polyethylene is replaced by either a single atom (chlorine) or a group of atoms, such as the methyl radical CH_3, the aromatic benzene ring C_6H_6 and the acetate radical $C_2H_3O_2$. These four polymers, polyvinyl chloride (PVC), polypropylene (PP), polystyrene (PS) and polyvinyl acetate (PVAc), are illustrated in Table 2.6.

Introduction of a different atom or group alongside the spine of the molecule makes certain alternative symmetries possible. For instance, when all the chloride atoms of the PVC molecule lie along the same side of each chain, the polymer is said to be isotactic. In the syndiotactic form, chlorine atoms are disposed symmetrically around and along the spine of the molecule. A fully-randomized arrangement of chlorine atoms is known as the atactic form. Like side-branching, tacticity can greatly influence molecular mobility. During addition polymerization, it is possible for two or more polymers to compete simultaneously during the joining of mers and thereby form a copolymer with its own unique properties. Thus, a vinyl copolymer is produced by combining mers of vinyl chloride and vinyl acetate in a random sequence.[1] In some copolymers, each type of constituent mer may form alternate blocks of considerable length within the copolymeric chains. Branching can, of course, occur in copolymers as well as in 'straight' polymers.

2.7.2 Elastomers

The development of a relatively small number of crosslinking chains between linear molecules can produce an elastomeric material which, according to an ASTM definition, can be stretched repeatedly at room temperature to at least twice its original length and which will, upon sudden release of the stress, return forcibly to its approximate original length. As shown in Figure 2.24, the constituent molecules are in a coiled and kinked condition when unstressed; during elastic strain, they rapidly uncoil. Segments of the structure are locally mobile but the crosslinks tend to prevent any gross relative movement of adjoining molecules, i.e. viscous deformation. However, under certain conditions it is possible for elastomers, like most polymers, to behave in a viscoelastic manner when stressed and to exhibit both viscous (time-dependent) and elastic (instantaneous) strain characteristics. These two effects can be broadly attributed, respectively, to the

[1]In the late 1940s this copolymer was chosen to provide the superior surface texture and durability required for the first long-play microgroove gramophone records. This $33\frac{1}{3}$ r.p.m. system, which quickly superseded 78 r.p.m. shellac records, has been replaced by compact discs made from polycarbonate thermoplastics of very high purity.

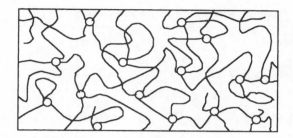

Figure 2.24 *Unstrained elastomeric structure showing entanglement, branching points, crosslinks, loops and free ends (after Young, 1991).*

relative movement and the uncoiling and/or unravelling of molecular segments.

Elastomers include natural polymers, such as polyisoprene and polybutadiene in natural rubbers, and synthetic polymers, such as polychloroprene (*Neoprene*), styrene-butadiene rubber (SBR) and silicone rubbers. The structural repeat units of some important elastomers are shown in Table 2.7. In the original vulcanization process, which was discovered by C. Goodyear in 1839 after much experimentation, isoprene was heated with a small amount of sulphur to a temperature of 140°C, causing primary bonds or crosslinks to form between adjacent chain molecules. Individual crosslinks take the form $C-(S)_n-C$, where n is equal to or greater than unity. Monosulphide links ($n = 1$) are preferred because they are less likely to break than longer links. They are also less likely to allow slow deformation under stress (creep). Examples of the potentially-reactive double bonds that open up and act as a branching points for crosslinking are shown in Table 2.7. Nowadays, the term vulcanization is applied to any crosslinking or curing process which improves elasticity and strength; it does not necessarily involve the use of sulphur. Hard rubber (*Ebonite*) contains 30–50% sulphur and is accordingly heavily crosslinked and no longer elastomeric. Its long-established use for electrical storage battery cases is now being challenged by polypropylene (PP).

The majority of polymers exhibit a structural change known as the glass transition point, T_g; this temperature value is specific to each polymer and is of great practical and scientific significance. (Its implications will be discussed more fully in Chapter 11.) In general terms, it marks a transition from hard, stiff and brittle behaviour (comparable to that of an inorganic glass) to soft, rubbery behaviour as the temperature increases. The previously-given ASTM definition described mechanical behaviour at room temperature; it follows that the elastomeric condition refers to temperatures well above T_g. Table 2.7 shows that typical values for most elastomers lie in the range −50° to −80°C. When an elastomeric structure is heated through T_g, the segments between the linkage or branching points are able to vibrate more vigorously.

Table 2.7 Repeat units of typical elastomers

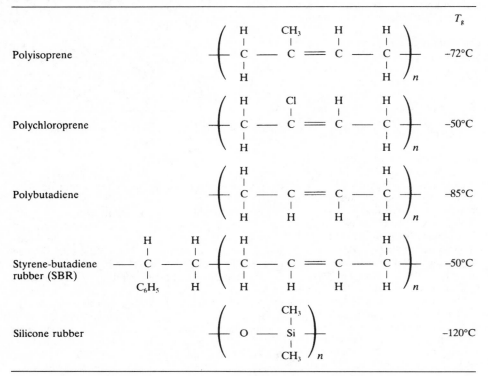

A simple linear equation expresses the temperature-dependence of an elastomer's response to shear stress:

$$\mu = NkT$$

where μ is the shear modulus, N is the number of segments per unit volume of structure (between successive points of crosslinking), k is the Boltzmann constant and T is absolute temperature. Segments are typically about 100 repeat units long. Clearly, for a given polymer, the stiffness under shear conditions is directly proportional to the absolute temperature. As temperature increases, deflection under load becomes less. This rather unusual feature has raised engineering problems in suspension systems.

At higher temperatures, well above T_g, the polymeric structure is likely to deform slowly under applied stress (creep) and ultimately to break down into smaller chemical entities, or degrade. As indicated in Figure 2.24, unstressed elastomers are disordered and non-crystalline. Interestingly, stressing to produce a high elastic strain, say 200% or more, will induce a significant amount of crystallinity. Stress aligns the chains and produces regions in which repeat units form ordered patterns. (This effect is readily demonstrated by projecting a monochromatic beam of X-rays through relaxed and stretched membranes of an elastomer and comparing the diffraction patterns formed upon photographic film.)

2.7.3 Thermosets

In the third and remaining category of polymers, known generally as thermosets or network polymers, the degree of crosslinking is highly developed. As a result, these structures contain many branching points. They are rigid and strong, being infinitely braced in three dimensions by numerous chain segments of relatively short length. Unlike thermoplastics, molecular mobility is virtually absent and T_g is accordingly high, usually being above 50°C. Thermosets are therefore regarded generally as being hard and brittle ('glassy') materials. Examples of thermoset resins in common use include phenol-formaldehyde (P-F resin; *Bakelite*), epoxy resins (structural adhesives; *Araldite*), urea formaldehyde (U-F resin; *Beetle*) and polyester resins. A resin is a partially-polymerized substance which requires further treatment.

The utilization of thermosets typically involves two stages of chemical reaction. In the first stage, a liquid or solid prepolymer form or precursor is produced which is physically suitable for casting or moulding. Resins are well-known examples of this intermediate state. Their structures consist mostly of linear molecules and they are potentially reactive, having a specifiable shelf-life. In the second stage, extensive crosslinking is promoted by heating, pressure application or addition of a hardening agent, depending upon the type of polymer. This stage is commonly referred to as curing. The resultant random network possesses the desired stiffness and strength. When heated, this structure does not exhibit viscoelastic flow and, being both chemically and physically stable, remains unaltered and hard until the decomposition temperature is reached. The formation of a thermoset may thus be regarded as an irreversible process. With phenolic resins, final crosslinking is induced by heating, as implied by the term thermoset. The latter term is also applied, in a looser sense, to polymers in which final crosslinking occurs as the result of adding a hardener; for example, in epoxy resin adhesives and in the polyester resin-based matrix of glass-reinforced polymer (GRP; *Fibreglass*). In these substances, an increase in the ratio of hardener to resin tends to increase the T_g value and the modulus.

Many thermoset structures develop by condensation polymerization. This process is quite different in chemical character to the process of addition polymerization by which linear molecules grow in an endwise manner in thermoplastics. Although it is one of the oldest synthetic polymers, having appeared as *Bakelite* in the early 1900s, phenolformaldehyde retains its industrial importance. It is widely used for injection-mouldings in the automotive and electrical industries, for surface coatings and as a binder for moulding sands in metal foundries. It is therefore appropriate to use phenolformaldehyde as an illustrative example of the condensation reaction, focusing upon the novolac resin route. In the first stage (Figure 2.25a), phenol and formaldehyde groups react to form an addition compound (methylol derivative). Figure 2.25b shows these derivatives joining with phenol groups in a condensation reaction. Methylene bridges (CH_2) begin to form between adjacent phenol groups and molecules of water are released. The two reactions shown diagrammatically in Figures 2.25a and 2.25b produce a

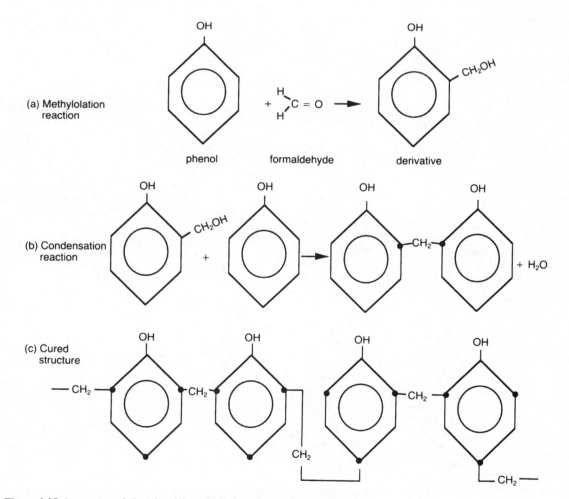

Figure 2.25 *Interaction of phenol and formaldehyde to form a thermoset structure.*

relatively unreactive novolac resin. (Control of the initial phenol/formaldehyde ratio above unity ensures that a deficiency of formaldehyde will inhibit crosslinking during this first stage.) After drying, grinding and the addition of fillers and colourants, the partly-condensed resin is treated with a catalysed curing agent which acts as a source of formaldehyde. Network formation then proceeds during hot-moulding at a temperature of 200–300°C. Each phenol group is said to be trifunctional because it can contribute three links to the three-dimensional random network (Figure 2.25c).

Another type of phenolic resin, the resole, is produced by using an initial phenol/formaldehyde ratio of less than unity and a different catalyst. Because of the excess of formaldehyde groups, it is then possible to form the network structure by heating without the addition of a curing agent.

2.7.4 Crystallinity in polymers

So far, we have regarded thermoplastic polymers as essentially disordered (non-crystalline) structures in which chain molecules of various lengths form a tangled mass. This image is quite appropriate for some polymers e.g. polystyrene and polymethyl methacrylate (*Perspex*). However, as indicated in the case of stressed elastomers (Section 2.7.2), it is possible for chain molecules to form regions in which repeat units are aligned in close-packed, ordered arrays. Crystalline regions in polymers are generally lamellar in form and often small, with their smallest dimension in the order of 10–20 nm. Inevitably, because of the complexity of the molecules, crystallized regions are associated with amorphous regions and defects. However, the degree of crystallinity attainable can approach 80–85% by volume of the structure. Thus, in polyethylene (PE), a simple 'linear' thermoplastic that has been the subject of much investigation, crystalline regions nucleate and grow extremely rapidly during polymerization, their formation being virtually unpreventable. Values of 50% and 80%, respectively, are quoted for the crystallinity of its low- and high-density forms, LDPE and HDPE. A medium-density form provides an appropriate balance of strength and flexibility and has been used for the yellow distribution pipes which convey natural gas in the UK. Crystalline regions are close-packed and act as barriers to the diffusion of gases and small molecules. This impermeability favours the use of LDPE for food wrappings. HDPE, being even more crystalline, has a lower permeability to gases and vapours than LDPE.

A typical chain molecule contains hundreds, often thousands, of single primary bonds along its zigzag-shaped backbone. Rotation about these bonds enables the molecule to bend, fold, coil and kink. Under certain circumstances, it is possible for it to adopt an overall shape, or conformation, and then interlock in a close-packed manner alongside other molecules (or even itself) to form a three-dimensional crystalline region. Figure 2.26 shows a small portion of a crystalline region in polyethylene. Segments of five chain molecules, composed of CH_2 groups, are packed together and said to be in extended conformation. (For convenience, the carbon and hydrogen atoms have been 'shrunk'.) The orthorhombic unit cell,[1] with its unequal axes, has been superimposed.

The criteria which determine the extent to which a polymer can crystallize are (1) the conditions under which the polymer forms or is processed, and (2) the structural character of its molecules. With regard to the conditions, crystallization is favoured by slow cooling rates. Industrially, rapid cooling usually prevails and there is relatively little time available for molecular packing and ordering. Application of stress can be used to induce crystallization either during or after polymerization. As the degree of crystallization increases, the polymer becomes denser; it also becomes more resistant to thermal degradation. This important feature is reflected in the crystalline melting point T_m.

[1]PE can also exist in a less stable monoclinic form. Cubic forms, with their high symmetry, do not appear in polymeric systems; consequently, crystalline polymers frequently exhibit pronounced anisotropy.

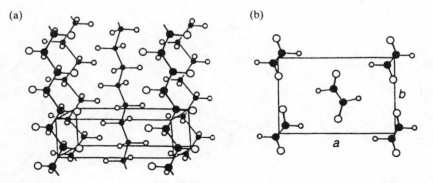

Figure 2.26 *Crystal structure of orthorhombic polyethylene. (a) General view of unit cell. (b) Projection of unit cell parallel to the chain direction, c. ● carbon atoms, ○ hydrogen atoms. a = 0.741 nm, b = 0.494 nm, c = 0.255 nm (from Young, 1991).*

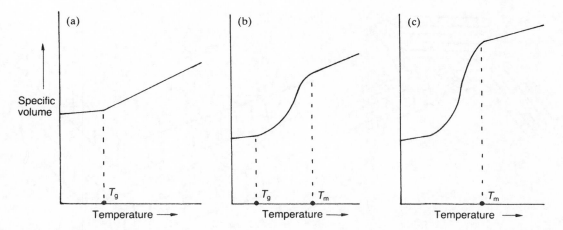

Figure 2.27 *Specific volume versus temperature plots for (a) 100% amorphous polymer, (b) partially-crystalline polymer, (c) 100% crystalline polymer.*

At this temperature, which is higher in value than T_g, the crystalline components of the structure break down completely on heating (Figure 2.27). For a given polymer, T_m increases with the degree of crystallinity; for example, its values for LDPE and HDPE are 110°C and 135°C, respectively. Holding a polymer at a temperature between T_g and T_m (annealing) is sometimes used as a method for increasing existing crystallinity.

Turning now to the matter of structural complexity, crystallization is less likely as molecules become longer and more complex. Thus the presence of side-branching is a steric hindrance to crystallization, particularly if the molecules are atactic in character. Iso- and syndio-tacticity are more readily accommodated; for example, isotactic polypropylene (iPP) is a material with useful engineering strength whereas atactic PP is a sticky gum. Scrutiny of the various repeat units shown earlier in Table 2.6 shows that they are usually asymmetrical and may be said to possess a 'head' and a 'tail'. Various combinations, or configurations, can result from addition polymerization, such as 'head to tail' (**XYXYXYXY**...) or 'head to head' (**XYXXYYXX**...), etc. If the molecules have a similar and consistent configuration, crystallization is favoured. For instance, the previously-mentioned polymer, isotactic PP, has a 'head to tail' configuration throughout and can develop a high degree of crystallinity. Matching configurations favour crystallinity. Crystallinity is therefore more likely in copolymers with regular block patterns of constituents than in random copolymers.

Summarizing, regular conformations and/or regular configurations favour crystallization. Each of these two characteristics is manipulated in a different way. Conformations are changed by physical means (annealing, application of stress): changes in configuration require the breaking of bonds and are achieved by chemical means.

The fine structure of crystalline regions in commercial polymers and their relation to associated amorphous regions have been the subject of much research. An important advance was made in the 1950s when single crystals of polyethylene were produced for the first time. A typical method of preparation is to dissolve <0.01% PE in xylene at a temperature of 135–138°C and then cool slowly to 70–80°C. The small PE crystals that precipitate are several microns across and only 10–20 nm thick. The thickness of these platey crystals (lamellae) is temperature-dependent. Diffraction studies by transmission electron microscope showed that, surprisingly, the axes of the chain molecules were approximately perpendicular to the two large faces of each lamella. In view of the smallness of the crystal thickness relative to the average length of molecules, it was deduced that multiple chain-folding had occurred during crystallization from the mother liquor. In other words, a molecule could exhibit an extended and/or folded conformation. The chain-folding model has been disputed but is now generally accepted. The exact nature of the fold surface has also been the subject of much debate; typical models for folding are shown in Figure 2.28. The measured density of these single crystals is less than the theoretical value; this feature indicates that irregular arrangements exist at fold surfaces and that the crystal itself contains defects. Most of the folds or loops are tight and each requires only two or three repeat units of molecular structure. 'Loose' loops of larger radius and chain ends (cilia) project from the surfaces of lamellae. These fundamental studies on single crystals had a far-reaching effect upon polymer physics, comparable in importance to that of single metal crystals upon metallurgical science.

In contrast to the relatively uncongested conditions that exist at the surface of an isolated crystal growing from a dilute solution, entanglement of chain molecules is more likely when a polymer

(a)   (b)

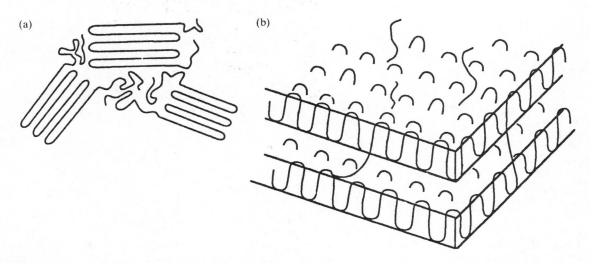

Figure 2.28 *Folded chain model for crystallinity in polymers shown in (a) two dimensions and (b) three dimensions (after Askeland, 1990, p. 534; by permission of Chapman and Hall, UK and PWS Publishers, USA).*

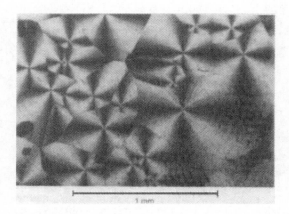

Figure 2.29 *Polarized light micrograph of two-dimensional spherulites grown in a thin film of polyethylene oxide (from Mills, 1986; by permission of Edward Arnold).*

crystallizes from the molten state. Consequently, melt-grown crystallites are more complex in physical character. Microscopical examination of thin sections of certain crystallizable polymers (PE, PS or *nylon*) can reveal a visually-striking spherulitic state of crystalline aggregation. In Figure 2.29, three-dimensional spherulites have grown in a radial manner from a number of nucleating points scattered throughout the melt. Nucleation can occur if a few molecular segments chance to order locally at a point (homogeneous nucleation). However, it is more likely that nucleation is heterogeneous, being initiated by the presence of particles of foreign matter or deliberately-added nucleating agents. Radial growth continues until spherulites impinge upon each other. Spherulites are much larger than the isolated single crystals previously

described and range in diameter from microns to millimetres, depending upon conditions of growth. Thus crystallization at a temperature just below T_m will proceed from relatively few nucleating points and will ultimately produce a coarse spherulitic structure. However, this prolonged 'annealing', or very slow cooling from the molten state, can produce cracks between the spherulites (over-crystallization).

Internally, spherulites consist of lamellae. During crystallization, the lamellae grow radially from the nucleus. As with solution-grown single crystals, these lamellae develop by chain-folding and are about 10 nm thick. Space-filling lamellar branches also form. Frequently, a chain molecule extends within one lamella and then leaves to enter another. The resultant inter-lamellar ties or links have an important role during deformation, as will be discussed later. Inevitably, the outward growth of lamellae traps amorphous material. Spherulites are about 70–80% crystalline if the constituent molecules are simple. The layered mixture of strong lamellae and weaker amorphous material is reminiscent of pearlite in steel and, as such, is sometimes regarded as a self-assembled (*in situ*) composite.

The distinctive patterns seen when spherulitic aggregates are examined between crossed polars in a light microscope (Figure 2.29) provide evidence that the lamellae radiating from the nucleus twist in synchronism. For orthorhombic PE, the *c*-axes of lamellae lie parallel to the length of the extended chain molecules and are tangential to the spherulite (Figure 2.30). The *a*- or *b*-axes are radial in direction. Lamellae twist as they grow and the *c*-axes remain normal to the growth direction. The refractive index gradually changes for each lamella, causing incident plane-polarized light to become elliptically polarized in four quadrants of each spherulite and to form the characteristic 'Maltese cross' figure. The grain boundary structures of

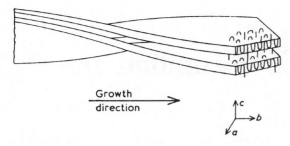

Growth direction

Figure 2.30 *Schematic representation of a possible model for twisted lamellae in spherulitic polyethylene showing chain-folds and intercrystalline links (from Young, 1991).*

polycrystalline metals and alloys are reminiscent of spherulitic structures in polymers. Indeed, control of spherulite size in partially-crystallized polymers and of grain (crystal) size in fully-crystallized metals and alloys are well-known means of manipulating strength and deformability.

The mechanism of molecular movement in a polymeric melt has presented a puzzling scientific problem. In particular, it was not known exactly how a molecule is able to move among entangled molecules and to participate in the progressive chain-folding action that is the outstanding feature of crystallization. Clearly, it is completely different from atomic and ionic diffusion in metals and ceramics. This long-standing problem

was convincingly resolved by de Gennes,[1] who pioneered the idea of reptation, a powerful concept that serves to explain various viscous and elastic effects in polymers. He proposed that a long-chain molecule, acting as an individual, is able to creep lengthwise in snake-like movements through the entangled mass of molecules. It moves within a constraining 'reptation tube' (Figure 2.31) which occupies free space between molecules; the diameter of this convoluted 'tube' is the minimum distance between two entangling molecules. The reptant motion enables a molecule to shift its centre of mass along a 'tube' and to progress through a tangled polymeric structure. Reptation is more difficult for the longer molecules. At the surface of a growing crystalline lamella, a molecule can be 'reeled in from its tube' and become part of the chain-folding process.

Further reading

Barrett, C. S. and Massalski, T. B. (1966). *Structure of Metals*, 3rd edn. McGraw-Hill, New York.

Brydson, J. A. (1989). *Plastics Materials*, 5th edn. Butterworths, London.

Evans, R. C. (1966) *An Introduction to Crystal Chemistry*, 2nd edn. Cambridge University Press, Cambridge.

Huheey, J. E. (1978). *Inorganic Chemistry: Principles of Structure and Reactivity*, 2nd edn. Harper and Row, New York.

Hume-Rothery, W., Smallman, R. E. and Haworth, C. W. (1988). *The Structure of Metals and Alloys*, revised 5th edn. Institute of Metals, London.

Kelly, A. and Groves, G. W. (1973). *Crystallography and Crystal Defects*. Longmans, Harlow.

Kingery, W. D., Bowen, H. K. and Uhlmann, D. R. (1976). *Introduction to Ceramics*, 2nd edn. John Wiley and Sons, Chichester.

Mills, N. J. (1986). *Plastics: Microstructure, Properties and Applications*. Edward Arnold, London.

Morton, M. (ed.) (1987). *Rubber Technology*, 3rd edn, Van Nostrand Reinhold, New York.

Young, R. J. and Lovell P. A. (1991). *Introduction to Polymers*. 2nd edn, Chapman and Hall, London.

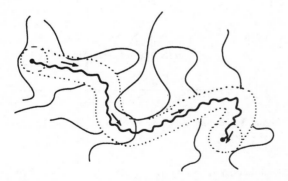

Figure 2.31 *Movement of a polymer molecule by reptation.*

[1] Pierre-Gilles de Gennes, physicist, was awarded the Nobel Prize for Physics (1991) for his theoretical work on liquid crystals and macromolecular motion in polymers; de Gennes acknowledged the stimulating influence of the ideas of Professor S. F. Edwards, University of Cambridge.

Chapter 3

Structural phases: their formation and transitions

3.1 Crystallization from the melt

3.1.1 Freezing of a pure metal

At some stage of production the majority of metals and alloys are melted and then allowed to solidify as a casting. The latter may be an intermediate product, such as a large steel ingot suitable for hotworking, or a complex final shape, such as an engine cylinder block of cast iron or a single-crystal gas-turbine blade of superalloy. Solidification conditions determine the structure, homogeneity and soundness of cast products and the governing scientific principles find application over a wide range of fields. For instance, knowledge of the solidification process derived from the study of conventional metal casting is directly relevant to many fusionwelding processes, which may be regarded as 'casting in miniature', and to the fusion-casting of oxide refractories. The liquid/solid transition is obviously of great scientific and technological importance.

First, in order to illustrate some basic principles, we will consider the freezing behaviour of a melt of like metal atoms. The thermal history of a slowly cooling metal is depicted in Figure 3.1; the plateau on the curve indicates the melting point (m.p.), which is pressure-dependent and specific to the metal. Its value relates to the bond strength of the metal. Thus, the drive to develop strong alloys for service at high temperatures has stimulated research into new and improved ways of casting high-m.p. alloys based upon iron, nickel or cobalt.

The transition from a highly-disordered liquid to an ordered solid is accompanied by a lowering in the energy state of the metal and the release of thermal energy (latent heat of solidification), forming the arrest on the cooling curve shown in Figure 3.1. This ordering has a marked and immediate effect upon other structure-sensitive properties of the metal; for instance, the volume typically decreases by 1–6%, the electrical

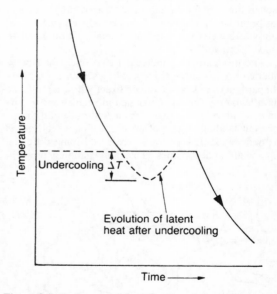

Figure 3.1 *Cooling curve for a pure metal showing possible undercooling.*

conductivity rises and the diffusivity, or ability of the atoms to migrate, falls.

Solidification is a classic example of a nucleation and growth process. In the general case of freezing within the bulk of pure molten metal, minute crystalline nuclei form independently at random points. After this homogeneous form of nucleation, continued removal of thermal energy from the system causes these small crystalline regions to grow independently at the expense of the surrounding melt. Throughout the freezing process, there is a tendency for bombardment by melt atoms to destroy embryonic crystals; only nuclei which exceed a critical size are able to survive. Rapid cooling of a pure molten metal reduces the time available for nuclei formation and delays the onset of freezing by a temperature interval of ΔT. This thermal undercooling (or supercooling), which is depicted in Figure 3.1, varies in extent, depending upon the metal and conditions, but can be as much as $0.1–0.3 \, T_m$, where T_m is the absolute melting point. However, commercial melts usually contain suspended insoluble particles of foreign matter (e.g. from the refractory crucible or hearth) which act as seeding nuclei for so-called heterogeneous nucleation. Undercooling is much less likely under these conditions; in fact, very pronounced undercooling is only obtainable when the melt is very pure and extremely small in volume. Homogeneous nucleation is not encountered in normal foundry practice.

The growing crystals steadily consume the melt and eventually impinge upon each other to form a structure of equiaxed (equal-sized) grains (Figures 3.2 and 3.3). Heterogeneous nucleation, by providing a larger population of nuclei, produces a smaller final grain size than homogeneous nucleation. The resultant grain (crystal) boundaries are several atomic diameters wide. The angle of misorientation between adjacent grains is usually greater than $10–15°$. Because of this misfit, such high-angle grain boundaries have a higher energy content than the bulk grains, and, on reheating, will tend to melt first. (During a grain-contrast etch of diamond-polished polycrystalline metal, the etchant attacks grain boundaries preferentially by an electrochemical process, producing a broad 'canyon' which scatters vertically incident light during normal microscopical examination. The boundary then appears as a black line.)

During the freezing of many metals (and alloys), nucleated crystals grow preferentially in certain directions, causing each growing crystal to assume a distinctive, non-faceted[1] tree-like form, known as a dendrite (Figure 3.2). In cubic crystals, the preferred axes of growth are $\langle 1\,0\,0 \rangle$ directions. As each dendritic spike grows, latent heat is transferred into the surrounding liquid, preventing the formation of other spikes in its immediate vicinity. The spacing of primary dendrites and of dendritic arms therefore tends to be regular. Ultimately, as the various crystals impinge upon each other, it is necessary for the interstices of the dendrites to be well-fed with melt if interdendritic shrinkage cavities are to be prevented from forming. Convection currents within the cooling melt are liable to disturb the delicate dendritic branches and produce slight angular misalignments in the final solidified structure (e.g. $5–10°$). These low-angle boundaries form a lineage (macromosaic) structure within the final grain, each surface of misfit being equivalent to an array of edge dislocations (Chapter 4). Convection currents can also

[1]Many metals and a few organic materials grow with non-faceted dendritic morphology, e.g. transparent succinonitrile-6% camphor has proved a valuable means of simulating dendrite growth on a hot-stage optical microscope. Most non-metals grow with a faceted morphology.

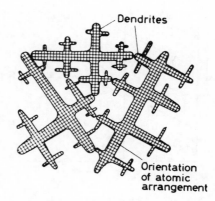

Figure 3.2 *Schematic diagram of three dendrites interlocking.*

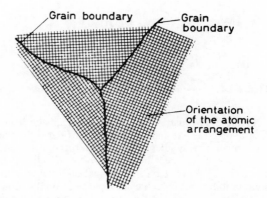

Figure 3.3 *Formation of grains from dendrites of Figure 3.2.*

provide thermal pulses which cause dendritic branch tips to melt off and enter the main body of the melt where they act as 'kindred nuclei'. Gentle stirring of the melt encourages this process, which is known as dendrite multiplication, and can be used to produce a fine-grained and equiaxed structure (e.g. electromagnetic stirring of molten steel). Dendrite multiplication is now recognised as an important source of crystals in castings and ingots.

3.1.2 Plane-front and dendritic solidification at a cooled surface

The previous section describes random, multidirectional crystallization within a cooling volume of pure molten metal. In practice, freezing often commences at the plane surface of a mould under more complex and constrained conditions, with crystals growing counter to the general direction of heat flow. The morphology of the interface, as well as the final grain structure of the casting, are then decided by thermal conditions at the solid/liquid interface.

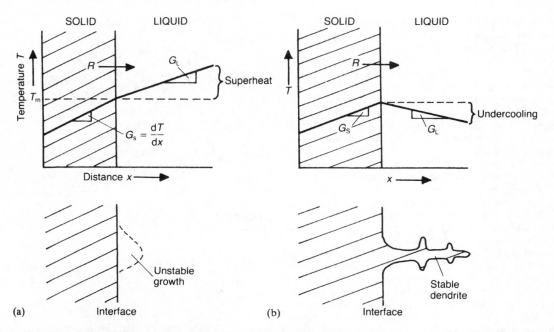

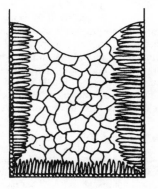

Figure 3.4 *Plane-front solidification (a) and dendritic solidification (b) of a pure metal, as determined by thermal conditions.*

Figure 3.4a illustrates the case where all the latent heat evolved at the interface flows into the solid and the temperature gradients in solid and liquid, G_S and G_L, are positive. The solidification front, which moves at a velocity R, is stable, isothermal and planar. Any solid protuberance which chances to form on this front will project into increasingly hotter, superheated liquid and will therefore quickly dissolve and be absorbed by the advancing front. Planar-front solidification is characterized by a high G_L/R ratio (e.g. slow cooling). If the solid is polycrystalline, emerging grain boundaries will form grooves in the stable planar front.

In the alternative scenario (Figure 3.4b), for which G_L/R has relatively low values, latent heat flows into both solid and liquid and G_L becomes negative. A planar interface becomes unstable. Dendritic protuberances (spikes) grow rapidly into the undercooled liquid, which quickly absorbs their evolved latent heat. Thermal undercooling is thus an essential prerequisite for dendritic growth; this form of growth becomes more and more likely as the degree of thermal undercooling increases. Melts almost invariably undercool slightly before solidification so that dendritic morphologies are very common. (The ability of dilute alloy melts to produce a cellular morphology as a result of constitutional undercooling will be described in Section 3.2.4.3.)

3.1.3 Forms of cast structure

Because of the interplay of a variety of physical and chemical factors during freezing, the as-cast grain structure is usually not as uniform and straightforward

as those discussed in the previous two sections. When solidification commences at the flat surface of a metallic ingot mould there is usually an extreme undercooling or chilling action which leads to the heterogeneous nucleation of a thin layer of small, randomly-oriented chill crystals (Figure 3.5). The size of these equiaxed crystals is strongly influenced by the texture of the mould surface. As the thickness of the zone of chill crystals increases, the temperature gradient G_L becomes less steep and the rate of cooling decreases. Crystal growth rather than the nucleation of new crystals now predominates and, in many metals and alloys, certain favourably-oriented crystals at the solid/liquid interface begin to grow into the melt. As in the case of the previously-described

Figure 3.5 *Chill-cast ingot structure.*

dendrites, the rapid growth directions are $\langle 1\,0\,0 \rangle$ for fcc and bcc crystals and lie along the direction of heat flow. Sideways growth is progressively hindered so that the crystals develop a preferred orientation and a characteristic columnar form. They therefore introduce directionality into the bulk structure; this effect will be most pronounced if the metal itself is strongly anisotropic (e.g. cph zinc). The preferred growth directions for cph crystals are $\langle 1\,0\,\bar{1}\,0 \rangle$. The growth form of the interface between the columnar crystals and the liquid varies from planar to dendritic, depending upon the particular metal (or alloy) and thermal conditions.

As the columnar zone thickens, the temperatures within the liquid become more shallow, undercooling more prominent and the presence of kindred nuclei from dendritic multiplication more likely. Under these conditions, independent nucleation (Section 3.1.1) is favoured and a central zone of equiaxed, randomly-oriented crystals can develop (Figure 3.5). Other factors such as a low pouring temperature (low superheat), moulds of low thermal conductivity and the presence of alloying elements also favour the development of this equiaxed zone. There is a related size effect, with the tendency for columnar crystals to form decreasing as the cross-section of the mould cavity decreases. However, in the absence of these influences, growth predominates over nucleation, and columnar zone may extend to the centre of the ingot (e.g. pure metals). The balance between the relative proportions of outer columnar crystals and inner equiaxed crystals is important and demands careful control. For some purposes, a completely fine-grained stucture is preferred, being stronger and more ductile. Furthermore, it will not contain the planes of weakness, shown in Figure 3.5, which form when columnar crystals impinge upon each other obliquely. (In certain specialized alloys, however, such as those for high-power magnets and creep-resistant alloys, a coarse grain size is prescribed.)

The addition of various 'foreign' nucleating agents, known as inoculants, is a common and effective method for providing centres for heterogeneous nucleation within the melt, inhibiting undercooling and producing a uniform fine-grained structure. Refining the grain structure disperses impurity elements over a greater area of grain boundary surface and generally benefits mechanical and founding properties (e.g. ductility, resistance to hot-tearing). However, the need for grain refinement during casting operations is often less crucial if the cast structure can be subsequently worked and/or heat-treated. Nucleating agents must remain finely dispersed, must survive and must be wetted by the superheated liquid. Examples of inoculants are titanium and/or boron (for aluminium alloys), zirconium or rare earth metals (for magnesium alloys) and aluminium (for steel). Zirconium is an extremely effective grain-refiner for magnesium and its alloys. The close similarity in lattice parameters between zirconium and magnesium suggests that the oriented overgrowth (epitaxy) of magnesium upon zirconium is an important factor; however, inoculants have largely been developed empirically.

3.1.4 Gas porosity and segregation

So far we have tended to concentrate upon the behaviour of pure metals. It is now appropriate to consider the general behaviour of dissimilar types of atoms which, broadly speaking, fall into two main categories: those that have been deliberately added for a specific purpose (i.e. alloying) and those that are accidentally present as undesirable impurities. Most metallic melts, when exposed to a furnace atmosphere, will readily absorb gases (e.g. oxygen, nitrogen, hydrogen). The solubility of gas in liquid metal can be expressed by Sievert's relation, which states that the concentration of dissolved gas is proportional to the square root of the partial pressure of the gas in the contacting atmosphere. Thus, for hydrogen, which is one of the most troublesome gases:

$$[H_{\text{solution}}] = K\{p(H_2)\}^{1/2} \tag{3.1}$$

The constant K is temperature-dependent. The solubility of gases decreases during the course of freezing, usually quite abruptly, and they are rejected in the form of gas bubbles which may become entrapped within and between the crystals, forming weakening blowholes. It follows from Sievert's relation that reducing the pressure of the contacting atmosphere will reduce the gas content of the melt; this principle is the basis of vacuum melting and vacuum degassing. Similarly, the passage of numerous bubbles of an inert, low-solubility gas through the melt will also favour gas removal (e.g. scavenging treatment of molten aluminium with chlorine). Conversely, freezing under high applied pressure, as in the die-casting process for light alloys, suppresses the precipitation of dissolved gas and produces a cast shape of high density.

Dissolved gas may precipitate as simple gas bubbles but may, like oxygen, react with melt constituents to form either bubbles of compound gas (e.g. CO_2, CO, SO_2, H_2O_{vap}) or insoluble non-metallic particles. The latter are potential inoculants. Although their presence may be accidental, as indicated previously, their deliberate formation is sometimes sought. Thus, a specific addition of aluminium, an element with a high chemical affinity for oxygen, is used to deoxidize molten steel in the ladle prior to casting; the resultant particles of alumina subsequently act as heterogeneous nucleants, refining the grain size.

Segregation almost invariably occurs during solidification; unfortunately, its complete elimination is impossible. Segregation, in its various forms, can seriously impair the physical, chemical and mechanical properties of a cast material. In normal segregation, atoms different to those which are crystallizing can be rejected into the melt as the solid/liquid interface advances. These atoms may be impurities or, as in the case of a solid solution alloy, solute atoms. Insoluble particles can also be pushed ahead of the interface.

Eventually, pronounced macro-segregation can be produced in the final regions to solidify, particularly if the volume of the cast mass is large. On a finer scale, micro-segregation can occur interdendritically within both equiaxed and columnar grains (coring) and at the surfaces of low- and high-angle grain boundaries. The modern analytical technique of Auger electron spectroscopy (AES) is capable of detecting monolayers of impurity atoms at grain boundary surfaces and has made it possible to study their very significant effect upon properties such as ductility and corrosion resistance (Chapter 5).

In the other main form of separation process,[1] which is known as inverse segregation, thermal contraction of the solidified outer shell forces a residual melt of low melting point outwards along intergranular channels until it freezes on the outside of the casting (e.g. 'tin sweat' on bronzes, 'phosphide sweat' on grey cast iron). The direction of this remarkable migration thus coincides with that of heat flow, in direct contrast to normal macro-segregation. Inverse segregation can be prevented by unidirectional solidification. Later, in Section 3.2.4.4, it will be shown how the process of zone-refining, as used in the production of high-purity materials for the electronics industry, takes positive advantage of segregation.

3.1.5 Directional solidification

The exacting mechanical demands made upon gas turbine blades have led to the controlled exploitation of columnar crystal growth and the development of directional solidification (DS) techniques for superalloys.[2] As the turbine rotor rotates, the hot blades are subject to extremely large centrifugal forces and to thermal excursions in temperature. Development has proceeded in two stages. First, a wholly columnar grain structure, without grain boundaries transverse to the major axis of the blade, has been produced during precision investment casting by initiating solidification at a watercooled copper chill plate in the mould base and then slowly withdrawing the vertically-positioned mould from the hot zone of an enclosing furnace. Most of the heat is removed by the chill plate. A restricted number of crystals is able to grow parallel to the major axis of the blade. Transverse grain boundaries, which have been the initiation sites for intergranular creep failures in equiaxed blade structures, are virtually eliminated. Grain shape is mainly dependent upon (1) the thermal gradient G_L extending into the melt from the melt/solid interface and (2) the growth rate R at which this interface moves. Graphical plots of G_L versus R

[1]Like all separation processes, even so-called chemical separations, it is essentially a physical process.

[2]Directional solidification of high-temperature alloys was pioneered by F. L. VerSnyder and R. W. Guard at the General Electric Company, Schenectady, USA, in the late 1950s; by the late 1960s, DS blades were being used in gas turbines of commercial aircraft.

Figure 3.6 *A single-crystal blade embodying a DS-starter block and helical constriction and a test plate facilitating orientation check by XRD.*

have provided a useful means for predicting the grain morphology of DS alloys.

In the second, logical stage of development, a single-crystal (SC) turbine blade is produced by placing a geometrical restriction between a chilled starter block and the mould proper (Figure 3.6). This spiral selector causes several changes in direction and ensures that only one of the upward-growing columnar crystals in the starter block survives. A typical production unit is illustrated in Figure 3.7. The orientation of every blade is checked by means of a computerized X-ray diffraction procedure (e.g. the SCORPIO system at Rolls-Royce). In the fcc nickel-based superalloys, the favoured [1 0 0] growth direction coincides with the major axis of the blade and fortunately offers the best overall mechanical properties. For instance, the modulus of elasticity is low in the ⟨1 0 0⟩ directions; consequently, thermal stresses are reduced and the resistance to thermal fatigue enhanced. If full three-dimensional control of crystal orientation is required, a seed crystal is precisely located close to the chill plate and gives the desired epitaxial, or oriented, overgrowth. The production of single-crystal turbine blades by directional solidification (DS) techniques has increased blade life

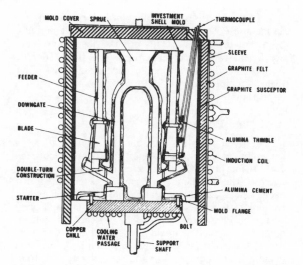

Figure 3.7 *Schematic diagram of directional solidification plant for turbine blades.*

substantially and has enabled operating temperatures to be raised by 30°C, thus improving engine efficiency. It has also had a far-reaching effect upon the philosophy of alloy design. Previously, certain elements, such as carbon, boron, hafnium and zirconium, were added in order to strengthen grain boundary surfaces. Unfortunately, their presence lowers the incipient melting point. The DS route dispenses with the need for them and permits beneficially higher temperatures to be used during subsequent heat-treatment. The DS approach requires to be carefully matched to alloy composition; it is possible for it to lower the creep strength of certain superalloys.

3.1.6 Production of metallic single crystals for research

Development of highly-specialized industrial-scale techniques, such as the directional solidification (DS) of turbine blades and the production of silicon, germanium and various compounds for semiconductors, owes much to expertise gained over many years in producing small single crystals for fundamental research. Several methods originally developed for metals have been adapted for ceramics. Experiments with single crystals of metals (and ceramics and polymers) have a special place in the history of materials science. The two basic methods for preparing single crystals involve (1) solidification from the melt and (2) grain growth in the solid state.

In the simplest version of the solidification method, the polycrystalline metal to be converted to a single crystal is supported in a horizontal, non-reactive boat (e.g. graphite) and made to to freeze progressively from one end by passing an electric furnace, with its peak temperature set about 10°C above the melting point, over the boat. Although several nuclei may

form during initial solidification, the sensitivity of the growth rate to orientation usually results in one of the crystals swamping the others and eventually forming the entire growth front. The method is particularly useful for seeding crystals of a predetermined orientation. A seed crystal is placed adjacent to the polycrystalline sample in the boat and the junction is melted before commencing the melting/solidification process. Wire specimens may be grown in silica or heat-resistant glass tubes internally coated with graphite (*Aquadag*). In a modern development of these methods, the sample is enclosed in an evacuated silica tube and placed in a water-cooled copper boat; passage through a high-frequency heating coil produces a melt zone.

Most solidification techniques for single crystals are derived from the Bridgman and Czochralski methods. In the former, a pure metal sample is loaded in a vertical mould of smooth graphite, tapered to a point at the bottom end. The mould is lowered slowly down a tubular furnace which produces a narrow melt zone. The crystal grows from the point of the mould. In the Czochralski method, often referred to as 'crystal-pulling', a seed crystal is withdrawn slowly from the surface of a molten metal, enabling the melt to solidify with the same orientation as the seed. Rotation of the crystal as it is withdrawn produces a cylindrical crystal. This technique is used for the preparation, *in vacuo*, of Si and Ge crystals.

Crystals may also be prepared by a 'floating-zone' technique (e.g. metals of high melting point such as W, Mo and Ta). A pure polycrystalline rod is gripped at the top and bottom in water-cooled grips and rotated in an inert gas or vacuum. A small melt zone, produced by either a water-cooled radio-frequency coil or electron bombardment from a circular filament, is passed up its length. High purity is possible because the specimen has no contact with any source of contamination and also because there is a zone-refining action (Section 3.2.4.4).

Methods involving grain growth in the solid state (2) depend upon the annealing of deformed samples. In the strain-anneal technique, a fine-grained polycrystalline metal is critically strained approximately 1–2% elongation in tension and then annealed in a moving-gradient furnace with a peak temperature set below the melting point or transformation temperature. Light straining produces very few nuclei for crystallization; during annealing, one favoured nucleus grows more rapidly than the other potential nuclei, which it consumes. The method has been applied to metals and alloys of high stacking-fault energy, e.g. Al, silicon-iron, (see Chapter 4). Single crystals of metals with low stacking-fault energy, such as Au and Ag, are difficult to grow because of the ease of formation of annealing twins which give multiple orientations. Hexagonal metals are also difficult to prepare because deformation twins formed during straining act as effective nucleation sites.

3.2 Principles and applications of phase diagrams

3.2.1 The concept of a phase

The term 'phase' refers to a separate and identifiable state of matter in which a given substance may exist. Being applicable to both crystalline and non-crystalline materials, its use provides a convenient way of expressing a material's structure. Thus, in addition to the three crystalline forms mentioned in Section 2.5.1, the element iron may exist as a liquid or vapour, giving five phases overall. Similarly, the important refractory oxide silica is able to exist as three crystalline phases, quartz, tridymite and cristobalite, as well as a non-crystalline phase, silica glass, and as molten silica. Under certain conditions, it is possible for two or more different phases to co-exist. The glass-reinforced polymer (GRP) known as *Fibreglass* is an example of a two-phase structure.

When referring to a particular phase in the structure of a material, we imply a region comprising a large number of atoms (or ions or molecules) and the existence of a bounding surface which separates it from contiguous phases. Local structural disturbances and imperfections are disregarded. Thus, a pure metal or an oxide solid solution are each described, by convention, as single-phase structures even though they may contain grain boundaries, concentration gradients (coring) and microdefects, such as vacancies, dislocations and voids (Chapter 4). Industrial practice understandably favours relatively rapid cooling rates, frequently causing phases to exist in a metastable condition. Some form of 'triggering' process, such as thermal activation, is needed before a metastable phase can adopt the stable, or equilibrium, state of lowest energy (e.g. annealing of metals and alloys). These two features, structural heterogeneity on a micro-scale and non-equilibrium, do not give rise to any untoward scientific difficulty.

The production of the thermal-shock resistant known as *Vycor* provides an interesting example of the potential of phase control.[1] Although glasses of very high silica content are eminently suitable for high-temperature applications, their viscosity is very high, making fabrication by conventional methods difficult and costly. This problem was overcome by taking advantage of phase separation in a workable silica containing a high proportion of boric oxide. After shaping, this glass is heat-treated at a temperature of 500–600°C in order to induce separation of two distinct and interpenetrating glassy phases. Electron microscopy reveals a wormlike boron-rich phase surrounded by a porous siliceous matrix. Leaching in hot acid dissolves away the former phase, leaving a porous

silica-rich structure. Consolidation heat-treatment at a temperature of 1000°C 'shrinks' this skeletal structure by a remarkable 30%. This product has a low linear coefficient of thermal expansion ($\alpha = 0.8 \times 10^{-60}$C; 0–300°C) and can withstand quenching into iced water from a temperature of 900°C, its maximum service temperature.

3.2.2 The Phase Rule

For a given metallic or ceramic material, there is a theoretical condition of equilibrium in which each constituent phase is in a reference state of lowest energy. The independent variables determining this energy state, which are manipulated by scientists and technologists, are composition, temperature and pressure. The Phase Rule derived by Willard Gibbs from complex thermodynamical theory provides a device for testing multi-phase (heterogeneous) equilibria and deciding the number of variables (degrees of freedom) necessary to define the energy state of a system. Its basic equation, $P + F = C + 2$, relates the number of phases present at equilibrium (P) and the number of degrees of freedom (F) to the number of components (C), which is the smallest number of substances of independently-variable composition making up the system. For metallic systems, the components are metallic elements: for ceramics, the components are frequently oxides (e.g. MgO, SiO_2, etc.).

Consider Figure 3.8, which is a single-component (unary) diagram representing the phase relations for a typical pure metal. Transitions such as melting, sublimation and vaporization occur as boundaries between the three single-phase fields are crossed. Suppose that conditions are such that three phases of a pure metal co-exist simultaneously, a unique condition represented by the triple-point O. Applying the Phase Rule: $P = 3$, $C = 1$ and $F = 0$. This system is said to be invariant with no degrees of freedom. Having stated that 'three phases co-exist', the energy state is fully defined and it is unnecessary to specify values for variables. One or two phases could be caused to disappear, but this would require temperature and/or pressure to be changed. Similarly, at a melting point on the line between single-phase fields, a solid and a liquid phase may co-exist (i.e. $P = 2$, $C = 1$ and $F = 1$). Specification of one variable (either temperature or pressure) will suffice to define the energy state of a particular two-phase equilibrium. A point within one of the single-phase fields represents a system that is bivariant ($F = 2$). Both temperature and pressure must be specified if its state is to be fully defined.

Let us consider some practical implications of Figure 3.8. The line for the solid/liquid transition is shown with a slight inclination to the right in this schematic diagram because, theoretically, one may reason from Le Chatelier's principle that a pressure increase on a typical melt, from P_1 to P_2, will favour a constraint-removing contraction and therefore lead to freezing. (Bismuth and gallium are exceptions: like

---

[1] *Vycor*, developed and patented by Corning Glass Works (USA) just before World War II, has a final composition of $96SiO_2–4B_2O_3$. It has been used for laboratory ware and for the outer windows of space vehicles.

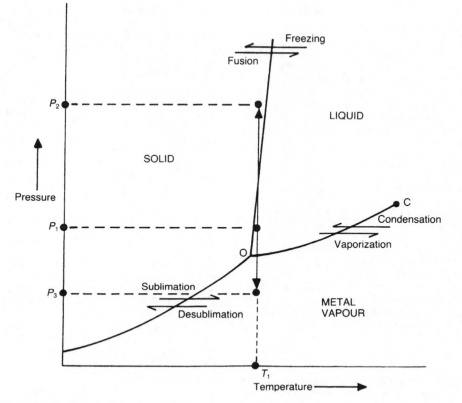

Figure 3.8 *Phase diagram for a typical pure metal.*

water, they expand on freezing and the solid/liquid line has an opposite slope.) The pressure at the critical point C, beyond which liquid and vapour are indistinguishable, is usually extremely high and considerably greater than one atmosphere. The diagram refers to an enclosed system. For instance, if molten metal is in a vacuum system, and the pressure is reduced isothermally from P_1 to P_3, then vaporization will be favoured.

For most metals, the pressure value at the triple point is far below one atmosphere (1 atm = 101.325 kN m^{-2} $\cong$ 1 bar). For copper and lead, it is in the order of 10^{-6} atm. The pressures associated with the solid/vapour line are obviously even lower. Thus, for most metals, vapour pressure is disregarded and it is customary to use isobaric (constant pressure) phase diagrams in which the composition (concentration) of component metals is plotted against temperature. As a consequence of this practice, metallurgists use a modified form of the Phase Rule equation i.e. $P + F = C + 1$.

Nevertheless, it is possible for vapour pressure to be a highly significant factor in alloy systems formed by volatile components such as mercury, zinc, cadmium and magnesium: in such cases, it is advisable to take

advantage of full pressure-temperature diagrams. Furthermore, the partial pressures of certain gases in a contacting atmosphere can be highly significant and at elevated temperatures can contribute to serious corrosion of metals (e.g. O_2, S_2 (Chapter 12)). Again, the pressure variable must be taken into account.

3.2.3 Stability of phases

3.2.3.1 The concept of free energy

Every material in a given state has a characteristic heat content or enthalpy, H, and the rate of change of heat content with temperature is equal to the specific heat of the material measured at constant pressure, $C_P = (\mathrm{d}H/\mathrm{d}T)_p$. A knowledge of the quantity H is clearly important to understand reactions but it does not provide a criterion for equilibrium, nor does it determine when a phase change occurs, as shown by the occurrence of both exothermic and endothermic reactions. To provide this criterion it is necessary to consider a second important property of state known as the entropy, S. In statistical terms S may be regarded as a measure of the state of disorder of the structure, but from classical thermodynamics it may be shown that for any material passing through a complete cycle of events

$$\oint \frac{dQ}{T} = 0$$

where dQ is the heat exchanged between the system and its surroundings during each infinitesimal step and T is the temperature at which the transfer takes place.

It is then convenient to define a quantity S such that $dS = dQ/T$, so that $\oint dS = 0$; entropy so defined is then a state property. At constant pressure, $dQ = dH$ and consequently

$$dS = dQ/T = C_p dT/T$$

which by integration gives

$$S = S_0 + \int_0^T (C_p/T)dT = S_0 + \int_0^T C_p \delta(\ln T)$$

where S is the entropy at T K usually measured in J K^{-1}. The integration constant S_0 represents the entropy at absolute zero, which for an ordered crystalline substance is taken to be zero; this is often quoted as the third law of thermodynamics. Clearly, any reaction or transformation within a system will be associated with a characteristic entropy change given by

$$dS = S_\beta - S_\alpha$$

where dS is the entropy of transformation and S_β and S_α are the entropy values of the new phase β and the old phase α, respectively. It is a consequence of this that any irreversible change which takes place in a system (e.g. the combustion of a metal) must be accompanied by an increase in the total entropy of the system. This is commonly known as the second law of thermodynamics.

The quantity entropy could be used for deciding the equilibrium state of a system, but it is much more convenient to work in terms of energy. Accordingly, it is usual to deal with the quantity TS, which has the units of energy, rather than just S, and to separate the total energy of the system H into two components according to the relation

$$H = G + TS$$

where G is that part of the energy of the system which causes the process to occur and TS is the energy requirement resulting from the change involved. The term G is known as Gibbs' free energy and is defined by the equation

$$G = H - TS$$

Every material in a given state will have a characteristic value of G. The change of free energy accompanying a change represents the 'driving force' of the change, and is given by the expression

$$dG = dH - TdS \equiv dE + PdV - TdS$$

All spontaneous changes in a system must be accompanied by a reduction of the total free energy of that system, and thus for a change to occur the free energy change ΔG must be negative. It also follows that the equilibrium condition of a reaction will correspond to the state where $dG = 0$, i.e. zero driving force.

For solids and liquids at atmospheric pressure the volume change accompanying changes of state is very small and hence PdV is also very small. It is therefore reasonable to neglect this term in the free energy equation and use as the criterion of equilibrium $dE - TdS = 0$. This is equivalent to defining the quantity $(E - TS)$ to be a minimum in the equilibrium state, for by differentiation

$$d(E - TS) = dE - TdS - SdT$$
$$= dE - TdS \text{ (since } T \text{ is constant)}$$
$$= 0 \text{ for the equilibrium state}$$

The quantity $(E - TS)$ thus defines the equilibrium state at constant temperature and volume, and is given the symbol F, the Helmholtz free energy $(F = E - TS)$, to distinguish it from the Gibbs free energy $(G = H - TS)$. In considering changes in the solid state it is thus a reasonable approximation to use F in place of G. The enthalpy H is the sum of the internal and external energies which reduces to $H \simeq E$ when the external energy PV is neglected.

3.2.3.2 Free energy and temperature

If a metal undergoes a structural change from phase α to phase β at a temperature T_t then it does so because above this temperature the free energy of the β phase, G_β, becomes lower than the free energy of the α phase, G_α. For this to occur the free energy curves must vary with temperature in the manner shown in Figure 3.9a. It can be seen that at T_t the free energy of the α-phase is equal to that of the β-phase so that ΔG is zero; T_t is, therefore, the equilibrium transformation point.

Figure 3.9a also shows that successive transformations occur in a given temperature range. The way in which the absolute value of the free energy of a crystal varies with temperature is shown in Figure 3.9b, where H and $-TS$ are plotted as a function of temperature. At the transformation temperature, T_t, the change in heat content ΔH is equal to the latent heat L, while the entropy change ΔS is equal to L/T_t. In consequence, a plot of $G = H - TS$ shows no sharp discontinuities at T_t (since $\Delta H = T_t \Delta S$) or T_m but merely a discontinuity of slope. A plot of G versus temperature for each of the three phases considered, α, β and liquid, would then be of the form shown in Figure 3.9(a). In any temperature range the phase with the lowest free energy is the stable phase.

3.2.3.3 Free energy and composition

Of considerable importance in metallurgical practice is the increase in entropy associated with the formation of

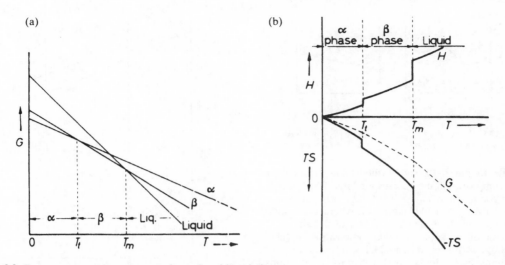

Figure 3.9 *Free energy-temperature curves for α, β and liquid phases.*

a disordered solid solution from the pure components. This arises because over and above the entropies of the pure components A and B, the solution of B in A has an extra entropy due to the numerous ways in which the two kinds of atoms can be arranged amongst each other. This entropy of disorder or mixing is of the form shown in Figure 3.10a.

As a measure of the disorder of a given state we can, purely from statistics, consider W the number of distributions which belong to that state. Thus, if the crystal contains N sites, n of which contain A-atoms and $(N - n)$ contain B-atoms, it can be shown that the total number of ways of distributing the A and B

atoms on the N sites is given by

$$W = \frac{N!}{n!(N - n)!}$$

This is a measure of the extra disorder of solution, since $W = 1$ for the pure state of the crystal because there is only one way of distributing N indistinguishable pure A or pure B atoms on the N sites. To ensure that the thermodynamic and statistical definitions of entropy are in agreement the quantity, W, which is a measure of the configurational probability of the system, is not used directly, but in the form

$$S = \mathbf{k} \ln W$$

where $\mathbf{k}$ is Boltzmann's constant. From this equation it can be seen that entropy is a property which measures the probability of a configuration, and that the greater the probability, the greater is the entropy. Substituting for W in the statistical equation of entropy and using Stirling's approximation[1] we obtain

$$S = \mathbf{k} \ln[N!/n!(N - n)!]$$
$$= \mathbf{k}[N \ln N - n \ln n - (N - n)\ln(N - n)]$$

for the entropy of disorder, or mixing. The form of this entropy is shown in Figure 3.10a, where $c = n/N$ is the atomic concentration of A in the solution. It is of particular interest to note the sharp increase in entropy for the addition of only a small amount of solute. This fact accounts for the difficulty of producing really pure metals, since the entropy factor, $-T\mathrm{d}S$, associated with impurity addition, usually outweighs the energy term, $\mathrm{d}H$, so that the free energy of the material is almost certainly lowered by contamination.

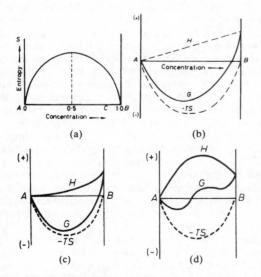

Figure 3.10 *Variation with composition of entropy (a) and free energy (b) for an ideal solid solution, and for non-ideal solid solutions (c) and (d).*

[1] Stirling's theorem states that if N is large

$$\ln N! = N \ln N - N$$

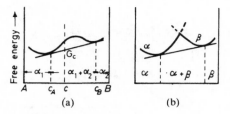

Figure 3.11 *Free energy curves showing extent of phase fields at a given temperature.*

While Figure 3.10a shows the change in entropy with composition the corresponding free energy versus composition curve is of the form shown in Figure 3.10b, c or d depending on whether the solid solution is ideal or deviates from ideal behaviour. The variation of enthalpy with composition, or heat of mixing, is linear for an ideal solid solution, but if A atoms prefer to be in the vicinity of B atoms rather than A atoms, and B atoms behave similarly, the enthalpy will be lowered by alloying (Figure 3.10c). A positive deviation occurs when A and B atoms prefer like atoms as neighbours and the free energy curve takes the form shown in Figure 3.10d. In diagrams 3.10b and 3.10c the curvature dG^2/dc^2 is everywhere positive whereas in 3.10d there are two minima and a region of negative curvature between points of inflexion[1] given by $dG^2/dc^2 = 0$. A free energy curve for which d^2G/dc^2 is positive, i.e. simple U-shaped, gives rise to a homogeneous solution. When a region of negative curvature exists, the stable state is a phase mixture rather than a homogeneous solid solution, as shown in Figure 3.11a. An alloy of composition c has a lower free energy G_c when it exists as a mixture of A-rich phase (α_1) of composition c_A and B-rich phase (α_2) of composition c_B in the proportions given by the Lever Rule, i.e. $\alpha_1/\alpha_2 = (c_B - c)/(c - c_A)$. Alloys with composition $c < c_A$ or $c > c_B$ exist as homogeneous solid solutions and are denoted by phases, α_1 and α_2 respectively. Partial miscibility in the solid state can also occur when the crystal structures of the component metals are different. The free energy curve then takes the form shown in Figure 3.11b, the phases being denoted by α and β.

3.2.4 Two-phase equilibria

3.2.4.1 Extended and limited solid solubility

Solid solubility is a feature of many metallic and ceramic systems, being favoured when the components have similarities in crystal structure and atomic (ionic) diameter. Such solubility may be either extended (continuous) or limited. The former case is illustrated by the binary phase diagram for the nickel–copper system (Figure 3.12) in which the solid solution (α) extends

<hr/>

[1] The composition at which $d^2G/dc^2 = 0$ varies with temperature and the corresponding temperature–composition curves are called spinodal lines.

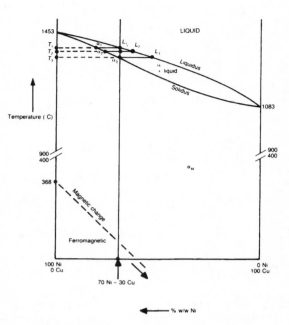

Figure 3.12 *Binary phase diagram for Ni–Cu system showing extended solid solubility.*

from component to component. In contrast to the abrupt (congruent) melting points of the pure metals, the intervening alloys freeze, or fuse, over a range of temperatures which is associated with a univariant two-phase (α + liquid) field. This 'pasty' zone is located between two lines known as the liquidus and solidus. (The phase diagrams for Ni–Cu and MgO–FeO systems are similar in form.)

Let us consider the very slow (equilibrating) solidification of a 70Ni–30Cu alloy. A commercial version of this alloy, *Monel*, also contains small amounts of iron and manganese. It is strong, ductile and resists corrosion by all forms of water, including sea-water (e.g. chemical and food processing, water treatment). An ordinate is erected from its average composition on the base line. Freezing starts at a temperature T_1. A horizontal tie-line is drawn to show that the first crystals of solid solution to form have a composition α_1. When the temperature reaches T_2, crystals of composition α_2 are in equilibrium with liquid of composition L_2. Ultimately, at temperature T_3, solidification is completed as the composition α_3 of the crystals coincides with the average composition of the alloy. It will be seen that the compositions of the α-phase and liquid have moved down the solidus and liquidus, respectively, during freezing.

Each tie-line joins two points which represent two phase compositions. One might visualize that a two-phase region in a binary diagram is made up of an infinite number of horizontal (isothermal) tie-lines. Using the average alloy composition as a fulcrum (x) and applying the Lever Rule, it is quickly possible to derive mass ratios and fractions. For instance, for equilibrium at temperature T_2 in Figure 3.12, the mass ratio of solid solution crystals to liquid is $xL_2/\alpha_2 x$. Similarly, the mass fraction of solid in the two-phase mixture at this temperature is $xL_2/L_2\alpha_2$. Clearly, the phase compositions are of greater structural significance than the average composition of the alloy. If the volumetric relations of two phases are required, these being what we naturally assess in microscopy, then the above values must be corrected for phase density.

In most systems, solid solubility is far more restricted and is often confined to the phase field adjacent to the end-component. A portion of a binary phase diagram for the copper–beryllium system, which contains a primary, or terminal, solid solution, is shown in Figure 3.13. Typically, the curving line known as the solvus shows an increase in the ability of the solvent copper to dissolve beryllium solute as the temperature is raised. If a typical 'beryllium–copper' containing 2% beryllium is first held at a temperature just below the solidus (solution-treated), water-quenched to preserve the α-phase and then aged at a temperature of 425°C, particles of a second phase (γ) will form within the α-phase matrix because the alloy is equilibrating in the ($\alpha + \gamma$) field of the diagram. This type of treatment, closely controlled, is known as precipitation-hardening; the mechanism of this important strengthening process will be discussed in detail in Chapter 8. Precipitation-hardening of a typical beryllium–copper, which also contains up to 0.5% cobalt or nickel, can raise the 0.1% proof stress to 1200 MN m^{-2} and the tensile strength to 1400 MN m^{-2}. Apart from being suitable for non-sparking tools, it is a valuable spring material, being principally used for electrically conductive brush springs and contact fingers in electrical switches. A curving solvus is an essential feature of phase diagrams for precipitation-hardenable alloys (e.g. aluminium–copper alloys (*Duralumin*)).

When solid-state precipitation takes place, say of β within a matrix of supersaturated α grains, this precipitation occurs in one or more of the following preferred locations: (1) at grain boundaries, (2) around dislocations and inclusions, and (3) on specific crystallographic planes. The choice of site for precipitation depends on several factors, of which grain size and rate of nucleation are particularly important. If the grain size is large, the amount of grain boundary surface is relatively small, and deposition of β-phase within the grains is favoured. When this precipitation occurs preferentially on certain sets of crystallographic planes within the grains, the etched structure has a mesh-like appearance which is known as a Widmanstätten-type structure.[1] Widmanstätten structures have been observed in many alloys (e.g. overheated steels).

[1] Named after Count Alois von Widmanstätten who discovered this morphology within an iron–nickel meteorite sample in 1808.

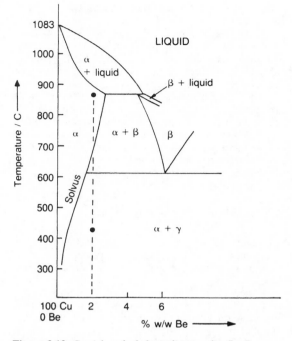

Figure 3.13 *Cu-rich end of phase diagram for Cu–Be system, showing field of primary solid solution (α).*

3.2.4.2 Coring

It is now possible to consider microsegregation, a phenomenon introduced in Section 3.1.4, in more detail. Referring again to the freezing process for a Ni–Cu alloy (Figure 3.12), it is clear that the composition of the α-phase becomes progressively richer in copper and, consequently, if equilibrium is to be maintained in the alloy, the two phases must continuously adjust their compositions by atomic migration. In the liquid phase such diffusion is relatively rapid. Under industrial conditions, the cooling rate of the solid phase is often too rapid to allow complete elimination of differences in composition by diffusion. Each grain of the α-phase will thus contain composition gradients between the core, which will be unduly rich in the metal of higher melting point, and the outer regions, which will be unduly rich in the metal of lower melting point. Such a non-uniform solid solution is said to be cored: etching of a polished specimen can reveal a pattern of dendritic segregation within each cored grain. The faster the rate of cooling, the more pronounced will be the degree of coring. Coring in chill-cast ingots is, therefore, quite extensive.

The physical and chemical hetereogeneity produced by non-equilibrium cooling rates impairs properties. Cored structures can be homogenized by annealing. For instance, an ingot may be heated to a temperature just below the solidus temperature where diffusion is rapid. The temperature must be selected with care because some regions might be rich enough in low melting point metal to cause localized fusion. However, when practicable, it is more effective to cold-work a cored structure before annealing. This treatment has three advantages. First, dendritic structures are broken up by deformation so that regions of different composition are intermingled, reducing the distances over which diffusion must take place. Second, defects introduced by deformation accelerate rates of diffusion during the subsequent anneal. Third, deformation promotes recrystallization during subsequent annealing, making it more likely that the cast structure will be completely replaced by a generation of new equiaxed grains. Hot-working is also capable of eliminating coring.

3.2.4.3 Cellular microsegregation

In the case of a solid solution, we have seen that it is possible for solvent atoms to tend to freeze before solute atoms, causing gradual solute enrichment of an alloy melt and, under non-equilibrium conditions, dendritic coring (e.g. Ni–Cu). When a very dilute alloy melt or impure metal freezes, it is possible for each crystal to develop a regular cell structure on a finer scale than coring. The thermal and compositional condition responsible for this cellular microsegregation is referred to as constitutional undercooling.

Suppose that a melt containing a small amount of lower-m.p. solute is freezing. The liquid becomes increasingly enriched in rejected solute atoms, particularly close to the moving solid/liquid interface. The variation of liquid composition with distance from

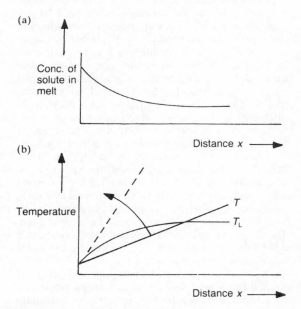

(a)

Conc. of solute in melt

Distance x ⟶

(b)

Temperature

T

T_L

Distance x ⟶

Figure 3.14 *Variation with distance from solid/liquid interface of (a) melt composition and (b) actual temperature T and freezing temperature T_L.*

the interface is shown in Figure 3.14a. There is a corresponding variation with distance of the temperature T_L at which the liquid will freeze, since solute atoms lower the freezing temperature. Consequently, for the positive gradient of melt temperature T shown in Figure 3.14b, there is a layer of liquid in which the actual temperature T is below the freezing temperature T_L: this layer is constitutionally undercooled. Clearly, the depth of the undercooled zone, as measured from the point of intersection, will depend upon the slope of the curve for actual temperature, i.e. $G_L = dT/dx$. As G_L decreases, the degree of constitutional undercooling will increase.

Suppose that we visualize a tie-line through the two-phase region of the phase diagram fairly close to the component of higher m.p. Assuming equilibrium, a partition or distribution coefficient k can be defined as the ratio of solute concentration in the solid to that in the liquid, i.e. c_S/c_L. For an alloy of average composition c_0, the solute concentration in the first solid to freeze is kc_0, where $k < 1$, and the liquid adjacent to the solid becomes richer in solute than c_0. The next solid to freeze will have a higher concentration of solute. Eventually, for a constant rate of growth of the solid/liquid interface, a steady state is reached for which the solute concentration at the interface reaches a limiting value of c_0/k and decreases exponentially within the liquid to the bulk composition. This concentration profile is shown in Figure 3.14a.

The following relation can be derived by applying Fick's second law of diffusion (Section 6.4.1):

$$c_L = c_0 \left[1 + \frac{1-k}{k} \exp\left(-\frac{Rx}{D}\right) \right] \qquad (3.2)$$

where x is the distance into the liquid ahead of the interface, c_L is the solute concentration in the liquid at point x, R is the rate of growth, and D is the diffusion coefficient of the solute in the liquid. The temperature distribution in the liquid can be calculated if it is assumed that k is constant and that the liquidus is a straight line of slope m. For the two curves of Figure 3.14b:

$$T = T_0 - mc_0/k + G_L x \qquad (3.3)$$

and

$$T_L = T_0 - mc_0 \left[1 + \frac{1-k}{k} \exp\left(-\frac{Rx}{D}\right) \right] \qquad (3.4)$$

where T_0 is the freezing temperature of pure solvent, T_L the liquidus temperature for the liquid of composition c_L and T is the actual temperature at any point x.

The zone of constitutional undercooling may be eliminated by increasing the temperature gradient G_L, such that:

$$G_L > dT_L/dx \qquad (3.5)$$

Substituting for T_L and putting $[1 - (Rx/D)]$ for the exponential gives the critical condition:

$$\frac{G_L}{R} > \frac{mc_0}{D} \left(\frac{1-k}{k}\right) \qquad (3.6)$$

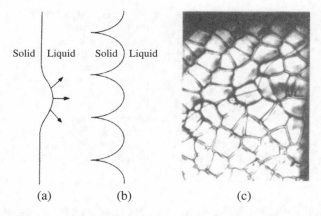

Solid | Liquid   Solid) Liquid

(a)   (b)   (c)

Figure 3.15 *The breakdown of a planar solid–liquid interface (a), (b) leading to the formation of a cellular structure of the form shown in (c) for Sn/0.5 at.% Sb × 140.*

This equation summarizes the effect of growth conditions upon the transition from planar to cellular growth and identifies the factors that stabilize a planar interface. Thus, a high G_L, low R and low c_0 will reduce the tendency for cellular (and dendritic) structures to form.

The presence of a zone of undercooled liquid ahead of a macroscopically planar solid/liquid interface (Section 3.1.2) makes it unstable and an interface with cellular morphology develops. The interface grows locally into the liquid from a regular array of points on its surface, forming dome-shaped cells. Figures 3.15a and 3.15b show the development of domes within a metallic melt. As each cell grows by rapid freezing, solute atoms are rejected into the liquid around its base which thus remains unfrozen. This solute-rich liquid between the cells eventually freezes at a much lower temperature and a crystal with a periodic columnar cell structure is produced. Solute or impurity atoms are concentrated in the cell walls. Decantation of a partly-solidified melt will reveal the characteristic surface structure shown in Figure 3.15c. The cells of metals are usually hexagonal in cross-section and about 0.05–1 mm across: for each grain, their major axes have the same crystallographic orientation to within a few minutes of arc. It is often found that a lineage or macromosaic structure (Section 3.1.1) is superimposed on the cellular structure; this other form of sub-structure is coarser in scale.

Different morphologies of a constitutionally-cooled surface, other than cellular, are possible. A typical overall sequence of observed growth forms is planar/cellular/cellular dendritic/dendritic. Substructures produced by constitutional undercooling have been observed in 'doped' single crystals and in ferrous and non-ferrous castings/weldments.[1] When

the extent of undercooling into the liquid is increased as, for example, by reducing the temperature gradient G_L, the cellular structure becomes unstable and a few cells grow rapidly as cellular dendrites. The branches of the dendrites are interconnected and are an extreme development of the dome-shaped bulges of the cell structure in directions of rapid growth. The growth of dendrites in a very dilute, constitutionally-undercooled alloy is slower than in a pure metal because solute atoms must diffuse away from dendrite/liquid surfaces and also because their growth is limited to the undercooled zone. Cellular impurity-generated substructures have also been observed in 'non-metals' as a result of constitutional undercooling. Unlike the dome-shaped cells produced with metals, non-metals produce faceted projections which relate to crystallographic planes. For instance, cells produced in a germanium crystal containing gallium have been reported in which cell cross-sections are square and the projection tips are pyramid-shaped, comprising four octahedral {1 1 1} planes.

3.2.4.4 Zone-refining

Extreme purification of a metal can radically improve properties such as ductility, strength and corrosion-resistance. Zone-refining was devised by W. G. Pfann, its development being 'driven' by the demands of the newly invented transistor for homogeneous and ultra-pure metals (e.g. Si, Ge). The method takes advantage of non-equilibrium effects associated with the 'pasty' zone separating the liquidus and solidus of impure metal. Considering the portion of Figure 3.12 where addition of solute lowers the liquidus temperature, the concentration of solute in the liquid, c_L, will always be greater than its concentration c_s in the solid phase; that is, the distribution coefficient $k = c_s/c_L$ is less than unity. If a bar of impure metal is threaded through a heating coil and the coil is slowly moved, a narrow zone of melt can be made to progress along the bar. The first solid to freeze is

[1] The geological equivalent, formed by very slowly cooling magma, is the hexagonal-columnar structure of the Giant's Causeway, Northern Ireland.

purer than the average composition by a factor of k, while that which freezes last, at the trailing interface, is correspondingly enriched in solute. A net movement of impurity atoms to one end of the bar takes place. Repeated traversing of the bar with a set of coils can reduce the impurity content well below the limit of detection (e.g. <1 part in 10^{10} for germanium). Crystal defects are also eliminated: Pfann reduced the dislocation density in metallic and semi-metallic crystals from about 3.5×10^6 cm^{-2} to almost zero. Zone-refining has been used to optimize the ductility of copper, making it possible to cold-draw the fine-gauge threads needed for interconnects in very large-scale integrated circuits.

3.2.5 Three-phase equilibria and reactions

3.2.5.1 The eutectic reaction

In many metallic and ceramic binary systems it is possible for two crystalline phases and a liquid to co-exist. The modified Phase Rule reveals that this unique condition is invariant; that is, the temperature and all phase compositions have fixed values. Figure 3.16 shows the phase diagram for the lead–tin system. It will be seen that solid solubility is limited for each of the two component metals, with α and β representing primary solid solutions of different crystal structure. A straight line, the eutectic horizontal, passes through three phase compositions (α_e, L_e and β_e) at the temperature T_e. As will become clear when ternary systems are discussed (Section 3.2.9), this line is a collapsed three-phase triangle: at any point on this line, three phases are in equilibrium. During slow cooling or heating, when the average composition of an alloy lies between its limits, α_e and β_e, a eutectic reaction takes place in accordance with the equation $L_e \rightleftharpoons \alpha_e + \beta_e$. The sharply-defined minimum in the liquidus, the eutectic (easy-melting) point, is a typical feature of the reaction.

Consider the freezing of a melt, average composition 37Pb–63Sn. At the temperature T_e of approximately 180°C, it freezes abruptly to form a mechanical mixture of two solid phases, i.e. Liquid $L_e \rightarrow \alpha_e + \beta_e$. From the Lever Rule, the α/β mass ratio is approximately 9:11. As the temperature falls further, slow cooling will allow the compositions of the two phases to follow their respective solvus lines. Tie-lines across this ($\alpha + \beta$) field will provide the mass ratio for any temperature. In contrast, a hypoeutectic alloy melt, say of composition 70Pb–30Sn, will form primary crystals of α over a range of temperature until T_e is reached. Successive tie-lines across the (α + Liquid) field show that the crystals and the liquid become enriched in tin as the temperature falls. When the liquid composition reaches the eutectic value L_e, all of the remaining liquid transforms into a two-phase mixture, as before. However, for this alloy, the final structure will comprise primary grains of α in a eutectic matrix of α and β. Similarly, one may deduce that the structure of a solidified hyper-eutectic alloy containing 30Pb–70Sn

will consist of a few primary β grains in a eutectic matrix of α and β.

Low-lead or low-tin alloys, with average compositions beyond the two ends of the eutectic horizontal,[1] freeze by transforming completely over a small range of temperature into a primary phase. (Changes in composition are similar in character to those described for Figure 3.12) When the temperature 'crosses' the relevant solvus, this primary phase becomes unstable and a small amount of second phase precipitates. Final proportions of the two phases can be obtained by superimposing a tie-line on the central two-phase field: there will be no signs of a eutectic mixture in the microstructure.

The eutectic (37Pb–63Sn) and hypo-eutectic (70Pb–30Sn) alloys chosen for the description of freezing represent two of the numerous types of solder[2] used for joining metals. Eutectic solders containing 60–65% tin are widely used in the electronics industry for making precise, high-integrity joints on a mass-production scale without the risk of damaging heat-sensitive components. These solders have excellent 'wetting' properties (contact angle <10°), a low liquidus and a negligible freezing range. The long freezing range of the 70Pb–30Sn alloy (plumbers' solder) enables the solder at a joint to be 'wiped' while 'pasty'.

The shear strength of the most widely-used solders is relatively low, say 25–55 MN m^{-2}, and mechanically-interlocking joints are often used. Fluxes (corrosive zinc chloride, non-corrosive organic resins) facilitate essential 'wetting' of the metal to be joined by dissolving thin oxide films and preventing re-oxidation. In electronic applications, minute solder preforms have been used to solve the problems of excess solder and flux.

Figure 3.16 shows the sequence of structures obtained across the breadth of the Pb–Sn system. Cooling curves for typical hypo-eutectic and eutectic alloys are shown schematically in Figure 3.17a. Separation of primary crystals produces a change in slope while heat is being evolved. Much more heat is evolved when the eutectic reaction takes place. The lengths (duration) of the plateaux are proportional to the amounts of eutectic structure formed, as summarized in Figure 3.17b. Although it follows that cooling curves can be used to determine the form of such a simple system, it is usual to confirm details by means of microscopical examination (optical, scanning electron) and X-ray diffraction analysis.

---

[1]Theoretically, the eutectic horizontal cannot cut the vertical line representing a pure component: some degree of solid solubility, however small, always occurs.

[2]Soft solders for engineering purposes range in composition from 20% to 65% tin; the first standard specifications for solders were produced in 1918 by the ASTM. The USA is currently contemplating the banning of lead-bearing products; lead-free solders are being sought.

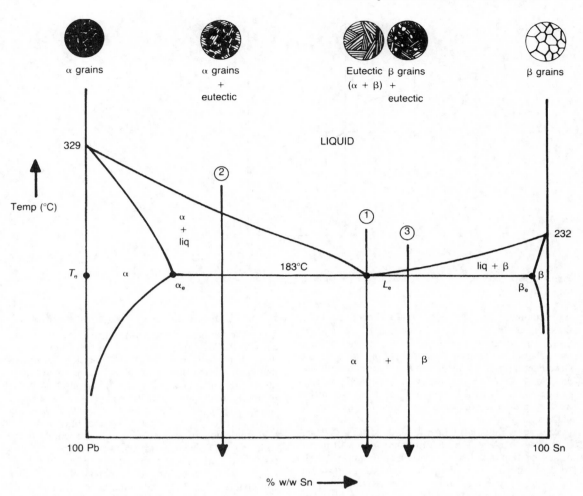

Figure 3.16 *Phase diagram for Pb–Sn system. Alloy 1: 63Sn–37Pb, Alloy 2: 70Pb–30Sn, Alloy 3: 70Sn–30Pb.*

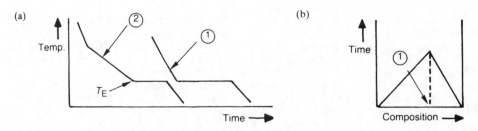

Figure 3.17 *(a) Typical cooling curves for hypo-eutectic alloy 2 and eutectic alloy 1 in Figure 3.16 and (b) dependence of duration of cooling arrest at eutectic temperature T_E on composition.*

3.2.5.2 The peritectic reaction

Whereas eutectic systems often occur when the melting points of the two components are fairly similar, the second important type of invariant three-phase condition, the peritectic reaction, is often found when the components have a large difference in melting points.

Usually they occur in the more complicated systems; for instance, there is a cascade of five peritectic reactions in the Cu–Zn system (Figure 3.20).

A simple form of peritectic system is shown in, Figure 3.18a; although relatively rare in practice (e.g. Ag–Pt), it can serve to illustrate the basic principles.

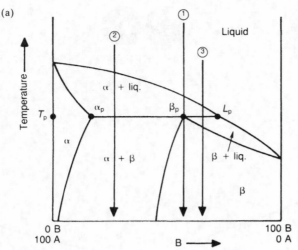

Figure 3.18 *(a) Simple peritectic system; (b) development of a peritectic 'wall'.*

A horizontal line, the key to the reaction, links three critical phase compositions; that is, α_p, β_p and liquid L_p. A peritectic reaction occurs if the average composition of the alloy crosses this line during either slow heating or cooling. It can be represented by the equation $\alpha_p + L_p \rightleftharpoons \beta_p$. Binary alloys containing less of component B than the point α_p will behave in the manner previously described for solid solutions. A melt of alloy 1, which is of peritectic composition, will freeze over a range of temperature, depositing crystals of primary α-phase. The melt composition will move down the liquidus, becoming richer in component B. At the peritectic temperature T_p, liquid of composition L_p will react with these primary crystals, transforming them completely into a new phase, β, of different crystal structure in accordance with the equation $\alpha_p + L_p \rightarrow \beta_p$. In the system shown, β remains stable during further cooling. Alloy 2 will aso deposit primary α, but the reaction at temperature T_p will not consume all these crystals and the final solid will consist of β formed by peritectic reaction and residual α. Initially, the α/β mass ratio will be approximately 2.5 to 1 but both phases will adjust their compositions during subsequent cooling. In the case of alloy 3, fewer primary crystals of α form: later, they are completely destroyed by the peritectic reaction. The amount of β in the resultant mixture of β and liquid increases until the liquid disappears and an entire structure of β is produced.

The above descriptions assume that equilibrium is attained at each stage of cooling. Although very slow cooling is unlikely in practice, the nature of the peritectic reaction introduces a further complication. The reaction product β tends to form a shell around the particles of primary α: its presence obviously inhibits the exchange of atoms by diffusion which equilibrium demands (Figure 3.18b).

3.2.5.3 Classification of three-phase equilibria
The principal invariant equilibria involving three condensed (solid, liquid) phases can be conveniently divided into eutectic- and peritectic-types and classified in the manner shown in Table 3.1. Interpretation of these reactions follows the methodology already set out for the more common eutectic and peritectic reactions.

The inverse relation between eutectic- and peritectic-type reactions is apparent from the line diagrams. Eutectoid and peritectoid reactions occur wholly in the solid state. (The eutectoid reaction $\gamma \rightleftharpoons \alpha + Fe_3C$ is the basis of the heat-treatment of steels.) In all the systems so far described, the components have been completely miscible in the liquid state. In monotectic and syntectic systems, the liquid phase field contains a region in which two different liquids (L_1 and L_2) are immiscible.

3.2.6 Intermediate phases
An intermediate phase differs in crystal structure from the primary phases and lies between them in a phase diagram. In Figure 3.19, which shows the diagram for the Mg–Si system, Mg_2Si is the intermediate phase. Sometimes intermediate phases have definite stoichiometric ratios of constituent atoms and appear as a single vertical line in the diagram. However, they frequently exist over a range of composition and it is therefore generally advisable to avoid the term 'compound'.

In some diagrams, such as Figure 3.19, they extend from room temperature to the liquidus and melt or freeze without any change in composition. Such a melting point is said to be congruent: the melting point of a eutectic alloy is incongruent. A congruently melting phase provides a convenient means to divide a complex phase diagram (binary or ternary) into more readily understandable parts. For instance, an

Table 3.1 Classification of three-phase equilibria

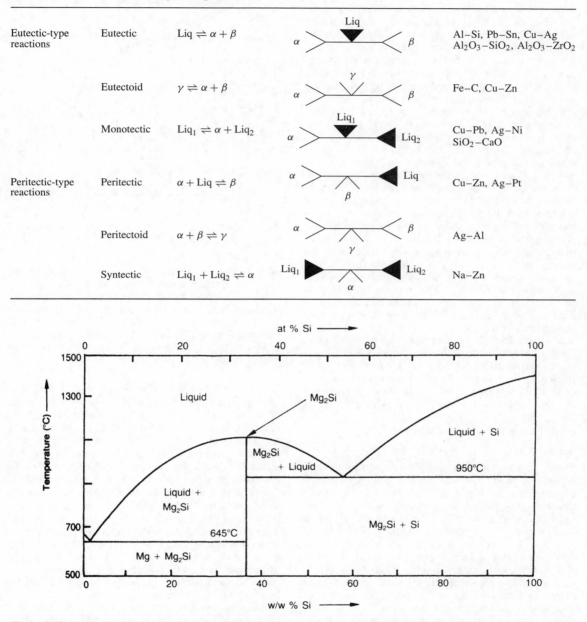

Eutectic-type reactions	Eutectic	$Liq \rightleftharpoons \alpha + \beta$		Al–Si, Pb–Sn, Cu–Ag, Al_2O_3–SiO_2, Al_2O_3–ZrO_2
	Eutectoid	$\gamma \rightleftharpoons \alpha + \beta$		Fe–C, Cu–Zn
	Monotectic	$Liq_1 \rightleftharpoons \alpha + Liq_2$		Cu–Pb, Ag–Ni, SiO_2–CaO
Peritectic-type reactions	Peritectic	$\alpha + Liq \rightleftharpoons \beta$		Cu–Zn, Ag–Pt
	Peritectoid	$\alpha + \beta \rightleftharpoons \gamma$		Ag–Al
	Syntectic	$Liq_1 + Liq_2 \rightleftharpoons \alpha$		Na–Zn

Figure 3.19 *Phase diagram for Mg–Si system showing intermediate phase Mg_2Si (after Brandes and Brook, 1992).*

ordinate through the vertex of the intermediate phase in Figure 3.19 produces two simple eutectic sub-systems. Similarly, an ordinate can be erected to pass through the minimum (or maximum) of the liquidus of a solid solution (Figure 3.38b).

In general, intermediate phases are hard and brittle, having a complex crystal structure (e.g. Fe_3C, $CuAl_2$ (θ)). For instance, it is advisable to restrict time and temperature when soldering copper alloys, otherwise it is possible for undesirable brittle layers of Cu_3Sn and Cu_6Sn_5 to form at the interface.

3.2.7 Limitations of phase diagrams

Phase diagrams are extremely useful in the interpretation of metallic and ceramic structures but they are subject to several restriction. Primarily, they identify which phases are likely to be present and provide compositional data. The most serious limitation is that they give no information on the structural form and distribution of phases (e.g. lamellae, spheroids, intergranular films, etc.). This is unfortunate, since these two features, which depend upon the surface energy effects between different phases and strain energy effects due to volume and shape changes during transformations, play an important role in the mechanical behaviour of materials. This is understood if we consider a two-phase $(\alpha + \beta)$ material containing only a small amount of β-phase. The β-phase may be dispersed evenly as particles throughout the α-grains, in which case the mechanical properties of the material would be largely governed by those of the α-phase. However, if the β-phase is concentrated at grain boundary surfaces of the α-phase, then the mechanical behaviour of the material will be largely dictated by the properties of the β-phase. For instance, small amounts of sulphide particles, such as grey manganese sulphide (MnS), are usually tolerable in steels but sulphide films at the grain boundaries cause unacceptable embrittlement.

A second limitation is that phase diagrams portray only equilibrium states. As indicated in previous sections, alloys are rarely cooled or heated at very slow rates. For instance, quenching, as practised in the heat-treatment of steels, can produce metastable phases known as martensite and bainite that will then remain unchanged at room temperature. Neither appears in phase diagrams. In such cases it is necessary to devise methods for expressing the rate at which equilibrium is approached and its temperature-dependency.

3.2.8 Some key phase diagrams

3.2.8.1 Copper–zinc system

Phase diagrams for most systems, metallic and ceramic, are usually more complex than the examples discussed so far. Figure 3.20 for the Cu–Zn system illustrates this point. The structural characteristics and mechanical behaviour of the industrial alloys known as brasses can be understood in terms of the copper-rich end of this diagram. Copper can dissolve up to 40% w/w of zinc and cooling of any alloy in this range will produce an extensive primary solid solution (fcc-α). By contrast, the other primary solid solution (η) is extremely limited. A special feature of the diagram is the presence of four intermediate phases (β, γ, δ, ε). Each is formed during freezing by peritectic reaction and each exists over a range of composition. Another notable feature is the order–disorder transformation which occurs in alloys containing about 50% zinc over the temperature range 450–470°C. Above this temperature range, bcc β-phase exists as a disordered solid solution. At lower temperatures, the zinc atoms are distributed regularly on the bcc lattice: this ordered phase is denoted by β'.

Suppose that two thin plates of copper and zinc are held in very close contact and heated at a temperature

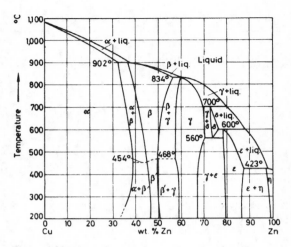

Figure 3.20 *Phase diagram for copper–zinc (from Raynor; courtesy of the Institute of Metals).*

of 400°C for several days. Transverse sectioning of the diffusion couple will reveal five phases in the sequence $\alpha/\beta/\gamma/\varepsilon/\eta$, separated from each other by a planar interface. The δ-phase will be absent because it is unstable at temperatures below its eutectoid horizontal (560°C). Continuation of diffusion will eventually produce one or two phases, depending on the original proportions of copper and zinc.

3.2.8.2 Iron–carbon system

The diagram for the part of the Fe–C system shown in Figure 3.21 is the basis for understanding the microstructures of the ferrous alloys known as steels and cast irons. Dissolved carbon clearly has a pronounced effect upon the liquidus, explaining why the difficulty of achieving furnace temperatures of 1600°C caused large-scale production of cast irons to predate that of steel. The three allotropes of pure iron are α-Fe (bcc), γ-Fe (fcc) and δ-Fe (bcc).[1] Small atoms of carbon dissolve interstitially in these allotropes to form three primary solid solutions: respectively, they are α-phase (ferrite), γ-phase (austenite) and δ-phase. At the other end of the diagram is the orthorhombic intermediate phase Fe_3C, which is known as cementite.

The large difference in solid solubility of carbon in austenite and ferrite, together with the existence of a eutectoid reaction, are responsible for the versatile behaviour of steels during heat-treatment. Ae_1, Ae_2, Ae_3 and A_{cm} indicate the temperatures at which phase changes occur: they are arrest points for equilibria detected during thermal analysis. For instance, slow cooling enables austenite (0.8% C) to decompose eutectoidally at the temperature Ae_1 and form the microconstituent pearlite, a lamellar composite of soft,

[1] The sequence omits β-Fe, a term once used to denote a non-magnetic form of α-Fe which exists above the Curie point.

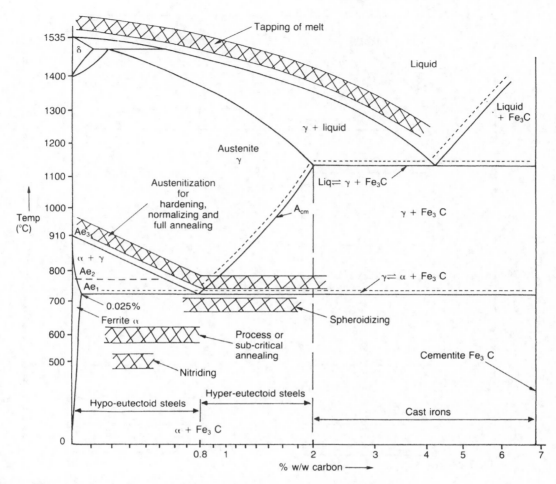

Figure 3.21 *Phase diagram for Fe–C system (dotted lines represent iron-graphite equilibrium).*

ductile ferrite (initially 0.025% C) and hard, brittle cementite (6.67% C). Quenching of austenite from a temperature above Ae₃ forms a hard metastable phase known as martensite. From the diagram one can see why a medium-carbon (0.4%) steel must be quenched from a higher Ae₃ temperature than a high-carbon (0.8%) steel. Temperature and composition 'windows' for some important heat-treatment operations have been superimposed upon the phase diagram.

3.2.8.3 Copper–lead system

The phase diagram for the Cu–Pb system (Figure 3.22) provides an interesting example of extremely limited solubility in the solid state and partial immiscibility in the liquid state. The two components differ greatly in density and melting point. Solid solutions, α and β, exist at the ends of the diagram. The 'miscibility gap' in the liquid phase takes the form of a dome-shaped two-phase $(L_1 + L_2)$ field. At temperatures above the top of the dome, the critical point,

liquid miscibility is complete. The upper isothermal represents a monotectic reaction, i.e. $L_1 \rightleftharpoons \alpha + L_2$.

On cooling, a hyper-monotectic 50Cu–50Pb melt will separate into two liquids of different composition. The degree of separation depends on cooling conditions. Like oil and water, the two liquids may form an emulsion of droplets or separate into layers according to density. At a temperature of 954°C, the copper-rich liquid L_1 disappears, forming α crystals and more of the lead-rich liquid L_2. This liquid phase gets richer in lead and eventually decomposes by eutectic reaction, i.e. $L_2 \rightleftharpoons \alpha + \beta$. (Tie-lines can be used for all two-phase fields, of course; however, because of density differences, mass ratios may differ greatly from observed volume ratios.)

The hypo-monotectic 70Cu–30Pb alloy, rapidly cast, has been used for steel-backed bearings: dispersed friction-reducing particles of lead-rich β are supported in a supporting matrix of copper-rich α. Binary combinations of conductive metal (Cu, Ag) and

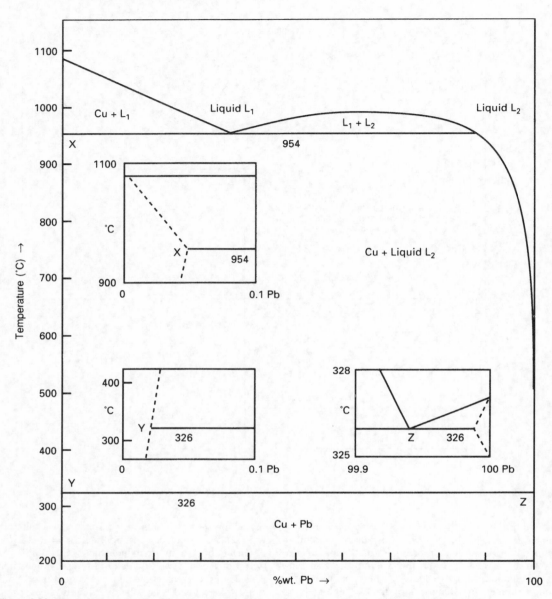

Figure 3.22 *Phase diagram for Cu–Pb system (by permission of the Copper Development Association, 1993).*

refractory arc-resistant metal (W, Mo, Ni) have been used for electrical contacts (e.g. 60Ag–40Ni). These particular monotectic systems, with their liquid immiscibility, are difficult to cast and are therefore made by powder metallurgy techniques.

3.2.8.4 Alumina–silica system

The binary phase diagram for alumina–silica (Figure 3.23) is of special relevance to the refractories industry, an industry which produces the bricks, slabs, shapes, etc. for the high-temperature plant that make steel-making, glass-making, heat-treatment, etc. possible. The profile of its liquidus shows a minimum and thus mirrors the refractoriness of aluminosilicate refractories (Figure 3.24). Refractoriness, the prime requirement of a refractory, is commonly determined by an empirical laboratory test. A sample cone of a given refractory is placed on a plaque and located at the centre of a ring of standard cones, each of which has a different softening or slumping temperature and is identified by a Pyrometric Cone Equivalent (PCE) number. All cones are then slowly heated until the

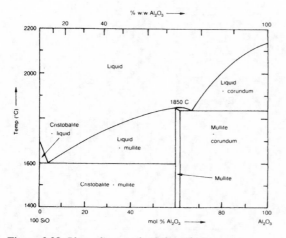

Figure 3.23 *Phase diagram for SiO₂–Al₂O₃ system.*

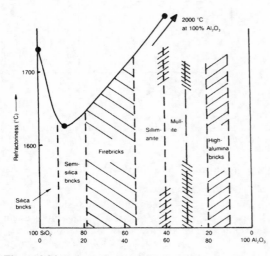

Figure 3.24 *Refractoriness of aluminosilicate ceramics.*

sample cone bends or slumps under gravity: the PCE of a standard cone that has behaved similarly is noted and taken to represent the refractoriness of the sample. It will be realized that the end-point of the PCE test is rather arbitrary, being a rising-temperature value. (Other requirements may include refractoriness-under-load, resistance to thermal shock, resistance to attack by molten slag, low thermal conductivity, etc.)

The steeply-descending liquidus shows the adverse effect of a few per cent of alumina on the refractoriness of silica bricks. (Sodium oxide, Na_2O, has an even more pronounced eutectic-forming effect and is commonly used to flux sand particles during glass-melting.) The discovery of this eutectic point led to immediate efforts to keep the alumina content as far below 5% as possible. Silica refractories are made by firing size-graded quartzite grains and a small amount

of lime (CaO) flux at a temperature of 1450°C: the final structure consists of tridymite, cristobalite and a minimal amount of unconverted quartz. Tridymite is preferred to cristobalite because of the large volume change (~1%) associated with the α/β cristobalite inversion. The lime forms an intergranular bond of SiO_2–CaO glass. Chequerwork assemblies of silica bricks are used in hot-blast stoves that regeneratively preheat combustion air for iron-making blast furnaces to temperatures of 1200–1300°C. Silica bricks have a surprisingly good refractoriness-under-load at temperatures only 50°C or so below the melting point of pure silica 1723°C). Apparently, the fired grains of tridymite and cristobalite interlock, being able to withstand a compressive stress of, say, 0.35 MN m⁻² at these high temperature levels.

Firebricks made from carefully-selected low-iron clays are traditionally used for furnace-building. These clays consist essentially of minute platey crystals of kaolinite, $Al_2(Si_2O_5)(OH)_4$: the (OH) groups are expelled during firing. The alumina content (46%) of fired kaolinite sets the upper limit of the normal composition range for firebricks. Refractoriness rises steeply with alumina content and aluminous fireclays containing 40% or more of alumina are therefore particularly valued. A fireclay suitable for refractories should have a PCE of at least 30 (equivalent to 1670°C): with aluminous clays the PCE can rise to 35 (1770°C). Firing the clay at temperatures of 1200–1400°C forms a glassy bond and an interlocking mass of very small lath-like crystals of mullite; this is the intermediate phase with a narrow range of composition which marks the edge of the important (mullite + corundum) plateau. High-alumina bricks, with their better refractoriness, have tended to replace firebricks. An appropriate raw material is obtained by taking clay and adding alumina (bauxite, artificial corundum) or a 'sillimanite-type' mineral, Al_2SiO_5 (andalusite, sillimanite, kyanite).

Phase transformations in ceramic systems are generally more sluggish than in metallic systems and steep concentration gradients can be present on a micro-scale. Thus tie-lines across the silica–mullite field usually only give approximate proportions of these two phases. The presence of traces of catalysing mineralizers, such as lime, can make application of the diagram nominal rather than rigorous. For instance, although silica bricks are fired at a temperature of 1450°C, which is within the stability range of tridymite (870–1470°C), cristobalite is able to form in quantity. However, during service, true stability is approached and a silica brick operating in a temperature gradient will develop clearly-defined and separate zones of tridymite and cristobalite.

By tradition, refractories are often said to be acid or basic, indicating their suitability for operation in contact with acid (SiO_2-rich) or basic (CaO- or FeO-rich) slags. For instance, suppose that conditions are reducing and the lower oxide of iron, FeO, forms in a basic steel-making slag (1600°C). 'Acid' silica refractory will be rapidly destroyed because this ferrous oxide reacts with silica to form fayalite, Fe_2SiO_4, which has a melting point of 1180°C. (The SiO_2-FeO phase diagram shows a sudden fall in the liquidus.) However, in certain cases, this approach is scientifically inadequate. For instance, 'acid' silica also has a surprising tolerance for basic CaO-rich slags. Reference to the SiO_2-CaO diagram reveals that there is a monotectic plateau at its silica-rich end, a feature that is preferable to a steeply descending liquidus. Its existence accounts for the slower rate of attack by molten basic slag and also, incidentally, for the feasibility of using lime as a bonding agent for silica grains during firing.

3.2.8.5 Nickel−sulphur−oxygen and chromium−sulphur−oxygen systems

The hot corrosion of superalloys based upon nickel, iron or cobalt by flue or exhaust gases from the combustion of sulphur-containing fuels is a problem common to a number of industries (e.g. power generation). These gases contain nitrogen, oxygen (excess to stoichiometric combustion requirements), carbon dioxide, water vapour, sulphur dioxide, sulphur trioxide, etc. In the case of a nickel-based alloy, the principal corrosive agents are sulphur and oxygen. They form nickel oxide and/or sulphide phases at the flue gas/alloy interface: their presence represents metal wastage. A phase diagram for the Ni−S−O system, which makes due allowance for the pressure variables, provides a valuable insight into the thermochemistry of attack of a Ni-based superalloy. Although disregarding kinetic factors, such as diffusion, a stability diagram of this type greatly helps understanding of underlying mechanisms. Primarily, it indicates which phases are likely to form. Application of these diagrams to hot corrosion phenomena is discussed in Chapter 12.

Under equilibrium conditions, the variables governing chemical reaction at a nickel/gas interface are temperature and the partial pressures p_{o_2} and p_{s_2} for the gas phase. For isothermal conditions, the general disposition of phases will be as shown schematically in Figure 3.25a. An isothermal section (900 K) is depicted in Figure 3.25b. A comprehensive three-dimensional representation, based upon standard free energy data for the various competing reactions, is given in Figure 3.26. Section **AA** is isothermal (1200 K): the full diagram may be regarded as a parallel stacking of an infinite number of such vertical sections. From the Phase Rule, $P + F = C + 2$, it follows that $F = 5 - P$. Hence, for equilibrium between gas and one condensed phase, there are three degrees of freedom and equilibrium is represented by a volume. Similarly, equilibrium between gas and three condensed phases is represented by a line. The bivariant and univariant equilibrium equations which form the basis of the three-dimensional stability diagram are given in Figure 3.26.

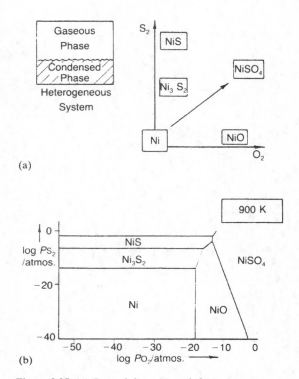

Figure 3.25 (a) General disposition of phases in Ni−S−O system; (b) isothermal section at temperature of 900 K (after Quets and Dresher, 1969, pp. 583−599).

3.2.9 Ternary phase diagrams

3.2.9.1 The ternary prism

Phase diagrams for three-component systems usually take the standard form of a prism which combines an equilateral triangular base (ABC) with three binary system 'walls' (A−B, B−C, C−A), as shown in Figure 3.27a. This three-dimensional form allows the three independent variables to be specified, i.e. two component concentrations and temperature. As the diagram is isobaric, the modified Phase Rule applies. The vertical edges can represent pure components of either metallic or ceramic systems. Isothermal contour lines are helpful means for indicating the curvature of liquidus and solidus surfaces.

Figure 3.27b shows some of the ways in which the base of the prism, the Gibbs triangle, is used. For instance, the recommended method for deriving the composition of a point P representing a ternary alloy is to draw two construction lines to cut the nearest of the three binary composition scales. In similar fashion, the composition of the phases at each end of the tie-line passing through P can be derived. Tie-triangles representing three-phase equilibria commonly appear in horizontal (isothermal) sections through the prism. The example in Figure 3.27b shows equilibrium between α, β and γ phases for an alloy of average

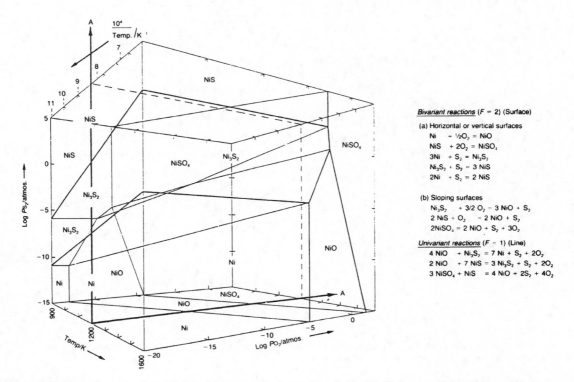

Figure 3.26 *Three-dimensional equilibrium diagram and basic reactions for the Ni–S–O system (after Quets and Dresher, 1969, pp. 583–99).*

(a)

(b)

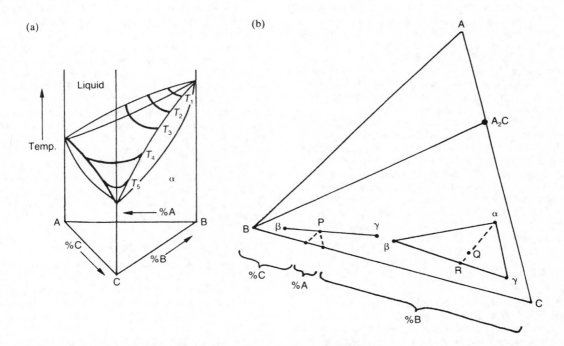

Figure 3.27 *(a) Ternary system with complete miscibility in solid and liquid phases and (b) the Gibbs triangle.*

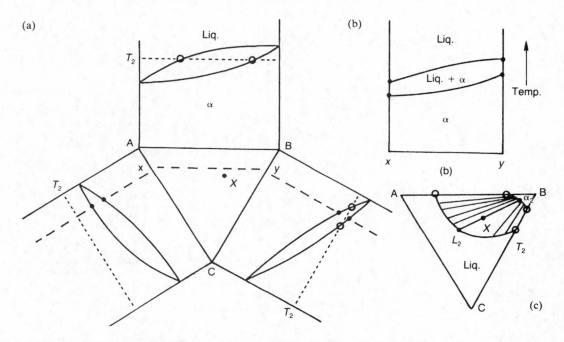

Figure 3.28 *Derivation of vertical and horizontal sections: (a) lines of construction, (b) vertical section (~20% C), (c) isothermal section at temperature T_2.*

composition Q. Each side of the tie-triangle is a tie-line. Applying the centre-of-gravity principle, the weight proportion of any phase in the mixture can be obtained by drawing a line and measuring two lengths (e.g. % $\alpha = QR/R\alpha \times 100$). Congruently-melting phases can be linked together by a join-line, thus simplifying interpretation. For example, component B may be linked to the intermediate phase A_2C. Any line originating at a pure component also provides a constant ratio between the other two components along its length.

Tie-lines can be drawn across two-phase regions of isothermal sections but, unlike equivalent tie-lines in binary systems, they do not necessarily point directly towards pure components: their exact disposition has to be determined by practical experiment. It follows that when vertical sections (isopleths) are taken through the prism, insertion of tie-lines across a two-phase field is not always possible because they may be inclined to the vertical plane. For related reasons, one finds that three-phase regions in vertical sections have slightly curved sides.

3.2.9.2 Complete solid miscibility

In the ternary diagram of Figure 3.27a the volume of continuous solid solubility (α) is separated from the liquid phase by a two-phase (α + Liquid) zone of convex-lens shape. Analysis of this diagram provides an insight into the compositional changes attending the freezing/melting of a ternary solid solution. It is convenient to fold down the binary 'walls' so that

construction lines for isothermal and vertical sections can be prepared (Figure 3.28a). A vertical section, which provides guidance on the freezing and melting of all alloys containing the same amount of component C, is shown in Figure 3.28b.

Assuming that an alloy of average composition X freezes over a temperature range of T_1 to T_4, Figure 3.28c shows an isothermal section for a temperature T_2 just below the liquidus. The two-phase field in this section is bivariant. As the proportions and compositions of α and liquid L gradually change, the tie-lines for the four temperatures change their orientation. This rotational effect, which is shown by projecting the tie-lines downward onto the basal triangle (Figure 3.29), occurs because the composition of the residual liquid L moves in the general direction of C, the component of lowest melting point, and the composition of α approaches the average composition X.

3.2.9.3 Three-phase equilibria

In a ternary system, a eutectic reaction such as Liquid $\rightleftharpoons \alpha + \beta$ is univariant ($F = 1$) and, unlike its binary equivalent, takes place over a range of temperature. Its characteristics can be demonstrated by considering the system shown in Figure 3.30a in which two binary eutectic reactions occur at different temperatures. The key feature of the diagram, a three-phase triangle ($\alpha + \beta$ + Liquid), evolves from the upper binary eutectic horizontal and then appears to move down

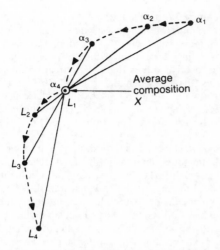

Figure 3.29 *Rotation of tie-lines during freezing of a ternary solid solution (α).*

three 'guide rails' until it finally degenerates into the lower binary eutectic horizontal. As the temperature

falls, the eutectic reaction Liquid → α + β takes place; hence the leading vertex of the triangle represents the composition of the liquid phase.

A vertical section between component A and the mid-point of the B–C 'wall' (Figure 3.30b) and an isothermal section taken when the tie-triangle is about halfway through its downward movement (Figure 3.30c) help to explain typical solidification sequences. The vertical section shows that when an alloy of composition X freezes, primary β formation is followed by eutectic reaction over a range of temperature. The final microstructure consists of primary β in a eutectic matrix of α and β. If we now concentrate on the immediate surroundings of the isothermal triangle, as depicted in Figure 3.30d, primary deposition of β is represented by the tie-line linking β and liquid compositions. (This type of tie-line rotates, as described previously.) The eutectic reaction (Liquid → α + β) starts when one leading edge of the triangle, in this case the tie-line L-β, cuts point X and is completed when its trailing edge, the tie-line α-β, cuts X. As X is traversed by tie-triangles, the relative amounts of the three phases can be derived for each isotherm. For alloy Y, lying on the valley line of the

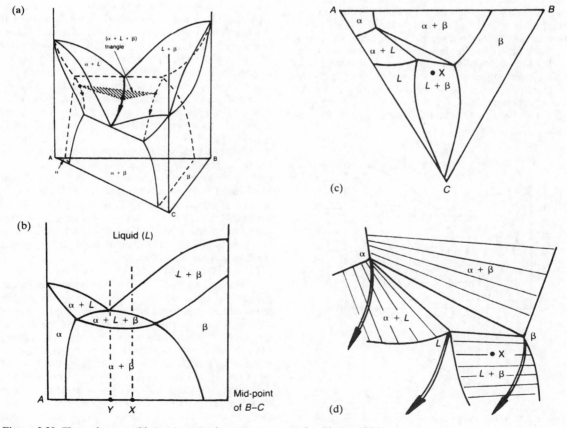

Figure 3.30 *Three-phase equilibrium in a simple ternary system (after Rhines, 1956).*

eutectic reaction, it can be seen from both sections that no primary β forms and freezing only produces a mixture of α and β phases.

These ideas can also be applied to three-phase peritectic reactions in ternary systems, i.e. α + Liquid $\rightleftharpoons$ β. Being the converse form of the eutectic reaction, its tie-triangle has the $(L + \alpha)$ tie-line as the leading side and a trailing vertex. The 'inflated' triangle in vertical sections is inverted.

3.2.9.4 Four-phase equilibria

The simplest form of four-phase equilibrium, referred to as Class I, is summarized by the ternary eutectic reaction, Liquid $\rightleftharpoons$ $\alpha + \beta + \gamma$. This invariant condition $(F = 0)$ is represented by the triangular plateau at the heart of Figure 3.31. The plateau itself is solid and can be regarded as a stack of three-phase triangles, i.e. $\alpha + \beta + \gamma$. Each of the three constituent binary systems is eutectiferous with limited solid solubility. It is necessary to visualize that, on cooling, each of the three eutectic horizontals becomes the independent source of a set of descending three-phase eutectic triangles which behave in the general manner described in the previous section. At the ternary eutectic level of the plateau, temperature T_6, these three triangles coalesce and the ternary reaction follows. (The lines where triangles meet are needed when sections are drawn.) The stack of tie-triangles associated with each binary eutectic reaction defines a beak-shaped volume, the upper edge being a 'valley' line. Figure 3.31 shows three valley lines descending to the ternary eutectic point.

A typical sequence of isothermal sections is shown in Figure 3.32 over the temperature range T_1–T_6. As the temperature falls, the three eutectic tie-triangles appear in succession. The liquid field shrinks until, at

temperature T_6, the three triangles coalesce to form the larger triangle of four-phase equilibrium. Below this level, after ternary eutectic reaction, the three solid phases adjust their composition in accordance with their respective solvus lines. A vertical section which includes all alloys containing about 30% component C is shown in Figure 3.33. The three-phase plateau is immediately apparent. This section also intersects three of the binary eutectic reaction 'beaks', i.e. Liquid + $\alpha + \beta$, Liquid + $\alpha + \gamma$ and Liquid + $\beta + \gamma$.

We will consider the solidification of four alloys which are superimposed on the T_6 isothermal section of Figure 3.32. The simplest case of solidification is the liquid ternary alloy W, which transforms to three phases immediately below the plateau temperature T_6. This type of alloy is the basis of fusible alloys which are used for special low m.p. applications (e.g. plugs for fire-extinguishing sprinkler systems). For example, a certain combination of lead (m.p. 327°C), tin (m.p. 232°C) and bismuth (m.p. 269°C) melts at a temperature of 93°C. On cooling, liquid alloy X will first decompose over a range of temperature to form a eutectic mixture $(\alpha + \beta)$ and then change in composition along a 'valley' line until its residue is finally consumed in the ternary eutectic reaction. Liquid alloy Y will first deposit primary β and then, with rotation of $(\beta +$ Liquid) tie-lines, become depleted in B until it reaches the nearby 'valley' line. Thereafter, it behaves like alloy X. In the special case of alloy Z, which lies on a construction line joining the ternary eutectic point to a corner of the ternary plateau, it will first form a primary phase (γ) and then undergo the ternary reaction: there will be no preliminary binary eutectic reaction.

We will now outline the nature of two other four-phase equilibria. Figure 3.34a illustrates the Class II reaction, Liquid + α $\rightleftharpoons$ $\beta + \gamma$. As the temperature is lowered, two sets of descending tie-triangles representing a peritectic-type reaction and a eutectic-type reaction, respectively, combine at the ternary reaction isotherm to form a kite-shaped plane. This plane then divides, as shown schematically in Figure 3.35a, forming a solid three-phase plateau and initiating a descending set of eutectic triangles. Unlike Class I equilibrium, the composition which reacts with α-phase lies outside the limits of the top of the plateau.

From Figure 3.34b, which illustrates Class III equilibrium, Liquid + $\alpha + \beta$ $\rightleftharpoons$ γ, a solid three-phase plateau is again a central feature. At the higher temperatures, a eutectic-type reaction (Liquid $\rightarrow$ $\alpha +$ β) generates a stack of tie-triangles. At the critical temperature of the ternary peritectic reaction, a large triangle breaks up and two sets of peritectic-type triangles are initiated (Figure 3.35b). If the Class II and Class III ternary reaction sequences are compared with that of Class I, it will be seen that there is an inverse relation between eutectic Class I and the peritectic Class III. Class II is intermediate in character to Classes I and III.

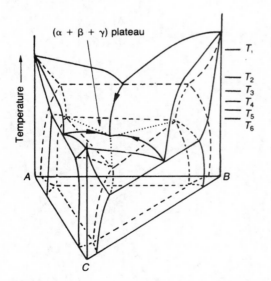

Figure 3.31 *Phase diagram for a ternary eutectic system (Class I four-phase equilibrium) (after Rhines, 1956).*

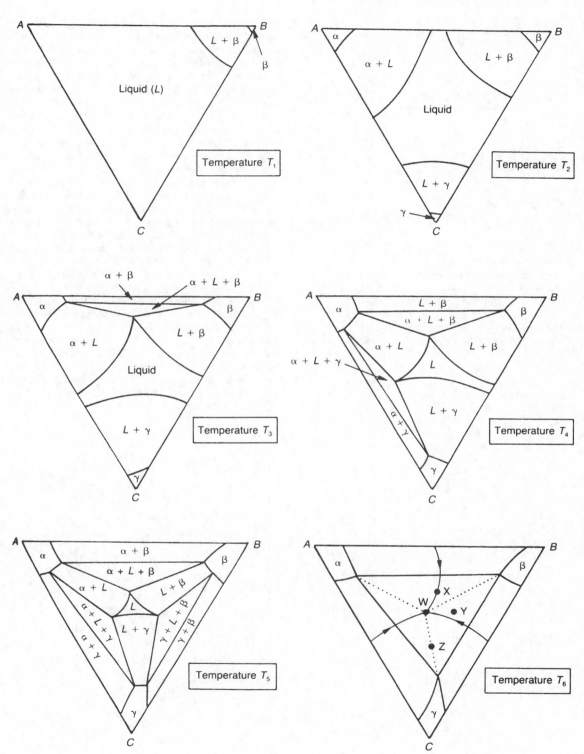

Figure 3.32 *Horizontal sections at six temperatures in phase diagram of Figure 3.31 (after Rhines, 1956).*

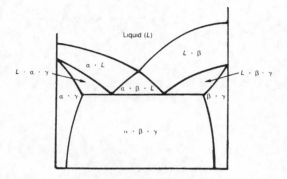

Figure 3.33 *Vertical section through ternary system shown in Figure 3.31.*

3.2.9.5 Application to dielectric ceramics

The phase diagram for the $MgO-Al_2O_3-SiO_2$ system (Figure 3.36) has proved extremely useful in providing guidance on firing strategies and optimum phase relations for important dielectric[1] ceramics. (The diagram is also relevant to basic steel-making refractories based on magnesia, i.e. periclase.) Its principal features are the straight join-lines, which link binary and/or ternary compounds, and the curving

---

[1] A dielectric is a material that contains few or no free electrons and has a lower electrical conductivity than a metal.

'valley' lines. Two junctions of these join-lines lie within the diagram, marking the ternary compounds sapphirine ($M_4A_5S_2$) and cordierite ($M_2A_2S_5$). The join-lines divide the projection into tie-triangles (sometimes termed compatibility triangles). These triangles enable the amounts and composition of stable phases to be calculated. The topology of the liquidus surface is always of prime interest. In this system, the lowest liquidus temperature (1345°C) is associated with the tridymite–protoenstatite–cordierite eutectic.

In the complementary Figure 3.37, compositional zones for four classes of dielectric ceramic have been superimposed, i.e. forsterite ceramics, low-loss-factor steatites, steatite porcelains and cordierite ceramics. These fired ceramics originate from readily-workable clay/talc mixtures. The nominal formula for talc is $Mg_3(Si_2O_5)_2(OH)_2$. Fired clay can be regarded as mullite (A_3S_2) plus silica, and fired talc as protoenstatite (MS) plus silica: accordingly, the zones are located toward the silica-rich corner of the diagram. Each fired product consists of small crystals in a glassy matrix (20–30%). The amount of glass must be closely controlled. Ideally, control of firing is facilitated when the amount of glass-forming liquid phase changes slowly with changing temperature. In this respect, the presence of a steeply sloping liquidus is favourable (e.g. forsterite (Mg_2SiO_4) ceramics). Unfortunately, the other three materials tend to liquefy rather abruptly and to form too much liquid, making firing a potentially difficult operation.

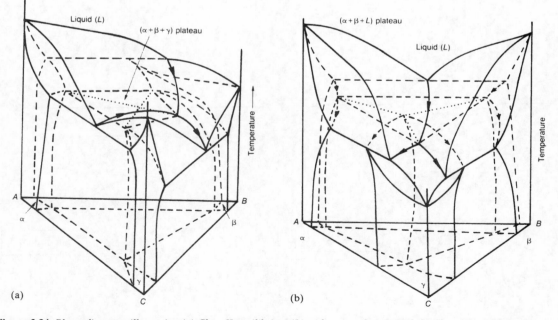

Figure 3.34 *Phase diagrams illustrating (a) Class II equilibrium (Liquid + $\alpha \rightleftharpoons \beta + \gamma$), and (b) Class III equilibrium (Liquid + $\alpha + \beta \rightleftharpoons \gamma$) (after Rhines, 1956).*

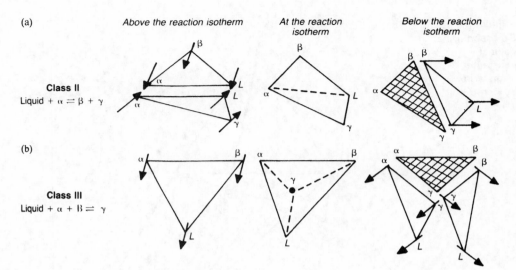

Figure 3.35 *Class II and Class III equilibria in ternary systems.*

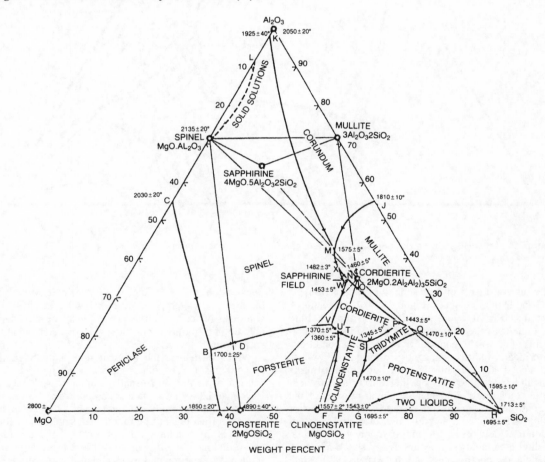

Figure 3.36 *Basal projection for MgO–Al₂O₃–SiO₂ system; regions of solid solution not shown (from Keith and Schairer, 1952; by permission of University of Chicago Press).*

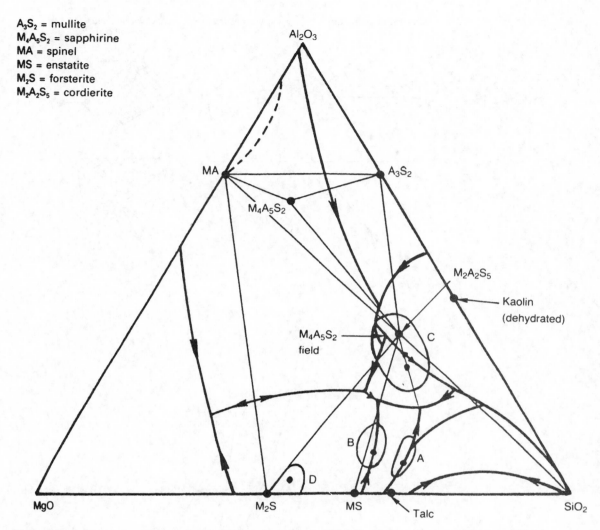

A$_3$S$_2$ = mullite
M$_4$A$_5$S$_2$ = sapphirine
MA = spinel
MS = enstatite
M$_2$S = forsterite
M$_2$A$_2$S$_5$ = cordierite

Figure 3.37 *Location of steatites, cordierite and forsterite in Figure 3.36 (after Kingery, Bowen and Uhlmann, 1976; by permission of Wiley-Interscience).*

Let us consider four representative compositions A, B, C and D in more detail (Figure 3.37). The steatite A is produced from a 90% talc–10% clay mixture. Pure talc liquefies very abruptly and clay is added to modify this undesirable feature; even then firing conditions are critical. Additional magnesia is used in low-loss steatites (B) for the same purpose but, again, firing is difficult to control. During cooling after firing, protoenstatite converts to clinoenstatite: small crystals of the latter are embedded in a glassy matrix. Low-loss-factor steatites also have a relatively high dielectric constant and are widely used for high-frequency insulators. Cordierite ceramics (C) are thermally shock-resistant, having a low coefficient of thermal expansion. Despite their restricted firing range, they find use as electrical resistor supports and burner tips. Fluxes can be used

to extend the freezing range of steatites and cordierites but the electrical properties are likely to suffer. In sharp contrast, forsterite ceramics (D), which are also suitable for high-frequency insulation, have a conveniently wide firing range.

When considering the application of phase diagrams to ceramics in general, it must be recognized that ceramic structures are usually complex in character. Raw materials often contain trace impurities which will shift boundaries in phase diagrams and influence rates of transformation. Furthermore, metastable glass formation is quite common. Determination of the actual phase diagrams is difficult and time-consuming; consequently, experimental work often focuses upon a specific problem or part of a system. Against this background, in circumstances where detailed information on phases is sought, it is advisable to refer back to the experimental conditions and data upon which the relevant phase diagram are based.

3.3 Principles of alloy theory

3.3.1 Primary substitutional solid solutions

3.3.1.1 The Hume-Rothery rules

The key phase diagrams outlined in Section 3.2.8 exhibit many common features (e.g. primary solid solutions, intermediate phases) and for systems based on simple metals some general rules[1] governing the formation of alloys have been formulated. These rules can form a useful basis for predicting alloying behaviour in other more complex systems.

In brief the rules for primary solid solubility are as follows:

1. *Atomic size factor* — If the atomic diameter of the solute atom differs by more than 15% from that of the solvent atom, the extent of the primary solid solution is small. In such cases it is said that the size-factor is unfavourable for extensive solid solution.

---

[1] These are usually called the Hume-Rothery rules because it was chiefly W. Hume-Rothery and his colleagues who formulated them.

2. *Electrochemical effect* — The more electropositive the one component and the more electronegative the other, the greater is the tendency for the two elements to form compounds rather than extensive solid solutions.
3. *Relative valency effect* — A metal of higher valency is more likely to dissolve to a large extent in one of lower valency than vice versa.

3.3.1.2 Size-factor effect

Two metals are able to form a continuous range of solid solutions only if they have the same crystal structure (e.g. copper and nickel). However, even when the crystal structure of the two elements is the same, the extent of the primary solubility is limited if the atomic size of the two metals, usually taken as the closest distance of approach of atoms in the crystal of the pure metal, is unfavourable. This is demonstrated in Figure 3.38 for alloy systems where rules 2 and 3 have been observed, i.e. the electrochemical properties of the two elements are similar and the solute is dissolved in a metal of lower valency. As the size difference between the atoms of the two component metals A and B approaches 15%, the equilibrium diagram changes from that of the copper–nickel type to one of a eutectic system with limited primary solid solubility.

The size-factor effect is due to the distortion produced in the parent lattice around the dissolved misfitting solute atom. In these localized regions the

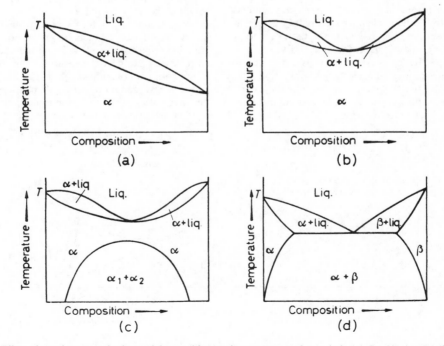

Figure 3.38 *Effect of size factor on the form of the equilibrium diagram; examples include (a) Cu–Ni, Au–Pt, (b) Ni–Pt, (c) Au–Ni, and (d) Cu–Ag.*

interatomic distance will differ from that given by the minimum in the $E-r$ curve of Figure 6.2, so that the internal energy and hence the free energy, G, of the system is raised. In the limit when the lattice distortion is greater than some critical value the primary solid solution becomes thermodynamically unstable relative to some other phase.

3.3.1.3 Electrochemical effect

This effect is best demonstrated by reference to the alloying behaviour of an electropositive solvent with solutes of increasing electronegativity. The electronegativity of elements in the Periodic Table increases from left to right in any period and from bottom to top in any group. Thus, if magnesium is alloyed with elements of Group IV the compounds formed, Mg_2 (Si, Sn or Pb), become more stable in the order lead, tin, silicon, as shown by their melting points, 550°C, 778°C and 1085°C, respectively. In accordance with rule 2 the extent of the primary solid solution is small (≈ 7.75 at.%, 3.35 at.%, and negligible, respectively, at the eutectic temperature) and also decreases in the order lead, tin, silicon. Similar effects are also observed with elements of Group V, which includes the elements bismuth, antimony and arsenic, when the compounds Mg_3 (Bi, Sb or As)$_2$ are formed.

The importance of compound formation in controlling the extent of the primary solid solution can be appreciated by reference to Figure 3.39, where the curves represent the free-energy versus composition relationship between the α-phase and compound at a temperature T. It is clear from Figure 3.39a that at this temperature the α-phase is stable up to a composition c_1, above which the phase mixture ($\alpha +$ compound) has the lower free energy. When the compound becomes more stable, as shown in Figure 3.39b, the solid solubility decreases, and correspondingly the phase mixture is now stable over a greater composition range which extends from c_3 to c_4.

The above example is an illustration of a more general principle that the solubility of a phase decreases with increasing stability, and may also be used to show that the concentration of solute in solution increases as the radius of curvature of the precipitate particle decreases. Small precipitate particles are less stable than large particles and the variation of solubility with particle size is recognized in classical thermodynamics by the Thomson–Freundlich equation

$$\ln[c(r)/c] = 2\gamma\Omega/kTr \qquad (3.7)$$

where $c(r)$ is the concentration of solute in equilibrium with small particles of radius r, c the equilibrium concentration, γ the precipitate/matrix interfacial energy and Ω the atomic volume (see Chapter 8).

3.3.1.4 Relative valency effect

This is a general rule for alloys of the univalent metals, copper, silver and gold, with those of higher valency. Thus, for example, copper will dissolve approximately 40% zinc in solid solution but the solution of copper in zinc is limited. For solvent elements of higher valencies the application is not so general, and in fact exceptions, such as that exhibited by the magnesium-indium system, occur.

3.3.1.5 The primary solid solubility boundary

It is not yet possible to predict the exact form of the α-solid solubility boundary, but in general terms the boundary may be such that the range of primary solid solution either (1) increases or (2) decreases with rise in temperature. Both forms arise as a result of the increase in entropy which occurs when solute atoms are added to a solvent. It will be remembered that this entropy of mixing is a measure of the extra disorder of the solution compared with the pure metal.

The most common form of phase boundary is that indicating that the solution of one metal in another increases with rise in temperature. This follows

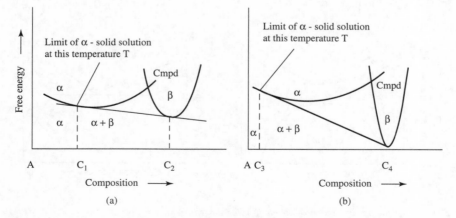

Figure 3.39 *Influence of compound stability on the solubility limit of the α phase at a given temperature.*

from thermodynamic reasoning since increasing the temperature favours the structure of highest entropy (because of the $-TS$ term in the relation $G = H - TS$) and in alloy systems of the simple eutectic type an α-solid solution has a higher entropy than a phase mixture $(\alpha + \beta)$. Thus, if the alloy exists as a phase mixture $(\alpha + \beta)$ at the lower temperatures, it does so because the value of H happens to be less for the mixture than for the homogeneous solution at that composition. However, because of its greater entropy term, the solution gradually becomes preferred at high temperatures. In more complex alloy systems, particularly those containing intermediate phases of the secondary solid solution type (e.g. copper–zinc, copper–gallium, copper–aluminium, etc.), the range of primary solid solution decreases with rise in temperature. This is because the β-phase, like the α-phase, is a disordered solid solution. However, since it occurs at a higher composition, it has a higher entropy of mixing, and consequently its free energy will fall more rapidly with rise in temperature. This is shown schematically in Figure 3.40. The point of contact on the free energy curve of the α-phase, determined by drawing the common tangent to the α and β curves, governs the solubility c at a given temperature T. The steep fall with temperature of this common tangent automatically gives rise to a decreasing solubility limit.

Many alloys of copper or silver reach the limit of solubility at an electron to atom ratio of about 1.4. The divalent elements zinc, cadmium and mercury have solubilities of approximately 40 at.%[1] (e.g. copper–zinc, silver–cadmium, silver–mercury), the trivalent elements approximately 20 at.% (e.g. copper–aluminium, copper–gallium, silver–aluminium, silver–indium) and the tetravalent elements about 13 at.% (e.g. copper–germanium, copper–silicon, silver–tin), respectively.

The limit of solubility has been explained by Jones in terms of the Brillouin zone structure (see Chapter 6). It is assumed that the density of states–energy curve for the two phases, α (the close-packed phase) and β (the more open phase), is of the form shown in Figure 3.41, where the $N(E)$ curve deviates from the parabolic relationship as the Fermi surface approaches the zone boundary.[2] As the solute is added to the solvent lattice and more electrons are added the top of the Fermi level moves towards A, i.e. where the

[1] For example, a copper–zinc alloy containing 40 at.% zinc has an e/a ratio of 1.4, i.e. for every 100 atoms, 60 are copper each contributing one valency electron and 40 are zinc each contributing two valency electrons, so that $e/a = (60 \times 1 + 40 \times 2)/100 = 1.4$.

[2] The shape of the Fermi surface may be determined from measurements of physical properties as a function of orientation in a single crystal. The surface resistance to a high-frequency current at low temperatures (the anomalous skin effect) shows that in copper the Fermi surface is distorted from the spherical shape but becomes more nearly spherical in copper alloys.

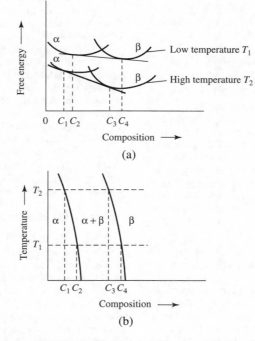

Figure 3.40 (a) The effect of temperature on the relative positions of the α- and β-phase free energy curves for an alloy system having a primary solid solubility of the form shown in (b).

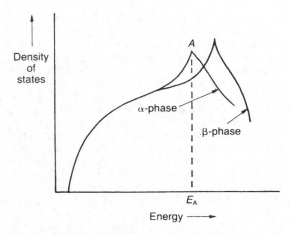

Figure 3.41 *Density of states versus energy diagram.*

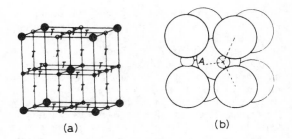

Figure 3.42 *(a) BCC lattice showing the relative positions of the main lattice sites, the octahedral interstices marked O, and the tetrahedral interstices marked T. (b) Structure cell of iron showing the distortions produced by the two different interstitial sites. Only three of the iron atoms surrounding the octahedral sites are shown; the fourth, centred at A, has been omitted for clarity (after Williamson and Smallman, 1953).*

density of states is high and the total energy E for a given electron concentration is low. Above this point the number of available energy levels decreases so markedly that the introduction of a few more electrons per atom causes a sharp increase in energy. Thus, just above this critical point the α structure becomes unstable relative to the alternative β structure which can accommodate the electrons within a smaller energy range, i.e. the energy of the Fermi level is lower if the β-phase curve is followed rather than the α-phase curve. The composition for which E_{max} reaches the point E_A is therefore a critical one, since the alloy will adopt that phase which has the lowest energy. It can be shown that this point corresponds to an electron-to-atom ratio of approximately 1.4.

3.3.2 Interstitial solid solutions

Interstitial solid solutions are formed when the solute atoms can fit into the interstices of the lattice of the solvent. However, an examination of the common crystal lattices shows that the size of the available interstices is restricted, and consequently only the small atoms, such as hydrogen, boron, carbon or nitrogen, with atomic radii very much less than one nanometre form such solutions. The most common examples occur in the transition elements and in particular the solution of carbon or nitrogen in iron is of great practical importance. In fcc iron (austenite) the largest interstice or 'hole' is at the centre of the unit cell (coordinates $\frac{1}{2}, \frac{1}{2}, \frac{1}{2}$) where there is space for an atom of radius 52 pm, i.e. $0.41r$ if r is the radius of the solvent atom. A carbon atom (80 pm (0.8 Å) diameter) or a nitrogen atom (70 pm diameter) therefore expands the lattice on solution, but nevertheless dissolves in quantities up to 1.7 wt% and 2.8 wt%, respectively. Although the bcc lattice is the more open structure the largest interstice is smaller than that in the fcc. In bcc iron (ferrite) the largest hole is at the position $(\frac{1}{2}, \frac{1}{4}, 0)$ and is a tetrahedral site where four iron atoms are

situated symmetrically around it; this can accommodate an atom of radius 36 pm, i.e. $0.29r$, as shown in Figure 3.42a. However, internal friction and X-ray diffraction experiments show that the carbon or nitrogen atoms do not use this site, but instead occupy a smaller site which can accommodate an atom only $0.154r$, or 19 pm. This position $(0, 0, \frac{1}{2})$ at the midpoints of the cell edges is known as the octahedral site since, as can be seen from Figure 3.42b, it has a distorted octahedral symmetry for which two of the iron atoms are nearer to the centre of the site than the other four nearest neighbours. The reason for the interstitial atoms preferring this small site is thought to be due to the elastic properties of the bcc lattice. The two iron atoms which lie above and below the interstice, and which are responsible for the smallness of the hole, can be pushed away more easily than the four atoms around the larger interstice. As a result, the solution of carbon in α-iron is extremely limited (0.02 wt%) and the structure becomes distorted into a body-centred tetragonal lattice. The c axis for each interstitial site is, however, disordered, so that this gives rise to a structure which is statistically cubic. The body-centred tetragonal structure forms the basis of martensite (an extremely hard metastable constituent of steel), since the quenching treatment given to steel retains the carbon in supersaturated solution (see Chapter 8).

3.3.3 Types of intermediate phases

3.3.3.1 Electrochemical compounds

The phases which form in the intermediate composition regions of the equilibrium diagram may be (1) electrochemical or full-zone compounds, (2) size-factor compounds, or (3) electron compounds. The term 'compound' still persists even though many of these phases do not obey the valency laws of chemistry and often exist over a wide composition range.

We have already seen that a strong tendency for compound formation exists when one element is electropositive and the other is electronegative. The magnesium-based compounds are probably the most common examples having the formula $Mg_2(Pb, Sn, Ge \text{ or } Si)$. These have many features in common with salt-like compounds since their compositions satisfy the chemical valency laws, their range of solubility is small, and usually they have high melting points. Moreover, many of these types of compounds have crystal structures identical to definite chemical compounds such as sodium chloride, NaCl, or calcium fluoride, CaF_2. In this respect the Mg_2X series are anti-isomorphous with the CaF_2 fluorspar structure, i.e. the magnesium metal atoms are in the position of the non-metallic fluoride atoms and the metalloid atoms such as tin or silicon take up the position of the metal atoms in calcium fluoride.

Even though these compounds obey all the chemical principles they may often be considered as special electron compounds. For example, the first Brillouin zone[1] of the CaF_2 structure is completely filled at $\frac{8}{3}$ electrons per atom, which significantly is exactly that supplied by the compound Mg_2Pb, Sn, ..., etc. Justification for calling these full-zone compounds is also provided by electrical conductivity measurements. In contrast to the behaviour of salt-like compounds which exhibit low conductivity even in the liquid state, the compound Mg_2Pb shows the normal conduction (which indicates the possibility of zone overlapping) while Mg_2Sn behaves like a semiconductor (indicating that a small energy gap exists between the first and second Brillouin zones).

In general, it is probable that both concepts are necessary to describe the complete situation. As we shall see in Section 3.3.3.3, with increasing electrochemical factor even true electron compounds begin to show some of the properties associated with chemical compounds, and the atoms in the structure take up ordered arrangements.

3.3.3.2 Size-factor compounds

When the atomic diameters of the two elements differ only slightly, electron compounds are formed, as discussed in the next section. However, when the difference in atomic diameter is appreciable, definite size-factor compounds are formed which may be of the (1) interstitial or (2) substitutional type.

A consideration of several interstitial solid solutions has shown that if the interstitial atom has an atomic radius 0.41 times that of the metal atom then it can fit into the largest available structural interstice without distortion. When the ratio of the radius of the interstitial atom to that of the metal atom is greater than 0.41 but less than 0.59, interstitial compounds are formed; hydrides, borides, carbides and nitrides of the transition metals are common examples. These compounds usually take up a simple structure of either the cubic

or hexagonal type, with the metal atoms occupying the normal lattice sites and the non-metal atoms the interstices. In general, the phases occur over a range of composition which is often centred about a simple formula such as M_2X and MX. Common examples are carbides and nitrides of titanium, zirconium, hafnium, vanadium, niobium and tantalum, all of which crystallize in the NaCl structure. It is clear, therefore, that these phases do not form merely as a result of the small atom fitting into the interstices of the solvent structure, since vanadium, niobium and tantalum are bcc, while titanium, zirconium and hafnium are cph. By changing their structure to fcc the transition metals allow the interstitial atom not only a larger 'hole' but also six metallic neighbours. The formation of bonds in three directions at right angles, such as occurs in the sodium chloride arrangement, imparts a condition of great stability to these MX carbides.

When the ratio $r_{\text{(interstitial)}}$ to $r_{\text{(metal)}}$ exceeds 0.59 the distortion becomes appreciable, and consequently more complicated crystal structures are formed. Thus, iron nitride, where $r_N/r_{Fe} = 0.56$, takes up a structure in which nitrogen lies at the centre of six atoms as suggested above, while iron carbide, i.e. cementite, Fe_3C, for which the ratio is 0.63, takes up a more complex structure.

For intermediate atomic size difference, i.e. about 20–30%, an efficient packing of the atoms can be achieved if the crystal structure common to the Laves phases is adopted (Table 3.2). These phases, classified by Laves and his co-workers, have the formula AB_2 and each A atom has 12 B neighbours and 4 A neighbours, while each B atom is surrounded by six like and six unlike atoms. The average coordination number of the structure (13.33) is higher, therefore, than that achieved by the packing of atoms of equal size. These phases crystallize in one of three closely related structures which are isomorphous with the compounds $MgCu_2$ (cubic), $MgNi_2$ (hexagonal) or $MgZn_2$ (hexagonal). The secret of the close relationship between these structures is that the small atoms are arranged on a space lattice of tetrahedra.

The different ways of joining such tetrahedra account for the different structures. This may be demonstrated by an examination of the $MgCu_2$ structure. The small B atoms lie at the corners of tetrahedra which are joined point-to-point throughout space, as shown in Figure 3.43a. Such an arrangement provides large holes of the type shown in Figure 3.43b and these are best filled when the atomic ratio $r_{\text{(large)}}/r_{\text{(small)}} = 1.225$. The complete cubic structure of $MgCu_2$ is shown in Figure 3.43c. The $MgZn_2$ structure is hexagonal, and in this case the tetrahedra are joined alternately point-to-point and base-to-base in long chains to form a wurtzite type of structure. The $MgNi_2$ structure is also hexagonal and although very complex it is essentially a mixture of both the $MgCu_2$ and $MgNi_2$ types.

The range of homogeneity of these phases is narrow. This limited range of homogeneity is not due to any

[1]Brillouin zones and electrical conductivity are dealt with in Chapter 6.

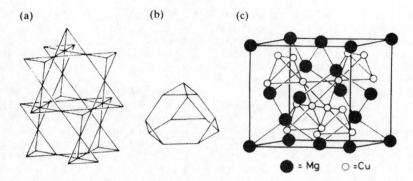

Figure 3.43 *(a) Framework of the MgCu$_2$ structure. (b) Shape of hole in which large Mg atom is accommodated. (c) Complete MgCu$_2$ structure (after Hume-Rothery, Smallman and Howorth, 1969; by courtesy of the Institute of Metals).*

Table 3.2 Compounds which exist in a Laves phase structure

MgCu$_2$ type	MgNi$_2$ type	MgZn$_2$ type	
AgBe$_2$			
BiAu$_2$	BaMg$_2$		
NbCo$_2$	Nb(Mn of Fe)$_2$	NbCo$_2$	with
TaCo$_2$	TaMn$_2$	TaCo$_2$	excess of
Ti (Be, Co, or Cr)$_2$	Ti (Mn or Fe)$_2$	TiCo$_2$	B metal
U (Al, Co, Fe or Mn)$_2$	UNi$_2$	ZrFe$_2$	
Zr (Co, Fe, or W)$_2$	Zr (Cr, Ir, Mn, Re, Ru, Os or V)$_2$		

ionic nature of the compound, since ionic compounds usually have low coordination numbers whereas. Laves phases have high coordination numbers, but because of the stringent geometrical conditions governing the structure. However, even though the chief reason for their existence is that the ratio of the radius of the large atom to that of the small is about 1.2, there are indications that electronic factors may play some small part. For example, provided the initial size-factor condition is satisfied then if the e/a ratio is high (e.g. 2), there is a tendency for compounds to crystallize in the MgZn$_2$ structure, while if the e/a ratio is low (e.g. $\frac{4}{3}$), then there is a tendency for the MgCu$_2$ type of structure to be formed. This electronic feature is demonstrated in the magnesium–nickel–zinc ternary system. Thus, even though the binary systems contain both the MgZn$_2$ and MgNi$_2$ phases the ternary compound MgNiZn has the MgCu$_2$ structure, presumably because its e/a ratio is $\frac{4}{3}$. Table 3.2 shows a few common examples of each type of Laves structure, from which it is evident that there is also a general tendency for transition metals to be involved.

3.3.3.3 Electron compounds

Alloys of copper, silver and gold with the B subgroup all possess the sequence α, β, γ, ϵ of structurally similar phases, and while each phase does not occur at the same composition when this is measured in weight per cent or atomic per cent, they do so if composition is expressed in terms of electron concentration. Hume-Rothery and his co-workers have pointed out that the e/a ratio is important not only in governing the limit of the α-solid solution but also in controlling the formation of certain intermediate phases; for this reason they have been termed 'electron compounds'.

In terms of those phases observed in the copper–zinc system (Figure 3.20), β-phases are found at an e/a ratio of $\frac{3}{2}$ and these phases are often either disordered bcc in structure or ordered CsCl-type, β'. In the copper–aluminium system for example, the β-structure is found at Cu$_3$Al, where the three valency electrons from the aluminium and the one from each copper atom make up a ratio of 6 electrons to 4 atoms, i.e. $e/a = \frac{3}{2}$. Similarly, in the copper–tin system the β-phase occurs at Cu$_5$Sn with 9 electrons to 6 atoms giving the governing e/a ratio. The γ-brass phase, Cu$_5$Zn$_8$, has a complex cubic (52 atoms per unit cell) structure, and is characterized by an e/a ratio of $\frac{21}{13}$, while the ϵ-brass phase, CuZn$_3$, has a cph structure and is governed by an e/a ratio of $\frac{7}{4}$. A list of some of these structurally-analogous phases is given in Table 3.3.

A close examination of this table shows that some of these phases, e.g. Cu$_5$Si and Ag$_3$Al, exist in different structural forms for the same e/a ratio. Thus, Ag$_3$Al is basically a $\frac{3}{2}$ bcc phase, but it only exists as such at high temperatures; at intermediate temperatures it is cph and at low temperatures β-Mn. It is also noticeable that to conform with the appropriate

Table 3.3 Some selected structurally-analogous phases

Electron-atom ratio 3:2			Electron-atom ratio 21:13 γ-brass (complex cubic)	Electron-atom ratio 7:4 ε-brass (cph)
β-brass (bcc)	β-manganese (complex cubic)	(cph)		
(Cu, Ag or Au)Zn		AgZn	(Cu, Ag or Au) (Zn or Cd)$_8$	(Cu, Ag or Au) (Zn or Cd)$_3$
CuBe	(Ag or Au)$_3$Al	AgCd		
	Cu$_5$Si		Cu$_9$Al$_4$	Cu$_3$Sn
(Ag or Au)Mg	CoZn$_3$	Ag$_3$Al		Cu$_3$Si
(Ag or Au)Cd		Au$_5$Sn	Cu$_{31}$Sn$_8$	Ag$_5$Al$_3$
(Cu or Ag)$_3$Al				
(Cu$_5$Sn or Si)				
(Fe, Co or Ni)Al			(Fe, Co, Ni, Pd or Pt)$_5$Zn$_{21}$	

electron-to-atom ratio the transition metals are credited with zero valency. The basis for this may be found in their electronic structure which is characterized by an incomplete d-band below an occupied outermost s-band. The nickel atom, for example, has an electronic structure denoted by (2) (8) (16) (2), i.e. two electrons in the first quantum shell, eight in the second, sixteen in the third and two in the fourth shells, and while this indicates that the free atom has two valency electrons, it also shows two electrons missing from the third quantum shell. Thus, if nickel contributes valency electrons, it also absorbs an equal number from other atoms to fill up the third quantum shell so that the net effect is zero.

Without doubt the electron concentration is the most important single factor which governs these compounds. However, as for the other intermediate phases, a closer examination shows an interplay of all factors. Thus, in general, the bcc $\frac{3}{2}$ compounds are only formed if the size factor is less than $\pm 18\%$, an increase in the valency of the solute tends to favour cph and β-Mn structures at the expense of the bcc structure, a high electrochemical factor leads to ordering up to the melting point, and an increase in temperature favours the bcc structure in preference to the cph or β-Mn structure.

3.3.4 Order–disorder phenomena

A substitutional solid solution can be one of two types, either ordered in which the A and B atoms are arranged in a regular pattern, or disordered in which the distribution of the A and B atoms is random. From the previous section it is clear that the necessary condition for the formation of a superlattice, i.e. an ordered solid solution, is that dissimilar atoms must attract each other more than similar atoms. In addition, the alloy must exist at or near a composition which can be expressed by a simple formula such as AB, A$_3$B or AB$_3$. The following are common structures:

1. *CuZn* While the disordered solution is bcc with equal probabilities of having copper or zinc atoms at each lattice point, the ordered lattice has copper atoms and zinc atoms segregated to cube corners (0, 0, 0) and centres ($\frac{1}{2}$, $\frac{1}{2}$, $\frac{1}{2}$), respectively. The superlattice in the β-phase therefore takes up the CsCl (also described as B2 or L2$_0$) structure as illustrated in Figure 3.44a. Other examples of the same type, which may be considered as being made up of two interpenetrating simple cubic lattices, are Ag(Mg, Zn or Cd), AuNi, NiAl, FeAl and FeCo.

2. *Cu$_3$Au* This structure, which occurs less frequently than the β-brass type, is based on the fcc structure with copper atoms at the centres of the faces (0, $\frac{1}{2}$, $\frac{1}{2}$) and gold atoms at the corners (0, 0, 0), as shown in Figure 3.44b. Other examples of the L1$_2$ structure include Ni$_3$Al, Ni$_3$Ti, Ni$_3$Si, Pt$_3$Al, Fe$_3$Ge, Zr$_3$Al.

3. *AuCu* The AuCu structure shown in Figure 3.44c is also based on the fcc lattice, but in this case alternate (0 0 1) layers are made up of copper and gold atoms, respectively. Hence, because the atomic sizes of copper and gold differ, the lattice is distorted into a tetragonal structure having an axial ratio $c/a = 0.93$. Other examples of the L1$_0$ include CoPt, FePt and TiAl.

4. *Fe$_3$Al* Like FeAl, the Fe$_3$Al structure is based on the bcc lattice but, as shown in Figure 3.44d, eight

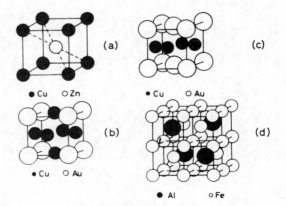

Figure 3.44 *Examples of ordered structures, (a) CuZn, (b) Cu$_3$Au, (c) CuAu, (d) Fe$_3$Al.*

simple cells are necessary to describe the complete ordered arrangement. In this structure any individual atom is surrounded by the maximum number of unlike atoms and the aluminium atoms are arranged tetrahedrally in the cell. Other examples of the DO_3 include Fe_3Si and Cu_3Al.

5. Mg_3Cd This ordered structure is based on the cph lattice. Other examples of the DO_{19} structure are Ti_3Al, $MgCd_3$ and Ni_3Sn.

Another important structure which occurs in certain intermetallics is the defect lattice. In the compound NiAl, as the composition deviates from stoichiometry towards pure aluminium, the electron to atom ratio becomes greater than $\frac{3}{2}$, but to prevent the compound becoming unstable the lattice takes up a certain proportion of vacancies to maintain the number of electrons per unit cell at a constant value of 3. Such defects obviously increase the entropy of the alloy, but the fact that these phases are stable at low temperatures, where the entropy factor is unimportant, demonstrates that their stability is due to a lowering of internal energy. Such defects produce an anomalous decrease in both the lattice parameter and the density above 50 at.% Al.

3.4 The mechanism of phase changes

3.4.1 Kinetic considerations

Changes of phase in the solid state involve a redistribution of the atoms in that solid and the kinetics of the change necessarily depend upon the rate of atomic migration. The transport of atoms through the crystal is more generally termed diffusion, and is dealt with in Section 6.4. This can occur more easily with the aid of vacancies, since the basic act of diffusion is the movement of an atom to an empty adjacent atomic site.

Let us consider that during a phase change an atom is moved from an α-phase lattice site to a more favourable β-phase lattice site. The energy of the atom should vary with distance as shown in Figure 3.45, where the potential barrier which has to be overcome arises from the interatomic forces between the moving atom and the group of atoms which adjoin it and the new site. Only those atoms (n) with an energy

greater than Q are able to make the jump, where $Q_{\alpha \to \beta} = H_m - H_\alpha$ and $Q_{\beta \to \alpha = H_m - H_\beta}$ are the activation enthalpies for heating and cooling, respectively. The probability of an atom having sufficient energy to jump the barrier is given, from the Maxwell–Boltzmann distribution law, as proportional to $\exp[-Q/kT]$ where $\mathbf{k}$ is Boltzmann's constant, T is the temperature and Q is usually expressed as the energy per atom in electron volts.[1]

The rate of reaction is given by

$$\text{Rate} = A \exp[-Q/\mathbf{k}T] \qquad (3.8)$$

where A is a constant involving n and v, the frequency of vibration. To determine Q experimentally, the reaction velocity is measured at different temperatures and, since

$$\ln(\text{Rate}) = \ln A - Q/\mathbf{k}T \qquad (3.9)$$

the slope of the ln (rate) versus $1/T$ curve gives $Q/\mathbf{k}$.

In deriving equation (3.8), usually called an Arrhenius equation after the Swedish chemist who first studied reaction kinetics, no account is taken of the entropy of activation, i.e. the change in entropy as a result of the transition. In considering a general reaction the probability expression should be written in terms of the free energy of activation per atom F or G rather than just the internal energy or enthalpy. The rate equation then becomes

$$\begin{aligned}\text{Rate} &= A \exp[-F/\mathbf{k}T] \\ &= A \exp[S/\mathbf{k}] \exp[-E/\mathbf{k}T] \qquad (3.10)\end{aligned}$$

The slope of the ln (rate) versus $1/T$ curve then gives the temperature-dependence of the reaction rate, which is governed by the activation energy or enthalpy, and the magnitude of the intercept on the ln (rate) axis depends on the temperature-independent terms and include the frequency factor and the entropy term.

During the transformation it is not necessary for the entire system to go from α to β at one jump and, in fact, if this were necessary, phase changes would practically never occur. Instead, most phase changes occur by a process of nucleation and growth (cf. solidification, Section 3.1.1). Chance thermal fluctuations provide a small number of atoms with sufficient activation energy to break away from the matrix (the old structure) and form a small nucleus of the new phase, which then grows at the expense of the matrix until the whole structure is transformed. By this mechanism, the amount of material in the intermediate configuration of higher free energy is kept to a minimum, as it is localized into atomically thin layers at the interface between the phases. Because of

[1]Q may also be given as the energy in J mol^{-1} in which case the rate equation becomes

$$\text{Rate of reaction} = A \exp[-Q/\mathbf{R}T]$$

where $\mathbf{R} = \mathbf{k}N$ is the gas constant, i.e. 8.314 J mol^{-1} K^{-1}.

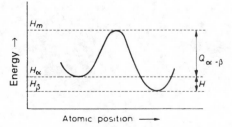

Figure 3.45 *Energy barrier separating structural states.*

this mechanism of transformation, the factors which determine the rate of phase change are: (1) the rate of nucleation, N (i.e. the number of nuclei formed in unit volume in unit time) and (2) the rate of growth, G (i.e. the rate of increase in radius with time). Both processes require activation energies, which in general are not equal, but the values are much smaller than that needed to change the whole structure from α to β in one operation.

Even with such an economical process as nucleation and growth transformation, difficulties occur and it is common to find that the transformation temperature, even under the best experimental conditions, is slightly higher on heating than on cooling. This sluggishness of the transformation is known as hysteresis, and is attributed to the difficulties of nucleation, since diffusion, which controls the growth process, is usually high at temperatures near the transformation temperature and is, therefore, not rate-controlling. Perhaps the simplest phase change to indicate this is the solidification of a liquid metal.

The transformation temperature, as shown on the equilibrium diagram, represents the point at which the free energy of the solid phase is equal to that of the liquid phase. Thus, we may consider the transition, as given in a phase diagram, to occur when the bulk or chemical free energy change, ΔG_v, is infinitesimally small and negative, i.e. when a small but positive driving force exists. However, such a definition ignores the process whereby the bulk liquid is transformed to bulk solid, i.e. nucleation and growth. When the nucleus is formed the atoms which make up the interface between the new and old phase occupy positions of compromise between the old and new structure, and as a result these atoms have rather higher energies than the other atoms. Thus, there will always be a positive free energy term opposing the transformation as a result of the energy required to create the surface of interface. Consequently, the transformation will occur only when the sum $\Delta G_v + \Delta G_s$ becomes negative, where ΔG_s arises from the surface energy of solid–liquid interface. Normally, for the bulk phase change, the number of atoms which form the interface is small and ΔG_s compared with ΔG_v can be ignored. However, during nucleation ΔG_v is small, since it is proportional to the amount

transformed, and ΔG_s, the extra free energy of the boundary atoms, becomes important due to the large surface area to volume ratio of small nuclei. Therefore before transformation can take place the negative term ΔG_v must be greater than the positive term ΔG_s and, since ΔG_v is zero at the equilibrium freezing point, it follows that undercooling must result.

3.4.2 Homogeneous nucleation

Quantitatively, since ΔG_v depends on the volume of the nucleus and ΔG_s is proportional to its surface area, we can write for a spherical nucleus of radius r

$$\Delta G = (4\pi r^3 \Delta G_v/3) + 4\pi r^2 \gamma \qquad (3.11)$$

where ΔG_v is the bulk free energy change involved in the formation of the nucleus of unit volume and γ is the surface energy of unit area. When the nuclei are small the positive surface energy term predominates, while when they are large the negative volume term predominates, so that the change in free energy as a function of nucleus size is as shown in Figure 3.46a. This indicates that a critical nucleus size exists below which the free energy increases as the nucleus grows, and above which further growth can proceed with a lowering of free energy; ΔG_{max} may be considered as the energy or work of nucleation W. Both r_c and W may be calculated since $d\Delta G/dr = 4\pi r^2 \Delta G_v + 8\pi r \gamma = 0$ when $r = r_c$ and thus $r_c = -2\gamma/\Delta G_v$. Substituting for r_c gives

$$W = 16\pi \gamma^3/3\Delta G_v^2 \qquad (3.12)$$

The surface energy factor γ is not strongly dependent on temperature, but the greater the degree of undercooling or supersaturation, the greater is the release of chemical free energy and the smaller the critical nucleus size and energy of nucleation. This can be shown analytically since $\Delta G_v = \Delta H - T\Delta S$, and at $T = T_e$, $\Delta G_v = 0$, so that $\Delta H = T_e \Delta S$. It therefore follows that

$$\Delta G_v = (T_e - T)\Delta S = \Delta T \Delta S$$

and because $\Delta G_v \propto \Delta T$, then

$$W \propto \gamma^3/\Delta T^2 \qquad (3.13)$$

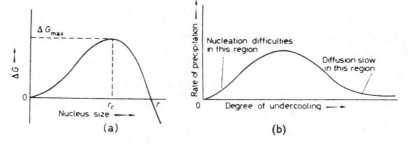

(a)   (b)

Figure 3.46 (a) Effect of nucleus size on the free energy of nucleus formation. (b) Effect of undercooling on the rate of precipitation.

Consequently, since nuclei are formed by thermal fluctuations, the probability of forming a smaller nucleus is greatly improved, and the rate of nucleation increases according to

$$\text{Rate} = A \exp\left[-Q/\mathbf{k}T\right] \exp\left[-\Delta G_{max}/\mathbf{k}T\right]$$

$$= A \exp\left[-(Q + \Delta G_{max})/\mathbf{k}T\right] \quad (3.14)$$

The term $\exp\left[-Q/\mathbf{k}T\right]$ is introduced to allow for the fact that rate of nucleus formation is in the limit controlled by the rate of atomic migration. Clearly, with very extensive degrees of undercooling, when $\Delta G_{max} \ll Q$, the rate of nucleation approaches exp $[-Q/\mathbf{k}T]$ and, because of the slowness of atomic mobility, this becomes small at low temperature (Figure 3.46b). While this range of conditions can be reached for liquid glasses the nucleation of liquid metals normally occurs at temperatures before this condition is reached. (By splat cooling, small droplets of the metal are cooled very rapidly (10^5 K s^{-1}) and an amorphous solid may be produced.) Nevertheless, the principles are of importance in metallurgy since in the isothermal transformation of eutectoid steel, for example, the rate of transformation initially increases and then decreases with lowering of the transformation temperature (*see* **TTT** curves, Chapter 8).

3.4.3 Heterogeneous nucleation

In practice, homogeneous nucleation rarely takes place and heterogeneous nucleation occurs either on the mould walls or on insoluble impurity particles. From equation (3.13) it is evident that a reduction in the interfacial energy γ would facilitate nucleation at small values of ΔT. Figure 3.47 shows how this occurs at a mould wall or pre-existing solid particle, where the nucleus has the shape of a spherical cap to minimize the energy and the 'wetting' angle θ is given by the balance of the interfacial tensions in the plane of the mould wall, i.e. $\cos\theta = (\gamma_{ML} - \gamma_{SM})/\gamma_{SL}$.

The formation of the nucleus is associated with an excess free energy given by

$$\Delta G = V\Delta G_v + A_{SL}\gamma_{SL} + A_{SM}\gamma_{SM} - A_{SM}\gamma_{ML}$$

$$= \pi/3(2 - 3\cos\theta + \cos^3\theta)r^3\Delta G_v$$

$$+ 2\pi(1 - \cos\theta)r^2\gamma_{SL}$$

$$+ \pi r^2 \sin^2\theta(\gamma_{SM} - \gamma_{LM}) \quad (3.15)$$

Differentiation of this expression for the maximum, i.e. $d\Delta G/dr = 0$, gives $r_c = -2\gamma_{SL}/\Delta G_v$ and

$$W = (16\pi\gamma^3/3\Delta G_v{}^2)[(1 - \cos\theta)^2(2 + \cos\theta)/4] \quad (3.16)$$

or

$$W_{\text{(heterogeneous)}} = W_{\text{(homogeneous)}}[S(\theta)]$$

The shape factor $S(\theta) \leq 1$ is dependent on the value of θ and the work of nucleation is therefore less for

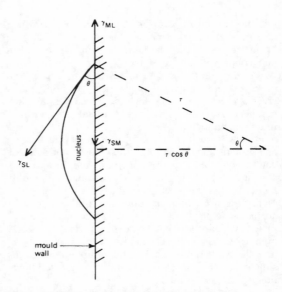

Figure 3.47 *Schematic geometry of heterogeneous nucleation.*

heterogeneous nucleation. When $\theta = 180°$, no wetting occurs and there is no reduction in W; when $\theta \to 0°$ there is complete wetting and $W \to 0$; and when $0 < \theta < 180°$ there is some wetting and W is reduced.

3.4.4 Nucleation in solids

When the transformation takes place in the solid state, i.e. between two solid phases, a second factor giving rise to hysteresis operates. The new phase usually has a different parameter and crystal structure from the old so that the transformation is accompanied by dimensional changes. However, the changes in volume and shape cannot occur freely because of the rigidity of the surrounding matrix, and elastic strains are induced. The strain energy and surface energy created by the nuclei of the new phase are positive contributions to the free energy and so tend to oppose the transition.

The total free energy change is

$$\Delta G = V\Delta G_v + A\gamma + V\Delta G_s \quad (3.17)$$

where A is the area of interface between the two phases and γ the interfacial energy per unit area, and ΔG_s is the misfit strain energy per unit volume of new phase. For a spherical nucleus of the second phase

$$\Delta G = \tfrac{4}{3}\pi r^3(\Delta G_v - \Delta G_s) + 4\pi r^2\gamma \quad (3.18)$$

and the misfit strain energy reduces the effective driving force for the transformation. Differentiation of equation (3.18) gives

$$r_c = -2\gamma/(\Delta G_v - \Delta G_s), \text{ and}$$

$$W = 16\pi\gamma^3/3(\Delta G_v - \Delta G_s)^2$$

The value of γ can vary widely from a few mJ/m^2 to several hundred mJ/m^2 depending on the coherency

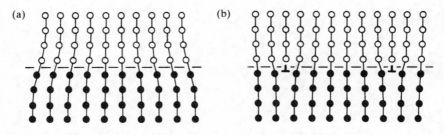

Figure 3.48 *Schematic representation of interface structures. (a) A coherent boundary with misfit strain and (b) a semi-coherent boundary with misfit dislocations.*

of the interface. A coherent interface is formed when the two crystals have a good 'match' and the two lattices are continuous across the interface. This happens when the interfacial plane has the same atomic configuration in both phases, e.g. $\{1\,1\,1\}$ in fcc and $\{0\,0\,0\,1\}$ in cph. When the 'match' at the interface is not perfect it is still possible to maintain coherency by straining one or both lattices, as shown in Figure 3.48a. These coherency strains increase the energy and for large misfits it becomes energetically more favourable to form a semi-coherent interface in which the mismatch is periodically taken up by misfit dislocations.[1] The coherency strains can then be relieved by a cross-grid of dislocations in the interface plane, the spacing of which depends on the Burgers vector b of the dislocation and the misfit ε, i.e. b/ε. The interfacial energy for semi-coherent interfaces arises from the change in composition across the interface or chemical contribution as for fully-coherent interfaces, plus the energy of the dislocations (see Chapter 4). The energy of a semi-coherent interface is $200-500$ mJ/m^2 and increases with decreasing dislocation spacing until the dislocation strain fields overlap. When this occurs, the discrete nature of the dislocations is lost and the interface becomes incoherent. The incoherent interface is somewhat similar to a high-angle grain boundary (see Figure 3.3) with its energy of 0.5 to 1 J/m^2 relatively independent of the orientation.

The surface and strain energy effects discussed above play an important role in phase separation. When there is coherence in the atomic structure across the interface between precipitate and matrix the surface energy term is small, and it is the strain energy factor which controls the shape of the particle. A plate-shaped particle is associated with the least strain energy, while a spherical-shaped particle is associated with maximum strain energy but the minimum surface energy. On the other hand, surface energy determines the crystallographic plane of the matrix on which a

plate-like precipitate forms. Thus, the habit plane is the one which allows the planes at the interface to fit together with the minimum of disregistry; the frequent occurrence of the Widmanstätten structures may be explained on this basis. It is also observed that precipitation occurs most readily in regions of the structure which are somewhat disarranged, e.g. at grain boundaries, inclusions, dislocations or other positions of high residual stress caused by plastic deformation. Such regions have an unusually high free energy and necessarily are the first areas to become unstable during the transformation. Also, new phases can form there with a minimum increase in surface energy. This behaviour is considered again in Chapter 7.

Further reading

Beeley, P. R. (1972). *Foundry Technology*. Butterworths, London.

Campbell, J. (1991). *Castings*. Butterworth-Heinemann, London.

Chadwick, G. A. (1972). *Metallography of Phase Transformations*. Butterworths, London.

Davies, G. J. (1973). *Solidification and Casting*. Applied Science, London.

Driver, D. (1985). Aero engine alloy development, Inst. of Metals Conf., Birmingham. 'Materials at their Limits' (25 September 1985).

Flemings, M. C. (1974). *Solidification Processing*. McGraw-Hill, New York.

Hume-Rothery, W., Smallman, R. E. and Haworth, C. (1969). *Structure of Metals and Alloys*, 5th edn. Institute of Metals, London.

Kingery, W. D., Bowen, H. K. and Uhlmann, D. R. (1976). *Introduction to Ceramics*, 2nd edn. Wiley-Interscience, New York.

Rhines, F. N. (1956). *Phase Diagrams in Metallurgy: their development and application*. McGraw-Hill, New York.

Quets, J. M. and Dresher, W. H. (1969). Thermo-chemistry of hot corrosion of superalloys. *Journal of Materials*, ASTM, JMSLA, **4**, 3, 583–599.

West, D. R. F. (1982). *Ternary Equilibrium Diagrams*, 2nd edn. Macmillan, London.

[1] A detailed treatment of dislocations and other defects is given in Chapter 4.

Chapter 4

Defects in solids

4.1 Types of imperfection

Real solids invariably contain structural discontinuities and localized regions of disorder. This heterogeneity can exist on both microscopic and macroscopic scales, with defects or imperfections ranging in size from missing or misplaced atoms to features that are visible to the naked eye. The majority of materials used for engineering components and structures are made up from a large number of small interlocking grains or crystals. It is therefore immediately appropriate to regard the grain boundary surfaces of such polycrystalline aggregates as a type of imperfection. Other relatively large defects, such as shrinkage pores, gas bubbles, inclusions of foreign matter and cracks, may be found dispersed throughout the grains of a metal or ceramic material. In general, however, these large-scale defects are very much influenced by the processing of the material and are less fundamental to the basic material. More attention will thus be given to the atomic-scale defects in materials. Within each grain, atoms are regularly arranged according to the basic crystal structure but a variety of imperfections, classified generally as crystal defects, may also occur. A schematic diagram of these basic defects is shown in Figure 4.1. These take the form of:

- Point defects, such as vacant atomic sites (or simply vacancies) and interstitial atoms (or simply interstitials) where an atom sits in an interstice rather than a normal lattice site
- Line defects, such as dislocations
- Planar defects, such as stacking faults and twin boundaries
- Volume defects, such as voids, gas bubbles and cavities.

In the following sections this type of classification will be used to consider the defects which can occur in metallic and ceramic crystals. Glasses already lack long-range order; we will therefore concentrate upon

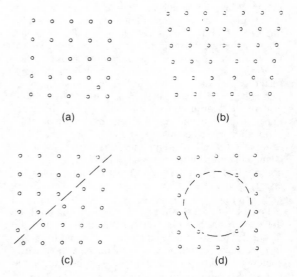

Figure 4.1 *(a) Vacancy–interstitial, (b) dislocation, (c) stacking fault, (d) void.*

crystal defects. Defects in crystalline macromolecular structures, as found in polymers, form a special subject and will be dealt with separately in Section 4.6.7.

4.2 Point defects

4.2.1 Point defects in metals

Of the various lattice defects the vacancy is the only species that is ever present in appreciable concentrations in thermodynamic equilibrium and increases exponentially with rise in temperature, as shown in Figure 4.2. The vacancy is formed by removing an atom from its lattice site and depositing it in a nearby atomic site where it can be easily accommodated. Favoured places are the free surface of the crystal, a

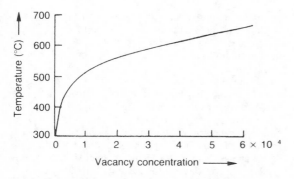

Figure 4.2 *Equilibrium concentration of vacancies as a function of temperature for aluminium (after Bradshaw and Pearson, 1957).*

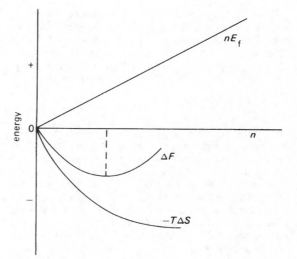

Figure 4.3 *Variation of the energy of a crystal with addition of n vacancies.*

grain boundary or the extra half-plane of an edge dislocation. Such sites are termed vacancy sources and the vacancy is created when sufficient energy is available (e.g. thermal activation) to remove the atom. If E_f is the energy required to form one such defect (usually expressed in electron volts per atom), the total energy increase resulting from the formation of n such defects is nE_f. The accompanying entropy increase may be calculated using the relations $S = k \ln W$, where W is the number of ways of distributing n defects and N atoms on $N + n$ lattice sites, i.e. $(N + n)!/n!N!$ Then the free energy, G, or strictly F of a crystal of n defects, relative to the free energy of the perfect crystal, is

$$F = nE_f - kT \ln [(N + n)!/n!N!] \tag{4.1}$$

which by the use of Stirling's theorem[1] simplifies to

$$F = nE_f - kT [N + n) \ln (N + n) - n \ln n - N \ln N] \tag{4.2}$$

The equilibrium value of n is that for which $dF/dn = 0$, which defines the state of minimum free energy as shown in Figure 4.3.[2] Thus, differentiating equation (4.2) gives

$$0 = E_f - kT [\ln (N + n) - \ln n]$$
$$= E_f - kT \ln [(N + n)/n]$$

so that

$$\frac{n}{N + n} = \exp [-E_f/kT]$$

Usually N is very large compared with n, so that the expression can be taken to give the atomic concentration, c, of lattice vacancies, $n/N = \exp [-E_f/kT]$. A more rigorous calculation of the concentration of vacancies in thermal equilibrium in a perfect lattice shows that although c is principally governed by

[1]Stirling's approximation states that $\ln N! = N \ln N$.
[2]dF/dn or dG/dn in known as the chemical potential.

the Boltzmann factor $\exp [-E_f/kT]$, the effect of the vacancy on the vibrational properties of the lattice also leads to an entropy term which is independent of temperature and usually written as $\exp [S_f/k]$. The fractional concentration may thus be written

$$c = n/N = \exp [S_f/k] \exp [-E_f/kT]$$
$$= A \exp [-E_f/kT] \tag{4.3}$$

The value of the entropy term is not accurately known but it is usually taken to be within a factor of ten of the value 10; for simplicity we will take it to be unity.

The equilibrium number of vacancies rises rapidly with increasing temperature, owing to the exponential form of the expression, and for most common metals has a value of about 10^{-4} near the melting point. For example, kT at room temperature (300 K) is $\approx 1/40$ eV and for aluminium $E_f = 0.7$ eV, so that at 900 K we have

$$c = \exp \left[-\frac{7}{10} \times \frac{40}{1} \times \frac{300}{900} \right]$$
$$= \exp [-9.3] = 10^{-[9.3/2.3]} \approx 10^{-4}$$

As the temperature is lowered, c should decrease in order to maintain equilibrium and to do this the vacancies must migrate to positions in the structure where they can be annihilated; these locations are then known as 'vacancy sinks' and include such places as the free surface, grain boundaries and dislocations.

The defect migrates by moving through the energy maxima from one atomic site to the next with a frequency

$$\nu = \nu_0 \exp \left(\frac{S_m}{K} \right) \exp \left(-\frac{E_m}{KT} \right)$$

where ν_0 is the frequency of vibration of the defect in the appropriate direction, S_m is the entropy increase and E_m is the internal energy increase associated with the process. The self-diffusion coefficient in a pure metal is associated with the energy to form a vacancy E_f and the energy to move it E_m, being given by the expression

$$E_{SD} = E_f + E_m$$

Clearly the free surface of a sample or the grain boundary interface are a considerable distance, in atomic terms, from the centre of a grain and so dislocations in the body of the grain or crystal are the most efficient 'sink' for vacancies. Vacancies are annihilated at the edge of the extra half-plane of atoms of the dislocation, as shown in Figure 4.4a and 4.4b. This causes the dislocation to climb, as discussed in Section 4.3.4. The process whereby vacancies are annihilated at vacancy sinks such as surfaces, grain boundaries and dislocations, to satisfy the thermodynamic equilibrium concentration at a given temperature is, of course, reversible. When a metal is heated the equilibrium concentration increases and, to produce this additional concentration, the surfaces, grain boundaries and dislocations in the crystal reverse their role and act as vacancy sources and emit vacancies; the extra half-plane of atoms climbs in the opposite sense (see Figures 4.4c and 4.4d).

Below a certain temperature, the migration of vacancies will be too slow for equilibrium to be maintained, and at the lower temperatures a concentration of vacancies in excess of the equilibrium number will be retained in the structure. Moreover, if the cooling rate of the metal or alloy is particularly rapid, as, for example, in quenching, the vast majority of the vacancies which exist at high temperatures can be 'frozen-in'.

Vacancies are of considerable importance in governing the kinetics of many physical processes. The industrial processes of annealing, homogenization, precipitation, sintering, surface-hardening, as well as oxidation and creep, all involve, to varying degrees, the transport of atoms through the structure with the help of vacancies. Similarly, vacancies enable dislocations to climb, since to move the extra half-plane of a dislocation up or down requires the mass transport of atoms. This mechanism is extremely important in the recovery stage of annealing and also enables dislocations to climb over obstacles lying in their slip plane; in this way materials can soften and lose their resistance to creep at high temperatures.

In metals the energy of formation of an interstitial atom is much higher than that for a vacancy and is of the order of 4 eV. At temperatures just below the melting point, the concentration of such point defects is only about 10^{-15} and therefore interstitials are of little consequence in the normal behaviour of metals. They are, however, more important in ceramics because of the more open crystal structure. They are also of importance in the deformation behaviour of solids when point defects are produced by the non-conservative motion of jogs in screw dislocation (see Section 4.3.4) and also of particular importance in materials that have been subjected to irradiation by high-energy particles.

4.2.2 Point defects in non-metallic crystals

Point defects in non-metallic, particularly ionic, structures are associated with additional features (e.g. the requirement to maintain electrical neutrality and the possibility of both anion-defects and cation-defects existing). An anion vacancy in NaCl, for example, will be a positively-charged defect and may trap an electron to become a neutral F-centre. Alternatively, an anion vacancy may be associated with either an anion interstitial or a cation vacancy. The vacancy-interstitial pair is called a Frenkel defect and the vacancy pair a Schottky defect, as shown in Figure 4.5. Interstitials are much more common in ionic structures than metallic structures because of the large 'holes' or interstices that are available.

In general, the formation energy of each of these two types of defect is different and this leads to different defect concentrations. With regard to vacancies, when $E_f^- > E_f^+$, i.e. the formation will initially produce more cation than anion vacancies from dislocations and boundaries as the temperature is raised. However, the electrical field produced will eventually oppose the production of further cations and promote the formation of anions such that of equilibrium there will be almost equal numbers of both types and the

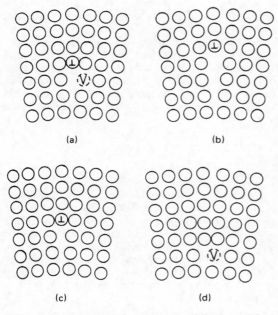

Figure 4.4 *Climb of a dislocation, (a) and (b) to annihilate, (c) and (d) to create a vacancy.*

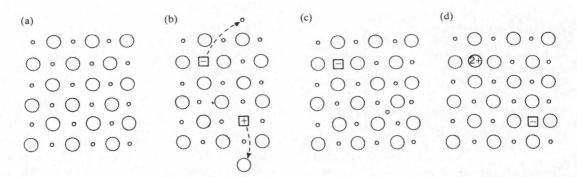

Figure 4.5 *Representation of point defects in two-dimensional ionic structure: (a) perfect structure and monovalent ions, (b) two Schottky defects, (c) Frenkel defect, and (d) substitutional divalent cation impurity and cation vacancy.*

combined or total concentration c of Schottky defects at high temperatures is $\sim 10^{-4}$.

Foreign ions with a valency different from the host cation may also give rise to point defects to maintain charge neutrality. Monovalent sodium ions substituting for divalent magnesium ions in MgO, for example, must be associated with an appropriate number of either cation interstitials or anion vacancies in order to maintain charge neutrality. Deviations from the stoichiometric composition of the non-metallic material as a result of excess (or deficiency) in one (or other) atomic species also results in the formation of point defects.

An example of excess-metal due to anion vacancies is found in the oxidation of silicon which takes place at the metal/oxide interface. Interstitials are more likely to occur in oxides with open crystal structures and when one atom is much smaller than the other as, for example, ZnO (Figure 4.6a). The oxidation of copper to Cu_2O, shown in Figure 4.6b, is an example of non-stoichiometry involving cation vacancies. Thus copper vacancies are created at the oxide surface and diffuse through the oxide layer and are eliminated at the oxide/metal interface.

Oxides which contain point defects behave as semiconductors when the electrons associated with the point defects either form positive holes or enter the conduction band of the oxide. If the electrons remain locally associated with the point defects, then charge can only be transferred by the diffusion of the charge carrying defects through the oxide. Both p- and n-type semiconductors are formed when oxides deviate from stoichiometry: the former arises from a deficiency of cations and the latter from an excess of cations.

Examples of p-type semiconducting oxides are NiO, PbO and Cu_2O while the oxides of Zn, Cd and Be are n-type semiconductors.

4.2.3 Irradiation of solids

There are many different kinds of high-energy radiation (e.g. neutrons, electrons, α-particles, protons, deuterons, uranium fission fragments, γ-rays, X-rays) and all of them are capable of producing some form of 'radiation damage' in the materials they irradiate. While all are of importance to some aspects of the solid state, of particular interest is the behaviour of materials under irradiation in a nuclear reactor. This is because the neutrons produced in a reactor by a fission reaction have extremely high energies of about 2 million electron volts (i.e. 2 MeV), and being electrically uncharged, and consequently unaffected by the electrical fields surrounding an atomic nucleus, can travel large distances through a structure. The resultant damage is therefore not localized, but is distributed throughout the solid in the form of 'damage spikes.'

The fast neutrons (they are given this name because 2 MeV corresponds to a velocity of 2×10^7 m s^{-1}) are slowed down, in order to produce further fission,

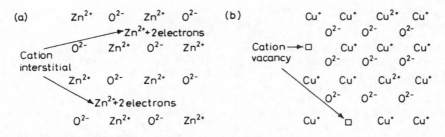

Figure 4.6 *Schematic arrangement of ions in two typical oxides. (a) $Zn_{>1}O$, with excess metal due to cation interstitials and (b) $Cu_{<2}O$, with excess non-metal due to cation vacancies.*

by the moderator in the pile until they are in thermal equilibrium with their surroundings. The neutrons in a pile will, therefore, have a spectrum of energies which ranges from about 1/40 eV at room temperature (thermal neutrons) to 2 MeV (fast neutrons). However, when non-fissile material is placed in a reactor and irradiated most of the damage is caused by the fast neutrons colliding with the atomic nuclei of the material.

The nucleus of an atom has a small diameter (e.g. 10^{-10} m), and consequently the largest area, or cross-section, which it presents to the neutron for collision is also small. The unit of cross-section is a barn, i.e. 10^{-28} m^2 so that in a material with a cross-section of 1 barn, an average of 10^9 neutrons would have to pass through an atom (cross-sectional area 10^{-19} m^2) for one to hit the nucleus. Conversely, the mean free path between collisions is about 10^9 atom spacings or about 0.3 m. If a metal such as copper (cross-section, 4 barns) were irradiated for 1 day (10^5 s) in a neutron flux of 10^{17} m^{-2} s^{-1} the number of neutrons passing through unit area, i.e. the integrated flux, would be 10^{22} n m^{-2} and the chance of a given atom being hit (=integrated flux × cross-section) would be 4×10^{-6}, i.e. about 1 atom in 250 000 would have its nucleus struck.

For most metals the collision between an atomic nucleus and a neutron (or other fast particle of mass m) is usually purely elastic, and the struck atom mass M will have equal probability of receiving any kinetic energy between zero and the maximum $E_{max} = 4E_n Mm/(M + m)^2$, where E_n is the energy of the fast neutron. Thus, the most energetic neutrons can impart an energy of as much as 200 000 eV, to a copper atom initially at rest. Such an atom, called a primary 'knock-on', will do much further damage on its subsequent passage through the structure often producing secondary and tertiary knock-on atoms, so that severe local damage results. The neutron, of course, also continues its passage through the structure producing further primary displacements until the energy transferred in collisions is less than the energy E_d (≈ 25 eV for copper) necessary to displace an atom from its lattice site.

The damage produced in irradiation consists largely of interstitials, i.e. atoms knocked into interstitial positions in the lattice, and vacancies, i.e. the holes they leave behind. The damaged region, estimated to contain about 60 000 atoms, is expected to be originally pear-shaped in form, having the vacancies at the centre and the interstitials towards the outside. Such a displacement spike or cascade of displaced atoms is shown schematically in Figure 4.7. The number of vacancy-interstitial pairs produced by one primary knock-on is given by $n \simeq E_{max}/4E_d$, and for copper is about 1000. Owing to the thermal motion of the atoms in the lattice, appreciable self-annealing of the damage will take place at all except the lowest temperatures, with most of the vacancies and interstitials

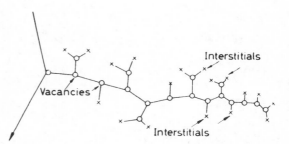

Figure 4.7 *Formation of vacancies and interstitials due to particle bombardment (after Cottrell, 1959; courtesy of the Institute of Mechanical Engineers).*

annihilating each other by recombination. However, it is expected that some of the interstitials will escape from the surface of the cascade leaving a corresponding number of vacancies in the centre. If this number is assumed to be 100, the local concentration will be 100/60 000 or $\approx 2 \times 10^{-3}$.

Another manifestation of radiation damage concerns the dispersal of the energy of the stopped atom into the vibrational energy of the lattice. The energy is deposited in a small region, and for a very short time the metal may be regarded as locally heated. To distinguish this damage from the 'displacement spike', where the energy is sufficient to displace atoms, this heat-affected zone has been called a 'thermal spike'. To raise the temperature by 1000°C requires about $3R \times 4.2$ kJ/mol or about 0.25 eV per atom. Consequently, a 25 eV thermal spike could heat about 100 atoms of copper to the melting point, which corresponds to a spherical region of radius about 0.75 nm. It is very doubtful if melting actually takes place, because the duration of the heat pulse is only about 10^{-11} to 10^{-12} s. However, it is not clear to what extent the heat produced gives rise to an annealing of the primary damage, or causes additional quenching damage (e.g. retention of high-temperature phases).

Slow neutrons give rise to transmutation products. Of particular importance is the production of the noble gas elements, e.g. krypton and xenon produced by fission in U and Pu, and helium in the light elements B, Li, Be and Mg. These transmuted atoms can cause severe radiation damage in two ways. First, the inert gas atoms are almost insoluble and hence in association with vacancies collect into gas bubbles which swell and crack the material. Second, these atoms are often created with very high energies (e.g. as α-particles or fission fragments) and act as primary sources of knock-on damage. The fission of uranium into two new elements is the extreme example when the fission fragments are thrown apart with kinetic energy ≈ 100 MeV. However, because the fragments carry a large charge their range is short and the damage restricted to the fissile material itself, or in materials which are in close proximity. Heavy ions can be accelerated to kilovolt energies in accelerators to produce heavy ion bombardment of materials being tested for reactor application. These moving particles have a short range and the damage is localized.

4.2.4 Point defect concentration and annealing

Electrical resistivity ρ is one of the simplest and most sensitive properties to investigate the point defect concentration. Point defects are potent scatterers of electrons and the increase in resistivity following quenching ($\Delta\rho$) may be described by the equation

$$\Delta\rho = A \exp\left[-E_f/\mathbf{k}T_Q\right] \tag{4.4}$$

where A is a constant involving the entropy of formation, E_f is the formation energy of a vacancy and T_Q the quenching temperature. Measuring the resistivity after quenching from different temperatures enables E_f to be estimated from a plot of $\Delta\rho_0$ versus $1/T_Q$. The activation energy, E_m, for the movement of vacancies can be obtained by measuring the rate of annealing of the vacancies at different annealing temperatures. The rate of annealing is inversely proportional to the time to reach a certain value of 'annealed-out' resistivity. Thus, $1/t_1 = A \exp\left[-E_m/\mathbf{k}T_1\right]$ and $1/t_2 = \exp\left[-E_m/\mathbf{k}T_2\right]$ and by eliminating A we obtain $\ln(t_2/t_1) = E_m[(1/T_2) - (1/T_1)]/\mathbf{k}$ where E_m is the only unknown in the expression. Values of E_f and E_m for different materials are given in Table 4.1.

At elevated temperatures the very high equilibrium concentration of vacancies which exists in the structure gives rise to the possible formation of divacancy and even tri-vacancy complexes, depending on the value of the appropriate binding energy. For equilibrium between single and di-vacancies, the total vacancy concentration is given by

$$c_v = c_{1v} + 2c_{2v}$$

and the di-vacancy concentration by

$$c_{2v} = Azc_{1v}^2 \exp\left[B_2/\mathbf{k}T\right]$$

where A is a constant involving the entropy of formation of di-vacancies, B_2 the binding energy for vacancy pairs estimated to be in the range 0.1–0.3 eV and z a configurational factor. The migration of di-vacancies is

an easier process and the activation energy for migration is somewhat lower than E_m for single vacancies.

Excess point defects are removed from a material when the vacancies and/or interstitials migrate to regions of discontinuity in the structure (e.g. free surfaces, grain boundaries or dislocations) and are annihilated. These sites are termed defect sinks. The average number of atomic jumps made before annihilation is given by

$$n = Azvt \exp\left[-E_m/\mathbf{k}T_a\right] \tag{4.5}$$

where A is a constant (≈ 1) involving the entropy of migration, z the coordination around a vacancy, v the Debye frequency ($\approx 10^{13}/s$), t the annealing time at the ageing temperature T_a and E_m the migration energy of the defect. For a metal such as aluminium, quenched to give a high concentration of retained vacancies, the annealing process takes place in two stages, as shown in Figure 4.8; stage I near room temperature with an activation energy ≈ 0.58 eV and $n \approx 10^4$, and stage II in the range 140–200°C with an activation energy of ~ 1.3 eV.

Assuming a random walk process, single vacancies would migrate an average distance ($\sqrt{n} \times$ atomic spacing b) ≈ 30 nm. This distance is very much less than either the distance to the grain boundary or the spacing of the dislocations in the annealed metal. In this case, the very high supersaturation of vacancies produces a chemical stress, somewhat analogous to an osmotic pressure, which is sufficiently large to create new dislocations in the structure which provide many new 'sinks' to reduce this stress rapidly.

The magnitude of this chemical stress may be estimated from the chemical potential, if we let dF represent the change of free energy when dn vacancies are added to the system. Then,

$$dF/dn = E_f + \mathbf{k}T \ln(n/N) = -\mathbf{k}T \ln c_0 + \mathbf{k}T \ln c$$
$$= \mathbf{k}T \ln(c/c_0)$$

where c is the actual concentration and c_0 the equilibrium concentration of vacancies. This may be rewritten as

Table 4.1 Values of vacancy formation (E_f) and migration (E_m) energies for some metallic materials together with the self-diffusion energy (E_{SD})

Energy (eV)	Cu	Al	Ni	Mg	Fe	W	NiAl
E_f	1.0–1.1	0.76	1.4	0.9	2.13	3.3	1.05
E_m	1.0–1.1	0.62	1.5	0.5	0.76	1.9	2.4
E_D	2.0–2.2	1.38	2.9	1.4	2.89	5.2	3.45

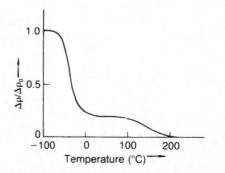

Figure 4.8 *Variation of quenched-in resistivity with temperature of annealing for aluminium (after Panseri and Federighi, 1958, 1223).*

$$dF/dV = \text{Energy/volume} \equiv \text{stress}$$

$$= (\mathbf{k}T/b^3)[\ln (c/c_0)] \qquad (4.6)$$

where dV is the volume associated with dn vacancies and b^3 is the volume of one vacancy. Inserting typical values, $\mathbf{K}T \simeq 1/40$ eV at room temperature, $b = 0.25$ nm, shows $\mathbf{K}T/b^3 \simeq 150$ MN/m^2. Thus, even a moderate 1% supersaturation of vacancies i.e. when $(c/c_0) = 1.01$ and $\ln (c/c_0) = 0.01$, introduces a chemical stress σ_c equivalent to 1.5 MN/m^2.

The equilibrium concentration of vacancies at a temperature T_2 will be given by $c_2 = \exp[-E_f/\mathbf{k}T_2]$ and at T_1 by $c_1 = \exp[-E_f/\mathbf{k}T_1]$. Then, since

$$\ln (c_2/c_1) = (E_f/\mathbf{k})\left[\frac{1}{T_1} - \frac{1}{T_2}\right]$$

the chemical stress produced by quenching a metal from a high temperature T_2 to a low temperature T_1 is

$$\sigma_c = (\mathbf{k}T/b^3)\ln (c_2/c_1) = (E_f/b^3)\left[1 - \frac{T_1}{T_2}\right]$$

For aluminium, E_f is about 0.7 eV so that quenching from 900 K to 300 K produces a chemical stress of about 3 GN/m^2. This stress is extremely high, several times the theoretical yield stress, and must be relieved in some way. Migration of vacancies to grain boundaries and dislocations will occur, of course, but it is not surprising that the point defects form additional vacancy sinks by the spontaneous nucleation of dislocations and other stable lattice defects, such as voids and stacking fault tetrahedra (see Sections 4.5.3 and 4.6).

When the material contains both vacancies and interstitials the removal of the excess point defect concentration is more complex. Figure 4.9 shows the 'annealing' curve for irradiated copper. The resistivity decreases sharply around 20 K when the interstitials start to migrate, with an activation energy $E_m \sim 0.1$ eV. In Stage I, therefore, most of the Frenkel (interstitial–vacancy) pairs anneal out. Stage II has been attributed to the release of interstitials from impurity traps as thermal energy supplies the necessary activation energy. Stage III is

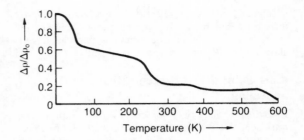

Figure 4.9 *Variation of resistivity with temperature produced by neutron irradiation for copper (after Diehl).*

around room temperature and is probably caused by the annihilation of free interstitials with individual vacancies not associated with a Frenkel pair, and also the migration of di-vacancies. Stage IV corresponds to the stage I annealing of quenched metals arising from vacancy migration and annihilation to form dislocation loops, voids and other defects. Stage V corresponds to the removal of this secondary defect population by self-diffusion.

4.3 Line defects

4.3.1 Concept of a dislocation

All crystalline materials usually contain lines of structural discontinuities running throughout each crystal or grain. These line discontinuities are termed dislocations and there is usually about 10^{10} to 10^{12} m of dislocation line in a metre cube of material.[1] Dislocations enable materials to deform without destroying the basic crystal structure at stresses below that at which the material would break or fracture if they were not present.

A crystal changes its shape during deformation by the slipping of atomic layers over one another. The theoretical shear strength of perfect crystals was first calculated by Frenkel for the simple rectangular-type lattice shown in Figure 4.10 with spacing a between

[1]This is usually expressed as the density of dislocations $\rho = 10^{10}$ to 10^{12} m^{-2}.

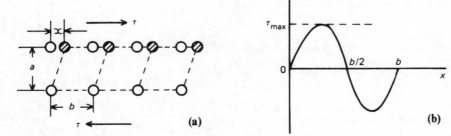

Figure 4.10 *Slip of crystal planes (a); shear stress versus displacement curve (b).*

the planes. The shearing force required to move a plane of atoms over the plane below will be periodic, since for displacements $x < b/2$, where b is the spacing of atoms in the shear direction, the lattice resists the applied stress but for $x > b/2$ the lattice forces assist the applied stress. The simplest function with these properties is a sinusoidal relation of the form

$$\tau = \tau_m \sin(2\pi x/b) \simeq \tau_m 2\pi x/b$$

where τ_m is the maximum shear stress at a displacement $= b/4$. For small displacements the elastic shear strain given by x/a is equal to τ/μ from Hooke's law, where μ is the shear modulus, so that

$$\tau_m = (\pi/2\pi)b/a \qquad (4.7)$$

and since $b \simeq a$, the theoretical strength of a perfect crystal is of the order of $\mu/10$.

This calculation shows that crystals should be rather strong and difficult to deform, but a striking experimental property of single crystals is their softness, which indicates that the critical shear stress to produce slip is very small (about $10^{-5}\ \mu$ or ≈ 50gf mm^{-2}). This discrepancy between the theoretical and experimental strength of crystals is accounted for if atomic planes do not slip over each other as rigid bodies but instead slip starts at a localized region in the structure and then spreads gradually over the remainder of the plane, somewhat like the disturbance when a pebble is dropped into a pond.

In general, therefore, the slip plane may be divided into two regions, one where slip has occurred and the other which remains unslipped. Between the slipped and unslipped regions the structure will be dislocated (Figure 4.11); this boundary is referred to as a dislocation line, or dislocation. Three simple properties of a dislocation are immediately apparent, namely: (1) it is a line discontinuity, (2) it forms a closed loop in the interior of the crystal or emerges at the surface and (3) the difference in the amount of slip across the dislocation line is constant. The last property is probably the most important, since a dislocation is characterized by the magnitude and direction of the slip movement associated with it. This is called the Burgers vector, b, which for any given dislocation line is the same all along its length.

4.3.2 Edge and screw dislocations

It is evident from Figure 4.11a that some sections of the dislocation line are perpendicular to b, others are parallel to b while the remainder lie at an angle to b. This variation in the orientation of the line with respect to the Burgers vector gives rise to a difference in the structure of the dislocation. When the dislocation line is normal to the slip direction it is called an edge dislocation. In contrast, when the line of the dislocations is parallel to the slip direction the dislocation line is known as a screw dislocation. From the diagram shown in Figure 4.11a it is evident that

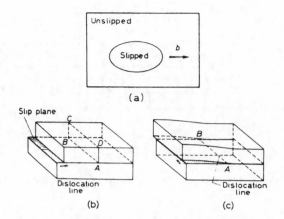

Figure 4.11 *Schematic representation of (a) a dislocation loop, (b) edge dislocation and (c) screw dislocation.*

the dislocation line is rarely pure edge or pure screw, but it is convenient to think of these ideal dislocations since any dislocation can be resolved into edge and screw components. The atomic structure of a simple edge and screw dislocation is shown in Figure 4.13 and 4.14.

4.3.3 The Burgers vector

It is evident from the previous sections that the Burgers vector b is an important dislocation parameter. In any deformation situation the Burgers vector is defined by constructing a Burgers circuit in the dislocated crystal as shown in Figure 4.12. A sequence of lattice vectors is taken to form a closed clockwise circuit around the dislocation. The same sequence of vectors is then taken in the perfect lattice when it is found that the circuit fails to close. The closure vector FS (finish-start) defines b for the dislocation. With this FS/RH (right-hand) convention it is necessary to choose one direction along the dislocation line as positive. If this direction is reversed the vector b is also reversed. The Burgers vector defines the atomic displacement produced as the dislocation moves across the slip plane. Its value is governed by the crystal structure because during slip it is necessary to retain an identical lattice structure both before and after the passage of the dislocation. This is assured if the dislocation has

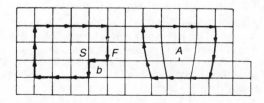

Figure 4.12 *Burgers circuit round a dislocation A fails to close when repeated in a perfect lattice unless completed by a closure vector FS equal to the Burgers vector b.*

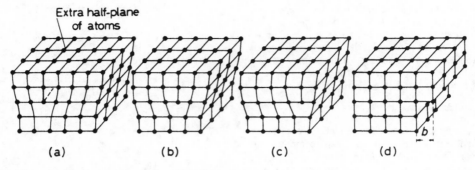

Figure 4.13 *Slip caused by the movement of an edge dislocation.*

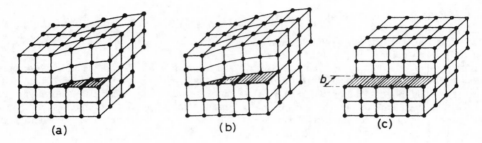

Figure 4.14 *Slip caused by the movement of a screw dislocation.*

a Burgers vector equal to one lattice vector and, since the energy of a dislocation depends on the square of the Burgers vector (see Section 4.3.5.2), its Burgers vector is usually the shortest available lattice vector. This vector, by definition, is parallel to the direction of closest packing in the structure, which agrees with experimental observation of the slip direction.

The Burgers vector is conveniently specified by its directional co-ordinates along the principal crystal axes. In the fcc lattice, the shortest lattice vector is associated with slip from a cube corner to a face centre, and has components $a/2$, $a/2$, 0. This is usually written $a/2[1\,1\,0]$, where a is the lattice parameter and $[1\,1\,0]$ is the slip direction. The magnitude of the vector, or the strength of the dislocation, is then given by $\sqrt{\{a^2(1^2 + 1^2 + 0^2)/4\}} = a/\sqrt{2}$. The corresponding slip vectors for the bcc and cph structures are $b = a/2[1\,1\,1]$ and $b = a/3[1\,1\,\bar{2}\,0]$ respectively.

4.3.4 Mechanisms of slip and climb

The atomic structure of an edge dislocation is shown in Figure 4.13a. Here the extra half-plane of atoms is above the slip plane of the crystal, and consequently the dislocation is called a positive edge dislocation and is often denoted by the symbol ⊥. When the half-plane is below the slip plane it is termed a negative dislocation. If the resolved shear stress on the slip plane is τ and the Burgers vector of the dislocation b, the force on the dislocation, i.e. force per unit length of dislocation, is $F = \tau b$. This can be seen by reference to

Figure 4.13 if the crystal is of side L. The force on the top face (stress × area) is $\tau \times L^2$. Thus, when the two halves of the crystal have slipped the relative amount b, the work done by the applied stress (force × distance) is $\tau L^2 b$. On the other hand, the work done in moving the dislocation (total force on dislocation FL × distance moved) is FL^2, so that equating the work done gives F (force per unit length of dislocation) = τb. Figure 4.13 indicates how slip is propagated by the movement of a dislocation under the action of such a force. The extra half-plane moves to the right until it produces the slip step shown at the surface of the crystal; the same shear will be produced by a negative dislocation moving from the right to left.[1]

The slip process as a result of a screw dislocation is shown in Figure 4.14. It must be recognized, however, that the dislocation is more usually a closed loop and slip occurs by the movement of all parts of the dislocation loop, i.e. edge, screw and mixed components, as shown in Figure 4.15.

A dislocation is able to glide in that slip plane which contains both the line of the dislocation and its Burgers vector. The edge dislocation is confined to glide in one plane only. An important difference between the motion of a screw dislocation and that of

---

[1] An obvious analogy to the slip process is the movement of a caterpillar in the garden or the propagation of a ruck in a carpet to move the carpet into place. In both examples, the effort to move is much reduced by this propagation process.

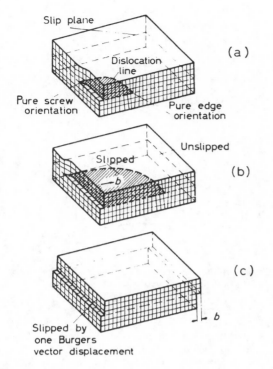

Figure 4.15 *Process of slip by the expansion of a dislocation loop in the slip plane.*

an edge dislocation arises from the fact that the screw dislocation is cylindrically symmetrical about its axis with its *b* parallel to this axis. To a screw dislocation all crystal planes passing through the axis look the same and, therefore, the motion of the screw dislocation is not restricted to a single slip plane, as is the case for a gliding edge dislocation. The process thereby a screw dislocation glides into another slip plane having a slip direction in common with the original slip plane, as shown in Figure 4.16, is called cross-slip. Usually, the cross-slip plane is also a close-packed plane, e.g. {1 1 1} in fcc crystals.

The mechanism of slip illustrated above shows that the slip or glide motion of an edge dislocation is restricted, since it can only glide in that slip plane

which contains both the dislocation line and its Burgers vector. However, movement of the dislocation line in a direction normal to the slip plane can occur under certain circumstances; this is called dislocation climb. To move the extra half-plane either up or down, as is required for climb, requires mass transport by diffusion and is a non-conservative motion. For example, if vacancies diffuse to the dislocation line it climbs up and the extra half-plane will shorten. However, since the vacancies will not necessarily arrive at the dislocation at the same instant, or uniformly, the dislocation climbs one atom at a time and some sections will lie in one plane and other sections in parallel neighbouring planes. Where the dislocation deviates from one plane to another it is known as a jog, and from the diagrams of Figure 4.17 it is evident that a jog in a dislocation may be regarded as a short length of dislocation not lying in the same slip plane as the main dislocation but having the same Burgers vector.

Jogs may also form when a moving dislocation cuts through intersecting dislocations, i.e. forest[1] dislocations, during its glide motion. In the lower range of temperature this will be the major source of jogs. Two examples of jogs formed from the crossings of dislocations are shown in Figure 4.18. Figure 4.18a shows a crystal containing a screw dislocation running from top to bottom which has the effect of 'ramping' all the planes in the crystal. If an edge dislocation moves through the crystal on a horizontal plane then the screw dislocation becomes jogged as the top half of the crystal is sheared relative to the bottom. In addition, the screw dislocation becomes jogged since one part has to take the upper ramp and the other part the lower ramp. The result is shown schematically in Figure 4.18b. Figure 4.18c shows the situation for a moving screw cutting through the vertical screw; the jog formed in each dislocation is edge in character since it is perpendicular to its Burgers vector which lies along the screw axis.

A jog in an edge dislocation will not impede the motion of the dislocation in its slip plane because it can, in general, move with the main dislocation line by glide, not in the same slip plane (see Figure 4.17b)

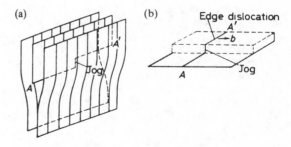

Figure 4.17 *Climb of an edge dislocation in a crystal.*

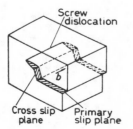

Figure 4.16 *Cross-slip of a screw dislocation in a crystal.*

[1] A number of dislocation lines may project from the slip plane like a forest, hence the term 'forest dislocation'.

(a)

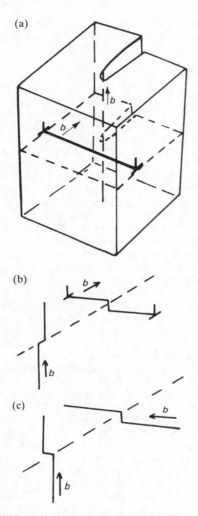

(b)

(c)

Figure 4.18 *Dislocation intersections. (a) and (b) screw–edge, (c) screw–screw.*

but in an intersecting slip plane that does contain the line of the jog and the Burgers vector. In the case of a jog in a screw dislocation the situation is not so clear, since there are two ways in which the jog can move. Since the jog is merely a small piece of edge dislocation it may move sideways, i.e. conservatively, along the screw dislocation and attach itself to an edge component of the dislocation line. Conversely, the jog may be dragged along with the screw dislocation. This latter process requires the jog to climb and, because it is a non-conservative process, must give rise to the creation of a row of point defects, i.e. either vacancies or interstitials depending on which way the jog is forced to climb. Clearly, such a movement is difficult but, nevertheless, may be necessary to give the dislocation sufficient mobility. The 'frictional' drag of

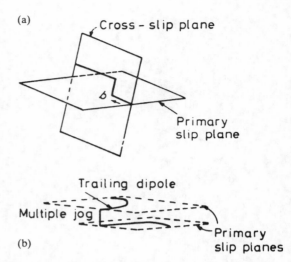

Figure 4.19 *(a) Formation of a multiple jog by cross-slip, and (b) motion of jog to produce a dipole.*

jogs will make a contribution to the work-hardening[1] of the material.

Apart from elementary jogs, or those having a height equal to one atomic plane spacing, it is possible to have multiple jogs where the jog height is several atomic plane spacings. Such jogs can be produced, for example, by part of a screw dislocation cross-slipping from the primary plane to the cross-slip plane, as shown in Figure 4.19a. In this case, as the screw dislocation glides forward it trails the multiple jog behind, since it acts as a frictional drag. As a result, two parallel dislocations of opposite sign are created in the wake of the moving screw, as shown in Figure 4.19b; this arrangement is called a dislocation dipole. Dipoles formed as debris behind moving dislocations are frequently seen in electron micrographs taken from deformed crystals (see Chapter 7). As the dipole gets longer the screw dislocation will eventually jettison the debris by cross-slipping and pinching off the dipole to form a prismatic loop, as shown in Figure 4.20. The loop is capable of gliding on the surface of a prism, the cross-sectional area of which is that of the loop.

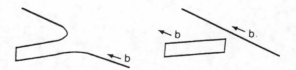

Figure 4.20 *Formation of prismatic dislocation loop from screw dislocation trailing a dipole.*

[1]When material is deformed by straining or working the flow stress increases with increase in strain (i.e. it is harder to deform a material which has been strained already). This is called strain- or work-hardening.

4.3.5 Strain energy associated with dislocations

4.3.5.1 Stress fields of screw and edge dislocations

The distortion around a dislocation line is evident from Figure 4.1 and 4.13. At the centre of the dislocation the strains are too large to be treated by elasticity theory, but beyond a distance r_0, equal to a few atom spacings Hooke's law can be applied. It is therefore necessary to define a core to the dislocation at a cut-off radius r_0 ($\approx b$) inside which elasticity theory is no longer applicable. A screw dislocation can then be considered as a cylindrical shell of length l and radius r contained in an elastically isotropic medium (Figure 4.21). A discontinuity in displacement exists only in the z-direction, i.e. parallel to the dislocation, such that $u = v = 0$, $w = b$. The elastic strain thus has to accommodate a displacement $w = b$ around a length $2\pi r$. In an elastically isotropic crystal the accommodation must occur equally all round the shell and indicates the simple relation $w = b\theta/2\pi$ in polar (r, θ, z) coordinates. The corresponding shear strain $\gamma_{\theta z}(= \gamma_{z\theta}) = b/2\pi r$ and shear stress $\tau_{\theta z}(= \tau_{z\theta}) = \mu b/2\pi r$ *which acts on the end faces of the cylinder with* σ_{rr} and $\tau_{r\theta}$ equal to zero.[1] Alternatively, the stresses are given in cartesian coordinates (x, y, z)

$$\tau_{xz}(= \tau_{zx}) = -\mu by/2\pi(x^2 + y^2)$$
$$\tau_{yz}(= \tau_{zy}) = -\mu bx/2\pi(x^2 + y^2) \quad (4.8)$$

with all other stresses equal to zero. The field of a screw dislocation is therefore purely one of shear, having radial symmetry with no dependence on θ. This mathematical description is related to the structure of a screw which has no extra half-plane of atoms and cannot be identified with a particular slip plane.

An edge dislocation has a more complicated stress and strain field than a screw. The distortion associated with the edge dislocation is one of plane strain, since there are no displacements along the z-axis, i.e. $w = 0$. In plane deformation the only stresses to be determined are the normal stresses σ_{xx}, σ_{yy} along the x- and y-axes respectively, and the shear stress τ_{xy} which acts in the direction of the y-axis on planes perpendicular to the x-axis. The third normal stress $\sigma_{zz} = \nu(\sigma_{xx} + \sigma_{yy})$ where ν is Poisson's ratio, and the other shear stresses τ_{yz} and τ_{zx} are zero. In polar coordinates r, θ and z, the stresses are σ_{rr}, $\sigma_{\theta\theta}$, and $\tau_{r\theta}$.

Even in the case of the edge dislocation the displacement b has to be accommodated round a ring of length $2\pi r$, so that the strains and the stresses must contain a term in $b/2\pi r$. Moreover, because the atoms in the region $0 < \theta < \pi$ are under compression and for $\pi < \theta < 2\pi$ in tension, the strain field must be of the form $(b/2\pi r)f(\theta)$, where $f(\theta)$ is a function such as $\sin\theta$ which changes sign when θ changes from 0 to 2π. It can be shown that the stresses are given by

$$\sigma_{rr} = \sigma_{\theta\theta} = -D\sin\theta/r; \quad \sigma_{r\theta} = D\cos\theta/r;$$

$$\sigma_{xz} = -D\frac{y(3x^2 + y^2)}{(x^2 + y^2)^2}; \quad \sigma_{yy} = D\frac{y(x^2 - y^2)}{(x^2 + y^2)^2} \quad (4.9)$$

$$\sigma_{xy} = D\frac{x(x^2 - y^2)}{(x^2 + y^2)^2}$$

where $D = \mu b/2\pi(1 - \nu)$. These equations show that the stresses around dislocations fall off as $1/r$ and hence the stress field is long-range in nature.

4.3.5.2 Strain energy of a dislocation

A dislocation is a line defect extending over large distances in the crystal and, since it has a strain energy per unit length (J m^{-1}), it possesses a total strain energy. An estimate of the elastic strain energy of screw dislocation can be obtained by taking the strain energy (i.e. $\frac{1}{2}$ × stress × strain per unit volume) in an annular ring around the dislocation of radius r and thickness dr to be $\frac{1}{2} \times (\mu b/2\pi r) \times (b/2\pi r) \times 2\pi r dr$. The total strain energy per unit length of dislocation is then obtained by integrating from r_0 the core radius, to r the outer radius of the strain field, and is

$$E = \frac{\mu b^2}{4\pi} \int_{r_0}^{r} \frac{dr}{r} = \frac{\mu b^2}{4\pi} \ln\left[\frac{r}{r_0}\right] \quad (4.10)$$

With an edge dislocation this energy is modified by the term $(1 - \nu)$ and hence is about 50% greater than a screw. For a unit dislocation in a typical crystal $r_0 \simeq 0.25$ nm, $r \simeq 2.5$ μm and $\ln[r/r_0] \simeq 9.2$, so that the energy is approximately μb^2 per unit length of dislocation, which for copper [taking $\mu = 40$ GNm^{-2}, $b = 0.25$ nm and 1 eV is about 1.6×10^{-19} J] is about 4 eV for every atom plane threaded by the dislocation.[2] If the reader prefers to think in terms of one metre of dislocation line, then this length is associated with about 2×10^{10} electron volts. We shall see later that heavily-deformed metals contain approximately 10^{16} m/m^3 of

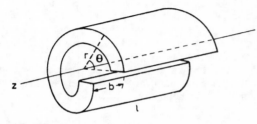

Figure 4.21 *Screw-dislocation in an elastic continuum.*

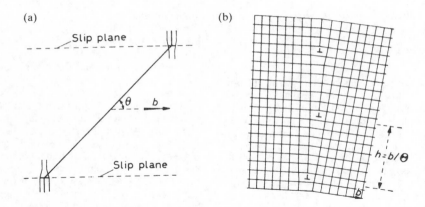

Figure 4.22 *Interaction between dislocations not on the same slip plane: (a) unlike dislocation, (b) like dislocations. The arrangement in (b) constitutes a small-angle boundary.*

dislocation line which leads to a large amount of energy stored in the lattice (i.e. ≈4J/g for Cu). Clearly, because of this high line energy a dislocation line will always tend to shorten its length as much as possible, and from this point of view it may be considered to possess a line tension, $T \approx \alpha \mu b^2$, analogous to the surface energy of a soap film, where $\alpha \approx \frac{1}{2}$.

4.3.5.3 Interaction of dislocations

The strain field around a dislocation, because of its long-range nature, is also important in influencing the behaviour of other dislocations in the crystal. Thus, it is not difficult to imagine that a positive dislocation will attract a negative dislocation lying on the same slip plane in order that their respective strain fields should cancel. Moreover, as a general rule it can be said that the dislocations in a crystal will interact with each other to take up positions of minimum energy to reduce the total strain energy of the lattice.

Two dislocations of the same sign will repel each other, because the strain energy of two dislocations on moving apart would be $2 \times b^2$ whereas if they combined to form one dislocation of Burgers vector $2b$, the strain energy would then be $(2b)^2 = 4b^2$; a force of repulsion exists between them. The force is, by definition, equal to the change of energy with position (dE/dr) and for screw dislocations is simply $F = \mu b^2/2\pi r$ where r is the distance between the two dislocations. Since the stress field around screw dislocations has cylindrical symmetry the force of interaction depends only on the distance apart, and the above expression for F applies equally well to parallel screw dislocations on neighbouring slip planes. For parallel edge dislocations the force–distance relationship is less simple. When the two edge dislocations lie in the same slip plane the relation is similar to that for two screws and has the form $F = \mu b^2/(1 - \nu)2\pi r$, but for edge dislocations with the same Burgers vector but not on the same slip plane the force also depends

on the angle θ between the Burgers vector and the line joining the two dislocations (Figure 4.22a).

Edge dislocations of the same sign repel and opposite sign attract along the line between them, but the component of force in the direction of slip, which governs the motion of a dislocation, varies with the angle θ. With unlike dislocations an attractive force is experienced for $\theta > 45°$ but a repulsive force for $\theta < 45°$, and in equilibrium the dislocations remain at an angle of $45°$ to each other. For like dislocations the converse applies and the position $\theta = 45°$ is now one of unstable equilibrium. Thus, edge dislocations which have the same Burgers vector but which do not lie on the same slip plane will be in equilibrium when $\theta = 90°$, and consequently they will arrange themselves in a plane normal to the slip plane, one above the other a distance h apart. Such a wall of dislocations constitutes a small-angle grain boundary as shown in Figure 4.22b, where the angle across the boundary is given by $\Theta = b/h$. This type of dislocation array is also called a sub-grain or low-angle boundary, and is important in the annealing of deformed metals.

By this arrangement the long-range stresses from the individual dislocations are cancelled out beyond a distance of the order of h from the boundary. It then follows that the energy of the crystal boundary will be given approximately by the sum of the individual energies, each equal to $\{\mu b^2/4\pi(1 - \nu)\} \ln (h/r_0)$ per unit length. There are $1/h$ or θ/b dislocations in a unit length, vertically, and hence, in terms of the misorientation across the boundary $\theta = b/h$, the energy γ_{gb} per unit area of boundary is

$$\gamma_{gb} = \frac{\mu b^2}{4\pi(1 - \nu)} \ln \left(\frac{h}{r_0} \right) \times \frac{\theta}{b}$$

$$= \left[\frac{\mu b}{4\pi(1 - \nu)} \right] \theta \ln \left(\frac{b}{\theta r} \right)$$

$$= E_0 \theta [A - \ln \theta] \tag{4.11}$$

where $E_0 = \mu b/4\pi(1 - \nu)$ and $A = \ln(b/r_0)$; this is known as the Read–Shockley formula. Values from it give good agreement with experimental estimates even up to relatively large angles. For $\theta \sim 25°$, $\gamma_{gb} \sim \mu b/25$ or ~ 0.4 J/m^2, which surprisingly is close to the value for the energy per unit area of a general large-angle grain boundary.

4.3.6 Dislocations in ionic structures

The slip system which operates in materials with NaCl structure is predominantly $a/2\langle 1\,\overline{1}\,0\rangle\{1\,1\,0\}$. The closest packed plane $\{1\,0\,0\}$ is not usually the preferred slip plane because of the strong electrostatic interaction that would occur across the slip plane during slip; like ions are brought into neighbouring positions across the slip plane for $(1\,0\,0)$ but not for the $(1\,1\,0)$. Dislocations in these materials are therefore simpler than fcc metals, but they may carry an electric charge (the edge dislocation on $\{1\,1\,0\}$, for example). Figure 4.23a has an extra 'half-plane' made up of a sheet of Na$^+$ ions and one of Cl$^-$ ions. The line as a whole can be charged up to a maximum of $e/2$ per atom length by acting as a source or sink for point defects. Figure 4.23b shows different jogs in the line which may either carry a charge or be uncharged. The jogs at B and C would be of charge $+e/2$ because the section BC has a net charge equal to e. The jog at D is uncharged.

4.4 Planar defects

4.4.1 Grain boundaries

The small-angle boundary described in Section 4.3.5.3 is a particular example of a planar defect interface in a crystal. Many such planar defects occur in materials ranging from the large-angle grain boundary, which is an incoherent interface with a relatively high energy of ~ 0.5 J/m^2, to atomic planes in the crystal across which there is a mis-stacking of the atoms, i.e. twin interfaces and stacking faults which retain the coherency of the packing and have much lower energies ($\leqslant 0.1$ J/m^2). Generally, all these planar defects are associated with dislocations in an extended form.

A small-angle tilt boundary can be described adequately by a vertical wall of dislocations. Rotation of one crystal relative to another, i.e. a twist boundary, can be produced by a crossed grid of two sets of screw dislocations as shown in Figure 4.24. These boundaries are of a particularly simple kind separating two crystals which have a small difference in orientation, whereas a general grain boundary usually separates crystals which differ in orientation by large angles. In this case, the boundary has five degrees of freedom, three of which arise from the fact that the adjoining crystals may be rotated with respect to each other about the three perpendicular axes, and the other two from the degree of freedom of the orientation of the boundary surface itself with respect to the crystals. Such a large-angle (30–40°) grain boundary may simply be regarded as a narrow region, about two atoms thick, across which the atoms change from the lattice orientation of the one grain to that of the other. Nevertheless, such a grain boundary may be described by an arrangement of dislocations, but their arrangement will be complex and the individual dislocations are not easily recognized or analysed.

The simplest extension of the dislocation model for low-angle boundaries to high-angle grain boundaries is to consider that there are islands of good atomic fit surrounded by non-coherent regions. In a low-angle boundary the 'good fit' is perfect crystal and the 'bad

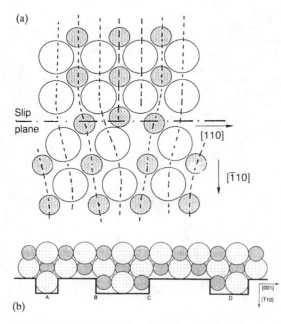

(a)

Slip plane

[110]

[$\overline{1}$10]

(b)

A   B   C   D

[001]
[$\overline{1}$10]

Figure 4.23 *Edge dislocation in NaCl, showing: (a) two extra half-sheets of ions: anions are open circles, cations are shaded; (b) charged and uncharged jogs (after Kelly and Groves, 1970).*

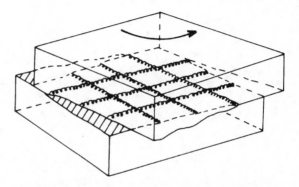

Figure 4.24 *Representation of a twist boundary produced by cross-grid of screw dislocations.*

fit' is accommodated by lattice dislocations, whereas for high-angle boundaries the 'good fit' could be an interfacial structure with low energy and the bad fit accommodated by dislocations which are not necessarily lattice dislocations. These dislocations are often termed intrinsic secondary grain boundary dislocations (gbds) and are essential to maintain the boundary at that mis-orientation.

The regions of good fit are sometimes described by the coincident site lattice (CSL) model, with its development to include the displacement shift complex (DSC) lattice. A CSL is a three-dimensional superlattice on which a fraction $1/\sum$ of the lattice points in both crystal lattices lie; for the simple structures there will be many such CSLs, each existing at a particular misorientation. One CSL is illustrated in Figure 4.25 but it must be remembered that the CSL is three-dimensional, infinite and interpenetrates both crystals; it does not in itself define an interface. However, an interface is likely to have a low energy if it lies between two crystals oriented such that they share a high proportion of lattice sites, i.e. preferred misorientations will be those with CSLs having low $\sum$ values. Such misorientations can be predicted from the expression

$$\theta = 2\tan^{-1}\left(\frac{b}{a}\sqrt{N}\right)$$

where b and a are integers and $N = h^2 + k^2 + l^2$; the $\sum$ value is then given by $a^2 + Nb^2$, divided by 2 until an odd number is obtained.

The CSL model can only be used to describe certain specific boundary misorientations but it can be extended to other misorientations by allowing the presence of arrays of dislocations which act to preserve a special orientation between them. Such intrinsic secondary dislocations must conserve the boundary structure and, generally, will have Burgers vectors smaller than those of the lattice dislocations.

When a polycrystalline specimen is examined in TEM other structural features apart from intrinsic grain boundary dislocations (gbds) may be observed in a grain boundary, such as 'extrinsic' dislocations which have probably run-in from a neighbouring grain, and interface ledges or steps which curve the boundary. At low temperatures the run-in lattice dislocation tends to retain its character while trapped in the interface, whereas at high temperatures it may dissociate into several intrinsic gbds resulting in a small change in misorientation across the boundary. The analysis of gbds in TEM is not easy, but information about them will eventually further our understanding of important boundary phenomena (e.g. migration of boundaries during recrystallization and grain growth, the sliding of grains during creep and superplastic flow and the way grain boundaries act as sources and sinks for point defects).

4.4.2 Twin boundaries

Annealing of cold-worked fcc metals and alloys, such as copper, α-brass and austenitic stainless steels usually causes many of the constituent crystals to form annealing twins. The lattice orientation changes at the twin boundary surface, producing a structure in which one part of the crystal or grain is the mirror-image of the other, the amount of atomic displacement being proportional to the distance from the twin boundary.

The surfaces of a sample within and outside an annealing twin have different surface energies, because of their different lattice orientations, and hence respond quite differently when etched with a chemical etchant (Figure 4.26). In this diagram, twins 1 and 2 both have two straight parallel sides which are coherent low-energy interfaces. The short end face of twin 2 is non-coherent and therefore has a higher energy content

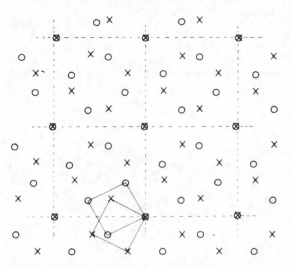

Figure 4.25 *Two-dimensional section of a CSL with $\sum 5$ 36.9° [100] twist orientation (courtesy of P. Goodhew).*

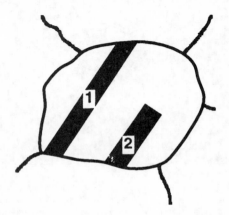

Figure 4.26 *Twinned regions within a single etched grain, produced by deformation and annealing.*

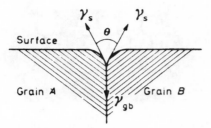

Figure 4.27 *Grain boundary/surface triple junction.*

per unit surface. Stacking faults are also coherent and low in energy content; consequently, because of this similarity in character, we find that crystalline materials which twin readily are also likely to contain many stacking faults (e.g. copper, α-brass).

Although less important and less common than slip, another type of twinning can take place during plastic deformation. These so-called deformation twins sometimes form very easily, e.g. during mechanical polishing of metallographic samples of pure zinc; this process is discussed in Chapter 7.

The free energies of interfaces can be determined from the equilibrium form of the triple junction where three interfaces, such as surfaces, grain boundaries or twins, meet. For the case of a grain boundary intersecting a free surface, shown in Figure 4.27,

$$\gamma_{gb} = 2\gamma_s \cos\theta/2 \qquad (4.12)$$

and hence γ_{gb} can be obtained by measuring the dihedral angle θ and knowing γ_s. Similarly, measurements can be made of the ratio of twin boundary energy to the average grain boundary energy and, knowing either γ_s or γ_{gb} gives an estimate of γ_T.

4.4.3 Extended dislocations and stacking faults in close-packed crystals

4.4.3.1 Stacking faults

Stacking faults associated with dislocations can be an extremely significant feature of the structure of many materials, particularly those with fcc and cph structure. They arise because to a first approximation there is little to choose electrostatically between the stacking sequence of the close-packed planes in the fcc metals **ABCABC**... and that in the cph metals **ABABAB**... Thus, in a metal like copper or gold, the atoms in a part of one of the close-packed layers may fall into the 'wrong' position relative to the atoms of the layers above and below, so that a mistake in the stacking sequence occurs (e.g. **ABCBCABC**...). Such an arrangement will be reasonably stable, but because some work will have to be done to produce it, stacking faults are more frequently found in deformed metals than annealed metals.

4.4.3.2 Dissociation into Shockley partials

The relationship between the two close-packed structures cph and fcc has been discussed in Chapter 2 where it was seen that both structures may be built up from stacking close-packed planes of spheres. The shortest lattice vector in the fcc structure joins a cube corner atom to a neighbouring face centre atom and defines the observed slip direction; one such slip vector $a/2[10\bar{1}]$ is shown as b_1 in Figure 4.28a which is for glide in the (111) plane. However, an atom which sits in a **B** position on top of the A plane would move most easily initially towards a **C** position and, consequently, to produce a macroscopical slip movement along $[10\bar{1}]$ the atoms might be expected to take a zigzag path of the type $\mathbf{B} \rightarrow \mathbf{C} \rightarrow \mathbf{B}$ following the vectors $b_2 = a/6[2\bar{1}\bar{1}]$ and $b_3 = a/6[11\bar{2}]$ alternately. It will be evident, of course, that during the initial part of the slip process when the atoms change from **B** positions to **C** positions, a stacking fault in the (111) layers is produced and the stacking sequence changes from **ABCABC**... to **ABCACABC**.... During the second part of the slip process the correct stacking sequence is restored.

To describe the atoms movement during slip, discussed above, Heidenreich and Shockley have pointed out that the unit dislocation must dissociate into two half dislocations,[1] which for the case of glide in the (111) plane would be according to the reaction:

$$a/2[10\bar{1}] \rightarrow a/6[2\bar{1}\bar{1}] + a/6[11\bar{2}]$$

Such a dissociation process is (1) algebraically correct, since the sum of the Burgers vector components of the two partial dislocations, i.e. $a/6[2+1]$, $a/6[1+1]$, $a/6[1+2]$, are equal to the components of the Burgers vector of the unit dislocation, i.e. $a/2$, 0, $\bar{a}/2$, and (2) energetically favourable, since the sum of the strain energy values for the pair of half dislocations is less than the strain energy value of the single unit dislocation, where the initial dislocation energy is proportional to $b_1^2 (= a^2/2)$ and the energy of the resultant partials to $b_2^2 + b_3^2 = a^2/3$. These half dislocations, or Shockley partial dislocations, repel each other by a force that is approximately $F = (\mu b_2 b_3 \cos 60)/2\pi d$, and separate as shown in Figure 4.28b. A sheet of stacking faults is then formed in the slip plane between the partials, and it is the creation of this faulted region, which has a higher energy than the normal lattice, that prevents the partials from separating too far. Thus, if γ J/m² is the energy per unit area of the fault, the force per unit length exerted on the dislocations by the fault is γ N/m and the equilibrium separation d is given by equating the repulsive force F between the two half dislocations to the force exerted by the fault, γ. The equilibrium separation of two partial dislocations is

---

[1] The correct indices for the vectors involved in such dislocation reactions can be obtained from Figure 4.37.

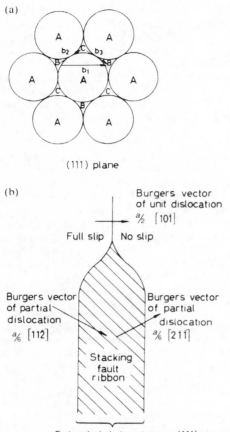

(a)

(111) plane

(b)

Burgers vector
of unit dislocation
$\frac{a}{2}$ [10$\bar{1}$]

Full slip | No slip

Burgers vector
of partial
dislocation
$\frac{a}{6}$ [112]

Burgers vector
of partial
dislocation
$\frac{a}{6}$ [21$\bar{1}$]

Stacking
fault
ribbon

Extended dislocation in (111) plane

Figure 4.28 *Schematic representation of slip in a (1 1 1) plane of a fcc crystal.*

then given by

$$d = \left(\frac{\mu b_2 b_3 \cos 60}{2\pi\gamma} \right)$$

$$= \left(\frac{\mu \dfrac{a}{\sqrt{6}} \dfrac{a}{\sqrt{6}} \dfrac{1}{2}}{2\pi\gamma} \right) = \frac{\mu a^2}{24\pi\gamma} \qquad (4.13)$$

from which it can be seen that the width of the stacking fault 'ribbon' is inversely proportional to the value of the stacking fault energy γ and also depends on the value of the shear modulus μ.

Figure 4.29a shows that the undissociated edge dislocation has its extra half-plane corrugated which may be considered as two (1 0 $\bar{1}$) planes displaced relative to each other and labelled a and b in Figure 4.29b. On dissociation, planes a and b are separated by a region of crystal in which across the slip plane the atoms are in the wrong sites (see Figure 4.29c). Thus the high

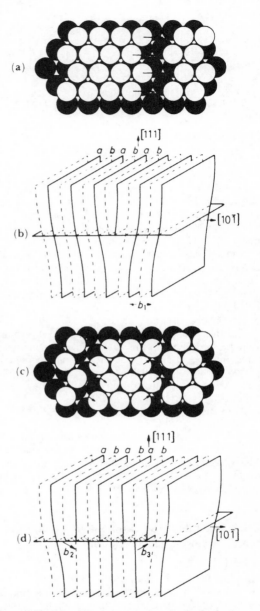

(a)

(b)

(c)

(d)

Figure 4.29 *Edge dislocation structure in the fcc lattice, (a) and (b) undissociated, (c) and (d) dissociated: (a) and (c) are viewed normal to the (1 1 1) plane (from Hume-Rothery, Smallman and Haworth, 1969; courtesy of the Institute of Metals).*

strain energy along a line through the crystal associated with an undissociated dislocation is spread over a plane in the crystal for a dissociated dislocation (see Figure 4.29d) thereby lowering its energy.

A direct estimate of γ can be made from the observation of extended dislocations in the electron microscope and from observations on other stacking fault defects (see Chapter 5). Such measurements show that the stacking fault energy for pure fcc metals ranges from about 16 mJ/m² for silver to ≈200 mJ/m² for nickel, with gold ≈30, copper ≈40 and aluminium 135 mJ/m², respectively. Since stacking faults are coherent interfaces or boundaries they have energies considerably lower than non-coherent interfaces such as free surfaces for which $\gamma_s \approx \mu b/8 \approx 1.5$ J/m² and grain boundaries for which $\gamma_{gb} \approx \gamma_s/3 \approx 0.5$ J/m².

The energy of a stacking fault can be estimated from twin boundary energies since a stacking fault **ABCBCABC** may be regarded as two overlapping twin boundaries **CBC** and **BCB** across which the next nearest neighbouring plane are wrongly stacked. In fcc crystals any sequence of three atomic planes not in the **ABC** or **CBA** order is a stacking violation and is accompanied by an increased energy contribution. A twin has one pair of second nearest neighbour planes in the wrong sequence, two third neighbours, one fourth neighbour and so on; an intrinsic stacking fault two second nearest neighbours, three third and no fourth nearest neighbour violations. Thus, if next-next nearest neighbour interactions are considered to make a relatively small contribution to the energy then an approximate relation $\gamma \simeq 2\gamma_T$ is expected.

The frequency of occurrence of annealing twins generally confirms the above classification of stacking fault energy and it is interesting to note that in aluminium, a metal with a relatively high value of γ, annealing twins are rarely, if ever, observed, while they are seen in copper which has a lower stacking fault energy. Electron microscope measurements of γ show that the stacking fault energy is lowered by solid solution alloying and is influenced by those factors which affect the limit of primary solubility. The reason for this is that on alloying, the free energies of the α-phase and its neighbouring phase become more nearly equal, i.e. the stability of the α-phase is decreased relative to some other phase, and hence can more readily tolerate mis-stacking. Figure 4.30 shows the reduction of γ for copper with addition of solutes such as Zn, Al, Sn and Ge, and is consistent with the observation that annealing twins occur more frequently in α-brass or Cu–Sn than pure copper. Substituting the appropriate values for μ, a and γ in equation (4.13) indicates that in silver

and copper the partials are separated to about 12 and 6 atom spacings, respectively. For nickel the width is about $2b$ since although nickel has a high γ its shear modulus is also very high. In contrast, aluminium has a lower $\gamma \approx 135$ mJ/m² but also a considerably lower value for μ and hence the partial separation is limited to about $1b$ and may be considered to be unextended. Alloying significantly reduces γ and very wide dislocations are produced, as found in the brasses, bronzes and austenitic stainless steels. However, no matter how narrow or how wide the partials are separated the two half-dislocations are bound together by the stacking fault, and consequently, they must move together as a unit across the slip plane.

The width of the stacking fault ribbon is of importance in many aspects of plasticity because at some stage of deformation it becomes necessary for dislocations to intersect each other; the difficulty which dislocations have in intersecting each other gives rise to one source of work-hardening. With extended dislocations the intersecting process is particularly difficult since the crossing of stacking faults would lead to a complex fault in the plane of intersection. The complexity may be reduced, however, if the half-dislocations coalesce at the crossing point, so that they intersect as perfect dislocations; the partials then are constricted together at their jogs, as shown in Figure 4.31a.

The width of the stacking fault ribbon is also important to the phenomenon of cross-slip in which a dislocation changes from one slip plane to another intersecting slip plane. As discussed previously, for glide to occur the slip plane must contain both the Burgers vector and the line of the dislocation, and, consequently, for cross-slip to take place a dislocation

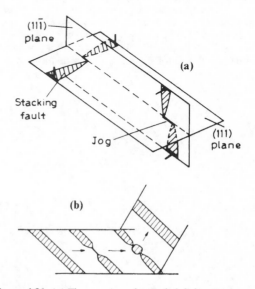

Figure 4.31 (a) The crossing of extended dislocations, (b) various stages in the cross-slip of a dissociated screw dislocation.

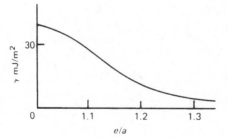

Figure 4.30 Decrease in stacking-fault energy γ for copper with alloying addition (e/a).

must be in an exact screw orientation. If the dislocation is extended, however, the partials have first to be brought together to form an unextended dislocation as shown in Figure 4.31b before the dislocation can spread into the cross-slip plane. The constriction process will be aided by thermal activation and hence the cross-slip tendency increases with increasing temperature. The constriction process is also more difficult, the wider the separation of the partials. In aluminium, where the dislocations are relatively unextended, the frequent occurrence of cross-slip is expected, but for low stacking fault energy metals (e.g. copper or gold) the activation energy for the process will be high. Nevertheless, cross-slip may still occur in those regions where a high concentration of stress exists, as, for example, when dislocations pile up against some obstacle, where the width of the extended dislocation may be reduced below the equilibrium separation. Often screw dislocations escape from the piled-up group by cross-slipping but then after moving a certain distance in this cross-slip plane return to a plane parallel to the original slip plane because the resolved shear stress is higher. This is a common method of circumventing obstacles in the structure.

4.4.3.3 Sessile dislocations

The Shockley partial dislocation has its Burgers vector lying in the plane of the fault and hence is glissile. Some dislocations, however, have their Burgers vector not lying in the plane of the fault with which they are associated, and are incapable of gliding, i.e. they are sessile. The simplest of these, the Frank sessile dislocation loop, is shown in Figure 4.32a. This dislocation is believed to form as a result of the collapse of the lattice surrounding a cavity which has been produced by the aggregation of vacancies on to a $(1\,1\,1)$ plane. As shown in Figure 4.32a, if the vacancies aggregate on the central A-plane the adjoining parts of the neighbouring B and C planes collapse to fit in close-packed formation. The Burgers vector of the dislocation line

bounding the collapsed sheet is normal to the plane with $b = \frac{a}{3}[1\,1\,1]$, where a is the lattice parameter, and such a dislocation is sessile since it encloses an area of stacking fault which cannot move with the dislocation. A Frank sessile dislocation loop can also be produced by inserting an extra layer of atoms between two normal planes of atoms, as occurs when interstitial atoms aggregate following high energy particle irradiation. For the loop formed from vacancies the stacking sequence changes from the normal **ABCABCA**... to **ABCBCA**..., whereas inserting a layer of atoms, e.g. an A-layer between B and C, the sequence becomes **ABCABACA**.... The former type of fault with one violation in the stacking sequence is called an intrinsic fault, the latter with two violations is called an extrinsic fault. The stacking sequence violations are conveniently shown by using the symbol $\triangle$ to denote any normal stacking sequence **AB, BC, CA** but $\triangledown$ for the reverse sequence **AC, CB, BA**. The normal fcc stacking sequence is then given by $\triangle\triangle\triangle\triangle$..., the intrinsic fault by $\triangle\triangle\triangledown\triangle\triangle$... and the extrinsic fault by $\triangle\triangle\triangledown\triangledown\triangle\triangle$.... The reader may verify that the fault discussed in the previous Section is also an intrinsic fault, and that a series of intrinsic stacking faults on neighbouring planes gives rise to a twinned structure **ABCABACBA** or $\triangle\triangle\triangle\triangle\triangledown\triangledown\triangledown\triangle$. Electron micrographs of Frank sessile dislocation loops are shown in Figures 4.38 and 4.39.

Another common obstacle is that formed between extended dislocations on intersecting $\{1\,1\,1\}$ slip planes, as shown in Figure 4.32b. Here, the combination of the leading partial dislocation lying in the $(1\,1\,1)$ plane with that which lies in the $(\overline{1}\,1\,\overline{1})$ plane forms another partial dislocation, often referred to as a 'stair-rod' dislocation, at the junction of the two stacking fault ribbons by the reaction

$$\frac{a}{6}[\overline{1}\,1\,2] + \frac{a}{6}[2\,\overline{1}\,\overline{1}] \rightarrow \frac{a}{6}[1\,0\,1]$$

The indices for this reaction can be obtained from Figure 4.37 and it is seen that there is a reduction in

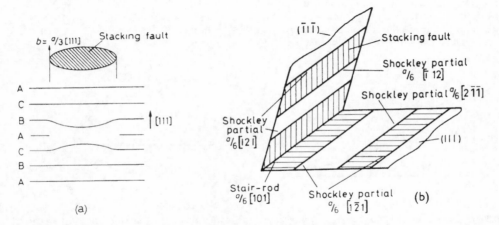

(a)   (b)

Figure 4.32 *Sessile dislocations: (a) a Frank sessile dislocation; (b) stair-rod dislocation as part of a Lomer-Cottrell barrier.*

energy from $[(a^2/6) + (a^2/6)]$ to $a^2/18$. This triangular group of partial dislocations, which bounds the wedge-shaped stacking fault ribbon lying in a $\langle 1\,0\,1 \rangle$ direction, is obviously incapable of gliding and such an obstacle, first considered by Lomer and Cottrell, is known as a Lomer-Cottrell barrier. Such a barrier impedes the motion of dislocations and leads to work-hardening, as discussed in Chapter 7.

4.4.3.4 Stacking faults in ceramics

Some ceramic oxides may be described in terms of fcc or cph packing of the oxygen anions with the cations occupying the tetrahedral or octahedral interstitial sites, and these are more likely to contain stacking faults. Sapphire, α-Al_2O_3, deforms at high temperatures on the $(0\,0\,0\,1)$ $\langle 1\,1\,\bar{2}\,0 \rangle$ basal systems and dissociated $\frac{1}{3}\langle 1\,0\,\bar{1}\,0 \rangle + \frac{1}{3}\langle 0\,1\,\bar{1}\,0 \rangle$ dislocations have been observed. Stacking faults also occur in spinels. Stoichiometric spinel ($MgAl_2O_4$ or $MgO \cdot nAl_2O_3$, $n = 1$) deforms predominantly on the $\{1\,\bar{1}\,1\}\langle 1\,1\,0 \rangle$ slip system at high temperature ($\sim 1800°C$) with dissociated dislocations. Non-stoichiometric crystals ($n > 1$) deform at lower temperatures when the $\{1\,\bar{1}\,0\}\langle 1\,1\,0 \rangle$ secondary slip system is also preferred. The stacking fault energy decreases with deviation from stoichiometry from a value around 0.2 J/m^2 for $n = 1$ crystals to around 0.02 J/m^2 for $n = 3.5$.

4.4.3.5 Stacking faults in semiconductors

Elemental semiconductors Si, Ge with the diamond cubic structure or III-V compounds InSb with the sphalerite (zinc blende) structure have perfect dislocation Burgers vectors similar to those in the fcc lattice. Stacking faults on $\{1\,1\,1\}$ planes associated with partial dislocations also exist. The $\{1\,1\,1\}$ planes are stacked in the sequence **AaBbCcAaBb** as shown in Figure 4.33 and stacking faults are created by the removal (intrinsic) or insertion (extrinsic) of pairs of layers such as **Aa** or **Bb**. These faults do not change the four nearest-neighbour covalent bonds and are low-energy

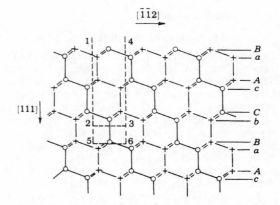

Figure 4.33 *A diamond cubic lattice projected normal to* $(1\,\bar{1}\,0)$. ◯ *represents atoms in the plane of the paper and* + *represents atoms in the plane below.* $(1\,1\,1)$ *is perpendicular to the plane of the paper and appears as a horizontal trace.*

faults ~ 50 mJ/m^2. Dislocations could slip between the narrowly spaced planes **Ba**, called the glide set, or between the widely spaced planes **bB**, called the shuffle set, but weak beam microscopy shows dissociation into Shockley partials occurs on the glide set. A 60° dislocation (i.e. 60° to its Burgers vector $\frac{1}{2}a\langle 1\,1\,0 \rangle$) of the glide set is formed by cutting out material bounded by the surface 1564 then joining the cut together. The extra plane of atoms terminates between **a** and **B** leaving a row of dangling bonds along its core which leads to the electrical effect of a half-filled band in the band gap; plastic deformation can make n-type Ge into p-type. The 60° dislocation **BC** and its dissociation into δ**C** and **B**δ is shown in Figure 4.34. The wurtzite (ZnS) structure has the hexagonal stacking sequence **AaBbAaBb** .... Similarly, stacking faults in the wurtzite structure are thin layers of sphalerite **BbAaBbCcAA** ... analogous to stacking faults in hexagonal metals.

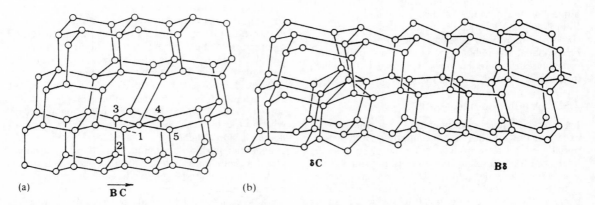

Figure 4.34 *(a) The 60° dislocation **BC**, (b) the dissociation of **BC** into δ**C** and **B**δ.*

4.5 Volume defects

4.5.1 Void formation and annealing

Defects which occupy a volume within the crystal may take the form of voids, gas bubbles and cavities. These defects may form by heat-treatment, irradiation or deformation and their energy is derived largely from the surface energy $(1-3 \text{ J/m}^2)$. In some materials with low stacking fault energy a special type of three-dimensional defect is formed, namely the defect tetrahedron. This consists of a tetrahedron made up from stacking faults on the four $\{1\,1\,1\}$ planes joined together by six low-energy stair-rod dislocations. This defect is discussed more fully in Section 4.6.2.3.

The growth of the original vacancy cluster into a three-dimensional aggregate, i.e. void, a collapsed vacancy disc, i.e. dislocation loop, should, in principle, depend on the relative surface to strain energy values for the respective defects. The energy of a three-dimensional void is mainly surface energy, whereas that of a Frank loop is mainly strain energy at small sizes. However, without a detailed knowledge of the surface energy of small voids and the core-energy of dislocations it is impossible to calculate, with any degree of confidence, the relative stability of these clustered vacancy defects.

The clustering of vacancies to form voids has now been observed in a number of metals with either fcc or cph structure. In as-quenched specimens the voids are not spherical but bounded by crystallographic faces (see Figure 4.58) and usually are about 50 nm radius in size. In fcc metals they are octahedral in shape with sides along $\langle 1\,1\,0 \rangle$, sometimes truncated by $\{1\,0\,0\}$ planes, and in cph metals bounded by prism and pyramidal planes. Void formation is favoured by slow quenching rates and high ageing temperatures, and the density of voids increases when gas is present in solid solution (e.g. hydrogen in copper, and either hydrogen or oxygen in silver). In aluminium and magnesium, void formation is favoured by quenching from a wet atmosphere, probably as a result of hydrogen production due to the oxidation reactions. It has been postulated that small clustered vacancy groups are stabilized by the presence of gas atoms and prevented from collapsing to a planar disc, so that some critical size for collapse can be exceeded. The voids are not conventional gas bubbles, however, since only a few gas atoms are required to nucleate the void after which it grows by vacancy adsorption.

4.5.2 Irradiation and voiding

Irradiation produces both interstitials and vacancies in excess of the equilibrium concentration. Both species initially cluster to form dislocation loops, but it is the interstitial loops formed from clustering of interstitials which eventually develop into a dislocation structure.

In general, interstitial loops grow during irradiation because the large elastic misfit associated with an interstitial causes dislocations to attract interstitials more strongly than vacancies. Interstitial loops are therefore intrinsically stable defects, whereas vacancy loops are basically unstable defects during irradiation. Thus interstitials attracted to a vacancy loop, i.e. a loop formed by clustering vacancies, will cause it to shrink as the interstitials are annihilated. Increasing the irradiation temperature results in vacancies aggregating to form voids. Voids are formed in an intermediate temperature range ≈ 0.3 to $0.6 T_m$, above that for long-range single vacancy migration and below that for thermal vacancy emission from voids. To create the excess vacancy concentration it is also necessary to build up a critical dislocation density from loop growth to bias the interstitial flow.

There are two important factors contributing to void formation. The first is the degree of bias the dislocation density (developed from the growth of interstitial loops) has for attracting interstitials, which suppresses the interstitial content compared to vacancies. The second factor is the important role played in void nucleation by gases, both surface-active gases such as oxygen, nitrogen and hydrogen frequently present as residual impurities, and inert gases such as helium which may be generated continuously during irradiation due to transmutation reactions. The surface-active gases such as oxygen in copper can migrate to embryo vacancy clusters and reduce the surface energy. The inert gas atoms can acquire vacancies to become gas molecules inside voids (when the gas pressure is not in equilibrium with the void surface tension) or gas bubbles when the gas pressure is considerable $(P \gtrsim 2\gamma_s/r)$. Voids and bubbles can give rise to irradiation swelling and embrittlement of materials.

4.5.3 Voiding and fracture

The formation of voids is an important feature in the ductile failure of materials. The fracture process involves three stages. First, small holes or cavities nucleate usually at weak internal interfaces (e.g. particle/matrix interfaces). These cavities then expand by plastic deformation and finally coalesce by localized necking of the metal between adjacent cavities to form a fibrous fracture. A scanning electron micrograph showing the characteristics of a typical ductile failure is shown in Figure 4.35. This type of fracture may be regarded as taking place by the nucleation of an internal plastic cavity, rather than a crack, which grows outwards to meet the external neck which is growing inwards. Experimental evidence suggests that nucleation occurs at foreign particles. For example, OFHC copper necks down to over 90% reduction in area, whereas tough-pitch copper shows only 70% reduction in area; a similar behaviour is noted for super-pure and commercial purity aluminium. Thus if no inclusions were present, failure should occur by the specimen pulling apart entirely by the inward growth of the external neck, giving nearly 100% reduction in area. Dispersion-hardened materials in general fail with a ductile fracture, the fibrous region often consisting

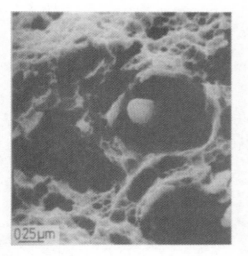

Figure 4.35 *SEM micrograph of a medium-carbon (0.4%) steel with a quenched and tempered martensite structure, showing large dimples associated with oxide inclusions and small dimples associated with small carbide precipitates (courtesy Dr L. Sidjanin).*

of many dimples arising from the dispersed particles nucleating holes and causing local ductile failure. Ductile failure is discussed further in Chapter 8.

4.6 Defect behaviour in some real materials

4.6.1 Dislocation vector diagrams and the Thompson tetrahedron

The classification of defects into point, line, planar and volume is somewhat restrictive in presenting an overview of defect behaviour in materials, since it is clear, even from the discussion so far, that these defects are interrelated and interdependent. In the following sections these features will be brought out as well as those which relate to specific structures.

In dealing with dislocation interactions and defects in real material it is often convenient to work with a vector notation rather than use the more conventional Miller indices notation. This may be illustrated by reference to the fcc structure and the Thompson tetrahedron.

All the dislocations common to the fcc structure, discussed in the previous sections, can be represented conveniently by means of the Thompson reference tetrahedron (Figure 4.36a), formed by joining the three nearest face-centring atoms to the origin D. Here ABCD is made up of four {1 1 1} planes (1 1 1), ($\bar{1}\,\bar{1}$ 1), ($\bar{1}$ 1 $\bar{1}$) and (1 $\bar{1}\,\bar{1}$) as shown by the stereogram given in Figure 4.36b, and the edges **AB**, **BC**, **CA** ... correspond to the $\langle 1\,1\,0 \rangle$ directions in these planes. Then, if the mid-points of the faces are labelled α, β, γ, δ, as shown in Figure 4.37a, all the dislocation Burgers vectors are represented. Thus, the edges (**AB**, **BC** ...) correspond to the normal slip vectors, $a/2\langle 1\,1\,0 \rangle$. The half-dislocations, or Shockley partials, into which these are dissociated have Burgers vectors of the $a/6\langle 1\,1\,2 \rangle$ type and are represented by the Roman–Greek symbols Aγ, Bγ, Dγ, Aδ, Bδ, etc, or Greek–Roman symbols γA, γB, γD, δA, δB, etc. The dissociation reaction given in the first reaction in Section 4.4.3.2 is then simply written

$$BC \rightarrow B\delta + \delta C$$

and there are six such dissociation reactions in each of the four {1 1 1} planes (see Figure 4.37). It is conventional to view the slip plane from outside the tetrahedron along the positive direction of the unit dislocation BC, and on dissociation to produce an intrinsic stacking fault arrangement; the Roman–Greek partial Bδ is on the right and the Greek–Roman partial δC on the left. A screw dislocation with Burgers vector BC which is normally dissociated in the δ-plane is capable of cross-slipping into the α-plane by first constricting B$\delta + \delta$C $\rightarrow$ BC and then redissociating in the α-plane BC $\rightarrow$ B$\alpha + \alpha$C.

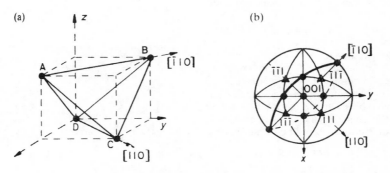

Figure 4.36 *(a) Construction and (b) orientation of the Thompson tetrahedron ABCD. The slip directions in a given {1 1 1} plane may be obtained from the trace of that plane as shown for the (1 1 1) plane in (b).*

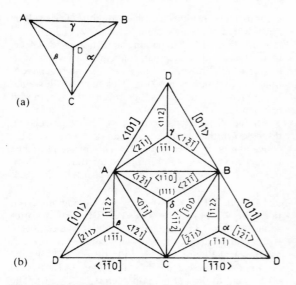

(a)

(b)

Figure 4.37 *A Thompson tetrahedron (a) closed and (b) opened out. In (b) the notation [1 1 0⟩ is used in place of the usual notation [1 1 0] to indicate the sense of the vector direction.*

4.6.2 Dislocations and stacking faults in fcc structures

4.6.2.1 Frank loops

A powerful illustration of the use of the Thompson tetrahedron can be made if we look at simple Frank loops in fcc metals (see Figure 4.32a). The Frank partial dislocation has a Burgers vector perpendicular to the (1 1 1) plane on which it lies and is represented by $A\alpha$, $B\beta$, $C\gamma$, $D\delta$, αA, etc. Such loops shown in the electron micrograph of Figure 4.38 have been produced in aluminium by quenching from about 600°C. Each loop arises from the clustering of vacancies into a disc-shaped cavity which then form a dislocation loop. To reduce their energy, the loops take up regular crystallographic forms with their edges parallel to the ⟨1 1 0⟩ directions in the loop plane. Along a ⟨1 1 0⟩ direction it can reduce its energy by dissociating on an intersecting {1 1 1} plane, forming a stair-rod at the junction of the two {1 1 1} planes, e.g. $A\alpha \rightarrow A\delta + \delta\alpha$ when the Frank dislocation lies along $[\bar{1}\,0\,1]$ common to both α- and δ-planes.

Some of the loops shown in Figure 4.38 are not Frank sessile dislocations as expected, but prismatic dislocations, since no contrast of the type arising from stacking faults, can be seen within the defects. The fault will be removed by shear if it has a high stacking fault energy thereby changing the sessile Frank loop into a glissile prismatic loop according to the reaction

$$a/3[1\,1\,1] + a/6[1\,1\,\bar{2}] \rightarrow a/2[1\,1\,0]$$

Stressing the foil while it is under observation in the microscope allows the unfaulting process to be

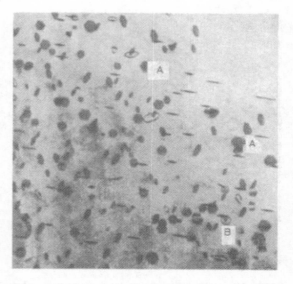

Figure 4.38 *Single-faulted, double-faulted (A) and unfaulted (B) dislocation loops in quenched aluminium (after Edington and Smallman, 1965; courtesy of Taylor and Francis).*

observed directly (see Figure 4.39). This reaction is more easily followed with the aid of the Thompson tetrahedron and rewritten as

$$D\delta + \delta C \rightarrow DC$$

Physically, this means that the disc of vacancies aggregated on a (1 1 1) plane of a metal with high stacking fault energy, besides collapsing, also undergoes a shear movement. The dislocation loops shown in Figure 4.39b are therefore unit dislocations with their Burgers vector $a/2[1\,1\,0]$ inclined at an angle to the original (1 1 1) plane. A prismatic dislocation loop lies on the surface of a cylinder, the cross-section of which is determined by the dislocation loop, and the axis of which is parallel to the [1 1 0] direction. Such a dislocation is not sessile, and under the action of a shear stress it is capable of movement by prismatic slip in the [1 1 0] direction.

Many of the large Frank loops in Figure 4.38 (for example, marked A) contain additional triangular-shaped loop contrast within the outer hexagonal loop.

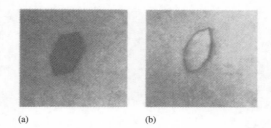

(a)   (b)

Figure 4.39 *Removal of the stacking fault from a Frank sessile dislocation by stress (after Goodhew and Smallman).*

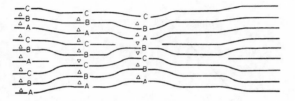

Figure 4.40 *The structure of a double dislocation loop in quenched aluminium (after Edington and Smallman, 1965; courtesy of Taylor and Francis).*

The stacking fault fringes within the triangle are usually displaced relative to those between the triangle and the hexagon by half the fringe spacing, which is the contrast expected from overlapping intrinsic stacking faults. The structural arrangement of those double-faulted loops is shown schematically in Figure 4.40, from which it can be seen that two intrinsic faults on next neighbouring planes are equivalent to an extrinsic fault. The observation of double-faulted loops in aluminium indicates that it is energetically more favourable to nucleate a Frank sessile loop on an existing intrinsic fault than randomly in the perfect lattice, and it therefore follows that the energy of a double or extrinsic fault is less than twice that of the intrinsic fault, i.e. $\gamma_E < 2\gamma_I$. The double loops marked B have the outer intrinsic fault removed by stress.

The addition of a third overlapping intrinsic fault would change the stacking sequence from the perfect **ABCABCABC** to **ABC ↓ B ↓ A ↓ CABC**, where the arrows indicate missing planes of atoms, and produce a coherent twinned structure with two coherent twin boundaries. This structure would be energetically favourable to form, since $\gamma_{twin} < \gamma_I < \gamma_E$. It is possible, however, to reduce the energy of the crystal even further by aggregating the third layer of vacancies between the two previously-formed neighbouring intrinsic faults to change the structure from an extrinsically faulted **ABC ↓ B ↓ ABC** to perfect **ABC ↓↓↓ ABC** structure. Such a triple-layer dislocation loop is shown in Figure 4.41.

4.6.2.2 Stair-rod dislocations

The stair-rod dislocation formed at the apex of a Lomer-Cottrell barrier can also be represented by the Thompson notation. As an example, let us take the interaction between dislocations on the δ- and α-planes. Two unit dislocations BA and DB, respectively, are dissociated according to

$$BA \rightarrow B\delta + \delta A \text{ (on the } \delta\text{-plane)}$$

$$\text{and } DB \rightarrow D\alpha + \alpha B \text{ (on the } \alpha\text{-plane)}$$

and when the two Shockley partials αB and Bδ interact, a stair-rod dislocation $\alpha\delta = a/6[1\,0\,1]$ is formed. This low-energy dislocation is pure edge and therefore sessile. If the other pair of partials interact then the resultant Burgers vector is $(\delta A + D\alpha) = a/3[1\,0\,1]$

(a)

Figure 4.41 *Triple-loop and Frank sessile loop in Al-0.65% Mg (after Kritzinger, Smallman and Dobson, 1969; courtesy of Pergamon Press).*

and of higher energy. This vector is written in Thompson's notation as δD/Aα and is a vector equal to twice the length joining the midpoints of δA and Dα.

4.6.2.3 Stacking-fault tetrahedra

In fcc metals and alloys, the vacancies may also cluster into a three-dimensional defect, forming a tetrahedral arrangement of stacking faults on the four $\{1\,1\,1\}$ planes with the six $\langle 1\,1\,0\rangle$ edges of the tetrahedron, where the stacking faults bend from one $\{1\,1\,1\}$ plane to another, consisting of stair-rod dislocations. The crystal structure is perfect inside and outside the tetrahedron, and the three-dimensional array of faults exhibits characteristic projected shape and contrast when seen in transmission electron micrographs as shown in Figure 4.44. This defect was observed originally in quenched gold but occurs in other materials with low stacking-fault energy. One mechanism for the formation of the defect tetrahedron by the dissociation of a Frank dislocation loop (see Figure 4.42) was first explained by Silcox and Hirsch. The Frank partial dislocation bounding a stacking fault has, because of its large Burgers vector, a high strain energy, and hence can lower its energy by dissociation according to a reaction of the type

$$a/3[1\,1\,1] \rightarrow a/6[1\,2\,1] + a/6[1\,0\,1]$$
$$(\tfrac{1}{3}) \qquad (\tfrac{1}{6}) \qquad (\tfrac{1}{18})$$

The figures underneath the reaction represent the energies of the dislocations, since they are proportional to the squares of the Burgers vectors. This reaction is, therefore, energetically favourable. This reaction can be seen with the aid of the Thompson tetrahedron, which shows that the Frank partial dislocation Aα can dissociate into a Shockley partial dislocation (Aβ, Aδ or Aγ) and a low energy stair-rod dislocation ($\beta\alpha$, $\delta\alpha$ or $\gamma\alpha$) for example A$\alpha \rightarrow$ A$\gamma + \gamma\alpha$.

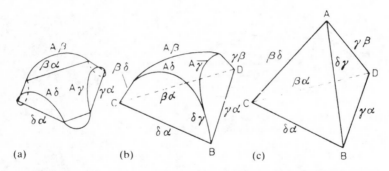

Figure 4.42 *Formation of defect tetrahedron: (a) dissociation of Frank dislocations. (b) formation of new stair-rod dislocations, and (c) arrangement of the six stair-rod dislocations.*

The formation of the defect tetrahedron of stacking faults may be envisaged as follows. The collapse of a vacancy disc will, in the first instance, lead to the formation of a Frank sessile loop bounding a stacking fault, with edges parallel to the $\langle 1\,1\,0\rangle$ directions. Each side of the loop then dissociates according to the above reaction into the appropriate stair-rod and partial dislocations and, as shown in Figure 4.42a, the Shockley dislocations formed by dissociation will lie on intersecting $\{1\,1\,1\}$ planes, above and below the plane of the hexagonal loop; the decrease in energy accompanying the dissociation will give rise to forces which tend to pull any rounded part of the loop into $\langle 1\,1\,0\rangle$. Moreover, because the loop will not in general be a regular hexagon, the short sides will be eliminated by the preferential addition of vacancies at the constricted site, and a triangular-shaped loop will form (Figure 4.42b). The partials $A\beta$, $A\gamma$ and $A\delta$ bow out on their slip plane as they are repelled by the stair-rods. Taking into account the fact that adjacent ends of the bowing loops are of opposite sign, the partials attract each other in pairs to form stair-rod dislocations along DA, BA and CA, according to the reactions

$$\gamma A + A\beta \rightarrow \gamma B,\; \delta A + A\gamma \rightarrow \delta\gamma,\; \beta A + A\delta \rightarrow \beta\delta$$

In vector notation the reactions are of the type

$$a/6[\bar{1}\,\bar{1}\,\bar{2}] + a/6[\bar{1}\,\bar{2}\,\bar{1}] \rightarrow a/6[0\,1\,\bar{1}]$$
$$\left(\tfrac{1}{6}\right) \qquad\qquad \left(\tfrac{1}{6}\right) \qquad\qquad \left(\tfrac{1}{6}\right)$$

(the reader may deduce the appropriate indices from Figure 4.37), and from the addition of the squares of the Burgers vectors underneath it is clear that this reaction is also energetically favourable. The final defect will therefore be a tetrahedron made up from the intersection of stacking faults on the four $\{1\,1\,1\}$ planes, so that the $\langle 1\,1\,0\rangle$ edges of the tetrahedron will consist of low-energy stair-rod dislocations (Figure 4.42c).

The tetrahedron of stacking faults formed by the above sequence of events is essentially symmetrical, and the same configuration would have been obtained if collapse had taken place originally on any other $(1\,1\,1)$ plane. The energy of the system of stair-rod dislocations in the final configuration is proportional

to $6 \times \frac{1}{18} = \frac{1}{3}$, compared with $3 \times \frac{1}{3} = 1$ for the original stacking fault triangle bounded by Frank partials. Considering the dislocation energies alone, the dissociation leads to a lowering of energy to one-third of the original value. However, three additional stacking fault areas, with energies of γ per unit area, have been newly created and if there is to be no net rise in energy these areas will impose an upper limit on the size of the tetrahedron formed. The student may wish to verify that a calculation of this maximum size shows the side of the tetrahedron should be around 50 nm.

De Jong and Koehler have proposed that the tetrahedra may also form by the nucleation and growth of a three-dimensional vacancy cluster. The smallest cluster that is able to collapse to a tetrahedron and subsequently grow by the absorption of vacancies is a hexa-vacancy cluster. Growth would then occur by the nucleation and propagation of jog lines across the faces of the tetrahedron, as shown in Figure 4.43. The hexavacancy cluster may form by clustering di-vacancies and is aided by impurities which have excess positive change relative to the matrix (e.g. Mg, Cd or Al in Au). Hydrogen in solution is also a potent nucleating agent because the di-vacancy/proton complex is mobile and attracted to 'free' di-vacancies. Figure 4.44 shows the increase in tetrahedra nucleation after preannealing gold in hydrogen.

4.6.3 Dislocations and stacking faults in cph structures

In a cph structure with axial ratio c/a, the most closely packed plane of atoms is the basal plane $(0\,0\,0\,1)$

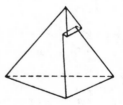

Figure 4.43 *Jog line forming a ledge on the face of a tetrahedron.*

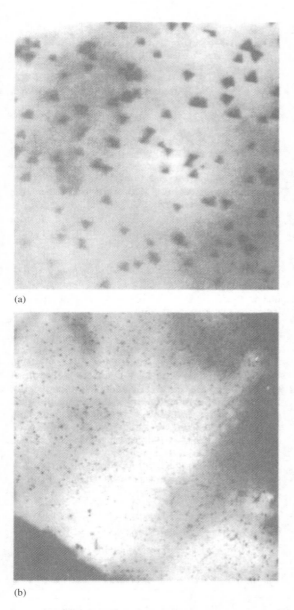

(a)

(b)

Figure 4.44 *Tetrahedra in gold (a) quenched and (b) preannealed in H_2-N_2 gas (after Johnston, Dobson and Smallman).*

and the most closely packed directions $\langle 1\,1\,\overline{2}\,0\rangle$. The smallest unit lattice vector is a, but to indicate the direction of the vector $\langle u, v, w\rangle$ in Miller–Bravais indices it is written as $a/3\langle 1\,1\,\overline{2}\,0\rangle$ where the magnitude of the vector in terms of the lattice parameters is given by $a[3(u^2 + uv + v^2) + (c/a)^2 w^2]^{1/2}$. The usual slip dislocation therefore has a Burgers vector $a/3\langle 1\,1\,\overline{2}\,0\rangle$ and glides in the $(0\,0\,0\,1)$ plane. This slip vector is $\langle a/3, a/3, 2(a/3), 0\rangle$ and has no

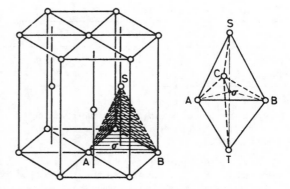

Figure 4.45 *Burgers vectors in the cph lattice (after Berghezan, Fourdeux and Amelinckx, 1961; courtesy of Pergamon Press).*

component along the c-axis and so can be written without difficulty as $a/3\langle 1\,1\,\overline{2}\,0\rangle$. However, when the vector has a component along the c-axis, as for example $\langle a/3, a/3, \overline{2}(a/3), 3c\rangle$, difficulty arises and to avoid confusion the vectors are referred to unit distances (a, a, a, c) along the respective axes (e.g. $1/3\langle 1\,1\,\overline{2}\,0\rangle$ and $1/3\langle 1\,1\,\overline{2}\,3\rangle$). Other dislocations can be represented in a notation similar to that for the fcc structure, but using a double-tetrahedron or bipyramid instead of the single tetrahedron previously adopted, as shown in Figure 4.45. An examination leads to the following simple types of dislocation:

1. Six perfect dislocations with Burgers vectors in the basal plane along the sides of the triangular base ABC. They are AB, BC, CA, BA, CB and AC and are denoted by a or $1/3\langle 1\,1\,\overline{2}\,0\rangle$.
2. Six partial dislocations with Burgers vectors in the basal plane represented by the vectors $A\sigma$, $B\sigma$, $C\sigma$ and their negatives. These dislocations arise from dissociation reactions of the type

    $$AB \rightarrow A\sigma + \sigma B$$

    and may also be written as p or $\frac{1}{3}\langle 1\,0\,\overline{1}\,0\rangle$.
3. Two perfect dislocations perpendicular to the basal plane represented by the vectors ST and TS of magnitude equal to the cell height c or $\langle 0\,0\,0\,1\rangle$.
4. Partial dislocations perpendicular to the basal plane represented by the vectors σS, σT, $S\sigma$, $T\sigma$ of magnitude $c/2$ or $\frac{1}{2}\langle 0\,0\,0\,1\rangle$.
5. Twelve perfect dislocations of the type $\frac{1}{3}\langle 1\,1\,\overline{2}\,3\rangle$ with a Burgers vector represented by SA/TB which is a vector equal to twice the join of the mid-points of SA and TB. These dislocations are more simply referred to as $(c + a)$ dislocations.
6. Twelve partial dislocations, which are a combination of the partial basal and non-basal dislocations, and represented by vectors AS, BS, CS, AT, BT and CT or simply $(c/2) + p$ equal to $\frac{1}{6}\langle 2\,0\,\overline{2}\,3\rangle$. Although these vectors represent a displacement

Table 4.2 Dislocations in structures

Type	AB, BC	$A\sigma$, $B\sigma$	ST, TS	σS, σT	AS, BS	SA/TB
Vector	$\frac{1}{3}\langle 1\,1\,\overline{2}\,0\rangle$	$\frac{1}{3}\langle 1\,0\,\overline{1}\,0\rangle$	$\langle 0\,0\,0\,1\rangle$	$\frac{1}{2}\langle 0\,0\,0\,1\rangle$	$\frac{1}{6}\langle 2\,0\,\overline{2}\,3\rangle$	$\frac{1}{3}\langle 1\,1\,\overline{2}\,3\rangle$
Energy	a^2	$a^2/3$	$c^2 = 8a^2/3$	$2a^2/3$	a^2	$11a^2/3$

from one atomic site to another the resultant dislocations are imperfect because the two sites are not identical.

The energies of the different dislocations are given in a relative scale in Table 4.2 assuming c/a is ideal.

There are many similarities between the dislocations in the cph and fcc structure and thus it is not necessary to discuss them in great detail. It is, however, of interest to consider the two basic processes of glide and climb.

4.6.3.1 Dislocation glide

A perfect slip dislocation in the basal plane $AB = \frac{1}{3}[\overline{1}\,2\,\overline{1}\,0]$ may dissociate into two Shockley partial dislocations separating a ribbon of intrinsic stacking fault which violates the two next-nearest neighbours in the stacking sequence. There are actually two possible slip sequences: either a B-layer slides over an A-layer, i.e. $A\sigma$ followed by σB (see Figure 4.46a) or an A-layer slides over a B-layer by the passage of a σB partial followed by an $A\sigma$ (see Figure 4.46b). The dissociation given by

$$AB \rightarrow A\sigma + \sigma B$$

may be written in Miller–Bravais indices as

$$\frac{1}{3}[\overline{1}\,2\,\overline{1}\,0] \rightarrow \frac{1}{3}[0\,1\,\overline{1}\,0] + \frac{1}{3}[\overline{1}\,1\,0\,0]$$

This reaction is similar to that in the fcc lattice and the width of the ribbon is again inversely proportional to the stacking fault energy γ. Dislocations dissociated in the basal planes have been observed in cobalt, which undergoes a phase transformation and for which γ is considered to be low (≈ 25 mJ/m^2). For the other common cph metals Zn, Cd, Mg, Ti, Be, etc. γ is high (250–300 mJ/m^2). No measurements of intrinsic faults with two next-nearest neighbour violations have been made, but intrinsic faults with one next-nearest neighbour violation have been measured and show that Mg ≈ 125 mJ/m^2, Zn ≈ 140 mJ/m^2, and Cd $\approx$ 150–175 mJ/m^2. It is thus reasonable to conclude that intrinsic faults associated with Shockley partials have somewhat higher energy. Dislocations in these metals are therefore not very widely dissociated. A screw dislocation lying along a $[\overline{1}\,2\,\overline{1}\,0]$ direction is capable of gliding in three different glide planes but the small extension in the basal plane will be sufficient to make basal glide easier than in either the pyramidal $(1\,0\,\overline{1}\,1)$ or prismatic $(1\,0\,\overline{1}\,0)$ glide (see Figure 7.19). Pyramidal and prismatic glide will be more favoured at high temperatures in metals with high stacking-fault energy when thermal activation aids the constriction of the dissociated dislocations.

4.6.3.2 Dislocation climb

Stacking faults may be produced in hexagonal lattices by the aggregation of point defects. If vacancies

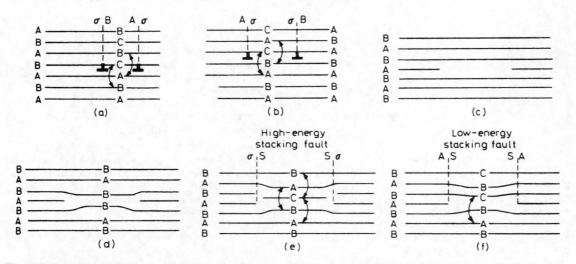

Figure 4.46 *Stacking faults in the cph lattice (after Partridge, 1967; by courtesy of the American Society for Metals).*

aggregate as a platelet, as shown in Figure 4.46c, the resultant collapse of the disc-shaped cavity (Figure 4.46d) would bring two similar layers into contact. This is a situation incompatible with the close-packing and suggests that simple Frank dislocations are energetically unfavourable in cph lattices. This unfavourable situation can be removed by either one of two mechanisms as shown in Figures 4.46e and 4.46f. In Figure 4.46e the B-layer is converted to a C-position by passing a pair of equal and opposite partial dislocations (dipole) over adjacent slip planes. The Burgers vector of the dislocation loop will be of the σS type and the energy of the fault, which is extrinsic, will be high because of the three next nearest neighbour violations. In Figure 4.46f the loop is swept by a $A\sigma$-type partial dislocation which changes the stacking of all the layers above the loop according to the rule $A \rightarrow B \rightarrow C \rightarrow A$. The Burgers vector of the loop is of the type AS, and from the dislocation reaction $A\sigma + \sigma S \rightarrow AS$ or $\frac{1}{3}[10\bar{1}0] + \frac{1}{2}[0001] \rightarrow \frac{1}{6}[20\bar{2}3]$ and the associated stacking fault, which is intrinsic, will have a lower energy because there is only one next-nearest neighbour violation in the stacking sequence. Faulted loops with $b = AS$ or $(\frac{1}{2}c + p)$ have

been observed in Zn, Mg and Cd (see Figure 4.47). Double-dislocation loops have also been observed when the inner dislocation loop encloses a central region of perfect crystal and the outer loop an annulus of stacking fault. The structure of such a double loop is shown in Figure 4.48. The vacancy loops on adjacent atomic planes are bounded by dislocations with non-parallel Burgers vectors, i.e. $b = (\frac{1}{2}c + p)$ and $b = (\frac{1}{2}c - p)$, respectively; the shear component of the second loop acts in such a direction as to eliminate the fault introduced by the first loop. There are six partial vectors in the basal plane p_1, p_2, p_3 and the negatives, and if one side of the loop is sheared by either p_1, p_2 or p_3 the stacking sequence is changed according to $A \rightarrow B \rightarrow C \rightarrow A$, whereas reverse shearing $A \rightarrow C \rightarrow B \rightarrow A$, results from either $-p_1$, $-p_2$ or $-p_3$. It is clear that the fault introduced by a positive partial shear can be eliminated by a subsequent shear brought about by any of the three negative partials. Three, four and more layered loops have also been observed in addition to the more common double loop. The addition of each layer of vacancies alternately introduces or removes stacking-faults, no matter whether the loops precipitate one above the other or on opposite sides of the original defect.

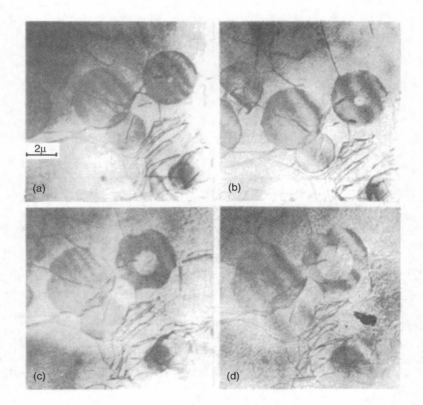

(a)   (b)

(c)   (d)

Figure 4.47 *Growth of single- and double-faulted loops in magnesium on annealing at 175 °C for (a) t = 0 min, (b) t = 5 min, (c) t = 15 min and (d) t = 25 min (after Hales, Smallman and Dobson).*

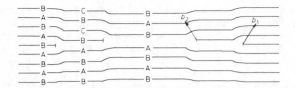

Figure 4.48 *Structure of double-dislocation loop in cph lattice.*

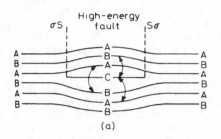

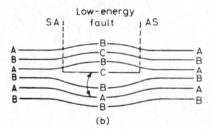

Figure 4.49 *Dislocation loop formed by aggregation of interstitials in a cph lattice with (a) high-energy and (b) low-energy stacking fault.*

As in fcc metals, interstitials may be aggregated into platelets on close-packed planes and the resultant structure, shown in Figure 4.49a, is a dislocation loop with Burgers vector $S\sigma$, containing a high-energy stacking fault. This high-energy fault can be changed to one with lower energy by having the loop swept by a partial as shown in Figure 4.49b.

All these faulted dislocation loops are capable of climbing by the addition or removal of point defects to the dislocation line. The shrinkage and growth of vacancy loops has been studied in some detail in Zn, Mg and Cd and examples, together with the climb analysis, are discussed in Section 4.7.1.

4.6.4 Dislocations and stacking faults in bcc structures

The shortest lattice vector in the bcc lattice is $a/2[1\,1\,1]$, which joins an atom at a cube corner to the one at the centre of the cube; this is the observed slip direction. The slip plane most commonly observed is $(1\,1\,0)$ which, as shown in Figure 4.50, has a distorted close-packed structure. The $(1\,1\,0)$ planes are packed

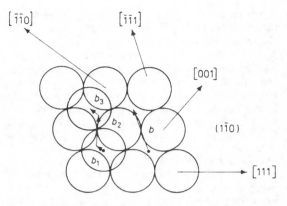

Figure 4.50 *The $(1\,1\,0)$ plane of the bcc lattice (after Weertman; by courtesy of Collier-Macmillan International).*

in an **ABABAB** sequence and three $\{1\,1\,0\}$ type planes intersect along a $\langle 1\,1\,1 \rangle$ direction. It therefore follows that screw dislocations are capable of moving in any of the three $\{1\,1\,0\}$ planes and for this reason the slip lines are often wavy and ill-defined. By analogy with the fcc structure it is seen that in moving the B-layer along the $[\bar{1}\,\bar{1}\,1]$ direction it is easier to shear in the directions indicated by the three vectors b_1, b_2 and b_3. These three vectors define a possible dissociation reaction

$$\frac{a}{2}[\bar{1}\,\bar{1}\,1] \rightarrow \frac{a}{8}[\bar{1}\,\bar{1}\,0] + \frac{a}{4}[\bar{1}\,\bar{1}\,2] + \frac{a}{8}[\bar{1}\,\bar{1}\,0]$$

The stacking fault energy of pure bcc metals is considered to be very high, however, and hence no faults have been observed directly. Because of the stacking sequence **ABABAB** of the $(1\,1\,0)$ planes the formation of a Frank partial dislocation in the bcc structure gives rise to a situation similar to that for the cph structure, i.e. the aggregation of vacancies or interstitials will bring either two A-layers or two B-layers into contact with each other. The correct stacking sequence can be restored by shearing the planes to produce perfect dislocations $a/2[1\,1\,1]$ or $a/2[1\,1\,\bar{1}]$.

Slip has also been observed on planes indexed as $(1\,\bar{1}\,2)$ and $(1\,2\,3)$ planes, and although some workers attribute this latter observation to varying amounts of slip on different $(1\,1\,0)$ planes, there is evidence to indicate that $(1\,1\,2)$ and $(1\,2\,3)$ are definite slip planes. The packing of atoms in a $(1\,1\,2)$ plane conforms to a rectangular pattern, the rows and columns parallel to the $[1\,\bar{1}\,0]$ and $[1\,1\,\bar{1}]$ directions, respectively, with the closest distance of approach along the $[1\,1\,\bar{1}]$ direction. The stacking sequence of the $(1\,1\,2)$ planes is **ABCDEFAB**... and the spacing between the planes $a/\sqrt{6}$. It has often been suggested that the unit dislocation can dissociate in the $(1\,1\,2)$ plane according to the reaction

$$\frac{a}{2}[1\,1\,\bar{1}] \rightarrow \frac{a}{3}[1\,1\,\bar{1}] + \frac{a}{6}[1\,1\,\bar{1}]$$

because the homogeneous shear necessary to twin the structure is $1/\sqrt{2}$ in a $\langle 1\,1\,1 \rangle$ on a $(1\,1\,2)$ and this shear can be produced by a displacement $a/6[1\,1\,\overline{1}]$ on every successive $(1\,1\,2)$ plane. It is therefore believed that twinning takes place by the movement of partial dislocations. However, it is generally recognized that the stacking fault energy is very high in bcc metals so that dissociation must be limited. Moreover, because the Burgers vectors of the partial dislocations are parallel, it is not possible to separate the partials by an applied stress unless one of them is anchored by some obstacle in the crystal.

When the dislocation line lies along the $[1\,1\,\overline{1}]$ direction it is capable of dissociating in any of the three $\{1\,1\,2\}$ planes, i.e. $(1\,1\,2)$, $(\overline{1}\,2\,1)$ and $(2\,\overline{1}\,1)$, which intersect along $[1\,1\,\overline{1}]$. Furthermore, the $a/2[1\,1\,\overline{1}]$ screw dislocation could dissociate according to

$$\frac{a}{2}[1\,1\,\overline{1}] \rightarrow \frac{a}{6}[1\,1\,\overline{1}] + \frac{a}{6}[1\,1\,\overline{1}] + \frac{a}{6}[1\,1\,\overline{1}]$$

to form the symmetrical fault shown in Figure 4.51.

The symmetrical configuration may be unstable, and the equilibrium configuration is one partial dislocation at the intersection of two $\{1\,1\,2\}$ planes and the other two lying equidistant, one in each of the other two

planes. At larger stresses this unsymmetrical configuration can be broken up and the partial dislocations induced to move on three neighbouring parallel planes, to produce a three-layer twin. In recent years an asymmetry of slip has been confirmed in many bcc single crystals, i.e. the preferred slip plane may differ in tension and compression. A yield stress asymmetry has also been noted and has been related to asymmetric glide resistance of screw dislocations arising from their 'core' structure.

An alternative dissociation of the slip dislocation proposed by Cottrell is

$$\frac{a}{2}[1\,1\,1] \rightarrow \frac{a}{3}[1\,1\,2] + \frac{a}{6}[1\,1\,\overline{1}]$$

The dissociation results in a twinning dislocation $a/6[1\,1\,\overline{1}]$ lying in the $(1\,1\,2)$ plane and a $a/3[1\,1\,2]$ partial dislocation with Burgers vector normal to the twin fault and hence is sessile. There is no reduction in energy by this reaction and is therefore not likely to occur except under favourable stress conditions.

Another unit dislocation can exist in the bcc structure, namely $a[0\,0\,1]$, but it will normally be immobile. This dislocation can form at the intersection of normal slip bands by the reaction.

$$\frac{a}{2}[\overline{1}\,\overline{1}\,1] + \frac{a}{2}[1\,1\,1] \rightarrow a[0\,0\,1]$$

with a reduction of strain energy from $3a^2/2$ to a^2. The new $a[0\,0\,1]$ dislocation lies in the $(0\,0\,1)$ plane and is pure edge in character and may be considered as a wedge, one lattice constant thick, inserted between the $(0\,0\,1)$ and hence has been considered as a crack nucleus. $a[0\,0\,1]$ dislocations can also form in networks of $a/2\langle 1\,1\,1 \rangle$ type dislocations.

4.6.5 Dislocations and stacking faults in ordered structures

When the alloy orders, a unit dislocation in a disordered alloy becomes a partial-dislocation in the superlattice with its attached anti-phase boundary interface, as shown in Figure 4.52a. Thus, when this dislocation moves through the lattice it will completely destroy the order across its slip plane. However, in an ordered alloy, any given atom prefers to have unlike atoms as

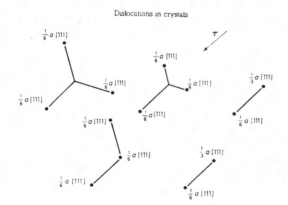

Dislocations in crystals

Figure 4.51 *Dissociated $a/2\,[1\,1\,1]$ dislocation in the bcc lattice (after Mitchell, Foxall and Hirsch, 1963; courtesy of Taylor and Francis).*

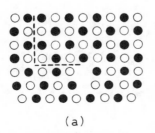

(a)

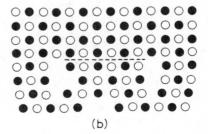

(b)

Figure 4.52 *Dislocations in ordered structures.*

its neighbours, and consequently such a process of slip would require a very high stress. To move a dislocation against the force γ exerted on it by the fault requires a shear stress $\tau = \gamma/b$, where b is the Burgers vector; in β-brass where γ is about 0.07 N/m this stress is 300 MN/m^2. In practice the critical shear stress of β-brass is an order of magnitude less than this value, and thus one must conclude that slip occurs by an easier process than the movement of unit dislocations. In consequence, by analogy with the slip process in fcc crystals, where the leading partial dislocation of an extended dislocation trails a stacking fault, it is believed that the dislocations which cause slip in an ordered lattice are not single dislocations but coupled pairs of dislocations, as shown in Figure 4.52b. The first dislocation of the pair, on moving across the slip plane, destroys the order and the second half of the couple completely restores it again, the third dislocation destroys it once more, and so on. In β-brass[1] and similar weakly-ordered alloys such as AgMg and FeCo the crystal structure is ordered bcc (or CsCl-type) and, consequently, deformation is believed to occur by the movement of coupled pairs of $a/2[1\,1\,1]$-type dislocations. The combined slip vector of the coupled pair of dislocations, sometimes called a super-dislocation, is then equivalent to $a[1\,1\,1]$, and, since this vector connects like atoms in the structure, long-range order will be maintained.

The separation of the super-partial dislocations may be calculated, as for Shockley partials, by equating the repulsive force between the two like $a/2\langle1\,1\,1\rangle$ dislocations to the surface tension of the anti-phase boundary. The values obtained for β-brass and FeCo are about 70 and 50 nm, respectively, and thus super-dislocations can be detected in the electron microscope using the weak beam technique (see Chapter 5). The separation is inversely proportional to the square of the ordering parameter and super-dislocation pairs ≈ 12.5 nm width have been observed more readily in partly ordered FeCo (S = 0.59).

In alloys with high ordering energies the antiphase boundaries associated with super-dislocations cannot be tolerated and dislocations with a Burgers vector equal to the unit lattice vector $a\langle1\,0\,0\rangle$ operate to produce slip in $\langle1\,0\,0\rangle$ directions. The extreme case of this is in ionic-bonded crystals such as CsBr, but strongly-ordered intermetallic compounds such as NiAl are also observed to slip in the $\langle1\,0\,0\rangle$ direction with dislocations having $b = a\langle1\,0\,0\rangle$.

Ordered A$_3$B-type alloys also give rise to super-dislocations. Figure 4.53a illustrates three $(1\,1\,1)$

[1]Chapter 3, Figure 3.40, shows the CsCl or L$_2$O structure. When disordered, the slip vector is $a/2[1\,1\,1]$, but this vector in the ordered structure moves an A atom to a B site. The slip vector to move an A atom to an A site in twice the length and equal to $a[1\,1\,1]$.

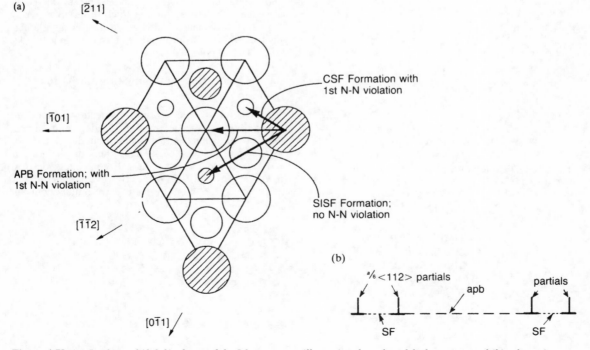

Figure 4.53 (a) Stacking of (1 1 1) planes of the L1$_2$ structure, illustrating the apb and fault vectors, and (b) schematic representation of super-dislocation structure.

layers of the Ll$_2$ structure, with different size atoms for each layer. The three vectors shown give rise to the formation of different planar faults; $a/2[\bar{1}\,0\,1]$ is a super-partial producing apb, $a/6[\bar{2}\,1\,1]$ produces the familiar stacking fault, and $a/3[\bar{1}\,\bar{1}\,2]$ produces a super-lattice intrinsic stacking fault (SISF). A $[\bar{1}\,0\,1]$ super-dislocation can therefore be composed of either

$$[\bar{1}\,0\,1] \to \frac{a}{2}[\bar{1}\,0\,1] + \text{ apb on } (1\,1\,1) + \frac{a}{2}[\bar{1}\,0\,1]$$

or

$$[\bar{1}\,0\,1] \to \frac{a}{3}[\bar{1}\,\bar{1}\,2] + \text{ SISF on } (1\,1\,1) + \frac{a}{3}[\bar{2}\,1\,1]$$

Each of the $a/2[\bar{1}\,0\,1]$ super-partials may also dissociate, as for fcc, according to

$$\frac{a}{2}[\bar{1}\,0\,1] \to \frac{a}{6}[\bar{2}\,1\,1] + \frac{a}{6}[\bar{1}\,\bar{1}\,2].$$

The resultant super-dislocation is schematically shown in Figure 4.53b. In alloys such as Cu$_3$Au, Ni$_3$Mn, Ni$_3$Al, etc., the stacking fault ribbon is too small to be observed experimentally but super-dislocations have been observed. It is evident, however, that the cross-slip of these super-dislocations will be an extremely difficult process. This can lead to a high work-hardening rate in these alloys, as discussed in Chapter 7.

In an alloy possessing short-range order, slip will not occur by the motion of super-dislocations since there are no long-range faults to couple the dislocations together in pairs. However, because the distribution of neighbouring atoms is not random the passage of a dislocation will destroy the short-range order between the atoms, across the slip plane. As before, the stress to do this will be large but in this case there is no mechanism, such as coupling two dislocations together, to make the process easier. The fact that, for instance, a crystal of AuCu$_3$ in the quenched state (short-range order) has nearly double the yield strength of the annealed state (long-range order) may be explained on this basis. The maximum strength is exhibited by a partially-ordered alloy with a critical domain size of about 6 nm. The transition from deformation by unit dislocations in the disordered state to deformation by super-dislocations in the ordered condition gives rise to a peak in the flow stress with change in degree of order (see Chapter 6).

4.6.6 Dislocations and stacking faults in ceramics

At room temperature, the primary slip system in the fcc structure of magnesia, MgO, is $\{1\,1\,0\}\langle1\,1\,0\rangle$. It is favoured because its Burgers vector is short and, most importantly, because this vector is parallel to rows of ions of like electrostatic charge, permitting the applied stress to shear the $\{1\,1\,0\}$ planes past each other. Slip in the $\langle1\,0\,0\rangle$ directions is resisted at room temperature because it involves forcing ions of like charge into close proximity. If we consider the slip geometry of ionic crystals in terms of the Thompson tetrahedron for cubic ionic crystals (Figure 4.54), six

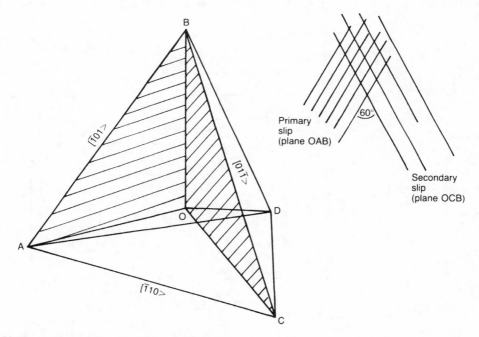

Figure 4.54 *Thompson tetrahedron for ionic crystals (cubic).*

primary {1 1 0} slip planes extend from the central point O of the tetrahedron to its ⟨1 1 0⟩ edges. There is no dissociation in the {1 1 1} faces and slip is only possible along the ⟨1 1 0⟩ edges of the tetrahedron. Thus, for each of the {1 1 0} planes, there is only one ⟨1 1 0⟩ direction available. This limiting 'one-to-one' relation for a cubic ionic crystal contrasts with the 'three-to-one' relation of the {1 1 1}⟨1 1 0⟩ slip system in cubic metallic crystals. As an alternative to the direct 'easy' translation $\overrightarrow{AC}$, we might postulate the route $\overrightarrow{AB} + \overrightarrow{BC}$ at room temperature. This process involves slip on plane OAB in the direction $\overrightarrow{AB}$ followed by slip on plane OCB in the direction $\overrightarrow{BC}$. It is not favoured because, apart from being unfavoured in terms of energy, it involves a 60° change in slip direction and the critical resolved shear stress for the second stage is likely to be much greater than that needed to activate the first set of planes. The two-stage route is therefore a difficult one. Finally, it will be noticed that the central point lies at the junction of the ⟨1 1 1⟩ directions which, being in a cubic system, are perpendicular to the four {1 1 1} faces. One can thus appreciate why raising the temperature of an ionic crystal often allows the {1 1 0}⟨1 1 1⟩ system to become active.

In a single crystal of alumina, which is rhombohedral-hexagonal in structure and highly anisotropic, slip is confined to the basal planes. At temperatures above 900°C, the slip system is {0 0 0 1}⟨1 1 $\bar{2}$ 0⟩. As seen from Figure 2.18, this resultant slip direction is not one of close-packing. If a unit translation of shear is to take place in a ⟨1 1 $\bar{2}$ 0⟩-type direction, the movement and re-registration of oxygen anions and aluminium cations must be in synchronism ('synchro-shear'). Figure 4.55 shows the Burgers vectors for slip in the [1 1 $\bar{2}$ 0] direction in terms of the two modes of dissociation proposed by M. Kronberg. These two routes are energetically favoured. The dissociation reaction for the oxygen anions is: $\frac{1}{3}[1 1 \bar{2} 0] \rightarrow \frac{1}{3}[1 0 \bar{1} 0] + \frac{1}{3}[0 1 \bar{1} 0]$. The vectors for these two half-partials lie in close-packed directions and enclose a stacking fault. In the case of the smaller aluminium cations, further dissociation of each of similar half-partials takes place (e.g. $\frac{1}{3}[1 0 \bar{1} 0] \rightarrow \frac{1}{9}[2 \bar{1} \bar{1} 0] + \frac{1}{9}[1 1 \bar{2} 0]$). These quarter-partials enclose three regions in which stacking is faulted. Slip involves a synchronized movement of both types of planar fault (single and triple) across the basal planes.

4.6.7 Defects in crystalline polymers

Crystalline regions in polymers are based upon long-chain molecules and are usually associated with at least some glassy (amorphous) regions. Although less intensively studied than defect structures in metals and ceramics, similar crystal defects, such as vacancies, interstitials and dislocations, have been observed in polymers. Their association with linear macromolecules, however, introduces certain special features. For instance, the chain ends of molecules can be regarded as point defects because they differ in chemical character from the chain proper. Vacancies, usually associated with chain ends, and foreign atoms, acting as interstitials, are also present. Edge and

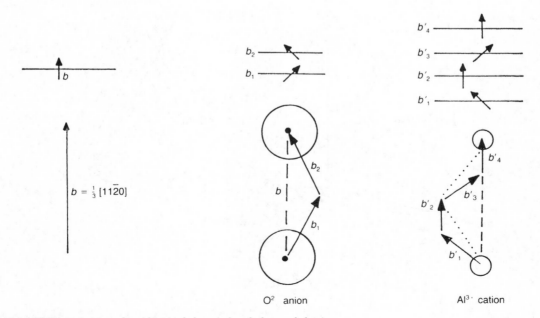

$b = \frac{1}{3}[1 1 \bar{2} 0]$

O^2 anion

Al^{3+} cation

Figure 4.55 *Dissociation and synchronized shear in basal planes of alumina.*

screw dislocations have been detected.[1] Moiré pattern techniques of electron microscopy are useful for revealing the presence of edge dislocations. Growth spirals, centred on screw dislocations, have frequently been observed on surfaces of crystalline polymers, with the Burgers vector, dislocation axis and chain directions lying in parallel directions (e.g. polyethylene crystals grown from a concentrated melt).

Within crystalline regions, such as spherulites composed of folded chain molecules, discrepancies in folding may be regarded as defects. The nature of the two-dimensional surface at the faces of spherulites where chains emerge, fold and re-enter the crystalline region is of particular interest. Similarly, the surfaces where spherulite edges impinge upon each other can be regarded as planar defects, being analogous to grain boundary surfaces.

X-ray diffraction studies of line-broadening effects and transmission electron microscopy have been used to elucidate crystal defects in polymers. In the latter case, the high energy of an electron beam can damage the polymer crystals and introduce artefacts. It is recognized that the special structural features found in polymer crystals such as the comparative thinness of many crystals, chain-folding, the tendency of the molecules to resist bending of bonds and the great difference between primary intramolecular bonding and secondary intermolecular bonding, make them unique and very different to metallic and ceramic crystals.

4.6.8 Defects in glasses

It is recognized that real glass structures are less homogeneous than the random network model might suggest. Adjacent glassy regions can differ abruptly in composition, giving rise to 'cords', and it has been proposed that extremely small micro-crystalline regions may exist within the glass matrix. Tinting of clear glass is evidence for the presence of trace amounts of impurity atoms (iron, chromium) dispersed throughout the structure. Modifying ions of sodium are relatively loosely held in the interstices and have been known to migrate through the structure and aggregate close to free surfaces. On a coarser scale, it is possible for bubbles ('seeds'), rounded by surface tension, to persist from melting/fining operations. Bubbles may contain gases, such as carbon dioxide, sulphur dioxide and sulphur trioxide. Solid inclusions ('stones') of crystalline matter, such as silica, alumina and silicates, may be present in the glass as a result of incomplete fusion, interaction with refractory furnace linings and localized crystallization (devitrification) during the final cooling.

---

[1] Transmission electron microscopy of the organic compound platinum phthalocyanine which has relatively large intermolecular spacing provided the first visual evidence for the existence of edge dislocations.

4.7 Stability of defects

4.7.1 Dislocation loops

During annealing, defects such as dislocation loops, stacking-fault tetrahedra and voids may exhibit shrinkage in size. This may be strikingly demonstrated by observing a heated specimen in the microscope. On heating, the dislocation loops and voids act as vacancy sources and shrink, and hence the defects annihilate themselves. This process occurs in the temperature range where self-diffusion is rapid, and confirms that the removal of the residual resistivity associated with Stage II is due to the dispersal of the loops, voids, etc.

The driving force for the emission of vacancies from a vacancy defect arises in the case of (1) a prismatic loop from the line tension of the dislocation, (2) a Frank loop from the force due to the stacking fault on the dislocation line since in intermediate and high γ-metals this force far outweighs the line tension contribution, and (3) a void from the surface energy γ_s. The annealing of Frank loops and voids in quenched aluminium is shown in Figures 4.56 and 4.58, respectively. In a thin metal foil the rate of annealing is generally controlled by the rate of diffusion of vacancies away from the defect to any nearby sinks, usually the foil surfaces, rather than the emission of vacancies at the defect itself. To derive the rate equation governing the annealing, the vacancy concentration at the surface of the defect is used as one boundary condition of a diffusion-controlled problem and the second boundary condition is obtained by assuming that the surfaces of a thin foil act as ideal sinks for vacancies. The rate then depends on the vacancy concentration gradient developed between the defect, where the vacancy concentration is given by

$$c = c_0 \exp \{(dF/dn)/kT\} \tag{4.14}$$

with (dF/dn) the change of free energy of the defect configuration per vacancy emitted at the temperature T, and the foil surface where the concentration is the equilibrium value c_0.

For a single, intrinsically-faulted circular dislocation loop of radius r the total energy of the defect F is given by the sum of the line energy and the fault energy, i.e.

$$F \simeq 2\pi r\{[\mu b^2/4\pi(1 - \nu)] \ln (r/r_0)\} + \pi r^2 \gamma$$

In the case of a large loop ($r > 50$ nm) in a material of intermediate or high stacking fault energy ($\gamma \gtrsim 60$ mJ/m^2) the term involving the dislocation line energy is negligible compared with the stacking fault energy term and thus, since $(dF/dn) = (dF/dr) \times (dr/dn)$, is given simply by γB^2, where B^2 is the cross-sectional area of a vacancy in the (1 1 1) plane. For large loops the diffusion geometry approximates to cylindrical diffusion[2] and a solution of the time-independent diffusion equation gives for the annealing rate,

---

[2] For spherical diffusion geometry the pre-exponential constant is D/b.

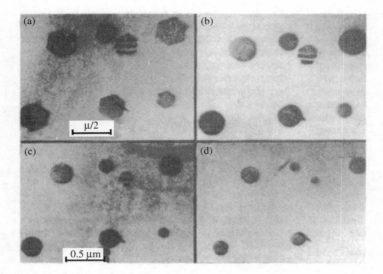

Figure 4.56 *Climb of faulted loops in aluminium at 140°C. (a) t = 0 min, (b) t = 12 min, (c) t = 24 min, (d) t = 30 min (after Dobson, Goodhew and Smallman, 1967; courtesy of Taylor and Francis).*

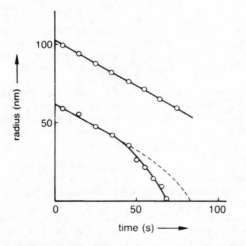

Figure 4.57 *Variation of loop radius with time of annealing for Frank dislocations in Al showing the deviation from linearity at small r.*

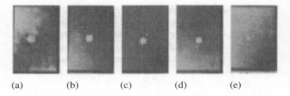

Figure 4.58 *Sequence of micrographs showing the shrinkage of voids in quenched aluminium during isothermal annealing at 170°C. (a) t = 3 min, (b) t = 8 min, (c) t = 21 min, (d) t = 46 min, (e) t = 98 min. In all micrographs the scale corresponds to 0.1 μm (after Westmacott, Smallman and Dobson, 1968, 117; courtesy of the Institute of Metals).*

$$dr/dt = -[2\pi D/b \ln (L/b)][\exp (\gamma B^2/kT) - 1]$$

$$= \text{const. } [\exp (\gamma B^2/kT) - 1] \qquad (4.15)$$

where $D = D_0 \exp (-U_D/kT)$ is the coefficient of self-diffusion and L is half the foil thickness. The annealing rate of a prismatic dislocation loop can be similarly determined, in this case dF/dr is determined solely by the line energy, and then

$$dr/dt = -[2\pi D/b \ln (L/b)](\alpha b/r)$$

$$= \text{const. } [\alpha b/r] \qquad (4.16)$$

where the term containing the dislocation line energy can be approximated to $\alpha b/r$. The annealing of Frank loops obeys the linear relation given by equation (4.15) at large r (Figure 4.57); at small r the curve deviates from linearity because the line tension term can no longer be neglected and also because the diffusion geometry changes from cylindrical to spherical symmetry. The annealing of prismatic loops is much slower, because only the line tension term is involved, and obeys an r^2 versus t relationship.

In principle, equation (4.15) affords a direct determination of the stacking fault energy γ by substitution, but since U_D is usually much bigger than γB^2 this method is unduly sensitive to small errors in U_D. This difficulty may be eliminated, however, by a comparative method in which the annealing rate of a faulted loop is compared to that of a prismatic one at the same temperature. The intrinsic stacking fault energy of aluminium has been shown to be 135 mJ/m^2 by this technique.

In addition to prismatic and single-faulted (Frank) dislocation loops, double-faulted loops have also been annealed in a number of quenched fcc metals. It is observed that on annealing, the intrinsic loop first shrinks until it meets the inner, extrinsically-faulted region, following which the two loops shrink together as one extrinsically-faulted loop. The rate of annealing of this extrinsic fault may be derived in a way similar to equation (4.15) and is given by

$$dr/dt = -[\pi D/b \ln (L/b)][\exp (\gamma_E B^2/kT) - 1]$$
$$= \text{const.} \{\exp (\gamma_E B^2/2kT) - 1\} \qquad (4.17)$$

from which the extrinsic stacking-fault energy may be determined. Generally γ_E is about 10–30% higher in value than the intrinsic energy γ

Loop growth can occur when the direction of the vacancy flux is towards the loop rather than away from it, as in the case of loop shrinkage. This condition can arise when the foil surface becomes a vacancy source, as, for example, during the growth of a surface oxide film. Loop growth is thus commonly found in Zn, Mg, Cd, although loop shrinkage is occasionally observed, presumably due to the formation of local cracks in the oxide film at which vacancies can be annihilated. Figure 4.47 shows loops growing in Mg as a result of the vacancy supersaturation produced by oxidation. For the double loops, it is observed that a stacking fault is created by vacancy absorption at the growing outer perimeter of the loop and is destroyed at the growing inner perfect loop. The perfect regions expand faster than the outer stacking fault, since the addition of a vacancy to the inner loop decreases the energy of the defect by γB^2 whereas the addition of a vacancy to the outer loop increases the energy by the same amount. This effect is further enhanced as the two loops approach each other due to vacancy transfer from the outer to inner loops. Eventually the two loops coalesce to give a perfect prismatic loop of Burgers vector $c = [0\,0\,0\,1]$ which continues to grow under the vacancy supersaturation. The outer loop growth rate is thus given by

$$\dot{r}_0 = -[2\pi D/B \ln (L/b)][(c_s/c_0) - \exp (\gamma B^2/kT)] \qquad (4.18)$$

when the vacancy supersaturation term (c_s/c_o) is larger than the elastic force term tending to shrink the loop. The inner loop growth rate is

$$\dot{r}_i = -[2\pi D/B \ln (L/b)][(c_s/c_0) - \exp (-\gamma B^2/kT)] \qquad (4.19)$$

where $\exp (-\gamma B^2/kT) \ll 1$, and the resultant prismatic loop growth rate is

$$\dot{r}_p = -[\pi D/B \ln (L/b)]\{(c_s/c_0) - [(\alpha b/r) + 1]\} \qquad (4.20)$$

where $(\alpha b/r) < 1$ and can be neglected. By measuring these three growth rates, values for γ, (c_s/c_o) and

D may be determined; Mg has been shown to have $\gamma = 125$ mJ/m^2 from such measurements.

4.7.2 Voids

Voids will sinter on annealing at a temperature where self-diffusion is appreciable. The driving force for sintering arises from the reduction in surface energy as the emission of vacancies takes place from the void surface. In a thin metal foil the rate of annealing is generally controlled by the rate of diffusion of vacancies away from the defect to any nearby sinks, usually the foil surfaces. The rate then depends on the vacancy concentration gradient developed between the defect (where the vacancy concentration is given by

$$c = c_0 \exp \{(dF/dn)/kT\} \qquad (4.21)$$

with (dF/dn) the change in free energy of the defect configuration per vacancy emitted at the temperature T) and the foil surface where the concentration is the equilibrium value c_o.

For a void in equilibrium with its surroundings the free energy $F \simeq 4\pi r^2 \gamma$, and since $(dF/dn) = (dF/dr)(dr/dn) = (8\pi r\gamma_s)(\Omega/4\pi r^2)$ where Ω is the atomic volume and n the number of vacancies in the void, equation (4.14), the concentration of vacancies in equilibrium with the void is

$$c_v = c_0 \exp (2\gamma_s \Omega/rkT)$$

Assuming spherical diffusion geometry, the diffusion equation may be solved to give the rate of shrinkage of a void as

$$dr/dt = -(D/r)\{\exp (2\Omega\gamma_s/rkT) - 1\} \qquad (4.22)$$

For large $r (>50$ nm$)$ the exponential term can be approximated to the first two terms of the series expansion and equation (4.22) may then be integrated to give

$$r^3 = r_i^3 - (6D\Omega\gamma_s/kT)t \qquad (4.23)$$

where r_i is the initial void radius at $t = 0$. By observing the shrinkage of voids as a function of annealing time at a given temperature (see Figure 4.58) it is possible to obtain either the diffusivity D or the surface energy γ_s. From such observations, γ_s for aluminium is shown to be 1.14 J/m^2 in the temperature range 150–200°C, and $D = 0.176 \times \exp (-1.31$ eV$/kT)$. It is difficult to determine γ_s for Al by zero creep measurements because of the oxide. This method of obtaining γ_s has been applied to other metals and is particularly useful since it gives a value of γ_s in the self-diffusion temperature range rather than near the melting point.

4.7.3 Nuclear irradiation effects

4.7.3.1 Behaviour of point defects and dislocation loops

Electron microscopy of irradiated metals shows that large numbers of small point defect clusters are formed

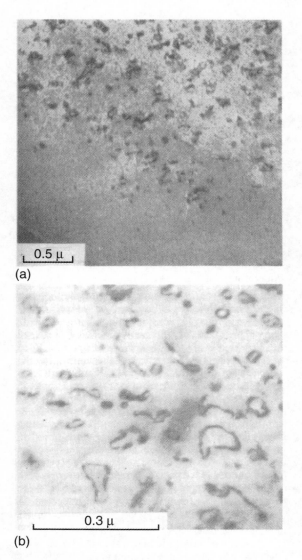

0.5 μ

(a)

0.3 μ

(b)

Figure 4.59 *A thin film of copper after bombardment with*
1.4×10^{21} α-particles m^{-2}. (a) Dislocation loops (~40 nm
dia) and small centres of strain (~4 nm dia); (b) after a
2-hour anneal at 350°C showing large prismatic loops
(after Barnes and Mazey, 1960).

loops, which eventually appear as dislocation tangles. Neutron bombardment produces similar effects to α-particle bombardment, but unless the dose is greater than 10^{21} neutrons/m^2 the loops are difficult to resolve. In copper irradiated at pile temperature the density of loops increases with dose and can be as high as 10^{14} m^{-2} in heavily bombarded metals.

The micrographs from irradiated metals reveal, in addition to the dislocation loops, numerous small centres of strain in the form of black dots somewhat less than 5 nm diameter, which are difficult to resolve (see Figure 4.59a). Because the two kinds of clusters differ in size and distribution, and also in their behaviour on annealing, it is reasonable to attribute the presence of one type of defect, i.e. the large loops, to the aggregation of interstitials and the other, i.e. the small dots, to the aggregation of vacancies. This general conclusion has been confirmed by detailed contrast analysis of the defects.

The addition of an extra (1 1 1) plane in a crystal with fcc structure (see Figure 4.60) introduces two faults in the stacking sequence and not one, as is the case when a plane of atoms is removed. In consequence, to eliminate the fault it is necessary for two partial dislocations to slip across the loop, one above the layer and one below, according to a reaction of the form

$$\frac{a}{3}[\bar{1}\,\bar{1}\,\bar{1}] + \frac{a}{6}[1\,1\,\bar{2}] + \frac{a}{6}[1\,\bar{2}\,1] \rightarrow \frac{a}{2}[0\,\bar{1}\,\bar{1}]$$

The resultant dislocation loop formed is identical to the prismatic loop produced by a vacancy cluster but has a Burgers vector of opposite sign. The size of the loops formed from interstitials increases with the irradiation dose and temperature, which suggests that small interstitial clusters initially form and subsequently grow by a diffusion process. In contrast, the vacancy clusters are much more numerous, and although their size increases slightly with dose, their number is approximately proportional to the dose and equal to the number of primary collisions which occur. This observation supports the suggestion that vacancy clusters are formed by the redistribution of vacancies created in the cascade.

Changing the type of irradiation from electron, to light charged particles such as protons, to heavy ions such as self-ions, to neutrons, results in a progressive increase in the mean recoil energy. This results in an increasingly non-uniform point defect generation due to the production of displacement cascades by primary knock-ons. During the creation of cascades, the interstitials are transported outwards (see Figure 4.7), most probably by focused collision sequences, i.e. along a close-packed row of atoms by a sequence of replacement collisions, to displace the last atom in this same crystallographic direction, leaving a vacancy-rich region at the centre of the cascade which can collapse to form vacancy loops. As the irradiation temperature increases, vacancies can also aggregate to form voids.

on a finer scale than in quenched metals, because of the high supersaturation and low diffusion distance. Bombardment of copper foils with 1.4×10^{21} 38 MeV α-particles m^{-2} produces about 10^{21} m^{-3} dislocation loops as shown in Figure 4.59a; a denuded region 0.8 μ m wide can also be seen at the grain boundary. These loops, about 40 nm diameter, indicate that an atomic concentration of $\approx 1.5 \times 10^{-4}$ point defects have precipitated in this form. Heavier doses of α-particle bombardment produce larger diameter

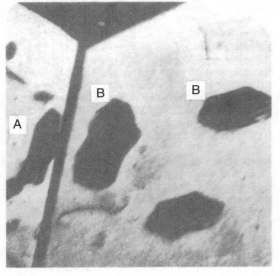

(a)

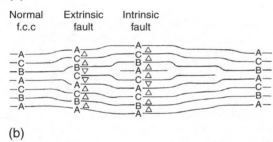

(b)

Figure 4.60 *(a) Single (A) and double (B) dislocation loops in proton-irradiated copper (×43 000). (b) Structure of a double-dislocation loop (after Mazey and Barnes, 1968; courtesy of Taylor and Francis).*

Frank sessile dislocation loops, double-faulted loops, tetrahedra and voids have all been observed in irradiated metals, but usually under different irradiation conditions. Results from Cu, Ag and Au show that cascades collapse to form Frank loops, some of which dissociate towards stacking fault tetrahedra. The fraction of cascades collapsing to form visible loops, defined as the defect yield, is high, ≈ 0.5 in Cu to 1.0 in Au irradiated with self-ions. Moreover, the fraction of vacancies taking part in the collapse process, expressed as the cascade efficiency, is also high (≈ 0.3 to 0.5). Vacancy loops have been observed on irradiation at R.T. in some bcc metals (e.g. Mo, Nb, W, α-Fe). Generally, the loops are perfect with $b = a/2\langle 1\,1\,1\rangle$ although they are thought to nucleate as $a/2\langle 1\,1\,0\rangle$ faulted loops on $\{1\,1\,0\}$ but unfault at an early stage because of the high stacking-fault energy. Vacancy loops have also been observed in some cph metals (e.g. Zr and Ti).

Interstitial defects in the form of loops are commonly observed in all metals. In fcc metals Frank loops containing extrinsic faults occur in Cu, Ag, Au, Ni, Al and austenitic steels. Clustering of interstitials on two neighbouring (1 1 1) planes to produce an intrinsically faulted defect may also occur, as shown in Figure 4.60. In bcc metals they are predominantly perfect $a/2\langle 1\,1\,1\rangle$.

The damage produced in cph metals by electron irradiation is very complex and for Zn and Cd ($c/a > 1.633$) several types of dislocation loops, interstitial in nature, nucleate and grow; thus $c/2$ loops, i.e. with $b = [c/2]$, c-loops, $(c/2 + p)$ loops, i.e. with $b = \frac{1}{6}\langle 2\,0\,\overline{2}\,3\rangle$, $[c/2] + [c/2]$ loops and $\langle c/2 + p\rangle + \langle c/2 - p\rangle$ loops are all formed; in the very early stages of irradiation most of the loops consist of $[c/2]$ dislocations, but as they grow a second loop of $b = [c/2]$ forms in the centre, resulting in the formation of a $[c/2] + [c/2]$ loop. The $\langle c/2 + p\rangle + \langle c/2 - p\rangle$ loops form either from the nucleation of a $\langle c/2 + p\rangle$ loop inside a $\langle c/2 - p\rangle$ loop or when a $[c/2] + [c/2]$ loop shears. At low dose rates and low temperatures many of the loops facet along $\langle 1\,1\,\overline{2}\,0\rangle$ directions.

In magnesium with c/a almost ideal the nature of the loops is very sensitive to impurities, and interstitial loops with either $b = \frac{1}{3}\langle 1\,1\,\overline{2}\,0\rangle$ on non-basal planes or basal loops with $b = (c/2 + p)$ have been observed in samples with different purity. Double loops with $b = (c/2 + p) + (c/2 - p)$ also form but no $c/2$-loops have been observed.

In Zr and Ti ($c/a < 1.633$) irradiated with either electrons or neutrons both vacancy and interstitial loops form on non-basal planes with $b = \frac{1}{3}\langle 1\,1\,\overline{2}\,0\rangle$. Loops with a c-component, namely $b = \frac{1}{3}\langle 1\,1\,\overline{2}\,3\rangle$ on $\{1\,0\,\overline{1}\,0\}$ planes and $b = c/2$ on basal planes have also been observed; voids also form in the temperature range $0.3\text{--}0.6T_{\mathrm{m}}$. The fact that vacancy loops are formed on electron irradiation indicates that cascades are not essential for the formation of vacancy loops. Several factors can give rise to the increased stability of vacancy loops in these metals. One factor is the possibility of stresses arising from oxidation or anisotropic thermal expansion, i.e. interstitial loops are favoured perpendicular to a tensile axis and vacancy loops parallel. A second possibility is impurities segregating to dislocations and reducing the interstitial bias.

4.7.3.2 Radiation growth and swelling

In non-cubic materials, partitioning of the loops on to specific habit planes can lead to an anisotropic dimensional change, known as irradiation growth. The aggregation of vacancies into a disc-shaped cavity which collapses to form a dislocation loop will give rise to a contraction of the material in the direction of the Burgers vector. Conversely, the precipitation of a plane of interstitials will result in the growth of the material. Such behaviour could account for the growth which takes place in α-uranium single crystals during neutron irradiation, since electron micrographs from thin films of irradiated uranium show the presence of clusters of point defects.

The energy of a fission fragment is extremely high ($\approx$200 MeV) so that a high concentration of both vacancies and interstitials might be expected. A dose of 10^{24} n m^{-2} at room temperature causes uranium to grow about 30% in the [0 1 0] direction and contract in the [1 0 0] direction. However, a similar dose at the temperature of liquid nitrogen produces ten times this growth, which suggests the preservation of about 10^4 interstitials in clusters for each fission event that occurs. Growth also occurs in textured polycrystalline α-uranium and to avoid the problem a random texture has to be produced during fabrication. Similar effects can be produced in graphite.

During irradiation vacancies may aggregate to form voids and the interstitials form dislocation loops. The voids can grow by acquiring vacancies which can be provided by the climb of the dislocation loops. However, because these loops are formed from interstitial atoms they grow, not shrink, during the climb process and eventually become a tangled dislocation network.

Interstitial point defects have two properties important in both interstitial loop and void growth. First, the elastic size interaction (see Chapter 7) causes dislocations to attract interstitials more strongly than vacancies and secondly, the formation energy of an interstitial E_f^i is greater than that of a vacancy E_f^v so that the dominant process at elevated temperatures is vacancy emission. The importance of these factors to loop stability is shown by the spherical diffusion-controlled rate equation

$$\frac{dr}{dt} = \frac{1}{b} \left\{ D_v c_v - Z_i D_i c_i \right.$$

$$\left. - D_v c_0 \exp\left[\frac{(F_{el} + \gamma)b^2}{kT}\right] \right\} \quad (4.24)$$

For the growth of voids during irradiation the spherical diffusion equation

$$\frac{dr}{dt} = \frac{1}{r} \left\{ \begin{array}{l} [1 + (\rho r)^{1/2}]D_v c_v - [1 + (Z_i \rho r)^{1/2}]D_i c_i \\ -[1 + (\rho r)^{1/2}]D_v c_0 \exp\left[\frac{[(2\gamma_s/r) - P]\Omega}{kT}\right] \end{array} \right\}$$

has been developed, where c_i and c_v are the interstitial and vacancy concentrations, respectively, D_v and D_i their diffusivities, γ_s the surface energy and Z_i is a bias term defining the preferred attraction of the loops for interstitials.

At low temperatures, voids undergo bias-driven growth in the presence of biased sinks, i.e. dislocation loops or network of density ρ. At higher temperatures when the thermal emission of vacancies becomes important, whether voids grow or shrink depends on the sign of $[(2\gamma_s/r) - P]$. During neutron irradiation when gas is being created continuously the gas pressure $P > 2\gamma_s/r$ and a flux of gas atoms can arrive at the voids causing gas-driven growth.

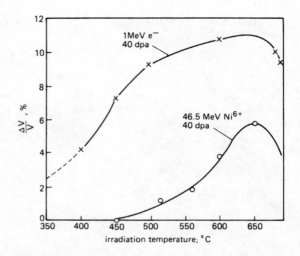

Figure 4.61 *Plots of void swelling versus irradiation temperature for 1050°C solution-treated Type 316 irradiated with 1 MeV electrons and 46 MeV Ni^{6+} to a dose of 40 dpa (after Nelson and Hudson, p. 19).*

The formation of voids leads to the phenomenon of void swelling and is of practical importance in the dimensional stability of reactor core components. The curves of Figure 4.61 show the variation in total void volume as a function of temperature for solution-treated Type 316 stainless steel; the upper cut-off arises when the thermal vacancy emission from the voids exceeds the net flow into them. Comparing the ion- and electron-irradiated curves shows that increasing the recoil energy moves the lower threshold to higher temperatures and is considered to arise from the removal of vacancies by the formation of vacancy loops in cascades; cascades are not created by electron irradiation.

Voids are formed in an intermediate temperature range $\approx$0.3 to 0.6T_m, above that for long-range single vacancy migration and below that for thermal vacancy emission from voids. To create the excess vacancy concentration it is also necessary to build up a critical dislocation density from loop growth to bias the interstitial flow. The sink strength of the dislocations, i.e. the effectiveness of annihilating point defects, is given by $K_i^2 = Z_i \rho$ for interstitials and $K_v^2 = Z_{v\rho}$ for vacancies where $(Z_i - Z_v)$ is the dislocation bias for interstitials $\approx$10% and ρ is the dislocation density. As voids form they also act as sinks, and are considered neutral to vacancies and interstitials, so that $K_i^2 = K_v^2 = 4\pi r_v C_v$, where r_v and C_v are the void radius and concentration, respectively.

The rate theory of void swelling takes all these factors into account and (1) for moderate dislocation densities as the dislocation structure is evolving, swelling is predicted to increase linearly with irradiation dose, (2) when ρ reaches a quasi-steady state the rate should increase as (dose)$^{3/2}$, and (3) when the void density is very high, i.e. the sink strength of the voids is greater than the sink strength of the dislocations ($K_v^2 \gg K_d^2$), the rate of swelling should again decrease. Results from electron irradiation of stainless steel show that the swelling rate is linear with dose up to 40 dpa (displacement per atom) and there is no tendency to a (dose)$^{3/2}$ law, which is consistent with dislocation structure continuing to evolve over the dose and temperature range examined.

In the fuel element itself, fission gas swelling can occur since uranium produces one atom of gas (Kr and Ze) for every five U atoms destroyed. This leads to ≈ 2 m^3 of gas (stp) per m^3 of U after a 'burnup' of only 0.3% of the U atoms.

In practice, it is necessary to keep the swelling small and also to prevent nucleation at grain boundaries when embrittlement can result. In general, variables which can affect void swelling include alloying elements together with specific impurities, and microstructural features such as precipitates, grain size and dislocation density. In ferritic steels, the interstitial solutes carbon and nitrogen are particularly effective in (1) trapping the radiation-induced vacancies and thereby enhancing recombination with interstitials, and (2) interacting strongly with dislocations and therefore reducing the dislocation bias for preferential annihilation of interstitials, and also inhibiting the climb rate of dislocations. Substitutional alloying elements with a positive misfit such as Cr, V and Mn with an affinity for C or N can interact with dislocations in combination with interstitials and are considered to have a greater influence than C and N alone.

These mechanisms can operate in fcc alloys with specific solute atoms trapping vacancies and also elastically interacting with dislocations. Indeed the inhibition of climb has been advanced to explain the low swelling of *Nimonic PE16* nickel-based alloys. In this case precipitates were considered to restrict dislocation climb. Such a mechanism of dislocation pinning is likely to be less effective than solute atoms since pinning will only occur at intervals along the dislocation line. Precipitates in the matrix which are coherent in nature can also aid swelling resistance by acting as regions of enhanced vacancy–interstitial recombination. TEM observations on θ' precipitates in Al–Cu alloys have confirmed that as these precipitates lose coherency during irradiation, the swelling resistance decreases.

4.7.3.3 Radiation-induced segregation, diffusion and precipitation

Radiation-induced segregation is the segregation under irradiation of different chemical species in an alloy towards or away from defect sinks (free surfaces, grain boundaries, dislocations, etc.). The segregation is caused by the coupling of the different types of atom with the defect fluxes towards the sinks. There are four different possible mechanisms, which fall into two pairs, one pair connected with size effects and the other with the Kirkendall effect.[1] With size effects, the point defects drag the solute atoms to the sinks because the size of the solute atoms differs from the other types of atom present (solvent atoms). Thus interstitials drag small solute atoms to sinks and vacancies drag large solute atoms to sinks. With Kirkendall effects, the faster diffusing species move in the opposite direction to the vacancy current, but in the same direction as the interstitial current. The former case is usually called the 'inverse Kirkendall effect', although it is still the Kirkendall effect, but solute atoms rather than the vacancies are of interest. The most important of these mechanisms, which are summarized in Figure 4.62, appear to be (1) the interstitial size effect mechanism–the dragging of small solute atoms to sinks by interstitials–and (2) the vacancy Kirkendall

[1]The Kirkendall effect is discussed in Chapter 6, Section 6.4.2.

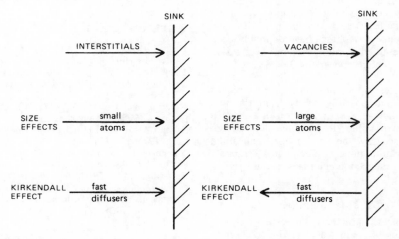

Figure 4.62 *Schematic representation of radiation-induced segregation produced by interstitial and vacancy flow to defect sinks.*

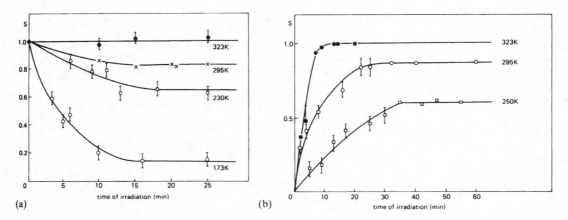

Figure 4.63 *Variation in the degree of long-range order S for initially (a) ordered and (b) disordered Cu₃Au for various irradiation temperatures as a function of irradiation time. Accelerating voltage 600 kV (after Hameed, Loretto and Smallman, 1982; by courtesy of Taylor and Francis).*

effect – the migration away from sinks of fast-diffusing atoms.

Radiation-induced segregation is technologically important in fast breeder reactors, where the high radiation levels and high temperatures cause large effects. Thus, for example, in Type 316 stainless steels, at temperatures in the range 350–650°C (depending on the position in the reactor) silicon and nickel segregate strongly to sinks. The small silicon atoms are dragged there by interstitials and the slow diffusing nickel stays there in increasing concentration as the other elements diffuse away by the vacancy inverse Kirkendall effect. Such diffusion (1) denudes the matrix of void-inhibiting silicon and (2) can cause precipitation of brittle phases at grain boundaries, etc.

Diffusion rates may be raised by several orders of magnitude because of the increased concentration of point defects under irradiation. Thus phases expected from phase diagrams may appear at temperatures where kinetics are far too slow under normal circumstances. Many precipitates of this type have been seen in stainless steels which have been in reactors. Two totally new phases have definitely been produced and identified in alloy systems (e.g. Pd_8W and Pd_8V) and others appear likely (e.g. Cu–Ni miscibility gap).

This effect relates to the appearance of precipitates after irradiation and possibly arises from the two effects described above, i.e. segregation or enhanced diffusion. It is possible to distinguish between these two causes by post-irradiation annealing, when the segregation-induced precipitates disappear but the diffusion-induced precipitates remain, being equilibrium phases.

4.7.3.4 Irradiation of ordering alloys

Ordering alloys have a particularly interesting response to the influence of point defects in excess of the eqilibrium concentration. Irradiation introduces point defects and their effect on the behaviour of ordered alloys depends on two competitive processes, i.e. radiation-induced ordering and radiation-induced disordering, which can occur simultaneously. The interstitials do not contribute significantly to ordering but the radiation-induced vacancies give rise to ordering by migrating through the crystal. Disordering is assumed to take place athermally by displacements. Figure 4.63 shows the influence of electron irradiation time and temperature on (a) initially ordered and (b) initially disordered Cu_3Au. The final state of the alloy at any irradiation temperature is independent of the initial condition. At 323 K, Cu_3Au is fully ordered on irradiation, whether it is initially ordered or not, but at low temperatures it becomes largely disordered because of the inability of the vacancies to migrate and develop order; the interstitials ($E_m^i \approx 0.1$ eV) can migrate at low temperatures.

Further reading

Hirth, J. P. and Lothe, J. (1984). *Theory of Dislocations.* McGraw-Hill, New York.

Hume-Rothery, W., Smallman, R. E. and Haworth, C. W. (1969). *Structure of Metals and Alloys*, Monograph No. 1. Institute of Metals.

Kelly, A. and Groves, G. W. (1970). *Crystallography and Crystal Defects.* Longman, London.

Loretto, M. H. (ed.) (1985). *Dislocations and Properties of Real Materials.* Institute of Metals, London.

Smallman, R. E. and Harris, J. E. (eds) (1976). *Vacancies.* The Metals Society, London.

Sprackling, M. T. (1976). *The Plastic Deformation of Simple Ionic Solids.* Academic Press, London.

Thompson, N. (1953). Dislocation nodes in fcc lattices. *Proc. Phys. Soc.*, **B66**, 481.

Chapter 5

The characterization of materials

5.1 Tools of characterization

Determination of the structural character of a material, whether massive in form or particulate, crystalline or glassy, is a central activity of materials science. The general approach adopted in most techniques is to probe the material with a beam of radiation or high-energy particles. The radiation is electromagnetic in character and may be monochromatic or polychromatic: the electromagnetic spectrum (Figure 5.1) conveniently indicates the wide choice of energy which is available. Wavelengths (λ) range from heat, through the visible range ($\lambda = 700$–400 nm) to penetrating X-radiation. Using de Broglie's important relation $\lambda = \mathbf{h}/mv$, which expresses the duality of radiation frequency and particle momentum, it is possible to apply the idea of wavelength to a stream of electrons.

The microscope, in its various forms, is the principal tool of the materials scientist. The magnification of the image produced by an electron microscope can be extremely high; however, on occasion, the

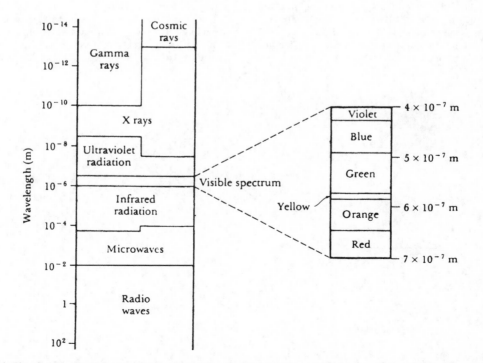

Figure 5.1 *The electromagnetic spectrum of radiation (from Askeland, 1990, p. 732; by permission of Chapman and Hall).*

modest magnification produced by a light stereo-microscope can be sufficient to solve a problem. In practical terms, the microscopist attaches more importance to resolution than magnification; that is, the ability of the microscope to distinguish fine detail. In a given microscope, increasing the magnification beyond a certain limit will fail to reveal further structural detail; such magnification is said to be 'empty'. Unaided, the human eye has a resolution of about 0.1 mm: the resolution of light microscopes and electron microscopes are, respectively, about 200 nm and 0.5 nm. In order to perceive or image a structural feature it is necessary that the wavelength of the probing radiation should be similar in size to that of the feature. In other words, and as will be enlarged upon later, resolution is a function of wavelength.

In this chapter we examine the principal ways in which light, X-rays, electrons, and neutrons are used to explore the structure of metals. Some degree of selectivity has been unavoidable. Although the prime purpose of microscopy is to provide qualitative information on structure, many complementary techniques are available that provide quantitative data on the chemical and physical attributes of a material.

5.2 Light microscopy

5.2.1 Basic principles

The light microscope provides two-dimensional representation of structure over a total magnification range of roughly $40\times$ to $1250\times$. Interpretation of such images is a matter of skill and experience and needs to allow for the three-dimensional nature of features observed. The main components of a bench-type microscope are (1) an illumination system comprising a light source and variable apertures, (2) an objective lens and an ocular lens (eyepiece) mounted at the ends of a cylindrical body-tube, and (3) a specimen stage (fixed or rotatable). Metallic specimens that are to be examined at high magnifications are successively polished with 6, 1 and sometimes 0.25 μM diamond grit. Examination in the as-polished condition, which is generally advisable, will reveal structural features such as shrinkage or gas porosity, cracks and inclusions of foreign matter. Etching with an appropriate chemical reagent is used to reveal the arrangement and size of grains, phase morphology, compositional gradients (coring), orientation-related etch pits and the effects of plastic deformation. Although actually only a few atomic diameters wide, grain boundaries are preferentially and grossly attacked by many etchants. In bright-field illumination, light is reflected back towards the objective from reflective surfaces, causing them to appear bright. Dark-field illumination reverses this effect, causing grain boundaries to appear bright. The degree of chemical attack is sensitive to crystal orientation and an etched polycrystalline aggregate will often display its grain structure clearly (Figure 5.2a). Preparation techniques for ceramics are essentially similar to those for metals and alloys. However, their porosity can cause two problems. First, there is a risk of entrapping diamond particles during polishing, making ultrasonic cleaning advisable. Second, it may be necessary to strengthen the structure by impregnating with liquid resin *in vacuo*, provided that pores are interconnected.

The objective, the most important and critical component in the optical train of the light microscope, is made up of a number of glass lenses and, sometimes, fluorite (CaF_2) lenses also. Lenses are subject to

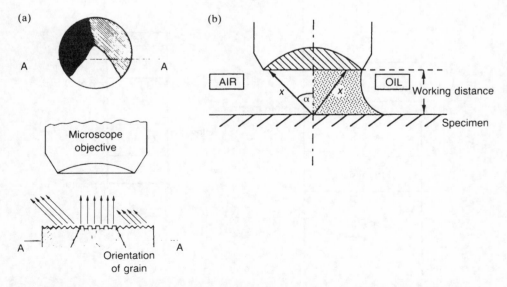

Figure 5.2 *(a) Reflection of light from etched specimen. (b) Use of oil to improve numerical aperture of objective.*

spherical and chromatic aberrations. Minimization and correction of these undesirable physical effects, greatly aided by modern computational techniques, is possible and objectives are classified according to the degree of correction, i.e. achromats, fluorites (semi-apochromats), apochromats. Lenses are usually coated in order to increase light transmission. As magnification is increased, the depth of field of the objective becomes smaller, typically falling from 250 μm at 15 × to 0.08 μm at 1200×, so that specimen flatness becomes more critical. The focal length and the working distance (separating its front lens from the specimen) of an objective differ. For instance, an $f\,2$ mm objective may have a working distance of 0.15 mm.

Resolution, rather than magnification, is usually the prime concern of the skilled microscopist. It is the smallest separating distance (δ) that can be discerned between two lines in the image. The unaided eye, at the least distance of comfortable vision (about 250 mm), can resolve 0.1 mm. Confusingly, the resolution value for a lens with a so-called high resolving power is small. Resolution is determined by (1) the wavelength (λ) of the radiation and (2) the numerical aperture (NA) of the objective and is expressed by the Abbe formula $\delta = \lambda/2\text{NA}$.

The numerical aperture value, which is engraved upon the side of the objective, indicates the light-gathering power of the compound lens system and is obtained from the relation $\text{NA} = n \sin \alpha$, where n is the refractive index of the medium between the front lens face of the objective and the specimen, and α is the semi-apex angle of the light cone defined by the most oblique rays collected by the lens. Numerical apertures range in typical value from 0.08 to 1.25. Despite focusing difficulties and the need for costly lenses, efforts have been made to use short-wavelength ultraviolet radiation: developments in electron microscopy have undermined the feasibility of this approach. Oil-immersion objectives enable the refractive index term to be increased (Figure 5.2b). Thus, by replacing air ($n = 1$) with a layer of cedar wood oil ($n = 1.5$) or monobromonaphthalene ($n = 1.66$), the number of rays of reflected light accepted by the front lens of the objective is increased and resolution and contrast are improved. The range of wavelengths for visible light is approximately 400–700 nm; consequently, using the Abbe formula, it can readily be shown that the resolution limit of the light microscope is in the order of 200 nm. The 'useful' range of magnification is approximately 500–1000 NA. The lower end of the range can be tiring to the eyes; at the top end, oil-immersion objectives are useful.

Magnification is a subjective term; for instance, it varies with the distance of an image or object from the eye. Hence, microscopists sometimes indicate this difficulty by using the more readily defined term 'scale of reproduction', which is the lineal size ratio of an image (on a viewing screen or photomicrograph) to the original object. Thus, strictly speaking, a statement such as 500× beneath a photomicrograph gives the scale of reproduction, not the magnification.

The ocular magnifies the image formed by the objective: the finally-observed image is virtual. It can also correct for certain objective faults and, in photomicrography, be used to project a real image. The ocular cannot improve the resolution of the system but, if inferior in quality, can worsen it. The most widely used magnifications for oculars are 8× and 12.5×.

Two dimensional features of a standard bench microscope, the mechanical tube length t_m and optical tube length t_o, are of special significance. The former is the fixed distance between the top of the body tube, on which the ocular rests, and the shoulder of the rotatable nosepiece into which several objectives are screwed. Objectives are designed for a certain t_m value. A value of 160 mm is commonly used. (In Victorian times, it was 250 mm, giving a rather unwieldy instrument.) The optical tube length t_o is the distance between the front focal point of the ocular and the rear focal plane of the objective. Parfocalization, using matched parfocal objectives and oculars, enables the specimen to remain in focus when objectives are step-changed by rotating the nosepiece. With each change, t_o changes but the image produced by the objective always forms in the fixed focal phase of the ocular. Thus the distance between the specimen and the aerial image is kept constant. Some manufacturers base their sequences of objective and ocular magnifications upon preferred numbers[1] rather than upon a decimal series. This device facilitates the selection of a basic set of lenses that is comprehensive and 'useful' (exempt from 'empty' magnification). For example, the Michel series of 6.3×, 8×, 10×, 12.5×, 16×, 20×, 25×, etc., a geometrical progression with a common ratio of approximately 1.25, provides a basis for magnification values for objectives and oculars. This rational approach is illustrated in Figure 5.3. Dot–dash lines represent oculars and thin solid lines represent objectives. The bold lines outline a box within which objective/ocular combinations give 'useful' magnifications. Thus, pairing of a 12.5× ocular with a 40× objective (NA = 0.65) gives a 'useful' magnification of 500×.

5.2.2 Selected microscopical techniques

5.2.2.1 Phase-contrast microscopy

Phase-contrast microscopy is a technique that enables special surface features to be studied even when

---

[1]The valuable concept of preferred numbers/sizes, currently described in document PD 6481 of the British Standards Institution, was devised by a French military engineer, Colonel Charles Renard (1847–1905). In 1879, during the development of dirigible (steerable) balloons, he used a geometrical progression to classify cable diameters. A typical Renard Series is 1.25, 1.6, 2.0, 2.5, 3.2, 4.0, 5.0, 6.4, 8.0, etc.

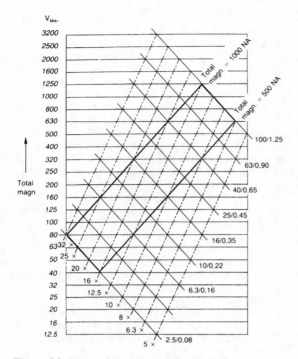

Figure 5.3 *Range of 'useful' magnification in light microscope (from Optical Systems for the Microscope, 1967, p. 15; by courtesy of Carl Zeiss, Germany).*

there is no colour or reflectivity contrast. The light reflected from a small depression in a metallographic specimen will be retarded in phase by a fraction of a light wavelength relative to that reflected from the surrounding matrix and, whereas in ordinary microscopy a phase difference in the light collected by the objective will not contribute to contrast in the final image, in phase-contrast microscopy small differences in phases are transformed into differences in brightness which the eye can detect.

General uses of the technique include the examination of multi-phased alloys after light etching, the detection of the early stages of precipitation, and the study of cleavage faces, twins and other deformation characteristics. The optimum range of differences in surface level is about 20–50 nm, although under favourable conditions these limits may be extended. A schematic diagram of the basic arrangement for phase contrast in the metallurgical microscope is shown in Figure 5.4a. A hollow cone of light produced by an annulus A, is reflected by the specimen and brought to an image in the back focal plane of the objective. A phase plate of suitable size should, strictly, be positioned in this plane but, for the ease of interchangeability of phase plates, the position Q in front of the eyepiece E is often preferred. This phase plate has an annulus, formed either by etching or deposition, such that the light it transmits is either advanced or retarded by a quarter of a wavelength relative to the light transmitted by the rest of the plate and, because the light reflected from a surface feature is also advanced or retarded by approximately $\lambda/4$, the beam is either in phase or approximately $\lambda/2$ or π out of phase with that diffracted by the surface features of the specimen. Consequently, reinforcement or cancellation occurs, and the image intensity at any point depends on the phase difference produced at the corresponding point on the specimen surface, and this in turn depends upon the height of this point relative to the adjacent parts of the surface. When the light passing through the annulus is advanced in phase, positive phase contrast results and areas of

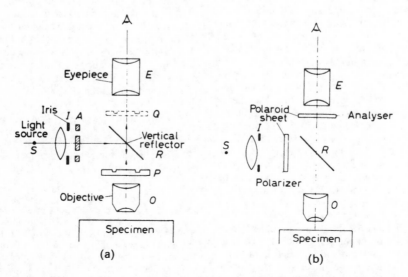

Figure 5.4 *Schematic arrangement of microscope system for (a) phase-contrast and (b) polarized light microscopy.*

the specimen which are proud of the matrix appear bright and depressions dark; when the phase is retarded negative contrast is produced and 'pits' appear bright and 'hills' dark.

5.2.2.2 Polarized-light microscopy

The basic arrangement for the use of polarized light is shown in Figure 5.4b. The only requirements of this technique are that the incident light on the specimen be plane-polarized and that the reflected light be analysed by a polarizing unit in a crossed relation with respect to the polarizer, i.e. the plane of polarization of the analyser is perpendicular to that of the polarizer.

The application of the technique depends upon the fact that plane-polarized light striking the surface of an optically isotropic metal is reflected unchanged if it strikes at normal incidence. If the light is not at normal incidence the reflected beam may still be unchanged but only if the angle of incidence is in, or at right angles to, the plane of polarization, otherwise it will be elliptically polarized. It follows that the unchanged reflected beam will be extinguished by an analyser in the crossed position whereas an elliptically polarized one cannot be fully extinguished by an analyser in any position. When the specimen being examined is optically anisotropic, the light incident normally is reflected with a rotation of the plane of polarization and as elliptically polarized light; the amount of rotation and of elliptical polarization is a property of the metal and of the crystal orientation.

If correctly prepared, as-polished specimens of anisotropic metals will 'respond' to polarized light and a grain contrast effect is observed under crossed polars as a variation of brightness with crystal orientation. Metals which have cubic structure, on the other hand, will appear uniformly dark under crossed polars, unless etched to invoke artificial anisotropy, by producing anisotropic surface films or well-defined pits. An etch pit will reflect the light at oblique incidence and elliptically-polarized light will be produced. However, because such a beam cannot be fully extinguished by the analyser in any position, it will produce a background illumination in the image which tends to mask the grain contrast effect.

Clearly, one of the main uses of polarized light is to distinguish between areas of varying orientation, since these are revealed as differences of intensity under crossed polars. The technique is, therefore, very useful for studying the effects of deformation, particularly the production of preferred orientation, but information on cleavage faces, twin bands and sub-grain boundaries can also be obtained. If a 'sensitive tint' plate is inserted between the vertical illuminator and the analyser each grain of a sample may be identified by a characteristic colour which changes as the specimen is rotated on the stage. This application is useful in the assessment of the degree of preferred orientation and in recrystallization studies. Other uses of polarized light include distinguishing and identifying phases in multi-phase alloys.

Near-perfect extinction occurs when the polars of a transmission microscope are crossed. If a thin section or slice of ceramic, mineral or rock is introduced and the stage slowly rotated, optically anisotropic crystals will produce polarization colours, developing maximum brilliance at 45° to any of the four symmetrical positions of extinction. The colour of a crystal depends upon its birefringence, or capacity for double-refraction, and thickness. By standardizing the thickness of the section at 30–50 µm and using a Michel-Lévy colour chart, it is possible to identify crystalline species. In refractory materials, it is relatively easy to identify periclase (MgO), chromite ($FeCrO_4$), tridymite (SiO_2) and zircon ($ZrSiO_4$) by their characteristic form and colour.

As birefringence occurs within the crystal, each incident ray forms ordinary and extraordinary rays which are polarized in different planes and travel through the crystal at different velocities. On leaving the analyser, these out-of-phase 'fast' and 'slow' rays combine to produce the polarization colour. This colour is complementary to colour cancelled by interference and follows Newton's sequence: yellow, orange, red, violet, blue, green. More delicate, higher-order colours are produced as the phase difference between the emergent rays increases. Anisotropic crystals are either uniaxial or biaxial, having one or two optic axes, respectively, along which birefringence does not occur. (Optic axes do not necessarily correspond with crystallographic axes.) It is therefore possible for quartz (uniaxial) and mica (biaxial) crystals to appear black because of their orientation in the slice. Uniaxial (tetragonal and hexagonal systems) can be distinguished from biaxial crystals (orthorhombic, triclinic and monoclinic systems) by introducing a Bertrand lens into the light train of the microscope to give a convergent beam, rotating the stage and comparing their interference figures: uniaxial crystals give a moving 'ring and brush' pattern, biaxial crystals give two static 'eyes'. Cubic crystals are isotropic, being highly symmetrical. Glassy phases are isotropic and also appear black between crossed polars; however, glass containing residual stresses from rapid cooling produces fringe patterns and polarization colours. The stress-anisotropic properties of plastics are utilized in photoelastic analyses of transparent models of engineering structures or components made from standard sheets of constant thickness and stress-optic coefficient (e.g. clear *Bakelite*, epoxy resin). The fringe patterns produced by monochromatic light and crossed polars in a polariscope reveal the magnitude and direction of the principal stresses that are developed when typical working loads are applied.

5.2.2.3 Hot-stage microscopy

The ability to observe and photograph phase transformations and structural changes in metals, ceramics and polymers at high magnifications while being heated holds an obvious attraction. Various designs of microfurnace cell are available for mounting in light microscope systems.

For studies at moderate temperatures, such as spherulite formation in a cooling melt of polypropylene, the sample can be placed on a glass slide, heated in a stage fitment, and viewed through crossed polars with transmitted light. For metals, which have an increasing tendency to vaporize as the temperature is raised, the polished sample is enclosed in a resistance-heated microfurnace and viewed by reflected light through an optically worked window of fused silica. The chamber can be either evacuated (10^{-6} torr) or slowly purged with inert gas (argon). The later must be dry and oxygen-free. A Pt:Pt–10Rh thermocople is inserted in the specimen. The furnace should have a low thermal inertia and be capable of heating or cooling the specimen at controlled rates; temperatures of up to 1800°C are possible in some designs. The presence of a window, and possibly cooling devices, drastically reduces the available working distance for the objective lens, particularly when a large numerical aperture or high magnification are desired. One common solution is to use a Burch-type reflecting objective with an internal mirror system which gives a useful working distance of 13–14 mm. The type of stage unit described has been used for studies of grain growth in austenite and the formation of bainite and martensite in steels, allotropic transformations, and sintering mechanisms in powder compacts.

When polished polycrystalline material is heated, individual grains tend to reduce their volume as a result of surface tension and grain boundaries appear as black lines, an effect referred to as thermal etching or grooving. If a grain boundary migrates, as in the grain growth stage of annealing, ghost images of former grooves act as useful markers. As the melting point is approached, there is often a noticeable tendency for grain boundary regions to fuse before the bulk grains; this liquation effect is due to the presence of impurities and the atomic misfit across the grain boundary surface. When interpreting the visible results of hot-stage microscopy, it is important to bear in mind that surface effects do not necessarily reflect what is happening within the bulk material beneath the surface. The technique can produce artefacts; the choice between evacuation and gas-purging can be crucial. For instance, heating *in vacuo* can favour decarburization and grain-coarsening in steel specimens.

The classic method for studying high-temperature phases and their equilibria in oxide systems was based upon rapid quenching (e.g. silicates). This indirect method is slow and does not always preserve the high-temperature phase(s). A direct microscopical technique uses the U-shaped notch of a thermocouple hot junction as the support for a small non-metallic sample. In the original design,[1] the junction was alternately connected by high-speed relay at a frequency of 50 Hz to a power circuit and a temperature-measuring circuit. The sample could be heated to temperatures of up to 2150°C and it was possible to observe crystallization from a melt directly through crossed polars. Although unsuitable for metals and highly volatile materials, the technique has been applied to glasses, slags, refractories, Portland cements, etc., providing information on phase changes, devitrification, sintering shrinkage, grain growth and the 'wetting' of solids by melts.

5.2.2.4 Microhardness testing

The measurement of hardness with a microscope attachment, comprising a diamond indentor and means for applying small loads, dates back more than 50 years. Initially used for small components (watch gears, thin wire, foils), microhardness testing was extended to research studies of individual phases, orientation effects in single crystals, diffusion gradients, ageing phenomena, etc. in metallic and ceramic materials. Nowadays, testing at temperatures up to 1000°C is possible. In Europe, the pyramidal Vickers-type (interfacial angle 136°) indentor, which produces a square impression, is generally favoured. Its counterpart in general engineering employs test loads of 5–100 kgf: in microhardness testing, typical test loads are in the range 1–100 gf (1 gf = 1 pond = 1 p = 9.81 mN). A rhombic-based Knoop indentor of American origin has been recommended for brittle and/or anisotropic material (e.g. carbides, oxides, glass) and for thin foils and coatings where a shallow depth of impression is desired. The kite-shaped Knoop impression is elongated, with a 7:1 axial ratio.

Microhardness tests need to be very carefully controlled and replicated, using as large a load as possible. The surface of the specimen should be strain-free (e.g. electropolished), plane and perpendicular to the indentor axis. The indentor is lowered slowly at a rate of <1 mm min^{-1} under vibration-free conditions, eventually deforming the test surface in a manner analogous to steady-state creep. This condition is achieved within 15 s, a test period commonly used.

The equations for Vickers hardness (H_V) and Knoop hardness (H_K) take the following forms:

$$H_V = 1854.4 (P/d^2) \quad \text{kgf mm}^{-2} \tag{5.1}$$

$$H_K = 14228 (P/d^2) \quad \text{kgf mm}^{-2} \tag{5.2}$$

In these equations, which have the dimensions of stress, load P and diagonal length d are measured in gf and μm, respectively. The first equation is based upon the surface area of the impression; the second is based upon its projected area and the length of the long diagonal.

---

[1] Developed by W. Gutt and co-workers at the Building Research Station, Watford.

The main potential difficulty concerns the possible dependence of microhardness values (H_m) upon test load. As the test load is reduced below a certain threshold, the measured microhardness value may tend to decrease or increase, depending upon the material. In these circumstances, when measuring absolute hardness rather than relative hardness, it is useful to consider the material's behaviour in terms of the Meyer equation which relates indenting force P to the diagonal length of the Vickers-type impression produced, d, as follows:

$$P = kd^n \qquad (5.3)$$

The Meyer exponent n expresses the strain-hardening characteristics of the material as it deforms plastically during the test; it increases in value with the degree of strain-hardening. For simple comparisons of relative microhardness, hardness values at a fixed load can be compared without allowance for load-dependence. On the other hand, if absolute values of hardness are required from low-load tests, it is advisable to determine the Meyer line for the particular material over a comparatively small load range by plotting P values against the corresponding d values, using log-log graph paper. (Extrapolations beyond the chosen load range are unwise because the profile of the Meyer line may change.) Figure 5.5 shows the Meyer line, slope n, for a material giving load-dependent microhardness values. The slope n is less than 2, which is usual. The H_m curve has a negative slope and microhardness values increase as the load increases. One way of reporting load-dependent microhardness results is to state three hardness numbers in terms of

a standard set of three diagonal d-values, as shown in Figure 5.5. The approximate values for the set shown are $H_{5\mu m} = 160$, $H_{10\mu m} = 140$, $H_{20\mu m} = 120$. When the anisotropy ratio for elastic moduli is high, microhardness values can vary greatly from grain to grain in polycrystalline material.

Combination of the Vickers equation with the Meyer equation gives the following expression:

$$H_V = \text{ constant } \times d^{n-2} \qquad (5.4)$$

Accordingly, if $n = 2$, which is true for the conventional Vickers macrohardness test, the gradient of the H_m line becomes zero and hardness values are conveniently load-independent.

5.2.2.5 Quantitative microscopy

Important standard methods for measuring grain size and contents of inclusions and phases have evolved in metallurgy and mineralogy. Grain size in metallic structures is commonly assessed by the ASTM (McQuaid-Ehn) method in which etched microstructures are compared with diagrams or standard sets of photographs at a standard magnification, say 100×. Numerous manual methods have been developed to assess the cleanliness of steels. They range from direct methods, in which particles beneath linearly-traversed crosswires are individually counted, to comparative methods based on photomicrographs or charts (e.g. Fox, Jernkontoret (JK), SAE-ASTM, Diergarten). Sometimes these methods attempt to identify inclusions according to type and form. Nowadays, automatic systems are available to convert microscopical information, from a light or electron microscope, to electronic signals that can be rapidly processed and evaluated. Although more objective than former methods, they must be set up correctly and their limitations appreciated. For instance, the plane of sectioning is critical when highly-directional features are present, such as filaments in composites and slag stringers in metals. Quantitative methods have made it possible to relate many microstructural characteristics, such as volume fractions, interparticle distances, grain boundary area per unit volume, interlamellar spacing, etc., to mechanical properties. In 1848, the French petrographer Delesse established mathematically that, in a fully random cross-section of a uniform aggregate, the area fraction of a given microconstituent (phase) in a field of view is equal to the corresponding volume fraction in the three-dimensional aggregate. There are three basic methods for measuring the area fraction, as the following exercise demonstrates.

Suppose that a certain field of view contains a dark phase and a light-coloured matrix, as shown in Figure 5.6a. Using the systematic notation for stereology given in Table 5.1, the total area occupied by the dark phase in the test area $A_T (= L^2)$ is A; it is the sum of i areas, each of area a. This areal fraction is A_A. Alternatively, the field of view may be systematically traversed with a random test line, length L_T, and a

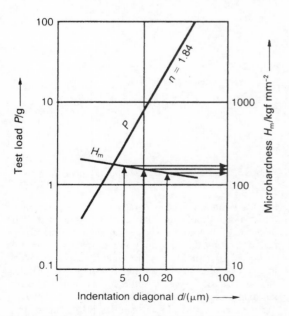

Figure 5.5 *Meyer line for material with load-independent hardness (by courtesy of Carl Zeiss, Germany).*

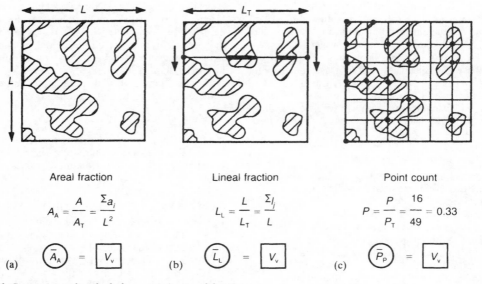

Figure 5.6 *Comparison of methods for measuring areal fraction.*

Table 5.1 Stereological notation (after Underwood)

Symbol	Dimensions	
P	Points of intersection	mm^0
N	Number of objects	mm^0
L	Lines	mm^1
A	Flat surfaces	mm^2
S	Curved surfaces	mm^2
V	Volumes	mm^3

$$P_P = L_L = A_A = V_V \qquad mm^0$$

$$2P_L = \frac{4}{\pi} L_A = S_V \qquad mm^{-1}$$

$$2P_A = L_V \qquad mm^{-2}$$

$$2P_A P_L = P_V \qquad mm^{-3}$$

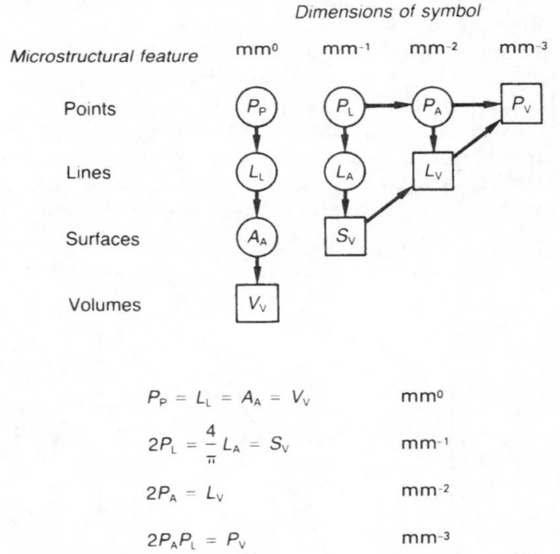

Figure 5.7 *Relationships between stereological quantities.*

length L derived from the sum of j intercepts, each of length l. The lineal fraction $L_L = L/L_T$ (Figure 5.6b). In the third method (Figure 5.6c), a regular grid of points is laid over the field and all coincidences of grid intersections with the dark phase counted to give a point count fraction P/P_T which is P_P. Only one field of view is shown in Figure 5.6; in practice, numerous different fields are tested in order to give statistically-significant average values, i.e. $\overline{A}_A$, $\overline{L}_L$, $\overline{P}_P$. The areal fraction method is chiefly of historical interest. The lineal method was used in the original Quantimet instruments which scanned[1] the images electronically with a raster. The point count method is the basis of modern instruments in which electronically-scanned fields are composed of thousands of pixels.

Certain three-dimensional properties of a structure have a great influence on a material's behaviour but cannot be measured directly by microscope. For instance, the amount of curved grain boundary surface

in unit volume S_V of a single-phase alloy is potentially more significant than average grain size. The diagram of Figure 5.7 shows some of the stereological equations that permit non-measurable quantities to be calculated. Measurable quantities are enclosed in circles, calculable quantities are enclosed in boxes and arrows show the possible routes in the triangular matrix. Important stereological equations providing the necessary links are shown below the matrix. Thus, the above-mentioned quantity S_V may be derived from intersections on test lines or traverses (P_L) since

---

[1]Contrary to popular belief, the term 'scan' implies accuracy and fine resolution.

$2P_L = S_V$. Similarly, length of dislocation lines, acicular precipitates or slag stringers per unit volume L_V may be obtained from point counts over an area, P_A.

5.3 X-ray diffraction analysis

5.3.1 Production and absorption of X-rays

The use of diffraction methods is of great importance in the analysis of crystalline solids. Not only can they reveal the main features of the structure, i.e. the lattice parameter and type of structure, but also other details such as the arrangement of different kinds of atoms in crystals, the presence of imperfections, the orientation, sub-grain and grain size, the size and density of precipitates.

X-rays are a form of electromagnetic radiation differing from light waves ($\lambda = 400-800$ nm) in that they have a shorter wavelength ($\lambda \approx 0.1$ nm). These rays are produced when a metal target is bombarded with fast electrons in a vacuum tube. The radiation emitted, as shown in Figure 5.8a, can be separated into two components, a continuous spectrum which is spread over a wide range of wavelengths and a superimposed line spectrum characteristic of the metal being bombarded. The energy of the 'white' radiation, as the continuous spectrum is called, increases as the atomic number of the target and approximately as the square of the applied voltage, while the characteristic radiation is excited only when a certain critical voltage is exceeded. The characteristic radiation is produced when the accelerated electrons have sufficient energy to eject one of the inner electrons (1s-level, for example) from its shell. The vacant 1s-level is then occupied by one of the other electrons from a higher energy level, and during the transition an emission of X-radiation takes place. If the electron falls from an adjacent shell then the radiation emitted is known as $K\alpha$-radiation, since the vacancy in the first K-shell $n = 1$ is filled by an electron from the second L-shell and the wavelength can be derived from the relation

$$\mathbf{h}v = E_L - E_K \tag{5.5}$$

However, if the K-shell vacancy is filled by an electron from an M-shell (i.e. the next highest quantum shell) then $K\beta$-radiation is emitted. Figure 5.8 shows that, in fact, one cannot be excited without the other, and the characteristic K-radiation emitted from a copper target is in detail composed of a strong $K\alpha$-doublet and a weaker $K\beta$-line.

In transversing a specimen, an X-ray beam loses intensity according to the equation

$$I = I_0 \exp[-\mu x] \tag{5.6}$$

where I_0 and I are the values of the initial and final intensities respectively, μ is a constant, known as the linear absorption coefficient which depends on the wavelength of the X-rays and the nature of

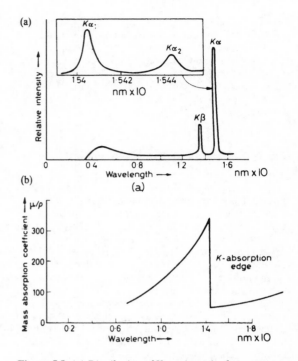

Figure 5.8 (a) Distribution of X-ray intensity from a copper target and (b) dependence of mass absorption coefficient on X-ray wavelength for nickel.

the absorber, and x is the thickness of the specimen.[1] The variation of mass absorption coefficient, i.e. linear absorption coefficient divided by density, μ/ρ, with wavelength is of particular interest, as shown in Figure 5.8b, which is the curve for nickel. It varies approximately as λ^3 until a critical value of λ(=0.148 nm) is reached, when the absorption decreases precipitously. The critical wavelength λ_K at which this decrease occurs is known as the K absorption edge, and is the value at which the X-ray beam has acquired just sufficient energy to eject an electron from the K-shell of the absorbing material. The value of λ_K is characteristic of the absorbing material, and similar L and M absorption edges occur at higher wavelengths.

This sharp variation in absorption with wavelength has many applications in X-ray practice, but its most common use is in filtering out unwanted $K\beta$-radiation. For example, if a thin piece of nickel foil is placed in a beam of X-rays from a copper target, absorption of some of the short wavelength 'white' radiation and most of the $K\beta$-radiation will result, but the

---

[1] This absorption equation is the basis of radiography, since a cavity, crack or similar defect will have a much lower μ-value than the sound metal. Such defects can be detected by the appearance of an intensity difference registered on a photographic film placed behind the X-irradiated object.

strong $K\alpha$-radiation will be only slightly attenuated. This filtered radiation is sufficiently monochromatic for many X-ray techniques, but for more specialized studies when a pure monochromatic beam is required, crystal monochromators are used. The X-ray beam is then reflected from a crystal, such as quartz or lithium fluoride, which is oriented so that only the desired wavelength is reflected according to the Bragg law (see below).

5.3.2 Diffraction of X-rays by crystals

The phenomena of interference and diffraction are commonplace in the field of light. The standard school physics laboratory experiment is to determine the spacing of a grating, knowing the wavelength of the light impinging on it, by measuring the angles of the diffracted beam. The only conditions imposed on the experiment are that (1) the grating be periodic, and (2) the wavelength of the light is of the same order of magnitude as the spacing to be determined. This experiment immediately points to the application of X-rays in determining the spacing and inter-atomic distances in crystals, since both are about $0.1-0.4$ nm in dimension. Rigorous consideration of diffraction from a crystal in terms of a three-dimensional diffraction grating is complex, but Bragg simplified the problem by showing that diffraction is equivalent to symmetrical reflection from the various crystal planes, provided certain conditions are fulfilled. Figure 5.9a shows a beam of X-rays of wavelength λ, impinging at an angle θ on a set of crystal planes of spacing d. The beam reflected at the angle θ can be real only if the rays from each successive plane reinforce each other. For this to be the case, the extra distance a ray, scattered from each successive plane, has to travel, i.e. the path difference, must be equal to an integral number of wavelengths, $n\lambda$. For example, the second ray shown in Figure 5.9a has to travel further than the first ray by the distance $PO + OQ$. The condition for reflection and reinforcement is then given by

$$n\lambda = PO + OQ = 2ON \sin\theta = 2d \sin\theta \qquad (5.7)$$

This is the well-known Bragg law and the critical angular values of θ for which the law is satisfied are known as Bragg angles.

The directions of the reflected beams are determined entirely by the geometry of the lattice, which in turn in governed by the orientation and spacing of the crystal planes. If for a crystal of cubic symmetry we are given the size of the structure cell, a, the angles at which the beam is diffracted from the crystal planes (hkl) can easily be calculated from the interplanar spacing relationship

$$d_{(hkl)} = a/\sqrt{(h^2 + k^2 + l^2)} \qquad (5.8)$$

It is conventional to incorporate the order of reflection, n, with the Miller index, and when this is done the Bragg law becomes

$$\lambda = 2a \sin\theta/\sqrt{(n^2h^2 + n^2k^2 + n^2l^2)}$$
$$= 2a \sin\theta/\sqrt{N} \qquad (5.9)$$

where N is known as the reflection or line number. To illustrate this let us take as an example the second-order reflection from $(1\,0\,0)$ planes. Then, since $n = 2$, $h = 1$, $k = 0$, and $l = 0$, this reflection is referred to either as the $2\,0\,0$ reflection or as line 4. The lattice planes which give rise to a reflection at the smallest Bragg angle are those which are most widely spaced, i.e. those with a spacing equal to the cell edge, d_{100}. The next planes in order of decreased spacing will be $\{1\,1\,0\}$ planes for which $d_{110} = a/\sqrt{2}$, while the octahedral $\{1\,1\,1\}$ planes will have a spacing equal to $a/\sqrt{3}$. The angle at which any of these planes in a crystal reflect an X-ray beam of wavelength λ may be calculated by inserting the appropriate value of d into the Bragg equation.

To ensure that Bragg's law is satisfied and that reflections from various crystal planes can occur, it is necessary to provide a range of either θ or λ values. The various ways in which this can be done leads to the standard methods of X-ray diffraction, namely: (1) the Laue method, and (2) the powder method.

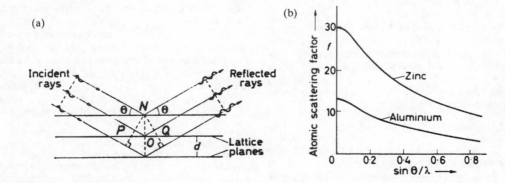

Figure 5.9 *(a) Diffraction from crystal planes. (b) Form of the atomic scattering curves for aluminium and zinc.*

5.3.3 X-ray diffraction methods

5.3.3.1 Laue method

In the Laue method, a stationary single crystal is bathed in a beam of 'white' radiation. Then, because the specimen is a fixed single crystal, the variable necessary to ensure that the Bragg law is satisfied for all the planes in the crystal has to be provided by the range of wavelengths in the beam, i.e. each set of crystal planes chooses the appropriate λ from the 'white' spectrum to give a Bragg reflection. Radiation from a target metal having a high atomic number (e.g. tungsten) is often used, but almost any form of 'white' radiation is suitable. In the experimental arrangement shown in Figure 5.10a, either a transmission photograph or a back-reflection photograph may be taken, and the pattern of spots which are produced lie on ellipses in the transmission case or hyperbolae in the back-reflection case. All spots on any ellipse or hyperbola are reflections from planes of a single zone (i.e. where all the lattice planes are parallel to a common direction, the zone axis) and, consequently, the Laue pattern is able to indicate the symmetry of the crystal. For example, if the beam is directed along a [1 1 1] or [1 0 0] direction in the crystal, the Laue pattern will show three- or fourfold symmetry, respectively. The Laue method is used extensively for the determination of the orientation of single crystals and, while charts are available to facilitate this determination, the method consists essentially of plotting the zones taken from the film on to a stereogram, and comparing the angles between them with a standard projection of that crystal structure. In recent years the use of the Laue technique has been extended to the study of imperfections resulting from crystal growth or deformation, because it is found that the Laue spots from perfect crystals are sharp, while those from deformed crystals are, as shown in Figure 5.10b, elongated. This elongated appearance of the diffraction spots is known as asterism and it arises in an analogous way to the reflection of light from curved mirrors.

5.3.3.2 Powder method

The powder method, devised independently by Debye and Scherrer, is probably the most generally useful of all the X-ray techniques. It employs monochromatic radiation and a finely-powdered, or fine-grained polycrystalline, wire specimen. In this case, θ is the variable, since the collection of randomly-oriented crystals will contain sufficient particles with the correct orientation to allow reflection from each of the possible reflecting planes, i.e. the powder pattern results from a series of superimposed rotating crystal patterns. The angle between the direct X-ray beam and the reflected ray is 2θ, and consequently each set of crystal planes gives rise to a cone of reflected rays of semi-angle 2θ, where θ is the Bragg angle for that particular set of reflecting planes producing the cone. Thus, if a film is placed around the specimen, as shown in Figure 5.11,

(a)

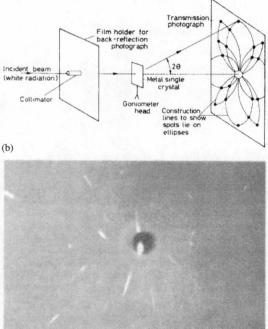

(b)

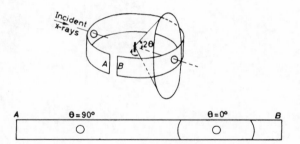

Figure 5.10 (a) Laue method of X-ray diffraction. (b) Asterisms on a Laue transmission photograph of deformed zinc (after Cahn, 1949).

the successive diffracted cones, which consist of rays from hundreds of grains, intersect the film to produce concentric curves around the entrance and exit holes. Figure 5.12 shows examples of patterns from bcc and fcc materials, respectively.

Precise measurement of the pattern of diffraction lines is required for many applications of the powder method, but a good deal of information can readily be obtained merely by inspection. One example of this is in the study of deformed metals, since after deformation the individual spots on the diffraction rings are blurred so much that line-broadening occurs, especially

Figure 5.11 Powder method of X-ray diffraction.

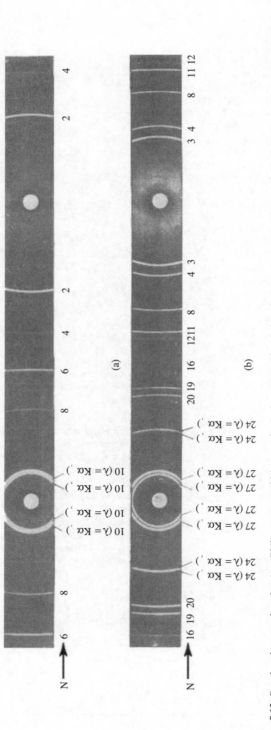

Figure 5.12 *Powder photographs taken in a Philips camera (114 mm radius) of (a) iron with cobalt radiation using an iron filter and (b) aluminium with copper radiation using a nickel filter. The high-angle lines are resolved and the separate reflections for $\lambda = K\alpha_1$ and $\lambda = K\alpha_2$ are observable..*

at high Bragg angles. On low-temperature annealing, the cold-worked material will tend to recover and this is indicated on the photograph by a sharpening of the broad diffraction lines. At higher annealing temperatures the metal will completely regain its softness by a process known as recrystallization (see Chapter 7) and this phenomenon is accompanied by the completion of the line-sharpening process. With continued annealing, the grains absorb each other to produce a structure with an overall coarser grain size and, because fewer reflections are available to contribute to the diffraction cones, the lines on the powder photograph take on a spotty appearance. This latter behaviour is sometimes used as a means of determining the grain size of a polycrystalline sample. In practice, an X-ray photograph is taken for each of a series of known grain sizes to form a set of standards, and with them an unknown grain size can be determined quite quickly by comparing the corresponding photograph with the set of standards. Yet a third use of the powder method as an inspection technique is in the detection of a preferred orientation of the grains of a polycrystalline aggregate. This is because a random orientation of the grains will produce a uniformly intense diffraction ring, while a preferred orientation, or texture, will concentrate the intensity at certain positions on the ring. The details of the texture require considerable interpretation and are discussed in Chapter 7.

5.3.3.3 X-ray diffractometry

In addition to photographic recording, the diffracted X-ray beam may be detected directly using a counter tube (either Geiger, proportional or scintillation type) with associated electrical circuitry. The geometrical arrangement of such an X-ray diffractometer is shown in Figure 5.13a. A divergent beam of filtered or monochromatized radiation impinges on the flat face of a powder specimen. This specimen is rotated at

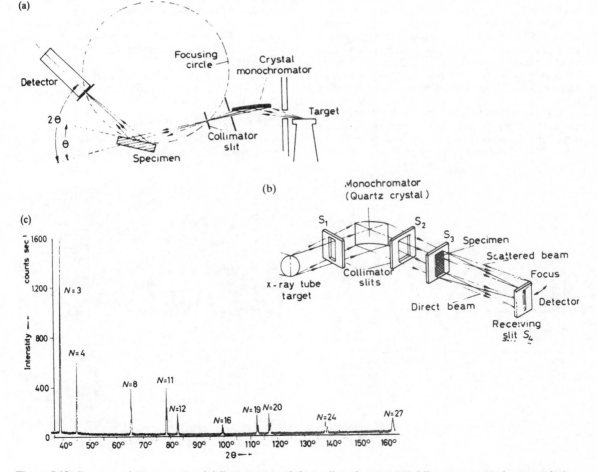

Figure 5.13 *Geometry of (a) conventional diffractometer and (b) small-angle scattering diffractometer, (c) chart record of diffraction pattern from aluminium powder with copper radiation using nickel filter.*

precisely one-half of the angular speed of the receiving slit so that a constant angle between the incident and reflected beams is maintained. The receiving slit is mounted in front of the counter on the counter tube arm, and behind it is usually fixed a scatter slit to ensure that the counter receives radiation only from the portion of the specimen illuminated by the primary beam. The intensity diffracted at the various angles is recorded automatically on a chart of the form shown in Figure 5.13c, and this can quickly be analysed for the appropriate θ and d values.

The technique is widely used in routine chemical analysis, since accurate intensity measurements allow a quantitative estimate of the various elements in the sample to be made. In research, the technique has been applied to problems such as the degree of order in alloys, the density of stacking faults in deformed alloys, elastic constant determination, the study of imperfections and preferred orientation.

5.3.3.4 X-ray topography

With X-rays it is possible to study individual crystal defects by detecting the differences in intensity diffracted by regions of the crystal near dislocations, for example, and more nearly perfect regions of the crystal. Figure 5.14a shows the experimental arrangement schematically in which collimated monochromatic $K\alpha$-radiation and photographic recording is used.

Any imperfections give rise to local changes in diffracted or transmitted X-ray intensities and, consequently, dislocations show up as bands of contrast, some 5 to 50 μm wide. No magnification is used in recording the diffraction image, but subsequent magnification of up to 500 times may be achieved with high-resolution X-ray emulsions. Large areas of the crystal to thicknesses of 10–100 μm can be mapped using scanning techniques provided the dislocation density is not too high ($\not> 10^{10}$ m^{-2}).

The X-ray method of detecting lattice defects suffers from the general limitations that the resolution is low and exposure times are long (12 h) although very high intensity X-ray sources are now available from synchrotrons and are being used increasingly with very short exposure times ($\sim$minutes). By comparison, the thin-film electron microscopy method (see Section 5.4.2) is capable of revealing dislocations with a much higher resolution because the dislocation image width is 10 nm or less and magnifications up to 100 000 times are possible. The X-ray method does, however, have the great advantage of being able to reveal dislocations in crystals which are comparatively thick ($\sim$1 mm, cf. 0.1 μm in foils suitable for transmission electron microscopy). The technique has been used for studying in detail the nature of dislocations in thick single crystals with very low dislocation densities, such as found in semiconducting materials; Figure 5.14b shows an example of an X-ray topograph revealing dislocations in magnesium by this technique.

5.3.4 Typical interpretative procedures for diffraction patterns

5.3.4.1 Intensity of diffraction

Many applications of the powder method depend on the accurate measurement of either line position or

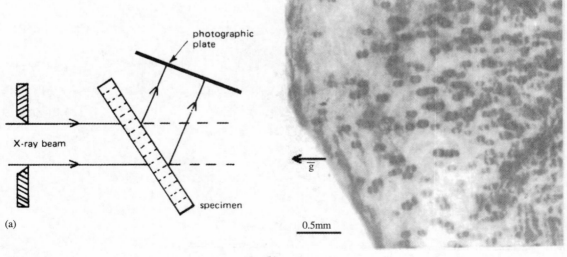

Figure 5.14 (a) Geometry of X-ray topographic technique, (b) topograph from a magnesium single crystal showing dislocation loops, $g = 0\ 1\ \bar{1}\ 0$ (after Vale and Smallman, 1977).

line intensity. The arrangement of the diffraction lines in any pattern is characteristic of the material being examined and, consequently, an important practical use of the method is in the identification of unknown phases. Thus, it will be evident that equation (5.9) can indicate the position of the reflected beams, as determined by the size and shape of the unit cell, but not the intensities of the reflected beams. These are determined not by the size of the unit cell but by the distribution of atoms within it, and while cubic lattices give reflections for every possible value of $(h^2 + k^2 + l^2)$ all other structures give characteristic absences. Studying the indices of the 'absent' reflections enables different structures to be distinguished.

In calculating the intensity scattered by a given atomic structure, we have first to consider the intensity scattered by one atom, and then go on to consider the contribution from all the other atoms in the particular arrangement which make up that structure. The efficiency of an atom in scattering X-rays is usually denoted by f, the atomic scattering factor, which is the ratio of amplitude scattered by an atom A_a to that by a single electron A_e. If atoms were merely points, their scattering factors would be equal to the number of electrons they contain, i.e. to their atomic numbers, and the relation $I_a = Z^2.I_e$ would hold since intensity is proportional to the square of amplitude. However, because the size of the atom is comparable to the wavelength of X-rays, scattering from different parts of the atom is not in phase, and the result is that $I_a \leq Z^2.I_e$. The scattering factor, therefore, depends both on angle θ and on the wavelength of X-rays used, as shown in Figure 5.9, because the path difference for the individual waves scattered from the various electrons in the atom is zero when $\theta = 0$ and increases with increasing θ. Thus, to consider the intensity scattered by a given structure, it is necessary to sum up the waves which come from all the atoms of one unit cell of that structure, since each wave has a different amplitude and a different phase angle due to the fact that it comes from a different part of the structure. The square of the amplitude of the resultant wave, F, then gives the intensity, and this may be calculated by using the f-values and the atomic coordinates of each atom in the unit cell. It can be shown that a general formula for the intensity is

$$
\begin{aligned}
I \propto |F|^2 = & [f_1 \cos 2\pi(hx_1 + ky_1 + lz_1) \\
& + f_2 \cos 2\pi(hx_2 + ky_2 + lz_2) + \cdots]^2 \\
& + [f_1 \sin 2\pi(hx_1 + ky_1 + lz_1) \\
& + f_2 \sin 2\pi(hx_2 + ky_2 + lz_2) + \cdots]^2
\end{aligned}
$$

$$(5.10)$$

where x_1, y_1, z_1; x_2, y_2, z_2, etc., are the coordinates of those atoms having scattering factors f_1, f_2, etc., respectively, and hkl are the indices of the reflection being computed. For structures having a centre of

symmetry, which includes most metals, the expression is much simpler because the sine terms vanish.

This equation may be applied to any structure, but to illustrate its use let us examine a pure metal crystallizing in the bcc structure. From Figure 2.11c it is clear that the structure has identical atoms (i.e. $f_1 = f_2$) at the coordinates $(0\,0\,0)$ and $\{\frac{1}{2}\,\frac{1}{2}\,\frac{1}{2}\}$ so that equation (5.10) becomes:

$$
\begin{aligned}
I \propto & f^2[\cos 2\pi.0 + \cos 2\pi(h/2 + k/2 + l/2)]^2 \\
& = f^2[1 + \cos \pi(h + k + l)]^2
\end{aligned}
$$

$$(5.11)$$

It then follows that I is equal to zero for every reflection having $(h + k + l)$ an odd number. The significance of this is made clear if we consider in a qualitative way the $1\,0\,0$ reflection shown in Figure 5.15a. To describe a reflection as the first-order reflection from $(1\,0\,0)$ planes implies that there is 1λ phase-difference between the rays reflected from planes A and those reflected from planes A'. However, the reflection from the plane B situated half-way between A and A' will be $\lambda/2$ out of phase with that from plane A, so that complete cancellation of the $1\,0\,0$ reflected ray will occur. The $1\,0\,0$ reflection is therefore absent, which agrees with the prediction made from equation (5.11) that the reflection is missing when $(h + k + l)$ is an odd number. A similar analysis shows that the $2\,0\,0$ reflection will be present (Figure 5.15b), since the ray from the B plane is now exactly 1λ out of phase with the rays from A and A'. In consequence, if a diffraction pattern is taken from a material having a bcc structure, because of the rule governing the sum of the indices, the film will show diffraction lines almost equally spaced with indices $N = 2$, $(1\,1\,0)$; 4, $(2\,0\,0)$; 6, $(2\,1\,1)$; 8, $(2\,2\,0)$; ..., as shown in Figure 5.12a. Application of equation (5.10) to a pure

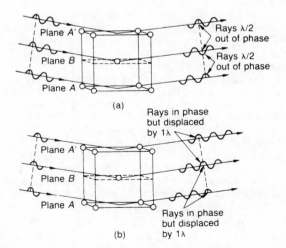

Figure 5.15 (a) $1\,0\,0$ reflection from bcc cell showing interference of diffracted rays, (b) $2\,0\,0$ reflection showing reinforcement (after Barrett, 1952; courtesy of McGraw-Hill).

metal with fcc structure shows that 'absent' reflections will occur when the indices of that reflection are mixed, i.e. when they are neither all odd nor all even. Thus, the corresponding diffraction pattern will contain lines according to $N = 3, 4, 8, 11, 12, 16, 19, 20$, etc; and the characteristic feature of the arrangement is a sequence of two lines close together and one line separated, as shown in Figure 5.12b.

Equation (5.10) is the basic equation used for determining unknown structures, since the determination of the atomic positions in a crystal is based on this relation between the coordinates of an atom in a unit cell and the intensity with which it will scatter X-rays.

5.3.4.2 Determination of lattice parameters

Perhaps the most common use of the powder method is in the accurate determination of lattice parameters. From the Bragg law we have the relation $a = \lambda\sqrt{N/2}\sin\theta$ which, because both λ and N are known and θ can be measured for the appropriate reflection, can be used to determine the lattice parameter of a material. Several errors are inherent in the method, however, and the most common include shrinkage of the film during processing, eccentricity of the specimen and the camera, and absorption of the X-rays in the sample. These errors affect the high-angle diffraction lines least and, consequently, the most accurate parameter value is given by determining a value of a from each diffraction line, plotting it on a graph against an angular function[1] of the $\cos^2\theta$-type and then extrapolating the curve to $\theta = 90°$.

The determination of precision lattice parameters is of importance in many fields of materials science, particularly in the study of thermal expansion coefficients, density determinations, the variation of properties with composition, precipitation from solid solution, and thermal stresses. At this stage it is instructive to consider the application of lattice parameter measurements to the determination of phase boundaries in equilibrium diagrams, since this illustrates the general usefulness of the technique. The diagrams shown in Figures 5.16a and 5.16b indicate the principle of the method. A variation of alloy composition within the single-phase field, α, produces a variation in the lattice parameter, a, since solute B, which has a different atomic size to the solvent A, is being taken into solution. However, at the phase boundary solvus this variation in a ceases, because at a given temperature the composition of the α-phase remains constant in the two-phase field, and the marked discontinuity in the plot of lattice parameter versus composition indicates the position of the phase boundary at that temperature. The change in solid solubility with temperature may then be obtained, either by taking diffraction photographs in a high-temperature

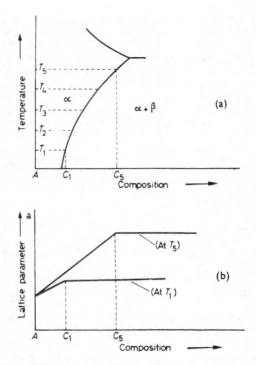

Figure 5.16 (a) and (b) Phase-boundary determination using lattice parameter measurements.

camera at various temperatures or by quenching the powder sample from the high temperature to room temperature (in order to retain the high temperature state of solid solution down to room temperature) and then taking a powder photograph at room temperature.

5.3.4.3 Line-broadening

Diffraction lines are not always sharp because of various instrumental factors such as slit size, specimen condition, and spread of wavelengths, but in addition the lines may be broadened as a result of lattice strain in the region of the crystal diffracting and also its limited dimension. Strain gives rise to a variation of the interplanar spacing Δd and hence diffraction occurs over a range $\Delta\theta$ and the breadth due to strain is then

$$\beta_s = \eta\tan\theta \tag{5.12}$$

where η is the strain distribution. If the dimension of the crystal diffracting the X-rays is small,[2] then this also gives rise to an appreciable 'particle-size' broadening given by the Scherrer formula

$$\beta_p = \lambda/t\cos\theta \tag{5.13}$$

---

[1] Nelson and Riley suggest the function

$$\left(\frac{\cos^2\theta}{\sin\theta} + \frac{\cos^2\theta}{\theta}\right)$$

[2] The optical analogue of this effect is the broadening of diffraction lines from a grating with a limited number of lines.

where t is the effective particle size. In practice this size is the region over which there is coherent diffraction and is usually defined by boundaries such as dislocation walls. It is possible to separate the two effects by plotting the experimentally measured broadening $\beta \cos \theta / \lambda$ against $\sin \theta / \lambda$, when the intercept gives a measure of t and the slope η.

5.3.4.4 Small-angle scattering

The scattering of intensity into the low-angle region ($\varepsilon = 2\theta < 10°$) arises from the presence of inhomogeneities within the material being examined (such as small clusters of solute atoms), where these inhomogeneities have dimensions only 10 to 100 times the wavelength of the incident radiation. The origin of the scattering can be attributed to the differences in electron density between the heterogeneous regions and the surrounding matrix,[1] so that precipitated particles afford the most common source of scattering; other heterogeneities such as dislocations, vacancies and cavities must also give rise to some small-angle scattering, but the intensity of the scattered beam will be much weaker than this from precipitated particles. The experimental arrangement suitable for this type of study is shown in Figure 5.13b.

Interpretation of much of the small-angle scatter data is based on the approximate formula derived by Guinier,

$$I = Mn^2 I_e \exp\left[-4\pi^2 \varepsilon^2 R^2 / 3\lambda^2\right] \quad (5.14)$$

where M is the number of scattering aggregates, or particles, in the sample, n represents the difference in number of electrons between the particle and an equal volume of the surrounding matrix, R is the radius of gyration of the particle, I_e is the intensity scattered by an electron, ε is the angle of scattering and λ is the wavelength of X-rays. From this equation it can be seen that the intensity of small-angle scattering is zero if the inhomogeneity, or cluster, has an electron density equivalent to that of the surrounding matrix, even if it has quite different crystal structure. On a plot of $\log_{10} I$ as a function of ε^2, the slope near the origin, $\varepsilon = 0$, is given by

$$P = -(4\pi^2 / 3\lambda^2) R^2 \log_{10} e$$

which for Cu $K\alpha$ radiation gives the radius of gyration of the scattering aggregate to be

$$R = 0.0645 \times P^{1/2} \text{nm} \quad (5.15)$$

It is clear that the technique is ideal for studying regions of the structure where segregation on too fine a scale to be observable in the light microscope has occurred, e.g. the early stages of phase precipitation

[1] The halo around the moon seen on a clear frosty night is the best example, obtained without special apparatus, of the scattering of light at small angles by small particles.

(see Chapter 8), and the aggregation of lattice defects (see Chapter 4).

5.3.4.5 The reciprocal lattice concept

The Bragg law shows that the conditions for diffraction depend on the geometry of sets of crystal planes. To simplify the more complex diffraction problems, use is made of the reciprocal lattice concept in which the sets of lattice planes are replaced by a set of points, this being geometrically simpler.

The reciprocal lattice is constructed from the real lattice by drawing a line from the origin normal to the lattice plane hkl under consideration of length, d^*, equal to the reciprocal of the interplanar spacing d_{hkl}. The construction of part of the reciprocal lattice from a face-centred cubic crystal lattice is shown in Figure 5.17.

Included in the reciprocal lattice are the points which correspond not only to the true lattice planes with Miller indices (hkl) but also to the fictitious planes (nh, nk, nl) which give possible X-ray reflections. The reciprocal lattice therefore corresponds to the diffraction spectrum possible from a particular crystal lattice and, since a particular lattice type is characterized by 'absent' reflections the corresponding spots in the reciprocal lattice will also be missing. It can be deduced that a fcc Bravais lattice is equivalent to a bcc reciprocal lattice, and vice versa.

A simple geometrical construction using the reciprocal lattice gives the conditions that correspond to Bragg reflection. Thus, if a beam of wavelength λ is incident on the origin of the reciprocal lattice, then a sphere of radius $1/\lambda$ drawn through the origin will intersect those points which correspond to the reflecting planes of a stationary crystal. This can be seen from Figure 5.18, in which the reflecting plane AB has a reciprocal point at d^*. If d^* lies on the surface of the sphere of radius $1/\lambda$ then

$$d^* = 1/d_{hkl} = 2 \sin \theta / \lambda \quad (5.16)$$

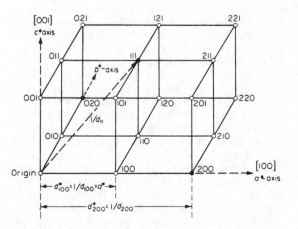

Figure 5.17 *fcc reciprocal lattice.*

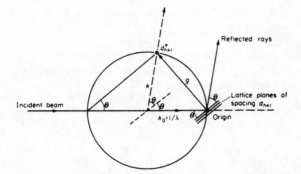

Figure 5.18 *Construction of the Ewald reflecting sphere.*

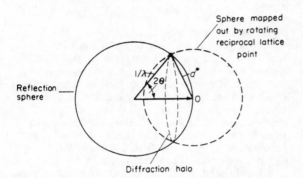

Figure 5.19 *Principle of the power method.*

and the Bragg law is satisfied; the line joining the origin to the operating reciprocal lattice spot is usually referred to as the g-vector. It will be evident that at any one setting of the crystal, few, if any, points will touch the sphere of reflection. This is the condition for a stationary single crystal and a monochromatic beam of X-rays, when the Bragg law is not obeyed except by chance. To ensure that the Bragg law is satisfied the crystal has to be rotated in the beam, since this corresponds to a rotation of the reciprocal lattice about the origin when each point must pass through the reflection surface. The corresponding reflecting plane reflects twice per revolution.

To illustrate this feature let us re-examine the powder method. In the powder specimen, the number of crystals is sufficiently large that all possible orientations are present and in terms of the reciprocal lattice construction we may suppose that the reciprocal lattice is rotated about the origin in all possible directions. The locus of any one lattice point during such a rotation is, of course, a sphere. This locus-sphere will intersect the sphere of reflection in a small circle about the axis of the incident beam as shown in Figure 5.19, and any line joining the centre of the reflection sphere to a point on this small circle is a possible direction for a diffraction maximum. This small circle corresponds to the powder halo discussed previously. From Figure 5.19 it can be seen that the radius of the sphere describing the locus of the reciprocal lattice point (hkl) is $1/d_{(hkl)}$ and that the angle of deviation of the diffracted beam 2θ is given by the relation

$$(2/\lambda)\sin\theta = 1/d_{(hkl)}$$

which is the Bragg condition.

5.4 Analytical electron microscopy

5.4.1 Interaction of an electron beam with a solid

When an electron beam is incident on a solid specimen a number of interactions take place which generate useful structural information. Figure 5.20 illustrates these interactions schematically. Some of the incident beam is back-scattered and some penetrates the sample. If the specimen is thin enough a significant amount is transmitted, with some electrons elastically scattered without loss of energy and some inelastically scattered. Interaction with the atoms in the specimen leads to the ejection of low-energy electrons and the creation of X-ray photons and Auger electrons, all of which can be used to characterize the material.

The two inelastic scattering mechanisms important in chemical analysis are (1) excitation of the electron gas plasmon scattering, and (2) single-electron scattering. In *plasmon scattering* the fast electron excites a ripple in the plasma of free electrons in the solid. The energy of this 'plasmon' depends only on the volume concentration of free electrons n in the solid and given by $E_p = [ne^2/m]^{1/2}$. Typically E_p, the energy loss suffered by the fast electron is ≈ 15 eV and the scattering intensity/unit solid angle has an angular half-width given by $\theta_E = E_p/2E_0$, where E_0 is the incident voltage; θ_E is therefore $\approx 10^{-4}$ radian. The energy

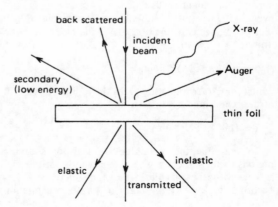

Figure 5.20 *Scattering of incident electrons by thin foil. With a bulk specimen the transmitted, elastic and inelastic scattered beams are absorbed.*

of the plasmon is converted very quickly into atom vibrations (heat) and the mean-free path for plasmon excitation is small, ≈ 50–150 nm. With *single-electron scattering* energy may be transferred to single electrons (rather than to the large number $\approx 10^5$ involved in plasmon excitation) by the incident fast electrons. Lightly-bound valency electrons may be ejected, and these electrons can be used to form secondary images in SEM; a very large number of electrons with energies up to ≈ 50 eV are ejected when a high-energy electron beam strikes a solid. The useful collisions are those where the single electron is bound. There is a minimum energy required to remove the single electron, i.e. ionization, but provided the fast electron gives the bound electron more than this minimum amount, it can give the bound electron any amount of energy, up to its own energy (e.g. $1\,0\,0$ keV). Thus, instead of the single-electron excitation process turning up in the energy loss spectrum of the fast electron as a peak, as happens with plasmon excitation, it turns up as an edge. Typically, the mean free path for inner shell ionization is several micrometres and the energy loss can be several keV. The angular half-width of scattering is given by $\Delta E/2E_0$. Since the energy loss ΔE can vary from ≈ 10 eV to tens of keV the angle can vary upwards from 10^{-4} radian (see Figure 5.36).

A plasmon, once excited, decays to give heat, which is not at all useful. In contrast, an atom which has had an electron removed from it decays in one of two ways, both of which turn out to be very useful in chemical analysis leading to the creation of X-rays and Auger electrons. The first step is the same for both cases. An electron from outer shell, which therefore has more energy than the removed electron, drops down to fill the hole left by the removal of the bound electron. Its extra energy, ΔE, equal to the difference in energy between the two levels involved and therefore absolutely characteristic of the atom, must be dissipated. This may happen in two ways: (1) by the creation of a photon whose energy, $h\nu$, equals the energy difference ΔE. For electron transitions of interest, ΔE, and therefore $h\nu$, is such that the photon is an X-ray, (2) by transferring the energy to a neighbouring electron, which is then ejected from the atom. This is an 'Auger' electron. Its energy when detected will depend on the original energy difference ΔE minus the binding energy of the ejected electron. Thus the energy of the Auger electron depends on three atomic levels rather than two as for emitted photons. The energies of the Auger electrons are sufficiently low that they escape from within only about 5 nm of the surface. This is therefore a surface analysis technique. The ratio of photon–Auger yield is called the fluorescence ratio ω, and depends on the atom and the shells involved. For the K-shell, ω is given by $\omega_K = X_K/(A_K + X_K)$, where X_K and A_K are, respectively, the number of X-ray photons and Auger electrons emitted. A_K is independent of atomic number Z, and X_K is proportional to Z^4 so that $\omega_K Z^4/(a + Z^4)$, where $a = 1.12 \times 10^6$. Light elements and outer shells (L-lines) have lower yields; for

K-series transitions ω_K varies from a few per cent for carbon up to $\geq 90\%$ for gold.

5.4.2 The transmission electron microscope (TEM)

Section 5.2.1 shows that to increase the resolving power of a microscope it is necessary to employ shorter wavelengths. For this reason the electron microscope has been developed to allow the observation of structures which have dimensions down to less than 1 nm. An electron microscope consists of an electron gun and an assembly of lenses all enclosed in an evacuated column. A very basic system for a transmission electron microscope is shown schematically in Figure 5.21. The optical arrangement is similar to that of the glass lenses in a projection-type light microscope, although it is customary to use several stages of magnification in the electron microscope. The lenses are usually of the magnetic type, i.e. current-carrying coils which are completely surrounded by a soft iron shroud except for a narrow gap in the bore, energized by d.c. and, unlike the lenses in a light microscope, which have fixed focal lengths, the focal length can be controlled by regulating the current through the coils of the lens.

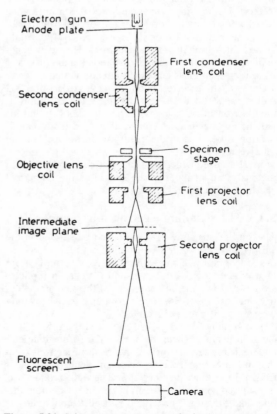

Figure 5.21 *Schematic arrangement of a basic transmission electron microscope system.*

This facility compensates for the fact that it is difficult to move the large magnetic lenses in the evacuated column of the electron microscope in an analogous manner to the glass lenses in a light microscope.

The condenser lenses are concerned with collimating the electron beam and illuminating the specimen which is placed in the bore of the objective lens. The function of the objective lens is to form a magnified image of up to about 40× in the object plane of the intermediate, or first projector lens. A small part of this image then forms the object for the first projector lens, which gives a second image, again magnified in the object plane of the second projector lens. The second projector lens is capable of enlarging this image further to form a final image on the fluorescent viewing screen. This image, magnified up to 100 000× may be recorded on a photographic film beneath the viewing screen. A stream of electrons can be assigned a wavelength λ given by the equation $\lambda = \mathbf{h}/mv$, where $\mathbf{h}$ is Planck's constant and mv is the and hence to the voltage applied to the electron gun, according to the approximate relation

$$\lambda = \sqrt{(1.5/V)}\,\text{nm} \qquad (5.17)$$

and, since normal operating voltages are between 50 and 100 kV, the value of λ used varies from 0.0054 nm to 0.0035 nm. With a wavelength of 0.005 nm if one could obtain a value of $(\mu \sin\alpha)$ for electron lenses comparable to that for optical lenses, i.e. 1.4, it would be possible to see the orbital electrons. However, magnetic lenses are more prone to spherical and chromatic aberration than glass lenses and, in consequence, small apertures, which correspond to α-values of about 0.002 radian, must be used. As a result, the resolution of the electron microscope is limited to about 0.2 nm. It will be appreciated, of course, that a variable magnification is possible in the electron microscope without relative movement of the lenses, as in a light microscope, because the depth of focus of each image, being inversely proportional to the square of the numerical aperture, is so great.

5.4.3 The scanning electron microscope

The surface structure of a metal can be studied in the TEM by the use of thin transparent replicas of the surface topography. Three different types of replica are in use, (1) oxide, (2) plastic, and (3) carbon replicas. However, since the development of the scanning electron microscope (SEM) it is very much easier to study the surface structure directly.

A diagram of the SEM is shown in Figure 5.22. The electron beam is focused to a spot ≈10 nm diameter and made to scan the surface in a raster. Electrons from the specimen are focused with an electrostatic electrode on to a biased scintillator. The light produced is transmitted via a *Perspex* light pipe to a photomultiplier and the signal generated is used to modulate the brightness of an oscilloscope spot which traverses a raster in exact synchronism with the electron beam at

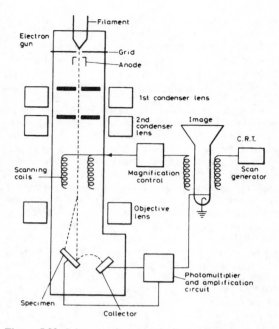

Figure 5.22 *Schematic diagram of a basic scanning electron microscope (courtesy of Cambridge Instrument Co.).*

the specimen surface. The image observed on the oscilloscope screen is similar to the optical image and the specimen is usually tilted towards the collector at a low angle (<30°) to the horizontal, for general viewing.

As initially conceived, the SEM used backscattered electrons (with $E \approx 30$ kV which is the incident energy) and secondary electrons ($E \approx 100$ eV) which are ejected from the specimen. Since the secondary electrons are of low energy they can be bent round corners and give rise to the topographic contrast. The intensity of backscattered (BS) electrons is proportional to atomic number but contrast from these electrons tends to be swamped because, being of higher energy, they are not so easily collected by the normal collector system used in SEMs. If the secondary electrons are to be collected a positive bias of ≈200 V is applied to the grid in front of the detector; if only the back-scattered electrons are to be collected the grid is biased negatively to ≈200 V.

Perhaps the most significant development in recent years has been the gathering of information relating to chemical composition. As discussed in Section 5.4.1, materials bombarded with high-energy electrons can give rise to the emissions of X-rays characteristic of the material being bombarded. The X-rays emitted when the beam is stopped on a particular region of the specimen may be detected either with a solid-state (Li-drifted silicon) detector which produces a voltage pulse proportional to the energy of the incident photons (energy-dispersive method) or with an X-ray spectrometer to measure the wavelength and intensity (wavelength-dispersive method). The microanalysis of materials is presented in Section 5.4.5. Alternatively, if the beam is scanned as usual and the intensity of the X-ray emission, characteristic of a particular element, is used to modulate the CRT, an image showing the distribution of that element in the sample will result. X-ray images are usually very 'noisy' because the X-ray production efficiency is low, necessitating exposures a thousand times greater than electron images.

Collection of the back-scattered (BS) electrons with a specially located detector on the bottom of the lens system gives rise to some exciting applications and opens up a completely new dimension for SEM from bulk samples. The BS electrons are very sensitive to atomic number Z and hence are particularly important in showing contrast from changes of composition, as illustrated by the image from a silver alloy in Figure 5.23. This atomic number contrast is particularly effective in studying alloys which normally are difficult to study because they cannot be etched. The intensity of back-scattered electrons is also sensitive to the orientation of the incident beam relative to the crystal. This effect will give rise to 'orientation' contrast from grain to grain in a polycrystalline specimen as the scan crosses several grains. In addition, the effect is also able to provide crystallographic information from bulk specimens by a process known as electron channelling. As the name implies, the electrons are channelled between crystal planes and the amount of channelling per plane depends on its packing and spacing. If the electron beam impinging on a crystal is rocked through a large angle then the amount of channelling will vary with angle and hence the BS image will exhibit contrast in the form of electron channelling patterns which can be used to provide crystallographic information. Figure 5.24 shows the 'orientation' or channelling contrast exhibited by a Fe–3%Si

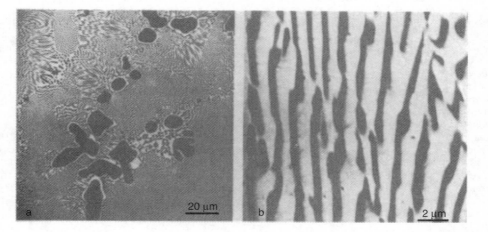

Figure 5.23 *Back-cattered electron image by atomic number contrast from 70Ag–30Cu alloy showing (a) α-dendrites + eutectic and (b) eutectic (courtesy of B. W. Hutchinson).*

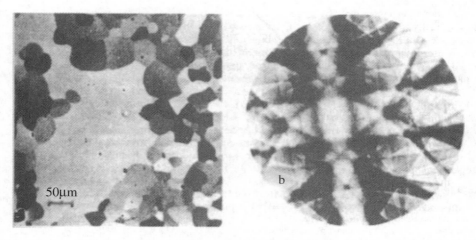

Figure 5.24 *(a) Back-scattered electron image and (b) associated channelling pattern, from secondary recrystallized Fe–3%Si (courtesy of B. W. Hutchinson).*

specimen during secondary recrystallization (a process used for transformer lamination production) and the channelling pattern can be analysed to show that the new grain possesses the Goss texture. Electron channelling occurs only in relatively perfect crystals and hence the degradation of electron channelling patterns may be used to monitor the level of plastic strain, for example to map out the plastic zone around a fatigue crack as it develops in an alloy.

The electron beam may also induce electrical effects which are of importance particularly in semiconductor materials. Thus a 30 kV electron beam can generate some thousand excess free electrons and the equivalent number of ions ('holes'), the vast majority of which recombine. In metals, this recombination process is very fast (1 ps) but in semiconductors may be a few seconds depending on purity. These excess current carriers will have a large effect on the limited conductivity. Also the carriers generated at one point will diffuse towards regions of lower carrier concentration and voltages will be established whenever the carriers encounter regions of different chemical composition (e.g. impurities around dislocations). The conductivity effect can be monitored by applying a potential difference across the specimen from an external battery and using the magnitude of the resulting current to modulate the CRT brightness to give an image of conductivity variation.

The voltage effect arising from different carrier concentrations or from accumulation of charge on an insulator surface or from the application of an external electromotive force can modify the collection of the emitted electrons and hence give rise to voltage contrast. Similarly, a magnetic field arising from ferromagnetic domains, for example, will affect the collection efficiency of emitted electrons and lead to magnetic field contrast.

The secondary electrons, i.e. lightly-bound electrons ejected from the specimen which give topographical information, are generated by the incident electrons, by the back-scattered electrons and by X-rays. The resolution is typically ≈ 10 nm at 20 kV for medium atomic weight elements and is limited by spreading of electrons as they penetrate into the specimen. The back-scattered electrons are also influenced by beam spreading and for a material of medium atomic weight the resolution is ≈ 100 nm. The specimen current mode is limited both by spreading of the beam and the noise of electronic amplification to a spatial resolution of 500 nm and somewhat greater values ≈ 1 µm apply to the beam-induced conductivity and X-ray modes.

5.4.4 Theoretical aspects of TEM

5.4.4.1 Imaging and diffraction

Although the examination of materials may be carried out with the electron beam impinging on the surface at a 'glancing incidence', most electron microscopes are aligned for the use of a transmission technique, since added information on the interior of the specimen may be obtained. In consequence, the thickness of the metal specimen has to be limited to below a micrometre, because of the restricted penetration power of the electrons. Three methods now in general use for preparing such thin films are (1) chemical thinning, (2) electropolishing, and (3) bombarding with a beam of ions at a potential of about 3 kV. Chemical thinning has the disadvantage of preferentially attacking either the matrix or the precipitated phases, and so the electropolishing technique is used extensively to prepare thin metal foils. Ion beam thinning is quite slow but is the only way of preparing thin ceramic and semiconducting specimens.

Transmission electron microscopy provides both image and diffraction information from the same small volume down to 1 µm in diameter. Ray diagrams for the two modes of operation, imaging and diffraction, are shown in Figure 5.25. Diffraction contrast[1] is the most common technique used and, as shown in Figure 5.25a, involves the insertion of an objective aperture in the back focal plane, i.e. in the plane in which the diffraction pattern is formed, to select either the directly-transmitted beam or a strong diffracted beam. Images obtained in this way cannot possibly contain information concerning the periodicity of

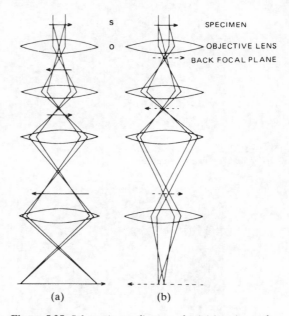

Figure 5.25 *Schematic ray diagrams for (a) imaging and (b) diffraction.*

[1] Another imaging mode does allow more than one beam to interfere in the image plane and hence crystal periodicity can be observed; the larger the collection angle, which is generally limited by lens aberrations, the smaller the periodicity that can be resolved. Interpretation of this direct imaging mode, while apparently straightforward, is still controversial, and will not be covered here.

the crystal, since this information is contained in the spacing of diffraction maxima and the directions of diffracted beams, information excluded by the objective aperture.

Variations in intensity of the selected beam is the only information provided. Such a mode of imaging, carried out by selecting one beam in TEM, is unusual and the resultant images cannot be interpreted simply as high-magnification images of periodic objects. In formulating a suitable theory it is necessary to consider what factors can influence the intensity of the directly-transmitted beam and the diffracted beams. The obvious factors are (1) local changes in scattering factor, e.g. particles of heavy metal in light metal matrix, (2) local changes in thickness, (3) local changes in orientation of the specimen, or (4) discontinuities in the crystal planes which give rise to the diffracted beams. Fortunately, the interpretation of any intensity changes is relatively straightforward if it is assumed that there is only one strong diffracted beam excited. Moreover, since this can be achieved quite easily experimentally, by orienting the crystal such that strong diffraction occurs from only one set of crystal planes, virtually all TEM is carried out with a two-beam condition: a direct and a diffracted beam. When the direct, or transmitted, beam only is allowed to contribute to the final image by inserting a small aperture in the back focal plane to block the strongly diffracted ray, then contrast is shown on a bright background and is known as bright-field imaging. If the diffracted ray only is allowed through the aperture by tilting the incident beam then contrast on a dark background is observed and is known as dark-field imaging. These two arrangements are shown in Figure 5.26.

A dislocation can be seen in the electron microscope because it locally changes the orientation of the crystal, thereby altering the diffracted intensity. This is illustrated in Figure 5.27. Any region of a grain or crystal which is not oriented at the Bragg angle, i.e. $\theta > \theta_B$, is not strongly diffracting electrons. However, in the vicinity of the dislocation the lattice planes are tilted such that locally the Bragg law is satisfied and then

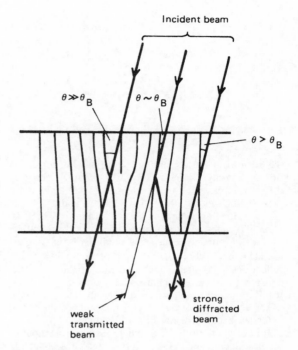

Figure 5.27 *Mechanism of diffraction contrast: the planes to the RHS of the dislocation are bent so that they closely approach the Bragg condition and the intensity of the direct beam emerging from the crystal is therefore reduced.*

strong diffraction arises from near the defect. These diffracted rays are blocked by the objective aperture and prevented from contributing to the final image. The dislocation therefore appears as a dark line (where electrons have been removed) on a bright background in the bright-field picture.

The success of transmission electron microscopy (TEM) is due, to a great extent, to the fact that it is possible to define the diffraction conditions which give rise to the dislocation contrast by obtaining a diffraction pattern from the same small volume of crystal (as small as 1 μm diameter) as that from which the electron micrograph is taken. Thus, it is possible to obtain the crystallographic and associated diffraction information necessary to interpret electron micrographs. To obtain a selected area diffraction pattern (SAD) an aperture is inserted in the plane of the first image so that only that part of the specimen which is imaged within the aperture can contribute to the diffraction pattern. The power of the diffraction lens is then reduced so that the back focal plane of the objective is imaged, and then the diffraction pattern, which is focused in this plane, can be seen after the objective aperture is removed.

The usual type of transmission electron diffraction pattern from a single crystal region is a cross-grating pattern of the form shown in Figure 5.28. The simple explanation of the pattern can be given by considering

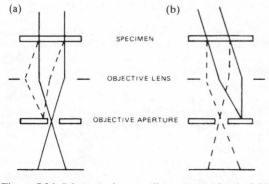

Figure 5.26 *Schematic diagram illustrating (a) bright-field and (b) dark-field image formation.*

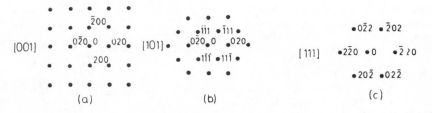

Figure 5.28 *fcc cross-grating patterns (a) [0 0 1], (b) [1 0 1] and (c) [1 1 1].*

the reciprocal lattice and reflecting sphere construction commonly used in X-ray diffraction. In electron diffraction, the electron wavelength is extremely short ($\lambda = 0.0037$ nm at 100 kV) so that the radius of the Ewald reflecting sphere is about 2.5 nm^{-1}, which is about 50 times greater than g, the reciprocal lattice vector. Moreover, because λ is small the Bragg angles are also small (about 10^{-2} radian or $\frac{1}{2}°$ for low-order reflections) and hence the reflection sphere may be considered as almost planar in this vicinity. If the electron beam is closely parallel to a prominent zone axis of the crystal then several reciprocal points (somewhat extended because of the limited thickness of the foil) will intersect the reflecting sphere, and a projection of the prominent zone in the reciprocal lattice is obtained, i.e. the SAD pattern is really a photograph of a reciprocal lattice section. Figure 5.28 shows some standard cross-grating for fcc crystals. Because the Bragg angle for reflection is small ($\approx \frac{1}{2}°$) only those lattice planes which are almost vertical, i.e. almost parallel to the direction of the incident electron beam, are capable of Bragg-diffracting the electrons out of the objective aperture and giving rise to image contrast. Moreover, because the foil is buckled or purposely tilted, only one family of the various sets of approximately vertical lattice planes will diffract strongly and the SAD pattern will then show only the direct beam spot and one strongly diffracted spot (see insert Figure 5.40). The indices g of the crystal planes (hkl) which are set at the Bragg angle can be obtained from the SAD. Often the planes are near to, but not exactly at, the Bragg angle and it is necessary to determine the precise deviation which is usually represented by the parameter s, as shown in the Ewald sphere construction in Figure 5.29. The deviation parameter s is determined from Kikuchi lines, observed in diffraction patterns obtained from somewhat thicker areas of the specimen, which form a pair of bright and dark lines associated with each reflection, spaced $|g|$ apart.

The Kikuchi lines arise from inelastically-scattered rays, originating at some point P in the specimen (see Figure 5.30), being subsequently Bragg-diffracted. Thus, for the set of planes in Figure 5.30a, those electrons travelling in the directions PQ and PR will be Bragg-diffracted at Q and R and give rise to rays in the directions QQ′ and RR′. Since the electrons in the beam RR′ originate from the scattered ray PR, this beam will be less intense than QQ′, which

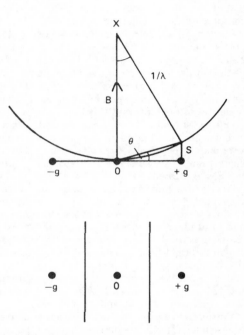

Figure 5.29 *Schematic diagram to illustrate the determination of s at the symmetry position, together with associated diffraction pattern.*

contains electrons scattered through a smaller angle at P. Because P is a spherical source this rediffraction at points such as Q and R gives rise to cones of rays which, when they intersect the film, approximate to straight lines.

The selection of the diffracting conditions used to image the crystal defects can be controlled using Kikuchi lines. Thus the planes (hkl) are at the Bragg angle when the corresponding pair of Kikuchi lines passes through 0 0 0 and g_{hkl}, i.e. $s = 0$. Tilting of the specimen so that this condition is maintained (which can be done quite simply, using modern double-tilt specimen stages) enables the operator to select a specimen orientation with a close approximation to two-beam conditions. Tilting the specimen to a particular orientation, i.e. electron beam direction, can also be selected using the Kikuchi lines as a 'navigational' aid. The series of Kikuchi lines make up a Kikuchi map, as shown in Figure 5.30b, which can be used to

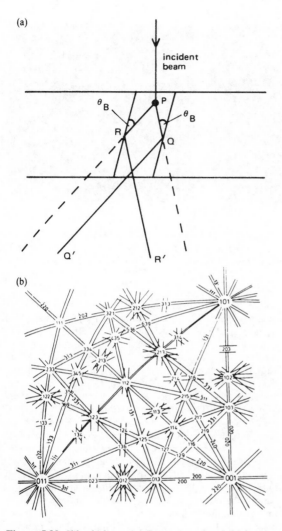

Figure 5.30 *Kikuchi lines. (a) Formation of and (b) from fcc crystal forming a Kikuchi map.*

tilt from one pole to another (as one would use an Underground map).

5.4.4.2 Convergent beam diffraction patterns

When a selected area diffraction pattern is taken with a convergent beam of electrons, the resultant pattern contains additional structural information. A ray diagram illustrating the formation of a convergent beam diffraction pattern (CBDP) is shown in Figure 5.31a. The discs of intensity which are formed in the back focal plane contain information which is of three types:

1. Fringes within discs formed by strongly diffracted beams. If the crystal is tilted to 2-beam conditions, these fringes can be used to determine the specimen thickness very accurately.

2. High-angle information in the form of fine lines (somewhat like Kikuchi lines) which are visible in the direct beam and in the higher-order Laue zones (HOLZ). These HOLZ are visible in a pattern covering a large enough angle in reciprocal space. The fine line pattern can be used to measure the lattice parameter to 1 in 10^4. Figure 5.31b shows an example of HOLZ lines for a silicon crystal centred [1 1 1]. Pairing a dark line through the zero-order disc with its corresponding bright line through the higher-order disc allows the lattice parameter to be determined, the distance between the pair being sensitive to the temperature, etc.

3. Detailed structure both within the direct beam and within the diffracted beams which show certain well-defined symmetries when the diffraction pattern is taken precisely along an important zone axis. The patterns can therefore be used to give crystal structure information, particularly the point group and space group. This information, together with the chemical composition from EELS or EDX, and the size of the unit cell from the indexed diffraction patterns can be used to define the specific crystal structure, i.e. the atomic positions. Figure 5.31c indicates the threefold symmetry in a CBDP from silicon taken along the [1 1 1] axis.

5.4.4.3 Higher-voltage electron microscopy

The most serious limitation of conventional transmission electron microscopes (CTEM) is the limited thickness of specimens examined (50–500 nm). This makes preparation of samples from heavy elements difficult, gives limited containment of particles and other structural features within the specimen, and restricts the study of dynamical processes such as deformation, annealing, etc., within the microscope. However, the usable specimen thickness is a function of the accelerating voltage and can be increased by the use of higher voltages. Because of this, higher-voltage microscopes (HVEM) have been developed.

The electron wavelength λ decreases rapidly with voltage and at 1000 kV the wavelength $\lambda \approx 0.001$ nm. The decrease in λ produces corresponding decreases in the Bragg angles θ, and hence the Bragg angles at 1000 kV are only about one third of their corresponding values at 100 kV. One consequence of this is that an additional projector lens is usually included in high-voltage microscope. This is often called the diffraction lens and its purpose is to increase the diffraction camera length so that the diffraction spots are more widely spaced on the photographic plate.

The principal advantages of HVEM are: (1) an increase in usable foil thickness and (2) a reduced ionization damage rate in ionic, polymer and biological specimens. The range of materials is therefore widened and includes (1) materials which are difficult to prepare as thin foils, such as tungsten and uranium and (2) materials in which the defect being studied is too large to be conveniently included within a 100 kV

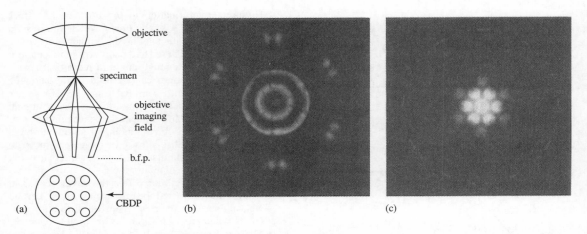

Figure 5.31 *(a) Schematic formation of convergent beam diffraction pattern in the backfocal plane of the objective lens, (b) and (c) ⟨1 1 1⟩ CBDPs from Si; (b) zero layer and HOLZ (Higher Order Laue Zones) in direct beam and (c) zero layer + FOLZ (First Order Laue Zones).*

specimen; these include large voids, precipitates and some dislocation structures such as grain boundaries.

Many processes such as recrystallization, deformation, recovery, martensitic transformation, etc. are dominated by the effects of the specimen surfaces in thin samples and the use of thicker foils enables these phenomena to be studied as they occur in bulk materials. With thicker foils it is possible to construct intricate stages which enable the specimen to be cooled, heated, strained and exposed to various chemical environments while it is being looked through.

A disadvantage of HVEM is that as the beam voltage is raised the energy transferred to the atom by the fast electron increases until it becomes sufficient to eject the atom from its site. The amount of energy transferred from one particle to another in a collision depends on the ratio of the two masses (see Chapter 4). Because the electron is very light compared with an atom, the transfer of energy is very inefficient and the electron needs to have several hundred keV before it can transmit the 25 eV or so necessary to displace an atom. To avoid radiation damage it is necessary to keep the beam voltage below the critical displacement value which is ≈100 kV for Mg and ≈1300 kV for Au. There is, however, much basic scientific interest in radiation damage for technological reasons and a HVEM enables the damage processes to be studied directly.

5.4.5 Chemical microanalysis

5.4.5.1 Exploitation of characteristic X-rays

Electron probe microanalysis (EPMA) of bulk samples is now a routine technique for obtaining rapid, accurate analysis of alloys. A small electron probe (≈100 nm diameter) is used to generate X-rays from a defined area of a polished specimen and the intensity of the various characteristic X-rays measured using either wavelength-dispersive spectrometers (WDS) or energy-dispersive spectrometers (EDS). Typically the accuracy of the analysis is ±0.1%. One of the limitations of EPMA of bulk samples is that the volume of the sample which contributes to the X-ray signal is relatively independent of the size of the electron probe, because high-angle elastic scattering of electrons within the sample generates X-rays (see Figure 5.32). The consequence of this is that the spatial resolution of EPMA is no better than ∼2 μm. In the last few years EDX detectors have been interfaced to transmission electron microscopes which are capable of operating with an electron probe as small as 2 nm. The combination of electron-transparent samples, in which high-angle elastic scattering is limited, and a small electron probe leads to a significant improvement in the potential spatial resolution of X-ray microanalysis. In addition, interfacing of energy loss spectrometers has enabled light elements to be detected and measured, so that electron microchemical analysis is now a powerful tool in the characterization of materials. With electron beam instrumentation it is required to measure (1) the wavelength or energies of emitted X-rays (WDX and EDX), (2) the energy losses of the fast electrons (EELS), and (3) the energies of emitted electrons (AES). Nowadays (1) and (2) can be carried out on the modern TEM using special detector systems, as shown schematically in Figure 5.33.

In a WDX spectrometer a crystal of known *d*-spacing is used which diffracts X-rays of a specific wavelength, λ, at an angle θ, given by the Bragg equation, $n\lambda = 2d \sin \theta$. Different wavelengths are selected by changing θ and thus to cover the necessary range of wavelengths, several crystals of different *d*-spacings are used successively in a spectrometer. The range of wavelength is 0.1–2.5 nm and the corresponding *d*-spacing for practicable values of θ, which

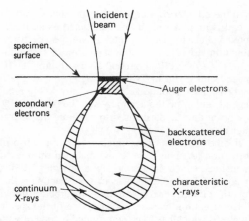

Figure 5.32 *Schematic diagram showing the generation of electrons and X-rays within the specimen.*

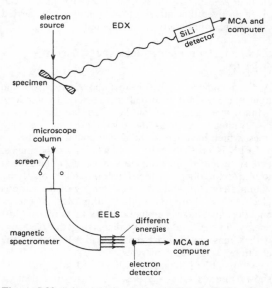

Figure 5.33 *Schematic diagram of EDX and EELS in TEM.*

lie between $\approx 15°$ and $65°$, is achieved by using crystals such as LiF, quartz, mica, etc. In a WDX spectrometer the specimen (which is the X-ray source), a bent crystal of radius $2r$ and the detector all lie on the focusing circle radius r and different wavelength X-rays are collected by the detector by setting the crystal at different angles, θ. The operation of the spectrometer is very time-consuming since only one particular X-ray wavelength can be focused on to the detector at any one time.

The resolution of WDX spectrometers is controlled by the perfection of the crystal, which influences the range of wavelengths over which the Bragg condition is satisfied, and by the size of the entrance slit to the X-ray detector; taking the resolution ($\Delta\lambda$) to ~ 0.001 nm

then $\lambda/\Delta\lambda$ is about 300 which, for a medium atomic weight sample, leads to a peak–background ratio of about 250. The crystal spectrometer normally uses a proportional counter to detect the X-rays, producing an electrical signal, by ionization of the gas in the counter, proportional to the X-ray energy, i.e. inversely proportional to the wavelength. The window of the counter needs to be thin and of low atomic number to minimize X-ray absorption. The output pulse from the counter is amplified and differentiated to produce a short pulse. The time constant of the electrical circuit is of the order of 1 μs which leads to possible count rates of at least $10^5/\text{s}$.

In recent years EDX detectors have replaced WDX detectors on transmission microscopes and are used together with WDX detectors on microprobes and on SEMs. A schematic diagram of a Si–Li detector is shown in Figure 5.34. X-rays enter through the thin Be window and produce electron-hole pairs in the Si–Li. Each electron-hole pair requires 3.8 eV, at the operating temperature of the detector, and the number of pairs produced by a photon of energy E_p is thus $E_p/3.8$. The charge produced by a typical X-ray photon is $\approx 10^{-16}$ C and this is amplified to give a shaped pulse, the height of which is then a measure of the energy of the incident X-ray photon. The data are stored in a multi-channel analyser. Provided that the X-ray photons arrive with a sufficient time interval between them, the energy of each incident photon can be measured and the output presented as an intensity versus energy display. The amplification and pulse shaping takes about 50 μs and if a second pulse arrives before the preceding pulse is processed, both pulses are rejected. This results in significant dead time for count rates $\geq 4000/\text{s}$.

The number of electron-hole pairs generated by an X-ray of a given energy is subject to normal statistical fluctuations and this, taken together with electronic noise, limits the energy resolution of a Si–Li detector to about a few hundred eV, which worsens with increase in photon energy. The main advantage of EDX detectors is that simultaneous collection of the whole range of X-rays is possible and the energy characteristics of all the elements $>Z = 11$ in the Periodic Table can be obtained in a matter of seconds. The main

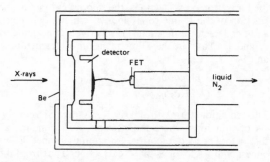

Figure 5.34 *Schematic diagram of Si–Li X-ray detector.*

disadvantages are the relatively poor resolution, which leads to a peak-background ratio of about 50, and the limited count rate.

The variation in efficiency of a Si–Li detector must be allowed for when quantifying X-ray analysis. At low energies (≤ 1 kV) the X-rays are mostly absorbed in the Be window and at high energies (≥ 20 kV), the X-rays pass through the detector so that the decreasing cross-section for electron-hole pair generation results in a reduction in efficiency. The Si–Li detector thus has optimum detection efficiency between about 1 and 20 kV.

5.4.5.2 Electron microanalysis of thin foils

There are several simplifications which arise from the use of thin foils in microanalysis. The most important of these arises from the fact that the average energy loss which electrons suffer on passing through a thin foil is only about 2%, and this small average loss means that the ionization cross-section can be taken as a constant. Thus the number of characteristic X-ray photons generated from a thin sample is given simply by the product of the electron path length and the appropriate cross-section Q, i.e. the probability of ejecting the electron, and the fluorescent yield ω. The intensity generated by element A is then given by

$$I_A = iQ\omega n$$

where Q is the cross-section per cm^2 for the particular ionization event, ω the fluorescent yield, n the number of atoms in the excited volume, and i the current incident on the specimen. Microanalysis is usually carried out under conditions where the current is unknown and interpretation of the analysis simply requires that the ratio of the X-ray intensities from the various elements be obtained. For the simple case of a very thin specimen for which absorption and X-ray fluorescence can be neglected, then the measured X-ray intensity from element A is given by

$$I_A \propto n_A Q_A \omega_A a_A \eta_A$$

and for element B by

$$I_B \propto n_B Q_B \omega_B a_B \eta_B$$

where n, Q, ω, a and η represent the number of atoms, the ionization cross-sections, the fluorescent yields, the fraction of the K line (or L and M) which is collected and the detector efficiencies, respectively, for elements A and B. Thus in the alloy made up of elements A and B

$$\frac{n_A}{n_B} \propto \frac{I_A Q_B \omega_B a_B \eta_B}{I_B Q_A \omega_A a_A \eta_A} = K_{AB} \frac{I_A}{I_B}$$

This equation forms the basis for X-ray microanalysis of thin foils where the constant K_{AB} contains all the factors needed to correct for atomic number differences, and is known as the Z-correction. Thus from the measured intensities, the ratio of the number of atoms

A to the number of atoms B, i.e. the concentrations of A and B in an alloy, can be calculated using the computed values for Q, ω, η, etc. A simple spectrum for stoichiometric NiAl is shown in Figure 5.35 and the values of I_K^{Al} and I_K^{Ni}, obtained after stripping the background, are given in Table 5.2 together with the final analysis. The absolute accuracy of any X-ray analysis depends either on the accuracy and the constants Q, ω, etc. or on the standards used to calibrate the measured intensities.

If the foil is too thick then an absorption correction (A) may have to be made to the measured intensities, since in traversing a given path length to emerge from the surface of the specimen, the X-rays of different energies will be absorbed differently. This correction involves a knowledge of the specimen thickness which has to be determined by one of various techniques but usually from CBDPs. Occasionally a fluorescence (F) correction is also needed since element $Z + 2$. This 'nostandards' Z(AF) analysis can given an overall accuracy of $\approx 2\%$ and can be carried out on-line with laboratory computers.

5.4.6 Electron energy loss spectroscopy (EELS)

A disadvantage of EDX is that the X-rays from the light elements are absorbed in the detector window. Windowless detectors can be used but have some disadvantages, such as the overlapping of spectrum lines, which have led to the development of EELS.

EELS is possible only on transmission specimens, and so electron spectrometers have been interfaced to TEMs to collect all the transmitted electrons lying within a cone of width α. The intensity of the various electrons, i.e. those transmitted without loss of energy and those that have been inelastically scattered and lost energy, is then obtained by dispersing the electrons with a magnetic prism which separates spatially the electrons of different energies.

A typical EELS spectrum illustrated in Figure 5.36 shows three distinct regions. The zero loss peak is made up from those electrons which have (1) not been scattered by the specimen, (2) suffered photon scattering ($\approx 1/40$ eV) and (3) elastically scattered. The energy width of the zero loss peak is caused by the energy spread of the electron source (up to ≈ 2 eV for a thermionic W filament) and the energy resolution of the spectrometer (typically a few eV). The second region of the spectrum extends up to about 50 eV loss and is associated with plasmon excitations corresponding to electrons which have suffered one, two, or more plasmon interactions. Since the typical mean free path for the generation of a plasmon is about 50 nm, many electrons suffer single-plasmon losses and only in specimens which are too thick for electron loss analysis will there be a significant third plasmon peak. The relative size of the plasmon loss peak and the zero loss peak can also be used to measure the foil thickness. Thus the ratio of the probability of

Table 5.2 Relationships between measured intensities and composition for a NiAl alloy

	Measured intensities	Cross-section Q, $(10^{-24}\ cm^2)$	Fluorescent yield ω	Detector efficiency η	Analysis at.%
NiK_α	16 250	297	0.392	0.985	50.6
AlK_α	7 981	2935	0.026	0.725	49.4

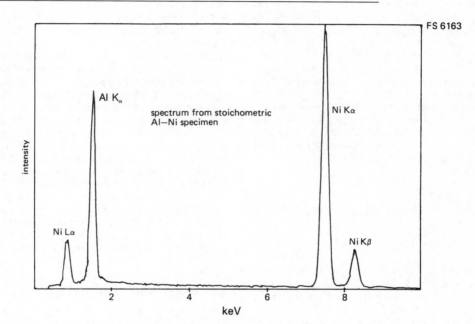

Figure 5.35 *EDX spectrum from a stoichiometric Ni–Al specimen.*

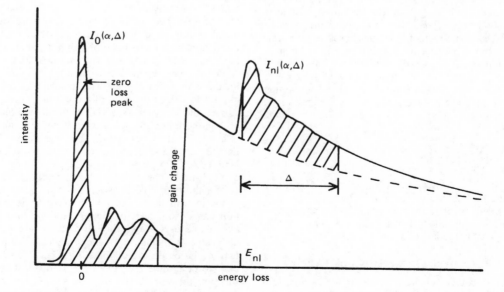

Figure 5.36 *Schematic energy-loss spectrum, showing the zero-loss and plasmon regions together with the characteristic ionization edge, energy E_{nl} and intensity I_{nl}.*

exciting a plasmon loss, P_1, to not exciting a plasmon, P_0, is given by $P_1/P_0 = t/L$, where t is the thickness, L the mean free path for plasmon excitation, and P_1 and P_0 are given by the relative intensities of the zero loss and the first plasmon peak. If the second plasmon peak is a significant fraction of the first peak this indicates that the specimen will be too thick for accurate microanalysis.

The third region is made up of a continuous background on which the characteristic ionization losses are superimposed. Qualitative elemental analysis can be carried out simply by measuring the energy of the edges and comparing them with tabulated energies. The actual shape of the edge can also help to define the chemical state of the element. Quantitative analysis requires the measurement of the ratios of the intensities of the electrons from elements A and B which have suffered ionization losses. In principle, this allows the ratio of the number of A atoms, N_A, and B atoms, N_B, to be obtained simply from the appropriate ionization cross-sections, Q_K. Thus the number of A atoms will be given by

$$N_A = (1/Q_K^A)[I_A^K/I_0]$$

and the number of B atoms by a similar expression, so that

$$N_A/N_B = I_K^A Q_K^B \big/ I_K^B Q_K^A$$

where I_K^A is the measured intensity of the K edge for element A, similarly for I_K^B and I_0 is the measured intensity of the zero loss peak. This expression is similar to the thin foil EDX equation.

To obtain I_K the background has to be removed so that only loss electrons remain. Because of the presence of other edges there is a maximum energy range over which I_K can be measured which is about 50–100 eV. The value of Q_K must therefore be replaced by $Q_K(\Delta)$ which is a partial cross-section calculated for atomic transition within an energy range Δ of the ionization threshold. Furthermore, only the loss electrons arising from an angular range of scatter α at the specimen are collected by the spectrometer so that a double partial cross-section $Q(\Delta, \alpha)$ is appropriate. Thus analysis of a binary alloy is carried out using the equation

$$\frac{N_A}{N_B} = \frac{Q_K^B(\Delta, \alpha)I_K^A(\Delta, \alpha)}{Q_K^A(\Delta, \alpha)I_K^B(\Delta, \alpha)}$$

Values of $Q(\Delta, \alpha)$ may be calculated from data in the literature for the specific value of ionization edge, Δ, α and incident accelerating voltage, but give an analysis accurate to only about 5%; a greater accuracy might be possible if standards are used.

5.4.7 Auger electron spectroscopy (AES)

Auger electrons originate from a surface layer a few atoms thick and therefore AES is a technique for studying the composition of the surface of a solid. It is obviously an important method for studying oxidation, catalysis and other surface chemical reactions, but has also been used successfully to determine the chemistry of fractured interfaces and grain boundaries (e.g. temper embrittlement of steels).

The basic instrumentation involves a focusable electron gun, an electron analyser and a sample support and manipulation system, all in an ultra-high-vacuum environment to minimize adsorption of gases onto the surface during analysis. Two types of analyser are in use, a cylindrical mirror analyser (CMA) and a hemispherical analyser (HSA), both of which are of the energy-dispersive type as for EELS, with the difference that the electron energies are much lower, and electrostatic rather than magnetic 'lenses' are used to separate out the electrons of different energies.

In the normal distribution the Auger electron peaks appear small on a large and often sloping background, which gives problems in detecting weak peaks since amplification enlarges the background slope as well as the peak. It is therefore customary to differentiate the spectrum so that the Auger peaks are emphasized as doublet peaks with a positive and negative displacement against a nearly flat background. This is achieved by electronic differentiation by applying a small a.c. signal of a particular frequency in the detected signal. Chemical analysis through the outer surface layers can be carried out by depth profiling with an argon ion gun.

5.5 Observation of defects

5.5.1 Etch pitting

Since dislocations are regions of high energy, their presence can be revealed by the use of an etchant which chemically attacks such sites preferentially. This method has been applied successfully in studying metals, alloys and compounds, and there are many fine examples in existence of etch-pit patterns showing small-angle boundaries and pile-ups. Figure 5.37a shows an etch-pit pattern from an array of piled-up dislocations in a zinc crystal. The dislocations are much closer together at the head of the pile-up, and an analysis of the array, made by Gilman, shows that their spacing is in reasonable agreement with the theory of Eshelby, Frank and Nabarro, who have shown that the number of dislocations n that can be packed into a length L of slip plane is $n = 2L\tau/\mu b$, where τ is the applied stress. The main disadvantage of the technique is its inability to reveal networks or other arrangements in the interior of the crystal, although some information can be obtained by taking sections through the crystal. Its use is also limited to materials with low dislocation contents ($<10^4$ mm^{-2}) because of the limited resolution. In recent years it has been successfully used to determine the velocity v of dislocations as a function of temperature and stress by measuring the distance travelled by a dislocation after the application of a stress for a known time (see Chapter 7).

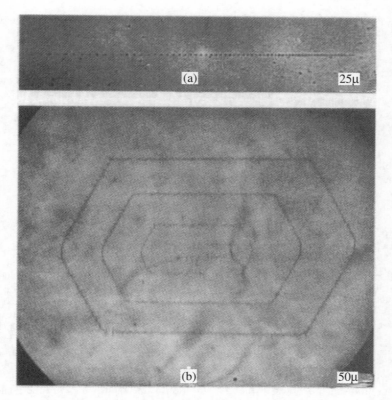

Figure 5.37 *Direct observation of dislocations. (a) Pile-up in a zinc single crystal (after Gilman, 1956, p. 1000).
(b) Frank-Read source in silicon (after Dash, 1957; courtesy of John Wiley and Sons).*

5.5.2 Dislocation decoration

It is well-known that there is a tendency for solute atoms to segregate to grain boundaries and, since these may be considered as made up of dislocations, it is clear that particular arrangements of dislocations and sub-boundaries can be revealed by preferential precipitation. Most of the studies in metals have been carried out on aluminium–copper alloys, to reveal the dislocations at the surface, but recently several decoration techniques have been devised to reveal internal structures. The original experiments were made by Hedges and Mitchell in which they made visible the dislocations in AgBr crystals with photographic silver. After a critical annealing treatment and exposure to light, the colloidal silver separates along dislocation lines. The technique has since been extended to other halides, and to silicon where the decoration is produced by diffusing copper into the crystal at 900°C so that on cooling the crystal to room temperature, the copper precipitates. When the silicon crystal is examined optically, using infrared illumination, the dislocation-free areas transmit the infrared radiation, but the dislocations decorated with copper are opaque. A fine example of dislocations observed using this technique is shown in Figure 5.37b.

The technique of dislocation decoration has the advantage of revealing internal dislocation networks but, when used to study the effect of cold-work on the dislocation arrangement, suffers the disadvantage of requiring some high-temperature heat-treatment during which the dislocation configuration may become modified.

5.5.3 Dislocation strain contrast in TEM

The most notable advance in the direct observation of dislocations in materials has been made by the application of transmission techniques to thin specimens. The technique has been used widely because the dislocation arrangements inside the specimen can be studied. It is possible, therefore, to investigate the effects of plastic deformation, irradiation, heat-treatment, etc. on the dislocation distribution and to record the movement of dislocations by taking cine-films of the images on the fluorescent screen of the electron microscope. One disadvantage of the technique is that the materials have to be thinned before examination and, because the surface-to-volume ratio of the resultant specimen is high, it is possible that some rearrangement of dislocations may occur.

A theory of image contrast has been developed which agrees well with experimental observations. The basic idea is that the presence of a defect in the lattice causes displacements of the atoms from their position in the perfect crystal and these lead to phase changes in the electron waves scattered by the atoms so that the amplitude diffracted by a crystal is altered. The image seen in the microscope represents the electron intensity distribution at the lower surface of the specimen. This intensity distribution has been calculated by a dynamical theory (see Section 5.5.7) which considers the coupling between the diffracted and direct beams but it is possible to obtain an explanation of many observed contrast effects using a simpler (kinematical) theory in which the interactions between the transmitted and scattered waves are neglected. Thus if an electron wave, represented by the function $\exp(2\pi i k_0.r)$ where k_0 is the wave vector of magnitude $1/\lambda$, is incident on an atom at position r there will be an elastically scattered wave $\exp(2\pi i k_1.r)$ with a phase difference equal to $2\pi r(k_1 - k_0)$ when k_1 is the wave vector of the diffracted wave. If the crystal is not oriented exactly at the Bragg angle the reciprocal lattice point will be either inside or outside the reflecting sphere and the phase difference is then $2\pi r(g + s)$ where g is the reciprocal lattice vector of the lattice plane giving rise to reflection and s is the vector indicating the deviation of the reciprocal lattice point from the reflection sphere (see Figure 5.39). To obtain the total scattered amplitude from a crystal it is necessary to sum all the scattered amplitudes from all the atoms in the crystal, i.e. take account of all the different path lengths for rays scattered by different atoms. Since most of the intensity is concentrated near the reciprocal lattice point it is only necessary to calculate the amplitude diffracted by a column of crystal in the direction of the diffracted by a column of crystal in the direction of the diffracted beam and not the whole crystal, as shown in Figure 5.38. The amplitude of the diffracted

beam ϕ_g for an incident amplitude $\phi_0 = 1$, is then

$$\phi_g = (\pi i/\xi_g) \int_0^t \exp[-2\pi i(g + s).r]dr$$

and since $r.s$ is small and $g.r$ is an integer this reduces to

$$\phi_g = (\pi i/\xi_g) \int_0^t \exp[-2\pi i s.r]dr$$

$$= (\pi i/\xi_g) \int_0^t \exp[-2\pi i s z]dz$$

where z is taken along the column. The intensity from such a column is

$$|\phi_g|^2 = I_g = [\pi^2/\xi_g^2](\sin^2 \pi t s/(\pi s)^2)$$

from which it is evident that the diffracted intensity oscillates with depth z in the crystal with a periodicity equal to $1/s$. The maximum wavelength of this oscillation is known as the extinction[1] distance ξ_g since the diffracted intensity is essentially extinguished at such positions in the crystal. This sinusoidal variation of intensity gives rise to fringes in the electron-optical image of boundaries and defects inclined to the foil surface, e.g. a stacking fault on an inclined plane is generally visible on an electron micrograph as a set of parallel fringes running parallel to the intersection of the fault plane with the plane of the foil (see Figure 5.43).

In an imperfect crystal, atoms are displaced from their true lattice positions. Consequently, if an atom at r_n is displaced by a vector **R**, the amplitude of the wave diffracted by the atom is multiplied by

[1] $\xi_g = \pi V \cos\theta/\lambda F$ where V is the volume of the unit cell, θ the Bragg angle and F the structure factor.

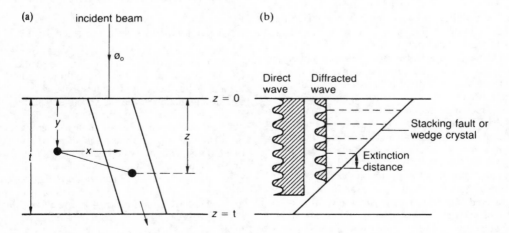

Figure 5.38 *(a) Column approximation used to calculate the amplitude of the diffracted beam ϕ_g on the bottom surface of the crystal. The dislocation is at a depth y and a distance x from the column. (b) Variation of intensity with depth in a crystal.*

an additional phase factor $\exp[2\pi i(k_1 - k_0).\mathbf{R}]$. Then, since $(k_1 - k_0) = g + s$ the resultant amplitude is

$$\phi_g = (\pi i/\xi_g)\int_0^t \exp[-2\pi i(g + s).(r + \mathbf{R})]dr$$

If we neglect $s.\mathbf{R}$ which is small in comparison with $g.\mathbf{R}$, and $g.r$ which gives an integer, then in terms of the column approximation

$$\phi_g = (\pi i/\xi_g)\int_0^t \exp(-2\pi i s z)\exp(-2\pi i g.\mathbf{R})dz$$

The amplitude, and hence the intensity, therefore may differ from that scattered by a perfect crystal, depending on whether the phase factor $\alpha = 2\pi g.\mathbf{R}$ is finite or not, and image contrast is obtained when $g.\mathbf{R} \neq 0$.

5.5.4 Contrast from crystals

In general, crystals observed in the microscope appear light because of the good transmission of electrons. In detail, however, the foils are usually slightly buckled so that the orientation of the crystal relative to the electron beam varies from place to place, and if one part of the crystal is oriented at the Bragg angle, strong diffraction occurs. Such a local area of the crystal then appears dark under bright-field illuminations, and is known as a bend or extinction contour. If the specimen is tilted while under observation, the angular conditions for strong Bragg diffraction are altered, and the extinction contours, which appear as thick dark bands, can be made to move across the specimen. To interpret micrographs correctly, it is essential to know the correct sense of both g and s. The g-vector is the line joining the origin of the diffraction pattern to the strong diffraction spot and it is essential that its sense is correct with respect to the micrograph, i.e. to allow for any image inversion or rotation by the electron optics. The sign of s can be determined from the position of the Kikuchi lines with respect to the diffraction spots, as discussed in Section 5.4.1.

5.5.5 Imaging of dislocations

Image contrast from imperfections arises from the additional phase factor $\alpha = 2\pi g.\mathbf{R}$ in the equation for the diffraction of electrons by crystals. In the case of dislocations the displacement vector $\mathbf{R}$ is essentially equal to b, the Burgers vector of the dislocation, since atoms near the core of the dislocation are displaced parallel to b. In physical terms, it is easily seen that if a crystal, oriented off the Bragg condition, i.e. $s \neq 0$, contains a dislocation then on one side of the dislocation core the lattice planes are tilted into the reflecting position, and on the other side of the dislocation the crystal is tilted away from the reflecting position. On the side of the dislocation in the reflecting position the transmitted intensity, i.e. passing through the objective aperture, will be

less and hence the dislocation will appear as a line in dark contrast. It follows that the image of the dislocation will lie slightly to one or other side of the dislocation core, depending on the sign of (g.b)s. This is shown in Figure 5.39 for the case where the crystal is oriented in such a way that the incident beam makes an angle greater than the Bragg angle with the reflecting planes, i.e. $s > 0$. The image occurs on that side of the dislocation where the lattice rotation brings the crystal into the Bragg position, i.e. rotates the reciprocal lattice point onto the reflection sphere. Clearly, if the diffracting conditions change, i.e. g or s change sign, then the image will be displaced to the other side of the dislocation core.

The phase angle introduced by a lattice defect is zero when $g.\mathbf{R} = 0$, and hence there is no contrast, i.e. the defect is invisible when this condition is satisfied. Since the scalar product $g.\mathbf{R}$ is equal to $gR\cos\theta$, where θ is the angle between g and $\mathbf{R}$, then $g.\mathbf{R} = 0$ when the displacement vector $\mathbf{R}$ is normal to g, i.e. parallel to the reflecting plane producing the image. If we think of the lattice planes which reflect the electrons as mirrors, it is easy to understand that no contrast results when $g.\mathbf{R} = 0$, because the displacement vector $\mathbf{R}$ merely moves the reflecting planes parallel to themselves without altering the intensity scattered from them. Only displacements which have a component perpendicular to the reflecting plane, i.e. tilting the planes, will produce contrast.

A screw dislocation only produces atomic displacements in the direction of its Burgers vector, and hence because $\mathbf{R} = b$ such a dislocation will be completely 'invisible' when b lies in the reflecting plane producing the image. A pure edge dislocation, however, produces some minor atomic displacements perpendicular to b, as discussed in Chapter 4, and the displacements give rise to a slight curvature of the lattice planes. An edge dislocation is therefore not completely invisible when b lies in the reflecting planes, but usually shows some evidence of faint residual contrast. In general,

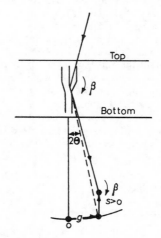

Figure 5.39 *Schematic diagram showing the dependence of the dislocation image position on diffraction conditions.*

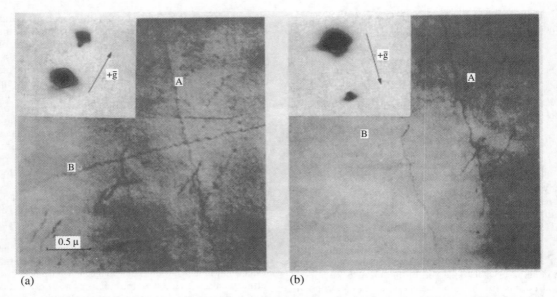

(a)   (b)

Figure 5.40 (a) Application of the g.b = 0 criterion. The effect of changing the diffraction condition (see diffraction pattern inserts) makes the long helical dislocation B in (a) disappear in (b) (after Hirsch, Howie and Whelan, 1960; courtesy of the Royal Society).

however, a dislocation goes out of contrast when the reflecting plane operating contains its Burgers vector, and this fact is commonly used to determine the Burgers vector. To establish b uniquely, it is necessary to tilt the foil so that the dislocation disappears on at least two different reflections. The Burgers vector must then be parallel to the direction which is common to these two reflecting planes. The magnitude of b is usually the repeat distance in this direction.

The use of the $g.b = 0$ criterion is illustrated in Figure 5.40. The helices shown in this micrograph have formed by the condensation of vacancies on to screw dislocations having their Burgers vector b parallel to the axis of the helix. Comparison of the two pictures in (a) and (b) shows that the effect of tilting the specimen, and hence changing the reflecting plane, is to make the long helix B in (a) disappear in (b). In detail, the foil has a [0 0 1] orientation and the long screws lying in this plane are $1/2[\bar{1} 1 0]$ and $1/2[1 1 0]$. In Figure 5.40a the insert shows the 0 2 0 reflection is operating and so $g.b \neq 0$ for either A or B, but in Figure 5.40b the insert shows that the $2\bar{2}0$ reflection is operating and the dislocation B is invisible since its Burgers vector b is normal to the g-vector, i.e. $g.b = 2\,\bar{2}\,0.1/2[1 1 0] = (\frac{1}{2} \times 1 \times 2) + (\frac{1}{2} \times 1 \times \bar{2}) + 0 = 0$ for the dislocation B, and is therefore invisible.

5.5.6 Imaging of stacking faults

Contrast at a stacking fault arises because such a defect displaces the reflecting planes relative to each other, above and below the fault plane, as illustrated in

Figure 5.41a. In general, the contrast from a stacking fault will not be uniformly bright or dark as would be the case if it were parallel to the foil surface, but in the form of interference fringes running parallel to the intersection of the foil surface with the plane containing the fault. These appear because the diffracted intensity oscillates with depth in the crystal as discussed. The stacking fault displacement vector **R**, defined as the shear parallel to the fault of the portion of crystal below the fault relative to that above the fault which is as fixed, gives rise to a phase difference $\alpha = 2\pi g.\mathbf{R}$ in the electron waves diffracted from either side of the fault. It then follows that stacking-fault contrast is absent with reflections for which $\alpha = 2\pi$, i.e. for which $g.\mathbf{R} = n$. This is equivalent to the $g.b = 0$ criterion for dislocations and can be used to deduce **R**.

The invisibility of stacking fault contrast when $g.\mathbf{R} = 0$ is exactly analogous to that of a dislocation when $g.b = 0$, namely that the displacement vector is parallel to the reflecting planes. The invisibility when $g.\mathbf{R} = 1, 2, 3, \ldots$ occurs because in these cases the vector **R** moves the imaging reflecting planes normal to themselves by a distance equal to a multiple of the spacing between the planes. From Figure 5.41b it can be seen that for this condition the reflecting planes are once again in register on either side of the fault, and, as a consequence, there is no interference between waves from the crystal above and below the fault.

5.5.7 Application of dynamical theory

The kinematical theory, although very useful, has limitations. The equations are only valid when the crystal is

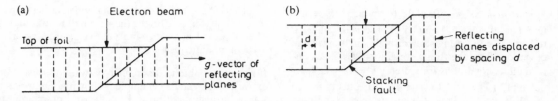

Figure 5.41 *Schematic diagram showing (a) displacement of reflecting planes by a stacking fault and (b) the condition for* $g.R = n$ *when the fault would be invisible.*

oriented far from the exact Bragg condition, i.e. when s is large. The theory is also only strictly applicable for foils whose thickness is less than about half an extinction distance ($\frac{1}{2}\xi_g$) and no account is taken of absorption. The dynamical theory has been developed to overcome these limitations.

The object of the dynamical theory is to take into account the interactions between the diffracted and transmitted waves. Again only two beams are considered, i.e. the transmitted and one diffracted beam, and experimentally it is usual to orient the specimen in a double-tilt stage so that only one strong diffracted beam is excited. The electron wave function is then considered to be made up of two plane waves –an incident or transmitted wave and a reflected or diffracted wave

$$\psi(r) = \phi_0 \exp(2\pi i k_0.r) + \phi_g \exp(2\pi i k_1.r)$$

The two waves can be considered to propagate together down a column through the crystal since the Bragg angle is small. Moreover, the amplitudes ϕ_0 and ϕ_g of the two waves are continually changing with depth z in the column because of the reflection of electrons from one wave to another. This is described by a pair of coupled first-order differential equations linking the wave amplitudes ϕ_0 and ϕ_g. Displacement of an atom R causes a phase change $\alpha = 2\pi g.R$ in the scattered wave, as before, and the two differential equations describing the dynamical equilibrium between incident and diffracted waves

$$\frac{d\phi_0}{dz} = \frac{\pi i}{\xi_g}\phi_g$$

$$\frac{d\phi_g}{dz} = \frac{\pi i}{\xi_g}\phi_0 + 2\pi\phi_g\left(s + g.\frac{dR}{dz}\right)$$

These describe the change in reflected amplitude ϕ_g because electrons are reflected from the transmitted wave (this change is proportional to ϕ_0, the transmitted wave amplitude, and contains the phase factor) and the reflection in the reverse direction.

These equations show that the effect of a displacement R is to modify s locally, by an amount proportional to the derivative of the displacement, i.e. dR/dz, which is the variation of displacement with depth z in the crystal. This was noted in the kinematical theory where dR/dz is equivalent to a local tilt of the lattice planes. The variation of the intensities $|\phi_0|^2$ and $|\phi_g|^2$

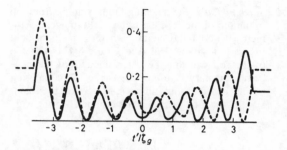

Figure 5.42 *Computed intensity profiles about the foil centre for a stacking fault with* $\alpha = +2\pi/3$. *The full curve is the B.F. and the broken curve the D.F. image (from Hirsch, Howie et al., 1965).*

for different positions of the column in the crystal, relative to the defect, then gives the bright and dark-field images respectively. Figure 5.42 shows the bright- and dark-field intensity profiles from a stacking fault on an inclined plane, in full and broken lines, respectively. A wide variety of defects have been computed, some of which are summarized below:

1. *Dislocations* In elastically isotropic crystals, perfect screw dislocations show no contrast if the condition $g.b = 0$ is satisfied. Similarly, an edge dislocation will be invisible if $g.b = 0$ and if $g.b \times u = 0$ where u is a unit vector along the dislocation line and $b \times u$ describes the secondary displacements associated with an edge dislocation normal to the dislocation line and b. The computations also show that for mixed dislocations and edge dislocations for which $g.b \times u < 0.64$ the contrast produced will be so weak as to render the dislocation virtually invisible. At higher values of $g.b \times u$ some contrast is expected. In addition, when the crystal becomes significantly anisotropic residual contrast can be observed even for $g.b = 0$.

   The image of a dislocation lies to one side of the core, the side being determined by $(g.b)s$. Thus the image of a dislocation loop lies totally outside the core when (using the appropriate convention) $(g.b)s$ is positive and inside when $(g.b)s$ is negative. Vacancy and interstitial loops can thus be distinguished by examining their size after changing from $+g$ to $-g$, since these loops differ only in the sign of b.

2. *Partial dislocations* Partials for which $g.b = \pm\frac{1}{3}$ (e.g. partial a/6[$\bar{1}\,\bar{1}\,2$] on (1 1 1) observed with 2 0 0 reflection) will be invisible at both small and large deviations from the Bragg condition. Partials examined under conditions for which $g.b = \pm\frac{2}{3}$ (i.e. partial a/6[$\bar{2}\,1\,1$] on (1 1 1) with 2 0 0 reflection) are visible except at large deviations from the Bragg condition. A partial dislocation lying above a similar stacking fault is visible for $g.b = \pm\frac{1}{3}$ and invisible for $g.b = \pm\frac{2}{3}$.

3. *Stacking faults* For stacking faults running from top to bottom of the foil, the bright-field image is symmetrical about the centre, whereas the dark-field image is asymmetrical (see Figure 5.42). The top of the foil can thus be determined from the non-complementary nature of the fringes by comparing bright- and dark-field images. Moreover, the intensity of the first fringe is determined by the sign

of the phase-factor α, such that when α is positive the first fringe is bright (corresponding to a higher transmitted intensity) and vice versa on a positive photographic print.

It is thus possible to distinguish between intrinsic and extrinsic faults and an example is shown in Figure 5.43 for an intrinsic fault on (1 1 1). The foil orientation is [1 1 0] and the non-complementary nature of the first fringe between B.F. and D.F. indicates the top of the foil, marked T. Furthermore, from the B.F. images the first fringe is bright with $\bar{1}\,1\,1$, and dark with $1\,\bar{1}\,\bar{1}$ and $\bar{1}\,1\,\bar{1}$.

5.5.8 Weak-beam microscopy

One of the limiting factors in the analysis of defects is the fact that dislocation images have widths of $\xi_g/3$, i.e. typically >10.0 nm. It therefore follows that dislocations closer together than about 20.0 nm are not

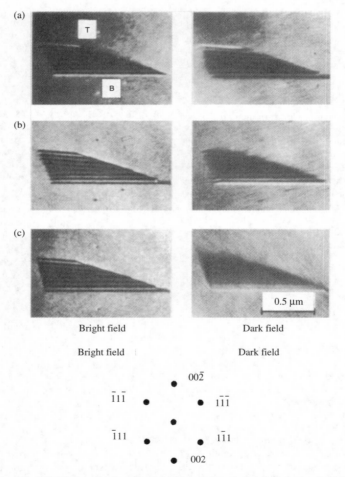

Figure 5.43 *Bright-field and dark-field micrographs of an intrinsic stacking fault in a copper–aluminium alloy; the operating diffraction vectors are (a) $\bar{1}\,1\,1$ (b) $1\,\bar{1}\,1$ and (c) $\bar{1}\,1\,\bar{1}$ (after Howie and Valdre, 1963; courtesy of Taylor and Francis).*

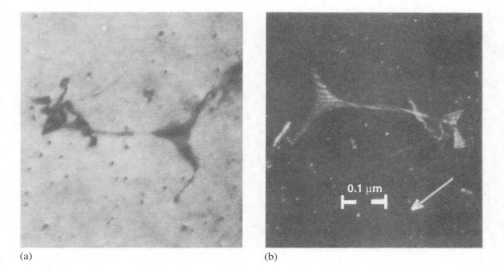

(a)   (b)

Figure 5.44 *Symmetrical node in Fe–21Cr–14Ni stainless steel with* $\gamma = 18 \pm 4\ mJ/m^2$, *(a) B.F. with g = 111 (b) weak beam with g(5g).*

generally resolved. With normal imaging techniques, the detail that can be observed is limited to a value about fifty to a hundred times greater than the resolution of the microscope. This limitation can be overcome by application of the weak-beam technique in which crystals are imaged in dark-field using a very large deviation parameter s. Under these conditions the background intensity is very low so that weak images are seen in very high contrast and the dislocation images are narrow, ($\approx$1.5 nm) as shown in Figure 5.44. At the large value of s used in weak-beam, the transfer of energy from the direct to the diffracted beam is very small, i.e. the crystal is a long way from the Bragg condition and there is negligible diffraction. Moreover, it is only very near the core of the dislocation that the crystal planes are sufficiently bent to cause the Bragg condition to be locally satisfied, i.e. g.(dR/dz) be large enough to satisfy the condition $[s + g.(dR/dz)] = 0$. Therefore diffraction takes place from only a small volume near the centre of the dislocation, giving rise to narrow images. The absolute intensity of these images is, however, very small even though the signal-to-background ratio is high and hence long exposures are necessary to record them.

5.6 Specialized bombardment techniques

5.6.1 Neutron diffraction

The advent of nuclear reactors stimulated the application of neutron diffraction to those problems of materials science which could not be solved satisfactorily by other diffraction techniques. In a conventional pile the fast neutrons produced by fission are slowed down by repeated collisions with a 'moderator' of graphite or heavy water until they are slow enough to produce further fission. If a collimator is inserted into the pile, some of these slow neutrons[1] will emerge from it in the form of a beam, and the equivalent wavelength λ of this neutron beam of energy E in electron-volts is given by $\lambda = 0.0081/E$. The equilibrium temperature in a pile is usually in the range 0–100°C, which corresponds to a peak energy of several hundredths of an electron-volt. The corresponding wavelength of the neutron beam is about 0.15 nm and since this is very similar to the wavelength of X-rays it is to be expected that thermal neutrons will be diffracted by crystals.

The properties of X-ray and neutron beams differ in many respects. The distribution of energy among the neutrons in the beam approximately follows the Maxwellian curve appropriate to the equilibrium temperature and, consequently, there is nothing which corresponds to characteristic radiation. The neutron beam is analogous to a beam of 'white' X-rays, and as a result it has to be monochromatized before it can be used in neutron crystallography. Then, because only about 1 in 10^3 of the neutrons in the originally weak collimated beam are reflected from the monochromator, it is necessary to employ very wide beams several inches in cross-section to achieve a sufficiently high counting rate on the boron trifluoride counter detector (photographic detection is possible but not generally useful). In consequence, neutron spectrometers, although similar in principle to X-ray diffractometers, have to be constructed on a massive scale.

---

[1] These may be called 'thermal' neutrons because they are in thermal equilibrium with their surroundings.

Neutron beams do, however, have advantages over X-rays or electrons, and one of these is the extremely low absorption of thermal neutrons by most elements. Table 5.3 shows that even in the most highly absorbent elements (e.g. lithium, boron, cadmium and gadolinium) the mass absorption coefficients are only of the same order as those for most elements for a comparable X-ray wavelength, and for other elements the neutron absorption is very much less indeed. This penetrative property of the neutron beam presents a wide scope for neutron crystallography, since the whole body of a specimen may be examined and not merely its surface. Problems concerned with preferred orientation, residual stresses, cavitation and structural defects are but a few of the possible applications, some of which are discussed more fully later.

Another difference concerns the intensity of scattering per atom, I_a. For X-rays, where the scattering is by electrons, the intensity I_a increases with atomic number and is proportional to the square of the atomic-form factor. For neutrons, where the scattering is chiefly by the nucleus, I_a appears to be quite unpredictable. The scattering power per atom varies not only apparently at random from atom to atom, but also from isotope to isotope of the same atom. Moreover, the nuclear component to the scattering does not decrease with increasing angle, as it does with X-rays, because the nucleus which causes the scattering is about 10^{-12} mm in size compared with 10^{-7} mm, which is the size of the electron cloud that scatters X-rays. Table 5.4 gives some of the scattering amplitudes for X-rays and thermal neutrons.

The fundamental difference in the origin of scattering between X-rays and neutrons affords a method of studying structures, such as hydrides and carbides, which contain both heavy and light atoms. When X-rays are used, the weak intensity contributions of the light atoms are swamped by those from the heavy atoms, but when neutrons are used, the scattering power of all atoms is roughly of the same order. Similarly, structures made up of atoms whose atomic numbers are nearly the same (e.g. iron and cobalt, or copper and zinc), can be studied more easily by using

Table 5.3 X-ray and neutron mass absorption coefficients

Element	At. no.	X-rays ($\lambda = 0.19$ nm)	Neutrons ($\lambda = 0.18$ nm)
Li	3	1.5	5.8
B	5	5.8	38.4
C	6	10.7	0.002
Al	13	92.8	0.005
Fe	26	72.8	0.026
Cu	29	98.8	0.03
Ag	47	402	0.3
Cd	48	417	13.0
Gd	61	199	183.0
Au	79	390	0.29
Pb	82	429	0.0006

Table 5.4 Scattering amplitudes for X-rays and thermal neutrons

Element	At. no.	Scattering amplitudes X-rays for $\sin\theta/\lambda = 0.5$ $\times 10^{-12}$	Neutrons* $\times 10^{-12}$	
H	1	0.02		-0.4
Li	3	0.28	Li^6	0.7
			Li^7	-0.25
C	6	0.48		0.64
N	7	0.54		0.85
O	8	0.62		0.58
Al	13	1.55		0.35
Ti	22	2.68		-0.38
Fe	26	3.27	Fe^{56}	1.0
			Fe^{57}	0.23
Co	27	3.42		0.28
Cu	29	3.75		0.76
Zn	30	3.92		0.59
Ag	47	6.71	Ag^{107}	0.83
			Ag^{109}	0.43
Au	79	12.37		0.75

*The negative sign indicates that the scattered and incident waves are in phase for certain isotopes and hence for certain elements. Usually the scattered wave from an atom is 180° out of phase with the incident wave.

neutrons. This aspect is discussed later in relation to the behaviour of ordered alloy phases.

The major contribution to the scattering power arises from the nuclear component, but there is also an electronic (magnetic spin) component to the scattering. This arises from the interaction between the magnetic moment of the neutron and any resultant magnetic moment which the atom might possess. As a result, the neutron diffraction pattern from paramagnetic materials, where the atomic moments are randomly directed (see Chapter 6), shows a broad diffuse background, due to incoherent (magnetic) scattering, superimposed on the sharp peaks which arise from coherent (nuclear) scattering. In ferromagnetic metals the atomic moments are in parallel alignment throughout a domain, so that this cause of incoherent scattering is absent. In some materials (e.g. NiO or FeO) an alignment of the spins takes place, but in this case the magnetization directions of neighbouring pairs of atoms in the structure are opposed and, in consequence, cancel each other out. For these materials, termed anti-ferromagnetic, there is no net spontaneous magnetization and neutron diffraction is a necessary and important tool for investigating their behaviour (see Chapter 6).

5.6.2 Synchrotron radiation studies

Very large electrical machines known as synchrotron radiation sources (SRS) provide a unique source of

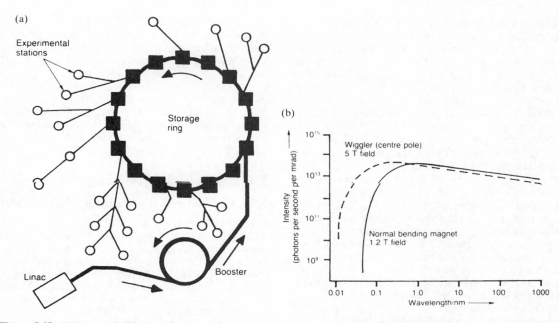

Figure 5.45 *(a) Layout of SRS, Daresbury, and (b) wavelength spectrum of synchrotron radiation (after Barnes, 1990, pp. 708–715; by permission of the Institute of Metals).*

electromagnetic radiation for materials characterisation.[1] Electrons from a hot cathode are accelerated in three stages by a linear accelerator (Linac), a booster synchrotron and an evacuated storage ring (Figure 5.45a). As bunches of electrons travel around the periphery of the storage ring they attain energies of up to 2 GeV and velocities approaching that of light. At these relativistic velocities, electron mass becomes 4000 times greater than the rest mass. Dipole and quadrupole magnets constrain the bunches into an approximately circular orbit and, by accelerating them centripetally, cause electromagnetic radiation to be produced. The spectrum of this synchrotron radiation is very wide, extending from short-wavelength ('hard') X-rays to the infrared range (Figure 5.45b). A wiggler magnet produces a strong (5 tesla) field and can extend the spectrum to even shorter wavelengths. Compared with more orthodox sources of electromagnetic radiation, the synchrotron offers the advantages of very high intensity, short wavelengths, precise collimation of the beam and a smooth, continuous spectrum. The high radiation intensity permits exposure times that are often several orders of magnitude shorter than those for comparable laboratory methods. The risk of beam damage to specimens by the flashes of radiation is accordingly lessened. Specimens of metals, ceramics, polymers, semiconductors, catalysts, etc.

[1] In 1980, the world's first totally radiation-dedicated SRS came into operation at Daresbury, England. Electrons are 'stored' in the main ring for 10–20 h, traversing its 96 m periphery more than 3×10^6 times per second.

are placed in independent experimental stations located around the periphery of the ring chamber and irradiated in order to produce spectroscopic, diffraction or imaging effects.

In the technique known as extended X-ray absorption fine-structure spectroscopy (EXAFS) attention is directed to the small discontinuities on the higher-energy flank beyond each vertical, characteristic 'edge' which appears in a plot of mass absorption versus X-ray wavelength. These 'finestructure' (FS) features derive from interference effects between electron waves from excited atoms and waves back-scattered from surrounding atoms. Mathematical treatment (using a Fourier transform) of the EXAFS spectra yields a radial distribution plot of surrounding atomic density versus distance from the excited atom. By selecting the 'edge' for a particular type of atom/ion and studying its fine structure, it is thus possible to obtain information on its local environment and coordination. This type of information is of great value in structural studies of materials, such as glasses, which only exhibit short-range order. For instance, the EXAFS technique has demonstrated that the network structure of $SiO_2 - Na_2O - CaO$ glass is threaded by percolation channels of modifier (sodium) cations.

5.6.3 Secondary ion mass spectrometry (SIMS)

This important and rapidly-developing technique, which enables material surfaces to be analysed with great chemical sensitivity and excellent resolution in depth, is based upon the well-known phenomenon of sputtering. The target surface is bombarded with a focused beam of primary ions that has been accelerated with a potential of 1–30 kV within a high-vacuum chamber (10^{-5}–10^{-10} torr). These ions generate a series of collision cascades in a shallow surface layer, 0.5–5 nm deep, causing neutral atoms and, to a much smaller extent, secondary ions to be ejected (sputtered) from the specimen surface. Thus, a metallic oxide (MO) sample may act as a source of M, O, M^+, O^+, M^-, O^-, MO^+ and MO^- species. The secondary ions, which are thus either monatomic or clustered, positive or negative, are directed into a mass spectrometer (analyser), wherein they are sorted and identified according to their mass/charge ratio. Exceptionally high elemental sensitivities, expressed in parts per million and even parts per billion, are achievable. All elements in the Periodic Table can be analysed and it is possible to distinguish between individual isotopes. Studies of the self-diffusion of oxygen and nitrogen have been hindered because these light elements have no isotopes that can be used as radioactive tracers. SIMS based on the stable isotope ^{18}O provides a rapid method for determining self-diffusion coefficients. The physical process whereby ions are ejected is difficult to express in rigorous theoretical terms, consequently SIMS is usually semiquantitative, with dependence upon calibration with standard samples of known composition. SIMS is a valuable complement to other methods of surface analysis.

The available range of beam diameter is 1 µm to several millimetres. Although various types of ion beam are available (e.g. Ar^-, $^{32}O_2{}^+$, $^{16}O^-$, Cs^+, etc.) positively-charged beams are a common choice. However, if the sample is insulating, positive charge tends to accumulate in the bombarded region, changing the effective value of the beam voltage and degrading the quality of signals. One partial remedy, applicable at low beam voltages, is to 'flood' the ion-bombarded area with a high-intensity electron beam. In some variants of SIMS laser beams are used instead of ion beams.

Of the large and growing variety of methods covered by the term SIMS, the dynamic, static and imaging modes are especially useful. Materials being investigated include metals, ceramics, polymers, catalysts, semiconductors and composites. Dynamic SIMS, which uses a relatively high beam current, is an important method for determining the distribution and very low concentration of dopants in semiconductors. The beam scans a raster, 100–500 µm in size, and slowly erodes the surface of the sample. Secondary ions from the central region of the crater are analysed to produce a precise depth profile of concentration. Static SIMS uses a much smaller beam current and the final spectra tend to be more informative, providing chemical data on the top few atomic layers of the sample surface. Little surface damage occurs and the method has been applied to polymers. The imaging version of SIMS has a resolution comparable to SEM

and provide 'maps' that show the lateral distribution of elements at grain boundaries and precipitated particles and hydrogen segregation in alloys. Imaging SIMS has been applied to transverse sections through the complex scale layers which form when alloys are exposed to hot oxidizing gases (e.g. O_2, CO_2). Its sensitivity is greater than that obtainable with conventional EDX in SEM analysis and has provided a better understanding of growth mechanisms and the special role of trace elements such as yttrium.

5.7 Thermal analysis

5.7.1 General capabilities of thermal analysis

Heating a material at a steady rate can produce chemical changes, such as oxidation and degradation, and/or physical changes, such as the glass transition in polymers, conversions/inversions in ceramics and phase changes in metals. Thermal analysis is used to complement X-ray diffraction analysis, optical and electron microscopy during the development of new materials and in production control. Sometimes it is used to define the temperature and energy change associated with a structural change; at other times it is used qualitatively to provide a characteristic 'fingerprint' trace of a particular material. The various techniques of thermal analysis measure one or more physical properties of a sample as a function of temperature. Figure 5.46 illustrates three basic methods of thermal analysis, namely thermogravimetric analysis (TGA), differential thermal analysis (DTA) and differential scanning calorimetry (DSC). Respectively, they measure change in mass (TGA) and energy flow (DTA, DSC). They can apply programmed heating and cooling, but usually operate with a slowly rising temperature. The sample chamber may contain air, oxygen, nitrogen, argon, etc. or be evacuated. A sample of a few tens of milligrams will often suffice.

Recently-developed methods have extended the range of thermal analysis and other aspects of behaviour can now be studied. For instance, using dynamic mechanical thermal analysis (DMTA), mechanical as well as structural information can be obtained on the viscoelastic response of a polymeric sample to tensile, bend or shear stresses during heating.

5.7.2 Thermogravimetric analysis

In a thermobalance the mass of a sample is continuously determined and recorded while the sample is being slowly heated (Figure 5.46a). Temperatures up to at least 1000°C are available. It has been applied to the decomposition of rubbers (Figure 5.47a), kinetic studies of metallic oxidation, glass transitions and softening in polymers. Equilibrium is not attained within the sample and the method is insensitive to the more subtle solid-state changes. When changes overlap, it can be helpful to plot the first derivative, $\delta m/\delta t$, of

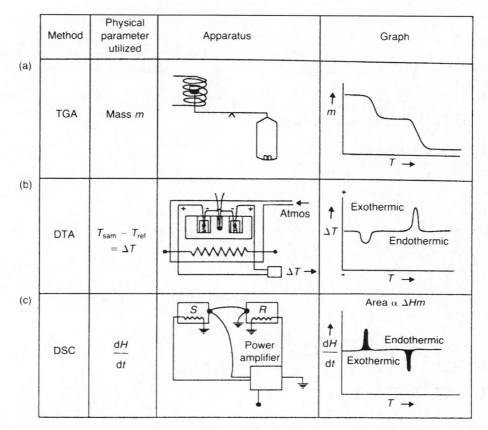

Figure 5.46 *Basic methods of thermal analysis. (a) Thermogravimetric analysis (TGA). (b) differential thermal analysis (DTA) and (c) differential scanning calorimetry (DSC).*

the graphical trace in a procedure known as derivative thermogravimetric analysis (DTGA).

5.7.3 Differential thermal analysis

DTA[1] reveals changes during the heating of a sample which involve evolution or absorption of energy. As shown diagrammatically in Figure 5.46b, a sample S and a chemically and thermally inert reference material R (sintered alumina or precipitated silica) are mounted in a recessed heating block and slowly heated. The thermocouples in S and R are connected in opposition; their temperature difference ΔT is amplified and plotted against temperature. Peak area on this trace is a function of the change in enthalpy (ΔH) as well as the mass and thermal characteristics of the sample S. Small samples can be used to give sharper, narrower peaks, provided that they are fully representative of the source

[1] Usually accredited to H. Le Chatelier (1887): improved version and forerunner of modern DTA used by W. C. Roberts-Austen (1899) in metallurgical studies of alloys.

material. Ideally, the specific heat capacities of S and R should be similar. DTA is generally regarded as a semi-quantitative or qualitative method. It has been used in studies of devitrification in oxide glasses and the glass transition in polymers. Figure 5.47b shows a comparison of the thermal response of high-alumina cement (HAC) and Portland cement. The amount of an undesirable weakening phase can be derived from the relative lengths of the ordinates X and Y in the HAC trace.

5.7.4 Differential scanning calorimetry

In this method, unlike DTA, the sample and reference body have separate resistive heaters (Figure 5.46c). When a difference in temperature develops between sample S and reference R, an automatic control loop heats the cooler of the two until the difference is eliminated. The electrical power needed to accomplish this equalizer is plotted against temperature. An endothermic change signifies that an enthalpy increase has occurred in S; accordingly, its peak is plotted upwards (unlike DTA traces). Differences in thermal conductivity and specific heat capacity have no effect

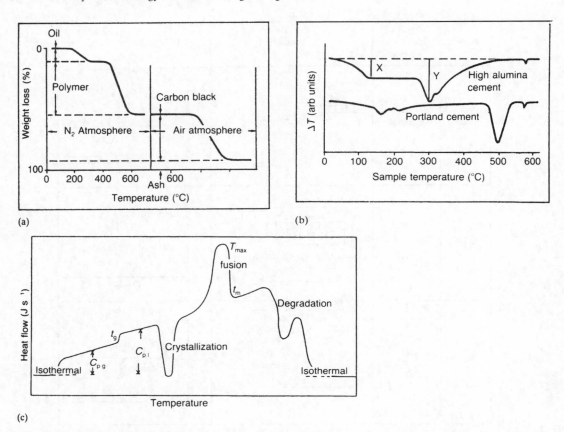

Figure 5.47 *Examples of thermal analysis (a) TGA curve for decomposition of rubber, showing decomposition of oil and polymer in N_2 up to 600°C and oxidation of carbon black in air above 600°C (Hill and Nicholas, 1989), (b) DTA curve for high-alumina cement and Portland cement (Hill and Nicholas, 1989) and (c) DTA curve for a quenched glassy polymer (Hay, 1982).*

and peak areas can be expressed as energy per unit mass. DSC has proved particularly valuable in polymer research, often being used in combination with other techniques, such as evolved gas analysis (EGA). DSC has been used in studies of the curing characteristics of rubbers and thermoset resins, transitions in liquid crystals and isothermal crystallization rates in thermoplastics. Figure 5.47c is a trace obtained for a quenched amorphous polymer. DSC has also been used in studies of the exothermic behaviour of cold-worked metals as they release 'stored energy' during annealing, energy absorption during eutectic melting of alloys, precipitation in aluminium-based alloys, relaxation transformations in metallic glasses and drying/firing transitions in clay minerals.

Further reading

Barnes, P. (1990). Synchrotron radiation for materials science research. *Metals and Materials*, November, 708–715, Institute of Materials.

Barrett, C. S. and Massalski, T. B. (1980). *Structure and Metals and Alloys*. McGraw-Hill, New York.

Cullity, B. D. (1978). *Elements of X-ray Diffraction*. Addison-Wesley, Reading, MA.

Dehoff, R. T. and Rhines, F. N. (eds) (1968). *Quantitative Microscopy*. McGraw-Hill, New York.

Gifkins, R. C. (1970). *Optical Microscopy of Metals*, Pitman, Melbourne.

Hay, J. N. (1982). Thermal methods of analysis of polymers. In *Analysis of Polymer Systems*, edited by L. S. Bark and N. S. Allen, Chap. 6. Applied Science, London.

Hill, M. and Nicholas, P. (1989). Thermal analysis in materials development. *Metals and Materials*, November, 639–642, Institute of Materials.

Jones, I. P. (1992). *Chemical Microanalysis using Electron Beams*. Institute of Materials, London.

Loretto, M. H. (1984). *Electron Beam Analysis of Materials*. Chapman and Hall, London.

Loretto, M. H. and Smallman, R. E. (1975). *Defect Analysis in Electron Microscopy*. Chapman and Hall, London.

Modin, H. and Modin, S. (1973). *Metallurgical Microscopy*. Butterworths, London.

Patzelt, W. J. (1974). *Polarised Light Microscopy: Principles, Instruments, Applications*, Ernst Leitz Wetzlar GmbH, Lahn-Wetzlar.

Pickering, F. B. (1976). *The Basis of Quantitative Metallography*, Inst. of Metallurgical Technicians Monograph No. 1.

Richardson, J. H. (1971). *Optical Microscopy for the Materials Sciences*. Marcell Dekker, New York.

Vickerman, J. C., Brown, A. and Reed, N. M. (eds) (1990). *Secondary Ion Mass Spectrometry: Principles and Applications*. Clarendon Press, Oxford.

Wendlandt, W. W. (1986). *Thermal Analysis*. 3rd edn. Wiley, New York.

Chapter 6

The physical properties of materials

6.1 Introduction

The ways in which any material interacts and responds to various forms of energy are of prime interest to scientists and, in the context of engineering, provide the essential base for design and innovation. The energy acting on a material may derive from force fields (gravitational, electric, magnetic), electromagnetic radiation (heat, light, X-rays), high-energy particles, etc. The responses of a material, generally referred to as its physical properties, are governed by the structural arrangement of atoms/ions/molecules in the material. The theme of the structure/property relation which has run through previous chapters is developed further. Special attention will be given to the diffusion of atoms/ions within materials because of the importance of thermal behaviour during manufacture and service. In this brief examination, which will range from density to superconductivity, the most important physical properties of materials are considered.

6.2 Density

This property, defined as the mass per unit volume of a material, increases regularly with increasing atomic numbers in each sub-group. The reciprocal of the density is the specific volume v, while the product of v and the relative atomic mass W is known as the atomic volume Ω. The density may be determined by the usual 'immersion' method, but it is instructive to show how X-rays can be used. For example, a powder photograph may give the lattice parameter of an fcc metal, say copper, as 0.36 nm. Then $1/(3.6 \times 10^{-10})^3$ or 2.14×10^{28} cells of this size (0.36 nm) are found in a cube 1 m edge length. The total number of atoms in 1 m^3 is then $4 \times 2.14 \times 10^{28} = 8.56 \times 10^{28}$ since an fcc cell contains four atoms. Furthermore, the mass of a copper atom is 63.57 times the mass of a hydrogen atom (which is 1.63×10^{-24} g) so

that the mass of 1 m^3 of copper, i.e. the density, is $8.56 \times 10^{28} \times 63.57 \times 1.63 \times 10^{-24} = 8900$ kg m^{-3}.

On alloying, the density of a metal changes. This is because the mass of the solute atom differs from that of the solvent, and also because the lattice parameter usually changes on alloying. The parameter change may often be deduced from Vegard's law, which assumes that the lattice parameter of a solid solution varies linearly with atomic concentration, but numerous deviations from this ideal behaviour do exist.

The density clearly depends on the mass of the atoms, their size and the way they are packed. Metals are dense because they have heavy atoms and close packing; ceramics have lower densities than metals because they contain light atoms, either C, N or O; polymers have low densities because they consist of light atoms in chains. Figure 6.1 shows the spread in density values for the different material classes. Such 'Material Property Charts', as developed by Ashby, are useful when selecting materials during engineering design.

6.3 Thermal properties

6.3.1 Thermal expansion

If we consider a crystal at absolute zero temperature, the ions sit in a potential well of depth E_{r_0} below the energy of a free atom (Figure 6.2). The effect of raising the temperature of the crystal is to cause the ions to oscillate in this asymmetrical potential well about their mean positions. As a consequence, this motion causes the energy of the system to rise, increasing with increasing amplitude of vibration. The increasing amplitude of vibration also causes an expansion of the crystal, since as a result of the sharp rise in energy below r_0 the ions as they vibrate to and fro do not approach much closer than the equilibrium separation, r_0, but separate more widely when moving apart. When

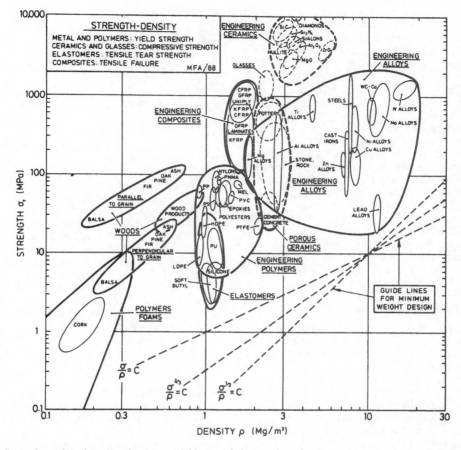

Figure 6.1 *Strength σ, plotted against density, ρ (yield strength for metals and polymers, compressive strength for ceramics, tear strength for elastomers and tensile strength for composites). The guide lines of constant σ/ρ, $\sigma^{2/3}/\rho$ and $\sigma^{1/2}/\rho$ are used in minimum weight, yield-limited, design (Ashby, 1989, pp. 1273–93, with permission of Elsevier Science Ltd.).*

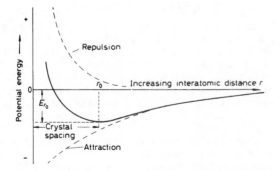

Figure 6.2 *Variation in potential energy with interatomic distance.*

the distance r is such that the atoms are no longer interacting, the material is transformed to the gaseous phase, and the energy to bring this about is the energy of evaporation.

The change in dimensions with temperature is usually expressed in terms of the linear coefficient of expansion α, given by $\alpha = (1/l)(\mathrm{d}l/\mathrm{d}T)$, where l is the original length of the specimen and T is the absolute temperature. Because of the anisotropic nature of crystals, the value of α usually varies with the direction of measurement and even in a particular crystallographic direction the dimensional change with temperature may not always be uniform.

Phase changes in the solid state are usually studied by *dilatometry*. The change in dimensions of a specimen can be transmitted to a sensitive dial gauge or electrical transducer by means of a fused silica rod. When a phase transformation takes place, because the new phase usually occupies a different volume to the old phase, discontinuities are observed in the coefficient of thermal expansion α versus T curve. Some of the 'nuclear metals' which exist in many allotropic forms, such as uranium and plutonium, show a negative coefficient along one of the crystallographic axes in certain of their allotropic modifications.

The change in volume with temperature is important in many metallurgical operations such as casting, welding and heat treatment. Of particular importance is the volume change associated with the melting or, alternatively, the freezing phenomenon since this is responsible for many of the defects, both of a macroscopic and microscopic size, which exist in crystals. Most metals increase their volume by about 3% on melting, although those metals which have crystal structures of lower coordination, such as bismuth, antimony or gallium, contract on melting. This volume change is quite small, and while the liquid structure is more open than the solid structure, it is clear that the liquid state resembles the solid state more closely than it does the gaseous phase. For the simple metals the latent heat of melting, which is merely the work done in separating the atoms from the close-packed structure of the solid to the more open liquid structure, is only about one thirtieth of the latent heat of evaporation, while the electrical and thermal conductivities are reduced only to three-quarters to one-half of the solid state values.

6.3.2 Specific heat capacity

The *specific heat* is another thermal property important in the processing operations of casting or heat treatment, since it determines the amount of heat required in the process. Thus, the specific heat (denoted by C_p, when dealing with the specific heat at constant pressure) controls the increase in temperature, dT, produced by the addition of a given quantity of heat, dQ, to one gram of matter so that $dQ = C_p dT$.

The specific heat of a metal is due almost entirely to the vibrational motion of the ions. However, a small part of the specific heat is due to the motion of the free electrons, which becomes important at high temperatures, especially in transition metals with electrons in incomplete shells.

The classical theory of specific heat assumes that an atom can oscillate in any one of three directions, and hence a crystal of N atoms can vibrate in $3N$ independent normal modes, each with its characteristic frequency. Furthermore, the mean energy of each normal mode will be kT, so that the total vibrational thermal energy of the metal is $E = 3NkT$. In solid and liquid metals, the volume changes on heating are very small and, consequently, it is customary to consider the specific heat at constant volume. If N, the number of atoms in the crystal, is equal to the number of atoms in a gram-atom (i.e. Avogadro number), the heat capacity per gram-atom, i.e. the atomic heat, at constant volume is given by

$$C_v \left(\frac{dQ}{dT} \right)_v = \frac{dE}{dT} = 3N\mathbf{k} = 24.95 \text{ J K}^{-1}$$

In practice, of course, when the specific heat is experimentally determined, it is the specific heat at constant pressure, C_p, which is measured, not C_v, and this is given by

$$C_p \left(\frac{dE + PdV}{dT} \right)_p = \frac{dH}{dT}$$

where $H = E + PV$ is known as the heat content or enthalpy, C_p is greater than C_v by a few per cent because some work is done against interatomic forces when the crystal expands, and it can be shown that

$$C_p - C_v = 9\alpha^2 VT/\beta$$

where α is the coefficient of linear thermal expansion, V is the volume per gram-atom and β is the compressibility.

Dulong and Petit were the first to point out that the specific heat of most materials, when determined at sufficiently high temperatures and corrected to apply to constant volume, is approximately equal to $3\mathbf{R}$, where $\mathbf{R}$ is the gas constant. However, deviations from the 'classical' value of the atomic heat occur at low temperatures, as shown in Figure 6.3a. This deviation is readily accounted for by the quantum theory, since the vibrational energy must then be quantized in multiples of $\mathbf{h}\nu$, where $\mathbf{h}$ is Planck's constant and ν is the characteristic frequency of the normal mode of vibration.

According to the quantum theory, the mean energy of a normal mode of the crystal is

$$E(\nu) = \tfrac{1}{2}\mathbf{h}\nu + \{\mathbf{h}\nu/ \exp{(\mathbf{h}\nu/kT)} - 1\}$$

where $\tfrac{1}{2}\mathbf{h}\nu$ represents the energy a vibrator will have at the absolute zero of temperature, i.e. the zero-point energy. Using the assumption made by Einstein (1907) that all vibrations have the same frequency (i.e. all atoms vibrate independently), the heat capacity is

$$C_v = (dE/dT)_v$$

$$= 3N\mathbf{k}(\mathbf{h}\nu/\mathbf{k}T)^2$$

$$[\exp{(\mathbf{h}\nu/\mathbf{k}T)}/\{\exp{(\mathbf{h}\nu/\mathbf{k}T)} - 1\}^2]$$

This equation is rarely written in such a form because most materials have different values of ν. It is more usual to express ν as an equivalent temperature defined by $\Theta_E = \mathbf{h}\nu/\mathbf{k}$, where Θ_E is known as the Einstein characteristic temperature. Consequently, when C_v is plotted against T/Θ_E, the specific heat curves of all pure metals coincide and the value approaches zero at very low temperatures and rises to the classical value of $3N\mathbf{k} = 3R \simeq 25.2$ J/g at high temperatures.

Einstein's formula for the specific heat is in good agreement with experiment for $T \gtrsim \Theta_E$, but is poor for low temperatures where the practical curve falls off less rapidly than that given by the Einstein relationship. However, the discrepancy can be accounted for, as shown by Debye, by taking account of the fact that the atomic vibrations are not independent of each other. This modification to the theory gives rise to a Debye characteristic temperature Θ_D, which is defined by

$$\mathbf{k}\Theta_D = \mathbf{h}\nu_D$$

where ν_D is Debye's maximum frequency. Figure 6.3b shows the atomic heat curves of Figure 6.3a plotted against T/Θ_D; in most metals for low temperatures ($T/\Theta_D \ll 1$) a T^3 law is obeyed, but at high temperatures the free electrons make a contribution to the atomic heat which is proportional to T and this causes a rise of C above the classical value.

6.3.3 The specific heat curve and transformations

The specific heat of a metal varies smoothly with temperature, as shown in Figure 6.3a, provided that no phase change occurs. On the other hand, if the metal undergoes a structural transformation the specific heat curve exhibits a discontinuity, as shown in Figure 6.4. If the phase change occurs at a fixed temperature, the metal undergoes what is known as a first-order transformation; for example, the α to γ, γ to δ and δ to liquid phase changes in iron shown in Figure 6.4a. At the transformation temperature the latent heat is absorbed without a rise in temperature, so that the specific heat ($\mathrm{d}Q/\mathrm{d}T$) at the transformation temperature is infinite. In some cases, known as transformations of the second order, the phase transition occurs over a range of temperature (e.g. the order–disorder transformation in alloys), and is associated with a specific heat peak of the form shown in Figure 6.4b. Obviously the narrower the temperature range $T_1 - T_c$, the sharper is the specific heat peak, and in the limit when the total change occurs at a single temperature, i.e. $T_1 = T_c$, the specific heat becomes infinite and equal to the latent heat of transformation. A second-order transformation also occurs in iron (see Figure 6.4a), and in this case is due to a change in ferromagnetic properties with temperature.

6.3.4 Free energy of transformation

In Section 3.2.3.2 it was shown that any structural changes of a phase could be accounted for in terms of the variation of free energy with temperature. The

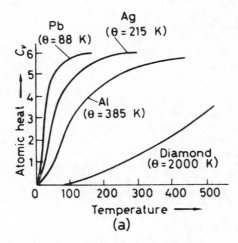
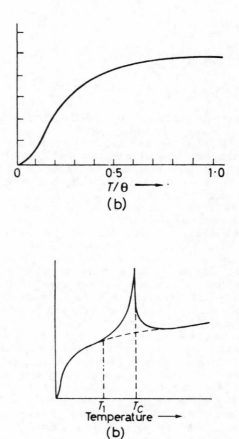

Figure 6.3 *The variation of atomic heat with temperature.*

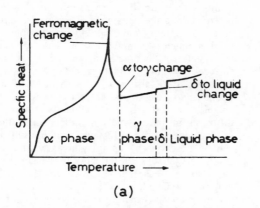

Figure 6.4 *The effect of solid state transformations on the specific heat–temperature curve.*

relative magnitude of the free energy value governs the stability of any phase, and from Figure 3.9a it can be seen that the free energy G at any temperature is in turn governed by two factors: (1) the value of G at 0 K, G_0, and (2) the slope of the G versus T curve, i.e. the temperature-dependence of free energy. Both of these terms are influenced by the vibrational frequency, and consequently the specific heat of the atoms, as can be shown mathematically. For example, if the temperature of the system is raised from T to $T + dT$ the change in free energy of the system dG is

$$dG = dH - TdS - SdT$$
$$= C_p dT - T(C_p dT/T) - SdT$$
$$= -SdT$$

so that the free energy of the system at a temperature T is

$$G = G_0 - \int_0^T SdT$$

At the absolute zero of temperature, the free energy G_0 is equal to H_0, and then

$$G = H_0 - \int_0^T SdT$$

which if S is replaced by $\int_0^T (C_p/T)dT$ becomes

$$G = H_0 - \int_0^T \left[\int_0^T (C_p/T)dT \right] dT \tag{6.1}$$

Equation (6.1) indicates that the free energy of a given phase decreases more rapidly with rise in temperature the larger its specific heat. The intersection of the free energy–temperature curves, shown in Figure 3.9a, therefore takes place because the low-temperature phase has a smaller specific heat than the higher-temperature phase.

At low temperatures the second term in equation (6.1) is relatively unimportant, and the phase that is stable is the one which has the lowest value of H_0, i.e. the most close-packed phase which is associated with a strong bonding of the atoms. However, the more strongly bound the phase, the higher is its elastic constant, the higher the vibrational frequency, and consequently the smaller the specific heat (see Figure 6.3a). Thus, the more weakly bound structure, i.e. the phase with the higher H_0 at low temperature, is likely to appear as the stable phase at higher temperatures. This is because the second term in equation (6.1) now becomes important and G decreases more rapidly with increasing temperature, for the phase with the largest value of $\int (C_p/T)dT$. From Figure 6.3b it is clear that a large $\int (C_p/T)dT$ is associated with a low characteristic temperature and hence, with a low vibrational frequency such as is displayed by a metal with a more open structure and small elastic strength. In general, therefore, when phase changes occur the more close-packed structure usually exists at the low temperatures and the more open structures at the high temperatures. From this viewpoint a liquid, which possesses no long-range structure, has a higher entropy than any solid phase so that ultimately all metals must melt at a sufficiently high temperature, i.e. when the TS term outweighs the H term in the free energy equation.

The sequence of phase changes in such metals as titanium, zirconium, etc. is in agreement with this prediction and, moreover, the alkali metals, lithium and sodium, which are normally bcc at ordinary temperatures, can be transformed to fcc at sub-zero temperatures. It is interesting to note that iron, being bcc (α-iron) even at low temperatures and fcc (γ-iron) at high temperatures, is an exception to this rule. In this case, the stability of the bcc structure is thought to be associated with its ferromagnetic properties. By having a bcc structure the interatomic distances are of the correct value for the exchange interaction to allow the electrons to adopt parallel spins (this is a condition for magnetism). While this state is one of low entropy it is also one of minimum internal energy, and in the lower temperature ranges this is the factor which governs the phase stability, so that the bcc structure is preferred.

Iron is also of interest because the bcc structure, which is replaced by the fcc structure at temperatures above 910°C, reappears as the δ-phase above 1400°C. This behaviour is attributed to the large electronic specific heat of iron which is a characteristic feature of most transition metals. Thus, the Debye characteristic temperature of γ-iron is lower than that of α-iron and this is mainly responsible for the α to γ transformation. However, the electronic specific heat of the α-phase becomes greater than that of the γ-phase above about 300°C and eventually at higher temperatures becomes sufficient to bring about the return to the bcc structure at 1400°C.

6.4 Diffusion

6.4.1 Diffusion laws

Some knowledge of diffusion is essential in understanding the behaviour of materials, particularly at elevated temperatures. A few examples include such commercially important processes as annealing, heat-treatment, the age-hardening of alloys, sintering, surface-hardening, oxidation and creep. Apart from the specialized diffusion processes, such as grain boundary diffusion and diffusion down dislocation channels, a distinction is frequently drawn between diffusion in pure metals, homogeneous alloys and inhomogeneous alloys. In a pure material self-diffusion can be observed by using radioactive tracer atoms. In a homogeneous alloy diffusion of each component can also be measured by a tracer method, but in an inhomogeneous alloy, diffusion can be determined by chemical analysis merely from the broadening of the interface between the two metals as a function of time.

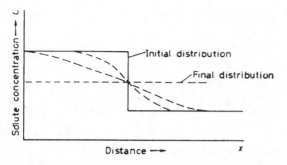

Figure 6.5 *Effect of diffusion on the distribution of solute in an alloy.*

Inhomogeneous alloys are common in metallurgical practice (e.g. cored solid solutions) and in such cases diffusion always occurs in such a way as to produce a macroscopic flow of solute atoms down the concentration gradient. Thus, if a bar of an alloy, along which there is a concentration gradient (Figure 6.5) is heated for a few hours at a temperature where atomic migration is fast, i.e. near the melting point, the solute atoms are redistributed until the bar becomes uniform in composition. This occurs even though the individual atomic movements are random, simply because there are more solute atoms to move down the concentration gradient than there are to move up. This fact forms the basis of Fick's law of diffusion, which is

$$dn/dt = -D dc/dx \qquad (6.2)$$

Here the number of atoms diffusing in unit time across unit area through a unit concentration gradient is known as the diffusivity or diffusion coefficient,[1] D. It is usually expressed as units of $cm^2 s^{-1}$ or $m^2 s^{-1}$ and depends on the concentration and temperature of the alloy.

To illustrate, we may consider the flow of atoms in one direction x, by taking two atomic planes A and B of unit area separated by a distance b, as shown in Figure 6.6. If c_1 and c_2 are the concentrations of diffusing atoms in these two planes ($c_1 > c_2$) the corresponding number of such atoms in the respective planes is $n_1 = c_1 b$ and $n_2 = c_2 b$. If the probability that any one jump in the $+x$ direction is p_x, then the number of jumps per unit time made by one atom is $p_x v$, where v is the mean frequency with which an atom leaves a site irrespective of directions. The number of diffusing atoms leaving A and arriving at B in unit time is $(p_x v c_1 b)$ and the number making the reverse transition is $(p_x v c_2 b)$ so that the net gain of atoms at B is

$$p_x v b(c_1 - c_2) = J_x$$

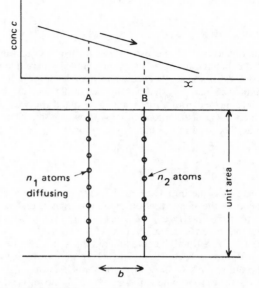

Figure 6.6 *Diffusion of atoms down a concentration gradient.*

with J_x the flux of diffusing atoms. Setting $c_1 - c_2 = -b(dc/dx)$ this flux becomes

$$J_x = -p_x v_v b^2 (dc/dx) = -\tfrac{1}{2} v b^2 (dc/dx)$$
$$= -D(dc/dx) \qquad (6.3)$$

In cubic lattices, diffusion is isotropic and hence all six orthogonal directions are equally likely so that $p_x = \tfrac{1}{6}$. For simple cubic structures $b = a$ and thus

$$D_x = D_y = D_z = \tfrac{1}{6} v a^2 = D \qquad (6.4)$$

whereas in fcc structures $b = a/\sqrt{2}$ and $D = \tfrac{1}{12} v a^2$, and in bcc structures $D = \tfrac{1}{24} v a^2$.

Fick's first law only applies if a steady state exists in which the concentration at every point is invariant, i.e. $(dc/dt) = 0$ for all x. To deal with nonstationary flow in which the concentration at a point changes with time, we take two planes A and B, as before, separated by unit distance and consider the rate of increase of the number of atoms (dc/dt) in a unit volume of the specimen; this is equal to the difference between the flux into and that out of the volume element. The flux across one plane is J_x and across the other $(J_x + 1) dJ/dx$ the difference being $-(dJ/dx)$. We thus obtain Fick's second law of diffusion

$$\frac{dc}{dt} = -\frac{dJ_x}{dx} = \frac{d}{dx}\left(D_x \frac{dc}{dx}\right) \qquad (6.5)$$

When D is independent of concentration this reduces to

$$\frac{dc_x}{dt} = D_x \frac{d^2 c}{dx^2} \qquad (6.6)$$

and in three dimensions becomes

$$\frac{dc}{dt} = \frac{d}{dx}\left(D_x\frac{dc}{dx}\right) + \frac{d}{dy}\left(D_y\frac{dc}{dy}\right) + \frac{d}{dz}\left(D_z\frac{dc}{dz}\right)$$

An illustration of the use of the diffusion equations is the behaviour of a diffusion couple, where there is a sharp interface between pure metal and an alloy. Figure 6.5 can be used for this example and as the solute moves from alloy to the pure metal the way in which the concentration varies is shown by the dotted lines. The solution to Fick's second law is given by

$$c = \frac{c_0}{2}\left[1 - \frac{2}{\sqrt{\pi}}\int_0^{x/[2\sqrt{(Dt)}]} \exp\left(-y^2\right)dy\right] \quad (6.7)$$

where c_0 is the initial solute concentration in the alloy and c is the concentration at a time t at a distance x from the interface. The integral term is known as the Gauss error function (erf (y)) and as $y \to \infty$, erf $(y) \to 1$. It will be noted that at the interface where $x = 0$, then $c = c_0/2$, and in those regions where the curvature $\partial^2c/\partial x^2$ is positive the concentration rises, in those regions where the curvature is negative the concentration falls, and where the curvature is zero the concentration remains constant.

This particular example is important because it can be used to model the depth of diffusion after time t, e.g. in the case-hardening of steel, providing the concentration profile of the carbon after a carburizing time t, or dopant in silicon. Starting with a constant composition at the surface, the value of x where the concentration falls to half the initial value, i.e. $1 - \text{erf}(y) = \frac{1}{2}$, is given by $x = \sqrt{(Dt)}$. Thus knowing D at a given temperature the time to produce a given depth of diffusion can be estimated.

The diffusion equations developed above can also be transformed to apply to particular diffusion geometries. If the concentration gradient has spherical symmetry about a point, c varies with the radial distance r and, for constant D,

$$\frac{dc}{dt} = D\left(\frac{d^2c}{dr^2} + \frac{2}{r}\frac{dc}{dr}\right) \quad (6.8)$$

When the diffusion field has radial symmetry about a cylindrical axis, the equation becomes

$$\frac{dc}{dt} = D\left(\frac{d^2c}{dr^2} + \frac{1}{r}\frac{dc}{dr}\right) \quad (6.9)$$

and the steady-state condition $(dc/dt) = 0$ is given by

$$\frac{d^2c}{dr^2} + \frac{1}{r}\frac{dc}{dr} = 0 \quad (6.10)$$

which has a solution $c = A\ln r + B$. The constants A and B may be found by introducing the appropriate boundary conditions and for $c = c_0$ at $r = r_0$ and $c = c_1$ at $r = r_1$ the solution becomes

$$c = \frac{c_0\ln(r_1/r) + c_1\ln(r/r_0)}{\ln(r_1/r_0)}$$

The flux through any shell of radius r is $-2\pi rD(dc/dr)$ or

$$J = -\frac{2\pi D}{\ln(r_1/r_0)}(c_1 - c_0) \quad (6.11)$$

Diffusion equations are of importance in many diverse problems and in Chapter 4 are applied to the diffusion of vacancies from dislocation loops and the sintering of voids.

6.4.2 Mechanisms of diffusion

The transport of atoms through the lattice may conceivably occur in many ways. The term 'interstitial diffusion' describes the situation when the moving atom does not lie on the crystal lattice, but instead occupies an interstitial position. Such a process is likely in interstitial alloys where the migrating atom is very small (e.g. carbon, nitrogen or hydrogen in iron). In this case, the diffusion process for the atoms to move from one interstitial position to the next in a perfect lattice is not defect-controlled. A possible variant of this type of diffusion has been suggested for substitutional solutions in which the diffusing atoms are only temporarily interstitial and are in dynamic equilibrium with others in substitutional positions. However, the energy to form such an interstitial is many times that to produce a vacancy and, consequently, the most likely mechanism is that of the continual migration of vacancies. With vacancy diffusion, the probability that an atom may jump to the next site will depend on: (1) the probability that the site is vacant (which in turn is proportional to the fraction of vacancies in the crystal), and (2) the probability that it has the required activation energy to make the transition. For self-diffusion where no complications exist, the diffusion coefficient is therefore given by

$$D = \frac{1}{6}a^2 f \nu \exp\left[(S_f + S_m)/k\right]$$
$$\times \exp\left[-E_f/kT\right]\exp\left[-E_m/kT\right]$$
$$= D_0\exp\left[-(E_f + E_m)/kT\right] \quad (6.12)$$

The factor f appearing in D_0 is known as a correlation factor and arises from the fact that any particular diffusion jump is influenced by the direction of the previous jump. Thus when an atom and a vacancy exchange places in the lattice there is a greater probability of the atom returning to its original site than moving to another site, because of the presence there of a vacancy; f is 0.80 and 0.78 for fcc and bcc lattices, respectively. Values for E_f and E_m are discussed in Chapter 4, E_f is the energy of formation of a vacancy, E_m the energy of migration, and the sum of the two energies, $Q = E_f + E_m$, is the activation energy for self-diffusion[1] E_d.

[1] The entropy factor $\exp\left[(S_f + S_m)/k\right]$ is usually taken to be unity.

In alloys, the problem is not so simple and it is found that the self-diffusion energy is smaller than in pure metals. This observation has led to the suggestion that in alloys the vacancies associate preferentially with solute atoms in solution; the binding of vacancies to the impurity atoms increases the effective vacancy concentration near those atoms so that the mean jump rate of the solute atoms is much increased. This association helps the solute atom on its way through the lattice, but, conversely, the speed of vacancy migration is reduced because it lingers in the neighbourhood of the solute atoms, as shown in Figure 6.7. The phenomenon of association is of fundamental importance in all kinetic studies since the mobility of a vacancy through the lattice to a vacancy sink will be governed by its ability to escape from the impurity atoms which trap it. This problem has been mentioned in Chapter 4.

When considering diffusion in alloys it is important to realize that in a binary solution of A and B the diffusion coefficients D_A and D_B are generally not equal. This inequality of diffusion was first demonstrated by Kirkendall using an α-brass/copper couple (Figure 6.8). He noted that if the position of the interfaces of the couple were marked (e.g. with fine W or Mo wires), during diffusion the markers move towards each other, showing that the zinc atoms diffuse out of the alloy more rapidly than copper atoms diffuse in. This being the case, it is not surprising that several workers have shown that porosity develops in such systems on that side of the interface from which there is a net loss of atoms.

The Kirkendall effect is of considerable theoretical importance since it confirms the vacancy mechanism of diffusion. This is because the observations cannot easily be accounted for by any other postulated mechanisms of diffusion, such as direct place-exchange, i.e. where neighbouring atoms merely change place with each other. The Kirkendall effect is readily explained in terms of vacancies since the lattice defect may interchange places more frequently with one atom than the other. The effect is also of

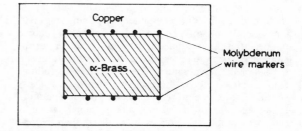

Figure 6.8 α-brass−copper couple for demonstrating the Kirkendall effect.

some practical importance, especially in the fields of metal-to-metal bonding, sintering and creep.

6.4.3 Factors affecting diffusion

The two most important factors affecting the diffusion coefficient D are temperature and composition. Because of the activation energy term the rate of diffusion increases with temperature according to equation (6.12), while each of the quantities D, D_0 and Q varies with concentration; for a metal at high temperatures $Q \approx 20RT_m$, D_0 is 10^{-5} to 10^{-3} m^2 s^{-1}, and $D \simeq 10^{-12}$ m^2 s^{-1}. Because of this variation of diffusion coefficient with concentration, the most reliable investigations into the effect of other variables necessarily concern self-diffusion in pure metals.

Diffusion is a structure-sensitive property and, therefore, D is expected to increase with increasing lattice irregularity. In general, this is found experimentally. In metals quenched from a high temperature the excess vacancy concentration $\approx 10^9$ leads to enhanced diffusion at low temperatures since $D = D_0 c_v \exp(-E_m/kT)$. Grain boundaries and dislocations are particularly important in this respect and produce enhanced diffusion. Diffusion is faster in the cold-worked state than in the annealed state, although recrystallization may take place and tend to mask the effect. The enhanced transport of material along dislocation channels has been demonstrated in aluminium where voids connected to a free surface by dislocations anneal out at appreciably higher rates than isolated voids. Measurements show that surface and grain boundary forms of diffusion also obey Arrhenius equations, with lower activation energies than for volume diffusion, i.e. $Q_{vol} \geq 2Q_{g.b} \geq 2Q_{surface}$. This behaviour is understandable in view of the progressively more open atomic structure found at grain boundaries and external surfaces. It will be remembered, however, that the relative importance of the various forms of diffusion does not entirely depend on the relative activation energy or diffusion coefficient values. The amount of material transported by any diffusion process is given by Fick's law and for a given composition gradient also depends on the effective area through which the atoms diffuse. Consequently, since the surface area (or grain boundary area) to volume

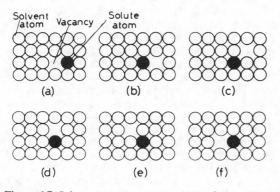

Figure 6.7 Solute atom−vacancy association during diffusion.

ratio of any polycrystalline solid is usually very small, it is only in particular phenomena (e.g. sintering, oxidation, etc.) that grain boundaries and surfaces become important. It is also apparent that grain boundary diffusion becomes more competitive, the finer the grain and the lower the temperature. The lattice feature follows from the lower activation energy which makes it less sensitive to temperature change. As the temperature is lowered, the diffusion rate along grain boundaries (and also surfaces) decreases less rapidly than the diffusion rate through the lattice. The importance of grain boundary diffusion and dislocation pipe diffusion is discussed again in Chapter 7 in relation to deformation at elevated temperatures, and is demonstrated convincingly on the deformation maps (see Figure 7.68), where the creep field is extended to lower temperatures when grain boundary (Coble creep) rather than lattice diffusion (Herring–Nabarro creep) operates.

Because of the strong binding between atoms, pressure has little or no effect but it is observed that with extremely high pressure on soft metals (e.g. sodium) an increase in Q may result. The rate of diffusion also increases with decreasing density of atomic packing. For example, self-diffusion is slower in fcc iron or thallium than in bcc iron or thallium when the results are compared by extrapolation to the transformation temperature. This is further emphasized by the anisotropic nature of D in metals of open structure. Bismuth (rhombohedral) is an example of a metal in which D varies by 10^6 for different directions in the lattice; in cubic crystals D is isotropic.

6.5 Anelasticity and internal friction

For an elastic solid it is generally assumed that stress and strain are directly proportional to one another, but in practice the elastic strain is usually dependent on time as well as stress so that the strain lags behind the stress; this is an anelastic effect. On applying a stress at a level below the conventional elastic limit, a specimen will show an initial elastic strain ε_e followed by a gradual increase in strain until it reaches an essentially constant value, $\varepsilon_e + \varepsilon_{an}$ as shown in Figure 6.9. When the stress is removed the strain will decrease, but a small amount remains which decreases slowly with time. At any time t the decreasing anelastic strain is given by the relation $\varepsilon = \varepsilon_{an} \exp(-t/\tau)$ where τ is known as the relaxation time, and is the time taken for the anelastic strain to decrease to $1/e \simeq 36.79\%$ of its initial value. Clearly, if τ is large, the strain relaxes very slowly, while if small the strain relaxes quickly.

In materials under cyclic loading this anelastic effect leads to a decay in amplitude of vibration and therefore a dissipation of energy by internal friction. Internal friction is defined in several different but related ways. Perhaps the most common uses the logarithmic decrement $\delta = \ln(A_n/A_{n+1})$, the natural logarithm of successive amplitudes of vibration. In a forced vibration experiment near a resonance, the factor $(\omega_2 - \omega_1)/\omega_0$

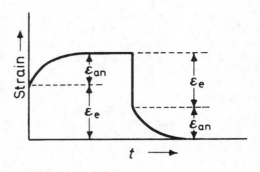

Figure 6.9 *Anelastic behaviour.*

is often used, where ω_1 and ω_2 are the frequencies on the two sides of the resonant frequency ω_0 at which the amplitude of oscillation is $1/\sqrt{2}$ of the resonant amplitude. Also used is the specific damping capacity $\Delta E/E$, where ΔE is the energy dissipated per cycle of vibrational energy E, i.e. the area contained in a stress–strain loop. Yet another method uses the phase angle α by which the strain lags behind the stress, and if the damping is small it can be shown that

$$\tan\alpha = \frac{\delta}{\pi} = \frac{1}{2\pi}\frac{\Delta E}{E} = \frac{\omega_2 - \omega_1}{\omega_0} = Q^{-1} \qquad (6.13)$$

By analogy with damping in electrical systems $\tan\alpha$ is often written equal to Q^{-1}.

There are many causes of internal friction arising from the fact that the migration of atoms, lattice defects and thermal energy are all time-dependent processes. The latter gives rise to thermoelasticity and occurs when an elastic stress is applied to a specimen too fast for the specimen to exchange heat with its surroundings and so cools slightly. As the sample warms back to the surrounding temperature it expands thermally, and hence the dilatation strain continues to increase after the stress has become constant.

The diffusion of atoms can also give rise to anelastic effects in an analogous way to the diffusion of thermal energy giving thermoelastic effects. A particular example is the stress-induced diffusion of carbon or nitrogen in iron. A carbon atom occupies the interstitial site along one of the cell edges slightly distorting the lattice tetragonally. Thus when iron is stretched by a mechanical stress, the crystal axis oriented in the direction of the stress develops favoured sites for the occupation of the interstitial atoms relative to the other two axes. Then if the stress is oscillated, such that first one axis and then another is stretched, the carbon atoms will want to jump from one favoured site to the other. Mechanical work is therefore done repeatedly, dissipating the vibrational energy and damping out the mechanical oscillations. The maximum energy is dissipated when the time per cycle is of the same order as the time required for the diffusional jump of the carbon atom.

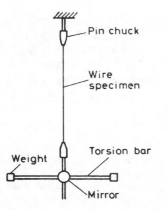

Figure 6.10 *Schematic diagram of a Kê torsion pendulum.*

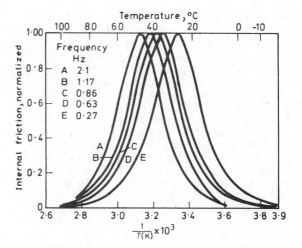

Figure 6.11 *Internal friction as a function of temperature for Fe with C in solid solution at five different pendulum frequencies (from Wert and Zener, 1949; by permission of the American Institute of Physics).*

The simplest and most convenient way of studying this form of internal friction is by means of a Kê torsion pendulum, shown schematically in Figure 6.10. The specimen can be oscillated at a given frequency by adjusting the moment of inertia of the torsion bar. The energy loss per cycle $\Delta E/E$ varies smoothly with the frequency according to the relation

$$\frac{\Delta E}{E} = 2 \left(\frac{\Delta E}{E} \right)_{max} \left[\frac{\omega\tau}{1 + (\omega\tau)^2} \right]$$

and has a maximum value when the angular frequency of the pendulum equals the relaxation time of the process; at low temperatures around room temperature this is interstitial diffusion. In practice, it is difficult to vary the angular frequency over a wide range and thus it is easier to keep ω constant and vary the relaxation time. Since the migration of atoms depends strongly on temperature according to an Arrhenius-type equation, the relaxation time $\tau_1 = 1/\omega_1$ and the peak occurs at a temperature T_1. For a different frequency value ω_2 the peak occurs at a different temperature T_2, and so on (see Figure 6.11). It is thus possible to ascribe an activation energy ΔH for the internal process producing the damping by plotting $\ln \tau$ versus $1/T$, or from the relation

$$\Delta H = R\frac{\ln(\omega_2/\omega_1)}{1/T_1 - 1/T_2}$$

In the case of iron the activation energy is found to coincide with that for the diffusion of carbon in iron. Similar studies have been made for other metals. In addition, if the relaxation time is τ the mean time an atom stays in an interstitial position is $(\frac{3}{2})\tau$, and from the relation $D = \frac{1}{24}a^2v$ for bcc lattices derived previously the diffusion coefficient may be calculated directly from

$$D = \frac{1}{36} \left(\frac{a^2}{\tau} \right)$$

Many other forms of internal friction exist in metals arising from different relaxation processes to those discussed above, and hence occurring in different frequency and temperature regions. One important source of internal friction is that due to stress relaxation across grain boundaries. The occurrence of a strong internal friction peak due to grain boundary relaxation was first demonstrated on polycrystalline aluminium at 300°C by Kê and has since been found in numerous other metals. It indicates that grain boundaries behave in a somewhat viscous manner at elevated temperatures and grain boundary sliding can be detected at very low stresses by internal friction studies. The grain boundary sliding velocity produced by a shear stress τ is given by $v = \tau d/\eta$ and its measurement gives values of the viscosity η which extrapolate to that of the liquid at the melting point, assuming the boundary thickness to be $d \simeq 0.5$ nm.

Movement of low-energy twin boundaries in crystals, domain boundaries in ferromagnetic materials and dislocation bowing and unpinning all give rise to internal friction and damping.

6.6 Ordering in alloys

6.6.1 Long-range and short-range order

An ordered alloy may be regarded as being made up of two or more interpenetrating sub-lattices, each containing different arrangements of atoms. Moreover, the term 'superlattice' would imply that such a coherent atomic scheme extends over large distances, i.e. the crystal possesses long-range order. Such a perfect arrangement can exist only at low temperatures, since the entropy of an ordered structure is much lower than that of a disordered one, and with increasing temperature the degree of long-range order, S, decreases until at a critical temperature T_c it becomes zero; the general form of the curve is shown in Figure 6.12. Partially-ordered structures are achieved by the formation of small regions (domains) of order, each of which are separated from each other by domain or anti-phase domain boundaries, across which the order changes phase (Figure 6.13). However, even when long-range order is destroyed, the tendency for unlike atoms to be neighbours still exists, and short-range order results above T_c. The transition from complete disorder to complete order is a nucleation and growth process and may be likened to the annealing of a cold-worked structure. At high temperatures well above T_c, there are more than the random number of AB atom pairs, and with the lowering of temperature small nuclei of order continually form and disperse in an otherwise disordered matrix. As the temperature, and hence thermal agitation, is lowered these regions of order become more extensive, until at T_c they begin to link together and the alloy consists of an interlocking mesh of small ordered regions. Below T_c these domains absorb each other (cf. grain growth) as a result of antiphase domain boundary mobility until long-range order is established.

Some order–disorder alloys can be retained in a state of disorder by quenching to room temperature while in others (e.g. β-brass) the ordering process occurs almost instantaneously. Clearly, changes in the degree of order will depend on atomic migration, so that the rate of approach to the equilibrium configuration will be governed by an exponential factor of the usual form, i.e. Rate $= Ae^{-Q/RT}$. However, Bragg

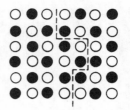

Figure 6.13 *An antiphase domain boundary.*

has pointed out that the ease with which interlocking domains can absorb each other to develop a scheme of long-range order will also depend on the number of possible ordered schemes the alloy possesses. Thus, in β-brass only two different schemes of order are possible, while in fcc lattices such as Cu_3Au four different schemes are possible and the approach to complete order is less rapid.

6.6.2 Detection of ordering

The determination of an ordered superlattice is usually done by means of the X-ray powder technique. In a disordered solution every plane of atoms is statistically identical and, as discussed in Chapter 5, there are reflections missing in the powder pattern of the material. In an ordered lattice, on the other hand, alternate planes become A-rich and B-rich, respectively, so that these 'absent' reflections are no longer missing but appear as extra superlattice lines. This can be seen from Figure 6.14: while the diffracted rays from the A planes are completely out of phase with those from the B planes their intensities are not identical, so that a weak reflection results.

Application of the structure factor equation indicates that the intensity of the superlattice lines is proportional to $|F^2| = S^2(f_A - f_B)^2$, from which it can be seen that in the fully-disordered alloy, where $S = 0$, the superlattice lines must vanish. In some alloys such as copper–gold, the scattering factor difference $(f_A - f_B)$ is appreciable and the superlattice lines are, therefore, quite intense and easily detectable. In other alloys, however, such as iron–cobalt, nickel–manganese, copper–zinc, the term $(f_A - f_B)$ is negligible for X-rays and the super-lattice lines are very weak; in copper–zinc, for

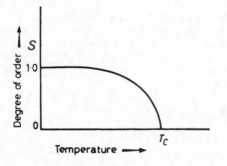

Figure 6.12 *Influence of temperature on the degree of order.*

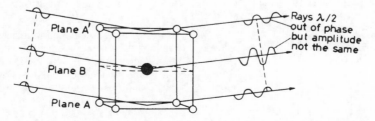

Figure 6.14 *Formation of a weak 100 reflection from an ordered lattice by the interference of diffracted rays of unequal amplitude.*

example, the ratio of the intensity of the superlattice lines to that of the main lines is only about 1:3500. In some cases special X-ray techniques can enhance this intensity ratio; one method is to use an X-ray wavelength near to the absorption edge when an anomalous depression of the f-factor occurs which is greater for one element than for the other. As a result, the difference between f_A and f_B is increased. A more general technique, however, is to use neutron diffraction since the scattering factors for neighbouring elements in the Periodic Table can be substantially different. Conversely, as Table 5.4 indicates, neutron diffraction is unable to show the existence of superlattice lines in Cu_3Au, because the scattering amplitudes of copper and gold for neutrons are approximately the same, although X-rays show them up quite clearly.

Sharp superlattice lines are observed as long as order persists over lattice regions of about 10^{-3} mm, large enough to give coherent X-ray reflections. When long-range order is not complete the superlattice lines become broadened, and an estimate of the domain size can be obtained from a measurement of the line breadth, as discussed in Chapter 5. Figure 6.15 shows variation of order S and domain size as determined from the intensity and breadth of powder diffraction lines. The domain sizes determined from the Scherrer line-broadening formula are in very good agreement with those observed by TEM. Short-range order is much more difficult to detect but nowadays direct measuring devices allow weak X-ray intensities to be measured more accurately, and as a result considerable information on the nature of short-range order has been obtained by studying the intensity of the diffuse background between the main lattice lines.

High-resolution transmission microscopy of thin metal foils allows the structure of domains to be examined directly. The alloy CuAu is of particular interest, since it has a face-centred tetragonal structure, often referred to as CuAu 1 below 380°C, but between 380°C and the disordering temperature of 410°C it has the CuAu 11 structures shown in Figure 6.16. The (002) planes are again alternately gold and copper, but halfway along the a-axis of the unit cell the copper atoms switch to gold planes and vice versa. The spacing between such periodic anti-phase domain boundaries is 5 unit cells or about 2 nm, so that the domains are easily resolvable in TEM, as seen in Figure 6.17a. The isolated domain boundaries in the simpler superlattice structures such as CuAu 1, although not in this case periodic, can also be revealed by electron microscope, and an example is shown in Figure 6.17b. Apart from static observations of these superlattice structures, annealing experiments inside the microscope also allow the effect of temperature on the structure to be examined directly. Such observations have shown that the transition from CuAu 1 to CuAu 11 takes place, as predicted, by the nucleation and growth of anti-phase domains.

6.6.3 Influence of ordering on properties

Specific heat The order–disorder transformation has a marked effect on the specific heat, since energy is necessary to change atoms from one configuration to another. However, because the change in lattice arrangement takes place over a range of temperature, the specific heat versus temperature curve will be of the form shown in Figure 6.4b. In practice the excess specific heat, above that given by Dulong and Petit's law, does not fall sharply to zero at T_c owing to the existence of short-range order, which also requires extra energy to destroy it as the temperature is increased above T_c.

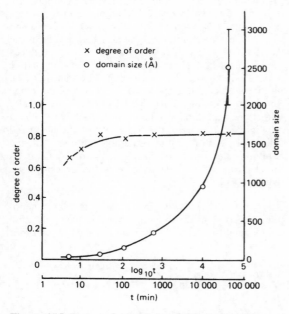

Figure 6.15 *Degree of order (×) and domain size (O) during isothermal annealing at 350°C after quenching from 465°C (after Morris, Besag and Smallman, 1974; courtesy of Taylor and Francis).*

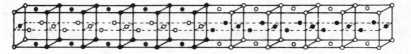

Figure 6.16 *One unit cell of the orthorhombic superlattice of CuAu, i.e. CuAu 11 (from J. Inst. Metals, 1958–9, courtesy of the Institute of Metals).*

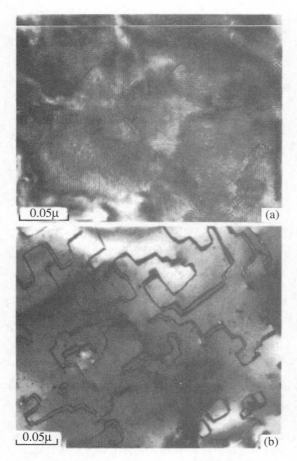

Figure 6.17 *Electron micrographs of (a) CuAu 11 and (b) CuAu 1 (from Pashley and Presland, 1958–9; courtesy of the Institute of Metals).*

Electrical resistivity As discussed in Chapter 4, any form of disorder in a metallic structure (e.g. impurities, dislocations or point defects) will make a large contribution to the electrical resistance. Accordingly, superlattices below T_c have a low electrical resistance, but on raising the temperature the resistivity increases, as shown in Figure 6.18a for ordered Cu_3Au. The influence of order on resistivity is further demonstrated by the measurement of resistivity as a function of composition in the copper–gold alloy system. As shown in Figure 6.18b, at composition near Cu_3Au and CuAu, where ordering is most complete, the resistivity is extremely low, while away from these stoichiometric compositions the resistivity increases; the quenched (disordered) alloys given by the dotted curve also have high resistivity values.

Mechanical properties The mechanical properties are altered when ordering occurs. The change in yield stress is not directly related to the degree of ordering, however, and in fact Cu_3Au crystals have a lower yield stress when well-ordered than when only partially-ordered. Experiments show that such effects can be accounted for if the maximum strength as a result of ordering is associated with critical domain size. In the alloy Cu_3Au, the maximum yield strength is exhibited by quenched samples after an annealing treatment of 5 min at 350°C which gives a domain size of 6 nm (see Figure 6.15). However, if the alloy is well-ordered and the domain size larger, the hardening is insignificant. In some alloys such as CuAu or CuPt, ordering produces a change of crystal structure and the resultant lattice strains can also lead to hardening. Thermal agitation is the most common means of destroying long-range order, but other methods (e.g. deformation) are equally effective. Figure 6.18c shows that cold work has a negligible effect upon the resistivity of the quenched (disordered) alloy but considerable influence on the well-annealed (ordered) alloy. Irradiation by neutrons or electrons also markedly affects the ordering (see Chapter 4).

Magnetic properties The order–disorder phenomenon is of considerable importance in the application of magnetic materials. The kind and degree of order

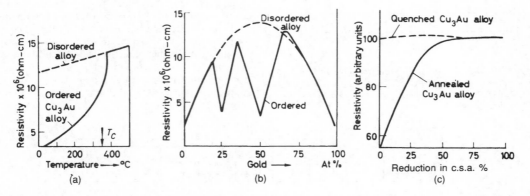

Figure 6.18 *Effect of (a) temperature, (b) composition, and (c) deformation on the resistivity of copper–gold alloys (after Barrett, 1952; courtesy of McGraw-Hill).*

affects the magnetic hardness, since small ordered regions in an otherwise disordered lattice induce strains which affect the mobility of magnetic domain boundaries (see Section 6.8.4).

6.7 Electrical properties

6.7.1 Electrical conductivity

One of the most important electronic properties of metals is the electrical conductivity, κ, and the reciprocal of the conductivity (known as the resistivity, ρ) is defined by the relation $R = \rho l / A$, where R is the resistance of the specimen, l is the length and A is the cross-sectional area.

A characteristic feature of a metal is its high electrical conductivity which arises from the ease with which the electrons can migrate through the lattice. The high thermal conduction of metals also has a similar explanation, and the Wiedmann−Franz law shows that the ratio of the electrical and thermal conductivities is nearly the same for all metals at the same temperature.

Since conductivity arises from the motion of conduction electrons through the lattice, resistance must be caused by the scattering of electron waves by any kind of irregularity in the lattice arrangement. Irregularities can arise from any one of several sources, such as temperature, alloying, deformation or nuclear irradiation, since all will disturb, to some extent, the periodicity of the lattice. The effect of temperature is particularly important and, as shown in Figure 6.19, the resistance increases linearly with temperature above about 100 K up to the melting point. On melting, the resistance increases markedly because of the exceptional disorder of the liquid state. However, for some metals such as bismuth, the resistance actually decreases, owing to the fact that the special zone structure which makes

bismuth a poor conductor in the solid state is destroyed on melting.

In most metals the resistance approaches zero at absolute zero, but in some (e.g. lead, tin and mercury) the resistance suddenly drops to zero at some finite critical temperature above 0 K. Such metals are called superconductors. The critical temperature is different for each metal but is always close to absolute zero; the highest critical temperature known for an element is 8 K for niobium. Superconductivity is now observed at much higher temperatures in some intermetallic compounds and in some ceramic oxides (see Section 6.7.4).

An explanation of electrical and magnetic properties requires a more detailed consideration of electronic structure than that briefly outlined in Chapter 1. There the concept of band structure was introduced and the electron can be thought of as moving continuously through the structure with an energy depending on the energy level of the band it occupies. The wave-like properties of the electron were also mentioned. For the electrons the regular array of atoms on the metallic lattice can behave as a three-dimensional diffraction grating since the atoms are positively-charged and interact with moving electrons. At certain wavelengths, governed by the spacing of the atoms on the metallic lattice, the electrons will experience strong diffraction effects, the results of which are that electrons having energies corresponding to such wavelengths will be unable to move freely through the structure. As a consequence, in the bands of electrons, certain energy levels cannot be occupied and therefore there will be energy gaps in the otherwise effectively continuous energy spectrum within a band.

The interaction of moving electrons with the metal ions distributed on a lattice depends on the wavelength of the electrons and the spacing of the ions in the direction of movement of the electrons. Since the ionic spacing will depend on the direction in the lattice, the wavelength of the electrons suffering diffraction by the ions will depend on their direction. The kinetic energy of a moving electron is a function of the wavelength according to the relationship

$$E = \mathbf{h}^2 / 2m\lambda^2 \tag{6.14}$$

Since we are concerned with electron energies, it is more convenient to discuss interaction effects in terms of the reciprocal of the wavelength. This quantity is called the wave number and is denoted by k.

In describing electron−lattice interactions it is usual to make use of a vector diagram in which the direction of the vector is the direction of motion of the moving electron and its magnitude is the wave number of the electron. The vectors representing electrons having energies which, because of diffraction effects, cannot penetrate the lattice, trace out a three-dimensional surface known as a Brillouin zone. Figure 6.20a shows such a zone for a face-centred cubic lattice. It is made up of plane faces which are, in fact, parallel to the most

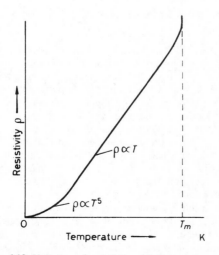

Figure 6.19 *Variation of resistivity with temperature.*

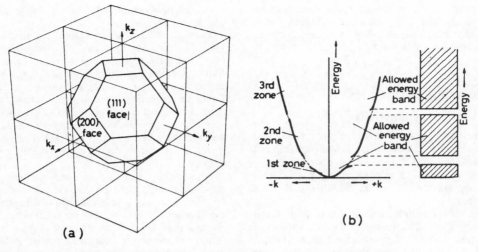

Figure 6.20 *Schematic representation of a Brillouin zone in a metal.*

widely-spaced planes in the lattice, i.e. in this case the {1 1 1} and {2 0 0} planes. This is a general feature of Brillouin zones in all lattices.

For a given direction in the lattice, it is possible to consider the form of the electron energies as a function of wave number. The relationship between the two quantities as given from equation (6.14) is

$$E = \mathbf{h}^2 k^2 / 2m \qquad (6.15)$$

which leads to the parabolic relationship shown as a broken line in Figure 6.20b. Because of the existence of a Brillouin zone at a certain value of k, depending on the lattice direction, there exists a range of energy values which the electrons cannot assume. This produces a distortion in the form of the E-k curve in the neighbourhood of the critical value of k and leads to the existence of a series of energy gaps, which cannot be occupied by electrons. The E-k curve showing this effect is given as a continuous line in Figure 6.20b.

The existence of this distortion in the E-k curve, due to a Brillouin zone, is reflected in the density of states versus energy curve for the free electrons. As previously stated, the density of states–energy curve is parabolic in shape, but it departs from this form at energies for which Brillouin zone interactions occur. The result of such interactions is shown in Figure 6.21a in which the broken line represents the $N(E)$-E curve for free electrons in the absence of zone effects and the full line is the curve where a zone exists. The total number of electrons needed to fill the zone of electrons delineated by the full line in Figure 6.21a is $2N$, where N is the total number of atoms in the metal. Thus, a Brillouin zone would be filled if the metal atoms each contributed two electrons to the band. If the metal atoms contribute more than two per atom, the excess electrons must be accommodated in the second or higher zones.

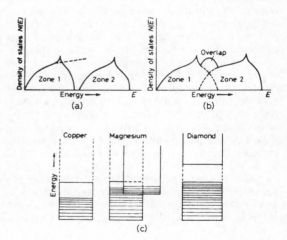

Figure 6.21 *Schematic representation of Brillouin zones.*

In Figure 6.21a the two zones are separated by an energy gap, but in real metals this is not necessarily the case, and two zones can overlap in energy in the $N(E)$-E curves so that no such energy gaps appear. This overlap arises from the fact that the energy of the forbidden region varies with direction in the lattice and often the energy level at the top of the first zone has a higher value in one direction than the lowest energy level at the bottom of the next zone in some other direction. The energy gap in the $N(E)$-E curves, which represent the summation of electronic levels in all directions, is then closed (Figure 6.21b).

For electrical conduction to occur, it is necessary that the electrons at the top of a band should be able to increase their energy when an electric field is applied to materials so that a net flow of electrons in the direction of the applied potential, which manifests itself as an electric current, can take place. If an energy gap between two zones of the type shown in Figure 6.21a occurs, and if the lower zone is just filled with electrons, then it is impossible for any electrons to increase their energy by jumping into vacant levels under the influence of an applied electric field, unless the field strength is sufficiently great to supply the electrons at the top of the filled band with enough energy to jump the energy gap. Thus metallic conduction is due to the fact that in metals the number of electrons per atom is insufficient to fill the band up to the point where an energy gap occurs. In copper, for example, the 4s valency electrons fill only one half of the outer s-band. In other metals (e.g. Mg) the valency band overlaps a higher energy band and the electrons near the Fermi level are thus free to move into the empty states of a higher band. When the valency band is completely filled and the next higher band, separated by an energy gap, is completely empty, the material is either an insulator or a semiconductor. If the gap is several electron volts wide, such as in diamond where it is 7 eV, extremely high electric fields would be necessary to raise electrons to the higher band and the material is an insulator. If the gap is small enough, such as 1–2 eV as in silicon, then thermal energy may be sufficient to excite some electrons into the higher band and also create vacancies in the valency band, the material is a semiconductor. In general, the lowest energy band which is not completely filled with electrons is called a conduction band, and the band containing the valency electrons the valency band. For a conductor the valency band is also the conduction band. The electronic state of a selection of materials of different valencies is presented in Figure 6.21c. Although all metals are relatively good conductors of electricity, they exhibit among themselves a range of values for their resistivities. There are a number of reasons for this variability. The resistivity of a metal depends on the density of states of the most energetic electrons at the top of the band, and the shape of the $N(E)$-E curve at this point.

In the transition metals, for example, apart from producing the strong magnetic properties, great strength and high melting point, the d-band is also responsible for the poor electrical conductivity and high electronic specific heat. When an electron is scattered by a lattice irregularity it jumps into a different quantum state, and it will be evident that the more vacant quantum states there are available in the same energy range, the more likely will be the electron to deflect at the irregularity. The high resistivities of the transition metals may, therefore, be explained by the ease with which electrons can be deflected into vacant d-states. Phonon-assisted s-d scattering gives rise to the non-linear variation of ρ with temperature observed at high temperatures. The high electronic specific heat is also due to the high density of states in the unfilled d-band, since this gives rise to a considerable number of electrons at the top of the Fermi distribution which can be excited by thermal activation. In copper, of course, there are no unfilled levels at the top of the d-band into which electrons can go, and consequently both the electronic specific heat and electrical resistance is low. The conductivity also depends on the degree to which the electrons are scattered by the ions of the metal which are thermally vibrating, and by impurity atoms or other defects present in the metal.

Insulators can also be modified either by the application of high temperatures or by the addition of impurities. Clearly, insulators may become conductors at elevated temperatures if the thermal agitation is sufficient to enable electrons to jump the energy gap into the unfilled zone above.

6.7.2 Semiconductors

Some materials have an energy gap small enough to be surmounted by thermal excitation. In such intrinsic semiconductors, as they are called, the current carriers are electrons in the conduction band and holes in the valency band in equal numbers. The relative position of the two bands is as shown in Figure 6.22. The motion of a hole in the valency band is equivalent to the motion of an electron in the opposite direction. Alternatively, conduction may be produced by the presence of impurities which either add a few electrons to an empty zone or remove a few from a full one. Materials which have their conductivity developed in this way are commonly known as semiconductors. Silicon and germanium containing small amounts of impurity have semiconducting properties at ambient temperatures and, as a consequence, they are frequently used in electronic transistor devices. Silicon normally has completely filled zones, but becomes conducting if some of the silicon atoms, which have four valency electrons, are replaced by phosphorus, arsenic or antimony atoms which have five valency electrons. The extra electrons go into empty zones, and as a

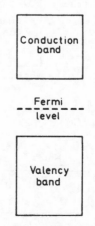

Figure 6.22 *Schematic diagram of an intrinsic semiconductor showing the relative positions of the conduction and valency bands.*

result silicon becomes an n-type semiconductor, since conduction occurs by negative carriers. On the other hand, the addition of elements of lower valency than silicon, such as aluminium, removes electrons from the filled zones leaving behind 'holes' in the valency band structure. In this case silicon becomes a p-type semiconductor, since the movement of electrons in one direction of the zone is accompanied by a movement of 'holes' in the other, and consequently they act as if they were positive carriers. The conductivity may be expressed as the product of (1) the number of charge carriers, n, (2) the charge carried by each (i.e. $e = 1.6 \times 10^{-19}$ C) and (3) the mobility of the carrier, μ.

A pentavalent impurity which donates conduction electrons without producing holes in the valency band is called a donor. The spare electrons of the impurity atoms are bound in the vicinity of the impurity atoms in energy levels known as the donor levels, which are near the conduction band. If the impurity exists in an otherwise intrinsic semiconductor the number of electrons in the conduction band become greater than the number of holes in the valency band and, hence, the electrons are the majority carriers and the holes the minority carriers. Such a material is an n-type extrinsic semiconductor (see Figure 6.23a).

Trivalent impurities in Si or Ge show the opposite behaviour leaving an empty electron state, or hole, in the valency band. If the hole separates from the so-called acceptor atom an electron is excited from the valency band to an acceptor level $\Delta E \approx 0.01$ eV. Thus, with impurity elements such as Al, Ga or In creating holes in the valency band in addition to those created thermally, the majority carriers are holes and the semiconductor is of the p-type extrinsic form (see Figure 6.23b). For a semiconductor where both electrons and holes carry current the conductivity is given by

$$\kappa = n_e e \mu_e + n_h e \mu_h \qquad (6.16)$$

where n_e and n_h are, respectively, the volume concentration of electrons and holes, and μ_e and μ_h the mobilities of the carriers, i.e. electrons and holes.

Semiconductor materials are extensively used in electronic devices such as the $p-n$ rectifying junction, transistor (a double-junction device) and the tunnel diode. Semiconductor regions of either p- or n-type can be produced by carefully controlling the distribution and impurity content of Si or Ge single crystals, and the boundary between p- and n-type extrinsic semiconductor materials is called a $p-n$ junction. Such a junction conducts a large current when the voltage is applied in one direction, but only a very small current when the voltage is reversed. The action of a $p-n$ junction as a rectifier is shown schematically in Figure 6.24. The junction presents no barrier to the flow of minority carriers from either side, but since the concentration of minority carriers is low, it is the flow of majority carriers which must be considered. When the junction is biased in the forward direction, i.e. n-type made negative and the p-type positive, the energy barrier opposing the flow of majority carriers from both sides of the junction is reduced. Excess majority carriers enter the p and n regions, and these recombine continuously at or near the junction to allow large currents to flow. When the junction is reverse-biased, the energy barrier opposing the flow of majority carriers is raised, few carriers move and little current flows.

A transistor is essentially a single crystal with two $p-n$ junctions arranged back to back to give either a $p-n-p$ or $n-p-n$ two-junction device. For a $p-n-p$ device the main current flow is provided by the positive holes, while for a $n-p-n$ device the electrons carry the current. Connections are made to the individual regions of the $p-n-p$ device, designated emitter, base

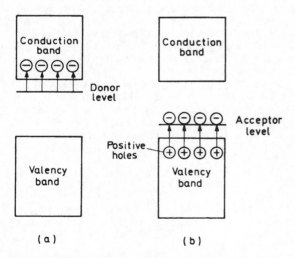

Figure 6.23 *Schematic energy band structure of (a) n-type and (b) p-type semiconductor.*

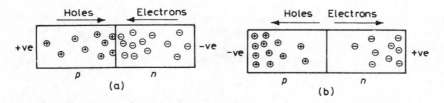

Figure 6.24 *Schematic illustration of p–n junction rectification with (a) forward bias and (b) reverse bias.*

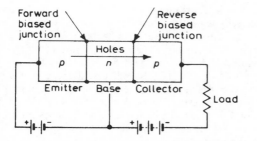

Figure 6.25 *Schematic diagram of a p–n–p transistor.*

and collector respectively, as shown in Figure 6.25, and the base is made slightly negative and the collector more negative relative to the emitter. The emitter-base junction is therefore forward-biased and a strong current of holes passes through the junction into the *n*-layer which, because it is thin (10^{-2} mm), largely reach the collector base junction without recombining with electrons. The collector-base junction is reverse-biased and the junction is no barrier to the passage of holes; the current through the second junction is thus controlled by the current through the first junction. A small increase in voltage across the emitter-base junction produces a large injection of holes into the base and a large increase in current in the collector, to give the amplifying action of the transistor.

Many varied semiconductor materials such as InSb and GaAs have been developed apart from Si and Ge. However, in all cases very high purity and crystal perfection is necessary for efficient semiconducting operations and to produce the material, zone-refining techniques are used. Semiconductor integrated circuits are extensively used in micro-electronic equipment and these are produced by vapour deposition through masks on to a single Si-slice, followed by diffusion of the deposits into the base crystal.

Doped ceramic materials are used in the construction of *thermistors*, which are semiconductor devices with a marked dependence of electrical resistivity upon temperature. The change in resistance can be quite significant at the critical temperature. Positive temperature coefficient (PTC) thermistors are used as switching devices, operating when a control temperature is reached during a heating process. PTC thermistors are commonly based on barium titanate. Conversely, NTC thermistors are based on oxide ceramics and can be used to signal a desired temperature change during cooling; the change in resistance is much more gradual and does not have the step-characteristic of the PTC types.

Doped zinc oxide does not exhibit the linear voltage/current relation that one expects from Ohm's Law. At low voltage, the resistivity is high and only a small current flows. When the voltage increases there is a sudden decrease in resistance, allowing a heavier current to flow. This principle is adopted in the *varistor*, a voltage-sensitive on/off switch. It is wired in parallel

with high-voltage equipment and can protect it from transient voltage 'spikes' or overload.

6.7.3 Superconductivity

At low temperatures (<20 K) some metals have zero electrical resistivity and become superconductors. This superconductivity disappears if the temperature of the metal is raised above a critical temperature T_c, if a sufficiently strong magnetic field is applied or when a high current density flows. The critical field strength H_c, current density J_c and temperature T_c are interdependent. Figure 6.26 shows the dependence of H_c on temperature for a number of metals; metals with high T_c and H_c values, which include the transition elements, are known as hard superconductors, those with low values such as Al, Zn, Cd, Hg, white-Sn are soft superconductors. The curves are roughly parabolic and approximate to the relation $H_c = H_0[1 - (T/T_c)^2]$ where H_0 is the critical field at 0 K; H_0 is about 1.6×10^5 A/m for Nb.

Superconductivity arises from conduction electron–electron attraction resulting from a distortion of the lattice through which the electrons are travelling; this is clearly a weak interaction since for most metals it is destroyed by thermal activation at very low temperatures. As the electron moves through the lattice it attracts nearby positive ions thereby locally causing a slightly higher positive charge density. A nearby electron may in turn be attracted by the net positive charge, the magnitude of the attraction depending on the electron density, ionic charge and lattice vibrational frequencies such that under favourable conditions the effect is slightly stronger than the electrostatic repulsion between electrons. The importance of the lattice ions in superconductivity is supported by the observation that different isotopes of the same metal (e.g. Sn and Hg) have different T_c values proportional to $M^{-1/2}$, where M is the atomic mass of the isotope. Since both the frequency of atomic vibrations and the velocity of elastic waves also varies as $M^{-1/2}$, the interaction between electrons and lattice vibrations

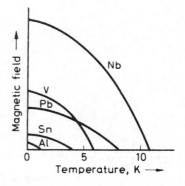

Figure 6.26 *Variation of critical field H_c as a function of temperature for several pure metal superconductors.*

(i.e. electron–phonon interaction) must be at least one cause of superconductivity.

The theory of superconductivity indicates that the electron–electron attraction is strongest between electrons in pairs, such that the resultant momentum of each pair is exactly the same and the individual electrons of each pair have opposite spin. With this particular form of ordering the total electron energy (i.e. kinetic and interaction) is lowered and effectively introduces a finite energy gap between this organized state and the usual more excited state of motion. The gap corresponds to a thin shell at the Fermi surface, but does not produce an insulator or semiconductor, because the application of an electric field causes the whole Fermi distribution, together with gap, to drift to an unsymmetrical position, so causing a current to flow. This current remains even when the electric field is removed, since the scattering which is necessary to alter the displaced Fermi distribution is suppressed.

At 0 K all the electrons are in paired states but as the temperature is raised, pairs are broken by thermal activation giving rise to a number of normal electrons in equilibrium with the superconducting pairs. With increasing temperature the number of broken pairs increases until at T_c they are finally eliminated together with the energy gap; the superconducting state then reverts to the normal conducting state. The superconductivity transition is a second-order transformation and a plot of C/T as a function of T^2 deviates from the linear behaviour exhibited by normal conducting metals, the electronic contribution being zero at 0 K. The main theory of superconductivity, due to Bardeen, Cooper and Schrieffer (BCS) attempts to relate T_c to the strength of the interaction potential, the density of states at the Fermi surface and to the average frequency of lattice vibration involved in the scattering, and provides some explanation for the variation of T_c with the e/a ratio for a wide range of alloys, as shown in Figure 6.27. The main effect is attributable to the change in density of states with e/a ratio. Superconductivity is thus favoured in compounds of polyvalent atoms with crystal structures having a high density of states at the Fermi surface. Compounds with high T_c values, such as Nb_3Sn (18.1 K), Nb_3Al (17.5 K), V_3Si (17.0 K), V_3Ga (16.8 K), all crystallize with the β-tungsten structure and have an e/a ratio close to 4.7; T_c is very sensitive to the degree of order and to deviation from the stoichiometric ratio, so values probably correspond to the non-stoichiometric condition.

The magnetic behaviour of superconductivity is as remarkable as the corresponding electrical behaviour, as shown in Figure 6.28 by the Meissner effect for an ideal (structurally perfect) superconductor. It is observed for a specimen placed in a magnetic field ($H < H_c$), which is then cooled down below T_c, that magnetic lines of force are pushed out. The specimen is a perfect diamagnetic material with zero inductance as well as zero resistance. Such a material is termed an ideal type I superconductor. An ideal type II superconductor behaves similarly at low field strengths, with $H < H_{c1} < H_c$, but then allows a gradual penetration of the field returning to the normal state when penetration is complete at $H > H_{c2} > H_c$. In detail, the field actually penetrates to a small extent in type I superconductors when it is below H_c and in type II superconductors when H is below H_{c1}, and decays away at a penetration depth $\approx$100–10 nm.

The observation of the Meissner effect in type I superconductors implies that the surface between the normal and superconducting phases has an effective positive energy. In the absence of this surface energy, the specimen would break up into separate fine regions of superconducting and normal material to reduce the work done in the expulsion of the magnetic flux. A negative surface energy exists between the normal and superconducting phases in a type II superconductor and hence the superconductor exists naturally in a state of finely-separated superconducting and normal regions. By adopting a 'mixed state' of normal and superconducting regions the volume of interface is maximized while at the same time keeping the volume

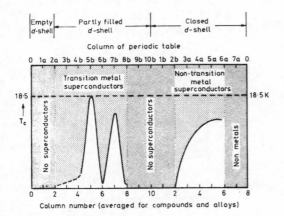

Figure 6.27 *The variation of T_C with position in the periodic table (from Mathias, 1959, p. 138; courtesy of North-Holland Publishing Co.).*

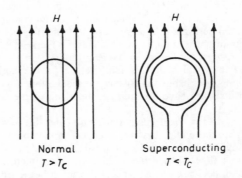

Figure 6.28 *The Meissner effect; shown by the expulsion of magnetic flux when the specimen becomes superconducting.*

of normal conduction as small as possible. The structure of the mixed state is believed to consist of lines of normal phases parallel to the applied field through which the field lines run, embedded in a superconducting matrix. The field falls off with distances from the centre of each line over the characteristic distance λ, and vortices or whirlpools of supercurrents flow around each line; the flux line, together with its current vortex, is called a fluxoid. At H_{c1}, fluxoids appear in the specimen and increase in number as the magnetic field is raised. At H_{c2}, the fluxoids completely fill the cross-section of the sample and type II superconductivity disappears. Type II superconductors are of particular interest because of their high critical fields which makes them potentially useful for the construction of high-field electromagnetics and solenoids. To produce a magnetic field of ≈ 10 T with a conventional solenoid would cost more than ten times that of a superconducting solenoid wound with Nb_3Sn wire. By embedding Nb wire in a bronze matrix it is possible to form channels of Nb_3Sn by interdiffusion. The conventional installation would require considerable power, cooling water and space, whereas the superconducting solenoid occupies little space, has no steady-state power consumption and uses relatively little liquid helium. It is necessary, however, for the material to carry useful currents without resistance in such high fields, which is not usually the case in annealed homogeneous type II superconductors. Fortunately, the critical current density is extremely sensitive to microstructure and is markedly increased by precipitation-hardening, cold work, radiation damage, etc., because the lattice defects introduced pin the fluxoids and tend to immobilize them. Figure 6.29 shows the influence of metallurgical treatment on the critical current density.

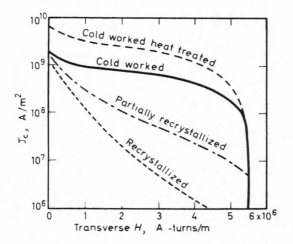

Figure 6.29 *The effect of processing on the J_c versus H curve of an Nb−25% Zr alloy wire which produces a fine precipitate and raises J_c (from Rose, Shepard and Wulff, 1966; courtesy of John Wiley and Sons).*

6.7.4 Oxide superconductors

In 1986 a new class of 'warm' superconductors, based on mixed ceramic oxides, was discovered by J. G. Bednorz and K. A. Müller. These lanthanum−copper oxide superconductors had a T_c around 35 K, well above liquid hydrogen temperature. Since then, three mixed oxide families have been developed with much higher T_c values, all around 100 K. Such materials give rise to optimism for superconductor technology; first, in the use of liquid nitrogen rather than liquid hydrogen and second, in the prospect of producing a room temperature superconductor.

The first oxide family was developed by mixing and heating the three oxides Y_2O_3, BaO and CuO. This gives rise to the mixed oxide $YBa_2Cu_3O_{7-x}$, sometimes referred to as 1−2−3 compound or YBCO. The structure is shown in Figure 6.30 and is basically made by stacking three perovskite-type unit cells one above the other; the top and bottom cells have barium ions at the centre and copper ions at the corners, the middle cell has yttrium at the centre. Oxygen ions sit half-way along the cell edges but planes, other than those containing barium, have some missing oxygen ions (i.e. vacancies denoted by x in the oxide formula). This structure therefore has planes of copper and oxygen ions containing vacancies, and copper−oxygen ion chains perpendicular to them. YBCO has a T_c value of about 90 K which is virtually unchanged when yttrium is replaced by other rare earth elements. The second family of oxides are $Bi−Ca−Sr−Cu−O_x$ materials with the metal ions in the ratio of 2111, 2122 or 2223, respectively. The 2111 oxide has only one copper−oxygen layer between the bismuth-oxygen layers, the 2122 two and the 2223 three giving rise to an increasing T_c up to about 105 K. The third family is based on $Tl−Ca−Ba−Cu−O$ with a 2223 structure having three copper−oxygen layers and a T_c of about 125 K.

While these oxide superconductors have high T_c values and high critical magnetic field (H_c)-values, they unfortunately have very low values of J_c, the critical current density. A high J_c is required if they are to be used for powerful superconducting magnets. Electrical applications are therefore unlikely until the J_c value can be raised by several orders of magnitude comparable to those of conventional superconductors, i.e. 10^6 A cm^{-2}. The reason for the low J_c is thought to be largely due to the grain boundaries in polycrystalline materials, together with dislocations, voids and impurity particles. Single crystals show J_c values around 10^5 A cm^{-2} and textured materials, produced by melt growth techniques, about 10^4 A cm^{-2}, but both processes have limited commercial application. Electronic applications appear to be more promising since it is in the area of thin (1 μm) films that high J_c values have been obtained. By careful deposition control, epitaxial and single-crystal films having $J_c \gg 10^6$ A cm^{-2} with low magnetic field dependence have been produced.

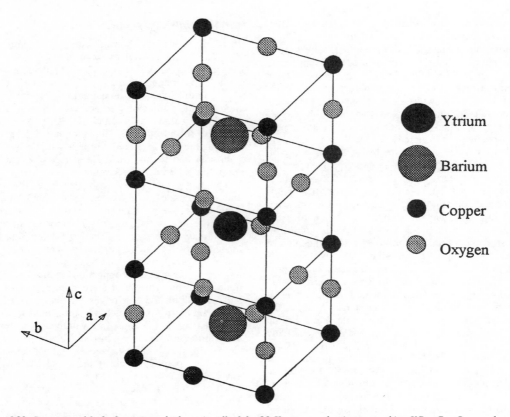

Figure 6.30 *Structure of 1–2–3 compound; the unit cell of the 90 K superconducting perovskite, $YBa_2Cu_3O_{7-x}$, where $x \sim 0$ (by courtesy of P. J. Hirst, Superconductivity Research Group, University of Birmingham, UK).*

6.8 Magnetic properties

6.8.1 Magnetic susceptibility

When a metal is placed in a magnetic field of strength H, the field induced in the metal is given by

$$B = H + 4\pi I \qquad (6.17)$$

where I is the intensity of magnetization. The quantity I is a characteristic property of the metal, and is related to the susceptibility per unit volume of the metal which is defined as

$$\kappa = I/H \qquad (6.18)$$

The susceptibility is usually measured by a method which depends upon the fact that when a metal specimen is suspended in a non-uniform transverse magnetic field, a force proportional to $\kappa V.H.\mathrm{d}H/\mathrm{d}x$, where V is the volume of the specimen and $\mathrm{d}H/\mathrm{d}x$ is the field gradient measured transversely to the lines of force, is exerted upon it. This force is easily measured by attaching the specimen to a sensitive balance, and one type commonly used is that designed by Sucksmith. In this balance the distortion of a copper–beryllium ring, caused by the force on the

specimen, is measured by means of an optical or electro-mechanical system. Those metals for which κ is negative, such as copper, silver, gold and bismuth, are repelled by the field and are termed diamagnetic materials. Most metals, however, have positive κ values (i.e. they are attracted by the field) and are either paramagnetic (when κ is small) or ferromagnetic (when κ is very large). Only four pure metals–iron, cobalt and nickel from the transition series, and gadolinium from the rare earth series–are ferromagnetic ($\kappa \approx 1000$) at room temperature, but there are several ferromagnetic alloys and some contain no metals which are themselves ferromagnetic. The Heusler alloy, which contains manganese, copper and aluminium, is one example; ferromagnetism is due to the presence of one of the transition metals.

The ability of a ferromagnetic metal to concentrate the lines of force of the applied field is of great practical importance, and while all such materials can be both magnetized and demagnetized, the ease with which this can be achieved usually governs their application in the various branches of engineering. Materials may be generally classified either as magnetically soft (temporary magnets) or as magnetically hard (permanent magnets), and the difference between the two

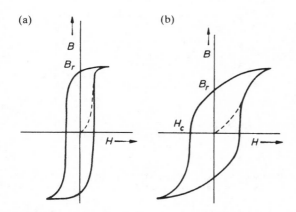

Figure 6.31 *B–H curves for (a) soft and (b) hard magnets.*

types of magnet may be inferred from Figure 6.31. Here, H is the magnetic field necessary to induce a field of strength B inside the material. Upon removal of the field H, a certain residual magnetism B_r, known as the remanence residual, is left in the specimen, and a field H_c, called the coercive force, must be applied in the opposite direction to remove it. A soft magnet is one which is easy both to magnetize and to demagnetize and, as shown in Figure 6.31a, a low value of H is sufficient to induce a large field B in the metal, while only a small field H_c is required to remove it; a hard magnet is a material that is magnetized and demagnetized with difficulty (Figure 6.31b).

6.8.2 Diamagnetism and paramagnetism

Diagmagnetism is a universal property of the atom since it arises from the motion of electrons in their orbits around the nucleus. Electrons moving in this way represent electrical circuits and it follows from Lenz's law that this motion is altered by an applied field in such a manner as to set up a repulsive force. The diamagnetic contribution from the valency electrons is small, but from a closed shell it is proportional to the number of electrons in it and to the square of the radius of the 'orbit'. In many metals this diamagnetic effect is outweighed by a paramagnetic contribution, the origin of which is to be found in the electron spin. Each electron behaves like a small magnet and in a magnetic field can take up one of two orientations, either along the field or in the other opposite direction, depending on the direction of the electron spin. Accordingly, the energy of the electron is either decreased or increased and may be represented conveniently by the band theory. Thus, if we regard the band of energy levels as split into two halves (Figure 6.32a), each half associated with electrons of opposite spin, it follows that in the presence of the field, some of the electrons will transfer their allegiance from one band to the other until the Fermi energy level is the same in both. It is clear, therefore, that in this state there will be a larger number of electrons which have their energy

lowered by the field than have their energy raised. This condition defines paramagnetism, since there will be an excess of unpaired spins which give rise to a resultant magnetic moment.

It is evident that an insulator will not be paramagnetic since the bands are full and the lowered half-band cannot accommodate those electrons which wish to 'spill over' from the raised half-band. On the other hand, it is not true, as one might expect, that conductors are always paramagnetic. This follows because in some elements the natural diamagnetic contribution outweighs the paramagnetic contribution; in copper, for example, the newly filled d-shell gives rise to a larger diamagnetic contribution.

6.8.3 Ferromagnetism

The theory of ferromagnetism is difficult and at present not completely understood. Nevertheless, from the electron theory of metals it is possible to build up a band picture of ferromagnetic materials which goes a long way to explain not only their ferromagnetic properties but also the associated high resistivity and electronic specific heat of these metals compared to copper. In recent years considerable experimental work has been done on the electronic behaviour of the transition elements, and this suggests that the electronic structure of iron is somewhat different to that of cobalt and nickel.

Ferromagnetism, like paramagnetism, has its origin in the electron spin. In ferromagnetic materials, however, permanent magnetism is obtained and this indicates that there is a tendency for electron spins to remain aligned in one direction even when the field has been removed. In terms of the band structure this means that the half-band associated with one spin is automatically lowered when the vacant levels at its top are filled by electrons from the top of the other

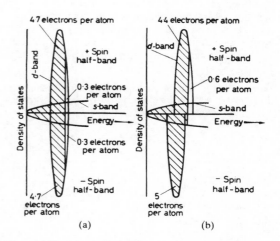

Figure 6.32 *Schematic representation of (a) paramagnetic nickel and (b) ferromagnetic nickel (after Raynor, 1958; by courtesy of Inst. of Materials).*

(Figure 6.32b); the change in potential energy associated with this transfer is known as the exchange energy. Thus, while it is energetically favourable for a condition in which all the spins are in the same direction, an opposing factor is the Pauli exclusion principle, because if the spins are aligned in a single direction many of the electrons will have to go into higher quantum states with a resultant increase in kinetic energy. In consequence, the conditions for ferro-magnetism are stringent, and only electrons from partially filled d or f levels can take part. This condition arises because only these levels have (1) vacant levels available for occupation, and (2) a high density of states which is necessary if the increase in kinetic energy accompanying the alignment of spins is to be smaller than the decrease in exchange energy. Both of these conditions are fulfilled in the transition and rare-earth metals, but of all the metals in the long periods only the elements iron, cobalt and nickel are ferromagnetic at room temperature, gadolinium just above RT ($T_c \approx 16°C$) and the majority are in fact strongly paramagnetic. This observation has led to the conclusion that the exchange interactions are most favourable, at least for the iron group of metals, when the ratio of the atomic radius to the radius of the unfilled shell, i.e. the d-shell, is somewhat greater than 3 (see Table 6.1). As a result of this condition it is hardly surprising that there are a relatively large number of ferromagnetic alloys and compounds, even though the base elements themselves are not ferromagnetic.

In ferromagnetic metals the strong interaction results in the electron spins being spontaneously aligned, even in the absence of an applied field. However, a specimen of iron can exist in an unmagnetized condition because such an alignment is limited to small regions, or domains, which statistically oppose each other. These domains are distinct from the grains of a polycrystalline metal and in general there are many domains in a single grain, as shown in Figure 6.33. Under the application of a magnetic field the favourably-oriented domains grow at the expense of the others by the migration of the domain boundaries until the whole specimen appears fully magnetized. At high

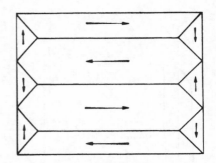

Figure 6.33 *Simple domain structure in a ferromagnetic material. The arrows indicate the direction of magnetization in the domains.*

field strengths it is also possible for unfavourably-oriented domains to 'snap-over' into more favourable orientations quite suddenly, and this process, which can often be heard using sensitive equipment, is known as the Barkhausen effect.

The state in which all the electron spins are in complete alignment is possible only at low temperatures. As the temperature is raised the saturation magnetization is reduced, falling slowly at first and then increasingly rapidly, until a critical temperature, known as the Curie temperature, is reached. Above this temperature, T_c, the specimen is no longer ferromagnetic, but becomes paramagnetic, and for the metals iron, cobalt, and nickel this transition occurs at 780°C, 1075°C and 365°C. respectively. Such a cooperative process may be readily understood from thermodynamic reasoning, since the additional entropy associated with the disorder of the electron spins makes the disordered (paramagnetic) state thermodynamically more stable at high temperatures. This behaviour is similar to that shown by materials which undergo the order–disorder transformation and, as a consequence, ferromagnetic metals exhibit a specific heat peak of the form previously shown (see Figure 6.4b).

A ferromagnetic crystal in its natural state has a domain structure. From Figure 6.33 it is clear that by dividing itself into domains the crystal is able to eliminate those magnetic poles which would otherwise occur at the surface. The width of the domain boundary or Bloch wall is not necessarily small, however, and in most materials is of the order of 100 atoms in thickness. By having a wide boundary the electron spins in neighbouring atoms are more nearly parallel, which is a condition required to minimize the exchange energy. On the other hand, within any one domain the direction of magnetization is parallel to a direction of easy magnetization (i.e. $\langle 100 \rangle$ in iron, $\langle 111 \rangle$ in nickel and $\langle 001 \rangle$ in cobalt) and as one passes across a boundary the direction of magnetization rotates away from one direction of easy magnetization to another. To minimize this magnetically-disturbed region the crystal will try to adopt a boundary which is as thin as possible. Consequently, the boundary width adopted is

Table 6.1 Radii (nm) of electronic orbits of atoms of transition metals of first long period (after Slater, *Quantum Theory of Matter*)

Element	3 d	4 s	Atomic radius in metal (nm)
Sc	0.061	0.180	0.160
Ti	0.055	0.166	0.147
V	0.049	0.152	0.136
Cr	0.045	0.141	0.128
Mn	0.042	0.131	0.128
Fe	0.039	0.122	0.128
Co	0.036	0.114	0.125
Ni	0.034	0.107	0.125
Cu	0.032	0.103	0.128

one of compromise between the two opposing effects, and the material may be considered to possess a magnetic interfacial or surface energy.

6.8.4 Magnetic alloys

The work done in moving a domain boundary depends on the energy of the boundary, which in turn depends on the magnetic anisotropy. The ease of magnetization also depends on the state of internal strain in the material and the presence of impurities. Both these latter factors affect the magnetic 'hardness' through the phenomenon of magnetostriction, i.e. the lattice constants are slightly altered by the magnetization so that a directive influence is put upon the orientation of magnetization of the domains. Materials with internal stresses are hard to magnetize or demagnetize, while materials free from stresses are magnetically soft. Hence, since internal stresses are also responsible for mechanical hardness, the principle which governs the design of magnetic alloys is to make permanent magnetic materials as mechanically hard and soft magnets as mechanically soft as possible.

Magnetically soft materials are used for transformer laminations and armature stampings where a high permeability and a low hysteresis are desirable: iron–silicon or iron–nickel alloys are commonly used for this purpose. In the development of magnetically soft materials it is found that those elements which form interstitial solid solutions with iron are those which broaden the hysteresis loop most markedly. For this reason, it is common to remove such impurities from transformer iron by vacuum melting or hydrogen annealing. However, such processes are expensive and, consequently, alloys are frequently used as 'soft' magnets, particularly iron–silicon and iron–nickel alloys (because silicon and nickel both reduce the amount of carbon in solution). The role of Si is to form a γ-loop and hence remove transformation strains and also improve orientation control. In the production of iron–silicon alloys the factors which are controlled include the grain size, the orientation difference from one grain to the next, and the presence of non-magnetic inclusions, since all are major sources of coercive force. The coercive force increases with decreasing grain size because the domain pattern in the neighbourhood of a grain boundary is complicated owing to the orientation difference between two adjacent grains. Complex domain patterns can also arise at the free surface of the metal unless these are parallel to a direction of easy magnetization. Accordingly, to minimize the coercive force, rolling and annealing schedules are adopted to produce a preferred oriented material with a strong 'cube-texture', i.e. one with two $\langle 100 \rangle$ directions in the plane of the sheet (see Chapter 7). This procedure is extremely important, since transformer material is used in the form of thin sheets to minimize eddy-current losses. Fe–Si–B in the amorphous state is finding increasing application in transformers.

The iron–nickel series, *Permalloys*, present many interesting alloys and are used chiefly in communication engineering where a high permeability is a necessary condition. The alloys in the range 40–55% nickel are characterized by a high permeability and at low field strengths this may be as high as 15 000 compared with 500 for annealed iron. The 50% alloy, *Hypernik*, may have a permeability which reaches a value of 70 000, but the highest initial and maximum permeability occurs in the composition range of the $FeNi_3$ superlattice, provided the ordering phenomenon is suppressed. An interesting development in this field is in the heat treatment of the alloys while in a strong magnetic field. By such a treatment the permeability of *Permalloy* 65 has been increased to about 260 000. This effect is thought to be due to the fact that during alignment of the domains, plastic deformation is possible and magnetostrictive strains may be relieved.

Magnetically hard materials are used for applications where a 'permanent magnetic field is required, but where electromagnets cannot be used, such as in electric clocks, meters, etc. Materials commonly used for this purpose include *Alnico* (Al–Ni–Co) alloys, *Cunico* (Cu–Ni–Co) alloys, ferrites (barium and strontium), samarium–cobalt alloys ($SmCo_5$ and $Sm_2(Co, Fe, Cu, Zr)_{17}$) and *Neomax* ($Nd_2Fe_{14}B$). The *Alnico* alloys have high remanence but poor coercivities, the ferrites have rather low remanence but good coercivities together with very cheap raw material costs. The rare-earth magnets have a high performance but are rather costly although the Nd-based alloys are cheaper than the Sm-based ones.

In the development of magnetically hard materials, the principle is to obtain, by alloying and heat treatment, a matrix containing finely divided particles of a second phase. These fine precipitates, usually differing in lattice parameter from the matrix, set up coherency strains in the lattice which affect the domain boundary movement. Alloys of copper–nickel–iron, copper–nickel–cobalt and aluminium–nickel–cobalt are of this type. An important advance in this field is to make the particle size of the alloy so small, i.e. less than a hundred nanometres diameter, that each grain contains only a single domain. Then magnetization can occur only by the rotation of the direction of magnetization *en bloc*. *Alnico* alloys containing 6–12% Al, 14–25% Ni, 0–35% Co, 0–8% Ti, 0–6% Cu in 40–70% Fe depend on this feature and are the most commercially important permanent magnet materials. They are precipitation-hardened alloys and are heat-treated to produce rod-like precipitates (30 nm × 100 nm) lying along $\langle 100 \rangle$ in the bcc matrix. During magnetic annealing the rods form along the $\langle 100 \rangle$ axis nearest to the direction of the field, when the remanence and coercivity are markedly increased, $Sm_2(Co, Fe, Cu, Zr)_{17}$ alloys also rely on the pinning of magnetic domains by fine precipitates. Clear correlation exists between mechanical hardness and intrinsic coercivity. $SmCo_5$ magnets depend on the very high magnetocrystalline anisotropy of this compound and the individual grains are single-domain particles. The big advantage of these magnets over the *Alnico* alloys is their much higher coercivities.

The Heusler alloys, copper–manganese–aluminium, are of particular interest because they are made up from non-ferromagnetic metals and yet exhibit ferromagnetic properties. The magnetism in this group of alloys is associated with the compound Cu_2MnAl, evidently because of the presence of manganese atoms. The compound has the Fe_3Al-type superlattice when quenched from 800°C, and in this state is ferromagnetic, but when the alloy is slowly cooled it has a γ-brass structure and is non-magnetic, presumably because the correct exchange forces arise from the lattice rearrangement on ordering. A similar behaviour is found in both the copper–manganese–gallium and the copper–manganese–indium systems.

The order–disorder phenomenon is also of magnetic importance in many other systems. As discussed previously, when ordering is accompanied by a structural change, i.e. cubic to tetragonal, coherency strains are set up which often lead to magnetic hardness. In FePt, for example, extremely high coercive forces are produced by rapid cooling. However, because the change in mechanical properties accompanying the transformation is found to be small, it has been suggested that the hard magnetic properties in this alloy are due to the small particle-size effect, which arises from the finely laminated state of the structure.

6.8.5 Anti-ferromagnetism and ferrimagnetism

Apart from the more usual dia-, para- and ferromagnetic materials, there are certain substances which are termed anti-ferromagnetic; in these, the net moments of neighbouring atoms are aligned in opposite directions, i.e. anti-parallel. Many oxides and chlorides of the transition metals are examples including both chromium and α-manganese, and also manganese–copper alloys. Some of the relevant features of anti-ferromagnetism are similar in many respects to ferromagnetism, and are summarized as follows:

1. In general, the magnetization directions are aligned parallel or anti-parallel to crystallographic axes, e.g. in MnI and CoO the moment of the Mn^{2+} and Co^{2+} ions are aligned along a cube edge of the unit cell. The common directions are termed directions of anti-ferromagnetism.
2. The degree of long-range anti-ferromagnetic ordering progressively decreases with increasing temperature and becomes zero at a critical temperature, T_n, known as the Néel temperature; this is the anti-ferromagnetic equivalent of the Curie temperature.
3. An anti-ferromagnetic domain is a region in which there is only one common direction of anti-ferromagnetism; this is probably affected by lattice defects and strain.

The most characteristic property of an anti-ferromagnetic material is that its susceptibility χ shows a maximum as a function of temperature, as shown in Figure 6.34a. As the temperature is raised from 0 K the interaction which leads to anti-parallel spin alignment becomes less effective until at T_n the spins are free.

Similar characteristic features are shown in the resistivity curves due to scattering as a result of spin disorder. However, the application of neutron diffraction techniques provides a more direct method of studying anti-ferromagnetic structures, as well as giving the magnetic moments associated with the ions of the metal. There is a magnetic scattering of neutrons in the case of certain magnetic atoms, and owing to the different scattering amplitude of the parallel and anti-parallel atoms, the possibility arises of the existence of superlattice lines in the anti-ferromagnetic state. In manganese oxide MnO, for example, the parameter of the magnetic unit cell is 0.885 nm, whereas the chemical unit cell (NaCl structure) is half this value, 0.443 nm. This atomic arrangement is analogous to the

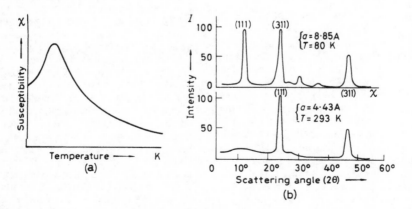

Figure 6.34 *(a) Variation of magnetic susceptibility with temperature for an anti-ferromagnetic material, (b) neutron diffraction pattern from the anti-ferromagnetic powder MnO above and below the critical temperature for ordering (after Shull and Smart, 1949).*

structure of an ordered alloy and the existence of magnetic superlattice lines below the Néel point (122 K) has been observed, as shown in Figure 6.34b.

Some magnetic materials have properties which are intermediate between those of anti-ferromagnetic and ferromagnetic. This arises if the moments in one direction are unequal in magnitude to those in the other, as, for example, in magnetite, Fe_3O_4, where the ferrous and ferric ions of the $FeO.Fe_2O_3$ compound occupy their own particular sites. Néel has called this state *ferrimagnetism* and the corresponding materials are termed ferrites. Such materials are of importance in the field of electrical engineering because they are ferromagnetic without being appreciably conducting; eddy current troubles in transformers are, therefore, not so great. Strontium ferrite is extensively used in applications such as electric motors, because of these properties and low material costs.

6.9 Dielectric materials

6.9.1 Polarization

Dielectric materials, usually those which are covalent or ionic, possess a large energy gap between the valence band and the conduction band. These materials exhibit high electrical resistivity and have important applications as insulators, which prevent the transfer of electrical charge, and capacitors which store electrical charge. Dielectric materials also exhibit piezoelectric and ferroelectric properties.

Application of an electric field to a material induces the formation of dipoles (i.e. atoms or groups of atoms with an unbalanced charge or moment) which become aligned with the direction of the applied field. The material is then polarized. This state can arise from several possible mechanisms–electronic, ionic or molecular, as shown in Figure 6.35a-c. With electronic polarization, the electron clouds of an atom are displaced with respect to the positively-charged ion core setting up an electric dipole with moment μ_e. For ionic solids in an electric field, the bonds between the ions are elastically deformed and the anion–cation distances become closer or further apart, depending on the direction of the field. These induced dipoles produce polarization and may lead to dimensional changes.

Molecular polarization occurs in molecular materials, some of which contain natural dipoles. Such materials are described as polar and for these the influence of an applied field will change the polarization by displacing the atoms and thus changing the dipole moment (i.e. atomic polarizability) or by causing the molecule as a whole to rotate to line up with the imposed field (i.e. orientation polarizability). When the field is removed these dipoles may remain aligned, leading to permanent polarization. Permanent dipoles exist in asymmetrical molecules such as H_2O, organic polymers with asymmetric structure and ceramic crystals without a centre of symmetry.

6.9.2 Capacitors and insulators

In a capacitor the charge is stored in a dielectric material which is easily polarized and has a high electrical resistivity $\sim 10^{11}$ V A^{-1} m to prevent the charge flowing between conductor plates. The ability of the material to polarize is expressed by the permittivity ε and the relative permittivity or dielectric constant κ is the ratio of the permittivity of the material and the permittivity of a vacuum ε_0. While a high κ is important for a capacitor, a high dielectric strength or breakdown voltage is also essential. The dielectric constant κ values for vacuum, water, polyethylene, *Pyrex* glass, alumina and barium titanate are 1, 78.3, 2.3, 4, 6.5 and 3000, respectively.

Structure is an important feature of dielectric behaviour. Glassy polymers and crystalline materials have a lower κ than their amorphous counterparts. Polymers with asymmetric chains have a high κ because of the strength of the associated molecular dipole; thus polyvinyl chloride (PVC) and polystyrene (PS) have larger κ's than polyethylene (PE). $BaTiO_3$ has an extremely high κ value because of its asymmetrical structure. Frequency response is also important in dielectric application, and depends on the mechanism of polarization. Materials which rely on electronic and ionic dipoles respond rapidly to frequencies of 10^{13}–10^{16} Hz but molecular polarization solids, which require groups of atoms to rearrange, respond less rapidly. Frequency is also important in governing dielectric loss due to heat and

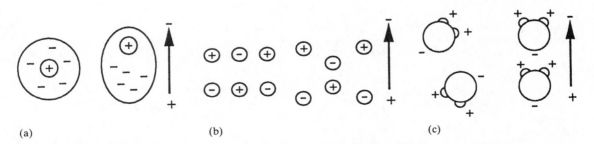

(a)   (b)   (c)

Figure 6.35 *Application of field to produce polarization by (a) electronic, (b) ionic and (c) molecular mechanisms.*

usually increases when one of the contributions to polarization is prevented. This behaviour is common in microwave heating of polymer adhesives; preferential heating in the adhesive due to dielectric losses starts the thermosetting reaction. For moderate increases, raising the voltage and temperature increases the polarizability and leads to a higher dielectric constant. Nowadays, capacitor dielectrics combine materials with different temperature dependence to yield a final product with a small linear temperature variation. These materials are usually titanates of Ba, Ca, Mg, Sr and rare-earth metals.

For an insulator, the material must possess a high electrical resistivity, a high dielectric strength to prevent breakdown of the insulator at high voltages, a low dielectric loss to prevent heating and small dielectric constant to hinder polarization and hence charge storage. Materials increasingly used are alumina, aluminium nitride, glass-ceramics, steatite porcelain and glasses.

6.9.3 Piezoelectric materials

When stress is applied to certain materials an electric polarization is produced proportional to the stress applied. This is the well-known piezoelectric effect. Conversely, dilatation occurs on application of an electric field. Common materials displaying this property are quartz, $BaTiO_3$, $Pb(Ti, Zr)O_3$ or PZT and Na or $LiNbO_3$. For quartz, the piezoelectric constant d relating strain ε to field strength $F (\varepsilon = d \times F)$ is 2.3×10^{-12} m V^{-1}, whereas for PZT it is 250×10^{-12} m V^{-1}. The piezoelectric effect is used in transducers which convert sound waves to electric fields, or vice versa. Applications range from microphones, where a few millivolts are generated, to military devices creating several kilovolts and from small sub-nanometre displacements in piezoelectrically-deformed mirrors to large deformations in power transducers.

6.9.4 Pyroelectric and ferroelectric materials

Some materials, associated with low crystal symmetry, are observed to acquire an electric charge when heated; this is known as pyroelectricity. Because of the low symmetry, the centre of gravity of the positive and negative charges in the unit cell are separated producing a permanent dipole moment. Moreover, alignment of individual dipoles leads to an overall dipole moment which is non-zero for the crystal. Pyroelectric materials are used as detectors of electromagnetic radiation in a wide band from ultraviolet to microwave, in radiometers and in thermometers sensitive to changes of temperature as small as $6 \times 10^{-6}°$C. Pyroelectric TV camera tubes have also been developed for long-wavelength infrared imaging and are useful in providing visibility through smoke. Typical materials are strontium barium niobate

and PZT with Pb_2FeNbO_6 additions to broaden the temperature range of operation.

Ferroelectric materials are those which retain a net polarization when the field is removed and is explained in terms of the residual alignment of permanent dipoles. Not all materials that have permanent dipoles exhibit ferroelectric behaviour because these dipoles become randomly-arranged as the field is removed so that no net polarization remains. Ferroelectrics are related to the pyroelectrics; for the former materials the direction of spontaneous polarization can be reversed by an electric field (Figure 6.36) whereas for the latter this is not possible. This effect can be demonstrated by a polarization versus field hysteresis loop similar in form and explanation to the B–H magnetic hysteresis loop (see Figure 6.31). With increasing positive field all the dipoles align to produce a saturation polarization. As the field is removed a remanent polarization P_r remains due to a coupled interaction between dipoles. The material is permanently polarized and a coercive field E_c has to be applied to randomize the dipoles and remove the polarization.

Like ferromagnetism, ferroelectricity depends on temperature and disappears above an equivalent Curie temperature. For $BaTiO_3$, ferroelectricity is lost at 120°C when the material changes crystal structure. By analogy with magnetism there is also a ferroelectric analogue of anti-ferromagnetism and ferrimagnetism. $NaNbO_3$, for example, has a T_c, of 640°C and anti-parallel electric dipoles of unequal moments characteristic of a ferrielectric material.

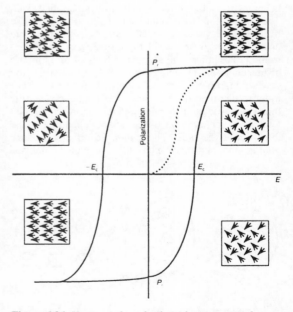

Figure 6.36 *Hysteresis loop for ferroelectric materials, showing the influence of electric field E on polarization P.*

6.10 Optical properties

6.10.1 Reflection, absorption and transmission effects

The optical properties of a material are related to the interaction of the material with electromagnetic radiation, particularly visible light. The electromagnetic spectrum is shown in Figure 5.1 from which it can be seen that the wavelength λ varies from 10^4 m for radio waves down to 10^{-14} m for γ-rays and the corresponding photon energies vary from 10^{-10} eV to 10^8 eV.

Photons incident on a material may be reflected, absorbed or transmitted. Whether a photon is absorbed or transmitted by a material depends on the energy gap between the valency and conduction bands and the energy of the photon. The band structure for metals has no gap and so photons of almost any energy are absorbed by exciting electrons from the valency band into a higher energy level in the conduction band. Metals are thus opaque to all electromagnetic radiation from radio waves, through the infrared, the visible to the ultraviolet, but are transparent to high-energy X-rays and γ-rays. Much of the absorbed radiation is reemitted as radiation of the same wavelength (i.e. reflected). Metals are both opaque and reflective and it is the wavelength distribution of the reflected light, which we see, that determines the colour of the metal. Thus copper and gold reflect only a certain range of wavelengths and absorb the remaining photons, i.e. copper reflects the longer-wavelength red light and absorbs the shorter-wavelength blue. Aluminium and silver are highly reflective over the complete range of the visible spectrum and appear silvery.

Because of the gaps in their band structure non-metals may be transparent. Thus if the photons have insufficient energy to excite electrons in the material to a higher energy level, they may be transmitted rather than absorbed and the material is transparent. In high-purity ceramics and polymers, the energy gap is large and these materials are transparent to visible light. In semiconductors, electrons can be excited into acceptor levels or out of donor levels and phonons having sufficient energy to produce these transitions will be absorbed. Semiconductors are therefore opaque to short wavelengths and transparent to long.[1] The band structure is influenced by crystallinity and hence materials such as glasses and polymers may be transparent in the amorphous state but opaque when crystalline.

High-purity non-metallics such as glasses, diamond or sapphire (Al_2O_3) are colourless but are changed by impurities. For example, small additions of Cr^{3+} ions (Cr_2O_3) to Al_2O_3 produces a ruby colour by introducing impurity levels within the band-gap of sapphire which give rise to absorption of specific wavelengths in the visible spectrum. Colouring of glasses and ceramics is produced by addition of transition metal impurities which have unfilled d-shells. The photons easily interact with these ions and are absorbed; Cr^{3+} gives green, Mn^{2+} yellow and Co^{2+} blue-violet colouring.

In photochromic sunglasses the energy of light quanta is used to produce changes in the ionic structure of the glass. The glass contains silver (Ag^+) ions as a dopant which are trapped in the disordered glass network of silicon and oxygen ions: these are excited by high-energy quanta (photons) and change to metallic silver, causing the glass to darken (i.e. light energy is absorbed). With a reduction in light intensity, the silver atoms re-ionize. These processes take a small period of time relying on absorption and non-absorption of light.

6.10.2 Optical fibres

Modern communication systems make use of the ability of optical fibres to transmit light signals over large distances. Optical guidance by a fibre is produced (see Figure 6.37) if a core fibre of refractive index n_1 is surrounded by a cladding of slightly lower index n_2 such that total internal reflection occurs confining the rays to the core; typically the core is about 100 μm and $n_1 - n_2 \approx 10^{-2}$. With such a simple optical fibre, interference occurs between different modes leading to a smearing of the signals. Later designs use a core in which the refractive index is graded, parabolically, between the core axis and the interface with the cladding. This design enables modulated signals to maintain their coherency. In vitreous silica, the refractive index can be modified by additions of dopants such as P_2O_5, GeO_2 which raise it and B_2O_5 and F which lower it. Cables are sheathed to give strength and environmental protection; PE and PVC are commonly used for limited fire-hazard conditions.

6.10.3 Lasers

A laser (Light Amplification by Stimulated Emission of Radiation) is a powerful source of coherent light (i.e. monochromatic and all in phase). The original laser material, still used, is a single crystal rod of ruby, i.e. Al_2O_3 containing dopant Cr^{3+} ions in solid solution. Nowadays, lasers can be solid, liquid or gaseous materials and ceramics, glasses and semiconductors. In all cases, electrons of the laser material are excited into a higher energy state by some suitable stimulus (see Figure 6.38). In a device this is produced by the photons from a flash tube, to give an intense

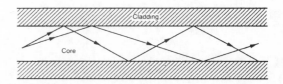

Figure 6.37 *Optical guidance in a multimode fibre.*

---

[1]Figure 5.37b shows a dislocation source in the interior of a silicon crystal observed using infrared light.

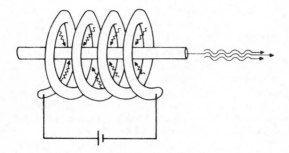

Figure 6.38 *Schematic diagram of a laser.*

source of light surrounding the rod to concentrate the energy into the laser rod. Alternatively an electrical discharge in a gas is used. The ends of the laser rod are polished flat and parallel and then silvered such that one end is totally-reflecting and the other partially-transmitting.

In the ruby laser, a xenon flash lamp excites the electrons of the Cr^{3+} ions into higher energy states. Some of these excited electrons decay back to their ground state directly and are not involved in the laser process. Other electrons decay into a metastable intermediate state before further emission returns them to the ground state. Some of the electrons in the metastable state will be spontaneously emitted after a short ($\sim$ms) rest period. Some of the resultant photons are retained in the rod because of the reflectivity of the silvered ends and stimulate the release of other electrons from the metastable state. Thus one photon releases another such that an avalanche of emissions is triggered all in phase with the triggering photon. The intensity of the light increases as more emissions are stimulated until a very high intense, coherent, collimated 'burst' of light is transmitted through the partially-silvered end lasting a few nanoseconds and with considerable intensity.

6.10.4 Ceramic 'windows'

Many ceramics, usually oxides, have been prepared in optically-transparent or translucent forms (translucent means that incident light is partly-reflected and partly-transmitted). Examples include aluminium oxide, magnesium oxide, their double oxide or spinel, and chalcogenides (zinc sulphide, zinc selenide). Very pure raw materials of fine particle size are carefully processed to eliminate voids and control grain sizes.

Thus, translucent alumina is used for the arc tube of high-pressure sodium lamps; a grain size of 25 µm gives the best balance of translucency and mechanical strength.

Ceramics are also available to transmit electromagnetic radiation with wavelengths which lie below or above the visible range of 400–700 nm (e.g. infrared, microwave, radar, etc.). Typical candidate materials for development include vitreous silica, cordierite glass-ceramics and alumina.

6.10.5 Electro-optic ceramics

Certain special ceramics combine electrical and optical properties in a unique manner. Lead lanthanum zirconium titanate, known as PLZT, is a highly-transparent ceramic which becomes optically birefringent when electrically charged. This phenomenon is utilized as a switching mechanism in arc welding goggles, giving protection against flash blindness. The PLZT plate is located between two 'crossed' sheets of polarizing material. A small impressed d.c. voltage on the PLZT plate causes it to split the incident light into two rays vibrating in different planes. One of these rays can pass through the inner polar sheet and enter the eye. A sudden flash of light will activate photodiodes in the goggles, reduce the impressed voltage and cause rapid darkening of the goggles.

Further reading

Anderson, J. C., Leaver, K. D., Rawlins, R. D. and Alexander, J. M. (1990). *Materials Science*. Chapman and Hall, London.

Braithwaite, N. and Weaver, G. (Eds) (1990). *Open University Materials in Action Series*. Butterworths, London.

Cullity, B. D. (1972). *Introduction to Magnetic Materials*. Addison-Wesley, Wokingham.

Hume-Rothery, W. and Coles, B. R. (1946, 1969). *Atomic Theory for Students of Metallurgy*. Institute of Metals, London.

Porter, D. A. and Easterling, K. E. (1992). *Phase Transformations in Metals and Alloys*, 2nd edn. Van Nostrand Reinhold, Wokingham.

Raynor, G. V. (1947, 1988). *Introduction to Electron Theory of Metals*. Institute of Metals, London.

Shewmon, P. G. (1989). *Diffusion in Solids*. Minerals, Metals and Materials Soc. Warrendale, USA.

Swalin, R. A. (1972). *Thermodynamics of Solids*. Wiley, Chichester.

Warn, J. R. W. (1969). *Concise Chemical Thermodynamics*. Van Nostrand, New York.

Chapter 7

Mechanical behaviour of materials

7.1 Mechanical testing procedures

7.1.1 Introduction

Real crystals, however carefully prepared, contain lattice imperfections which profoundly affect those properties sensitive to structure. Careful examination of the mechanical behaviour of materials can give information on the nature of these atomic defects. In some branches of industry the common mechanical tests, such as tensile, hardness, impact, creep and fatigue tests, may be used, not to study the 'defect state' but to check the quality of the product produced against a standard specification. Whatever its purpose, the mechanical test is of importance in the development of both materials science and engineering properties. It is inevitable that a large number of different machines for performing the tests are in general use. This is because it is often necessary to know the effect of temperature and strain rate at different levels of stress depending on the material being tested. Consequently, no attempt is made here to describe the details of the various testing machines. The elements of the various tests are outlined below.

7.1.2 The tensile test

In a tensile test the ends of a test piece are fixed into grips, one of which is attached to the load-measuring device on the tensile machine and the other to the straining device. The strain is usually applied by means of a motor-driven crosshead and the elongation of the specimen is indicated by its relative movement. The load necessary to cause this elongation may be obtained from the elastic deflection of either a beam or proving ring, which may be measured by using hydraulic, optical or electromechanical methods. The last method (where there is a change in the resistance of strain gauges attached to the beam) is, of course, easily adapted into a system for autographically recording the load–elongation curve.

The load–elongation curves for both polycrystalline mild steel and copper are shown in Figures 7.1a and 7.1b. The corresponding stress (load per unit area, P/A) versus strain (change in length per unit length, dl/l) curves may be obtained knowing the dimensions of the test piece. At low stresses the deformation is elastic, reversible and obeys Hooke's law with stress linearly proportional to strain. The proportionality constant connecting stress and strain is known as the elastic modulus and may be either (a) the elastic or Young's modulus, E, (b) the rigidity or shear modulus μ, or (c) the bulk modulus K, depending on whether the strain is tensile, shear or hydrostatic compressive, respectively. Young's modulus, bulk modulus, shear modulus and Poisson's ratio ν, the ratio of lateral contractions to longitudinal extension in uniaxial tension, are related according to

$$K = \frac{E}{2(1-2\nu)}, \quad \mu = \frac{E}{2(1+\nu)}, \quad E = \frac{9K\mu}{3K+\mu}$$

(7.1)

In general, the elastic limit is an ill-defined stress, but for impure iron and low carbon steels the onset of plastic deformation is denoted by a sudden drop in load indicating both an upper and lower yield point.[1] This yielding behaviour is characteristic of many metals, particularly those with bcc structure containing small amounts of solute element (see Section 7.4.6). For materials not showing a sharp yield point, a conventional definition of the beginning of plastic flow is the 0.1% proof stress, in which a line is drawn parallel

---

[1]Load relaxations are obtained only on 'hard' beam Polanyi-type machines where the beam deflection is small over the working load range. With 'soft' machines, those in which the load-measuring device is a soft spring, rapid load variations are not recorded because the extensions required are too large, while in dead-loading machines no load relaxations are possible. In these latter machines sudden yielding will show as merely an extension under constant load.

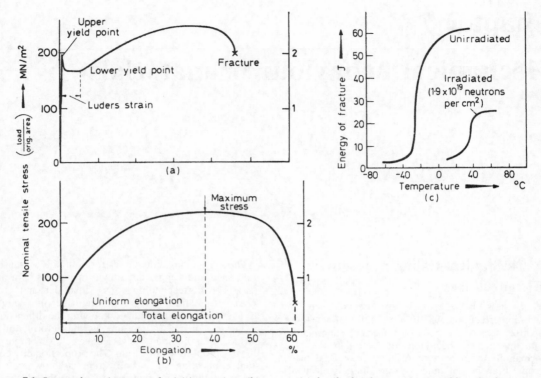

Figure 7.1 *Stress–elongation curves for (a) impure iron, (b) copper, (c) ductile–brittle transition in mild steel (after Churchman, Mogford and Cottrell, 1957).*

to the elastic portion of the stress–strain curve from the point of 0.1% strain.

For control purposes the tensile test gives valuable information on the tensile strength (TS = maximum load/original area) and ductility (percentage reduction in area or percentage elongation) of the material. When it is used as a research technique, however, the exact shape and fine details of the curve, in addition to the way in which the yield stress and fracture stress vary with temperature, alloying additions and grain size, are probably of greater significance.

The increase in stress from the initial yield up to the TS indicates that the specimen hardens during deformation (i.e. work-hardens). On straining beyond the TS the metal still continues to work-harden, but at a rate too small to compensate for the reduction in cross-sectional area of the test piece. The deformation then becomes unstable, such that as a localized region of the gauge length strains more than the rest, it cannot harden sufficiently to raise the stress for further deformation in this region above that to cause further strain elsewhere. A neck then forms in the gauge length, and further deformation is confined to this region until fracture. Under these conditions, the reduction in area $(A_0 - A_1)/A_0$ where A_0 and A_1 are the initial and final areas of the neck gives a measure of the localized strain, and is a better indication

than the strain to fracture measured along the gauge length.

True stress–true strain curves are often plotted to show the work hardening and strain behaviour at large strains. The true stress σ is the load P divided by the area A of the specimen at that particular stage of strain and the total true strain in deforming from initial length l_0 to length l_1 is $\varepsilon = \int_{l_0}^{l_1} (\mathrm{d}l/l) = \ln (l_1/l_0)$. The true stress–strain curves often fit the Ludwig relation $\sigma = k\varepsilon^n$ where n is a work-hardening coefficient ≈ 0.1–0.5 and k the strength coefficient. Plastic instability, or necking, occurs when an increase in strain produces no increase in load supported by the specimen, i.e. $\mathrm{d}P = 0$, and hence since $P = \sigma A$, then

$$\mathrm{d}P = A\mathrm{d}\sigma + \sigma\mathrm{d}A = 0$$

defines the instability condition. During deformation, the specimen volume is essentially constant (i.e. $\mathrm{d}V = 0$) and from

$$\mathrm{d}V = \mathrm{d}(lA) = A\mathrm{d}l + l\mathrm{d}A = 0$$

we obtain

$$\frac{\mathrm{d}\sigma}{\sigma} = -\frac{\mathrm{d}A}{A} = \frac{\mathrm{d}l}{l} = \mathrm{d}\varepsilon \qquad (7.2)$$

Thus, necking occurs at a strain at which the slope of the true stress–true strain curve equals the true

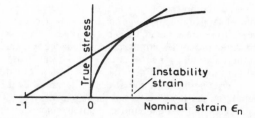

Figure 7.2 *Considère's construction.*

stress at that strain, i.e. $d\sigma/d\varepsilon = \sigma$. Alternatively, since $k\varepsilon^n = \sigma = d\sigma/d\varepsilon = nk\varepsilon^{n-1}$ then $\varepsilon = n$ and necking occurs when the true strain equals the strain-hardening exponent. The instability condition may also be expressed in terms of the conventional (nominal strain)

$$\frac{d\sigma}{d\varepsilon} = \frac{d\sigma}{d\varepsilon_n}\frac{d\varepsilon_n}{d\varepsilon} = \frac{d\sigma}{d\varepsilon_n}\left(\frac{dl/l_0}{dl/l}\right) = \frac{d\sigma}{d\varepsilon_n}\frac{l}{l_0}$$

$$= \frac{d\sigma}{d\varepsilon_n}(1 + \varepsilon_n) = \sigma \qquad (7.3)$$

which allows the instability point to be located using Considère's construction (see Figure 7.2), by plotting the true stress against nominal strain and drawing the tangent to the curve from $\varepsilon_n = -1$ on the strain axis. The point of contact is the instability stress and the tensile strength is $\sigma/(1 + \varepsilon_n)$.

Tensile specimens can also give information on the type of fracture exhibited. Usually in polycrystalline metals transgranular fractures occur (i.e. the fracture surface cuts through the grains) and the 'cup and cone' type of fracture is extremely common in really ductile metals such as copper. In this, the fracture starts at the centre of the necked portion of the test piece and at first grows roughly perpendicular to the tensile axis, so forming the 'cup', but then, as it nears the outer surface, it turns into a 'cone' by fracturing along a surface at about 45° to the tensile axis. In detail the 'cup' itself consists of many irregular surfaces at about 45° to the tensile axis, which gives the fracture a fibrous appearance. Cleavage is also a fairly common type of transgranular fracture, particularly in materials of bcc structure when tested at low temperatures. The fracture surface follows certain crystal planes (e.g. {1 0 0} planes), as is shown by the grains revealing large bright facets, but the surface also appears granular with 'river lines' running across the facets where cleavage planes have been torn apart. Intercrystalline fractures sometimes occur, often without appreciable deformation. This type of fracture is usually caused by a brittle second phase precipitating out around the grain boundaries, as shown by copper containing bismuth or antimony.

7.1.3 Indentation hardness testing

The hardness of a metal, defined as the resistance to penetration, gives a conveniently rapid indication of its deformation behaviour. The hardness tester forces a small sphere, pyramid or cone into the surface of the metals by means of a known applied load, and the hardness number (Brinell or Vickers diamond pyramid) is then obtained from the diameter of the impression. The hardness may be related to the yield or tensile strength of the metal, since during the indentation, the material around the impression is plastically deformed to a certain percentage strain. The Vickers hardness number (VPN) is defined as the load divided by the pyramidal area of the indentation, in kgf/mm², and is about three times the yield stress for materials which do not work harden appreciably. The Brinell hardness number (BHN) is defined as the stress P/A, in kgf/mm² where P is the load and A the surface area of the spherical cap forming the indentation. Thus

$$\text{BHN} = P \bigg/ \left(\frac{\pi}{2}D^2\right)\{1 - [1 - (d/D)^2]^{1/2}\}$$

where d and D are the indentation and indentor diameters respectively. For consistent results the ratio d/D should be maintained constant and small. Under these conditions soft materials have similar values of BHN and VPN. Hardness testing is of importance in both control work and research, especially where information on brittle materials at elevated temperatures is required.

7.1.4 Impact testing

A material may have a high tensile strength and yet be unsuitable for shock loading conditions. To determine this the impact resistance is usually measured by means of the notched or un-notched Izod or Charpy impact test. In this test a load swings from a given height to strike the specimen, and the energy dissipated in the fracture is measured. The test is particularly useful in showing the decrease in ductility and impact strength of materials of bcc structure at moderately low temperatures. For example, carbon steels have a relatively high ductile–brittle transition temperature (Figure 7.1c) and, consequently, they may be used with safety at sub-zero temperatures only if the transition temperature is lowered by suitable alloying additions or by refining the grain size. Nowadays, increasing importance is given to defining a fracture toughness parameter K_c for an alloy, since many alloys contain small cracks which, when subjected to some critical stress, propagate; K_c defines the critical combination of stress and crack length. Brittle fracture is discussed more fully in Chapter 8.

7.1.5 Creep testing

Creep is defined as plastic flow under constant stress, and although the majority of tests are carried out under constant load conditions, equipment is available for reducing the loading during the test to compensate for the small reduction in cross-section of the specimen. At relatively high temperatures creep appears to

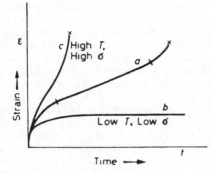

Figure 7.3 *Typical creep curves.*

occur at all stress levels, but the creep rate increase with increasing stress at a given temperature. For the accurate assessment of creep properties, it is clear that special attention must be given to the maintenance of the specimen at a constant temperature, and to the measurement of the small dimensional changes involved. This latter precaution is necessary, since in many materials a rise in temperature by a few tens of degrees is sufficient to double the creep rate. Figure 7.3, curve *a*, shows the characteristics of a typical creep curve and following the instantaneous strain caused by the sudden application of the load, the creep process may be divided into three stages, usually termed primary or transient creep, second or steady-state creep and tertiary or accelerating creep. The characteristics of the creep curve often vary, however, and the tertiary stage of creep may be advanced or retarded if the temperature and stress at which the test is carried out is high or low respectively (see Figure 7.3, curves *b* and *c*). Creep is discussed more fully in Section 7.9.

7.1.6 Fatigue testing

The fatigue phenomenon is concerned with the premature fracture of metals under repeatedly applied low stresses, and is of importance in many branches of engineering (e.g. aircraft structures). Several different types of testing machines have been constructed in which the stress is applied by bending, torsion, tension or compression, but all involve the same principle of subjecting the material to constant cycles of stress. To express the characteristics of the stress system, three properties are usually quoted: these include (1) the maximum range of stress, (2) the mean stress, and (3) the time period for the stress cycle. Four different arrangements of the stress cycle are shown in Figure 7.4, but the reverse and the repeated cycle tests (e.g. 'push–pull') are the most common, since they are the easiest to achieve in the laboratory.

The standard method of studying fatigue is to prepare a large number of specimens free from flaws, and to subject them to tests using a different range of stress, *S*, on each group of specimens. The number of stress cycles, *N*, endured by each specimen at a given stress level is recorded and plotted, as shown in Figure 7.5. This *S–N* diagram indicates that some metals can withstand indefinitely the application of a large number of stress reversals, provided the applied stress is below a limiting stress known as the endurance limit. For certain ferrous materials when they are used in the absence of corrosive conditions the assumption of a safe working range of stress seems justified, but for non-ferrous materials and for steels when they are used in corrosive conditions a definite endurance limit cannot be defined. Fatigue is discussed in more detail in Section 7.11.

7.1.7 Testing of ceramics

Direct tensile testing of ceramics is not generally favoured, mainly because of the extreme sensitivity of ceramics to surface flaws. First, it is difficult to apply a truly uniaxial tensile stress: mounting the specimen in the machine grips can seriously damage the surface and any bending of the specimen during the test will cause premature failure. Second, suitable waisted specimens with the necessary fine and flawless finish are expensive to produce. It is therefore common practice to use bend tests for engineering ceramics and glasses. (They have long been used for other non-ductile materials such as concretes and grey cast iron.) In the three- and four-point bend methods portrayed in Figure 7.6, a beam specimen is placed between rollers and carefully loaded at a constant strain rate. The flexural strength at failure, calculated from the standard formulae, is

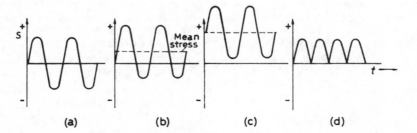

Figure 7.4 *Alternative forms of stress cycling: (a) reversed; (b) alternating (mean stress ≠ zero), (c) fluctuating and (d) repeated.*

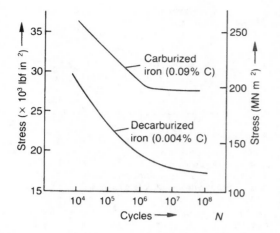

Figure 7.5 *S–N curve for carburized and decarburized iron.*

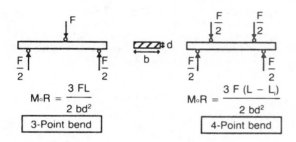

Figure 7.6 *Bend test configurations. MoR = modulus of rupture, F = applied force, L = outer span, L_i = inner span, b = breadth of specimen, d = depth of specimen.*

known as the modulus of rupture (MoR) and expresses the maximum tensile stress which develops on the convex face of the loaded beam. Strong ceramics, such as silicon carbide and hot-pressed silicon nitride, have very high MoR values. The four-point loading method is often preferred because it subjects a greater volume and area of the beam to stress and is therefore more searching. MoR values from four-point tests are often substantially lower than those from three-point tests on the same material. Similarly, strength values tend to decrease as the specimen size is increased. To provide worthwhile data for quality control and design activities, close attention must be paid to strain rate and environment, and to the size, edge finish and surface texture of the specimen. With oxide ceramics and silica glasses, a high strain rate will give an appreciably higher flexural strength value than a low strain rate, which leads to slow crack growth and delayed fracture (Section 10.7).

The bend test has also been adapted for use at high temperatures. In one industrial procedure, specimens of magnesia (basic) refractory are fed individually from a magazine into a three-point loading zone at the centre of an electric furnace heated by SiC elements. A similar type of hot-bend test has been used for the routine testing of graphite electrode samples and gives a useful indication of their ability to withstand accidental lateral impact during service in steel melting furnaces.

Proof-testing is a long-established method of testing certain engineering components and structures. In a typical proof test, each component is held at a certain proof stress for a fixed period of time; loading and unloading conditions are standardized. In the case of ceramics, it may involve bend-testing, internal pressurization (for tubes) or rotation at high speed ('overspeeding' of grinding wheels). Components that withstand the proof test are, in the simplest analysis, judged to be sound and suitable for long-term service at the lower design stress. The underlying philosophy has been often questioned, not least because there is a risk that the proof test itself may cause incipient cracking. Nevertheless, proof-testing now has an important role in the statistical control of strength in ceramics.

7.2 Elastic deformation

7.2.1 Elastic deformation of metals

It is well known that metals deform both elastically and plastically. Elastic deformation takes place at low stresses and has three main characteristics, namely (1) it is reversible, (2) stress and strain are linearly proportional to each other according to Hooke's Law and (3) it is usually small (i.e. <1% elastic strain).

The stress at a point in a body is usually defined by considering an infinitesimal cube surrounding that point and the forces applied to the faces of the cube by the surrounding material. These forces may be resolved into components parallel to the cube edges and when divided by the area of a face give the nine stress components shown in Figure 7.7. A given component σ_{ij} is the force acting in the j-direction per unit area of face normal to the i-direction. Clearly, when $i = j$ we have normal stress components (e.g. σ_{xx}) which may be either tensile (conventionally positive) or compressive (negative), and when $i \neq j$ (e.g. σ_{xy}) the stress components are shear. These shear stresses exert couples on the cube and to prevent rotation of the cube the couples on opposite faces must balance and hence $\sigma_{ij} = \sigma_{ji}$.[1] Thus, stress has only six independent components.

When a body is strained, small elements in that body are displaced. If the initial position of an element

---

[1] The nine components of stress σ_{ij} form a second-rank tensor usually written

$$\begin{matrix} \sigma_{xx} & \sigma_{xy} & \sigma_{xz} \\ \sigma_{yx} & \sigma_{yy} & \sigma_{yz} \\ \sigma_{zx} & \sigma_{zy} & \sigma_{zz} \end{matrix}$$

and is known as the stress tensor.

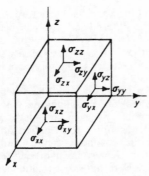

Figure 7.7 *Normal and shear stress components.*

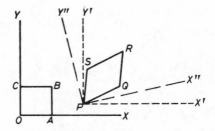

Figure 7.8 *Deformation of a square OABC to a parallelogram PQRS involving (i) a rigid body translation OP allowed for by redefining new axes X'Y', (ii) a rigid body rotation allowed for by rotating the axes to X"Y", and (iii) a change of shape involving both tensile and shear strains.*

is defined by its coordinates (x, y, z) and its final position by $(x + u, y + v, z + w)$ then the displacement is (u, v, w). If this displacement is constant for all elements in the body, no strain is involved, only a rigid translation. For a body to be under a condition of strain the displacements must vary from element to element. A uniform strain is produced when the displacements are linearly proportional to distance. In one dimension then $u = ex$ where $e = du/dx$ is the coefficient of proportionality or nominal tensile strain. For a three-dimensional uniform strain, each of the three components u, v, w is made a linear function in terms of the initial elemental coordinates, i.e.

$$u = e_{xx}x + e_{xy}y + e_{xz}z$$

$$v = e_{yx}x + e_{yy}y + e_{yz}z$$

$$w = e_{zx}x + e_{zy}y + e_{zz}z$$

The strains $e_{xx} = du/dx$, $e_{yy} = dv/dy$, $e_{zz} = dw/dz$ are the tensile strains along the x, y and z axes, respectively. The strains e_{xy}, e_{yz}, etc., produce shear strains and in some cases a rigid body rotation. The rotation produces no strain and can be allowed for by rotating the reference axes (see Figure 7.8). In general, therefore, $e_{ij} = \varepsilon_{ij} + \omega_{ij}$ with ε_{ij} the strain components and ω_{ij} the rotation components. If, however, the shear strain is defined as the angle of shear, this is twice the corresponding shear strain component, i.e. $\gamma = 2\varepsilon_{ij}$. The strain tensor, like the stress tensor, has nine components which are usually written as:

$$\begin{matrix} \varepsilon_{xx} & \varepsilon_{xy} & \varepsilon_{xz} \\ \varepsilon_{yx} & \varepsilon_{yy} & \varepsilon_{yz} \\ \varepsilon_{zx} & \varepsilon_{zy} & \varepsilon_{zz} \end{matrix} \text{ or } \begin{matrix} \varepsilon_{xx} & \frac{1}{2}\gamma_{xy} & \frac{1}{2}\gamma_{xz} \\ \frac{1}{2}\gamma_{yx} & \varepsilon_{yy} & \frac{1}{2}\gamma_{yz} \\ \frac{1}{2}\gamma_{zx} & \frac{1}{2}\gamma_{zy} & \varepsilon_{zz} \end{matrix}$$

where ε_{xx} etc. are tensile strains and γ_{xy}, etc. are shear strains. All the simple types of strain can be produced from the strain tensor by setting some of the components equal to zero. For example, a pure dilatation (i.e. change of volume without change of shape) is obtained when $\varepsilon_{xx} = \varepsilon_{yy} = \varepsilon_{zz}$ and all other components are zero. Another example is a uniaxial tensile test when the tensile strain along the x-axis

is simply $e = \varepsilon_{xx}$. However, because of the strains introduced by lateral contraction, $\varepsilon_{yy} = -ve$ and $\varepsilon_{zz} = -ve$, where v is Poisson's ratio; all other components of the strain tensor are zero.

At small elastic deformations, the stress is linearly proportional to the strain. This is Hooke's law and in its simplest form relates the uniaxial stress to the uniaxial strain by means of the modulus of elasticity. For a general situation, it is necessary to write Hooke's law as a linear relationship between six stress components and the six strain components, i.e.

$$\sigma_{xx} = c_{11}\varepsilon_{xx} + c_{12}\varepsilon_{yy} + c_{13}\varepsilon_{zz} + c_{14}\gamma_{yz} + c_{15}\gamma_{zx} + c_{16}\gamma_{xy}$$

$$\sigma_{yy} = c_{21}\varepsilon_{xx} + c_{22}\varepsilon_{yy} + c_{23} + c_{24}\gamma_{yz} + c_{25}\gamma_{zx} + c_{26}\gamma_{xy}$$

$$\sigma_{zz} = c_{31}\varepsilon_{xx} + c_{32}\varepsilon_{yy} + c_{33}\varepsilon_{zz} + c_{34}\gamma_{yz} + c_{35}\gamma_{zx} + c_{36}\gamma_{xy}$$

$$\tau_{yz} = c_{41}\varepsilon_{xx} + c_{42}\varepsilon_{yy} + c_{43}\varepsilon_{zz} + c_{44}\gamma_{yz} + c_{45}\gamma_{zx} + c_{46}\gamma_{xy}$$

$$\tau_{zx} = c_{51}\varepsilon_{xx} + c_{52}\varepsilon_{yy} + c_{53}\varepsilon_{zz} + c_{54}\gamma_{yz} + c_{55}\gamma_{zx} + c_{56}\gamma_{xy}$$

$$\tau_{xy} = c_{61}\varepsilon_{xx} + c_{62}\varepsilon_{yy} + c_{63}\varepsilon_{zz} + c_{64}\gamma_{yz} + c_{65}\gamma_{zx} + c_{66}\gamma_{xy}$$

The constants $c_{11}, c_{12}, \ldots, c_{ij}$ are called the elastic stiffness constants.[1]

Taking account of the symmetry of the crystal, many of these elastic constants are equal or become zero. Thus in cubic crystals there are only three independent elastic constants c_{11}, c_{12} and c_{44} for the three independent modes of deformation. These include the application of (1) a hydrostatic stress p to produce a dilatation Θ given by

$$p = -\tfrac{1}{3}(c_{11} + 2c_{12})\Theta = -\kappa$$

where κ is the bulk modulus, (2) a shear stress on a cube face in the direction of the cube axis defining the shear modulus $\mu = c_{44}$, and (3) a rotation about a cubic axis defining a shear modulus $\mu_1 = \tfrac{1}{2}(c_{11} - c_{12})$. The ratio μ/μ_1 is the elastic anisotropy factor and in elastically isotropic crystals it is unity with $2c_{44} =$

---

[1] Alternatively, the strain may be related to the stress, e.g. $\varepsilon_x = s_{11}\sigma_{xx} + s_{12}\sigma_{yy} + s_{13}\sigma_{zz} + \ldots$, in which case the constants $s_{11}, s_{12}, \ldots, s_{ij}$ are called elastic compliances.

Table 7.1 Elastic constants of cubic crystals (GN/m^2)

Metal	c_{11}	c_{12}	c_{44}	$2c_{44}/(c_{11} - c_{12})$
Na	006.0	004.6	005.9	8.5
K	004.6	003.7	002.6	5.8
Fe	237.0	141.0	116.0	2.4
W	501.0	198.0	151.0	1.0
Mo	460.0	179.0	109.0	0.77
Al	108.0	62.0	28.0	1.2
Cu	170.0	121.0	75.0	3.3
Ag	120.0	90.0	43.0	2.9
Au	186.0	157.0	42.0	3.9
Ni	250.0	160.0	118.0	2.6
β-brass	129.1	109.7	82.4	8.5

$c_{11} - c_{12}$; the constants are all interrelated with $c_{11} = \kappa + \frac{4}{3}\mu$, $c_{12} = \kappa - \frac{2}{3}\mu$ and $c_{44} = \mu$.

Table 7.1 shows that most metals are far from isotropic and, in fact, only tungsten is isotropic; the alkali metals and β-compounds are mostly anisotropic. Generally, $2c_{44} > (c_{11} - c_{12})$ and hence, for most elastically anisotropic metals E is maximum in the $\langle 1\,1\,1 \rangle$ and minimum in the $\langle 1\,0\,0 \rangle$ directions. Molybdenum and niobium are unusual in having the reverse anisotropy when E is greatest along $\langle 1\,0\,0 \rangle$ directions. Most commercial materials are polycrystalline, and consequently they have approximately isotropic properties. For such materials the modulus value is usually independent of the direction of measurement because the value observed is an average for all directions, in the various crystals of the specimen. However, if during manufacture a preferred orientation of the grains in the polycrystalline specimen occurs, the material will behave, to some extent, like a single crystal and some 'directionality' will take place.

7.2.2 Elastic deformation of ceramics

At ambient temperatures the profile of the stress versus strain curve for a conventional ceramic is similar to that of a non-ductile metal and can be described as linear-elastic, remaining straight until the point of fracture is approached. The strong interatomic bonding of engineering ceramics confers mechanical stiffness. Moduli of elasticity (elastic, shear) can be much higher than those of metallic materials. In the case of single ceramic crystals, these moduli are often highly anisotropic (e.g. alumina). However, in their polycrystalline forms, ceramics are often isotropic as a result of the randomizing effect of processing (e.g. isostatic pressing); nevertheless, some processing routes preserve anisotropic tendencies (e.g. extrusion). (Glasses are isotropic, of course.) Moduli are greatly influenced by the presence of impurities, second phases and porosity; for instance, the elastic modulus of a ceramic is lowered as porosity is increased. As the temperature of testing is raised, elastic moduli usually show a decrease, but there are exceptions.

7.3 Plastic deformation

7.3.1 Slip and twinning

The limit of the elastic range cannot be defined exactly but may be considered to be that value of the stress below which the amount of plasticity (irreversible deformation) is negligible, and above which the amount of plastic deformation is far greater than the elastic deformation. If we consider the deformation of a metal in a tensile test, one or other of two types of curve may be obtained. Figure 7.1a shows the stress–strain curve characteristic of iron, from which it can be seen that plastic deformation begins abruptly at A and continues initially with no increase in stress. The point A is known as the yield point and the stress at which it occurs is the yield stress. Figure 7.1b shows a stress–strain curve characteristic of copper, from which it will be noted that the transition to the plastic range is gradual. No abrupt yielding takes place and in this case the stress required to start macroscopic plastic flow is known as the flow stress.

Once the yield or flow stress has been exceeded plastic or permanent deformation occurs, and this is found to take place by one of two simple processes, slip (or glide) and twinning. During slip, shown in Figure 7.9a, the top half of the crystal moves over the bottom half along certain crystallographic planes, known as slip planes, in such a way that the atoms move forward by a whole number of lattice vectors; as a result the continuity of the lattice is maintained. During twinning (Figure 7.9b) the atomic movements are not whole lattice vectors, and the lattice generated in the deformed region, although the same as the parent lattice, is oriented in a twin relationship to it. It will also be observed that in contrast to slip, the sheared region in twinning occurs over many atom planes, the atoms in each plane being moved forward by the same amount relative to those of the plane below them.

7.3.2 Resolved shear stress

All working processes such as rolling, extrusion, forging etc. cause plastic deformation and, consequently, these operations will involve the processes of slip or twinning outlined above. The stress system applied during these working operations is often quite complex, but for plastic deformation to occur the presence of a shear stress is essential. The importance of shear

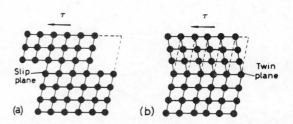

Figure 7.9 *Slip and twinning in a crystal.*

stresses becomes clear when it is realized that these stresses arise in most processes and tests even when the applied stress itself is not a pure shear stress. This may be illustrated by examining a cylindrical crystal of area A in a conventional tensile test under a uniaxial load P. In such a test, slip occurs on the slip plane, shown shaded in Figure 7.10, the area of which is $A/\cos\phi$, where ϕ is the angle between the normal to the plane OH and the axis of tension. The applied force P is spread over this plane and may be resolved into a force normal to the plane along OH, $P\cos\phi$, and a force along OS, $P\sin\phi$. Here, OS is the line of greatest slope in the slip plane and the force $P\sin\phi$ is a shear force. It follows that the applied stress (force/area) is made up of two stresses, a normal stress $(P/A)\cos^2\phi$ tending to pull the atoms apart, and a shear stress $(P/A)\cos\phi\sin\phi$ trying to slide the atoms over each other.

In general, slip does not take place down the line of greatest slope unless this happens to coincide with the crystallographic slip of direction. It is necessary, therefore, to know the resolved shear stress on the slip plane and in the slip direction. Now, if OT is taken to represent the slip direction the resolved shear stress will be given by

$$\sigma = P\cos\phi\sin\phi\cos\chi/A$$

where χ is the angle between OS and OT. Usually this formula is written more simply as

$$\sigma = P\cos\phi\cos\lambda/A \qquad (7.4)$$

where λ is the angle between the slip direction OT and the axis of tension. It can be seen that the resolved shear stress has a maximum value when the slip plane is inclined at $45°$ to the tensile axis, and becomes smaller for angles either greater than or less than $45°$. When the slip plane becomes more nearly perpendicular to the tensile axis ($\phi > 45°$) it is easy to imagine that the applied stress has a greater tendency to pull the atoms apart than to slide them. When the slip plane becomes more nearly parallel to the tensile axis ($\phi < 45°$) the shear stress is again small but in this case it is because the area of the slip plane, $A/\cos\phi$, is correspondingly large.

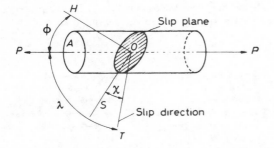

Figure 7.10 *Relation between the slip plane, slip direction and the axis of tension for a cylindrical crystal.*

A consideration of the tensile test in this way shows that it is shear stresses which lead to plastic deformation, and for this reason the mechanical behaviour exhibited by a material will depend, to some extent, on the type of test applied. For example, a ductile material can be fractured without displaying its plastic properties if tested in a state of hydrostatic or triaxial tension, since under these conditions the resolved shear stress on any plane is zero. Conversely, materials which normally exhibit a tendency to brittle behaviour in a tensile test will show ductility if tested under conditions of high shear stresses and low tension stresses. In commercial practice, extrusion approximates closely to a system of hydrostatic pressure, and it is common for normally brittle materials to exhibit some ductility when deformed in this way (e.g. when extruded).

7.3.3 Relation of slip to crystal structure

An understanding of the fundamental nature of plastic deformation processes is provided by experiments on single crystals only, because if a polycrystalline sample is used the result obtained is the average behaviour of all the differently oriented grains in the material. Such experiments with single crystals show that, although the resolved shear stress is a maximum along lines of greatest slope in planes at $45°$ to the tensile axis, slip occurs preferentially along certain crystal planes and directions. Three well-established laws governing the slip behaviour exist, namely: (1) the direction of slip is almost always that along which the atoms are most closely packed, (2) slip usually occurs on the most closely packed plane, and (3) from a given set of slip planes and directions, the crystal operates on that system (plane and direction) for which the resolved shear stress is largest. The slip behaviour observed in fcc metals shows the general applicability of these laws, since slip occurs along $\langle 1\,1\,0\rangle$ directions in $\{1\,1\,1\}$ planes. In cph metals slip occurs along $\langle 1\,1\,\overline{2}\,0\rangle$ directions, since these are invariably the closest packed, but the active slip plane depends on the value of the axial ratio. Thus, for the metals cadmium and zinc, c/a is 1.886 and 1.856, respectively, the planes of greatest atomic density are the $\{0\,0\,0\,1\}$ basal planes and slip takes place on these planes. When the axial ratio is appreciably smaller than the ideal value of $c/a = 1.633$ the basal plane is not so closely packed, nor so widely spaced, as in cadmium and zinc, and other slip planes operate. In zirconium ($c/a = 1.589$) and titanium ($c/a = 1.587$), for example, slip takes place on the $\{1\,0\,\overline{1}\,0\}$ prism planes at room temperature and on the $\{1\,0\,\overline{1}\,1\}$ pyramidal planes at higher temperatures. In magnesium the axial ratio ($c/a = 1.624$) approximates to the ideal value, and although only basal slip occurs at room temperature, at temperatures above 225°C slip on the $\{1\,0\,\overline{1}\,1\}$ planes has also been observed. Bcc metals have a single well-defined close-packed $\langle 1\,1\,1\rangle$ direction, but several planes of equally high density of packing, i.e. $\{1\,1\,2\}$, $\{1\,1\,0\}$ and $\{1\,2\,3\}$.

The choice of slip plane in these metals is often influenced by temperature and a preference is shown for {1 1 2} below $T_m/4$, {1 1 0} from $T_m/4$ to $T_m/2$ and {1 2 3} at high temperatures, where T_m is the melting point. Iron often slips on all the slip planes at once in a common $\langle 1 1 1 \rangle$ slip direction, so that a slip line (i.e. the line of intersection of a slip plane with the outer surface of a crystal) takes on a wavy appearance.

7.3.4 Law of critical resolved shear stress

This law states that slip takes place along a given slip plane and direction when the shear stress reaches a critical value. In most crystals the high symmetry of atomic arrangement provides several crystallographic equivalent planes and directions for slip (i.e. cph crystals have three systems made up of one plane containing three directions, fcc crystals have twelve systems made up of four planes each with three directions, while bcc crystals have many systems) and in such cases slip occurs first on that plane and along that direction for which the maximum stress acts (law 3 above). This is most easily demonstrated by testing in tension a series of zinc single crystals. Then, because zinc is cph in structure only one plane is available for the slip process and the resultant stress–strain curve will depend on the inclination of this plane to the tensile axis. The value of the angle ϕ is determined by chance during the process of single-crystal growth, and consequently all crystals will have different values of ϕ, and the corresponding stress–strain curves will have different values of the flow stress as shown in Figure 7.11a. However, because of the criterion of a critical resolved shear stress, a plot of resolved shear stress (i.e. the stress on the glide plane in the glide direction) versus strain should be a common curve, within experimental error, for all the specimens. This plot is shown in Figure 7.11b.

The importance of a critical shear stress may be demonstrated further by taking the crystal which has its basal plane oriented perpendicular to the tensile axis, i.e. $\phi = 0°$, and subjecting it to a bend test. In contrast to its tensile behaviour, where it is brittle it will now appear ductile, since the shear stress on the slip plane is only zero for a tensile test and not for a bend test. On the other hand, if we take the crystal with its basal plane oriented parallel to the tensile axis (i.e. $\phi = 90°$) this specimen will appear brittle whatever stress system is applied to it. For this crystal, although the shear force is large, owing to the large area of the slip plane, $A/\cos \phi$, the resolved shear stress is always very small and insufficient to cause deformation by slipping.

7.3.5 Multiple slip

The fact that slip bands, each consisting of many slip lines, are observed on the surface of deformed crystals shows that deformation is inhomogeneous, with extensive slip occurring on certain planes, while the crystal planes lying between them remain practically undeformed. Figures 7.12a and 7.12b show such a crystal in which the set of planes shear over each other in the slip direction. In a tensile test, however, the ends of a crystal are not free to move 'sideways' relative to each other, since they are constrained by the grips of the tensile machine. In this case, the central portion of the crystal is altered in orientation, and rotation of both the slip plane and slip direction into the axis of tension occurs, as shown in Figure 7.12c. This behaviour is more conveniently demonstrated on a stereographic projection of the crystal by considering the rotation of the tensile axis relative to the crystal rather than vice versa. This is illustrated in Figure 7.13a for the deformation of a crystal with fcc structure. The tensile axis, P, is shown in the unit triangle and the angles between P and $[\bar{1} 0 1]$, and P and $(1 1 1)$ are equal to λ and ϕ, respectively. The active slip system is the $(1 1 1)$ plane and the $[\bar{1} 0 1]$ direction, and as deformation proceeds the change in orientation is represented by the point, P,

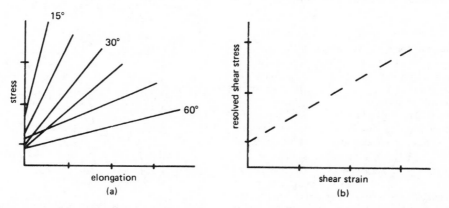

Figure 7.11 *Schematic representation of (a) variation of stress versus elongation with orientation of basal plane and (b) constancy of revolved shear stress.*

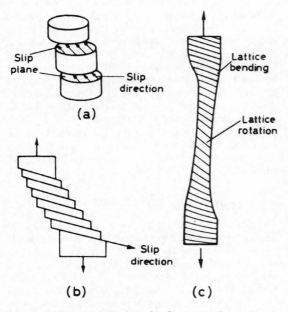

Figure 7.12 *(a) and (b) show the slip process in an unconstrained single crystal; (c) illustrates the plastic bending in a crystal gripped at its ends.*

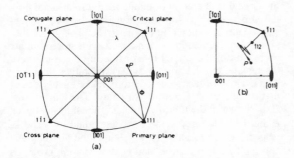

Figure 7.13 *Stereographic representation of (a) slip systems in fcc crystals and (b) overshooting of the primary slip system.*

moving along the zone, shown broken in Figure 7.13a, towards [$\bar{1}$0 1], i.e. λ decreasing and ϕ increasing.

As slip occurs on the one system, the primary system, the slip plane rotates away from its position of maximum resolved shear stress until the orientation of the crystal reaches the [0 0 1] − [$\bar{1}$1 1] symmetry line. Beyond this point, slip should occur equally on both the primary system and a second system (the conjugate system) ($\bar{1}\,\bar{1}$ 1) [0 1 1], since these two systems receive equal components of shear stress. Subsequently, during the process of multiple or duplex slip the lattice will rotate so as to keep equal stresses on the two active systems, and the tensile axis moves along the symmetry line towards [$\bar{1}$ 1 2]. This behaviour agrees with early

observations on virgin crystals of aluminium and copper, but not with those made on certain alloys, or pure metal crystals given special treatments (e.g. quenched from a high temperature or irradiated with neutrons). Results from the latter show that the crystal continues to slip on the primary system after the orientation has reached the symmetry line, causing the orientation to overshoot this line, i.e. to continue moving towards [$\bar{1}$ 0 1], in the direction of primary slip. After a certain amount of this additional primary slip the conjugate system suddenly operates, and further slip concentrates itself on this system, followed by overshooting in the opposite direction. This behaviour, shown in Figure 7.13b, is understandable when it is remembered that slip on the conjugate system must intersect that on the primary system, and to do this is presumably more difficult than to 'fit' a new slip plane in the relatively undeformed region between those planes on which slip has already taken place. This intersection process is more difficult in materials which have a low stacking fault energy (e.g. α-brass).

7.3.6 Relation between work-hardening and slip

The curves of Figure 7.1 show that following the yield phenomenon a continual rise in stress is required to continue deformation, i.e. the flow stress of a deformed metal increases with the amount of strain. This resistance of the metal to further plastic flow as the deformation proceeds is known as work-hardening. The degree of work-hardening varies for metals of different crystal structure, and is low in hexagonal metal crystals such as zinc or cadmium, which usually slip on one family of planes only. The cubic crystals harden rapidly on working but even in this case when slip is restricted to one slip system (see the curve for specimen *A*, Figure 7.14) the coefficient of hardening, defined as the slope of the plastic portion of the stress−strain curve, is small. Thus this type of hardening, like overshoot, must be associated with the interaction which results from slip on intersecting families of planes. This interaction will be dealt with more fully in Section 7.6.2.

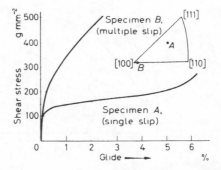

Figure 7.14 *Stress−strain curves for aluminium deformed by single and multiple slip (after Lücke and Lange, 1950).*

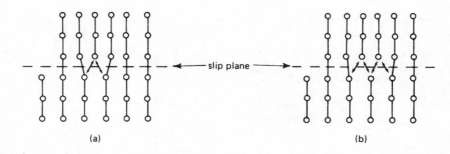

Figure 7.15 *Diagram showing structure of edge dislocation during gliding from (a) equilibrium to (b) metastable position.*

7.4 Dislocation behaviour during plastic deformation

7.4.1 Dislocation mobility

The ease with which crystals can be plastically deformed at stresses many orders of magnitude less than the theoretical strength ($\tau_t = \mu b/2\pi a$) is quite remarkable, and due to the mobility of dislocations. Figure 7.15a shows that as a dislocation glides through the lattice it moves from one symmetrical lattice position to another and at each position the dislocation is in neutral equilibrium, because the atomic forces acting on it from each side are balanced. As the dislocation moves from these symmetrical lattice positions some imbalance of atomic forces does exist, and an applied stress is required to overcome this lattice friction. As shown in Figure 7.15b, an intermediate displacement of the dislocation also leads to an approximately balanced force system.

The lattice friction depends rather sensitively on the dislocation width w and has been shown by Peierls and Nabarro to be given by

$$\tau \simeq \mu \exp\left[-2\pi w/b\right] \qquad (7.5)$$

for the shear of a rectangular lattice of interplanar spacing a with $w = \mu b/2\pi(1 - \nu)\tau_t = a(1 - \nu)$. The friction stress is therefore often referred to as the Peierls–Nabarro stress. The two opposing factors affecting w are (1) the elastic energy of the crystal, which is reduced by spreading out the elastic strains, and (2) the misfit energy, which depends on the number of misaligned atoms across the slip plane. Metals with close-packed structures have extended dislocations and hence w is large. Moreover, the close-packed planes are widely spaced with weak alignment forces between them (i.e. have a small b/a factor). These metals have highly mobile dislocations and are intrinsically soft. In contrast, directional bonding in crystals tends to produce narrow dislocations, which leads to intrinsic hardness and brittleness. Extreme examples are ionic and ceramic crystals and the covalent materials such as diamond and silicon. The bcc transition metals display intermediate behaviour (i.e. intrinsically ductile above room temperatures but brittle below).

Direct measurements of dislocation velocity v have now been made in some crystals by means of the etch pitting technique; the results of such an experiment are shown in Figure 7.16. Edge dislocations move faster than screws, because of the frictional drag of jogs on screws, and the velocity of both varies rapidly with

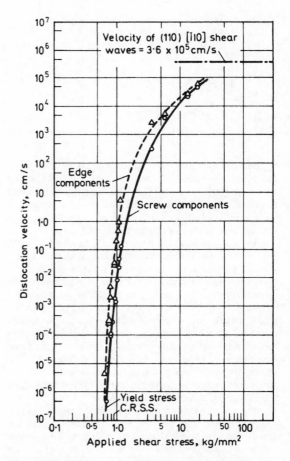

Figure 7.16 *Stress dependence of the velocity of edge and screw dislocations in lithium fluoride (from Johnston and Gilman, 1959; courtesy of the American Institute of Physics).*

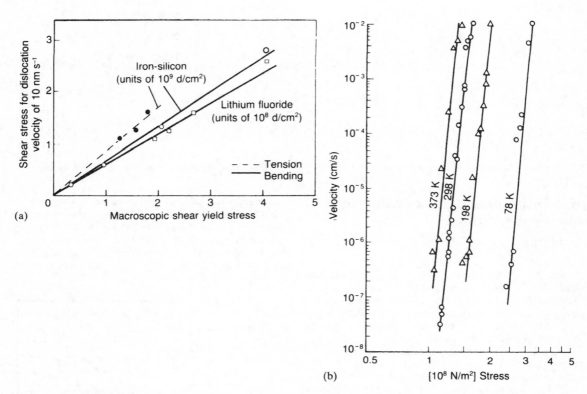

Figure 7.17 *(a) Correlation between stress to cause dislocation motion and the macro-yield stresses of crystals. (b) Edge dislocation motions in Fe-3% Si crystals (after Stein and Low, 1960; courtesy of the American Institute of Physics).*

applied stress τ according to an empirical relation of the form $v = (\tau/\tau_0)^n$, where τ_0 is the stress for unit speed and n is an index which varies for different materials. At high stresses the velocity may approach the speed of elastic waves $\approx 10^3$ m/s. The index n is usually low (<10) for intrinsically hard, covalent crystals such as Ge, ≈ 40 for bcc crystals, and high (≈ 200) for intrinsically soft fcc crystals. It is observed that a critical applied stress is required to start the dislocations moving and denotes the onset of microplasticity. A macroscopic tensile test is a relatively insensitive measure of the onset of plastic deformation and the yield or flow stress measured in such a test is related not to the initial motion of an individual dislocation but to the motion of a number of dislocations at some finite velocity, e.g. ~ 10 nm/s as shown in Figure 17.17a. Decreasing the temperature of the test or increasing the strain-rate increases the stress level required to produce the same finite velocity (see Figure 7.17b), i.e. displacing the velocity–stress curve to the right. Indeed, hardening the material by any mechanism has the same effect on the dislocation dynamics. This observation is consistent with the increase in yield stress with decreasing temperature or increasing strain-rate. Most metals and alloys are hardened by cold working or by placing obstacles (e.g. precipitates) in the path

of moving dislocations to hinder their motion. Such strengthening mechanisms increase the stress necessary to produce a given finite dislocation velocity in a similar way to that found by lowering the temperature.

7.4.2 Variation of yield stress with temperature and strain rate

The high Peierls–Nabarro stress, which is associated with materials with narrow dislocations, gives rise to a short-range barrier to dislocation motion. Such barriers are effective only over an atomic spacing or so, hence thermal activation is able to aid the applied stress in overcoming them. Thermal activation helps a portion of the dislocation to cross the barrier after which glide then proceeds by the sideways movement of kinks. (This process is shown in Figure 7.29, Section 7.4.8.) Materials with narrow dislocations therefore exhibit a significant temperature-sensitivity; intrinsically hard materials rapidly lose their strength with increasing temperature, as shown schematically in Figure 7.18a. In this diagram the (yield stress/modulus) ratio is plotted against T/T_m to remove the effect of modulus which decreases with temperature. Figure 7.18b shows that materials which exhibit a strong temperature-dependent yield stress also exhibit a high strain-rate

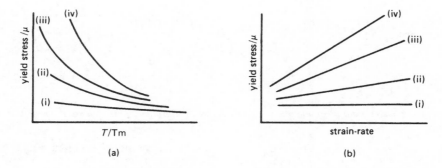

Figure 7.18 *Variation of yield stress with (a) temperature, (b) strain-rate for crystals with (i) fcc, (ii) bcc, (iii) ionic-bonded, (iv) covalent-bonded structure.*

sensitivity, i.e. the higher the imposed strain rate, the higher the yield stress. This arises because thermal activation is less effective at the faster rate of deformation.

In bcc metals a high lattice friction to the movement of a dislocation may arise from the dissociation of a dislocation on several planes. As discussed in Chapter 4, when a screw dislocation with Burgers vector $a/2[1\,1\,1]$ lies along a symmetry direction it can dissociate on three crystallographically equivalent planes. If such a dissociation occurs, it will be necessary to constrict the dislocation before it can glide in any one of the slip planes. This constriction will be more difficult to make as the temperature is lowered so that the large temperature dependence of the yield stress in bcc metals, shown in Figure 7.18a and also Figure 7.30, may be due partly to this effect. In fcc metals the dislocations lie on $\{1\,1\,1\}$ planes, and although a dislocation will dissociate in any given $(1\,1\,1)$ plane, there is no direction in the slip plane along which the dislocation could also dissociate on other planes; the temperature-dependence of the yield stress is small as shown in Figure 7.18a. In cph metals the dissociated dislocations moving in the basal plane will also have a small Peierls force and be glissile with low temperature-dependence. However, screw dislocations moving on non-basal planes (i.e. prismatic and pyramidal planes) may have a high Peierls force because they are able to extend in the basal plane as shown in Figure 7.19. Hence, constrictions will once again have to be made before the screw dislocations can advance on non-basal planes. This effect contributes to the high critical shear stress and strong temperature-dependence of non-basal glide observed in this crystal system, as mentioned in Chapter 4.

7.4.3 Dislocation source operation

When a stress is applied to a material the specimen plastically deforms at a rate governed by the strain rate of the deformation process (e.g. tensile testing, rolling, etc.) and the strain rate imposes a particular velocity on the mobile dislocation population. In a crystal of

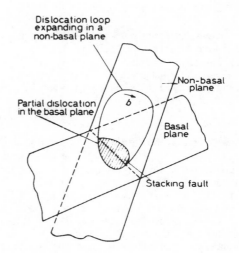

Figure 7.19 *Dissociation in the basal plane of a screw dislocation moving on a non-basal glide plane.*

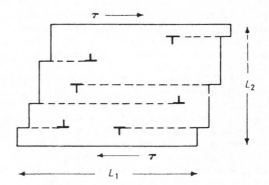

Figure 7.20 *Shear produced by gliding dislocations.*

dimensions $L_1 \times L_2 \times 1$ cm shown in Figure 7.20 a dislocation with velocity v moves through the crystal in time $t = L_1/v$ and produces a shear strain b/L_2, i.e.

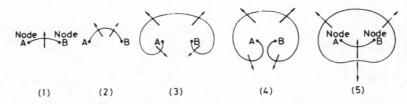

Figure 7.21 *Successive stages in the operation of a Frank–Read source. The plane of the paper is assumed to be the slip plane.*

the strain rate is bv/L_1L_2. If the density of glissible dislocations is ρ, the total number of dislocations which become mobile in the crystal is ρL_1L_2 and the overall strain rate is thus given by

$$\gamma = \frac{b}{L_2}\frac{v}{L_1}\rho L_1L_2 = \rho bv \qquad (7.6)$$

At conventional strain rates (e.g. 1 s^{-1}) the dislocations would be moving at quite moderate speeds of a few cm/s if the mobile density $\approx 10^7/\text{cm}^2$. During high-speed deformation the velocity approaches the limiting velocity. The shear strain produced by these dislocations is given by

$$\gamma = \rho b\overline{x} \qquad (7.7)$$

where $\overline{x}$ is the average distance a dislocation moves. If the distance $x \simeq 10^{-4}$ cm (the size of an average sub-grain) the maximum strain produced by $\rho \approx 10^7$ is about $(10^7 \times 3 \times 10^{-8} \times 10^{-4})$ which is only a fraction of 1%. In practice, shear strains >100% can be achieved, and hence to produce these large strains many more dislocations than the original ingrown dislocations are required. To account for the increase in number of mobile dislocations during straining the concept of a dislocation source has been introduced. The simplest type of source is that due to Frank and Read and accounts for the regenerative multiplication of dislocations. A modified form of the Frank–Read source is the multiple cross-glide source, first proposed by Koehler, which, as the name implies, depends on the cross-slip of screw dislocations and is therefore more common in metals of intermediate and high stacking fault energy.

Figure 7.21 shows a Frank–Read source consisting of a dislocation line fixed at the nodes A and B (fixed, for example, because the other dislocations that join the nodes do not lie in slip planes). Because of its high elastic energy ($\approx$4 eV per atom plane threaded by a dislocation) the dislocation possesses a line tension tending to make it shorten its length as much as possible (position 1, Figure 7.21). This line tension T is roughly equal to $\alpha\mu b^2$, where μ is the shear modulus, b the Burgers vector and α a constant usually taken to be about $\frac{1}{2}$. Under an applied stress the dislocation line will bow out, decreasing its radius of curvature until it reaches an equilibrium position in which the line tension balances the force due to the applied stress. Increasing the applied stress causes the

line to decrease its radius of curvature further until it becomes semi-circular (position 2). Beyond this point it has no equilibrium position so it will expand rapidly, rotating about the nodes and taking up the succession of forms indicated by 3, 4 and 5. Between stages 4 and 5 the two parts of the loop below AB meet and annihilate each other to form a complete dislocation loop, which expands into the slip plane and a new line source between A and B. The sequence is then repeated and one unit of slip is produced by each loop that is generated.

To operate the Frank–Read source the force applied must be sufficient to overcome the restoring force on the dislocation line due to its line tension. Referring to Figure 7.22 this would be $2T\text{d}\theta/2 > \tau bl\text{d}\theta/2$, and if $T \sim \mu b^2/2$ the stress to do this is about $\mu b/l$, where μ and b have their usual meaning and l is the length of the Frank–Read source; the substitution of typical values ($\mu = 4 \times 10^{10}$ N m^{-2}, $b = 2.5 \times 10^{-10}$ m, and $l = 10^{-6}$ m) into this estimate shows that a critical shear stress of about 100 gf mm^{-2} is required. This value is somewhat less than but of the same order as that observed for the yield stress of virgin pure metal single crystals. Another source mechanism involves multiple cross-slip as shown in Figure 7.23. It depends

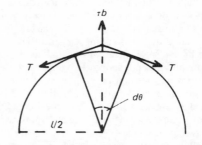

Figure 7.22 *Geometry of Frank–Read source used to calculate the stress to operate.*

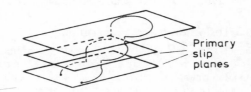

Figure 7.23 *Cross-slip multiplication source.*

on the Frank–Read principle but does not require a dislocation segment to be anchored by nodes. Thus, if part of a moving screw dislocation undergoes double cross-slip the two pieces of edge dislocation on the cross-slip plane effectively act as anchoring points for a new source. The loop expanding on the slip plane parallel to the original plane may operate as a Frank–Read source and any loops produced may in turn cross slip and become a source. This process therefore not only increases the number of dislocations on the original slip plane but also causes the slip band to widen.

The concept of the dislocation source accounts for the observation of slip bands on the surface of deformed metals. The amount of slip produced by the passage of a single dislocation is too small to be observable as a slip line or band under the light microscope. To be resolved it must be at least 300 nm in height and hence $\approx$1000 dislocations must have operated in a given slip band. Moreover, in general, the slip band has considerable width, which tends to support the operation of the cross-glide source as the predominant mechanism of dislocation multiplication during straining.

7.4.4 Discontinuous yielding

In some materials the onset of macroscopic plastic flow begins in an abrupt manner with a yield drop in which the applied stress falls, during yielding, from an upper to a lower yield point. Such yield behaviour is commonly found in iron containing small amounts of carbon or nitrogen as impurity. The main characteristics of the yield phenomenon in iron may be summarized as follows.

Yield point A specimen of iron during tensile deformation (Figure 7.24a, curve 1) behaves elastically up to a certain high load A, known as the upper yield point, and then it suddenly yields plastically. The important feature to note from this curve is that the stress required to maintain plastic flow immediately after yielding has started is lower than that required to start it, as shown by the fall in load from A to B (the lower yield point). A yield point elongation to C then occurs after which the specimen work hardens and the curve rises steadily and smoothly.

Overstraining The yield point can be removed temporarily by applying a small preliminary plastic strain to the specimen. Thus, if after reaching the point D, for example, the specimen is unloaded and a second test is made fairly soon afterwards, a stress–strain curve of type 2 will be obtained. The specimen deforms elastically up to the unloading point, D, and the absence of a yield point at the beginning of plastic flow is characteristic of a specimen in an overstrained condition.

Strain-age hardening If a specimen which has been overstrained to remove the yield point is allowed to rest, or age, before retesting, the yield point returns as shown in Figure 7.24a, curve 3. This process, which is accompanied by hardening (as shown by the increased stress, EF, to initiate yielding) is known as strain-ageing or, more specifically, strain-age hardening. In iron, strain-ageing is slow at room temperature but is greatly speeded up by annealing at a higher temperature. Thus, a strong yield point returns after an ageing treatment of only a few seconds at 200°C, but the same yield point will take many hours to develop if ageing is carried out at room temperature.

Lüders band formation Closely related to the yield point is the formation of Lüders bands. These bands are markings on the surface of the specimen which

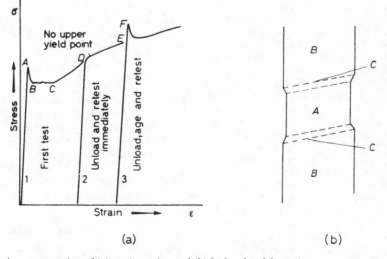

(a)   (b)

Figure 7.24 *Schematic representation of (a) strain ageing and (b) Lüders band formation.*

distinguish those parts of the specimen that have yielded, A, from those which have not, B. Arrival at the upper yield point is indicated by the formation of one or more of these bands and as the specimen passes through the stage of the yield point elongation these bands spread along the specimen and coalesce until the entire gauge length has been covered. At this stage the whole of the material within the gauge length has been overstrained, and the yield point elongation is complete. The growth of a Lüders band is shown diagrammatically in Figure 7.24b. It should be noted that the band is a macroscopic band crossing all the grains in the cross-section of a polycrystalline specimen, and thus the edges of the band are not necessarily the traces of individual slip planes. A second point to observe is that the rate of plastic flow in the edges of a band can be very high even in an apparently slow test; this is because the zones, marked C in Figure 7.24b, are very narrow compared with the gauge length.

These Lüders bands frequently occur in drawing and stamping operations when the surface markings in relief are called stretcher strains. These markings are unsightly in appearance and have to be avoided on many finished products. The remedy consists in overstraining the sheet prior to pressing operations, by means of a temper roll, or roller levelling, pass so that the yield phenomenon is eliminated. It is essential, once this operation has been performed, to carry out pressing before the sheet has time to strain-age; the use of a 'non-ageing' steel is an alternative remedy.

These yielding effects are influenced by the presence of small amounts of carbon or nitrogen atoms interacting with dislocations. The yield point can be removed by annealing at 700°C in wet-hydrogen atmosphere, and cannot subsequently be restored by any strain-ageing treatment. Conversely, exposing the decarburized specimen to an atmosphere of dry hydrogen containing a trace of hydrocarbon at 700°C for as little as one minute restores the yield point. The carbon and nitrogen atoms can also be removed from solution in other ways: for example, by adding to the iron such elements as molybdenum, manganese, chromium, vanadium, niobium or titanium which have a strong affinity for forming carbides or nitrides in steels. For this reason, these elements are particularly effective in removing the yield point and producing a non-strain ageing steel.

The carbon/nitrogen atoms are important in yielding process because they interact with the dislocations and immobilize them. This locking of the dislocations is brought about because the strain energy due to the distortion of a solute atom can be relieved if it fits into a structural region where the local lattice parameter approximates to that of the natural lattice parameter of the solute. Such a condition will be brought about by the segregation of solute atoms to the dislocations, with large substitutional atoms taking up lattice positions in the expanded region, and small ones in the compressed region; small interstitial atoms will tend to segregate to

interstitial sites below the half-plane. Thus, where both dislocations and solute atoms are present in the lattice, interactions of the stress field can occur, resulting in a lowering of the strain energy of the system. This provides a driving force tending to attract solute atoms to dislocations and if the necessary time for diffusion is allowed, a solute atom 'atmosphere' will form around each dislocation.

When a stress is applied to a specimen in which the dislocations are locked by carbon atoms the dislocations are not able to move at the stress level at which free dislocations are normally mobile. With increasing stress yielding occurs when dislocations suddenly become mobile either by breaking away from the carbon atmosphere or by nucleating fresh dislocations at stress concentrations. At this high stress level the mobile dislocation density increases rapidly. The lower yield stress is then the stress at which free dislocations continue to move and produce plastic flow. The overstrained condition corresponds to the situation where the mobile dislocations, brought to rest by unloading the specimen, are set in motion again by reloading before the carbon atmospheres have time to develop by diffusion. If, however, time is allowed for diffusion to take place, new atmospheres can re-form and immobilize the dislocations again. This is the strain-aged condition when the original yield characteristics reappear.

The upper yield point in conventional experiments on polycrystalline materials is the stress at which initially yielded zones trigger yield in adjacent grains. As more and more grains are triggered the yield zones spread across the specimen and form a Lüders band.

The propagation of yield is thought to occur when a dislocation source operates and releases an avalanche of dislocations into its slip plane which eventually pile up at a grain boundary or other obstacle. The stress concentration at the head of the pile-up acts with the applied stress on the dislocations of the next grain and operates the nearest source, so that the process is repeated in the next grain. The applied shear stress σ_y at which yielding propagates is given by

$$\sigma_y = \sigma_i + (\sigma_c r^{1/2})d^{-1/2} \qquad (7.8)$$

where r is the distance from the pile-up to the nearest source, $2d$ is the grain diameter and σ_c is the stress required to operate a source which involves unpinning a dislocation τ_c at that temperature. Equation (7.8) reduces to the Hall–Petch equation $\sigma_y = \sigma_i + k_y d^{-1/2}$, where σ_i is the 'friction' stress term and k_y the grain size dependence parameter ($= m^2 \tau_c r^{1/2}$) discussed in Section 7.4.11.

7.4.5 Yield points and crystal structure

The characteristic feature of discontinuous yielding is that at the yield point the specimen goes from a condition where the availability of mobile dislocations is limited to one where they are in abundance, the increase in mobile density largely arising from dislocation multiplication at the high stress level. A further feature is that not all the dislocations have to be immobilized to observe a yield drop. Indeed, this is not usually possible because specimen handling, non-axial loading, scratches, etc. give rise to stress concentrations that provide a small local density of mobile dislocations (i.e. pre-yield microtrain).

For materials with a high Peierls–Nabarro (P–N) stress, yield drops may be observed even when they possess a significant mobile dislocation density. A common example is that observed in silicon; this is an extremely pure material with no impurities to lock dislocations, but usually the dislocation density is quite modest (10^7 m/m^3) and possesses a high P–N stress.

When these materials are pulled in a tensile test the overall strain rate $\dot{\gamma}$ imposed on the specimen by the machine has to be matched by the motion of dislocations according to the relation $\dot{\gamma} = \rho b v$. However, because ρ is small the individual dislocations are forced to move at a high speed v, which is only attained at a high stress level (the upper yield stress) because of the large P–N stress. As the dislocations glide at these high speeds, rapid multiplication occurs and the mobile dislocation density increases rapidly. Because of the increased value of the term ρv, a lower average velocity of dislocations is then required to maintain a constant strain rate, which means a lower glide stress. The stress that can be supported by the specimen thus drops during initial yielding to the lower yield point, and does not rise again until the dislocation–dislocation interactions caused by the increased ρ produce a significant work-hardening.

In the fcc metals, the P–N stress is quite small and the stress to move a dislocation is almost independent of velocity up to high speeds. If such metals are to show a yield point, the density of mobile dislocations must be reduced virtually to zero. This can be achieved as shown in Figure 7.25 by the tensile testing of whisker crystals which are very perfect. Yielding begins at the stress required to create dislocations in the perfect lattice, and the upper yield stress approaches the theoretical yield strength. Following multiplication, the stress for glide of these dislocations is several orders of magnitude lower.

Bcc transition metals such as iron are intermediate in their plastic behaviour between the fcc metals and diamond cubic Si and Ge. Because of the significant P–N stress these bcc metals are capable of exhibiting a sharp yield point even when the initial mobile dislocation density is not zero, as shown by the calculated curves of Figure 7.26. However, in practice, the dislocation density of well-annealed pure metals is about 10^{10} m/m^3 and too high for any significant yield drop without an element of dislocation locking by carbon atoms.

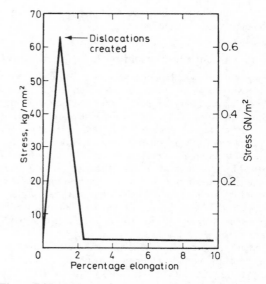

Figure 7.25 *Yield point in a copper whisker.*

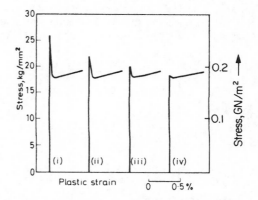

Figure 7.26 *Calculated stress–strain curves showing influence of initial dislocation density on the yield drop in iron for n = 35 with (i) 10^1 cm^{-2}, (ii) 10^3 cm^{-2}, (iii) 10^5 cm^{-2} and (iv) 10^7 cm^{-2} (after Hahn, 1962; courtesy of Pergamon Press).*

It is evident that discontinuous yielding can be produced in all the common metal structures provided the appropriate solute elements are present, and correct testing procedure adopted. The effect is particularly strong in the bcc metals and has been observed in α-iron, molybdenum, niobium, vanadium and β-brass each containing a strongly interacting interstitial solute element. The hexagonal metals (e.g. cadmium and zinc) can also show the phenomenon provided interstitial nitrogen atoms are added. The copper- and aluminium-based fcc alloys also exhibit yielding behaviour but often to a lesser degree. In this case it is substitutional atoms (e.g. zinc in α-brass and copper in aluminium alloys) which are responsible for the phenomenon (see Section 7.4.7).

7.4.6 Discontinuous yielding in ordered alloys

Discontinuous yield points have been observed in a wide variety of A_3B-type alloys. Figure 7.27 shows the development of the yield point in Ni_3Fe on ageing. The addition of Al speeds up the kinetics of ordering and therefore the onset of the yield point. Ordered materials deform by superdislocation motion and the link between yield points and superdislocations is confirmed by the observation that in Cu_3Au, for example, a transition from groups of single dislocations to more randomly arranged superdislocation pairs takes place at $\sim S = 0.7$ (see Chapter 4) and this coincides with the onset of a large yield drop and rapid rise in work hardening.

Sharp yielding may be explained by at least two mechanisms, namely (1) cross-slip of the superdislocation onto the cube plane to lower the APB energy effectively pinning it and (2) dislocation locking by rearrangement of the APB on ageing. The shear APB between a pair of superdislocations is likely to be energetically unstable since there are many like bonds across the interface and thermal activation will modify this sharp interface by atomic rearrangement. This

APB-locking model will give rise to sharp yielding because the energy required by the lead dislocation in creating sharp APB is greater than that released by the trailing dislocation initially moving across diffuse APB. Experimental evidence favours the APB-model and weak-beam electron microscopy (see Figure 7.28) shows that the superdislocation separation for a shear APB corresponds to an energy of 48 ± 5 mJ/m^2, whereas a larger dislocation separation corresponding to an APB energy of 25 ± 3 mJ/m^2 was observed for a strained and aged Cu_3Au.

7.4.7 Solute–dislocation interaction

Iron containing carbon or nitrogen shows very marked yield point effects and there is a strong elastic interaction between these solute atoms and the dislocations. The solute atoms occupy interstitial sites in the lattice and produce large tetragonal distortions as well as large-volume expansions. Consequently, they can interact with both shear and hydrostatic stresses and can lock screw as well as edge dislocations. Strong yielding behaviour is also expected in other bcc metals, provided they contain interstitial solute elements. On the other hand, in the case of fcc metals the arrangement of lattice positions around either interstitial or substitutional sites is too symmetrical to allow a solute atom to produce an asymmetrical distortion, and

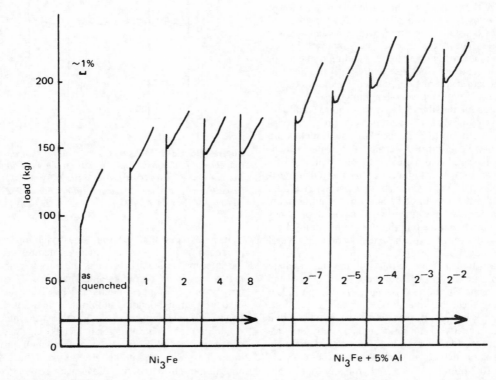

Figure 7.27 *Development of a yield point with ageing at 490°C for the times indicated. (a) Ni_3Fe, (b) $Ni_3Fe + 5\%$ Al; the tests are at room temperature.*

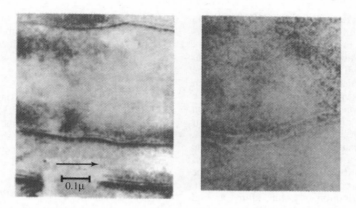

Figure 7.28 *Weak-beam micrographs showing separation of superdislocation partials in Cu_3Au. (a) As deformed, (b) after ageing at 225°C (after Morris and Smallman, 1975).*

the atmosphere locking of screw dislocations, which requires a shear stress interaction, would appear to be impossible. Then by this argument, since the screw dislocations are not locked, a drop in stress at the yield point should not be observed. Nevertheless, yield points are observed in fcc materials and one reason for this is that unit dislocations in fcc metals dissociate into pairs of partial dislocations which are elastically coupled by a stacking fault. Moreover, since their Burgers vectors intersect at 120° there is no orientation of the line of the pair for which both can be pure screws. At least one of them must have a substantial edge component, and a locking of this edge component by hydrostatic interactions should cause a locking of the pair although it will undoubtedly be weaker.

In its quantitative form the theory of solute atom locking has been applied to the formation of an atmosphere around an edge dislocation due to hydrostatic interaction. Since hydrostatic stresses are scalar quantities, no knowledge is required in this case of the orientation of the dislocation with respect to the interacting solute atom, but it is necessary in calculating shear stresses interactions.[1] Cottrell and Bilby have shown that if the introduction of a solute atom causes a volume change Δv at some point in the lattice where the hydrostatic pressure of the stress field is p, the interaction energy is

$$V = p\Delta v = K\Theta\Delta v \tag{7.9}$$

where K is the bulk modulus and Θ is the local dilatation strain. The dilatation strain at a point (R, θ) from a positive edge dislocation is $b(1 - 2v) \times \sin\theta/2\pi R(1 -$

$v)$, and substituting $K = 2\mu(1 + v)/3(1 - 2v)$, where μ is the shear modulus and v Poisson's ratio, we get the expression

$$V_{(R,\theta)} = b(1 + v)\mu\Delta v \sin\theta/3\pi R(1 - v)$$
$$= A\sin\theta/R \tag{7.10}$$

This is the interaction energy at a point whose polar coordinates with respect to the centre of the dislocation are R and θ. We note that V is positive on the upper side $(0 < \theta < \pi)$ of the dislocation for a large atom $(\Delta v > 0)$, and negative on the lower side, which agrees with the qualitative picture of a large atom being repelled from the compressed region and attracted into the expanded one.

It is expected that the site for the strongest binding energy V_{max} will be at a point $\theta = 3\pi/2$, $R = r_0 \simeq b$; and using known values of μ, v and Δv in equation (7.10) we obtain $A \simeq 3 \times 10^{-29}$ N m^2 and $V_{max} \simeq 1$ eV for carbon or nitrogen in α-iron. This value is almost certainly too high because of the limitations of the interaction energy equation in describing conditions near the centre of a dislocation, and a more realistic value obtained from experiment (e.g. internal friction experiments) is $V_{max} \simeq \frac{1}{2}$ to $\frac{3}{4}$ eV. For a substitutional solute atom such as zinc in copper Δv is not only smaller but also easier to calculate from lattice parameter measurements. Thus, if r and $r(1 + \varepsilon)$ are the atomic radii of the solvent and solute, respectively, where ε is the misfit value, the volume change Δv is $4\pi r^3\varepsilon$ and equation (7.10) becomes

$$V = 4(1 + v)\mu b\varepsilon r^3 \sin\theta/3(1 - v)R$$
$$= A\sin\theta/R \tag{7.11}$$

Taking the known values $\mu = 40$ GN/m^2, $v = 0.36$, $b = 2.55 \times 10^{-10}$ m, r_0 and $\varepsilon = 0.06$, we find $A \simeq 5 \times 10^{-30}$ N m^2, which gives a much lower binding energy, $V_{max} = \frac{1}{8}$ eV.

The yield phenomenon is particularly strong in iron because an additional effect is important; this concerns

[1]To a first approximation a solute atom does not interact with a screw dislocation since there is no dilatation around the screw; a second-order dilatation exists however, which gives rise to a non-zero interaction falling off with distance from the dislocation according to $1/r^2$. In real crystals, anisotropic elasticity will lead to first-order size effects even with screw dislocations and hence a substantial interaction is to be expected.

the type of atmosphere a dislocation gathers round itself which can be either condensed or dilute. During the strain-ageing process migration of the solute atoms to the dislocation occurs and two important cases arise. First, if all the sites at the centre of the dislocation become occupied the atmosphere is then said to be condensed; each atom plane threaded by the dislocation contains one solute atom at the position of maximum binding together with a diffuse cloud of other solute atoms further out. If, on the other hand, equilibrium is established before all the sites at the centre are saturated, a steady state must be reached in which the probability of solute atoms leaving the centre can equal the probability of their entering it. The steady-state distribution of solute atoms around the dislocations is then given by the relation

$$C_{(R,\theta)} = c_0 \exp\left[V_{(R,\theta)}/\mathbf{k}T\right]$$

where c_0 is the concentration far from a dislocation, $\mathbf{k}$ is Boltzmann's constant, T is the absolute temperature and c the local impurity concentration at a point near the dislocation where the binding energy is V. This is known as the dilute or Maxwellian atmosphere. Clearly, the form of an atmosphere will be governed by the concentration of solute atoms at the sites of maximum binding energy, V_{max} and for a given alloy (i.e. c_0 and V_{max} fixed) this concentration will be

$$c_{V_{max}} = c_0 \exp\left(V_{max}/\mathbf{k}T\right) \tag{7.12}$$

as long as $c_{V_{max}}$ is less than unity. The value of $c_{V_{max}}$ depends only on the temperature, and as the temperature is lowered $c_{V_{max}}$ will eventually rise to unity. By definition the atmosphere will then have passed from a dilute to a condensed state. The temperature at which this occurs is known as the condensation temperature T_c, and can be obtained by substituting the value $c_{V_{max}} = 1$ in equation (7.12) when

$$T_c = V_{max}/\mathbf{k}\ln(1/c_0) \tag{7.13}$$

Substituting the value of V_{max} for iron, i.e. $\frac{1}{2}$ eV in this equation we find that only a very small concentration of carbon or nitrogen is necessary to give a condensed atmosphere at room temperature, and with the usual concentration strong yielding behaviour is expected up to temperatures of about 400°C.

In the fcc structure although the locking between a solute atom and a dislocation is likely to be weaker, condensed atmospheres are still possible if this weakness can be compensated for by sufficiently increasing the concentration of the solution. This may be why examples of yielding in fcc materials have been mainly obtained from alloys. Solid solution alloys of aluminium usually contain less than 0.1 at. % of solute element, and these show yielding in single crystals only at low temperature (e.g. liquid nitrogen temperature, −196°C) whereas supersaturated alloys show evidence of strong yielding even in polycrystals at room temperature; copper dissolved in aluminium has a misfit value $\varepsilon \simeq 0.12$ which corresponds to $V_{max} = \frac{1}{4}$ eV,

and from equation (7.13) it can be shown that a 0.1 at. % alloy has a condensation temperature $T_c = 250$ K. Copper-based alloys, on the other hand, usually form extensive solid solutions, and, consequently, concentrated alloys may exhibit strong yielding phenomena.

The best-known example is α-brass and, because $V_{max} \simeq \frac{1}{8}$ eV, a dilute alloy containing 1 at. % zinc has a condensation temperature $T_c \simeq 300$ K. At low zinc concentrations (1–10%) the yield point in brass is probably solely due to the segregation of zinc atoms to dislocations. At higher concentrations, however, it may also be due to short-range order.

7.4.8 Dislocation locking and temperature

The binding of a solute atom to a dislocation is short range in nature, and is effective only over an atomic distance or so (Figure 7.29). Moreover, the dislocation line is flexible and this enables yielding to begin by throwing forward a small length of dislocation line, only a few atomic spacings long, beyond the position marked x_2. The applied stress then separates the rest of the dislocation line from its anchorage by pulling the sides of this loop outward along the dislocation line, i.e. by double kink movement. Such a breakaway process would lead to a yield stress which depends sensitively on temperature, as shown in Figure 7.30a. It is observed, however, that k_y, the grain-size dependence parameter in the Hall–Petch equation, in most annealed bcc metals is almost independent of temperature down to the range (<100 K) where twinning occurs, and that practically all the large temperature-dependence is due to σ_i (see Figure 7.30b). To explain this observation it is argued that when locked dislocations exist initially in the material, yielding starts by unpinning them if they are weakly locked (this corresponds to the condition envisaged by Cottrell–Bilby), but if they are strongly locked it starts instead by

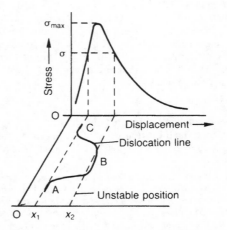

Figure 7.29 *Stress–displacement curve for the breakaway of a dislocation from its atmosphere (after Cottrell, 1957; courtesy of the Institution of Mechanical Engineers).*

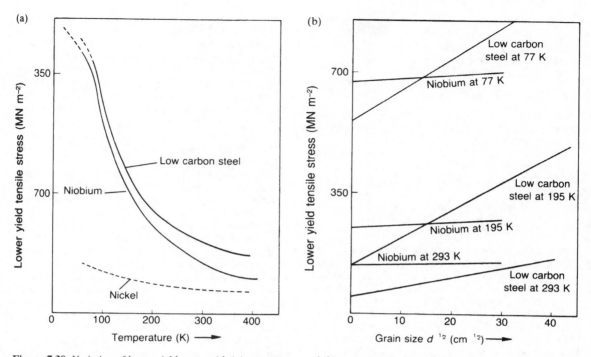

Figure 7.30 *Variation of lower yield stress with (a) temperature and (b) grain size, for low-carbon steel and niobium; the curve for nickel is shown in (a) for comparison (after Adams, Roberts and Smallman, 1960; Hull and Mogford, 1958).*

the creation of new dislocations at points of stress concentration. This is an athermal process and thus k_y is almost independent of temperature. Because of the rapid diffusion of interstitial elements the conventional annealing and normalizing treatments should commonly produce strong locking. In support of this theory, it is observed that k_y is dependent on temperature in the very early stages of ageing following either straining or quenching but on subsequent ageing k_y becomes temperature-independent. The interpretation of k_y therefore depends on the degree of ageing.

Direct observations of crystals that have yielded show that the majority of the strongly anchored dislocations remain locked and do not participate in the yielding phenomenon. Thus large numbers of dislocations are generated during yielding by some other mechanism than breaking away from Cottrell atmospheres, and the rapid dislocation multiplication, which can take place at the high stress levels, is now considered the most likely possibility. Prolonged ageing tends to produce coarse precipitates along the dislocation line and unpinning by bowing out between them should easily occur before grain boundary creation. This unpinning process would also give k_y independent of temperature.

7.4.9 Inhomogeneity interaction

A different type of elastic interaction can exist which arises from the different elastic properties of the solvent matrix and the region near a solute. Such an inhomogeneity interaction has been analysed for both a rigid and a soft spherical region; the former corresponds to a relatively hard impurity atom and the latter to a vacant lattice site. The results indicate that the interaction energy is of the form B/r^2 where B is a constant involving elastic constants and atomic size. It is generally believed that the inhomogeneity effect is small for solute–dislocation interactions but dominates the size effect for vacancy–dislocation interaction. The kinetics of ageing support this conclusion.

7.4.10 Kinetics of strain-ageing

Under a force F an atom migrating by thermal agitation acquires a steady drift velocity $v = DF/\mathbf{k}T$ (in addition to its random diffusion movements) in the direction of the F, where D is the coefficient of diffusion. The force attracting a solute atom to a dislocation is the gradient of the interaction energy dV/dr and hence $v = (D/\mathbf{k}T)(A/r^2)$. Thus atoms originally at a distance r from the dislocation reach it in a time given approximately by

$$t = r/v = r^3\mathbf{k}T/AD$$

After this time t the number of atoms to reach unit length of dislocation is

$$n(t) = \pi r^2 c_0 = \pi c_0[(AD/\mathbf{k}T)t]^{2/3}$$

where c_0 is the solute concentration in uniform solution in terms of the number of atoms per unit volume. If ρ is the density of dislocations (cm/cm^3) and f the fraction of the original solute which has segregated to the dislocation in time t then,

$$f = \pi\rho[(AD/\mathbf{k}T)t]^{2/3} \tag{7.14}$$

This expression is valid for the early stages of ageing, and may be modified to fit the later stages by allowing for the reduction in the matrix concentration as ageing proceeds, such that the rate of flow is proportional to the amount left in the matrix,

$$\mathrm{d}f/\mathrm{d}t = \pi\rho(AD/\mathbf{k}T)^{2/3}(2/3)t^{-2/3}(1-f)$$

which when integrated gives

$$f = 1 - \exp\{-\pi\rho[(AD/\mathbf{k}T)t]^{1/3}\} \tag{7.15}$$

This reduces to the simpler equation (7.14) when the exponent is small, and is found to be in good agreement with the process of segregation and precipitation on dislocations in several bcc metals. For carbon in α-Fe, Harper determined the fraction of solute atom still in solution using an internal friction technique and showed that $\log(1-f)$ is proportional to $t^{2/3}$; the slope of the line is $\pi\rho(AD/\mathbf{k}T)$ and evaluation of this slope at a series of temperatures allows the activation energy for the process to be determined from an Arrhenius plot. The value obtained for α-iron is 84 kJ/mol which is close to that for the diffusion of carbon in ferrite.

The inhomogeneity interaction is considered to be the dominant effect in vacancy–dislocation interactions, with $V = -B/r^2$ where B is a constant; this compares with the size effect for which $V = -A/r$ would be appropriate for the interstitial–dislocation interaction. It is convenient, however, to write the interaction energy in the general form $V = -A/r^n$ and hence, following the treatment previously used for the kinetics of strain-ageing, the radial velocity of a point defect towards the dislocation is

$$V = (D/\mathbf{k}T)(nA/r^{n+1}) \tag{7.16}$$

The number of a particular point defect specie that reach the dislocation in time t is

$$n(t) = \pi r^2 c_0$$
$$= \pi c_0[ADn(n+2)/\mathbf{k}T]^{2/(n+2)}t^{2/(n+2)} \tag{7.17}$$

and when $n=2$ then $n(t) \propto t^{1/2}$, and when $n=1$, $n(t) \propto t^{2/3}$. Since the kinetics of ageing in quenched copper follow $t^{1/2}$ initially, the observations confirm the importance of the inhomogeneity interaction for vacancies.

7.4.11 Influence of grain boundaries on plasticity

It might be thought that when a stress is applied to a polycrystalline metal, every grain in the sample deforms as if it were an unconstrained single crystal. This is not the case, however, and the fact that the aggregate does not deform in this manner is indicated by the high yield stress of polycrystals compared with that of single crystals. This increased strength of polycrystals immediately poses the question–is the hardness of a grain caused by the presence of the grain boundary or by the orientation difference of the neighbouring grains? It is now believed that the latter is the case but that the structure of the grain boundary itself may be of importance in special circumstances such as when brittle films, due to bismuth in copper or cementite in steel, form around the grains or when the grains slip past each other along their boundaries during high-temperature creep. The importance of the orientation change across a grain boundary to the process of slip has been demonstrated by experiments on 'bamboo'-type specimens, i.e. where the grain boundaries are parallel to each other and all perpendicular to the axis of tension. Initially, deformation occurs by slip only in those grains most favourably oriented, but later spreads to all the other grains as those grains which are deformed first, work harden. It is then found that each grain contains wedge-shaped areas near the grain boundary, as shown in Figure 7.31a, where slip does not operate, which indicates that the continuance of slip from one grain to the next is difficult. From these observations it is natural to enquire what happens in a completely polycrystalline metal where the slip planes must in all cases make contact with a grain boundary. It will be clear that the polycrystalline aggregate must be stronger because, unlike the deformation of bamboo-type samples where it is not necessary to raise the stress sufficiently high to operate those slip planes which made contact with a grain boundary, all the slip planes within any grain of a polycrystalline aggregate make contact with a grain boundary, but, nevertheless, have to be operated. The importance of the grain size

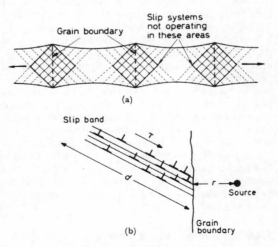

Figure 7.31 *(a) Grain-boundary blocking of slip. (b) Blocking of a slip band by a grain boundary.*

on a strength is emphasized by Figure 7.30b, which shows the variation in lower yield stress, σ_y, with grain diameter, $2d$, for low-carbon steel. The smaller the grain size, the higher the yield strength according to a relation of the form

$$\sigma_y = \sigma_i + kd^{-1/2} \qquad (7.18)$$

where σ_i is a lattice friction stress and k a constant usually denoted k_y to indicate yielding. Because of the difficulties experienced by a dislocation in moving from one grain to another, the process of slip in a polycrystalline aggregate does not spread to each grain by forcing a dislocation through the boundary. Instead, the slip band which is held up at the boundary gives rise to a stress concentration at the head of the pile-up group of dislocations which acts with the applied stress and is sufficient to trigger off sources in neighbouring grains. If τ_i is the stress a slip band could sustain if there were no resistance to slip across the grain boundary, i.e. the friction stress, and τ the higher stress sustained by a slip band in a polycrystal, then $(\tau - \tau_i)$ represents the resistance offered by the boundary, which reaches a limiting value when slip is induced in the next grain. The influence of grain size can be explained if the length of the slip band is proportional to d as shown in Figure 7.31(b). Thus, since the stress concentration a short distance r from the end of the slip band is proportional to $(d/4r)^{1/2}$, the maximum shear stress at a distance r ahead of a slip band carrying an applied stress τ in a polycrystal is given by $(\tau - \tau_i)[d/4r]^{1/2}$ and lies in the plane of the slip band. If this maximum stress has to reach a value $\tau_{\max}$ to operate a new source at a distance r then

$$(\tau - \tau_i)[d/4r]^{1/2} = \tau_{\max}$$

or, rearranging,

$$\tau = \tau_i + (\tau_{\max} 2r^{1/2})d^{-1/2}$$

which may be written as

$$\tau = \tau_i + k_s d^{-1/2}$$

It then follows that the tensile flow curve of a polycrystal is given by

$$\sigma = m(\tau_i + k_s d^{-1/2}) \qquad (7.19)$$

where m is the orientation factor relating the applied tensile stress σ to the shear stress, i.e. $\sigma = m\tau$. For a single crystal the m-factor has a minimum value of 2 as discussed, but in polycrystals deformation occurs in less favourably oriented grains and sometimes (e.g. hexagonal, intermetallics, etc.) on 'hard' systems, and so the m-factor is significantly higher. From equation (7.18) it can be seen that $\sigma_i = m\tau_i$ and $k = mk_s$.

While there is an orientation factor on a macroscopic scale in developing the critical shear stress within the various grains of a polycrystal, so there is a local orientation factor in operating a dislocation source ahead of a blocked slip band. The slip plane of the sources will not, in general, lie in the plane of maximum shear stress, and hence $\tau_{\max}$ will need to be such that the shear stress, τ_c required to operate the new source must be generated in the slip plane of the source. In general, the local orientation factor dealing with the orientation relationship of adjacent grains will differ from the macroscopic factor of slip plane orientation relative to the axis of stress, so that $\tau_{\max} = \frac{1}{2}m'\tau_c$. For simplicity, however, it will be assumed $m' = m$ and hence the parameter k in the Petch equation is given by $k = m^2 \tau_c r^{1/2}$.

It is clear from the above treatment that the parameter k depends essentially on two main factors. The first is the stress to operate a source dislocation, and this depends on the extent to which the dislocations are anchored or locked by impurity atoms. Strong locking implies a large τ_c and hence a large k; the converse is true for weak locking. The second factor is contained in the parameter m which depends on the number of available slip systems. A multiplicity of slip systems enhances the possibility for plastic deformation and so implies a small k. A limited number of slip systems available would imply a large value of k. It then follows, as shown in Figure 7.32, that for (1) fcc metals, which have weakly locked dislocations and a multiplicity of slip systems, k will generally be small, i.e. there is only a small grain size dependence of the flow stress, (2) cph metals, k will be large because of the limited slip systems, and (3) bcc metals, because of the strong locking, k will be large.

Each grain does not deform as a single crystal in simple slip, since, if this were so, different grains would then deform in different directions with the result that voids would be created at the grain boundaries. Except in high-temperature creep, where the grains slide past each other along their boundaries, this does not happen and each grain deforms in coherence with its neighbouring grains. However, the fact that

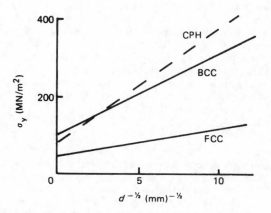

Figure 7.32 *Schematic diagram showing the grain size-dependence of the yield stress for crystals of different crystal structure.*

the continuity of the metal is maintained during plastic deformation must mean that each grain is deformed into a shape that is dictated by the deformation of its neighbours. Such behaviour will, of course, require the operation of several slip systems, and von Mises has shown that to allow this unrestricted change of shape of a grain requires at least five independent shear modes. The deformation of metal crystals with cubic structure easily satisfies this condition so that the polycrystals of these metals usually exhibit considerable ductility, and the stress–strain curve generally lies close to that of single crystals of extreme orientations deforming under multiple slip conditions. The hexagonal metals do, however, show striking differences between their single crystal and polycrystalline behaviour. This is because single crystals of these metals deform by a process of basal plane slip, but the three shear systems (two independent) which operate do not provide enough independent shear mechanisms to allow unrestricted changes of shape in polycrystals. Consequently, to prevent gaps opening up at grain boundaries during the deformation of polycrystals, some additional shear mechanisms, such as non-basal slip and mechanical twinning, must operate. Hence, because the resolved stress for non-basal slip and twinning is greater than that for basal-plane slip, yielding in a polycrystal is prevented until the applied stress is high enough to deform by these mechanisms.

7.4.12 Superplasticity

A number of materials, particularly two-phase eutectic or eutectoid alloys, have been observed to exhibit large elongations ($\approx 1000\%$) without fracture, and such behaviour has been termed superplasticity. Several metallurgical factors have been put forward to explain superplastic behaviour and it is now generally recognized that the effect can be produced in materials either (1) with a particular structural condition or (2) tested under special test conditions. The particular structural condition is that the material has a very fine grain size and the presence of a two-phase structure is usually of importance in maintaining this fine grain size during testing. Materials which exhibit superplastic behaviour under special test conditions are those for which a phase boundary moves through the strained material during the test (e.g. during temperature cycling).

In general, the superplastic material exhibits a high strain-rate sensitivity. Thus the plastic flow of a solid may be represented by the relation

$$\sigma = K\dot{\varepsilon}^m \tag{7.20}$$

where σ is the stress, $\dot{\varepsilon}$ the strain-rate and m an exponent generally known as the strain-rate sensitivity. When $m = 1$ the flow stress is directly proportional to strain rate and the material behaves as a Newtonian viscous fluid, such as hot glass. Superplastic materials are therefore characterized by high m-values, since this leads to increased stability against necking in a tensile test. Thus, for a tensile specimen length l

with cross-sectional area A under an applied load P then $dl/l = -dA/A$ and introducing the time factor we obtain

$$\dot{\varepsilon} = -(1/A)dA/dt$$

and if during deformation the equation $\sigma = K\dot{\varepsilon}^m$ is obeyed, then

$$dA/dt = (P/K)^{1/m} A^{\{1-(1/m)\}} \tag{7.21}$$

For most metals and alloys $m \approx 0.1-0.2$ and the rate at which A changes is sensitively dependent on A, and hence once necking starts the process rapidly leads to failure. When $m = 1$, the rate of change of area is independent of A and, as a consequence, any irregularities in specimen geometry are not accentuated during deformation. The resistance to necking therefore depends sensitively on m, and increases markedly when $m \gtrsim 0.5$. Considering, in addition, the dependence of the flow stress on strain, then

$$\sigma = K^1 \varepsilon^n \dot{\varepsilon}^m \tag{7.22}$$

and in this case, the stability against necking depends on a factor $(1 - n - m)/m$, but n-values are not normally very high. Superplastic materials such as Zn–Al eutectoid, Pb–Sn eutectic, Al–Cu eutectic, etc. have m values approaching unity at elevated temperatures.

The total elongation increases as m increases and with increasing microstructural fineness of the material (grain-size or lamella spacing) the tendency for superplastic behaviour is increased. Two-phase structures are advantageous in maintaining a fine grain size during testing, but exceptionally high ductilities have been produced in several commercially pure metals (e.g. Ni, Zn and Mg), for which the fine grain size was maintained during testing at a particular strain-rate and temperature.

It follows that there must be several possible conditions leading to superplasticity. Generally, it is observed metallographically that the grain structure remains remarkably equiaxed during extensive deformation and that grain boundary sliding is a common deformation mode in several superplastic alloys. While grain boundary sliding can contribute to the overall deformation by relaxing the five independent mechanisms of slip, it cannot give rise to large elongations without bulk flow of material (e.g. grain boundary migration). In polycrystals, triple junctions obstruct the sliding process and give rise to a low m-value. Thus to increase the rate sensitivity of the boundary shear it is necessary to lower the resistance to sliding from barriers, relative to the viscous drag of the boundary; this can be achieved by grain boundary migration. Indeed, it is observed that superplasticity is controlled by grain boundary diffusion.

The complete explanation of superplasticity is still being developed, but it is already clear that during deformation individual grains or groups of grains with suitably aligned boundaries will tend to slide. Sliding

Table 7.2 Twinning elements for some common metals

Structure	Plane	Direction	Metals
cph	$\{10\bar{1}2\}$	$\langle 10\bar{1}\bar{1}\rangle$	Zn, Cd, Be, Mg
bcc	$\{112\}$	$\langle 111\rangle$	Fe, β-brass, W, Ta, Nb, V, Cr, Mo
fcc	$\{111\}$	$\langle 112\rangle$	Cu, Ag, Au, Ag–Au, Cu–Al
Tetragonal	$\{331\}$	—	Sn
Rhombohedral	$\{001\}$	—	Bi, As, Sb

continues until obstructed by a protrusion in a grain boundary, when the local stress generates dislocations which slip across the blocked grain and pile up at the opposite boundary until the back stress prevents further generation of dislocations and thus further sliding. At the temperature of the test, dislocations at the head of the pile-up can climb into and move along grain boundaries to annihilation sites. The continual replacement of these dislocations would permit grain boundary sliding at a rate governed by the rate of dislocation climb, which in turn is governed by grain boundary diffusion. It is important that any dislocations created by local stresses are able to traverse yielded grains and this is possible only if the 'dislocation cell size' is larger than, or at least of the order of, the grain size, i.e. a few microns. At high strain-rates and low temperatures the dislocations begin to tangle and form cell structures and superplasticity then ceases.

The above conditions imply that any metal in which the grain size remains fine during deformation could behave superplastically; this conclusion is borne out in practice. The stability of grain size can, however, be achieved more readily with a fine micro-duplex structure as observed in some Fe–20Cr–6Ni alloys when hot worked to produce a fine dispersion of austenite and ferrite. Such stainless steels have an attractive combination of properties (strength, toughness, fatigue strength, weldability and corrosion resistance) and unlike the usual range of two-phase stainless steels have good hot workability if 0.5Ti is added to produce a random distribution of TiC rather than $Cr_{23}C_6$ at ferrite-austenite boundaries.

Superplastic forming is now an established and growing industry largely using vacuum forming to produce intricate shapes with high draw ratios. Two alloys which have achieved engineering importance are *Supral* (containing Al–6Cu–0.5Zr) and *IMI 318* (containing Ti–6Al–4V). *Supral* is deformed at 460°C and *IMI 318* at 900°C under argon. Although the process is slow, the loads required are also low and the process can be advantageous in the press-forming field to replace some of the present expensive and complex forming technology.

7.5 Mechanical twinning

7.5.1 Crystallography of twinning

Mechanical twinning plays only a minor part in the deformation of the common metals such as copper or aluminium, and its study has consequently been neglected. Nevertheless, twinning does occur in all the common crystal structures under some conditions of deformation. Table 7.2 shows the appropriate twinning elements for the common structures.

The geometrical aspects of twinning can be represented with the aid of a unit sphere shown in Figure 7.33. The twinning plane k_1 intersects the plane of the drawing in the shear direction η_1. On twinning the unit sphere is distorted to an ellipsoid of equal volume, and the shear plane k_1 remains unchanged during twinning, while all other planes become tilted. Distortion of planes occurs in all cases except k_1 and k_2. The shear strain, s, at unit distance from the twinning plane is related to the angle between k_1 and k_2. Thus the amount of shear is fixed by the crystallographic nature of the two undistorted planes. In the bcc lattice, the two undistorted planes are the (112) and $(11\bar{2})$ planes, displacement occurring in a $[111]$ direction a distance of 0.707 lattice vectors. The twinning elements are thus:

k_1	k_2	η_1	η_2	*Shear*
(112)	$(11\bar{2})$	$[11\bar{1}]$	$[111]$	0.707

where k_1 and k_2 denote the first and second undistorted planes, respectively, and η_1 and η_2 denote directions lying in k_1 and k_2, respectively, perpendicular to the line of intersection of these planes. k_1 is also called the composition or twinning plane, while η_1 is called the shear direction. The twins consist of regions of crystal

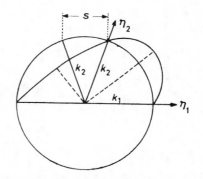

Figure 7.33 *Crystallography of twinning.*

in which a particular set of {1 1 2} planes (the k_1 set of planes) is homogeneously sheared by 0.707 in a $\langle 1\,1\,1 \rangle$ direction (the η_1 direction). The same atomic arrangement may be visualized by a shear of 1.414 in the *reverse* $\langle 1\,1\,1 \rangle$ direction, but this larger shear has never been observed.

7.5.2 Nucleation and growth of twins

During the development of mechanical twins, thin lamellae appear very quickly ($\approx$ speed of sound) and these thicken with increasing stress by the steady movement of the twin interface. New twins are usually formed in bursts and are sometimes accompanied by a sharp audible click which coincides with the appearance of irregularities in the stress–strain curve, as shown in Figure 7.34. The rapid production of clicks is responsible for the so-called twinning cry (e.g. in tin).

Although most metals show a general reluctance to twin, when tested under suitable conditions they can usually be made to do so. As mentioned in Section 7.3.1, the shear process involved in twinning must occur by the movement of partial dislocations and, consequently, the stress to cause twinning will depend not only on the line tension of the source dislocation, as in the case of slip, but also on the surface tension of the twin boundary. The stress to cause twinning is, therefore, usually greater than that required for slip and at room temperature deformation will nearly always occur by slip in preference to twinning. As the

deformation temperature is lowered the critical shear stress for slip increases, and then, because the general stress level will be high, the process of deformation twinning is more likely.

Twinning is most easily achieved in metals of cph structure where, because of the limited number of slip systems, twinning is an essential and unavoidable mechanism of deformation in polycrystalline specimens (see Section 7.4.11), but in single crystals the orientation of the specimen, the stress level and the temperature of deformation are all important factors in the twinning process. In metals of the bcc structure twinning may be induced by impact at room temperature or with more normal strain rates at low temperature where the critical shear stress for slip is very high. In contrast, only a few fcc metals have been made to twin, even at low temperatures.

In zinc single crystals it is observed that there is no well-defined critical resolved shear stress for twinning such as exists for slip, and that a very high stress indeed is necessary to nucleate twins. In most crystals, slip usually occurs first and twin nuclei are then created by means of the very high stress concentration which exists at dislocation pile-ups. Once formed, the twins can propagate provided the resolved shear stress is higher than a critical value, because the stress to propagate a twin is much lower than that to nucleate it. This can be demonstrated by deforming a crystal oriented in such a way that basal slip is excluded,

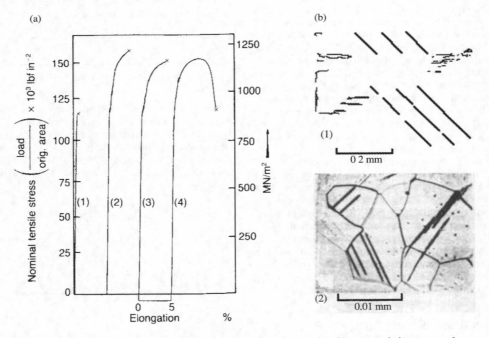

Figure 7.34 (a) Effect of grain size on the stress–strain curves of specimens of niobium extended at a rate of 2.02×10^{-4} s^{-1} at 20 K; (1) grain size $2d = 1.414$ mm, (2) grain size $2d = 0.312$ mm, (3) grain size $2d = 0.0951$ mm, (4) grain size $2d = 0.0476$ mm. (b) Deformation twins in specimen 1 and specimen 3 extended to fracture. Etched in 95% $HNO_3 + 5\%$ HF (after Adams, Roberts and Smallman, 1960).

i.e. when the basal planes are nearly parallel to the specimen axis. Even in such an oriented crystal it is found that the stress to cause twinning is higher than that for slip on non-basal planes. In this case, non-basal slip occurs first so that when a dislocation pile-up arises and a twin is formed, the applied stress is so high that an avalanche or burst of twins results.

It is also believed that in the bcc metals twin nucleation is more difficult than twin propagation. One possible mechanism is that nucleation is brought about by the stress concentration at the head of a piled-up array of dislocations produced by a burst of slip as a Frank–Read source operates. Such a behaviour is favoured by impact loading, and it is well known that twin lamellae known as Neumann bands are produced this way in α-iron at room temperature. At normal strain rates, however, it should be easier to produce a slip burst suitable for twin nucleation in a material with strongly locked dislocations, i.e. one with a large k value (as defined by equation (7.19)) than one in which the dislocation locking is relatively slight (small k values). In this context it is interesting to note that both niobium and tantalum have a small k value and, although they can be made to twin, do so with reluctance compared, for example, with α-iron.

In all the bcc metals the flow stress increases so rapidly with decreasing temperature (see Figure 7.30), that even with moderate strain rates (10^{-4} s^{-1}) α-iron will twin at 77 K, while niobium with its smaller value of k twins at 20 K. The type of stress–strain behaviour for niobium is shown in Figure 7.34a. The pattern of behaviour is characterized by small amounts of slip interspersed between extensive bursts of twinning in the early stages of deformation. Twins, once formed, may themselves act as barriers, allowing further dislocation pile-up and further twin nucleation. The action of twins as barriers to slip dislocations could presumably account for the rapid work-hardening observed at 20 K.

Fcc metals do not readily deform by twinning but it can occur at low temperatures, and even at 0°C, in favourably oriented crystals. The apparent restriction of twinning to certain orientations and low temperatures may be ascribed to the high shear stress attained in tests on crystals with these orientations, since the stress necessary to produce twinning is high. Twinning has been confirmed in heavily rolled copper. The exact mechanism for this twinning is not known, except that it must occur by the propagation of a half-dislocation and its associated stacking fault across each plane of a set of parallel (1 1 1) planes. For this process the half-dislocation must climb onto successive twin planes, as below for bcc iron.

7.5.3 Effect of impurities on twinning

It is well established that solid solution alloying favours twinning in fcc metals. For example, silver–gold alloys twin far more readily than the pure metals. Attempts have been made to correlate this effect with stacking fault energy and it has been shown that the twinning stress of copper-based alloys increases with increasing stacking fault energy. Twinning is also favoured by solid solution alloying in bcc metals, and alloys of Mo–Re, W–Re and Nb–V readily twin at room temperature. In this case it has been suggested that the lattice frictional stress is increased and the ability to cross-slip reduced by alloying, thereby confining slip dislocations to bands where stress multiplication conducive to twin nucleation occurs.

7.5.4 Effect of prestrain on twinning

Twinning can be suppressed in most metals by a certain amount of prestrain; the ability to twin may be restored by an ageing treatment. It has been suggested that the effect may be due to the differing dislocation distribution produced under different conditions. For example, niobium will normally twin at −196°C, when a heterogeneous arrangement of elongated screw dislocations capable of creating the necessary stress concentrations are formed. Room temperature prestrain, however, inhibits twin formation as the regular network of dislocations produced provides more mobile dislocations and homogenizes the deformation.

7.5.5 Dislocation mechanism of twinning

In contrast to slip, the shear involved in the twinning process is homogeneous throughout the entire twinning region, and each atom plane parallel to the twinning plane moves over the one below it by only a fraction of a lattice spacing in the twinning direction. Nevertheless, mechanical twinning is thought to take place by a dislocation mechanism for the same reasons as slip but the dislocations that cause twinning are partial and not unit dislocations. From the crystallography of the process it can be shown that twinning in the cph lattice, in addition to a simple shear on the twinning plane, must be accompanied by a localized rearrangement of the atoms, and furthermore, only in the bcc lattice does the process of twinning consist of a simple shear on the twinning plane (e.g. a twinned structure in this lattice can be produced by a shear of $1/\sqrt{2}$ in a $\langle 1 1 1 \rangle$ direction on a $\{1 1 2\}$ plane).

An examination of Figure 7.9 shows that the main problem facing any theory of twinning is to explain how twinning develops homogeneously through successive planes of the lattice. This could be accomplished by the movement of a single twinning (partial) dislocation successively from plane to plane. One suggestion, similar in principle to the crystal growth mechanism, is the pole-mechanism proposed by Cottrell and Bilby illustrated in Figure 7.35a. Here, OA, OB, and OC are dislocation lines. The twinning dislocation is OC, which produces the correct shear as it sweeps through the twin plane about its point of emergence O, and OA and OB form the pole dislocation, being partly or wholly of screw character with a pitch

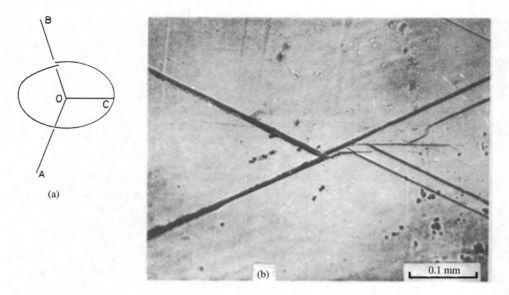

Figure 7.35 (a) Diagram illustrating the pole mechanism of twinning. (b) The formation of a crack at a twin intersection in silicon–iron (after Hull, 1960).

equal to the spacing of the twinning layers. The twinning dislocation rotates round the pole dislocation and in doing so, not only produces a monolayer sheet of twinned crystal but also climbs up the 'pole' to the next layer. The process is repeated and a thick layer of twin is built up.

The dislocation reaction involved is as follows. The line AOB represents a unit dislocation with a Burgers vector $a/2[1\,1\,1]$ and that part OB of the line lies in the $(1\,1\,2)$ plane. Then, under the action of stress dissociation of this dislocation can occur according to the reaction

$$a/2[1\,1\,1] \longrightarrow a/3[1\,1\,2] + a/6[1\,1\,\bar{1}]$$

The dislocation with vector $a/6[1\,1\,\bar{1}]$ forms a line OC lying in one of the other $\{1\,1\,2\}$ twin planes (e.g. the $(\bar{1}\,2\,1)$ plane) and produces the correct twinning shear. The line OB is left with a Burgers vector $a/3[1\,1\,2]$ which is of pure edge type and sessile in the $(1\,1\,2)$ plane.

7.5.6 Twinning and fracture

It has been suggested that a twin, like a grain boundary, may present a strong barrier to slip and that a crack can be initiated by the pile-up of slip dislocations at the twin interface (see Figure 8.32). In addition, cracks may be initiated by the intersection of twins, and examples are common in molybdenum, silicon–iron (bcc) and zinc (cph). Figure 7.35b shows a very good example of crack nucleation in 3% silicon–iron; the crack has formed along an $\{0\,0\,1\}$ cleavage plane at the intersection of two $\{1\,1\,2\}$ twins, and part of the

crack has developed along one of the twins in a zigzag manner while still retaining $\{0\,0\,1\}$ cleavage facets.

In tests at low temperature on bcc and cph metals both twinning and fracture readily occur, and this has led to two conflicting views. First, that twins are nucleated by the high stress concentrations associated with fracture, and second, that the formation of twins actually initiates the fracture. It is probable that both effects occur.

7.6 Strengthening and hardening mechanisms

7.6.1 Point defect hardening

The introduction of point defects into materials to produce an excess concentration of either vacancies or interstitials often gives rise to a significant change in mechanical properties (Figures 7.36 and 7.37). For aluminium the shape of the stress–strain curve is very dependent on the rate of cooling and a large increase in the yield stress may occur after quenching. We have already seen in Chapter 4 that quenched-in vacancies result in clustered vacancy defects and these may harden the material. Similarly, irradiation by high-energy particles may produce irradiation-hardening (see Figure 7.37). Information on the mechanisms of hardening can be obtained from observation of the dependence of the lower yield stress on grain size. The results, reproduced in Figure 7.37b, show that the relation $\sigma_y = \sigma_i + k_y d^{-1/2}$, which is a general relation describing the propagation of yielding in materials, is obeyed.

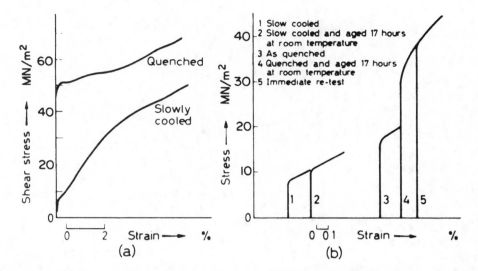

Figure 7.36 *Effect of quenching on the stress–strain curves from (a) aluminium (after Maddin and Cottrell, 1955), and (b) gold (after Adams and Smallman, unpublished).*

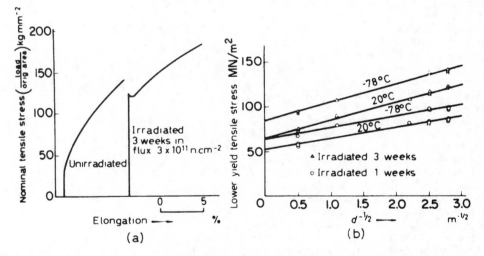

Figure 7.37 *(a) Stress–strain curves for unirradiated and irradiated fine-grained polycrystalline copper, tested at 20°C; (b) variation of yield stress with grain size and neutron dose (after Adams and Higgins, 1959).*

This dependence of the yield stress, σ_y, on grain size indicates that the hardening produced by point defects introduced by quenching or irradiation, is of two types: (1) an initial dislocation source hardening and (2) a general lattice hardening which persists after the initial yielding. The k_y term would seem to indicate that the pinning of dislocations may be attributed to point defects in the form of coarsely spaced jogs, and the electron-microscope observations of jogged dislocations would seem to confirm this.

The lattice friction term σ_i is clearly responsible for the general level of the stress–strain curve after yielding and arises from the large density of dislocation defects. However, the exact mechanisms

whereby loops and tetrahedra give rise to an increased flow stress is still controversial. Vacancy clusters are believed to be formed *in situ* by the disturbance introduced by the primary collision, and hence it is not surprising that neutron irradiation at 4 K hardens the material, and that thermal activation is not essential.

Unlike dispersion-hardened alloys, the deformation of irradiated or quenched metals is characterized by a low initial rate of work hardening (see Figure 7.36). This has been shown to be due to the sweeping out of loops and defect clusters by the glide dislocations, leading to the formation of cleared channels. Diffusion-controlled mechanisms are not thought to be important since defect-free channels are produced by deformation at 4 K. The removal of prismatic loops both unfaulted and faulted and tetrahedra can occur as a result of the strong coalescence interactions with screws to form helical configurations and jogged dislocations when the gliding dislocations and defects make contact. Clearly, the sweeping-up process occurs only if the helical and jogged configurations can glide easily. Resistance to glide will arise from jogs not lying in slip planes and also from the formation of sessile jogs (e.g. Lomer–Cottrell dislocations in fcc crystals).

7.6.2 Work-hardening

7.6.2.1 Theoretical treatment

The properties of a material are altered by cold-working, i.e. deformation at a low temperature relative to its melting point, but not all the properties are improved, for although the tensile strength, yield strength and hardness are increased, the plasticity and general ability to deform decreases. Moreover, the physical properties such as electrical conductivity, density and others are all lowered. Of these many changes in properties, perhaps the most outstanding are those that occur in the mechanical properties; the yield stress of mild steel, for example, may be raised by cold work from 170 up to 1050 MN/m^2.

Such changes in mechanical properties are, of course, of interest theoretically, but they are also of great importance in industrial practice. This is because the rate at which the material hardens during deformation influences both the power required and the method of working in the various shaping operations, while the magnitude of the hardness introduced governs the frequency with which the component must be annealed (always an expensive operation) to enable further working to be continued.

Since plastic flow occurs by a dislocation mechanism the fact that work-hardening occurs means that it becomes difficult for dislocations to move as the strain increases. All theories of work-hardening depend on this assumption, and the basic idea of hardening, put forward by Taylor in 1934, is that some dislocations become 'stuck' inside the crystal and act as sources of internal stress which oppose the motion of other gliding dislocations.

One simple way in which two dislocations could become stuck is by elastic interaction. Thus, two parallel edge dislocations of opposite sign moving on parallel slip planes in any sub-grain may become stuck, as a result of the interaction discussed in Chapter 4. G. I. Taylor assumed that dislocations become stuck after travelling an average distance, L, while the density of dislocations reaches ρ, i.e. work-hardening is due to the dislocations getting in each other's way. The flow stress is then the stress necessary to move a dislocation in the stress field of those dislocations surrounding it. This stress τ is quite generally given by

$$\tau = \alpha \mu b / l \qquad (7.23)$$

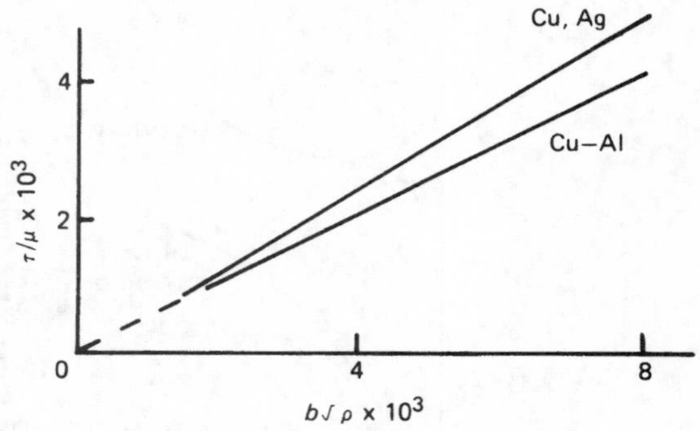

Figure 7.38 *Dependence of flow stress on (dislocation density)$^{1/2}$ for Cu, Ag and Cu–Al.*

where μ is the shear modulus, b the Burgers vector, l the mean distance between dislocations which is $\approx \rho^{-1/2}$, and α a constant; in the Taylor model $\alpha = 1/8\pi(1 - \nu)$. Figure 7.38 shows such a relationship for Cu–Al single crystals and polycrystalline Ag and Cu.

In his theory Taylor considered only a two-dimensional model of a cold-worked metal. However, because plastic deformation arises from the movement of dislocation loops from a source, it is more appropriate to assume that when the plastic strain is γ, N dislocation loops of side L (if we assume for convenience that square loops are emitted) have been given off per unit volume. The resultant plastic strain is then given by

$$\gamma = NL^2 b \qquad (7.24)$$

and l by

$$l \simeq [1/\rho^{1/2}] = [1/4LN]^{1/2} \qquad (7.25)$$

Combining these equations, the stress–strain relation

$$\tau = \text{const.} \ (b/L)^{1/2} \gamma^{1/2} \qquad (7.26)$$

is obtained. Taylor assumed L to be a constant, i.e. the slip lines are of constant length, which results in a parabolic relationship between τ and γ.

Taylor's assumption that during cold work the density of dislocations increases has been amply verified, and indeed the parabolic relationship between stress and strain is obeyed, to a first approximation, in many polycrystalline aggregates where deformation in all grains takes place by multiple slip. Experimental work on single crystals shows, however, that the work- or strain-hardening curve may deviate considerably from parabolic behaviour, and depends not only on crystal structure but also on other variables such as crystal orientation, purity and surface conditions (see Figures 7.39 and 7.40).

The crystal structure is important (see Figure 7.39) in that single crystals of some hexagonal metals slip only on one family of slip planes, those parallel to the basal plane, and these metals show a low rate of

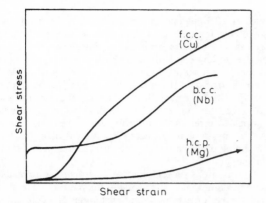

Figure 7.39 *Stress–strain curves of single crystals (after Hirsch and Mitchell, 1967; courtesy of the National Research Council of Canada).*

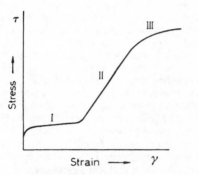

Figure 7.40 *Stress–strain curve showing the three stages of work hardening.*

work-hardening. The plastic part of the stress–strain curve is also more nearly linear than parabolic with a slope which is extremely small: this slope $(d\tau/d\gamma)$ becomes even smaller with increasing temperature of deformation. Cubic crystals, on the other hand, are capable of deforming in a complex manner on more than one slip system, and these metals normally show a strong work-hardening behaviour. The influence of temperature depends on the stress level reached during deformation and on other factors which must be considered in greater detail. However, even in cubic crystals the rate of work-hardening may be extremely small if the crystal is restricted to slip on a single slip system. Such behaviour points to the conclusion that strong work-hardening is caused by the mutual interference of dislocations gliding on intersecting slip planes.

Many theories of work-hardening similar to that of Taylor exist but all are oversimplified, since work-hardening depends not so much on individual dislocations as on the group behaviour of large numbers of them. It is clear, therefore, that a theoretical treatment which would describe the complete stress–strain relationship is difficult, and consequently the present-day approach is to examine the various stages of hardening and then attempt to explain the mechanisms likely to give rise to the different stages. The work-hardening behaviour in metals with a cubic structure is more complex than in most other structures because of the variety of slip systems available, and it is for this reason that much of the experimental evidence is related to these metals, particularly those with fcc structures.

7.6.2.2 Three-stage hardening

The stress–strain curve of a fcc single crystal is shown in Figure 7.40 and three regions of hardening are experimentally distinguishable. Stage I, or the easy glide region, immediately follows the yield point and is characterized by a low rate of work hardening θ_1 up to several per cent glide; the length of this region depends on orientation, purity and size of the crystals. The hardening rate $(\theta_1/\mu) \sim 10^{-4}$ and is of the same order as for hexagonal metals. Stage II, or the linear hardening region, shows a rapid increase in work-hardening rate. The ratio $(\theta_{11}/\mu) = (d\tau/d\gamma)/\mu$ is of the same order of magnitude for all fcc metals, i.e. 1/300 although this is $\approx 1/150$ for orientations at the corners of the stereographic triangle. In this stage short slip lines are formed during straining quite suddenly, i.e. in a short increment of stress $\Delta\tau$, and thereafter do not grow either in length or intensity. The mean length of the slip lines, $L \approx 25$ μm decreases with increasing strain. Stage III, or the parabolic hardening region, the onset of which is markedly dependent on temperature, exhibits a low rate of work-hardening, θ_{111}, and the appearance of coarse slip bands. This stage sets in at a strain which increases with decreasing temperature and is probably associated with the annihilation of dislocations as a consequence of cross-slip.

The low stacking fault energy metals exhibit all three work-hardening stages at room temperature, but metals with a high stacking fault energy often show only two stages of hardening. It is found, for example, that at 78 K aluminium behaves like copper at room temperature and exhibits all three stages, but at room temperature and above, stage II is not clearly developed and stage III starts before stage II becomes at all predominant. This difference between aluminium and the noble metals is not due only to the difference in melting point but also to the difference in stacking fault energies which affects the width of extended dislocations. The main effect of a change of temperature of deformation is, however, a change in the onset of stage III; the lower the temperature of deformation, the higher is the stress τ_{111} corresponding to the onset of stage III.

Because the flow stress of a metal may be affected by a change of temperature or strain-rate, it has been found convenient to think of the stress as made up of two parts according to the relation

$$\tau = \tau_s + \tau_g$$

where τ_s is that part of the flow stress which is dependent on temperature apart from the variation of the elastic modulus μ with temperature, and τ_g is a temperature-independent contribution. The relative importance of τ_s and τ_g can be studied conveniently by measuring the dependence of flow stress on temperature or strain rate, i.e. the change in flow stress $\Delta\tau$ on changing the temperature or strain rate, as function of deformation.

The Stage I easy glide region in cubic crystals, with its small linear hardening, corresponds closely to the hardening of cph crystals where only one glide plane operates. It occurs in crystals oriented to allow only one glide system to operate, i.e. for orientations near the [0 1 1] pole of the unit triangle (Figure 7.13). In this case the slip distance is large, of the order of the specimen diameter, with the probability of dislocations slipping out of the crystal. Electron microscope observations have shown that the slip lines on the surface are very long ($\approx$1 mm) and closely spaced, and that the slip steps are small corresponding to the passage of only a few dislocations. This behaviour obviously depends on such variables as sample size and oxide films, since these influence the probability of dislocations passing out of the crystal. It is also to be expected that the flow stress in easy glide will be governed by the ease with which sources begin to operate, since there is no slip on a secondary slip system to interfere with the movement of primary glide dislocations.

As soon as another glide system becomes activated there is a strong interaction between dislocations on the primary and secondary slip systems, which gives rise to a steep increase in work-hardening. It is reasonable to expect that easy glide should end, and turbulent flow begin, when the crystal reaches an orientation for which two or more slip systems are equally stressed, i.e. for orientations on the symmetry line between [0 0 1] and [1 1 1]. However, easy glide generally ends before symmetrical orientations are reached and this is principally due to the formation of deformation bands to accommodate the rotation of the glide plane in fixed grips during tensile tests. This rotation leads to a high resolved stress on the secondary slip system, and its operation gives rise to those lattice irregularities which cause some dislocations to become 'stopped' in the crystal. The transformation to Stage II then occurs.

The characteristic feature of deformation in Stage II is that slip takes place on both the primary and secondary slip systems. As a result, several new lattice irregularities may be formed which will include (1) forest dislocations, (2) Lomer–Cottrell barriers, and (3) jogs produced either by moving dislocations cutting through forest dislocations or by forest dislocations cutting through source dislocations. Consequently, the flow stress τ may be identified, in general terms, with a stress which is sufficient to operate a source and then move the dislocations against (1) the internal elastic stresses from the forest dislocations, (2) the long-range stresses from groups of dislocations piled up behind barriers, and (3) the frictional resistance due to jogs. In a cold-worked metal all these factors may exist to some extent, but because a linear hardening law can be derived by using any one of the various contributory factors, there have been several theories of Stage II-hardening, namely (1) the pileup theory, (2) the forest theory and (3) the jog theory. All have been shown to have limitations in explaining various features of the deformation process, and have given way to a more phenomenological theory based on direct observations of the dislocation distribution during straining.

Observations on fcc and bcc crystals have revealed several common features of the microstructure which include the formation of dipoles, tangles and cell structures with increasing strain. The most detailed observations have been made for copper crystals, and these are summarized below to illustrate the general pattern of behaviour. In Stage I, bands of dipoles are formed (see Figure 7.41a) elongated normal to the primary Burgers vector direction. Their formation is associated with isolated forest dislocations and individual dipoles are about 1 μm in length and $\geqslant$10 nm wide. Different patches are arranged at spacings of about 10 μm along the line of intersection of a secondary slip plane. With increasing strain in Stage I the size of the gaps between the dipole clusters decreases and therefore the stress required to push dislocations through these gaps increases. Stage II begins (see Figure 7.41b) when the applied stress plus internal stress resolved on the secondary systems is sufficient to activate secondary sources near the dipole clusters. The resulting local secondary slip leads to local interactions between primary and secondary dislocations both in the gaps and in the clusters of dipoles, the gaps being filled with secondary dislocations and short lengths of other dislocations formed by interactions (e.g. Lomer–Cottrell dislocations in fcc crystals and $a\langle 1 0 0 \rangle$ type dislocations in bcc crystals). Dislocation barriers are thus formed surrounding the original sources.

In Stage II (see Figure 7.41c) it is proposed that dislocations are stopped by elastic interaction when they pass too close to an existing tangled region with high dislocation density. The long-range internal stresses due to the dislocations piling up behind are partially relieved by secondary slip, which transforms the discrete pile-up into a region of high dislocation density containing secondary dislocation networks and dipoles. These regions of high dislocation density act as new obstacles to dislocation glide, and since every new obstacle is formed near one produced at a lower strain, two-dimensional dislocation structures are built up forming the walls of an irregular cell structure. With increasing strain the number of obstacles increases, the distance a dislocation glides decreases and therefore the slip line length decreases in Stage II. The structure remains similar throughout Stage II but is reduced in scale. The obstacles are in the form of ribbons of high densities of dislocations which, like pile-ups, tend to form sheets. The work-hardening rate depends mainly

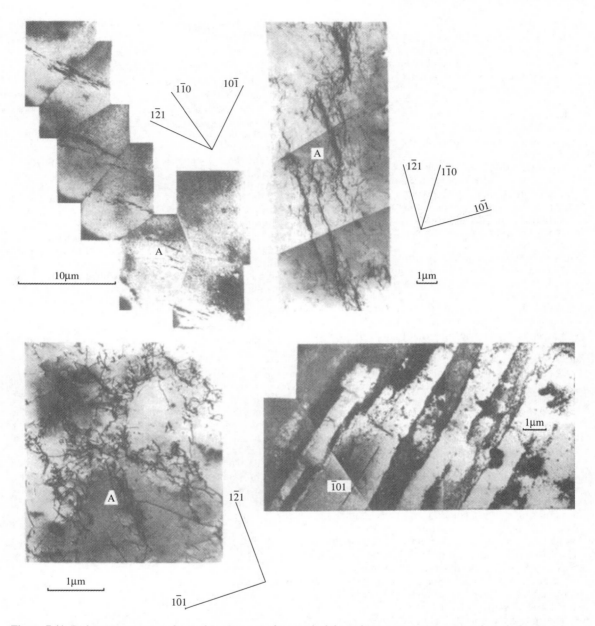

Figure 7.41 *Dislocation structure observed in copper single crystals deformed in tension to (a) stage I, (b) end of easy-glide and beginning of stage II, (c) top of stage II, and (d) stage III (after Steeds, 1963; Crown copyright; reproduced by permission of the Controller, H.M. Stationery Office).*

on the effective radius of the obstacles, and this has been considered in detail by Hirsch and co-workers and shown to be a constant fraction k of the discrete pile-up length on the primary slip system. In general, the work-hardening rate is given by $\theta_{11} = k\mu/3\pi$ and for an fcc crystal the small variation in k with orientation and alloying element is able to account for the variation of θ_{11} with those parameters.

The dislocation arrangement in metals with other structures is somewhat similar to that of copper with differences arising from stacking fault energy. In Cu–Al alloys the dislocations tend to be confined more to the active slip planes, the confinement increasing with decreasing γ_{SF}. In Stage I dislocation multipoles are formed as a result of dislocations of opposite sign on parallel nearby slip planes 'pairing up' with one another. Most of these dislocations are primaries. In Stage II the density of secondary dislocations is much less ($\approx\frac{1}{3}$) than that of the primary dislocations. The secondary slip occurs in bands and in each band slip on one particular secondary plane predominates. In niobium, a metal with high γ_{SF}, the dislocation distribution is rather similar to copper. In Mg, typical of cph metals, Stage I is extensive and the dislocations are mainly in the form of primary edge multipoles, but forest dislocations threading the primary slip plane do not appear to be generated.

From the curve shown in Figure 7.40 it is evident that the rate of work-hardening decreases in the later stages of the test. This observation indicates that at a sufficiently high stress or temperature the dislocations held up in Stage II are able to move by a process which at lower stresses and temperature had been suppressed. The onset of Stage III is accompanied by cross-slip, and the slip lines are broad, deep and consist of segments joined by cross-slip traces. Electron metallographic observations on sections of deformed crystal inclined to the slip plane (see Figure 7.41d) show the formation of a cell structure in the form of boundaries, approximately parallel to the primary slip plane of spacing about 1–3 μm plus other boundaries extending normal to the slip plane as a result of cross-slip.

The simplest process which is in agreement with the experimental observations is that the screw dislocations held up in Stage II, cross-slip and possibly return to the primary slip plane by double cross-slip. By this mechanism, dislocations can bypass the obstacles in their glide plane and do not have to interact strongly with them. Such behaviour leads to an increase in slip distance and a decrease in the accompanying rate of work-hardening. Furthermore, it is to be expected that screw dislocations leaving the glide plane by cross-slip may also meet dislocations on parallel planes and be attracted by those of opposite sign. Annihilation then takes place and the annihilated dislocation will be replaced, partly at least, from the original source. This process if repeated can lead to slip-band formation, which is also an important experimental feature of Stage III. Hardening in Stage III is then due to the edge parts of the loops which remain in the crystal and increase in density as the source continues to operate.

The importance of the value of the stacking fault energy, γ, on the stress–strain curve is evident from its importance to the process of cross-slip. Low values of γ give rise to wide stacking fault 'ribbons', and consequently cross-slip is difficult at reasonable stress levels. Thus, the screws cannot escape from their slip plane, the slip distance is small, the dislocation density is high and the transition from Stage II to Stage III is delayed. In aluminium the stacking fault ribbon width is very small because γ has a high value, and cross-slip occurs at room temperature. Stage II is, therefore, poorly developed unless testing is carried out at low temperatures. These conclusions are in agreement with the observations of dislocation density and arrangement.

7.6.2.3 Work-hardening in polycrystals

The dislocation structure developed during the deformation of fcc and bcc polycrystalline metals follows the same general pattern as that in single crystals; primary dislocations produce dipoles and loops by interaction with secondary dislocations, which give rise to local dislocation tangles gradually developing into three-dimensional networks of sub-boundaries. The cell size decreases with increasing strain, and the structural differences that are observed between various metals and alloys are mainly in the sharpness of the sub-boundaries. In bcc metals, and fcc metals with high stacking-fault energy, the tangles rearrange into sharp boundaries but in metals of low stacking fault energy the dislocations are extended, cross-slip is restricted, and sharp boundaries are not formed even at large strains. Altering the deformation temperature also has the effect of changing the dislocation distribution; lowering the deformation temperature reduces the tendency for cell formation, as shown in Figure 7.42. For a given dislocation distribution the dislocation density is simply related to the flow stress τ by an equation of the form

$$\tau = \tau_0 + \alpha\mu b\rho^{1/2} \tag{7.27}$$

where α is a constant at a given temperature ≈ 0.5; τ_0 for fcc metals is zero (see Figure 7.38). The workhardening rate is determined by the ease with which tangled dislocations rearrange themselves and is high in materials with low γ, i.e. brasses, bronzes and austenitic steels compared to Al and bcc metals. In some austenitic steels, work-hardening may be increased and better sustained by a strain-induced phase transformation (see Chapter 8).

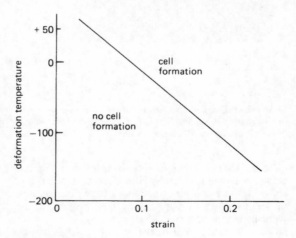

Figure 7.42 *Influence of deformation strain and temperature on the formation of a cell structure in α-iron.*

Grain boundaries affect work-hardening by acting as barriers to slip from one grain to the next. In addition, the continuity criterion of polycrystals enforces complex slip in the neighbourhood of the boundaries which spreads across the grains with increasing deformation. This introduces a dependence of work-hardening rate on grain size which extends to several per cent elongation. After this stage, however, the work-hardening rate is independent of grain size and for fcc polycrystals is about $\mu/40$, which, allowing for the orientation factors, is roughly comparable with that found in single crystals deforming in multiple slip. Thus from the relations $\sigma = m\tau$ and $\varepsilon = \gamma/m$ the average resolved shear stress on a slip plane is rather less than half the applied tensile stress, and the average shear strain parallel to the slip plane is rather more than twice the tensile elongation. The polycrystal work-hardening rate is thus related to the single-crystal work-hardening rate by the relation

$$d\sigma/d\varepsilon = m^2 d\tau/d\gamma \qquad (7.28)$$

For bcc metals with the multiplicity of slip systems and the ease of cross-slip m is more nearly 2, so that the work-hardening rate is low. In polycrystalline cph metals the deformation is complicated by twinning, but in the absence of twinning $m \approx 6.5$, and hence the work-hardening rate is expected to be more than an order of magnitude greater than for single crystals, and also higher than the rate observed in fcc polycrystals for which $m \approx 3$.

7.6.2.4 Dispersion-hardened alloys

On deforming an alloy containing incoherent, non-deformable particles the rate of work-hardening is much greater than that shown by the matrix alone (see Figure 8.10). The dislocation density increases very rapidly with strain because the particles produce a turbulent and complex deformation pattern around them. The dislocations gliding in the matrix leave loops around particles either by bowing between the particles or by cross-slipping around them; both these mechanisms are discussed in Chapter 8. The stresses in and around particles may also be relieved by activating secondary slip systems in the matrix. All these dislocations spread out from the particle as strain proceeds and, by intersecting the primary glide plane, hinder primary dislocation motion and lead to intense work-hardening. A dense tangle of dislocations is built up at the particle and a cell structure is formed with the particles predominantly in the cell walls.

At small strains ($\lesssim 1\%$) work-hardening probably arises from the back-stress exerted by the few Orowan loops around the particles, as described by Fisher, Hart and Pry. The stress–strain curve is reasonably linear with strain ε according to

$$\sigma = \sigma_i + \alpha\mu f^{3/2}\varepsilon$$

with the work-hardening depending only on f, the volume fraction of particles. At larger strains the 'geometrically necessary' dislocations stored to accommodate the strain gradient which arises because one component deforms plastically more than the other, determine the work-hardening. A determination of the average density of dislocations around the particles with which the primary dislocations interact allows an estimate of the work-hardening rate, as initially considered by Ashby. Thus, for a given strain ε and particle diameter d the number of loops per particle is

$$n \sim \varepsilon d/b$$

and the number of particles per unit volume

$$N_v = 3f/4\pi r^2, \text{ or } 6f/\pi d^3$$

The total number of loops per unit volume is nN_v and hence the dislocation density $\rho = nN_v\pi d = 6f\varepsilon/db$. The stress–strain relationship from equation (7.27) is then

$$\sigma = \sigma_i + \alpha\mu(fb/d)^{1/2}\varepsilon^{1/2} \qquad (7.29)$$

and the work-hardening rate

$$d\sigma/d\varepsilon = \alpha'\mu(f/d)^{1/2}(b/\varepsilon)^{1/2} \qquad (7.30)$$

Alternative models taking account of the detailed structure of the dislocation arrays (e.g. Orowan, prismatic and secondary loops) have been produced to explain some of the finer details of dispersion-hardened materials. However, this simple approach provides a useful working basis for real materials. Some additional features of dispersion-strengthened alloys are discussed in Chapter 8.

7.6.2.5 Work-hardening in ordered alloys

A characteristic feature of alloys with long-range order is that they work-harden more rapidly than in the disordered state. θ_{11} for Fe-Al with a B2 ordered structure is $\approx\mu/50$ at room temperature, several times greater than a typical fcc or bcc metal. However, the density of secondary dislocations in Stage II is relatively low and only about 1/100 of that of the primary dislocations. One mechanism for the increase in work-hardening rate is thought to arise from the generation of antiphase domain boundary (apb) tubes. A possible geometry is shown in Figure 7.43a; the superdislocation partials shown each contain a jog produced, for example, by intersection with a forest dislocation, which are nonaligned along the direction of the Burgers vector. When the dislocation glides and the jogs move nonconservatively a tube of apbs is generated. Direct evidence for the existence of tubes from weak-beam electron microscopy studies was first reported for Fe-30 at.% Al. The micrographs show faint lines along $\langle 1\,1\,1 \rangle$, the Burgers vector direction, and are about 3 nm in width. The images are expected to be weak, since the contrast arises from two closely spaced overlapping faults, the second effectively cancelling the displacement caused by the first, and are visible only when superlattice reflections

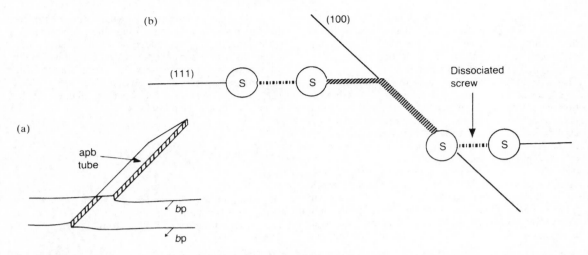

Figure 7.43 *Schematic diagram of superdislocation (a) with non-aligned jogs, which, after glide, produce an apb-tube and (b) cross-slipped onto the cube plane to form a Kear–Wilsdorf (K–W) lock.*

are excited. APB tubes have since been observed in other compounds.

Theory suggests that jogs in superdislocations in screw orientations provide a potent hardening mechanism, estimated to be about eight times as strong as that resulting from pulling out of apb tubes on non-aligned jogs on edge dislocations. The major contributions to the stress to move a dislocation are (1) τ_s, the stress to generate point defects or tubes, and (2) the interaction stress τ_i with dislocations on neighbouring slip planes, and $\tau_s + \tau_i = \frac{3}{4}\alpha_s\mu(\rho_f/\rho_p)\varepsilon$. Thus, with $\alpha_s = 1.3$ and provided ρ_f/ρ_p is constant and small, linear hardening with the observed rate is obtained.

In crystals with A_3B order only one rapid stage of hardening is observed compared with the normal three-stage hardening of fcc metals. Moreover, the temperature-dependence of θ_{11}/μ increases with temperature and peaks at $\sim 0.4T_m$. It has been argued that the apb tube model is unable to explain why anomalously high work-hardening rates are observed for those single crystal orientations favourable for single slip on {1 1 1} planes alone. An alternative model to apb tubes has been proposed based on cross-slip of the leading unit dislocation of the superdislocation. If the second unit dislocation cannot follow exactly in the wake of the first, both will be pinned.

For alloys with $L1_2$ structure the cross-slip of a screw superpartial with $b = \frac{1}{2}[\bar{1}0\,1]$ from the primary (1 1 1) plane to the (0 1 0) plane was first proposed by Kear and Wilsdorf. The two $\frac{1}{2}[\bar{1}0\,1]$ superpartials, one on the (1 1 1) plane and the other on the (0 1 0) plane, are of course dissociated into $\langle 1\,1\,2\rangle$-type partials and the whole configuration is sessile. This dislocation arrangement is known as a Kear–Wilsdorf (K–W) lock and is shown in Figure 7.43b. Since cross-slip is thermally activated, the number of locks and therefore the resistance to (1 1 1) glide will increase with

increasing temperature. This could account for the increase in yield stress with temperature, while the onset of cube slip at elevated temperatures could account for the peak in the flow stress.

Cube cross-slip and cube slip has now been observed in a number of $L1_2$ compounds by TEM. There is some TEM evidence that the apb energy on the cube plane is lower than that on the (1 1 1) plane (see Chapter 9) to favour cross-slip which would be aided by the torque, arising from elastic anisotropy, exerted between the components of the screw dislocation pair.

7.6.3 Development of preferred orientation

7.6.3.1 Crystallographic aspects

When a polycrystalline metal is plastically deformed the individual grains tend to rotate into a common orientation. This preferred orientation is developed gradually with increasing deformation, and although it becomes extensive above about 90% reduction in area, it is still inferior to that of a good single crystal. The degree of texture produced by a given deformation is readily shown on a monochromatic X-ray transmission photograph, since the grains no longer reflect uniformly into the diffraction rings but only into certain segments of them. The results are usually described in terms of an ideal orientation, such as [u, v, w] for the fibre texture developed by drawing or swaging, and $\{hkl\}\langle uvw\rangle$ for a rolling texture for which a plane of the form (hkl) lies parallel to the rolling plane and a direction of the type $\langle uvw\rangle$ is parallel to the rolling direction. However, the scatter about the ideal orientation can only be represented by means of a pole-figure which describes the spread of orientation about the ideal orientation for a particular set of (hkl) poles (see Figure 7.44).

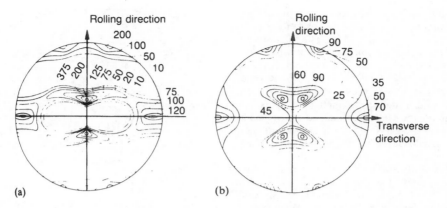

Figure 7.44 *(1 1 1) pole figures from (a) copper, and (b) α-brass after 95% deformation (intensities in arbitrary units).*

In tension, the grains rotate in such a way that the movement of the applied stress axis is towards the operative slip direction as discussed in Section 7.3.5 and for compression the applied stress moves towards the slip plane normal. By considering the deformation process in terms of the particular stresses operating and applying the appropriate grain rotations it is possible to predict the stable end-grain orientation and hence the texture developed by extensive deformation. Table 7.3 shows the predominant textures found in different metal structures for both wires and sheet.

For fcc metals a marked transition in deformation texture can be effected either by lowering the deformation temperature or by adding solid solution alloying elements which lower the stacking fault energy. The transition relates to the effect on deformation modes of reducing stacking fault energy or thermal energy, deformation banding and twinning becoming more prevalent and cross-slip less important at lower temperatures and stacking fault energies. This texture transition can be achieved in most fcc metals by alloying additions and by altering the rolling temperature. Al, however, has a high fault energy and because of the limited solid solubility it is difficult to lower by alloying. The extreme types of rolling texture, shown by copper and 70/30 brass, are given in Figures 7.44a and 7.44b.

In bcc metals there are no striking examples of solid solution alloying effects on deformation texture, the preferred orientation developed being remarkably insensitive to material variables. However, material variables can affect cph textures markedly. Variations in c/a ratio alone cause alterations in the orientation

developed, as may be appreciated by consideration of the twinning modes, and it is also possible that solid solution elements alter the relative values of critical resolved shear stress for different deformation modes. Processing variables are also capable of giving a degree of control in hexagonal metals. No texture, stable to further deformation, is found in hexagonal metals and the angle of inclination of the basal planes to the sheet plane varies continuously with deformation. In general, the basal plane lies at a small angle (<45°) to the rolling plane, tilted either towards the rolling direction (Zn, Mg) or towards the transverse direction (Ti, Zr, Be, Hf).

The deformation texture cannot, in general, be eliminated by an annealing operation even when such a treatment causes recrystallization. Instead, the formation of a new annealing texture usually results, which is related to the deformation texture by standard lattice rotations.

7.6.3.2 Texture-hardening

The flow stress in single crystals varies with orientation according to Schmid's law and hence materials with a preferred orientation will also show similar plastic anisotropy, depending on the perfection of the texture. The significance of this relationship is well illustrated by a crystal of beryllium which is cph and capable of slip only on the basal plane, a compressive stress approaching ≈2000 MN/m² applied normal to the basal plane produces negligible plastic deformation. Polycrystalline beryllium sheet, with a texture such that the basal planes lie in the plane of the sheet,

Table 7.3 Deformation textures in metals with common crystal structures

Structure	Wire (fibre texture)	Sheet (rolling texture)
bcc	[1 1 0]	{1 1 2} ⟨1 $\bar{1}$ 0⟩ to {1 0 0} ⟨0 1 1⟩
fcc	[1 1 1], [1 0 0] double fibre	{1 1 0} ⟨1 1 2⟩ to {3 $\bar{5}$ 1} ⟨1 1 2⟩
cph	[2 1 0]	{0 0 0 1} ⟨1 0 0 0⟩

shows a correspondingly high strength in biaxial tension. When stretched uniaxially the flow stress is also quite high, when additional (prismatic) slip planes are forced into action even though the shear stress for their operation is five times greater than for basal slip. During deformation there is little thinning of the sheet, because the $\langle 1\,1\,\bar{2}0\rangle$ directions are aligned in the plane of the sheet. Other hexagonal metals, such as titanium and zirconium, show less marked strengthening in uniaxial tension because prismatic slip occurs more readily, but resistance to biaxial tension can still be achieved. Applications of texture-hardening lie in the use of suitably textured sheet for high biaxial strength, e.g. pressure vessels, dent resistance, etc. Because of the multiplicity of slip systems, cubic metals offer much less scope for texture-hardening. Again, a consideration of single crystal deformation gives the clue; for whereas in a hexagonal crystal m can vary from 2 (basal planes at 45° to the stress axis) to infinity (when the basal planes are normal), in an fcc crystal m can vary only by a factor of 2 with orientation, and in bcc crystals the variation is rather less. In extending this approach to polycrystalline material certain assumptions have to be made about the mutual constraints between grains. One approach gives $m = 3.1$ for a random aggregate of fcc crystals and the calculated orientation dependence of σ/τ for fibre texture shows that a rod with $\langle 1\,1\,1\rangle$ or $\langle 1\,1\,0\rangle$ texture ($\sigma/\tau = 3.664$) is 20% stronger than a random structure; the cube texture ($\sigma/\tau = 2.449$) is 20% weaker.

If conventional mechanical properties were the sole criterion for texture-hardened materials, then it seems unlikely that they would challenge strong precipitation-hardened alloys. However, texture-hardening has more subtle benefits in sheet metal forming in optimizing fabrication performance. The variation of strength in the plane of the sheet is readily assessed by tensile tests carried out in various directions relative to the rolling direction. In many sheet applications, however, the requirement is for through-thickness strength (e.g. to resist thinning during pressing operations). This is more difficult to measure and is often assessed from uniaxial tensile tests by measuring the ratio of the strain in the width direction to that in the thickness direction of a test piece. The strain ratio R is given by

$$R = \varepsilon_w/\varepsilon_t = \ln(w_0/w)/\ln(t_0/t)$$
$$= \ln(w_0/w)/\ln(wL/w_0L) \tag{7.31}$$

where w_0, L_0, t_0 are the original dimensions of width, length and thickness and w, L and t are the corresponding dimensions after straining, which is derived assuming no change in volume occurs. The average strain ratio $\bar{R}$, for tests at various angles in the plane of the sheet, is a measure of the normal anisotropy, i.e. the difference between the average properties in the plane of the sheet and that property in the direction normal to the sheet surface. A large value of R means that there is a lack of deformation modes oriented to provide

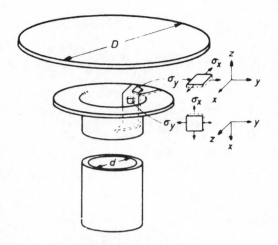

Figure 7.45 *Schematic diagram of the deep-drawing operations indicating the stress systems operating in the flange and the cup wall. Limiting drawing ratio is defined as the ratio of the diameter of the largest blank which can satisfactorily complete the draw (D_{max}) to the punch diameter (d) (after Dillamore, Smallman and Wilson, 1969; courtesy of the Canadian Institute of Mining and Metallurgy).*

strain in the through-thickness direction, indicating a high through-thickness strength.

In deep-drawing, schematically illustrated in Figure 7.45, the dominant stress system is radial tension combined with circumferential compression in the drawing zone, while that in the base and lower cup wall (i.e. central stretch-forming zone) is biaxial tension. The latter stress is equivalent to a through-thickness compression, plus a hydrostatic tension which does not affect the state of yielding. Drawing failure occurs when the central stretch-forming zone is insufficiently strong to support the load needed to draw the outer part of the blank through the die. Clearly differential strength levels in these two regions, leading to greater ease of deformation in the drawing zone compared with the stretching zone, would enable deeper draws to be made: this is the effect of increasing the $\bar{R}$ value, i.e. high through-thickness strength relative to strength in the plane of the sheet will favour drawability. This is confirmed in Figure 7.46, where deep drawability as determined by limiting drawing ratio (i.e. ratio of maximum drawable blank diameter to final cup diameter) is remarkably insensitive to ductility and, by inference from the wide range of materials represented in the figure, to absolute strength level. Here it is noted that for hexagonal metals slip occurs readily along $\langle 1\,1\,\bar{2}0\rangle$ thus contributing no strain in the c-direction, and twinning only occurs on the $\{1\,0\,\bar{1}2\}$ when the applied stress nearly parallel to the c-axis is compressive for $c/a > \sqrt{3}$ and tensile for $c/a < \sqrt{3}$. Thus titanium, $c/a < \sqrt{3}$, has a high strength in through-thickness compression, whereas

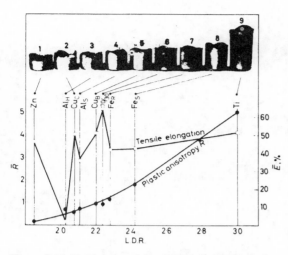

Figure 7.46 *Limiting draw ratios (LDR) as a function of average values of R and of elongation to fracture measured in tensile tests at 0°, 45° and 90° to the rolling direction (after Wilson, 1966; courtesy of the Institute of Metals).*

Zn with $c/a < \sqrt{3}$ has low through-thickness strength when the basal plane is oriented parallel to the plane of the sheet. In contrast, hexagonal metals with $c/a > \sqrt{3}$ would have a high R for $\{10\bar{1}0\}$ parallel to the plane of the sheet.

Texture-hardening is much less in the cubic metals, but fcc materials with $\{111\}\langle110\rangle$ slip system and bcc with $\{110\}\langle111\rangle$ are expected to increase R when the texture has component with $\{111\}$ and $\{110\}$ parallel to the plane of the sheet. The range of values of $\bar{R}$ encountered in cubic metals is much less. Face-centred cubic metals have $\bar{R}$ ranging from about 0.3 for cube-texture, $\{100\}\langle001\rangle$, to a maximum, in textures so far attained, of just over 1.0. Higher values are sometimes obtained in body-centred cubic metals. Values of $\bar{R}$ in the range $1.4 \sim 1.8$ obtained in aluminium-killed low-carbon steel are associated with significant improvements in deep-drawing performance compared with rimming steel, which has $\bar{R}$-values between 1.0 and 1.4. The highest values of R in steels are associated with texture components with $\{111\}$ parallel to the surface, while crystals with $\{100\}$ parallel to the surface have a strongly depressing effect on $\bar{R}$.

In most cases it is found that the R values vary with testing direction and this has relevance in relation to the strain distribution in sheet metal forming. In particular, ear formation on pressings generally develops under a predominant uniaxial compressive stress at the edge of the pressing. The ear is a direct consequence of the variation in strain ratio for different directions of uniaxial stressing, and a large variation in R value, where $\Delta R = (R_{max} - R_{min})$ generally correlates with a tendency to form pronounced ears. On this basis we could write a simple recipe for good deep-drawing properties in terms of strain ratio measurements made in a uniaxial tensile test as high R and low ΔR. Much research is aimed at improving forming properties through texture control.

7.7 Macroscopic plasticity

7.7.1 Tresca and von Mises criteria

In dislocation theory it is usual to consider the flow stress or yield stress of ductile metals under simple conditions of stressing. In practice, the engineer deals with metals under more complex conditions of stressing (e.g. during forming operations) and hence needs to correlate yielding under combined stresses with that in uniaxial testing. To achieve such a yield stress criterion it is usually assumed that the metal is mechanically isotropic and deforms plastically at constant volume, i.e. a hydrostatic state of stress does not affect yielding. In assuming plastic isotropy, macroscopic shear is allowed to take place along lines of maximum shear stress and crystallographic slip is ignored, and the yield stress in tension is equal to that in compression, i.e. there is no Bauschinger effect.

A given applied stress state in terms of the principal stresses $\sigma_1, \sigma_2, \sigma_3$ which act along three principal axes, X_1, X_2 and X_3, may be separated into the hydrostatic part (which produces changes in volume) and the deviatoric components (which produce changes in shape). It is assumed that the hydrostatic component has no effect on yielding and hence the more the stress state deviates from pure hydrostatic, the greater the tendency to produce yield. The stresses may be represented on a stress–space plot (see Figure 7.47a), in which a line equidistant from the three stress axes represents a pure hydrostatic stress state. Deviation from this line will cause yielding if the deviation is sufficiently large, and define a yield surface which has sixfold symmetry about the hydrostatic line. This arises because the conditions of isotropy imply equal yield stresses along all three axes, and the absence of the Bauschinger effect implies equal yield stresses along σ_1 and $-\sigma_1$. Taking a section through stress space, perpendicular to the hydrostatic line gives the two simplest yield criteria satisfying the symmetry requirements corresponding to a regular hexagon and a circle.

The hexagonal form represents the Tresca criterion (see Figure 7.47c) which assumes that plastic shear takes place when the maximum shear stress attains a critical value k equal to shear yield stress in uniaxial tension. This is expressed by

$$\tau_{max} = \frac{\sigma_1 - \sigma_3}{2} = k \tag{7.32}$$

where the principal stresses $\sigma_1 > \sigma_2 > \sigma_3$. This criterion is the isotropic equivalent of the law of resolved shear stress in single crystals. The tensile yield stress $Y = 2k$ is obtained by putting $\sigma_1 = Y$, $\sigma_2 = \sigma_3 = 0$.

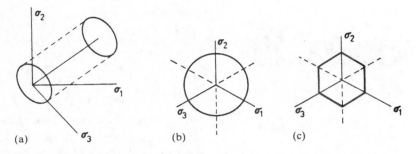

Figure 7.47 *Schematic representation of the yield surface with (a) principal stresses σ_1, σ_2 and σ_3, (b) von Mises yield criterion and (c) Tresca yield criterion.*

The circular cylinder is described by the equation

$$(\sigma_1 - \sigma_2)^2 + (\sigma_2 - \sigma_3)^2 + (\sigma_3 - \sigma_1)^2 = \text{constant}$$

$$(7.33)$$

and is the basis of the von Mises yield criterion (see Figure 7.47b). This criterion implies that yielding will occur when the shear energy per unit volume reaches a critical value given by the constant. This constant is equal to $6k^2$ or $2Y^2$ where k is the yield stress in simple shear, as shown by putting $\sigma_2 = 0$, $\sigma_1 = \sigma_3$, and Y is the yield stress in uniaxial tension when $\sigma_2 = \sigma_3 = 0$. Clearly $Y = 3k$ compared to $Y = 2k$ for the Tresca criterion and, in general, this is found to agree somewhat closer with experiment.

In many practical working processes (e.g. rolling), the deformation occurs under approximately plane strain conditions with displacements confined to the X_1X_2 plane. It does not follow that the stress in this direction is zero, and, in fact, the deformation conditions are satisfied if $\sigma_3 = \frac{1}{2}(\sigma_1 + \sigma_2)$ so that the tendency for one pair of principal stresses to extend the metal along the X_3 axis is balanced by that of the other pair to contract it along this axis. Eliminating σ_3 from the von Mises criterion, the yield criterion becomes

$$(\sigma_1 - \sigma_2) = 2k$$

and the plane strain yield stress, i.e. when $\sigma_2 = 0$, given when

$$\sigma_1 = 2k = 2Y/\sqrt{3} = 1.15Y$$

For plane strain conditions, the Tresca and von Mises criteria are equivalent and two-dimensional flow occurs when the shear stress reaches a critical value. The above condition is thus equally valid when written in terms of the deviatoric stresses σ_1', σ_2', σ_3' defined by equations of the type $\sigma_1' = \sigma_1 - \frac{1}{3}(\sigma_1 + \sigma_2 + \sigma_3)$.

Under plane stress conditions, $\sigma_3 = 0$ and the yield surface becomes two-dimensional and the von Mises criterion becomes

$$\sigma_1^2 + \sigma_1\sigma_2 + \sigma_2^2 = 3k^2 = Y^2 \qquad (7.34)$$

which describes an ellipse in the stress plane. For the Tresca criterion the yield surface reduces to a hexagon

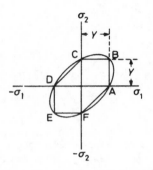

Figure 7.48 *The von Mises yield ellipse and Tresca yield hexagon.*

inscribed in the ellipse as shown in Figure 7.48. Thus, when σ_1 and σ_2 have opposite signs, the Tresca criterion becomes $\sigma_1 - \sigma_2 = 2k = Y$ and is represented by the edges of the hexagon CD and FA. When they have the same sign then $\sigma_1 = 2k = Y$ or $\sigma_2 = 2k = Y$ and defines the hexagon edges AB, BC, DE and EF.

7.7.2 Effective stress and strain

For an isotropic material, a knowledge of the uniaxial tensile test behaviour together with the yield function should enable the stress–strain behaviour to be predicted for any stress system. This is achieved by defining an effective stress–effective strain relationship such that if $\sigma = K\varepsilon^n$ is the uniaxial stress–strain relationship then we may write.

$$\bar{\sigma} = K\bar{\varepsilon}^n \qquad (7.35)$$

for any state of stress. The stress–strain behaviour of a thin-walled tube with internal pressure is a typical example, and it is observed that the flow curves obtained in uniaxial tension and in biaxial torsion coincide when the curves are plotted in terms of effective stress and effective strain. These quantities are defined by:

$$\bar{\sigma} = \frac{\sqrt{2}}{2}[(\sigma_1 - \sigma_2)^2 + (\sigma_2 - \sigma_3)^2 + (\sigma_3 - \sigma_1)^2]^{1/2}$$

$$(7.36)$$

and

$$\bar{\varepsilon} = \frac{\sqrt{2}}{3}[(\varepsilon_1 - \varepsilon_2)^2 + (\varepsilon_2 - \varepsilon_3)^2 + (\varepsilon_3 - \varepsilon_1)^2]^{1/2}$$

(7.37)

where ε_1, ε_2 and ε_3 are the principal strains, both of which reduce to the axial normal components of stress and strain for a tensile test. It should be emphasized, however, that this generalization holds only for isotropic media and for constant loading paths, i.e. $\sigma_1 = \alpha\sigma_2 = \beta\sigma_3$ where α and β are constants independent of the value of σ_1.

7.8 Annealing

7.8.1 General effects of annealing

When a metal is cold-worked, by any of the many industrial shaping operations, changes occur in both its physical and mechanical properties. While the increased hardness and strength which result from the working treatment may be of importance in certain applications, it is frequently necessary to return the metal to its original condition to allow further forming operations (e.g. deep drawing) to be carried out of for applications where optimum physical properties, such as electrical conductivity, are essential. The treatment given to the metal to bring about a decrease of the hardness and an increase in the ductility is known as annealing. This usually means keeping the deformed metal for a certain time at a temperature higher than about one-third the absolute melting point.

Cold working produces an increase in dislocation density; for most metals ρ increases from the value of 10^{10}–10^{12} lines m^{-2} typical of the annealed state, to 10^{12}–10^{13} after a few per cent deformation, and up to 10^{15}–10^{16} lines m^{-2} in the heavily deformed state. Such an array of dislocations gives rise to a substantial strain energy stored in the lattice, so that the cold-worked condition is thermodynamically unstable relative to the undeformed one. Consequently, the deformed metal will try to return to a state of lower free energy, i.e. a more perfect state. In general, this return to a more equilibrium structure cannot occur spontaneously but only at elevated temperatures where thermally activated processes such as diffusion, cross-slip and climb take place. Like all non-equilibrium processes the rate of approach to equilibrium will be governed by an Arrhenius equation of the form

Rate $= A \exp[-Q/\boldsymbol{k}T]$

where the activation energy Q depends on impurity content, strain, etc.

The formation of atmospheres by strain-ageing is one method whereby the metal reduces its excess lattice energy but this process is unique in that it usually leads to a further increase in the structure-sensitive properties rather than a reduction to the value characteristic of the annealed condition. It is necessary,

therefore, to increase the temperature of the deformed metal above the strain-ageing temperature before it recovers its original softness and other properties.

The removal of the cold-worked condition occurs by a combination of three processes, namely: (1) recovery, (2) recrystallization and (3) grain growth. These stages have been successfully studied using light microscopy, transmission electron microscopy, or X-ray diffraction; mechanical property measurements (e.g. hardness); and physical property measurements (e.g. density, electrical resistivity and stored energy). Figure 7.49 shows the change in some of these properties on annealing. During the recovery stage the decrease in stored energy and electrical resistivity is accompanied by only a slight lowering of hardness, and the greatest simultaneous change in properties occurs during the primary recrystallization stage. However, while these measurements are no doubt striking and extremely useful, it is necessary to understand them to correlate such studies with the structural changes by which they are accompanied.

7.8.2 Recovery

This process describes the changes in the distribution and density of defects with associated changes in physical and mechanical properties which take place in worked crystals before recrystallization or alteration of orientation occurs. It will be remembered that the structure of a cold-worked metal consists of dense dislocation networks, formed by the glide and interaction of dislocations, and, consequently, the recovery stage of annealing is chiefly concerned with the rearrangement of these dislocations to reduce the lattice energy and does not involve the migration of large-angle boundaries. This rearrangement of the dislocations is

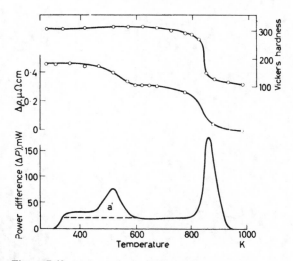

Figure 7.49 (a) Rate of release of stored energy (ΔP), increment in electrical resistivity ($\Delta\rho$) and hardness (VPN) for specimens of nickel deformed in torsion and heated at 6 k/min (Clareborough, Hargreaves and West, 1955).

assisted by thermal activation. Mutual annihilation of dislocations is one process.

When the two dislocations are on the same slip plane, it is possible that as they run together and annihilate they will have to cut through intersecting dislocations on other planes, i.e. 'forest' dislocations. This recovery process will, therefore, be aided by thermal fluctuations since the activation energy for such a cutting process is small. When the two dislocations of opposite sign are not on the same slip plane, climb or cross-slip must first occur, and both processes require thermal activation.

One of the most important recovery processes which leads to a resultant lowering of the lattice strain energy is rearrangement of the dislocations into cell walls. This process in its simplest form was originally termed polygonization and is illustrated schematically in Figure 7.50, whereby dislocations all of one sign align themselves into walls to form small-angle or sub-grain boundaries. During deformation a region of the lattice is curved, as shown in Figure 7.50a, and the observed curvature can be attributed to the formation of excess edge dislocations parallel to the axis of bending. On heating, the dislocations form a sub-boundary by a process of annihilation and rearrangement. This is shown in Figure 7.50b, from which it can be seen that it is the excess dislocations of one sign which remain after the annihilation process that align themselves into walls.

Polygonization is a simple form of sub-boundary formation and the basic movement is climb whereby the edge dislocations change their arrangement from a horizontal to a vertical grouping. This process involves the migration of vacancies to or from the edge of the half-planes of the dislocations (see Section 4.3.4). The removal of vacancies from the lattice, together with the reduced strain energy of dislocations which results, can account for the large change in both electrical resistivity and stored energy observed during this stage, while the change in hardness can be attributed to the rearrangement of dislocations and to the reduction in the density of dislocations.

The process of polygonization can be demonstrated using the Laue method of X-ray diffraction. Diffraction from a bent single crystal of zinc takes the form of continuous radial streaks. On annealing, these asterisms (see Figure 5.10) break up into spots as shown in Figure 7.50c, where each diffraction spot originates from a perfect polygonized sub-grain, and the distance between the spots represents the angular misorientation across the sub-grain boundary. Direct evidence for this process is observed in the electron microscope, where, in heavily deformed polycrystalline aggregates at least, recovery is associated with the formation of sub-grains out of complex dislocation networks by a process of dislocation annihilation and rearrangement. In some deformed metals and alloys the dislocations are already partially arranged in sub-boundaries forming diffuse cell structures by dynamical recovery (see Figure 7.41). The conventional recovery process is then one in which these cells sharpen and grow. In other metals, dislocations are more uniformly distributed after deformation, with hardly any cell structure discernible, and the recovery process then involves formation, sharpening and growth of sub-boundaries. The sharpness of the cell structure formed by deformation depends on the stacking fault energy of the metal, the deformation temperature and the extent of deformation (see Figure 7.42).

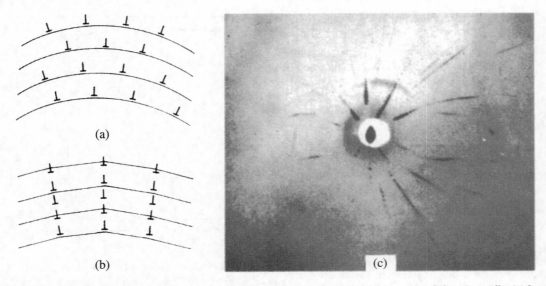

Figure 7.50 *(a) Random arrangement of excess parallel edge dislocations and (b) alignment into dislocation walls; (c) Laue photograph of polygonized zinc (after Cahn, 1949).*

7.8.3 Recrystallization

The most significant changes in the structure-sensitive properties occur during the primary recrystallization stage. In this stage the deformed lattice is completely replaced by a new unstrained one by means of a nucleation and growth process, in which practically stress-free grains grow from nuclei formed in the deformed matrix. The orientation of the new grains differs considerably from that of the crystals they consume, so that the growth process must be regarded as incoherent, i.e. it takes place by the advance of large-angle boundaries separating the new crystals from the strained matrix.

During the growth of grains, atoms get transferred from one grain to another across the boundary. Such a process is thermally activated as shown in Figure 7.51, and by the usual reaction-rate theory the frequency of atomic transfer one way is

$$\nu \exp \left(-\frac{\Delta F}{kT} \right) s^{-1} \tag{7.38}$$

and in the reverse direction

$$\nu \exp \left(-\frac{\Delta F^* + \Delta F}{kT} \right) s^{-1} \tag{7.39}$$

where ΔF is the difference in free energy per atom between the two grains, i.e. supplying the driving force for migration, and ΔF^* is an activation energy. For each net transfer the boundary moves forward a distance b and the velocity ν is given by

$$\nu = M \Delta F \tag{7.40}$$

where M is the mobility of the boundary, i.e. the velocity for unit driving force, and is thus

$$M = \frac{b\nu}{kT} \exp \left(\frac{\Delta S^*}{k} \right) \exp \left(-\frac{\Delta E^*}{kT} \right) \tag{7.41}$$

Generally, the open structure of high-angle boundaries should lead to a high mobility. However they are susceptible to the segregation of impurities, low concentrations of which can reduce the boundary mobility by orders of magnitude. In contrast, special boundaries which are close to a CSL are much less affected by impurity segregation and hence can lead to higher relative mobility.

It is well known that the rate of recrystallization depends on several important factors, namely: (1) the amount of prior deformation (the greater the degree of cold work, the lower the recrystallization temperature and the smaller the grain size), (2) the temperature of the anneal (as the temperature is lowered the time to attain a constant grain size increases exponentially[1]) and (3) the purity of the sample (e.g. zone-refined aluminium recrystallizes below room temperature, whereas aluminium of commercial purity must be heated several hundred degrees). The role these variables play in recrystallization will be evident once the mechanism of recrystallization is known. This mechanism will now be outlined.

Measurements, using the light microscope, of the increase in diameter of a new grain as a function of time at any given temperature can be expressed as shown in Figure 7.52. The diameter increases linearly with time until the growing grains begin to impinge on one another, after which the rate necessarily decreases. The classical interpretation of these observations is that nuclei form spontaneously in the matrix after a so-called nucleation time, t_0, and these nuclei then proceed to grow steadily as shown by the linear relationship. The driving force for the process is provided by the stored energy of cold work contained in the strained grain on one side of the boundary relative to that on the other side. Such an interpretation would suggest that the recrystallization process occurs in two distinct stages, i.e. first nucleation and then growth.

During the linear growth period the radius of a nucleus is $R = G(t - t_0)$, where G, the growth rate, is

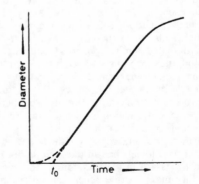

Figure 7.52 *Variation of grain diameter with time at a constant temperature.*

[1] The velocity of linear growth of new crystals usually obeys an exponential relationship of the form $\nu = \nu_0 \exp [-Q/RT]$.

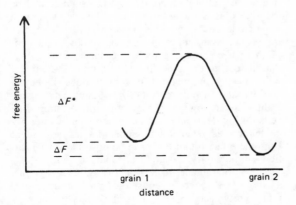

Figure 7.51 *Variation in free energy during grain growth.*

dR/dt and, assuming the nucleus is spherical, the volume of the recrystallized nucleus is $4\pi/3G^3(t-t_0)^3$. If the number of nuclei that form in a time increment dt is $N\,dt$ per unit volume of unrecrystallized matrix, and if the nuclei do not impinge on one another, then for unit total volume

$$f = \frac{4}{3}\pi N G^3 \int_0^t (t-t_0)^3 dt$$

or

$$f = \frac{\pi}{3} G^3 t^4 \qquad (7.42)$$

This equation is valid in the initial stages when $f \ll 1$. When the nuclei impinge on one another the rate of recrystallization decreases and is related to the amount untransformed $(1-f)$ by

$$f = 1 - \exp\left(-\frac{\pi}{3}N G^3 t^4\right) \qquad (7.43)$$

where, for short times, equation (7.43) reduces to equation (7.42). This Johnson–Mehl equation is expected to apply to any phase transformation where there is random nucleation, constant N and G and small t_0. In practice, nucleation is not random and the rate not constant so that equation (7.43) will not strictly apply. For the case where the nucleation rate decreases exponentially, Avrami developed the equation

$$f = 1 - \exp\left(-kt^n\right) \qquad (7.44)$$

where k and n are constants, with $n \approx 3$ for a fast and $n \approx 4$ for a slow, decrease of nucleation rate. Provided there is no change in the nucleation mechanism, n is independent of temperature but k is very sensitive to temperature T; clearly from equation (7.43), $k = \pi N G^3/3$ and both N and G depend on T.

An alternative interpretation is that the so-called incubation time t_0 represents a period during which small nuclei, of a size too small to be observed in the light microscope, are growing very slowly. This latter interpretation follows from the recovery stage of annealing. Thus, the structure of a recovered metal consists of sub-grain regions of practically perfect crystal and, thus, one might expect the 'active' recrystallization nuclei to be formed by the growth of certain sub-grains at the expense of others.

The process of recrystallization may be pictured as follows. After deformation, polygonization of the bent lattice regions on a fine scale occurs and this results in the formation of several regions in the lattice where the strain energy is lower than in the surrounding matrix; this is a necessary primary condition for nucleation. During this initial period when the angles between the sub-grains are small and less than one degree, the sub-grains form and grow quite rapidly. However, as the sub-grains grow to such a size that the angles between them become of the order of a few degrees, the growth of any given sub-grain at the expense of the others is very slow. Eventually one of the sub-grains

will grow to such a size that the boundary mobility begins to increase with increasing angle. A large angle boundary, $\theta \approx 30$–$40°$, has a high mobility because of the large lattice irregularities or 'gaps' which exist in the boundary transition layer. The atoms on such a boundary can easily transfer their allegiance from one crystal to the other. This sub-grain is then able to grow at a much faster rate than the other sub-grains which surround it and so acts as the nucleus of a recrystallized grain. The further it grows, the greater will be the difference in orientation between the nucleus and the matrix it meets and consumes, until it finally becomes recognizable as a new strain-free crystal separated from its surroundings by a large-angle boundary.

The recrystallization nucleus therefore has its origin as a sub-grain in the deformed microstructure. Whether it grows to become a strain-free grain depends on three factors: (1) the stored energy of cold work must be sufficiently high to provide the required driving force, (2) the potential nucleus should have a size advantage over its neighbours, and (3) it must be capable of continued growth by existing in a region of high lattice curvature (e.g. transition band) so that the growing nucleus can quickly achieve a high-angle boundary. In situ experiments in the HVEM have confirmed these factors. Figure 7.53a shows the as-deformed substructure in the transverse section of rolled copper, together with the orientations of some selected areas. The sub-grains are observed to vary in width from 50 to 500 nm, and exist between regions 1 and 8 as a transition band across which the orientation changes sharply. On heating to 200°C, the sub-grain region 2 grows into the transition region (Figure 7.53b) and the orientation of the new grain well developed at 300°C is identical to the original sub-grain (Figure 7.53c).

With this knowledge of recrystallization the influence of several variables known to affect the recrystallization behaviour of a metal can now be understood. Prior deformation, for example, will control the extent to which a region of the lattice is curved. The larger the deformation, the more severely will the lattice be curved and, consequently, the smaller will be the size of a growing sub-grain when it acquires a large-angle boundary. This must mean that a shorter time is necessary at any given temperature for the sub-grain to become an 'active' nucleus, or, conversely, that the higher the annealing temperature, the quicker will this stage be reached. In some instances, heavily cold-worked metals recrystallize without any significant recovery owing to the formation of strain-free cells during deformation. The importance of impurity content on recrystallization temperature is also evident from the effect impurities have on obstructing dislocation sub-boundary and grain boundary mobility.

The intragranular nucleation of strain-free grains, as discussed above, is considered as abnormal sub-grain growth, in which it is necessary to specify that some sub-grains acquire a size advantage and

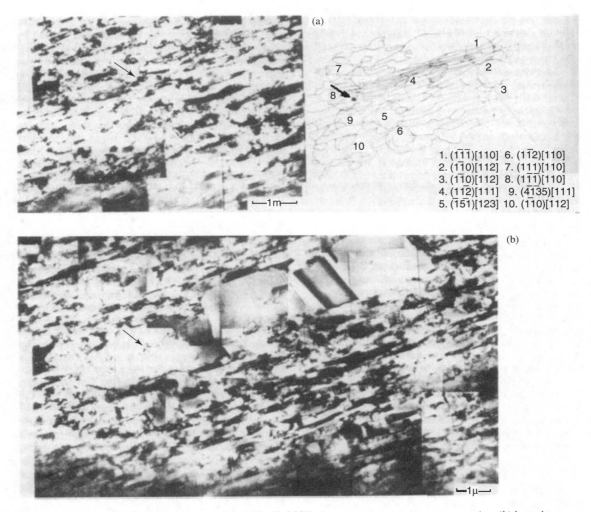

1. ($\bar{1}\bar{1}\bar{1}$)[110] 6. (1$\bar{1}$2)[110]
2. ($\bar{1}$10)[112] 7. (111)[110]
3. ($\bar{1}\bar{1}$0)[112] 8. (1$\bar{1}\bar{1}$)[110]
4. (1$\bar{1}$2)[111] 9. ($\bar{4}$135)[111]
5. ($\bar{1}$51)[123] 10. ($\bar{1}$10)[112]

Figure 7.53 *Electron micrographs of copper. (a) Cold-rolled 95% at room temperature, transverse section, (b) heated to 300°C in the HVEM.*

are able to grow at the expense of the normal sub-grains. It has been suggested that nuclei may also be formed by a process involving the rotation of individual cells so that they coalesce with neighbouring cells to produce larger cells by volume diffusion and dislocation rearrangement.

In some circumstances, intergranular nucleation is observed in which an existing grain boundary bows out under an initial driving force equal to the difference in free energy across the grain boundary. This strain-induced boundary migration is irregular and is from a grain with low strain (i.e. large cell size) to one of larger strain and smaller cell size. For a boundary to grow in this way the strain energy difference per unit volume across the boundary must be sufficient to supply the energy increase to bow out a length of boundary ≈ 1 μm.

Segregation of solute atoms to, and precipitation on, the grain boundary tends to inhibit intergranular nucleation and gives an advantage to intragranular nucleation, provided the dispersion is not too fine. In general, the recrystallization behaviour of two-phase alloys is extremely sensitive to the dispersion of the second phase. Small, finely dispersed particles retard recrystallization by reducing both the nucleation rate and the grain boundary mobility, whereas large coarsely dispersed particles enhance recrystallization by increasing the nucleation rate. During deformation, zones of high dislocation density and large misorientations are formed around non-deformable particles, and on annealing, recrystallization nuclei are created within these zones by a process of polygonization by sub-boundary migration. Particle-stimulated nucleation occurs above a critical particle size which decreases with increasing deformation. The finer dispersions tend to homogenize the microstructure (i.e. dislocation distribution) thereby minimizing local lattice curvature and reducing nucleation.

The formation of nuclei becomes very difficult when the spacing of second-phase particles is so small that each developing sub-grain interacts with a particle before it becomes a viable nucleus. The extreme case of this is SAP (sintered aluminium powder) which contains very stable, close-spaced oxide particles. These particles prevent the rearrangement of dislocations into cell walls and their movement to form high-angle boundaries, and hence SAP must be heated to a temperature very close to the melting point before it recrystallizes.

7.8.4 Grain growth

When primary recrystallization is complete (i.e. when the growing crystals have consumed all the strained material) the material can lower its energy further by reducing its total area of grain surface. With extensive annealing it is often found that grain boundaries straighten, small grains shrink and larger ones grow. The general phenomenon is known as grain growth, and the most important factor governing the process is the surface tension of the grain boundaries. A grain boundary has a surface tension, T (= surface-free energy per unit area) because its atoms have a higher free energy than those within the grains. Consequently, to reduce this energy a polycrystal will tend to minimize the area of its grain boundaries and when this occurs the configuration taken up by any set of grain boundaries (see Figure 7.54) will be governed by the condition that

$$T_A/\sin A = T_B/\sin B = T_C/\sin C \qquad (7.45)$$

Most grain boundaries are of the large-angle type with their energies approximately independent of orientation, so that for a random aggregate of grains $T_A = T_B = T_C$ and the equilibrium grain boundary angles are each equal to 120°. Figure 7.54b shows an idealized grain in two dimensions surrounded by others

of uniform size, and it can be seen that the equilibrium grain shape takes the form of a polygon of six sides with 120° inclusive angles. All polygons with either more or less than this number of sides cannot be in equilibrium. At high temperatures where the atoms are mobile, a grain with fewer sides will tend to become smaller, under the action of the grain boundary surface tension forces, while one with more sides will tend to grow.

Second-phase particles have a major inhibiting effect on boundary migration and are particularly effective in the control of grain size. The pinning process arises from surface tension forces exerted by the particle–matrix interface on the grain boundary as it migrates past the particle. Figure 7.55 shows that the drag exerted by the particle on the boundary, resolved in the forward direction, is

$$F = \pi r \gamma \sin 2\theta$$

where γ is the specific interfacial energy of the boundary; $F = F_{max} = \pi r \gamma$ when $\theta = 45°$. Now if there are N particles per unit volume, the volume fraction is $4\pi r^3 N/3$ and the number n intersecting unit area of boundary is given by

$$n = 3f/2\pi r^2 \qquad (7.46)$$

For a grain boundary migrating under the influence of its own surface tension the driving force is $2\gamma/R$, where R is the minimum radius of curvature and as the grains grow, R increases and the driving force decreases until it is balanced by the particle-drag, when growth stops. If $R \sim d$ the mean grain diameter, then the critical grain diameter is given by the condition

$$nF \approx 2\gamma/d_{crit}$$

or

$$d_{crit} \approx 2\gamma(2\pi r^2/3f\pi r\gamma) = 4r/3f \qquad (7.47)$$

This Zener drag equation overestimates the driving force for grain growth by considering an isolated

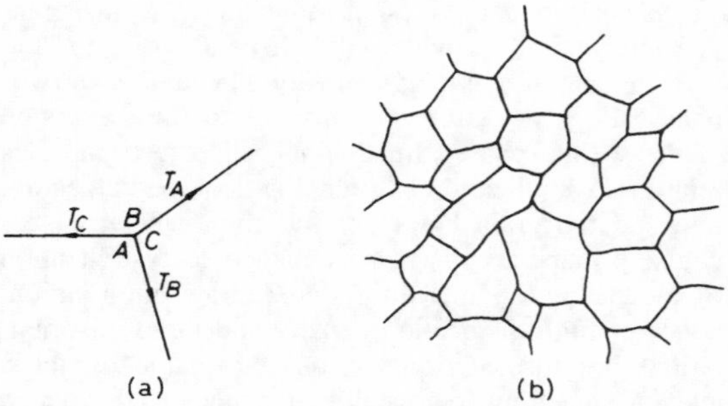

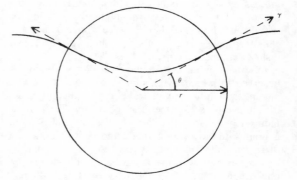

Figure 7.54 (a) Relation between angles and surface tensions at a grain boundary triple point; (b) idealized polygonal grain structure.

Figure 7.55 Diagram showing the drag exerted on a boundary by a particle.

spherical grain. A heterogeneity in grain size is necessary for grain growth and taking this into account gives a revised equation:

$$d_{\text{crit}} \approx \frac{\pi r}{3f} \left[\frac{3}{2} - \frac{2}{Z} \right] \tag{7.48}$$

where Z is the ratio of the diameters of growing grains to the surrounding grains. This treatment explains the successful use of small particles in refining the grain size of commercial alloys.

During the above process growth is continuous and a uniform coarsening of the polycrystalline aggregate usually occurs. Nevertheless, even after growth has finished the grain size in a specimen which was previously severely cold-worked remains relatively small, because of the large number of nuclei produced by the working treatment. Exaggerated grain growth can often be induced, however, in one of two ways, namely: (1) by subjecting the specimen to a critical strain-anneal treatment or (2) by a process of secondary recrystallization. By applying a critical deformation (usually a few per cent strain) to the specimen the number of nuclei will be kept to a minimum, and if this strain is followed by a high-temperature anneal in a thermal gradient some of these nuclei will be made more favourable for rapid growth than others. With this technique, if the conditions are carefully controlled, the whole of the specimen may be turned into one crystal, i.e. a single crystal. The term secondary recrystallization describes the process whereby a specimen which has been given a primary recrystallization treatment at a low temperature is taken to a higher temperature to enable the abnormally rapid growth of a few grains to occur. The only driving force for secondary recrystallization is the reduction of grain boundary-free energy, as in normal grain growth, and consequently, certain special conditions are necessary for its occurrence. One condition for this 'abnormal' growth is that normal continuous growth is impeded by the presence of inclusions, as is indicated by the exaggerated grain growth of tungsten wire containing thoria, or the sudden coarsening of deoxidized steel at about 1000°C. A possible explanation for the phenomenon is that in some regions the grain boundaries become free (e.g. if the inclusions slowly dissolve or the boundary tears away) and as a result the grain size in such regions becomes appreciably larger than the average (Figure 7.56a). It then follows that the grain

boundary junction angles between the large grain and the small ones that surround it will not satisfy the condition of equilibrium discussed above. As a consequence, further grain boundary movement to achieve 120° angles will occur, and the accompanying movement of a triple junction point will be as shown in Figure 7.56b. However, when the dihedral angles at each junction are approximately 120° a severe curvature in the grain boundary segments between the junctions will arise, and this leads to an increase in grain boundary area. Movement of these curved boundary segments towards their centres of curvature must then take place and this will give rise to the configuration shown in Figure 7.56c. Clearly, this sequence of events can be repeated and continued growth of the large grains will result.

The behaviour of the dispersed phase is extremely important in secondary recrystallization and there are many examples in metallurgical practice where the control of secondary recrystallization with dispersed particles has been used to advantage. One example is in the use of Fe–3% Si in the production of strip for transformer laminations. This material is required with $(1\,1\,0)\,[0\,0\,1]$ 'Goss' texture because of the $[0\,0\,1]$ easy direction of magnetization, and it is found that the presence of MnS particles favours the growth of secondary grains with the appropriate Goss texture. Another example is in the removal of the pores during the sintering of metal and ceramic powders, such as alumina and metallic carbides. The sintering process is essentially one of vacancy creep involving the diffusion of vacancies from the pore of radius r to a neighbouring grain boundary, under a driving force $2\gamma_s/r$ where γ_s is the surface energy. In practice, sintering occurs fairly rapidly up to about 95% full density because there is a plentiful association of boundaries and pores. When the pores become very small, however, they are no longer able to anchor the grain boundaries against the grain growth forces, and hence the pores sinter very slowly, since they are stranded within the grains some distance from any boundary. To promote total sintering, an effective dispersion is added. The dispersion is critical, however, since it must produce sufficient drag to slow down grain growth, during which a particular pore is crossed by several migrating boundaries, but not sufficiently large to give rise to secondary recrystallization when a given pore would be stranded far from any boundary.

The relation between grain-size, temperature and strain is shown in Figure 7.57 for commercially pure aluminium. From this diagram it is clear that either a critical strain-anneal treatment or a secondary recrystallization process may be used for the preparation of perfect strain-free single crystals.

7.8.5 Annealing twins

A prominent feature of the microstructures of most annealed fcc metals and alloys is the presence of many straight-sided bands that run across grains. These

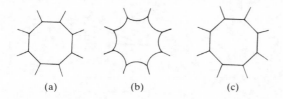

(a)   (b)   (c)

Figure 7.56 *Grain growth during secondary recrystallization.*

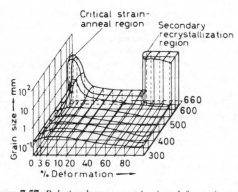

Figure 7.57 *Relation between grain size, deformation and temperature for aluminium (after Buergers, courtesy of Akademie-Verlags-Gesellschaft).*

bands have a twinned orientation relative to their neighbouring grain and are referred to as annealing twins (see Chapter 4). The parallel boundaries usually coincide with a (1 1 1) twinning plane with the structure coherent across it, i.e. both parts of the twin hold a single (1 1 1) plane in common.

As with formation of deformation twins, it is believed that a change in stacking sequence is all that is necessary to form an annealing twin. Such a change in stacking sequence may occur whenever a properly oriented grain boundary migrates. For example, if the boundary interface corresponds to a (1 1 1) plane, growth will proceed by the deposition of additional (1 1 1) planes in the usual stacking sequence **ABCABC** . . .. If, however, the next newly deposited layer falls into the wrong position, the sequence **ABCABCB** is produced which constitutes the first layer of a twin. Once a twin interface is formed, further growth may continue with the sequence in reverse order, **ABCABC|BACB** . . . until a second accident in the stacking sequence completes the twin band, **ABCABCBACBACBABC**. When a stacking error, such as that described above, occurs the number of nearest neighbours is unchanged, so that the ease of formation of a twin interface depends on the relative value of the interface energy. If this interface energy is low, as in copper where $\gamma_{twin} < 20$ mJ/m^2 twinning occurs frequently while if it is high, as in aluminium, the process is rare.

Annealing twins are rarely (if ever) found in cast metals because grain boundary migration is negligible during casting. Worked and annealed metals show considerable twin band formation; after extensive grain growth a coarse-grained metal often contains twins which are many times wider than any grain that was present shortly after recrystallization. This indicates that twin bands grow in width, during grain growth, by migration in a direction perpendicular to the (1 1 1) composition plane, and one mechanism whereby this can occur is illustrated schematically in Figure 7.58. This shows that a twin may form at the corner of a

Figure 7.58 *Formation and growth of annealing twins (from Burke and Turnbull, 1952; courtesy of Pergamon Press).*

grain, since the grain boundary configuration will then have a lower interfacial energy. If this happens the twin will then be able to grow in width because one of its sides forms part of the boundary of the growing grain. Such a twin will continue to grow in width until a second mistake in the positioning of the atomic layers terminates it; a complete twin band is then formed. In copper and its alloys $\gamma_{twin}/\gamma_{gb}$ is low and hence twins occur frequently, whereas in aluminium the corresponding ratio is very much higher and so twins are rare.

Twins may develop according to the model shown in Figure 7.59 where during grain growth a grain contact is established between grains C and D. Then if the orientation of grain D is close to the twin orientation of grain C, the nucleation of an annealing twin at the grain boundary, as shown in Figure 7.60d, will lower the total boundary energy. This follows because the twin/D interface will be reduced to about 5% of the normal grain boundary energy, the energies of the C/A and twin/A interface will be approximately the same, and the extra area of interface C/twin has only a

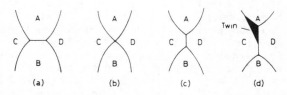

Figure 7.59 *Nucleation of an annealing twin during grain growth.*

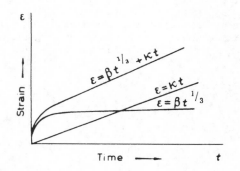

Figure 7.60 *Combination of transient and steady-state creep.*

very low energy. This model indicates that the number of twins per unit grain boundary area only depends on the number of new grain contacts made during grain growth, irrespective of grain size and annealing temperature.

7.8.6 Recrystallization textures

The preferred orientation developed by cold work often changes on recrystallization to a totally different preferred orientation. To explain this observation, Barrett and (later) Beck have put forward the 'oriented growth' theory of recrystallization textures in which it is proposed that nuclei of many orientations initially form but, because the rate of growth of any given nucleus depends on the orientation difference between the matrix and growing crystal, the recrystallized texture will arise from those nuclei which have the fastest growth rate in the cold-worked matrix, i.e. those bounded by large-angle boundaries. It then follows that because the matrix has a texture, all the nuclei which grow will have orientations that differ by $30-40°$ from the cold-worked texture. This explains why the new texture in fcc metals is often related to the old texture, by a rotation of approximately $30-40°$ around $\langle 1\,1\,1 \rangle$ axes, in bcc metals by $30°$ about $\langle 1\,1\,0 \rangle$ and in hcp by $30°$ about $\langle 0\,0\,0\,1 \rangle$. However, while it is undoubtedly true that oriented growth provides a selection between favourable and unfavourable oriented nuclei, there are many observations to indicate that the initial nucleation is not entirely random. For instance, because of the crystallographic symmetry one would expect grains appearing in a fcc texture to be related to rotations about all four $\langle 1\,1\,1 \rangle$ axes, i.e. eight orientations arising from two possible rotations about each of the four $\langle 1\,1\,1 \rangle$ axes. All these possible orientations are rarely (if ever) observed.

To account for such observations, and for those cases where the deformation texture and the annealing texture show strong similarities, oriented nucleation is considered to be important. The oriented nucleation theory assumes that the selection of orientations is determined in the nucleation stage. It is generally accepted that all recrystallization nuclei pre-exist in the deformed matrix, as sub-grains, which become more perfect through recovery processes prior to recrystallization. It is thus most probable that there is some selection of nuclei determined by the representation of the orientations in the deformation texture, and that the oriented nucleation theory should apply in some cases. In many cases the orientations which are strongly represented in the annealing texture are very weakly represented in the deformed material. The most striking example is the 'cube' texture, $(1\,0\,0)\,[0\,0\,1]$, found in most fcc pure metals which have been annealed following heavy rolling reductions. In this texture, the cube axes are extremely well aligned along the sheet axes, and its behaviour resembles that of a single crystal. It is thus clear that cube-oriented grains or sub-grains must have a very high initial growth rate in order to

form the remarkably strong quasi-single-crystal cube texture. The percentage of cubically aligned grains increases with increased deformation, but the sharpness of the textures is profoundly affected by alloying additions. The amount of alloying addition required to suppress the texture depends on those factors which affect the stacking fault energy, such as the lattice misfit of the solute atom in the solvent lattice, valency etc., in much the same way as that described for the transition of a pure metal deformation texture.

In general, however, if the texture is to be altered a distribution of second-phase must either be present before cold rolling or be precipitated during annealing. In aluminium, for example, the amount of cube texture can be limited in favour of retained rolling texture by limiting the amount of grain growth with a precipitate dispersion of Si and Fe. By balancing the components, earing can be minimized in drawn aluminium cups. In aluminium-killed steels AlN precipitation prior to recrystallization produces a higher proportion of grains with $\{1\,1\,1\}$ planes parallel to the rolling plane and a high $\overline{R}$ value suitable for deep drawing. The AlN dispersion affects sub-grain growth, limiting the available nuclei and increasing the orientation-selectivity, thereby favouring the high-energy $\{1\,1\,1\}$ grains. Improved $\overline{R}$-values in steels in general are probably due to the combined effect of particles in homogenizing the deformed microstructure and in controlling the subsequent sub-grain growth. The overall effect is to limit the availability of nuclei with orientations other than $\{1\,1\,1\}$.

7.9 Metallic creep

7.9.1 Transient and steady-state creep

Creep is the process by which plastic flow occurs when a constant stress is applied to a metal for a prolonged period of time. After the initial strain ε_0 which follows the application of the load, creep usually exhibits a rapid transient period of flow (stage 1) before it settles down to the linear steady-stage stage 2 which eventually gives way to tertiary creep and fracture. Transient creep, sometimes referred to as β-creep, obeys a $t^{1/3}$ law. The linear stage of creep is often termed steady-state creep and obeys the relation

$$\varepsilon = \kappa t \tag{7.49}$$

Consequently, because both transient and steady-state creep usually occur together during creep at high temperatures, the complete curve (Figure 7.60) during the primary and secondary stages of creep fits the equation

$$\varepsilon = \beta t^{1/3} + \kappa t \tag{7.50}$$

extremely well. In contrast to transient creep, steady-state creep increases markedly with both temperature and stress. At constant stress the dependence on temperature is given by

$$\dot{\varepsilon}_{ss} = d\varepsilon/dt = \text{const.} \exp\left[-Q/\mathbf{k}T\right] \qquad (7.51)$$

where Q is the activation energy for steady-state creep, while at constant temperature the dependence on stress σ (compensated for modulus E) is

$$\dot{\varepsilon}_{ss} = \text{const.} (\sigma/E)^n \qquad (7.52)$$

Steady-state creep is therefore described by the equation

$$\dot{\varepsilon}_{ss} = A(\sigma/E)^n \exp\left(-Q/\mathbf{k}T\right) \qquad (7.53)$$

The basic assumption of the mechanism of steady-state creep is that during the creep process the rate of recovery r (i.e. decrease in strength $(d\sigma/dt)$) is sufficiently fast to balance the rate of work hardening $h = (d\sigma/d\varepsilon)$. The creep rate $(d\varepsilon/dt)$ is then given by

$$d\varepsilon/dt = (d\sigma/dt)/(d\sigma/d\varepsilon) = r/h \qquad (7.54)$$

To prevent work hardening, both the screw and edge parts of a glissile dislocation loop must be able to escape from tangled or piled-up regions. The edge dislocations will, of course, escape by climb, and since this process requires a higher activation energy than cross-slip, it will be the rate-controlling process in steady-state creep. The rate of recovery is governed by the rate of climb, which depends on diffusion and stress such that

$$r = A(\sigma/E)^p D = A(\sigma/E)^p D_0 \exp\left[-Q/\mathbf{k}T\right]$$

where D is a diffusion coefficient and the stress term arises because recovery is faster, the higher the stress level and the closer dislocations are together. The work-hardening rate decreases from the initial rate h_0 with increasing stress, i.e. $h = h_0(E/\sigma)^q$, thus

$$\dot{\varepsilon}_{ss} = r/h = B(\sigma/E)^n D \qquad (7.55)$$

where $B(= A/h_0)$ is a constant and $n(= p + q)$ is the stress exponent.

The structure developed in creep arises from the simultaneous work-hardening and recovery. The dislocation density ρ increases with ε and the dislocation network gets finer, since dislocation spacing is proportional to $\rho^{-1/2}$. At the same time, the dislocations tend to reduce their strain energy by mutual annihilation and rearrange to form low-angle boundaries and this increases the network spacing. Straining then proceeds at a rate at which the refining action just balances the growth of the network by recovery; the equilibrium network size being determined by the stress. Although dynamical recovery can occur by cross-slip, the rate-controlling process in steady-state creep is climb whereby edge dislocations climb out of their glide planes by absorbing or emitting vacancies; the activation energy is therefore that of self-diffusion. Structural observations confirm the importance of the recovery process to steady-state creep. These show that sub-grains form within the original grains and, with increasing deformation, the sub-grain angle increases while the dislocation density within them remains constant.[1] The climb process may, of course, be important in several different ways. Thus, climb may help a glissile dislocation to circumvent different barriers in the structure such as a sessile dislocation, or it may lead to the annihilation of dislocations of opposite sign on different glide planes. Moreover, because creep-resistant materials are rarely pure metals, the climb process may also be important in allowing a glissile dislocation to get round a precipitate or move along a grain boundary. A comprehensive analysis of steady-state creep, based on the climb of dislocations, has been given by Weertman.

The activation energy for creep Q may be obtained experimentally by plotting $\ln \dot{\varepsilon}_{ss}$ versus $1/T$, as shown in Figure 7.61. Usually above $0.5T_m$, Q corresponds

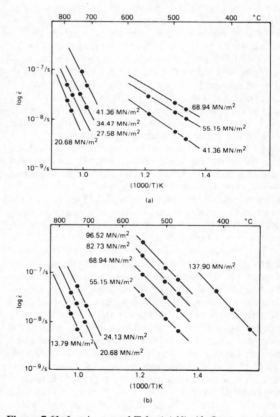

Figure 7.61 *Log $\dot{\varepsilon}$ versus 1/T for (a) Ni–Al$_2$O$_3$, (b) Ni–67Co–Al$_2$O$_3$, showing the variation in activation energy above and below $0.5T_m$ (after Hancock, Dillamore and Smallman, 1972).*

---

[1] Sub-grains do not always form during creep and in some metallic solid solutions where the glide of dislocations is restrained due to the dragging of solute atoms, the steady-state substructure is essentially a uniform distribution of dislocations.

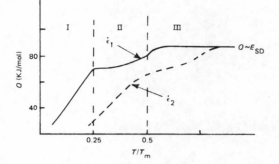

Figure 7.62 *Variation in activation energy Q with temperature for aluminium.*

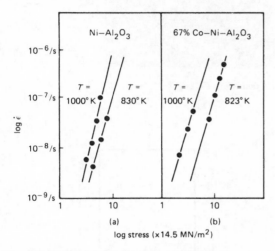

Figure 7.63 *Log $\dot{\varepsilon}$ versus log σ for (a) Ni–Al$_2$O$_3$, (b) Ni–67 Co–Al$_2$O$_3$ (after Hancock, Dillamore and Smallman, 1972).*

to the activation energy for self-diffusion E_{SD}, in agreement with the climb theory, but below $0.5T_m$, $Q < E_{SD}$, possibly corresponding to pipe diffusion. Figure 7.62 shows that three creep regimes may be identified and the temperature range where $Q = E_{SD}$ can be moved to higher temperatures by increasing the strain rate. Equation (7.55) shows that the stress exponent n can be obtained experimentally by plotting $\ln \dot{\varepsilon}_{ss}$ versus $\ln \sigma$, as shown in Figure 7.63, where $n \approx 4$. While n is generally about 4 for dislocation creep, Figure 7.64 shows that n may vary considerably from this value depending on the stress regime; at low stresses (i.e. regime I) creep occurs not by dislocation glide and climb but by stress-directed flow of vacancies.

7.9.2 Grain boundary contribution to creep

In the creep of polycrystals at high temperatures the grain boundaries themselves are able to play an important part in the deformation process due to the fact that

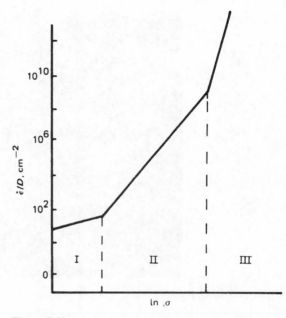

Figure 7.64 *Schematic diagram showing influence of stress on diffusion-compensated steady-state creep.*

they may (1) slide past each other or (2) create vacancies. Both processes involve an activation energy for diffusion and therefore may contribute to steady-state creep.

Grain boundary sliding during creep was inferred initially from the observation of steps at the boundaries, but the mechanism of sliding can be demonstrated on bi-crystals. Figure 7.65 shows a good example of grain boundary movement in a bi-crystal of tin, where the displacement of the straight grain boundary across its middle is indicated by marker scratches. Grain boundaries, even when specially produced for bi-crystal experiments, are not perfectly straight, and after a small amount of sliding at the boundary interface, movement will be arrested by protuberances. The grains are then locked, and the rate of slip will be determined by the rate of plastic flow in the protuberances. As a result, the rate of slip along a grain boundary is not constant with time, because the dislocations first form into piled-up groups, and later these become relaxed. Local relaxation may be envisaged as a process in which the dislocations in the pile-up climb towards the boundary. In consequence, the activation energy for grain boundary slip may be identified with that for steady-state creep. After climb, the dislocations are spread more evenly along the boundary, and are thus able to give rise to grain boundary migration, when sliding has temporarily ceased, which is proportional to the overall deformation.

A second creep process which also involves the grain boundaries is one in which the boundary acts

Figure 7.65 *Grain boundary sliding on a bi-crystal tin (after Puttick and King, 1952).*

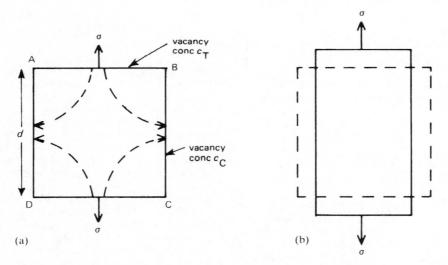

Figure 7.66 *Schematic representation of Herring–Nabarro creep; with $c_T > c_c$ vacancies flow from the tensile faces to the longitudinal faces (a) to produce creep as shown in (b).*

as a source and sink for vacancies. The mechanism depends on the migration of vacancies from one side of a grain to another, as shown in Figure 7.66, and is often termed Herring–Nabarro creep, after the two workers who originally considered this process. If, in a grain of sides d under a stress σ, the atoms are transported from faces BC and AD to the faces AB and DC the grain creeps in the direction of the stress. To transport atoms in this way involves creating vacancies on the tensile faces AB and DC and destroying them on the other compressive faces by diffusion along the paths shown.

On a tensile face AB the stress exerts a force σb^2 (or $\sigma \Omega^{2/3}$) on each surface atom and so does work $\sigma b^2 \times b$ each time an atom moves forward one atomic spacing b (or $\Omega^{1/3}$) to create a vacancy. The energy of vacancy formation at such a face is thus reduced to $(E_f - \sigma b^3)$ and the concentration of vacancies in equilibrium correspondingly increased to $c_\tau = \exp\left[(-E_f + \sigma b^3)/kT\right] = c_0 \exp \sigma b^3/kT$). The vacancy concentration on the compressive faces will be reduced to $c_c = c_0 \exp\left(-\sigma b^3/kT\right)$. Vacancies will therefore flow down the concentration gradient, and the number crossing a face under tension to one under compression will be given by Fick's law as

$$\phi = -D_v d^2 (c_T - c_c)/\alpha d$$

where D_v is the vacancy diffusivity and α relates to the diffusion length. Substituting for c_T, c_c and $D = (D_v c_0 b^3)$ leads to

$$\phi = 2dD \sinh(\sigma b^3/\mathbf{k}T)/\alpha b^3$$

Each vacancy created on one face and annihilated on the other produces a strain $\varepsilon = b^3/d^3$, so that the creep strain-rate $\dot{\varepsilon} = \phi(b^3/d^3)$. At high temperatures and low stresses this reduces to

$$\dot{\varepsilon}_{H-N} = 2D\sigma b^3/\alpha d^2 \mathbf{k}T = B_{H-N} D\sigma\Omega/d^2 \mathbf{k}T \quad (7.56)$$

where the constant $B_{H-N} \sim 10$.

In contrast to dislocation creep, Herring–Nabarro creep varies linearly with stress and occurs at $T \approx 0.8T_m$ with $\sigma \approx 10^6$ N/m^2. The temperature range over which vacancy-diffusion creep is significant can be extended to much lower temperatures (i.e. $T \approx 0.5T_m$) if the vacancies flow down the grain boundaries rather than through the grains. Equation (7.56) is then modified for Coble or grain boundary diffusion creep, and is given by

$$\dot{\varepsilon}_{Coble} = B_c D_{gb} \sigma\Omega\omega/\mathbf{k}T d^3 \quad (7.57)$$

where ω is the width of the grain boundary. Under such conditions (i.e. $T \approx 0.5$ to $0.6T_m$ and low stresses) diffusion creep becomes an important creep mechanism in a number of high-technology situations, and has been clearly identified in magnesium-based canning materials used in gas-cooled reactors.

7.9.3 Tertiary creep and fracture

Tertiary creep and fracture are logically considered together, since the accelerating stage represents the initiation of conditions which lead to fracture. In many cases the onset of accelerating creep is an indication that voids or cracks are slowly but continuously forming in the material, and this has been confirmed by metallography and density measurements. The type of fracture resulting from tertiary creep is not transcrystalline but grain boundary fracture. Two types of grain boundary fracture have been observed. The first occurs principally at the triple point formed where three grain boundaries meet, and sliding along boundaries on which there is a shear stress produces stress concentrations at the point of conjunction sufficiently high to start cracks. However, under conditions of slow strain rate for long times, which would be expected to favour recovery, small holes form on grain boundaries, especially those perpendicular to the tensile axis, and these gradually grow and coalesce.

Second-phase particles play an important part in the nucleation of cracks and cavities by concentrating stress in sliding boundaries and at the intersection of slip bands with particles but these stress concentrations are greatly reduced by plastic deformation by power-law creep and by diffusional processes. Cavity formation and early growth is therefore intimately linked to the creep process itself and the time-to-fracture correlates well with the minimum creep rate for many structural materials. Fracture occurs when the larger, more closely spaced cavities coalesce. Creep fracture is discussed further in Chapter 8.

7.9.4 Creep-resistant alloy design

The problem of the design of engineering creep-resistant alloys is complex, and the optimum alloy for a given service usually contains several constituents in various states of solution and precipitation. Nevertheless, it is worth considering some of the principles underlying creep-resistant behaviour in the light of the preceding theories.

First, let us consider the strengthening of the solid solution by those mechanisms which cause dislocation locking and those which contribute to lattice friction hardening. The former include solute atoms interacting with (1) the dislocation or (2) the stacking fault. Friction hardening can arise from (1) the stress fields around individual atoms (i.e. the Mott–Nabarro effect), (2) clusters of solute atoms in solid solutions, (3) by increasing the separation of partial dislocations and so making climb, cross-slip and intersection more difficult, (4) by the solute atoms becoming attached to jogs and thereby impeding climb, and (5) by influencing the energies of formation and migration of vacancies. The alloy can also be hardened by precipitation, and it is significant that many of the successful industrial creep-resistant alloys are of this type (e.g. the nickel alloys, and both ferritic and austenitic steels).

The effectiveness of these various methods of conferring strength on the alloy will depend on the conditions of temperature and stress during creep. All the effects should play some part during fast primary creep, but during the slow secondary creep stage the impeding of dislocation movement by solute locking effects will probably be small. This is because modern creep-resistant alloys are in service up to temperatures of about two-thirds the absolute melting point $(T/T_m \simeq \frac{2}{3})$ of the parent metal, whereas above about $T/T_m \simeq \frac{1}{2}$ solute atoms will migrate as fast as dislocations. Hardening which relies on clusters will be more difficult to remove than that which relies upon single atoms and should be effective up to higher temperatures. However, for any hardening mechanism to be really effective, whether it is due to solute atom clusters or actual precipitation, the rate of climb and cross-slip past the barriers must be slow. Accordingly, the most probable role of solute alloying elements in modern creep-resistant alloys is in reducing the rate of climb and cross-slip processes. The three hardening mechanisms listed as 3, 4 and 5 above are all effective in this way. From this point of view, it is clear that the best parent metals on which to base creep-resistant alloys will be those in which climb and cross-slip is difficult; these include the fcc and cph metals of low stacking-fault energy, for which the slip dislocations readily dissociate. Generally, the creep rate is described by the empirical relation

$$\dot{\varepsilon} = A(\sigma/E)^n (\gamma)^m D \qquad (7.58)$$

where A is a constant, n, m stress and fault energy exponents, respectively, and D the diffusivity; for fcc materials $m \approx 3$ and $n \approx 4$. The reason for the good creep strength of austenitic and Ni-base materials containing Co, Cr, etc. arises from their low fault energy and also because of their relatively high melting point when D is small.

From the above discussion it appears that a successful creep-resistant material would be an alloy, the composition of which gives a structure with a hardened solid–solution matrix containing a sufficient number of precipitated particles to force glissile partial dislocations either to climb or to cross-slip to circumvent them. The constitution of the *Nimonic* alloys, which consist of a nickel matrix containing dissolved chromium, titanium, aluminium and cobalt, is in accordance with these principles, and since no large atomic size factors are involved it appears that one of the functions of these additions is to lower the stacking-fault energy and thus widen the separation of the partial dislocations. A second object of the titanium and aluminium alloy additions[1] is to produce precipitation, and in the *Nimonic* alloys much of the precipitate is Ni_3Al. This precipitate is isomorphous with the matrix, and while it has a parameter difference ($\approx \frac{1}{2}\%$) small enough to give a low interfacial energy, it is, nevertheless, sufficiently large to give a source of hardening. Thus, since the energy of the interface provides the driving force for particle growth, this low-energy interface between particle and matrix ensures a low rate of particle growth and hence a high service temperature.

Grain boundary precipitation is advantageous in reducing grain boundary sliding. Alternatively, the weakness of the grain boundaries may be eliminated altogether by using single-crystal material. *Nimonic* alloys used for turbine blades have been manufactured in single-crystal form by directional solidification (see Chapters 3 and 10).

Dispersions are effective in conferring creep strength by two mechanisms. First the particle will hinder a dislocation and force it to climb and cross-slip. Second, and more important, is the retarding effect on recovery as shown by some dispersions, $Cu-Al_2O_3$ (extruded), SAP (sintered alumina powder), and $Ni-ThO_2$ which retain their hardness almost to the melting point. A comparison of SAP with a 'conventional' complex aluminium alloy shows that at 250°C there is little to choose between them but at 400°C SAP is several times stronger. Generally, the dislocation network formed by strain-hardening interconnects the particles and is thereby anchored by them. To do this effectively, the particle must be stable at the service temperature and remain finely dispersed. This depends on the solubility C, diffusion coefficient D and interfacial energy γ_1, since the time to dissolve the particle is $t = r^4 kT/DC\gamma_1 R^2$. In precipitation-hardening alloys, C is appreciable, and D offers little scope for adjustment; great importance is therefore placed on γ_1 as for the Ni_3 (TiAl) phase in *Nimonics* where it is very low.

Figure 7.63 shows that $n \approx 4$ both above and below $0.5T_m$ for the $Ni-Al_2O_3$ and $Ni-Co-Al_2O_3$ alloys that were completely recrystallized, which contrasts with values very much greater than 4 for extruded TD nickel and other dispersion-strengthened alloys[1] containing a dislocation substructure. This demonstrates the importance of substructure and probably indicates that in completely recrystallized alloys containing a dispersoid, the particles control the creep behaviour, whereas in alloys containing a substructure the dislocation content is more important. Since $n \approx 4$ for the Ni– and $Ni-Co-Al_2O_3$ alloys in both temperature regimes, the operative deformation mechanism is likely to be the same, but it is clear from the activation energies, listed in Table 7.4, that the rate-controlling thermally activated process changes with temperature. The activation energy is greater at the higher temperature when it is also, surprisingly, composition (or stacking-fault energy) independent.

Such behaviour may be explained, if it is assumed that the particles are bypassed by cross-slip (see Chapter 8) and this process is easy at all temperatures, but it is the climb of the edge segments of the cross-slipped dislocations that is rate-controlling. At low temperatures, climb would proceed by pipe-diffusion

[1] The chromium forms a spinel with NiO and hence improves the oxidation resistance.

[1] To analyse these it is generally necessary to introduce a threshold (or friction) stress σ_0, so that the effective stress is $(\sigma - \sigma_0)$.

Table 7.4 Experimentally-determined parameters from creep of $Ni-Al_2O_3$ and $Ni-Co-Al_2O_3$ alloys

Test temperature Alloy	773 K		1000 K		
	$Q(kJ\,mol^{-1})$	$A(s^{-1})$	$Q(kJ\,mol^{-1})$	$A(s^{-1})$	A/D_0
Ni	85	1.67×10^{16}	276	1.1×10^{28}	5.5×10^{28}
Ni–67% Co	121	9.95×10^{19}	276	2.2×10^{28}	5.8×10^{28}

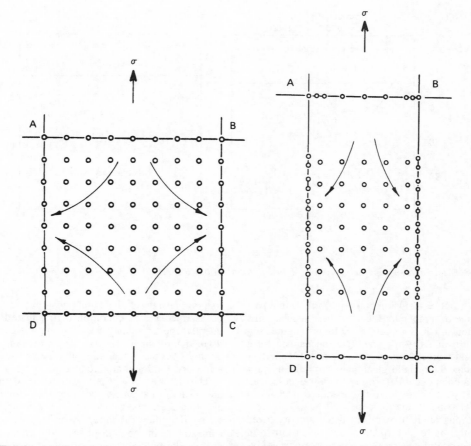

Figure 7.67 *Schematic diagram showing the distribution of second-phase particles before and after diffusion creep.*

so that the composition-dependence relates to the variation in the ease of pipe-diffusion along dislocations of different widths. At high temperatures, climb occurs by bulk diffusion and the absence of any composition-dependence is due to the fact that in these alloys the jog distribution is determined mainly by dislocation/particle interactions and not, as in single-phase alloys and in dispersion-strengthened alloys containing a substructure, by the matrix stacking-fault energy. The optimum creep resistance of dispersion-strengthened alloys is produced when a uniform dislocation network in a fibrous grain structure is anchored by the particles and recovery is minimized. Such a structure can reduce the creep rate by several orders of magnitude from that given in Figure 7.63, but it depends critically upon the working and heat-treatment used in fabricating the alloy.

Second-phase particles can also inhibit diffusion creep. Figure 7.67 shows the distribution of particles before and after diffusion creep and indicates that the longitudinal boundaries tend to collect precipitates as vacancies are absorbed and the boundaries migrate inwards, while the tensile boundaries acquire

a PFZ. Such a structural change has been observed in Mg−0.5%Zr (*Magnox ZR55*) at 400°C and is accompanied by a reduced creep rate. It is not anticipated that diffusion is significantly affected by the presence of particles and hence the effect is thought to be due to the particles affecting the vacancy-absorbing capabilities of the grain boundaries. Whatever mechanism is envisaged for the annihilation of vacancies at a grain boundary, the climb-glide of grain boundary dislocations is likely to be involved and such a process will be hindered by the presence of particles.

7.10 Deformation mechanism maps

The discussion in this chapter has emphasized that over a range of stress and temperature an alloy is capable of deforming by several alternative and independent mechanisms, e.g. dislocation creep with either pipe diffusion at low temperatures and lattice diffusion at high temperatures being the rate-controlling mechanism, and diffusional creep with either grain-boundary

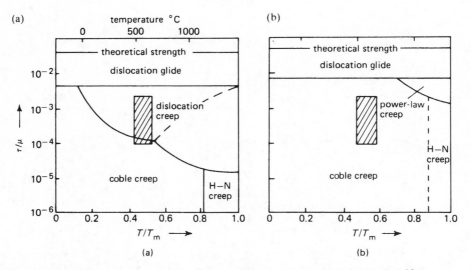

Figure 7.68 *Deformation-mechanism maps for (a) nickel, (b) nickel-based superalloy (after M. F. Ashby).*

diffusion or lattice diffusion being important. In a particular range of temperature, one of these mechanisms is dominant and it is therefore useful in engineering application to identify the operative mechanism for a given stress–temperature condition, since it is ineffective to change the metallurgical factors to influence, for example, a component deforming by power-law creep controlled by pipe diffusion if the operative mechanism is one of Herring–Nabarro creep.

The various alternative mechanisms are displayed conveniently on a deformation-mechanism map in which the appropriate stress, i.e. shear stress or equivalent stress, compensated by modulus on a log scale, is plotted against homologous temperature T/T_m as shown in Figure 7.68 for nickel and a nickel-based superalloy with a grain size of 100 μm. By comparing the diagrams it is evident that solid solution strengthening and precipitation-hardening have raised the yield stress and reduced the dislocation creep field. The shaded boxes shown in Figure 7.68 indicate the typical stresses and temperatures to which a turbine blade would be subjected; it is evident that the mechanism of creep during operation has changed and, indeed, the creep rate is reduced by several orders of magnitude.

7.11 Metallic fatigue

7.11.1 Nature of fatigue failure

The term fatigue applies to the behaviour of a metal which, when subjected to a cyclically variable stress of sufficient magnitude (often below the yield stress) produces a detectable change in mechanical properties. In practice, a large number of service failures are due to fatigue, and so engineers are concerned mainly with fatigue failure where the specimen is actually separated into two parts. Some of these failures can be attributed

to poor design of the component, but in some can be ascribed to the condition of the material. Consequently, the treatment of fatigue may be conveniently divided into three aspects: (1) engineering considerations, (2) gross metallurgical aspects, and (3) fine-scale structural and atomic changes.

The fatigue conditions which occur in service are usually extremely complex. Common failures are found in axles where the eccentric load at a wheel or pulley produces a varying stress which is a maximum in the skin of the axle. Other examples, such as the flexure stresses produced in aircraft wings and in undercarriages during ground taxi-ing, do, however, emphasize that the stress system does not necessarily vary in a regular sinusoidal manner. The series of aircraft disasters attributed to pressurized-cabin failures is perhaps the most spectacular example of this type of fatigue failure.

7.11.2 Engineering aspects of fatigue

In laboratory testing of materials the stress system is usually simplified, and both the Woehler and push-pull type of test are in common use. The results are usually plotted on the familiar $S–N$ curve (i.e. stress versus the number of cycles to failure, usually plotted on a logarithmic scale). Ferritic steels may be considered to exhibit a genuine fatigue limit with a fatigue ratio $S/\text{TS} \approx 0.5$. However, other materials, such as aluminium or copper-based alloys, certainly those of the age-hardening variety, definitely do not show a sharp discontinuity in the $S–N$ curve. For these materials no fatigue limit exists and all that can be specified is the endurance limit at N cycles. The importance of the effect is illustrated by the behaviour of commercial aluminium-based alloys containing zinc, magnesium and copper. Such an alloy may have a TS of

617 MN/m^2 but the fatigue stress for a life of 10^8 cycles is only 154 MN/m^2 (i.e. a fatigue ratio at 10^8 cycles of 0.25).

The amplitude of the stress cycle to which the specimen is subjected is the most important single variable in determining its life under fatigue conditions, but the performance of a material is also greatly affected by various other conditions, which may be summarized as follows:

1. *Surface preparation* Since fatigue cracks frequently start at or near the surface of the component, the surface condition is an important consideration in fatigue life. The removal of machining marks and other surface irregularities invariably improves the fatigue properties. Putting the surface layers under compression by shot peening or surface treatment improves the fatigue life.

2. *Effect of temperature* Temperature affects the fatigue properties in much the same way as it does the tensile strength (TS); the fatigue strength is highest at low temperatures and decreases gradually with rising temperature. For mild steel the ratio of fatigue limit to TS remains fairly constant at about 0.5, while the ratio of fatigue limit to yield stress varies over much wider limits. However, if the temperature is increased above about 100°C, both the tensile strength and the fatigue strength of mild steel show an increase, reaching a maximum value between 200°C and 400°C. This increase, which is not commonly found in other materials, has been attributed to strain-ageing.

3. *Frequency of stress cycle* In most metals the frequency of the stress cycle has little effect on the fatigue life, although lowering the frequency usually results in a slightly reduced fatigue life. The effect becomes greater if the temperature of the fatigue test is raised, when the fatigue life tends to depend on the total time of testing rather than on the number of cycles. With mild steel, however, experiments show that the normal speed effect is reversed in a certain temperature range and the number of cycles to failure increases with decrease in the frequency of the stress cycle. This effect may be correlated with the influence of temperature and strain-rate on the TS. The temperature at which the tensile strength reaches a maximum depends on the rate of strain, and it is, therefore, not surprising that the temperature at which the fatigue strength reaches a maximum depends on the cyclic frequency.

4. *Mean stress* For conditions of fatigue where the mean stress, i.e.

$$\Delta\sigma N_f^a = (\sigma_{max} + \sigma_{min})/2$$

does not exceed the yield stress σ_y, then the relationship

$$\Delta\sigma N_f^a = \text{const.} \qquad (7.59)$$

known as Basquin's law, holds over the range 10^2 to $\approx 10^5$ cycles, i.e. N less than the knee of the S–N curve, where $a \approx \frac{1}{10}$ and N_f the number of cycles to failure. For low cycle fatigue with $\Delta\sigma > \sigma_y$ then Basquin's law no longer holds, but a reasonable relationship

$$\Delta\varepsilon_p N_f^b = D^b = \text{const.} \qquad (7.60)$$

known as the Coffin–Manson law, is found where $\Delta\varepsilon_p$ is the plastic strain range, $b \approx 0.6$, and D is the ductility of the material. If the mean stress becomes tensile a lowering of the fatigue limit results. Several relationships between fatigue limit and mean stress have been suggested, as illustrated in Figure 7.69a. However, there is no theoretical reason why a material should follow any given relationship and the only safe rule on which to base design is to carry out prior tests on the material concerned to determine its behaviour under conditions similar to those it will meet in service. Another common engineering relationship frequently used, known as Miner's concept

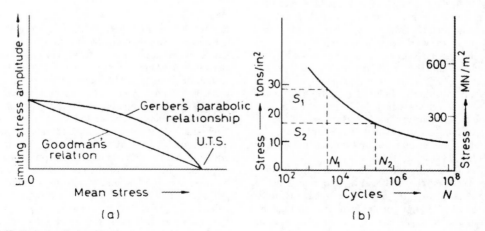

Figure 7.69 *Fatigue relationships.*

of cumulative damage, is illustrated in Figure 7.69b. This hypothesis states that damage can be expressed in terms of the number of cycles applied divided by the number to produce failure at a given stress level. Thus, if a maximum stress of value S_1 is applied to a specimen for n_1 cycles which is less than the fatigue life N_1, and then the maximum stress is reduced to a value equal to S_2, the specimen is expected to fail after n_2 cycles, since according to Miner the following relationship will hold

$$n_1/N_1 + n_2/N_2 + \ldots = \Sigma n/N = 1 \qquad (7.61)$$

5. *Environment* Fatigue occurring in a corrosive environment is usually referred to as corrosion fatigue. It is well known that corrosive attack by a liquid medium can produce etch pits which may act as notches, but when the corrosive attack is simultaneous with fatigue stressing, the detrimental effect is far greater than just a notch effect. Moreover, from microscopic observations the environment appears to have a greater effect on crack propagation than on crack initiation. For most materials even atmospheric oxygen decreases the fatigue life by influencing the speed of crack propagation, and it is possible to obtain a relationship between fatigue life and the degree of vacuum in which the specimen has been held.

It is now well established that fatigue starts at the surface of the specimen. This is easy to understand in the Woehler test because, in this test, it is there that the stress is highest. However, even in push–pull fatigue, the surface is important for several reasons: (1) slip is easier at the surface than in the interior of the grains, (2) the environment is in contact with the surface, and (3) any specimen misalignment will always give higher stresses at the surface. Accordingly, any alteration in surface properties must bring about a change in the fatigue properties. The best fatigue resistance occurs in materials with a worked surface layer produced by polishing with emery, shot-peening or skin-rolling the surface. This beneficial effect of a worked surface layer is principally due to the fact that the surface is put into compression, but the increased TS as a result of work hardening also plays a part. Electropolishing the specimen by removing the surface layers usually has a detrimental effect on the fatigue properties, but other common surface preparations such as nitriding and carburizing, both of which produce a surface layer which is in compression, may be beneficial. Conversely, such surface treatments as the decarburizing of steels and the cladding of aluminium alloys with pure aluminium, increase their susceptibility to fatigue.

The alloy composition and thermal and mechanical history of the specimen are also of importance in the fatigue process. Any treatment which increases the hardness or yield strength of the material will increase the level of the stress needed to produce slip and, as we shall see later, since the fundamental processes

of fatigue are largely associated with slip, this leads directly to an increase in fatigue strength. It is also clear that grain size is a relevant factor: the smaller the grain size, the higher is the fatigue strength at a given temperature.

The fatigue processes in stable alloys are essentially the same as those of pure metals but there is, of course, an increase in fatigue strength. However, the processes in unstable alloys and in materials exhibiting a yield point are somewhat different. In fatigue, as in creep, structural instability frequently leads to enhancement of the fundamental processes. In all cases the approach to equilibrium is more complete, so that in age-hardening materials, solution-treated specimens become harder and fully aged specimens become softer. The changes which occur are local rather than general, and are associated with the enhanced diffusion brought about by the production of vacancies during the fatigue test. Clearly, since vacancy mobility is a thermally activated process such effects can be suppressed at sufficiently low temperatures.

In general, non-ferrous alloys do not exhibit the type of fatigue limit shown by mild steel. One exception to this generalization is the alloy aluminium 2–7% magnesium, 0.5% manganese, and it is interesting to note that this alloy also has a sharp yield point and shows Lüders markings in an ordinary tensile test. Accordingly, it has been suggested that the fatigue limit occupies a similar place in the field of alternating stresses to that filled by the yield point in unidirectional stressing. Stresses above the fatigue limit readily unlock the dislocations from their solute atom atmospheres, while below the fatigue limit most dislocations remain locked. In support of this view, it is found that when the carbon and nitrogen content of mild steel is reduced, by annealing in wet hydrogen, striking changes take place in the fatigue limit (Figure 7.5) as well as in the sharp yield point.

7.11.3 Structural changes accompanying fatigue

Observations of the structural details underlying fatigue hardening show that in polycrystals large variations in slip-band distributions and the amount of lattice misorientation exist from one grain to another. Because of such variations it is difficult to typify structural changes, so that in recent years this structural work has been carried out more and more on single crystals; in particular, copper has received considerable attention as being representative of a typical metal. Such studies have now established that fatigue occurs as a result of slip, the direction of which changes with the stress cycle, and that the process continues throughout the whole of the test (shown, for example, by interrupting a test and removing the slip bands by polishing; the bands reappear on subsequent testing).

Moreover, four stages in the fatigue life of a specimen are distinguishable; these may be summarized as follows. In the early stages of the test, the whole of the specimen hardens. After about 5% of the life, slip becomes localized and persistent slip bands appear; they are termed persistent because they reappear and are not permanently removed by electropolishing. Thus, reverse slip does not continue throughout the whole test in the bulk of the metal (the matrix). Electron microscope observations show that metal is extruded from the slip bands and that fine crevices called intrusions are formed within the band. During the third stage of the fatigue life the slip bands grow laterally and become wider, and at the same time cracks develop in them. These cracks spread initially along slip bands, but in the later stages of fracture the propagation of the crack is often not confined to certain crystallographic directions and catastrophic rupture occurs. These two important crack growth stages, i.e. stage I in the slip band and stage II roughly perpendicular to the principal stress, are shown in Figure 7.70 and are influenced by the formation of localized (persistent) slip bands (i.e. PSBs). However, PSBs are not clearly defined in low stacking fault energy, solid solution alloys.

Cyclic stressing therefore produces plastic deformation which is not fully reversible and the build-up of dislocation density within grains gives rise to fatigue hardening with an associated structure which is characteristic of the strain amplitude and the ability of the dislocations to cross-slip, i.e. temperature and SFE. The non-reversible flow at the surface leads to intrusions, extrusions and crack formation in PSBs. These two aspects will now be considered separately and in greater detail.

Fatigue hardening If a single or polycrystalline specimen is subjected to many cycles of alternating stress, it becomes harder than a similar specimen extended uni-directionally by the same stress applied only once. This may be demonstrated by stopping the fatigue test and performing a static tensile test on the specimen when, as shown in Figure 7.71, the yield stress is increased. During the process, persistent slip bands appear on the surface of the specimen and it is in such bands that cracks eventually form. The behaviour of a fatigue-hardened specimen has

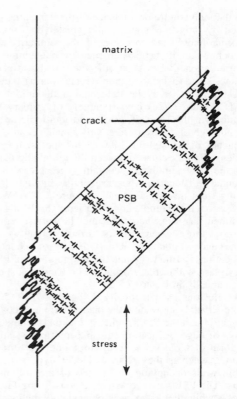

Figure 7.70 *Persistent slip band (PSB) formation in fatigue, and stage I and stage II crack growth.*

two unusual features when compared with an ordinary work-hardened material. The fatigue-hardened material, having been stressed symmetrically, has the same yield stress in compression as in tension, whereas the work-hardened specimen (e.g. prestrained in tension) exhibits a Bauschinger effect, i.e. weaker in compression than tension. It arises from the fact that the obstacles behind the dislocation are weaker than those resisting further dislocation motion, and the pile-up stress causes it to slip back under a reduced load in the reverse direction. The other important feature is

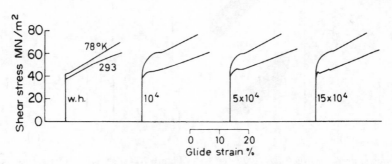

Figure 7.71 *Stress–strain curves for copper after increasing amounts of fatigue testing (after Broom and Ham, 1959).*

that the temperature-dependence of the hardening produced by fatigue is significantly greater than that of work-hardening and, because of the similarity with the behaviour of metals hardened by quenching and by irradiation, it has been attributed to the effect of vacancies and dislocation loops created during fatigue.

At the start of cyclic deformation the initial slip bands (Figure 7.72a) consist largely of primary dislocations in the form of dipole and multipole arrays; the number of loops is relatively small because the frequency of cross-slip is low. As the specimen work-hardens slip takes place between the initial slip bands, and the new slip bands contain successively more secondary dislocations because of the internal stress arising from nearby slip bands (Figure 7.72b). When the specimen is completely filled with slip bands, the specimen has work-hardened and the softest regions are now those where slip occurred originally since these bands contain the lowest density of secondary dislocations. Further slip and the development of PSBs takes place within these original slip bands, as shown schematically in Figure 7.72c.

As illustrated schematically in Figure 7.73, TEM of copper crystals shows that the main difference between the matrix and the PSBs is that in the matrix the dense arrays of edge dislocation (di- and multipoles) are in the form of large veins occupying about 50% of the volume, whereas they form a 'ladder'-type structure within walls occupying about 10% of the volume in PSBs. The PSBs are the active regions in the fatigue process while the matrix is associated with the inactive parts of the specimen between the PSBs. Steady-state deformation then takes place by the to-and-fro glide of the same dislocations in the matrix, whereas an equilibrium between dislocation multiplication and

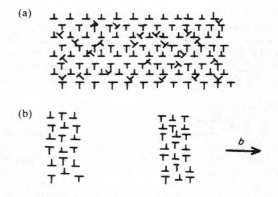

Figure 7.73 *Schematic diagram showing (a) vein structure of matrix and (b) ladder structure of PSBs.*

annihilation exists in the PSBs. Multiplication occurs by bowing-out of the walls and annihilation takes place by interaction with edge dislocations of opposite sign ($\approx 75b$ apart) on glide planes in the walls and of screw dislocations ($\approx 200b$ apart) on glide planes in the low-dislocation channels, the exact distance depending on the ease of cross-slip.

7.11.4 Crack formation and fatigue failure

Extrusions, intrusions and fatigue cracks can be formed at temperatures as low as 4 K where thermally activated movement of vacancies does not take place. Such observations indicate that the formation of intrusions and cracks cannot depend on either chemical or thermal action and the mechanism must be a purely geometrical process which depends on cyclic stressing.

Two general mechanisms have been suggested. The first, the Cottrell 'ratchet' mechanism, involves the use of two different slip systems with different directions and planes of slip, as is shown schematically in Figure 7.74. The most favoured source (e.g. S_1 in Figure 7.74a) produces a slip step on the surface at P during a tensile half-cycle. At a slightly greater stress in the same half-cycle, the second source S_2 produces

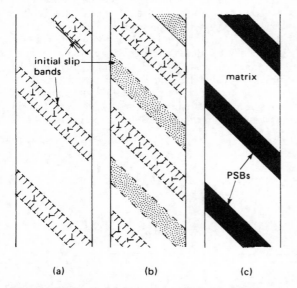

Figure 7.72 *Formation of persistent slip bands (PSBs) during fatigue.*

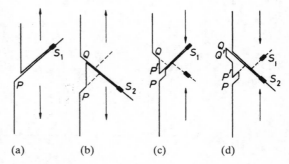

Figure 7.74 *Formation of intrusions and extrusions (after Cottrell; courtesy of John Wiley and Sons).*

a second step at Q (Figure 7.74b). During the compression half-cycle, the source S_1 produces a surface step of opposite sign at P' (Figure 7.74c), but, owing to the displacing action of S_2, this is not in the same plane as the first and thus an intrusion is formed. The subsequent operation of S_2 produces an extrusion at QQ' (Figure 7.74d) in a similar manner. Such a mechanism requires the operation of two slip systems and, in general, predicts the occurrence of intrusions and extrusions with comparable frequency, but not in the same slip band.

The second mechanism, proposed by Mott, involves cross-slip resulting in a column of metal extruded from the surface and a cavity is left behind in the interior of the crystal. One way in which this could happen is by the cyclic movement of a screw dislocation along a closed circuit of crystallographic planes, as shown in Figure 7.75. During the first half-cycle the screw dislocation glides along two faces $ABCD$ and $BB'C'C$ of the band, and during the second half-cycle returns along the faces $B'C'A'D$ and $A'D'DA$. Unlike the Cottrell mechanism this process can be operated with a single slip direction, provided cross-slip can occur.

Neither mechanism can fully explain all the experimental observations. The interacting slip mechanism predicts the occurrence of intrusions and extrusions with comparable frequency but not, as is often found, in the same slip band. With the cross-slip mechanism, there is no experimental evidence to show that cavities exist beneath the material being extruded. It may well be that different mechanisms operate under different conditions.

In a polycrystalline aggregate the operation of several slip modes is necessary and intersecting slip unavoidable. Accordingly, the widely differing fatigue behaviour of metals may be accounted for by the relative ease with which cross-slip occurs. Thus, those factors which affect the onset of stage III in the work-hardening curve will also be important in fatigue, and conditions suppressing cross-slip would, in general, increase the resistance to fatigue failure, i.e. low stacking-fault energy and low temperatures. Aluminium would be expected to have poor fatigue properties on this basis but the unfavourable fatigue characteristics of the high-strength aluminium alloys

is probably also due to the unstable nature of the alloy and to the influence of vacancies.

In pure metals and alloys, transgranular cracks initiate at intrusions in PSBs or at sites of surface roughness associated with emerging planar slip bands in low SFE alloys. Often the microcrack forms at the PSB-matrix interface where the stress concentration is high. In commercial alloys containing inclusions or second-phase particles, the fatigue behaviour depends on the particle size. Small particles ≈ 0.1 μm can have beneficial effects by homogenizing the slip pattern and delaying fatigue-crack nucleation. Larger particles reduce the fatigue life by both facilitating crack nucleation by slip band/particle interaction and increasing crack growth rates by interface decohesion and voiding within the plastic zone at the crack tip. The formation of voids at particles on grain boundaries can lead to intergranular separation and crack growth. The preferential deformation of 'soft' precipitate-free zones (PFZs) associated with grain boundaries in age-hardened alloys also provides a mechanism of intergranular fatigue-crack initiation and growth. To improve the fatigue behaviour it is therefore necessary to avoid PFZs and obtain a homogeneous deformation structure and uniform precipitate distribution by heat-treatment; localized deformation in PFZs can be restricted by a reduction in grain size.

From the general appearance of a typical fatigue fracture, shown in Figure 7.76, one can distinguish two distinct regions. The first is a relatively smooth area, through which the fatigue crack has spread slowly. This area usually has concentric marks about the point of origin of the crack which correspond to the positions at which the crack was stationary for some period. The remainder of the fracture surface shows a typically rough transcrystalline fracture where the failure has been catastrophic. Electron micrographs of the relatively smooth area show that this surface is covered with more or less regular contours perpendicular to the direction of the propagation front. These fatigue striations represent the successive positions of the propagation front and are spaced further apart the higher the

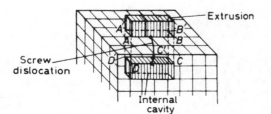

Figure 7.75 *Formation of an extrusion and associated cavity by the Mott mechanism.*

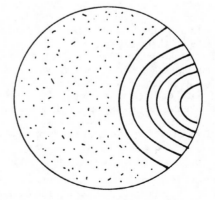

Figure 7.76 *A schematic fatigue fracture.*

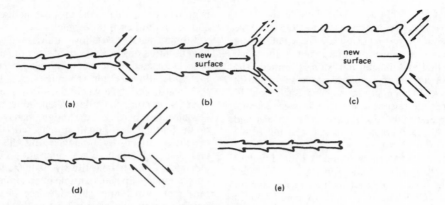

Figure 7.77 *Schematic illustration of the formation of fatigue striations.*

velocity of propagation. They are rather uninfluenced by grain boundaries and in metals where cross-slip is easy (e.g. mild steel or aluminium) may be wavy in appearance. Generally, the lower the ductility of the material, the less well defined are the striations.

Stage II growth is rate-controlling in the fatigue failure of most engineering components, and is governed by the stress intensity at the tip of the advancing crack. The striations seen on the fracture surface may form by a process of plastic blunting at the tip of the crack, as shown in Figure 7.77. In (a) the crack under the tensile loading part of the cycle generates shear stresses at the tip. With increasing tensile load the crack opens up and new surface is created (b), separation occurs in the slip band and 'ears' are formed at the end of the crack. The plastic deformation causes the crack to be both extended and blunted (c). On the compressive part of the cycle the crack begins to close (d), the shear stresses are reversed and with increasing load the crack almost closes (e). In this part of the cycle the new surface folds and the ears correspond to the new striations on the final fracture surface. A one-to-one correlation therefore exists between the striations and the opening and closing with ear formation. Crack growth continues in this manner until it is long enough to cause the final instability when either brittle or ductile (due to the reduced cross-section not being able to carry the load) failure occurs. In engineering alloys, rather than pure metals, which contain inclusions or second-phase particles, cracking or voiding occurs ahead of the original crack tip rather than in the ears when the tensile stress or strain reaches a critical value. This macroscopic stage of fracture is clearly of importance to engineers in predicting the working life of a component and has been successfully treated by the application of fracture mechanics as discussed in Chapter 8.

7.11.5 Fatigue at elevated temperatures

At ambient temperature the fatigue process involves intracrystalline slip and surface initiation of cracks, followed by transcrystalline propagation. When fatigued at elevated temperatures $\geqslant 0.5T_m$, pure metals and solid solutions show the formation of discrete cavities on grain boundaries, which grow, link up and finally produce failure. It is probable that vacancies produced by intracrystalline slip give rise to a supersaturation which causes the vacancies to condense on those grain boundaries that are under a high shear stress where the cavities can be nucleated by a sliding or ratchet mechanism. It is considered unlikely that grain boundary sliding contributes to cavity growth, increasing the grain size decreases the cavity growth because of the change in boundary area. *Magnox* (Mg) and alloys used in nuclear reactors up to $0.75T_m$ readily form cavities, but the high-temperature nickel-base alloys do not show intergranular cavity formation during fatigue at temperatures within their normal service range, because intracrystalline slip is inhibited by γ' precipitates. Above about $0.7T_m$, however, the γ' precipitates coarsen or dissolve and fatigue then produce cavities and eventually cavity failure.

Further reading

Argon, A. (1969). *The Physics of Strength and Plasticity*. MIT Press, Cambridge, MA.

Cottrell, A. H. (1964). *Mechanical Properties of Matter*. John Wiley, Chichester.

Cottrell, A. H. (1964). *The Theory of Crystal Dislocations*. Blackie, Glasgow.

Dislocations and Properties of Real Metals (1984). Conf. Metals Society.

Evans, R. W. and Wilshire, B. (1993). *Introduction to Creep*. Institute of Materials, London.

Freidel, J. (1964). *Dislocations*. Pergamon Press, London.

Hirsch, P. B. (ed.). (1975). *The Physics of Metals. 2. Defects*. Cambridge University Press, Cambridge.

Hirth, J. P. and Lothe, J. (1984). *Theory of Dislocations*. McGraw-Hill, New York.

Chapter 8

Strengthening and toughening

8.1 Introduction

The production of materials which possess considerable strength at both room and elevated temperatures is of great practical importance. We have already seen how alloying, solute-dislocation interaction, grain size control and cold-working can give rise to an increased yield stress. Of these methods, refining the grain size is of universal application to materials in which the yield stress has a significant dependence upon grain size. In certain alloy systems, it is possible to produce an additional increase in strength and hardness by heat-treatment alone. Such a method has many advantages, since the required strength can be induced at the most convenient stage of production or fabrication; moreover, the component is not sent into service in a highly stressed, plastically deformed state. The basic requirement for such a special alloy is that it should undergo a phase transformation in the solid state. One type of alloy satisfying this requirement, already considered, is that which can undergo an order–disorder reaction; the hardening accompanying this process (similar in many ways to precipitation-hardening) is termed order-hardening. However, conditions for this form of hardening are quite stringent, so that the two principal hardening methods, commonly used for alloys, are based upon (1) precipitation from a supersaturated solid solution and (2) eutectoid decomposition.

In engineering applications, strength is, without doubt, an important parameter. However, it is by no means the only important one and usually a material must provide a combination of properties. Some ductility is generally essential, enabling the material to relieve stress concentrations by plastic deformation and to resist fracture. The ability of materials to resist crack propagation and fracture, known generally as toughness, will be discussed in this chapter. Fracture can take many forms; some special forms, such as brittle fracture by cleavage, ductile fracture by microvoid coalescence, creep fracture by triple-point cracking and fatigue cracking, will be examined.

This chapter primarily concerns alloy behaviour, partly because of the inherent versatility of alloy systems and partly because the research background to much of the current understanding of strength, toughness and fracture is essentially metallurgical. However, it is often possible to extend the basic principles to non-metallic materials, particularly in the case of fracture processes. This will be apparent later, in Chapter 10, when we describe how the unique transformation characteristics of zirconia can be used to inhibit crack propagation in a brittle ceramic such as alumina. Methods for toughening glasses are described in the same chapter. In Chapter 11 we consider the strengthening and toughening effects produced when plastics, metals and ceramics are reinforced with filaments to form composite materials.

8.2 Strengthening of non-ferrous alloys by heat-treatment

8.2.1 Precipitation-hardening of Al–Cu alloys

8.2.1.1 Precipitation from supersaturated solid solution

The basic requirements of a precipitation-hardening alloy system is that the solid solubility limit should decrease with decreasing temperature as shown in Figure 8.1 for the Al–Cu system. During the precipitation-hardening heat-treatment procedure the alloy is first solution heat-treated at the high temperature and then rapidly cooled by quenching into water or some other cooling medium. The rapid cooling suppresses the separation of the θ-phase so that the alloy exists at the low temperature in an unstable supersaturated state. If, however, after quenching, the alloy is allowed to 'age' for a sufficient length of time, the second phase precipitates out. This precipitation occurs by a nucleation and growth process, fluctuations in solute concentration providing small clusters of atoms in the lattice which act as nuclei for the precipitate. However, the size of the precipitate becomes finer as the temperature at which precipitation occurs is lowered, and extensive hardening of the alloy is associated with a critical dispersion of the precipitate. If, at any given temperature, ageing is allowed to proceed too far, coarsening of the particles occurs (i.e. the small ones tend to redissolve, and the large ones to grow still larger as discussed in Section 8.2.6) and the numerous finely dispersed, small particles are gradually replaced by a smaller number of more widely dispersed, coarser particles. In this state the alloy becomes softer, and it is then said to be in the over-aged condition (see Figure 8.2).

8.2.1.2 Changes in properties accompanying precipitation

The actual quenching treatment gives rise to small changes in many of the mechanical and physical properties of alloys because both solute atoms and point defects in excess of the equilibrium concentration are retained during the process, and because the quench itself often produces lattice strains. Perhaps the property most markedly affected is the electrical resistance and this is usually considerably increased. In contrast, the mechanical properties are affected relatively much less.

On ageing, the change in properties in a quenched material is more marked and, in particular, the mechanical properties often show striking modifications. For example, the tensile strength of *Duralumin* (i.e. an aluminium−4% copper alloy containing magnesium, silicon and manganese) may be raised from 0.21 to 0.41 GN/m^2 while that of a Cu−2Be alloy may be increased from 0.46 to

1.23 GN/m^2. The structure-sensitive properties such as hardness, yield stress, etc. are, of course, extremely dependent on the structural distribution of the phases and, consequently, such alloys usually exhibit softening as the finely dispersed precipitates coarsen.

A simple theory of precipitation, involving the nucleation and growth of particles of the expected new equilibrium phase, leads one to anticipate that the alloy would show a single hardening peak, the electrical resistivity a decrease, and the lattice parameter an increase (assuming the solute atom is smaller than the solvent atom) as the solute is removed from solution. Such property changes are found in practice, but only at low supersaturations and high ageing temperatures. At higher supersaturations and lower ageing temperatures the various property changes are not consistent with such a simple picture of precipitation; the alloy may show two or more age-hardening peaks, and the electrical resistivity and lattice parameter may not change in the anticipated manner. A hardening process which takes place in two stages is shown in aluminium−copper alloys (Figure 8.2a) where the initial hardening occurs without any attendant precipitation being visible in the light microscope and, moreover, is accompanied by a decrease in conductivity and no change in lattice parameter. Such behaviour may be accounted for if precipitation is a process involving more than one stage. The initial stage of precipitation, at the lower ageing temperatures, involves a clustering of solute atoms on the solvent lattice sites to form zones or clusters, coherent with the matrix; the zones cannot be seen in the light microscope and for this reason this stage was at one time termed pre-precipitation. At a later stage of the ageing process these clusters break away from the matrix lattice to form distinct particles with their own crystal structure and a definite interface. These hypotheses were confirmed originally by structural studies using X-ray diffraction techniques but nowadays the so-called pre-precipitation effects can be observed directly in the electron microscope.

Even though clustering occurs, the general kinetic behaviour of the precipitation process is in agreement with that expected on thermodynamic grounds. From Figure 8.2 it is evident that the rate of ageing increases markedly with increasing temperature while the peak hardness decreases. Two-stage hardening takes place at low ageing temperatures and is associated with high maximum hardness, while single-stage hardening occurs at higher ageing temperatures, or at lower ageing temperatures for lower solute contents.

Another phenomenon commonly observed in precipitation-hardening alloys is reversion or retrogression. If an alloy hardened by ageing at low temperature is subsequently heated to a higher ageing temperature it softens temporarily, but becomes harder again on more prolonged heating. This temporary softening, or reversion of the hardening process, occurs because the very small nuclei or zones precipitated at the low temperature are unstable when raised to the higher ageing temperature, and consequently they redissolve and the

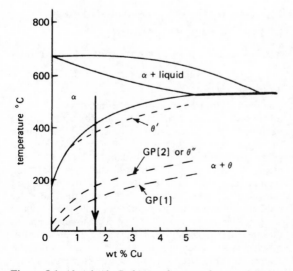

Figure 8.1 *Al-rich Al−Cu binary diagram showing GP [1], θ″ and θ′ solvus lines (dotted).*

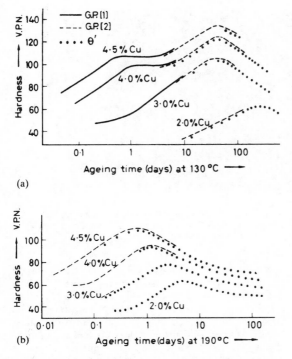

(a)

(b)

Figure 8.2 *The ageing of aluminium–copper alloys at (a) 130°C and (b) at 190°C (after Silcock, Heal and Hardy, 1953–4).*

alloy becomes softer; the temperature above which the nuclei or zones dissolve is known as the solvus temperature; Figure 8.1 shows the solvus temperatures for GP zones, θ'', θ' and θ. On prolonged ageing at the higher temperature larger nuclei, characteristic of that temperature, are formed and the alloy again hardens. Clearly, the reversion process is reversible, provided re-hardening at the higher ageing temperature is not allowed to occur.

8.2.1.3 Structural changes during precipitation

Early metallographic investigations showed that the microstructural changes which occur during the initial stages of ageing are on too fine a scale to be resolved by the light microscope, yet it is in these early stages that the most profound changes in properties are found. Accordingly, to study the process, it is necessary to employ the more sensitive and refined techniques of X-ray diffraction and electron microscopy.

The two basic X-ray techniques, important in studying the regrouping of atoms during the early stages of ageing, depend on the detection of radiation scattered away from the main diffraction lines or spots (see Chapter 5). In the first technique, developed independently by Guinier and Preston in 1938, the Laue method is used. They found that the single-crystal diffraction pattern of an aluminium–copper

alloy developed streaks extending from an aluminium lattice reflection along $\langle 100 \rangle_{Al}$ directions. This was attributed to the formation of copper-rich regions of plate-like shape on $\{100\}$ planes of the aluminium matrix (now called Guinier–Preston zones or GP zones). The net effect of the regrouping is to modify the scattering power of, and spacing between, very small groups of $\{100\}$ planes throughout the crystal. However, being only a few atomic planes thick, the zones produce the diffraction effect typical of a two-dimensional lattice, i.e. the diffraction spot becomes a diffraction streak. In recent years the Laue method has been replaced by a single-crystal oscillation technique employing monochromatic radiation, since interpretation is made easier if the wavelength of the X-rays used is known. The second technique makes use of the phenomenon of scattering of X-rays at small angles (see Chapter 5). Intense small-angle scattering can often be observed from age-hardening alloys (as shown in Figures 8.3 and 8.5) because there is usually a difference in electron density between the precipitated zone and the surrounding matrix. However, in alloys such as aluminium–magnesium or aluminium–silicon the technique is of no value because in these alloys the small difference in scattering power between the aluminium and silicon or magnesium atoms, respectively, is insufficient to give rise to appreciable scattering at small angles.

With the advent of the electron microscope the ageing of aluminium alloys was one of the first subjects to be investigated with the thin-foil transmission method. Not only can the detailed structural changes which occur during the ageing process be followed, but electron diffraction pictures taken from selected areas of the specimen while it is still in the microscope enable further important information on the structure of the precipitated phase to be obtained. Moreover, under some conditions the interaction of moving dislocations and precipitates can be observed. This naturally leads to a more complete understanding of the hardening mechanism.

Both the X-ray and electron-microscope techniques show that in virtually all age-hardening systems the initial precipitate is not the same structure as the equilibrium phase. Instead, an ageing sequence: zones → intermediate precipitates → equilibrium precipitate is followed. This sequence occurs because the equilibrium precipitate is incoherent with the matrix, whereas the transition structures are either fully coherent, as in the case of zones, or at least partially coherent. Then, because of the importance of the surface energy and strain energy of the precipitate to the precipitation process, the system follows such a sequence in order to have the lowest free energy in all stages of precipitation. The surface energy of the precipitates dominates the process of nucleation when the interfacial energy is large (i.e. when there is a discontinuity in atomic structure, somewhat like a grain boundary, at the interface between the nucleus and the matrix), so that for the incoherent type of precipitate the nuclei must exceed a

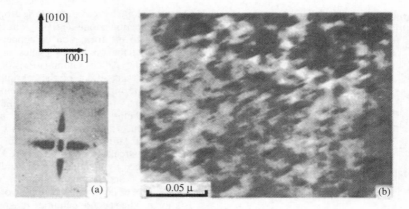

Figure 8.3 *(a) Small-angle X-ray pattern from aluminium–4% copper single crystal taken with molybdenum Kα radiation at a sample to film distance of 4 cm (after Guinier and Fournet, 1955; courtesy of John Wiley and Sons). (b) Electron micrograph of aluminium–4% copper aged 16 hours at 130°C, showing GP [1] zones (after Nicholson, Thomas and Nutting, 1958–9).*

certain minimum size before they can nucleate a new phase. To avoid such a slow mode of precipitation a coherent type of precipitate is formed instead, for which the size effect is relatively unimportant. The condition for coherence usually requires the precipitate to strain its equilibrium lattice to fit that of the matrix, or to adopt a metastable lattice. However, in spite of both a higher volume free energy and a higher strain energy, the transition structure is more stable in the early stages of precipitation because of its lower interfacial energy.

When the precipitate does become incoherent the alloy will, nevertheless, tend to reduce its surface energy as much as possible, by arranging the orientation relationship between the matrix and the precipitate so that the crystal planes which are parallel to, and separated by, the bounding surface have similar atomic spacings. Clearly, for these habit planes, as they are called, the better the crystallographic match, the less will be the distortion at the interface and the lower the surface energy. This principle governs the precipitation of many alloy phases, as shown by the frequent occurrence of the Widmanstätten structure, i.e. plate-shaped precipitates lying along prominent crystallographic planes of the matrix. Most precipitates are plate-shaped because the strain energy factor is least for this form.

The existence of a precipitation sequence is reflected in the ageing curves and, as we have seen in Figure 8.2, often leads to two stages of hardening. The zones, by definition, are coherent with the matrix, and as they form the alloy becomes harder. The intermediate precipitate may be coherent with the matrix, in which case a further increase of hardness occurs, or only partially coherent, when either hardening or softening may result. The equilibrium precipitate is incoherent and its formation always leads to softening. These features are best illustrated by a consideration of some actual age-hardening systems.

Precipitation reactions occur in a wide variety of alloy systems as shown in Table 8.1. The aluminium–copper alloy system exhibits the greatest number of intermediate stages in its precipitation process, and consequently is probably the most widely studied. When the copper content is high and the ageing temperature low, the sequence of stages followed is GP [1], GP [2], θ' and θ (CuAl$_2$). On ageing at higher temperatures, however, one or more of these intermediate stages may be omitted and, as shown in Figure 8.2, corresponding differences in the hardness curves can be detected. The early stages of ageing are due to GP [1] zones, which are interpreted as plate-like clusters of copper atoms segregated onto {1 0 0} planes of the aluminium matrix. A typical small-angle X-ray scattering pattern and thin-foil transmission electron micrograph from GP [1] zones are shown in Figure 8.3. The plates are only a few atomic planes thick (giving rise to the ⟨1 0 0⟩ streaks in the X-ray pattern), but are about 10 nm long, and hence appear as bright or dark lines on the electron micrograph.

GP [2] is best described as a coherent intermediate precipitate rather than a zone, since it has a definite crystal structure; for this reason the symbol θ'' is often preferred. These precipitates, usually of maximum thickness 10 nm and up to 150 nm diameter, have a tetragonal structure which fits perfectly with the aluminium unit cell in the a and b directions but not in the c. The structure postulated has a central plane which consists of 100% copper atoms, the next two planes a mixture of copper and aluminium and the other two basal planes of pure aluminium, giving an overall composition of CuAl$_2$. Because of their size, θ'' precipitates are easily observed in the electron microscope, and because of the ordered arrangements of copper and aluminium atoms within the structure, their presence gives rise to intensity maxima on the diffraction streaks in an X-ray photograph.

Table 8.1 Some common precipitation-hardening systems

Base metal	Solute	Transition structure	Equilibrium precipitate
Al	Cu	(i) Plate-like solute rich GP [1] zones on $\{1\,0\,0\}_{Al}$; (ii) ordered zones of GP [2]; (iii) θ'-phase (plates).	θ-CuAl$_2$
	Ag	(i) Spherical solute-rich zones; (ii) platelets of hexagonal γ' on $\{1\,1\,1\}_{Al}$.	γ-Ag$_2$Al
	Mg, Si	(i) GP zones rich in Mg and Si atoms on $\{1\,0\,0\}_{Al}$ planes; (ii) ordered zones of β'.	β-Mg$_2$Si (plates)
	Mg, Cu	(i) GP zones rich in Mg and Cu atoms on $\{1\,0\,0\}_{Al}$ planes; (ii) S$'$ platelets on $\{0\,2\,1\}_{Al}$ planes.	S-Al$_2$CuMg (laths)
	Mg, Zn	(i) Spherical zones rich in Mg and Zn; (ii) platelets of η' phase on $\{1\,1\,1\}_{Al}$.	η-MgZn$_2$ (plates)
Cu	Be	(i) Be-rich regions on $\{1\,0\,0\}_{Cu}$ planes; (ii) γ'.	γ-CuBe
	Co	Spherical GP zones.	β-Co plates
Fe	C	(i) Martensite (α'); (ii) martensite (α''); (iii) ε-carbide.	Fe$_3$C plates cementite
	N	(i) Nitrogen martensite (α'); (ii) martensite (α'') discs.	Fe$_4$N
Ni	Al, Ti	γ' cubes	γ-Ni$_3$(AlTi)

Since the c parameter 0.78 nm differs from that of aluminium 0.404 nm the aluminium planes parallel to the plate are distorted by elastic coherency strains. Moreover, the precipitate grows with the c direction normal to the plane of the plate, so that the strain fields become larger as it grows and at peak hardness extend from one precipitate particle to the next (see Figure 8.4a). The direct observation of coherency strains confirms the theories of hardening based on the development of an elastically strained matrix (see next section).

The transition structure θ' is tetragonal; the true unit cell dimensions are $a = 0.404$ and $c = 0.58$ nm and the axes are parallel to $\langle 1\,0\,0 \rangle_{Al}$ directions. The strains around the θ' plates can be relieved, however, by the formation of a stable dislocation loop around the precipitate and such a loop has been observed around small θ' plates in the electron microscope as shown in Figure 8.4b. The long-range strain fields of the precipitate and its dislocation largely cancel. Consequently, it is easier for glide dislocations to move through the lattice of the alloy containing an incoherent precipitate such as θ' than a coherent precipitate such as θ'', and the hardness falls.

The θ structure is also tetragonal, with $a = 0.606$ and $c = 0.487$ nm. This equilibrium precipitate is incoherent with the matrix and its formation always leads to softening, since coherency strains disappear.

8.2.2 Precipitation-hardening of Al–Ag alloys

Investigations using X-ray diffraction and electron microscopy have shown the existence of three distinct stages in the age-hardening process, which may be summarized: silver-rich clusters $\rightarrow$ intermediate

hexagonal $\gamma' \rightarrow$ equilibrium hexagonal γ. The hardening is associated with the first two stages in which the precipitate is coherent and partially coherent with the matrix, respectively.

During the quench and in the early stages of ageing, silver atoms cluster into small spherical aggregates and a typical small-angle X-ray picture of this stage, shown in Figure 8.5a, has a diffuse ring surrounding the trace of the direct beam. The absence of intensity in the centre of the ring (i.e. at $(0\,0\,0)$) is attributed to the fact that clustering takes place so rapidly that there is left a shell-like region surrounding each cluster which is low in silver content. On ageing, the clusters grow in size and decrease in number, and this is characterized by the X-ray pattern showing a gradual decrease in ring diameter. The concentration and size of clusters can be followed very accurately by measuring the intensity distribution across the ring as a function of ageing time. This intensity may be represented (see Chapter 5) by an equation of the form

$$l(\varepsilon) = Mn^2[\exp{(-2\pi^2 R^2 \varepsilon^2 / 3\lambda^2)}$$
$$- \exp{(-2\pi^2 R_1^2 \varepsilon^2 / 3\lambda^2)}]^2 \qquad (8.1)$$

and for values of ε greater than that corresponding to the maximum intensity, the contribution of the second term, which represents the denuded region surrounding the cluster, can be neglected. Figure 8.5b shows the variation in the X-ray intensity, scattered at small angles (SAS) with cluster growth, on ageing an aluminium–silver alloy at 120°C. An analysis of this intensity distribution, using equation (8.1), indicates that the size of the zones increases from 2 to 5 nm in just a few hours at 120°C. These zones may, of course, be seen in the electron microscope and Figure 8.6a

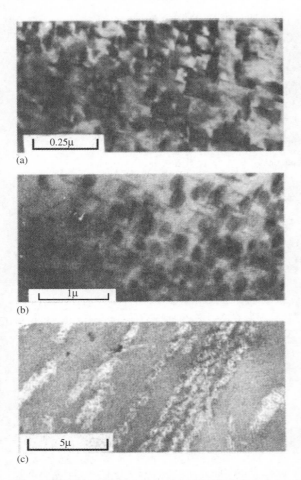

(a)

(b)

(c)

Figure 8.4 *Electron micrographs from Al–4Cu (a) aged 5 hours at 160°C showing θ″ plates, (b) aged 12 hours at 200°C showing a dislocation ring round θ″ plates, (c) aged 3 days at 160°C showing θ″ precipitated on helical dislocations (after Nicholson, Thomas and Nutting, 1958–9).*

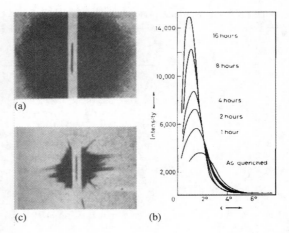

(a)

(c)

(b)

Figure 8.5 *Small-angle scattering of Cu Kα radiation by polycrystalline Al–Ag. (a) After quenching from 520°C (after Guinier and Walker, 1953). (b) The change in ring intensity and ring radius on ageing at 120°C (after Smallman and Westmacott, unpublished). (c) After ageing at 140°C for 10 days (after Guinier and Walker, 1953).*

is an electron micrograph showing spherical zones in an aluminium–silver alloy aged 5 hours at 160°C; the diameter of the zones is about 10 nm in good agreement with that deduced by X-ray analysis. The zone shape is dependent upon the relative diameters of solute and solvent atoms. Thus, solute atoms such as silver and zinc which have atomic sizes similar to aluminium give rise to spherical zones, whereas solute atoms such as copper which have a high misfit in the solvent lattice form plate-like zones.

With prolonged annealing, the formation and growth of platelets of a new phase, γ', occur. This is characterized by the appearance in the X-ray pattern of short streaks passing through the trace of the direct beam (Figure 8.5c). The γ' platelet lies parallel to the {1 1 1} planes of the matrix and its structure has lattice parameters very close to that of aluminium. However, the

structure is hexagonal and, consequently, the precipitates are easily recognizable in the electron microscope by the stacking fault contrast within them, as shown in Figure 8.6b. Clearly, these precipitates are never fully coherent with the matrix, but, nevertheless, in this alloy system, where the zones are spherical and have little or no coherency strain associated with them, and where no coherent intermediate precipitate is formed, the partially coherent γ' precipitates do provide a greater resistance to dislocation movement than zones and a second stage of hardening results.

The same principles apply to the constitutionally more complex ternary and quaternary alloys as to the binary alloys. Spherical zones are found in aluminium–magnesium–zinc alloys as in aluminium–zinc, although the magnesium atom is some 12% larger than the aluminium atom. The intermediate precipitate forms on the {1 1 1}$_{Al}$ planes, and is partially coherent with the matrix with little or no strain field associated with it. Hence, the strength of the alloy is due purely to dispersion hardening, and the alloy softens as the precipitate becomes coarser. In nickel-based alloys the hardening phase is the ordered γ'-Ni$_3$Al; this γ' is an equilibrium phase in the Ni–Al and Ni–Cr–Al systems and a metastable phase in Ni–Ti and Ni–Cr–Ti. These systems form the basis of the 'superalloys' (see Chapter 9) which owe their properties to the close matching of the γ' and the fcc matrix. The two phases have very similar lattice parameters (($\lesssim 0.25\%$), depending on composition) and the coherency (interfacial energy $\gamma_1 \approx$ 10–20 mJ/m^2) confers a very low coarsening rate on the precipitate so that the alloy overages extremely slowly even at $0.7T_m$.

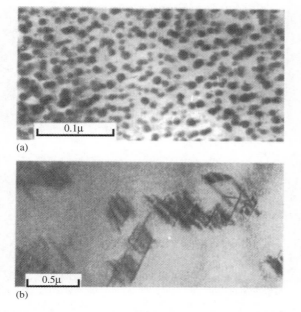

(a)

(b)

Figure 8.6 *Electron micrographs from Al–Ag alloy (a) aged 5 hours at 160°C showing spherical zones, and (b) aged 5 days at 160°C showing γ′ precipitate (after Nicholson, Thomas and Nutting, 1958–9).*

8.2.3 Mechanisms of precipitation-hardening

8.2.3.1 The significance of particle deformability

The strength of an age-hardening alloy is governed by the interaction of moving dislocations and precipitates. The obstacles in precipitation-hardening alloys which hinder the motion of dislocations may be either (1) the strains around GP zones, (2) the zones or precipitates themselves, or both. Clearly, if it is the zones themselves which are important, it will be necessary for the moving dislocations either to cut through them or go round them. Thus, merely from elementary reasoning, it would appear that there are at least three causes of hardening, namely: (1) coherency strain hardening, (2) chemical hardening, i.e. when the dislocation cuts through the precipitate, or (3) dispersion hardening, i.e. when the dislocation goes round or over the precipitate.

The relative contributions will depend on the particular alloy system but, generally, there is a critical dispersion at which the strengthening is a maximum, as shown in Figure 8.7. In the small-particle regime the precipitates, or particles, are coherent and deformable as the dislocations cut through them, while in the larger-particle regime the particles are incoherent and non-deformable as the dislocations bypass them. For deformable particles, when the dislocations pass through the particle, the intrinsic properties of the particle are of importance and alloy strength varies only weakly with particle size. For non-deformable

particles, when the dislocations bypass the particles, the alloy strength is independent of the particle properties but is strongly dependent on particle size and dispersion strength decreasing as particle size or dispersion increases. The transition from deformable to non-deformable particle-controlled deformation is readily recognized by the change in microstructure, since the 'laminar' undisturbed dislocation flow for the former contrasts with the turbulent plastic flow for non-deformable particles. The latter leads to the production of a high density of dislocation loops, dipoles and other debris which results in a high rate of work-hardening. This high rate of work-hardening is a distinguishing feature of all dispersion-hardened systems.

8.2.3.2 Coherency strain-hardening

The precipitation of particles having a slight misfit in the matrix gives rise to stress fields which hinder the movement of gliding dislocations. For the dislocations to pass through the regions of internal stress the applied stress must be at least equal to the average internal stress, and for spherical particles this is given by

$$\tau = 2\mu\varepsilon f \tag{8.2}$$

where μ is the shear modulus, ε is the misfit of the particle and f is the volume fraction of precipitate. This suggestion alone, however, cannot account for the critical size of dispersion of a precipitate at which the hardening is a maximum, since equation (8.2) is independent of L, the distance between particles. To explain this, Mott and Nabarro consider the extent to which a dislocation can bow round a particle under the action of a stress τ. Like the bowing stress of a Frank–Read source this is given by

$$r = \alpha\mu b/\tau \tag{8.3}$$

where r is the radius of curvature to which the dislocation is bent which is related to the particle spacing. Hence, in the hardest age-hardened alloys where the

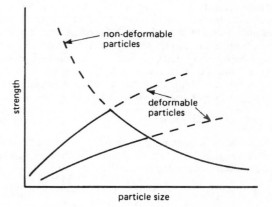

Figure 8.7 *Variation of strength with particle size, defining the deformable and non-deformable particle regimes.*

yield strength is about $\mu/100$, the dislocation can bend to a radius of curvature of about 100 atomic spacings, and since the distance between particles is of the same order it would appear that the dislocation can avoid the obstacles and take a form like that shown in Figure 8.8a. With a dislocation line taking up such a configuration, in order to produce glide, each section of the dislocation line has to be taken over the adverse region of internal stress without any help from other sections of the line — the alloy is then hard. If the precipitate is dispersed on too fine a scale (e.g. when the alloy has been freshly quenched or lightly aged) the dislocation is unable or bend sufficiently to lie entirely in the regions of low internal stress. As a result, the internal stresses acting on the dislocation line largely cancel and the force resisting its movement is small — the alloy then appears soft. When the dispersion is on a coarse scale, the dislocation line is able to move between the particles, as shown in Figure 8.8b, and the hardening is again small.

For coherency strain hardening the flow stress depends on the ability of the dislocation to bend and thus experience more regions of adverse stress than of aiding stress. The flow stress therefore depends on the treatment of averaging the stress, and recent attempts separate the behaviour of small and large coherent particles. For small coherent particles the flow stress is given by

$$\tau = 4.1\mu\varepsilon^{3/2} f^{1/2}(r/b)^{1/2} \tag{8.4}$$

which predicts a greater strengthening than the simple arithmetic average of equation (8.2). For large coherent particles

$$\tau = 0.7\mu f^{1/2}(\varepsilon b^3/r^3)^{1/4} \tag{8.5}$$

8.2.3.3 Chemical hardening

When a dislocation actually passes through a zone as shown in Figure 8.9 a change in the number of solvent–solute near-neighbours occurs across the slip plane. This tends to reverse the process of clustering and, hence, additional work must be done by the applied stress to bring this about. This process, known as chemical hardening, provides a short-range interaction between dislocations and precipitates and arises from three possible causes: (1) the energy required to create an additional particle/matrix interface with energy γ_1 per unit area which is provided by a stress

$$\tau \simeq \alpha\gamma_1^{3/2}(fr)^{1/2}/\mu b^2 \tag{8.6}$$

where α is a numerical constant, (2) the additional work required to create an antiphase boundary inside the particle with ordered structure, given by

$$\tau \simeq \beta\gamma_{apb}^{3/2}(fr)^{1/2}/\mu b^2 \tag{8.7}$$

where β is a numerical constant, and (3) the change in width of a dissociated dislocation as it passes

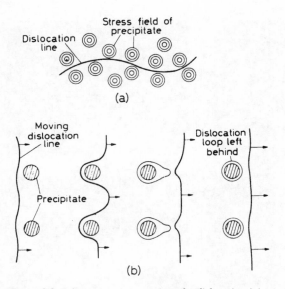

Figure 8.8 *Schematic representation of a dislocation (a) curling round the stress fields from precipitates and (b) passing between widely spaced precipitates (Orowan looping).*

through the particle where the stacking fault energy differs from the matrix (e.g. Al–Ag where $\Delta\gamma_{SF} \sim 100$ mJ/m^2 between Ag zones and Al matrix) so that

$$\tau \simeq \Delta\gamma_{SF}/b \tag{8.8}$$

Usually $\gamma_1 < \gamma_{apb}$ and so γ_1 can be neglected, but the ordering within the particle requires the dislocations to glide in pairs. This leads to a strengthening given by

$$\tau = (\gamma_{apb}/2b)[4\gamma_{apb}rf/\pi T)^{1/2} - f] \tag{8.9}$$

where T is the dislocation line tension.

8.2.3.4 Dispersion-hardening

In dispersion-hardening it is assumed that the precipitates do not deform with the matrix and that the yield stress is the stress necessary to expand a loop of dislocation between the precipitates. This will be given by the Orowan stress

$$\tau = \alpha\mu b/L \tag{8.10}$$

where L is the separation of the precipitates. As discussed above, this process will be important in the later stages of precipitation when the precipitate becomes incoherent and the misfit strains disappear. A moving dislocation is then able to bypass the obstacles, as shown in Figure 8.8b, by moving in the clean pieces of crystal between the precipitated particles. The yield stress decreases as the distance between the obstacles increases in the over-aged condition. However, even when the dispersion of the precipitate is coarse a greater applied stress is necessary to force a dislocation past the obstacles than would be the case if the

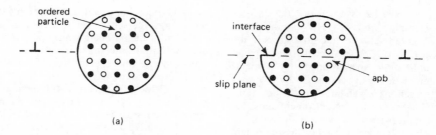

Figure 8.9 *Ordered particle (a) cut by dislocations in (b) to produce new interface and apb.*

obstruction were not there. Some particle or precipitate strengthening remains but the majority of the strengthening arises from the dislocation debris left around the particles giving rise to high work-hardening.

8.2.3.5 Hardening mechanisms in Al–Cu alloys

The actual hardening mechanism which operates in a given alloy will depend on several factors, such as the type of particle precipitated (e.g. whether zone, intermediate precipitate or stable phase), the magnitude of the strain and the testing temperature. In the earlier stages of ageing (i.e. before over-ageing) the coherent zones are cut by dislocations moving through the matrix and hence both coherency strain hardening and chemical hardening will be important, e.g. in such alloys as aluminium–copper, copper-beryllium and iron–vanadium–carbon. In alloys such as aluminium–silver and aluminium–zinc, however, the zones possess no strain field, so that chemical hardening will be the most important contribution. In the important high-temperature creep-resistant nickel alloys the precipitate is of the Ni_3Al form which has a low particle/matrix misfit and hence chemical hardening due to dislocations cutting the particles is again predominant. To illustrate that more than one mechanism of hardening is in operation in a given alloy system, let us examine the mechanical behaviour of an aluminium–copper alloy in more detail.

Figure 8.10 shows the deformation characteristics of single crystals of an aluminium–copper (nominally 4%) alloy in various structural states. The curves were obtained by testing crystals of approximately the same orientation, but the stress–strain curves from crystals containing GP [1] and GP [2] zones are quite different from those for crystals containing θ' or θ precipitates. When the crystals contain either GP [1] or GP [2] zones, the stress–strain curves are very similar to those of pure aluminium crystals, except that there is a two- or threefold increase in the yield stress. In contrast, when the crystals contain either θ' or θ precipitates the yield stress is less than for crystals containing zones, but the initial rate of work-hardening is extremely rapid. In fact, the stress–strain curves bear no similarity to those of a pure aluminium crystal. It is also observed that when θ' or θ is present as a precipitate, deformation does not take place on a single slip system but on several systems; the crystal then deforms, more nearly as a polycrystal does and the X-ray pattern develops extensive asterism. These factors are consistent with the high rate of work-hardening observed in crystals containing θ' or θ precipitates.

The separation of the precipitates cutting any slip plane can be deduced from both X-ray and electron-microscope observations. For the crystals, relating to Figure 8.10, containing GP [1] zones this value is 15 nm and for GP [2] zones it is 25 nm. It then follows from equation (8.3) that to avoid these precipitates the dislocations would have to bow to a radius of curvature of about 10 nm. To do this requires a stress several times greater than the observed flow stress and,

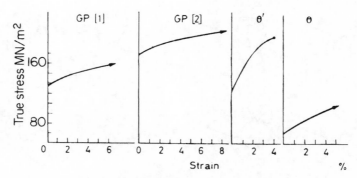

Figure 8.10 *Stress–strain curves from single crystals of aluminium–4% copper containing GP [1] zones, GP [2], zones, θ'-precipitates and θ-precipitates respectively (after Fine, Bryne and Kelly).*

in consequence, it must be assumed that the dislocations are forced through the zones. Furthermore, if we substitute the observed values of the flow stress in the relation $\mu b/\tau = L$, it will be evident that the bowing mechanism is unlikely to operate unless the particles are about 60 nm apart. This is confirmed by electron-microscope observations which show that dislocations pass through GP zones and coherent precipitates, but bypass non-coherent particles. Once a dislocation has cut through a zone, however, the path for subsequent dislocations on the same slip plane will be easier, so that the work-hardening rate of crystals containing zones should be low, as shown in Figure 8.10. The straight, well-defined slip bands observed on the surfaces of crystals containing GP [1] zones also support this interpretation.

If the zones possess no strain field, as in aluminium–silver or aluminium-zinc alloys, the flow stress would be entirely governed by the chemical hardening effect. However, the zones in aluminium copper alloys do possess strain fields, as shown in Figure 8.4, and, consequently, the stresses around a zone will also affect the flow stress. Each dislocation will be subjected to the stresses due to a zone at a small distance from the zone.

It will be remembered from Chapter 7 that temperature profoundly affects the flow stress if the barrier which the dislocations have to overcome is of a short-range nature. For this reason, the flow stress of crystals containing GP [1] zones will have a larger dependence on temperature than that of those containing GP [2] zones. Thus, while it is generally supposed that the strengthening effect of GP [2] zones is greater than that of GP [1], and this is true at normal temperatures (see Figure 8.10), at very low temperatures it is probable that GP [1] zones will have the greater strengthening effect due to the short-range interactions between zones and dislocations.

The θ' and θ precipitates are incoherent and do not deform with the matrix, so that the critical resolved shear stress is the stress necessary to expand a loop of dislocation between them. This corresponds to the over-aged condition and the hardening to dispersion-hardening. The separation of the θ particles is greater than that of the θ', being somewhat greater than 1 μm and the initial flow stress is very low. In both cases, however, the subsequent rate of hardening is high because, as suggested by Fisher, Hart and Pry, the gliding dislocation interacts with the dislocation loops in the vicinity of the particles (see Figure 8.8b). The stress–strain curves show, however, that the rate of work-hardening falls to a low value after a few per cent strain, and these authors attribute the maximum in the strain-hardening curve to the shearing of the particles. This process is not observed in crystals containing θ precipitates at room temperature and, consequently, it seems more likely that the particles will be avoided by cross-slip. If this is so, prismatic loops of dislocation will be formed at the particles, by the mechanism shown in Figure 8.11, and these will give approximately the same mean internal stress as that calculated by Fisher, Hart and Pry, but a reduced stress on the particle. The maximum in the work-hardening curve would then correspond to the stress necessary to expand these loops; this stress will be of the order of $\mu b/r$ where r is the radius of the loop which is somewhat greater than the particle size. At low temperatures cross-slip is difficult and the stress may be relieved either by initiating secondary slip or by fracture.

8.2.4 Vacancies and precipitation

It is clear that because precipitation is controlled by the rate of atomic migration in the alloy, temperature will have a pronounced effect on the process. Moreover, since precipitation is a thermally activated process, other variables such as time of annealing, composition, grain size and prior cold work are also important. However, the basic treatment of age-hardening alloys is solution treatment followed by quenching, and the introduction of vacancies by the latter process must play an important role in the kinetic behaviour.

It has been recognized that near room temperature, zone formation in alloys such as aluminium–copper and aluminium–silver occurs at a rate many orders of magnitude greater than that calculated from the

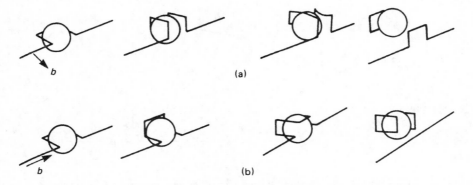

Figure 8.11 *Cross-slip of (a) edge and (b) screw dislocation over a particle producing prismatic loops in the process.*

diffusion coefficient of the solute atoms. In aluminium–copper, for example, the formation of zones is already apparent after only a few minutes at room temperature, and is complete after an hour or two, so that the copper atoms must therefore have moved through several atomic spacings in that time. This corresponds to an apparent diffusion coefficient of copper in aluminium of about 10^{-20}–10^{-22} m^2 s^{-1}, which is many orders of magnitude faster than the value of 5×10^{-29} m^2 s^{-1} obtained by extrapolation of high-temperature data. Many workers have attributed this enhanced diffusion to the excess vacancies retained during the quenching treatment. Thus, since the expression for the diffusion coefficient at a given temperature contains a factor proportional to the concentration of vacancies at that temperature, if the sample contains an abnormally large vacancy concentration then the diffusion coefficient should be increased by the ratio c_Q/c_o, where c_Q is the quenched-in vacancy concentration and c_o is the equilibrium concentration. The observed clustering rate can be accounted for if the concentration of vacancies retained is about 10^{-3}–10^{-4}.

The observation of loops by transmission electron microscopy allows an estimate of the number of excess vacancies to be made, and in all cases of rapid quenching the vacancy concentration in these alloys is somewhat greater than 10^{-4}, in agreement with the predictions outlined above. Clearly, as the excess vacancies are removed, the amount of enhanced diffusion diminishes, which agrees with the observations that the isothermal rate of clustering decreases continuously with increasing time. In fact, it is observed that D decreases rapidly at first and then remains at a value well above the equilibrium value for months at room temperature; the process is therefore separated into what is called the fast and slow reactions. A mechanism proposed to explain the slow reaction is that some of the vacancies quenched-in are trapped temporarily and then released slowly. Measurements show that the activation energy in the fast reaction (≈ 0.5 eV) is smaller than in the slow reaction (≈ 1 eV) by an amount which can be attributed to the binding energy between vacancies and trapping sites. These traps are very likely small dislocation loops or voids formed by the clustering of vacancies. The equilibrium matrix vacancy concentration would then be greater than that for a well-annealed crystal by a factor $\exp[\gamma\Omega/rkT]$, where γ is the surface energy, Ω the atomic volume and r the radius of the defect (see Chapter 4). The experimental diffusion rate can be accounted for if $r \approx 2$ nm, which is much smaller than the loops and voids usually seen, but they do exist. The activation energy for the slow reaction would then be $E_D - (\gamma\Omega/r)$ or approximately 1 eV for $r \approx 2$ nm.

Other factors known to affect the kinetics of the early stages of ageing (e.g. altering the quenching rate, interrupted quenching and cold work) may also be rationalized on the basis that these processes lead to different concentrations of excess vacancies. In general, cold working the alloy prior to ageing causes a decrease in the rate of formation of zones, which must mean that the dislocations introduced by cold work are more effective as vacancy sinks than as vacancy sources. Cold working or rapid quenching therefore have opposing effects on the formation of zones. Vacancies are also important in other aspects of precipitation-hardening. For example, the excess vacancies, by condensing to form a high density of dislocation loops, can provide nucleation sites for intermediate precipitates. This leads to the interesting observation in aluminium–copper alloys that cold working or rapid quenching, by producing dislocations for nucleation sites, have the same effect on the formation of the θ' phase but, as we have seen above, the opposite effect on zone formation. It is also interesting to note that screw dislocations, which are not normally favourable sites for nucleation, can also become sites for preferential precipitation when they have climbed into helical dislocations by absorbing vacancies, and have thus become mainly of edge character. The long arrays of θ' phase observed in aluminium–copper alloys, shown in Figure 8.4c, have probably formed on helices in this way. In some of these alloys, defects containing stacking faults are observed, in addition to the dislocation loops and helices, and examples have been found where such defects nucleate an intermediate precipitate having a hexagonal structure. In aluminium–silver alloys it is also found that the helical dislocations introduced by quenching absorb silver and degenerate into long narrow stacking faults on {1 1 1} planes; these stacking-fault defects then act as nuclei for the hexagonal γ' precipitate.

Many commercial alloys depend critically on the interrelation between vacancies, dislocations and solute atoms and it is found that trace impurities significantly modify the precipitation process. Thus trace elements which interact strongly with vacancies inhibit zone formation, e.g. Cd, In, Sn prevent zone formation in slowly quenched Al–Cu alloys for up to 200 days at 30°C. This delays the age-hardening process at room temperature which gives more time for mechanically fabricating the quenched alloy before it gets too hard, thus avoiding the need for refrigeration. On the other hand, Cd increases the density of θ' precipitate by increasing the density of vacancy loops and helices which act as nuclei for precipitation and by segregating to the matrix-θ' interfaces thereby reducing the interfacial energy.

Since grain boundaries absorb vacancies in many alloys there is a grain boundary zone relatively free from precipitation. The Al–Zn–Mg alloy is one commercial alloy which suffers grain boundary weakness but it is found that trace additions of Ag have a beneficial effect in refining the precipitate structure and removing the precipitate free grain boundary zone. Here it appears that Ag atoms stabilize vacancy clusters near the grain boundary and also increase the stability of the GP zone thereby raising the GP zone solvus temperature. Similarly, in the 'Concorde' alloy, *RR58* (basically Al–2.5Cu–1.2Mg with additions), Si addition (0.25Si) modifies the as-quenched dislocation distribution inhibiting the nucleation and growth of dislocation loops and reducing the diameter of helices. The S-precipitate (Al$_2$CuMg) is homogeneously nucleated in the presence of Si rather than heterogeneously nucleated at dislocations, and the precipitate grows directly from zones, giving rise to improved and more uniform properties.

Apart from speeding up the kinetics of ageing, and providing dislocations nucleation sites, vacancies may play a structural role when they precipitate cooperatively with solute atoms to facilitate the basic atomic arrangements required for transforming the parent crystal structure to that of the product phase. In essence, the process involves the systematic incorporation of excess vacancies, produced by the initial quench or during subsequent dislocation loop annealing, in a precipitate zone or plate to change the atomic stacking. A simple example of θ' formation in Al–Cu is shown schematically in Figure 8.12. Ideally, the structure of the θ'' phase in Al–Cu consists of layers of copper on {1 0 0} separated by three layers of aluminium atoms. If a next-nearest neighbour layer of aluminium atoms from the copper layer is removed by condensing a vacancy loop, an embryonic θ' unit cell with Al in the correct **AAA**... stacking sequence is formed (Figure 8.12b). Formation of the final CuAl$_2$ θ' fluorite structure requires only shuffling half of the copper atoms into the newly created next-nearest neighbour space and concurrent relaxation of the Al atoms to the correct θ' interplanar distances (Figure 8.12c).

The structural incorporation of vacancies in a precipitate is a non-conservative process since atomic sites are eliminated. There exist equivalent conservative processes in which the new precipitate structure is created from the old by the nucleation and expansion of partial dislocation loops with predominantly shear character. Thus, for example, the **BABAB** {1 0 0} plane stacking sequence of the fcc structure can be changed to **BAABA** by the propagation of an $a/2\langle 1\,0\,0\rangle$ shear loop in the {1 0 0} plane, or to **BAAAB** by the propagation of a pair of $a/2\langle 1\,0\,0\rangle$ partials of opposite sign on adjacent planes. Again, the **AAA** stacking resulting from the double shear is precisely that required for the embryonic formation of the fluorite structure from the fcc lattice.

In visualizing the role of lattice defects in the nucleation and growth of plate-shaped precipitates, a simple analogy with Frank and Shockley partial dislocation loops is useful. In the formation of a Frank loop, a layer of hcp material is created from the fcc lattice by the (non-conservative) condensation of a layer of vacancies in {1 1 1}. Exactly the same structure is formed by the (conservative) expansion of a Shockley partial loop on a {1 1 1} plane. In the former case a semi-coherent 'precipitate' is produced bounded by an $a/3\langle 1\,1\,1\rangle$ dislocation, and in the latter a coherent one bounded by an $a/6\langle 1\,1\,2\rangle$. Continued growth of precipitate plates occurs by either process or a combination of processes. Of course, formation of the final precipitate structure requires, in addition to these structural rearrangements, the long-range diffusion of the correct solute atom concentration to the growing interface.

The growth of a second-phase particle with a disparate size or crystal structure relative to the matrix is controlled by two overriding principles–the accommodation of the volume and shape change, and the optimized use of the available deformation mechanisms. In general, volumetric transformation strains are accommodated by vacancy or interstitial condensation, or prismatic dislocation loop punching, while deviatoric strains are relieved by shear loop propagation. An example is shown in Figure 8.13. The formation of semi-coherent Cu needles in Fe–1%Cu is accomplished by the generation of shear loops in

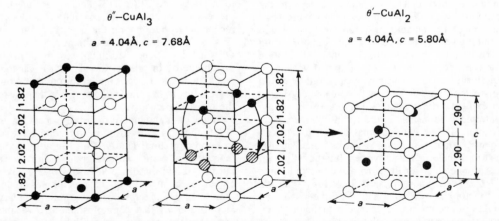

θ''–CuAl$_3$

$a = 4.04\text{Å}, c = 7.68\text{Å}$

θ'–CuAl$_2$

$a = 4.04\text{Å}, c = 5.80\text{Å}$

Figure 8.12 *Schematic diagram showing the transition of θ'' to θ' in Al–Cu by the vacancy mechanism. Vacancies from annealing loops are condensed on a next-nearest Al plane from the copper layer in θ'' to form the required AAA Al stacking. Formation of the θ' fluorite structure then requires only slight redistribution of the copper atom layer and relaxation of the Al layer spacings (courtesy of K. H. Westmacott).*

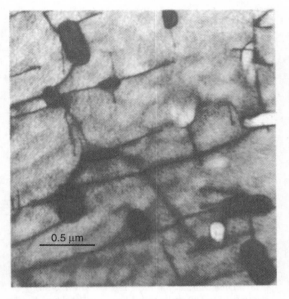

Figure 8.13 *The formation of semicoherent Cu needles in Fe–1% Cu (courtesy of K. H. Westacott).*

the precipitate/matrix interface. Expansion of the loops into the matrix and incorporation into nearby precipitate interfaces leads to a complete network of dislocations interconnecting the precipitates.

8.2.5 Duplex ageing

In non-ferrous heat-treatment there is considerable interest in double (or duplex) ageing treatments to obtain the best microstructure consistent with optimum properties. It is now realized that it is unlikely that the optimum properties will be produced in alloys of the precipitation-hardening type by a single quench and ageing treatment. For example, while the interior of grains may develop an acceptable precipitate size and density, in the neighbourhood of efficient vacancy sinks, such as grain boundaries, a precipitate-free zone (PFZ) is formed which is often associated with over-ageing in the boundary itself. This heterogeneous structure gives rise to poor properties, particularly under stress corrosion conditions.

Duplex ageing treatments have been used to overcome this difficulty. In Al–Zn–Mg, for example, it was found that storage at room temperature before heating to the ageing temperature leads to the formation of finer precipitate structure and better properties. This is just one special example of two-step or multiple ageing treatments which have commercial advantages and have been found to be applicable to several alloys. Duplex ageing gives better competitive mechanical properties in Al-alloys (e.g. Al–Zn–Mg alloys) with much enhanced corrosion resistance since the grain boundary zone is removed. It is possible to obtain strengths of $267–308$ MN/m^2 in Mg–Zn–Mn alloys

which have very good strength/weight ratio applications, and nickel alloys also develop better properties with multiple ageing treatments.

The basic idea of all heat-treatments is to 'seed' a uniform distribution of stable nuclei at the low temperature which can then be grown to optimum size at the higher temperature. In most alloys, there is a critical temperature T_c above which homogeneous nucleation of precipitate does not take place, and in some instances has been identified with the GP zone solvus. On ageing above T_c there is a certain critical zone size above which the zones are able to act as nuclei for precipitates and below which the zones dissolve.

In general, the ageing behaviour of Al–Zn–Mg alloys can be divided into three classes which can be defined by the temperature ranges involved:

1. Alloys quenched and aged above the GP zone solvus (i.e. the temperature above which the zones dissolve, which is above ~155°C in a typical Al–Zn–Mg alloy). Then, since no GP zones are ever formed during heat treatment, there are no easy nuclei for subsequent precipitation and a very coarse dispersion of precipitates results with nucleation principally on dislocations.
2. Alloys quenched and aged below the GP zone solvus. GP zones form continuously and grow to a size at which they are able to transform to precipitates. The transformation will occur rather more slowly in the grain boundary regions due to the lower vacancy concentration there but since ageing will always be below the GP zone solvus, no PFZ is formed other than a very small (~30 nm) solute-denuded zone due to precipitation in the grain boundary.
3. Alloys quenched below the GP zone solvus and aged above it (e.g. quenched to room temperature and aged at 180°C for Al–Zn–Mg). This is the most common practical situation. The final dispersion of precipitates and the PFZ width are controlled by the nucleation treatment below 155°C where GP zone size distribution is determined. A long nucleation treatment gives a fine dispersion of precipitates and a narrow PFZ.

It is possible to stabilize GP zones by addition of trace elements. These have the same effect as raising T_c, so that alloys are effectively aged below T_c. One example is Ag to Al–Zn–Mg which raises T_c from 155°C to 185°C, another is Si to Al–Cu–Mg, another Cu to Al–Mg–Si and yet another Cd or Sn to Al–Cu alloys. It is then possible to get uniform distribution and optimum properties by single ageing, and is an example of achieving by chemistry what can similarly be done with physics during multiple ageing. Whether it is best to alter the chemistry or to change the physics for a given alloy usually depends on other factors (e.g. economics).

8.2.6 Particle-coarsening

With continued ageing at a given temperature, there is a tendency for the small particles to dissolve and the resultant solute to precipitate on larger particles causing them to grow, thereby lowering the total interfacial energy. This process is termed particle-coarsening, or sometimes Ostwald ripening. The driving force for particle growth is the difference between the concentration of solute (S_r) in equilibrium with small particles of radius r and that in equilibrium with larger particles. The variation of solubility with surface curvature is given by the Gibbs–Thomson or Thomson–Freundlich equation

$$\ln(S_r/S) = 2\gamma\Omega/\mathbf{k}Tr \qquad (8.11)$$

where S is the equilibrium concentration, γ the particle/matrix interfacial energy and Ω the atomic volume; since $2\gamma\Omega \ll \mathbf{k}Tr$ then $S_r = S[1 + 2\gamma\Omega/\mathbf{k}Tr]$.

To estimate the coarsening rate of a particle it is necessary to consider the rate-controlling process for material transfer. Generally, the rate-limiting factor is considered to be diffusion through the matrix and the rate of change of particle radius is then derived from the equation

$$4\pi r^2 (\mathrm{d}r/\mathrm{d}t) = D4\pi R^2 (\mathrm{d}S/\mathrm{d}R)$$

where $\mathrm{d}S/\mathrm{d}R$ is the concentration gradient across an annulus at a distance R from the particle centre. Rewriting the equation after integration gives

$$\mathrm{d}r/\mathrm{d}t = -D(S_r - S_a)/r \qquad (8.12)$$

where S_a is the average solute concentration a large distance from the particle and D is the solute diffusion coefficient. When the particle solubility is small, the total number of atoms contained in particles may be assumed constant, independent of particle size distribution. Further consideration shows that

$$(S_a - S_r) = \{2\gamma\Omega S/\mathbf{k}T\}[(1/\bar{r}) - (1/r)]$$

and combining with equation (8.11) gives the variation of particle growth rate with radius according to

$$\mathrm{d}r/\mathrm{d}t = \{2DS\gamma\Omega/\mathbf{k}Tr\}[(1/\bar{r}) - (1/r)] \qquad (8.13)$$

This function is plotted in Figure 8.14, from which it is evident that particles of radius less than $\bar{r}$ are dissolving at increasing rates with decreasing values of r. All particles of radius greater than $\bar{r}$ are growing but the graph shows a maximum for particles twice the mean radius. Over a period of time the number of particles decreases discontinuously when particles dissolve, and ultimately the system would tend to form one large particle. However, before this state is reached the mean radius $\bar{r}$ increases and the growth rate of the whole system slows down.

A more detailed theory than that, due to Greenwood, outlined above has been derived by Lifshitz and Slyozov, and by Wagner taking into consideration the

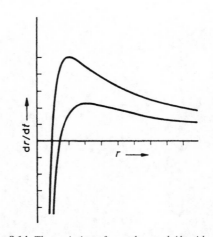

Figure 8.14 *The variation of growth rate dr/dt with particle radius r for diffusion-controlled growth, for two values of r. The value of r for the lower curve is 1.5 times that for the upper curve. Particles of radius equal to the mean radius of all particles in the system at any instant are neither growing nor dissolving. Particles of twice this radius are growing at the fastest rate. The smallest particles are dissolving at a rate approximately proportional $\bar{r}^2$ (after Greenwood, 1968; courtesy of the Institute of Metals).*

initial particle size distribution. They show that the mean particle radius varies with time according to

$$\bar{r}^3 - \bar{r}_0^3 = Kt \qquad (8.14)$$

where $\bar{r}_0$ is the mean particle radius at the onset of coarsening and K is a constant given by

$$K = 8DS\gamma\Omega/9\mathbf{k}T$$

This result is almost identical to that obtained by integrating equation (8.13) in the elementary theory and assuming that the mean radius is increasing at half the rate of that of the fastest-growing particle.

Coarsening rate equations have also been derived assuming that the most difficult step in the process is for the atom to enter into solution across the precipitate/matrix interface; the growth is then termed interface-controlled. The appropriate rate equation is

$$\mathrm{d}r/\mathrm{d}t = -C(S_r - S_a)$$

and leads to a coarsening equation of the form

$$\bar{r}^2 - \bar{r}_0^2 = (64CS\gamma\Omega t/8\mathbf{k}T) \qquad (8.15)$$

where C is some interface constant.

Measurements of coarsening rates so far carried out support the analysis basis on diffusion control of the particle growth. The most detailed results have been obtained for nickel-based systems, particularly the coarsening of $\gamma'(\mathrm{Ni_3Al-Ti}$ or Si), which show a good $\bar{r}^3$ versus t relationship over a wide range of temperatures. Strains due to coherency and the fact that γ' precipitates are cube-shaped do not seriously affect the

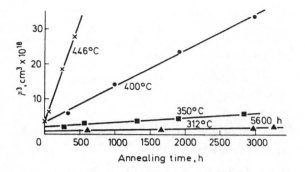

Figure 8.15 *The variation of $\bar{r}^3$ with time of annealing for manganese precipitates in a magnesium matrix (after Smith, 1967; courtesy of Pergamon Press).*

analysis in these systems. Concurrent measurements of $\bar{r}$ and the solute concentration in the matrix during coarsening have enabled values for the interfacial energy ≈ 13 mJ/m^2 to be determined. In other systems the agreement between theory and experiment is generally less precise, although generally the cube of the mean particle radius varies linearly with time, as shown in Figure 8.15 for the growth of Mn precipitates in a Mg–Mn alloy.

Because of the ease of nucleation, particles may tend to concentrate on grain boundaries, and hence grain boundaries may play an important part in particle growth. For such a case, the Thomson–Freundlich equation becomes

$$\ln(S_r/S) = (2\gamma - \gamma_g)\Omega/\mathbf{k}Tx$$

where γ_g is the grain boundary energy per unit area and $2x$ the particle thickness, and their growth follows a law of the form

$$r_f^4 - r_0^4 = Kt \tag{8.16}$$

where the constant K includes the solute diffusion coefficient in the grain boundary and the boundary width. The activation energy for diffusion is lower in the grain boundary than in the matrix and this leads to a less strong dependence on temperature for the growth of grain boundary precipitates. For this reason their preferential growth is likely to be predominant only at relatively low temperature.

8.2.7 Spinodal decomposition

For any alloy composition where the free energy curve has a negative curvature, i.e. $(d^2G/dc^2) < 0$, small fluctuations in composition that produce A-rich and B-rich regions will bring about a lowering of the total free energy. At a given temperature the alloy must lie between two points of inflection (where $d^2G/dc^2 = 0$) and the locus of these points at different temperatures is depicted on the phase diagram by the chemical spinodal line (see Figure 8.16).

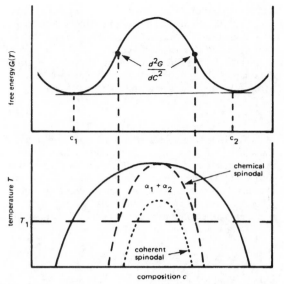

Figure 8.16 *Variation of chemical and coherent spinodal with composition.*

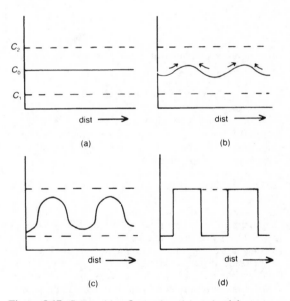

Figure 8.17 *Composition fluctuations in a spinodal system.*

For an alloy c_0 quenched inside this spinodal, composition fluctuations increase very rapidly with time and have a time constant $\tau = -\lambda/4\pi^2 D$, where λ is the wavelength of composition modulations in one dimension and D is the interdiffusion coefficient. For such a kinetic process, shown in Figure 8.17, 'uphill' diffusion takes place, i.e. regions richer in solute than the average become richer, and poorer become poorer until the equilibrium compositions c_1 and c_2 of the A-rich and B-rich regions are formed. As for normal precipitation, interfacial energy and strain energy influency the decomposition. During the early stages of decomposition the interface between A-rich and B-rich regions is diffuse and the interfacial energy becomes a gradient energy which depends on the composition gradient across the interface according to

$$\Delta G_{int} = K(\Delta c/\lambda)^2 \tag{8.17}$$

where λ is the wavelength and Δc the amplitude of the sinusoidal composition modulation, and K depends on the difference in bond energies between like and unlike atom pairs. The coherency strain energy term is related to the misfit ε between regions A and B, where $\varepsilon = (1/a)da/dc$, the fractional change in lattice parameter a per unit composition change, and is given for an elastically isotropic solid, by

$$\Delta G_{strain} = \varepsilon^2 \Delta c^2 EV/(1 - \nu) \tag{8.18}$$

with E Young's modulus, ν Poisson's ratio and V the molar volume. The total free energy change arising from a composition fluctuation is therefore

$$\Delta G = \left[\frac{d^2 G}{dc^2} + \frac{2K}{\lambda^2} + (2\varepsilon^2 EV/(1 - \nu)) \right] \Delta c^2/2 \tag{8.19}$$

and a homogeneous solid solution will decompose spinodally provided

$$-(d^2 G/dc^2) > (2K/\lambda^2) + (2\varepsilon^2 EV/1 - \nu) \tag{8.20}$$

For $\lambda = \infty$, the condition $[(d^2 G/dc^2) + (2\varepsilon^2 EV/1 - \nu)] = 0$ is known as the coherent spinodal, as shown in Figure 8.16. The λ of the composition modulations has to satisfy the condition

$$\lambda^2 > 2K/[d^2 G/dc^2 + (2\varepsilon^2 EV/(1 - \nu)] \tag{8.21}$$

and decreases with increasing degree of undercooling below the coherent spinodal line. A λ-value of 5–10 nm is favoured, since shorter λ's have too sharp a concentration gradient and longer λ's have too large a diffusion distance. For large misfit values, a large undercooling is required to overcome the strain energy effect. In cubic crystals, E is usually smaller along $\langle 1\,0\,0 \rangle$ directions and the high strain energy is accommodated more easily in the elastically soft directions, with composition modulations localized along this direction.

Spinodal decompositions have now been studied in a number of systems such as Cu–Ni–Fe, Cu–Ni–Si, Ni–12Ti, Cu–5Ti exhibiting 'side-bands' in X-ray small-angle scattering, satellite spots in electron diffraction patterns and characteristic modulation of structure along $\langle 1\,0\,0 \rangle$ in electron micrographs. Many of the alloys produced by splat cooling might be expected to exhibit spinodal decomposition, and it has

been suggested that in some alloy systems GP zones form in this way at high supersaturations, because the GP zone solvus (see Figure 8.1) gives rise to a metastable coherent miscibility gap.

The spinodally decomposed microstructure is believed to have unusually good mechanical stability under fatigue conditions.

8.3 Strengthening of steels by heat-treatment

8.3.1 Time–temperature–transformation diagrams

Eutectoid decomposition occurs in both ferrous (e.g. iron–carbon) and non-ferrous (e.g. copper–aluminium, copper–tin) alloy systems, but it is of particular importance industrially in governing the hardening of steels. In the iron–carbon system (see Figure 3.18) the γ-phase, austenite, which is a solid solution of carbon in fcc iron, decomposes on cooling to give a structure known as pearlite, composed of alternate lamellae of cementite (Fe_3C) and ferrite. However, when the cooling conditions are such that the alloy structure is far removed from equilibrium, an alternative transformation may occur. Thus, on very rapid cooling, a metastable phase called martensite, which is a supersaturated solid solution of carbon in ferrite, is produced. The microstructure of such a transformed steel is not homogeneous but consists of plate-like needles of martensite embedded in a matrix of the parent austenite. Apart from martensite, another structure known as bainite may also be formed if the formation of pearlite is avoided by cooling the austenite rapidly through the temperature range above 550°C, and then holding the steel at some temperature between 250°C and 550°C. A bainitic structure consists of platelike grains of ferrite, somewhat like the plates of martensite, inside which carbide particles can be seen.

The structure produced when austenite is allowed to transform isothermally at a given temperature can be conveniently represented by a diagram of the type shown in Figure 8.18, which plots the time necessary at a given temperature to transform austenite of eutectoid composition to one of the three structures: pearlite, bainite or martensite. Such a diagram, made up from the results of a series of isothermal-decomposition experiments, is called a TTT curve, since it relates the transformation product to the time at a given temperature. It will be evident from such a diagram that a wide variety of structures can be obtained from the austenite decomposition of a particular steel; the structure may range from 100% coarse pearlite, when the steel will be soft and ductile, to fully martensitic, when the steel will be hard and brittle. It is because this wide range of properties can be produced by the transformation of a steel that it remains a major constructional material for engineering purposes.

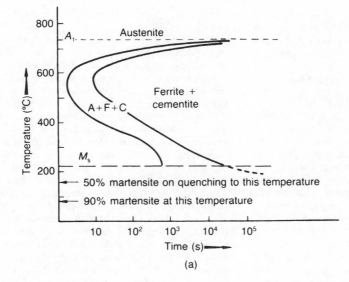

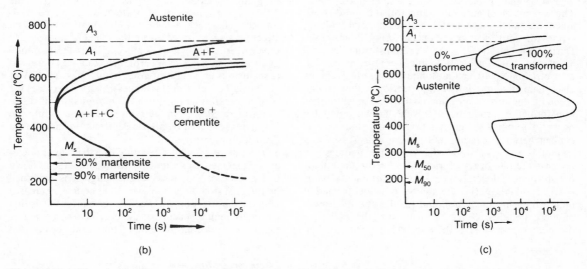

Figure 8.18 *TTT curves for (a) eutectoid, (b) hypo-eutectoid and (c) low alloy (e.g. Ni/Cr/Mo) steels (after ASM Metals Handbook).*

From the TTT curve it can be seen that just below the critical temperature, A_1, the rate of transformation is slow even though the atomic mobility must be high in this temperature range. This is because any phase change involving nucleation and growth (e.g. the pearlite transformation) is faced with nucleation difficulties, which arise from the necessary surface and strain energy contributions to the nucleus. Of course, as the transformation temperature approaches the temperature corresponding to the knee of the curve, the transformation rate increases. The slowness of the transformation below the knee of the TTT curve, when bainite is formed, is also readily understood, since atomic migration is slow at these lower temperatures and the bainite transformation depends on diffusion. The lower part of the TTT curve below about 250–300°C indicates, however, that the transformation speeds up again and takes place exceedingly fast, even though atomic mobility in this temperature range must be very low. For this reason, it is concluded that the martensite transformation does not depend on the speed of migration of carbon atoms and, consequently, it is often referred to as a diffusionless transformation. The austenite only starts transforming to martensite when the temperature falls below a critical temperature, usually denoted by M_s. Below M_s the percentage of austenite transformed to martensite is indicated on the diagram by a series of horizontal lines.

The M_s temperature may be predicted for steels containing various alloying elements in weight per cent by the formula, due to Steven and Haynes, given by $M_s(°C) = 561 - 474C - 33Mn - 17Ni - 17Cr - 21Mo$.

8.3.2 Austenite–pearlite transformation

8.3.2.1 Nucleation and growth of pearlite

If a homogeneous austenitic specimen of eutectoid composition were to be transferred quickly to a bath held at some temperature between 720°C and 550°C, decomposition curves of the form shown in Figure 8.19a would be obtained. These curves, typical of a nucleation and growth process, indicate that the transformation undergoes an incubation period, an accelerating stage and a decelerating stage; the volume transformed into pearlite has the time-dependence described by the Avrami equation (7.44). When the transformation is in its initial stage the austenite contains a few small pearlite nodules each of which grow during the period A to B (see curve obtained at 690°C) and, at the same time, further nuclei form. The percentage of austenite transformed is quite small, since the nuclei are small and their total volume represents only a fraction of the original austenite. During the B to C stage the transformation rate accelerates, since as each nodule increases in size the area of contact between austenite and pearlite regions also increases: the larger the pearlite volumes, the greater is the surface area upon which to deposit further transformation products. At C, the growing nodules begin to impinge on each other, so that the area of contact between pearlite and austenite decreases and from this stage onwards, the larger the nodules the lower is the rate of transformation. Clearly, the rate of transformation depends on (1) the rate of nucleation of pearlite nodules, N (i.e. the number of nuclei formed in unit volume in unit time), and (2) the rate of growth of these nodules, G (i.e. the rate that the radius of the nodule increases with time). The variation of N

and G with temperature for a eutectoid steel is shown in Figure 8.19b.

The rate of nucleation increases with decreasing temperature down to the knee of the curve and in this respect is analogous to other processes of phase precipitation where hysteresis occurs (see Chapter 3). In addition, the nucleation rate is very structure sensitive so that nucleation occurs readily in regions of high energy where the structure is distorted. In homogeneous austenite the nucleation of pearlite occurs almost exclusively at grain boundaries and, for this reason, the size of the austenite grains, prior to quenching, has an important effect on hardenability (a term which denotes the depth in a steel to which a fully martensitic structure can be obtained). Coarse-grained steels can be hardened more easily than fine-grained steels because to obtain maximum hardening in a steel, the decomposition of austenite to pearlite should be avoided, and this is more easily accomplished if the grain boundary area, or the number of potential pearlite nucleation sites, is small. Thus, an increase in austenite grain size effectively pushes the upper part of the TTT curve to longer times, so that, with a given cooling rate, the knee can be avoided more easily. The structure-sensitivity of the rate of nucleation is also reflected in other ways. For example, if the austenite grain is heterogeneous, pearlite nucleation is observed at inclusions as well as at grain boundaries. Moreover, plastic deformation during transformation increases the rate of transformation, since the introduction of dislocations provides extra sites for nucleation, while the vacancies produced by plastic deformation enhance the diffusion process.

The rate of growth of pearlite, like the rate of nucleation, also increases with decreasing temperature down to the knee of the curve, even though it is governed by the diffusion of carbon, which, of course, decreases with decreasing temperature. The reason for this is that the interlamellar spacing of the pearlite also decreases rapidly with decreasing temperature, and because the

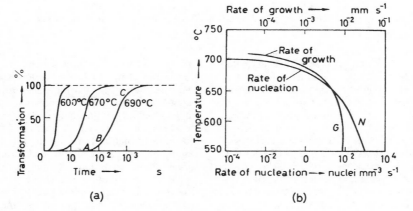

Figure 8.19 *Effect of temperature on (a) amount of pearlite formed with time and (b) rate of nucleation and rate of growth of pearlite (after Mehl and Hagel, 1956; courtesy of Pergamon Press).*

carbon atoms do not have to travel so far, the carbon supply is easily maintained. In contrast to the rate of nucleation, however, the rate of growth of pearlite is quite structure-insensitive and, therefore, is indifferent to the presence of grain boundaries or inclusions. These two factors are important in governing the size of the pearlite nodules produced. If, for instance, the steel is transformed just below A_1, where the rate of nucleation is very low in comparison with the rate of growth (i.e. the ratio N/G is small), very large nodules are developed. Then, owing to the structure-insensitivity of the growth process, the few nodules formed are able to grow across grain boundaries, with the result that pearlite nodules larger than the original austenite grain size are often observed. By comparison, if the steel is transformed at a lower temperature, just above the knee of the TTT curve where N/G is large, the rate of nucleation is high and the pearlite nodule size is correspondingly small.

8.3.2.2 Mechanism and morphology of pearlite formation

The growth of pearlite from austenite clearly involves two distinct processes: (1) a redistribution of carbon (since the carbon concentrates in the cementite and avoids the ferrite) and (2) a crystallographic change (since the structure of both ferrite and cementite differs from that of austenite). Of these two processes it is generally agreed that the rate of growth is governed by the diffusion of carbon atoms, and the crystallographic change occurs as readily as the redistribution of carbon will allow. The active nucleus of the pearlite nodule may be either a ferrite or cementite platelet, depending on the conditions of temperature and composition which prevail during the transformation, but usually it is assumed to be cementite. The nucleus may form at a grain boundary as shown in Figure 8.20a, and after its formation the surrounding matrix is depleted of carbon, so that conditions favour the nucleation of ferrite plates adjacent to the cementite nucleus (Figure 8.20b). The ferrite plates in turn reject carbon atoms into the surrounding austenite and this favours the formation of cementite nuclei, which then continue to grow. At the same time as the pearlite nodule grows sideways, the ferrite and cementite lamellae advance into the austenite, since the carbon atoms rejected ahead of the advancing ferrite diffuse into the path of the growing cementite (Figure 8.20c). Eventually, a cementite plate of different orientation forms and this acts as a new nucleus as shown in Figures 8.20d and 8.20e.

Homogeneous austenite, when held at a constant temperature, produces pearlite at a constant rate and with a constant interlamellar spacing. However, the interlamellar spacing decreases with decreasing temperature, and becomes irresolvable in the optical microscope as the temperature approaches that corresponding to the knee of the curve. An increase in hardness occurs as the spacing decreases. Zener explains the dependence of interlamellar spacing

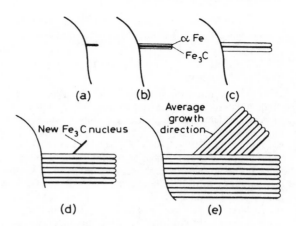

Figure 8.20 *Nucleation and growth of pearlite nodules. (a) Initial Fe_3C nucleus; (b) Fe_3C plate full grown, α-Fe now nucleated; (c) α-Fe plates now full grown, new Fe_3C plates nucleated; (d) Fe_3C nucleus of different orientation forms and original nodule grows; (e) new nodule at advanced stage of growth (after Mehl and Hagel; courtesy of Pergamon Press).*

on temperature in the following way. If the interlamellar spacing is large, the diffusion distance of the carbon atoms in order to concentrate in the cementite is also large, and the rate of carbon redistribution is correspondingly slow. Conversely, if the spacing is small the area, and hence energy, of the ferrite-cementite interfaces become large. In consequence, such a high proportion of the free energy released in the austenite to pearlite transformation is needed to provide the interfacial energy that little will remain to provide the 'driving force' for the change. Thus, a balance between these two opposing conditions is necessary to allow the formation of pearlite to proceed, and at a constant temperature the interlamellar spacing remains constant. However, because the free energy change, ΔG, accompanying the transformation increases with increasing degree of undercooling, larger interfacial areas can be tolerated as the temperature of transformation is lowered, with the result that the interlamellar spacing decreases with decreasing temperature.

The majority of commercial steels are not usually of the eutectoid composition (0.8% carbon), but hypo-eutectoid (i.e. <0.8% carbon). In such steels, pro-eutectoid ferrite is first formed before the pearlite reaction begins and this is shown in the TTT curve by a third decomposition line. From Figure 8.18b it can be seen that the amount of pro-eutectoid ferrite decreases as the isothermal transformation temperature is lowered. The morphology of the precipitated ferrite depends on the usual precipitation variables (i.e. temperature, time, carbon content and grain size) and growth occurs preferentially at grain boundaries and on certain crystallographic planes. The Widmanstätten pattern with ferrite growing along {1 1 1} planes of the parent austenite is a familiar structure of these steels.

8.3.2.3 Influence of alloying elements on pearlite formation

With the exception of cobalt, all alloying elements in small amounts retard the transformation of austenite to pearlite. These elements decrease both the rate of nucleation, N, and the rate of growth, G, so that the top part of the TTT curve is displaced towards longer times. This has considerable technological importance since in the absence of such alloying elements, a steel can only transform into the harder constituents of bainite or martensite if it is in the form of very thin sections so that the cooling rate will be fast enough to avoid crossing the knee of the TTT curve during the cooling process and hence avoid pearlite transformation. For this reason, most commercially heat-treatable steels contain one or more of the elements chromium, nickel, manganese, vanadium, molybdenum or tungsten. Cobalt increases both N and G, and its effect on the pearlite interlamellar spacing is contrary to the other elements in that it decreases the spacing.

With large additions of alloying elements, the simple form of TTT curve often becomes complex, as shown in Figure 8.18c. Thus to obtain any desired structure by heat treatment a detailed knowledge of the TTT curve is essential.

8.3.3 Austenite–martensite transformation

8.3.3.1 Crystallography of martensite formation

Martensite, the hardening constituent in quenched steels, is formed at temperatures below about 200°C. The regions of the austenite which have transformed to martensite are lenticular in shape and may easily be recognized by etching, or from the distortion they produce on the polished surface of the alloy. These relief effects, shown schematically in Figure 8.21, indicate that the martensite needles have been formed not with the aid of atomic diffusion but by a shear process, since if atomic mobility were allowed the large strain energy associated with the transformed volume would then be largely avoided. The lenticular shape of a martensite needle is a direct consequence of the stresses produced in the surrounding matrix by the shear mechanism of the transformation and is exactly analogous to the similar effect found in mechanical twinning. The strain energy associated with martensite is tolerated because the growth of such sheared regions does not depend on diffusion, and since the regions are coherent with the matrix they are able to spread at great speed through the crystal. The large free energy change associated with the rapid formation of the new phase outweighs the strain energy, so that there is a net lowering of free energy.

Direct TEM observations of martensite plates have shown that there are two main types of martensite, one with a twinned structure (see Figure 8.22), known as

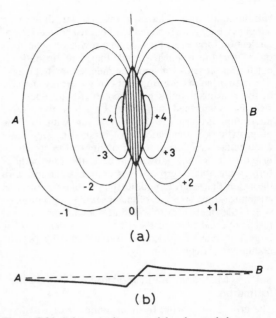

Figure 8.21 *Schematic diagram of the observed shape deformation produced by a martensite plate. (a) Contour lines on an originally flat surface; (b) section of the surface through AB. Vertical scale much exaggerated (after Bilby and Christian, 1956; courtesy of the Institute of Metals).*

acicular martensite, and the other with a high density of dislocations but few or no twins, called massive martensite.

In contrast to the pearlite transformation, which involves both a redistribution of carbon atoms and a structural change, the martensite transformation involves only a change in crystal structure.

The structure cell of martensite is body-centred tetragonal, which is a distorted form of a body-centred cubic structure, and hence may be regarded as a supersaturated solution of carbon in α-iron. X-ray examination shows that while the c/a ratio of the bct structure of martensite increases with increasing carbon content, the curve of c/a ratio against composition extrapolates back to $c/a = 1$ for zero carbon content, and the lattice parameter is equal to that of pure α-iron (Figure 8.23).

From the crystallographic point of view the most important experimental data in any martensite transformation are the orientation relations of the two phases and the habit plane. In steel, there are three groups of orientations often quoted; those due to Kurdjumov and Sachs, Nishiyama, and Greninger and Troiano, respectively. According to the Kurdjumov–Sachs relation, in iron–carbon alloys with 0.5–1.4% carbon, a $\{1\,1\,1\}_\gamma$ plane of the austenite lattice is parallel to the $\{1\,1\,0\}_\alpha$ plane of the martensite, with a $\langle 1\,1\,0\rangle_\gamma$ axis of the former parallel to a $\langle 1\,1\,1\rangle_\alpha$ axis of the latter; the associated habit plane is $\{2\,2\,5\}_\gamma$. In any one crystal there are 24 possible variants of the Kurdjumov–Sachs relationship, consisting of 12 twin pairs, both orientations

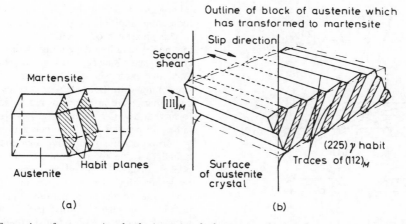

Figure 8.22 *(a) Formation of a martensite platelet in a crystal of austenite; (b) the inhomogeneous twinning shear within the platelet (after Kelly and Nutting, 1960; courtesy of the Royal Society).*

of a pair having the same habit plane. However, for general discussion it is usual to choose one relation which may be written

$$(1\,1\,1)_\gamma\ (1\,0\,1)_\alpha \text{ with } [1\,\bar{1}\,0]_\gamma\ [1\,1\,\bar{1}]_\alpha$$

In the composition range 1.5–1.8% carbon the habit plane changes to $\approx \{2\,5\,9\}_\gamma$ with the orientation relationship unspecified. This latter type of habit plane has also been reported by Nishiyama for iron–nickel alloys (27–34% nickel) for which the orientation relationship is of the form

$$(1\,1\,1)_\gamma\ (1\,0\,1)_\alpha \text{ with } [1\,\bar{2}\,1]_\gamma\ [1\,0\,\bar{1}]_\alpha$$

However, Greninger and Troiano have shown by precision orientation determinations that irrational relationships are very probable, and that in a ternary iron–nickel–carbon alloy (0.8% carbon, 22% nickel), $(1\,1\,1)_\gamma$ is approximately 1° from $(1\,0\,1)_\alpha$ with $[1\,\bar{2}\,1]_\gamma$ approximately 2° from $[1\,0\,1]_\alpha$, and is associated with a habit plane about 5° from $(2\,5\,9)$.

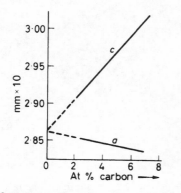

Figure 8.23 *Variation of c and a parameters with carbon content in martensite (after Kurdjumov, 1948).*

8.3.3.2 Mechanism of martensite formation

The martensite transformation is diffusionless, and therefore martensite forms without any interchange in the position of neighbouring atoms. Accordingly, the observed orientation relationships are a direct consequence of the atom movements that occur during the transformation. The first suggestion of a possible transformation mechanism was made by Bain in 1934. He suggested that since austenite may be regarded as a body-centred tetragonal structure of axial ratio $\sqrt{2}$, the transformation merely involves a compression of the c-axis of the austenite unit cell and expansion of the α-axis. The interstitially dissolved carbon atoms prevent the axial ratio from going completely to unity, and, depending on composition, the c/a ratio will be between 1.08 and 1.0. Clearly, such a mechanism can only give rise to three martensite orientations whereas, in practice, 24 result. To account for this, Kurdjumov and Sachs proposed that the transformation takes place not by one shear process but by a sequence of two shears (Figure 8.24), first along the elements $(1\,1\,1)_\gamma\langle 1\,1\,2\rangle_\gamma$, and then a minor shear along the elements $(1\,1\,2)_\alpha\langle 1\,1\,1\rangle_\alpha$; these elements are the twinning elements of the fcc and bcc lattice, respectively. This mechanism predicts the correct orientation relations, but not the correct habit characteristics or relief effects. Accordingly, Greninger and Troiano in 1941 proposed a different two-stage transformation, consisting of an initial shear on the irrational habit plane which produces the relief effects, together with a second shear along the twinning elements of the martensite lattice. If slight adjustments in spacing are then allowed, the mechanism can account for the relief effects, habit plane, the orientation relationship and the change of structure.

Further additions to these theories have been made in an effort to produce the ideal general theory of the crystallography of martensite transformation. Bowles, for example, replaces the first shear of the

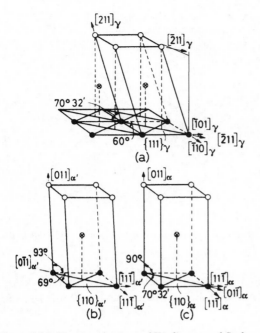

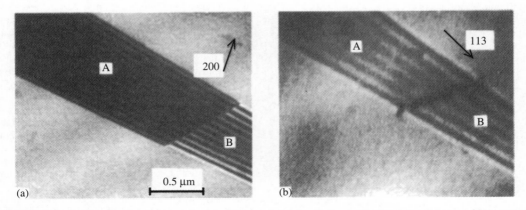

Figure 8.24 *Shear mechanisms of Kurdjumov and Sachs. (a) Face-centred austenite with* $\{1\,1\,1\}_\gamma$ *in horizontal plane, (b) body-centred tetragonal martensite* (α')*, (c) cubic ferrite* (α) *(after Bowles and Barrett, 1952; courtesy of Pergamon Press).*

Greninger–Troiano mechanism by the general type of homogeneous deformation in which the habit plane remains invariant, i.e. all directions in this plane are unrotated and unchanged in length. However, in all such cases the problem resolves itself into one of determining whether a homogeneous strain can transform the γ-lattice into the α-lattice, while preserving coherency at the boundary between them. The

homogeneous strain does not do this, so that some reasonable additional type of strain has to be added.

This shear can occur either by twinning or by slip, the mode prevailing depending on the composition and cooling rate. Between carbon contents of 0.2% and 0.5% the martensite changes from dislocated martensite arranged in thin lathes or needles to twinned acicular martensite arranged in plates. In the martensite formed at low C contents (e.g. Fe–Ni alloys) the thin lathes lie parallel to each other, with a $\{1\,1\,1\}_\gamma$ habit, to form pockets of massive martensite with jagged boundaries due to the impingement of other nearby pockets of lathes. The inhomogeneous shear produced by deformation twinning occurs on $\{1\,1\,2\}$ planes in the martensite, so that each martensite plate is made up of parallel twin plates of thickness 2–50 nm. By operation of such a complex transformation mode with a high index habit plan the system maintains an invariant interfacial plane.

Because of the shears involved and the speed of the transformation it is attractive to consider that dislocations play an important role in martensite formation. Some insight into the basic dislocation mechanisms has been obtained by *in situ* observations during either cooling below M_s or by straining, but unfortunately only for Ni–Cr austenitic steels with low stacking-fault energy (i.e. $\gamma \approx 20$ mJ/m^2). For these alloys it has been found that stacking faults are formed either by emitting partial dislocations with $b = a/6\langle1\,1\,2\rangle$ from grain boundaries or by the dissociation of unit dislocations with $b = a/2\langle1\,1\,0\rangle$. In regions of the grain where on cooling or deformation a high density of stacking faults developed, the corresponding diffraction pattern revealed cph ε-martensite. On subsequent deformation or cooling, regions of ε-martensite transform rapidly into bcc α-martensite, and indeed, the only way in which α-martensite was observed to form was from an ε nucleus.

Figure 8.25 *Electron micrographs showing (a) contrast from overlapping faults on* (1 1 1)*; A is extrinsic and B is intrinsic in nature. (b) Residual contrast arising from a supplementary displacement across the faults which is intrinsic in nature for both faults A and B (after Brooks, Loretto and Smallman, 1979).*

Because straining or cooling can be interrupted during the *in situ* experiments it was possible to carry out a detailed analysis of the defect structure formed prior to a region becoming recognizably (from diffraction patterns) martensitic. In this way it has been shown that the interplanar spacing across the individual stacking faults in the austenite decreased to the (0 0 0 1) spacing appropriate to ε-martensite. Figure 8.25 shows micrographs which reveal this change of spacing; no contrast is expected in Figure 8.25b if the faulted {1 1 1} planes remained at the fcc spacing, since the condition of invisibility $g.R = n$ is obeyed. The residual contrast observed arises from the supplementary displacement ΔR across the faults which, from the white outer fringe, is positive (intrinsic) in nature for both faults and $\approx$ 2% of the {1 1 1} spacing. The formation of regions of α from ε could also be followed although in this case the speed of the transformation precluded detailed analysis. Figure 8.26 shows a micrograph taken after the formation of α-martensite and this, together with continuous observations, show that the martensite/matrix interface changes from {1 1 1} to the well-known {2 2 5} as it propagates. Clearly, one of the important roles that the formation of ε-martensite plays in acting as a precursor for the formation of α-martensite is in the generation of close-packed planes with **ABAB** stacking so that atomic shuffles can subsequently transform these planes to {1 1 0} bcc which are, of course, stacked **ABAB** (see Figure 8.27). The α-martensite forms in dislocation pile-ups where the $a/6\langle 1 1 2\rangle$ partials are forced closer together by the applied stress. The volume of effective bcc material increases as more dislocations join the pile-up until the nucleus formed by this process reaches a critical size and rapid growth takes place. The martensite initially grows perpendicular to, and principally on, one side of the {1 1 1}$_\gamma$ slip plane associated with the nucleus, very likely corresponding to the side of the dislocations with missing half-planes since α-martensite is less dense than austenite.

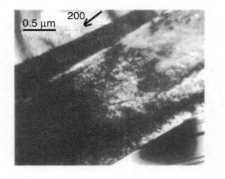

Figure 8.26 *Electron micrograph showing an α-martensite plate, the austenite–martensite interface, and the faults in the austenite matrix.*

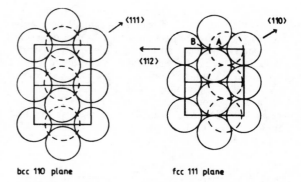

bcc 110 plane   fcc 111 plane

Figure 8.27 *A shear of $a/6\langle 1 1 2\rangle$ moves atoms in the fcc structure from A sites to B sites, and after half this shear the structure has pseudo-bcc packing.*

8.3.3.3 Kinetics of martensite formation

One of the most distinctive features of the martensite transformation is that in most systems martensite is formed only when the specimen is cooling, and that the rate of martensite formation is negligible if cooling is stopped. For this reason, the reaction is often referred to as an athermal[1] transformation, and the percentage of austenite transformed to martensite is indicated on the TTT curve by a series of horizontal lines. The transformation begins at a temperature M_s, which is not dependent on cooling rate, but is dependent on prior thermal and mechanical history, and on composition. For example, it is well established that the M_s temperature decreases approximately linearly with increasing concentration of solutes such as carbon, nickel or manganese.

Speed of formation The observation that martensite plates form rapidly and at a rate which is temperature-independent shows that thermal activation is not required for the growth process. Electronic methods show that the martensite needles form, in iron–nickel–carbon alloys, for example, in about 10^{-7} s and, moreover, that the linear growth velocity is about 10^3 m s^{-1} even at very low temperatures. Such observations show that the activation energy for the growth of a martensite plate is virtually zero, and that the velocity of growth approaches the speed of sound in the matrix. Sometimes a 'burst phenomenon' is exhibited, as, for example, in iron–nickel alloys, when the stresses produced by one martensite plate assist in the nucleation of others. The whole process is autocatalytic and about 25% of the transformation can occur in the time interval of an audible click.

The effect of applied stress Since the formation of martensite involves a homogeneous distortion of the parent structure, it is expected that externally applied

---

[1]In some alloys, such as iron–manganese–carbon and iron–manganese–nickel, the martensitic transformation occurs isothermally. For these systems, growth is still very rapid but the nuclei are formed by thermal activation.

stresses will be of importance. Plastic deformation is effective in forming martensite above the M_s temperature, provided the temperature does not exceed a critical value usually denoted by M_d. However, cold work above M_d may either accelerate or retard the transformation on subsequent cooling. Even elastic stresses, when applied above the M_s temperature and maintained during cooling, can affect the transformation; uniaxial compression or tensile stresses raise the M_s temperature while hydrostatic stresses lower the M_s temperature.

Stabilization When cooling is interrupted below M_s, stabilization of the remaining austenite often occurs. Thus, when cooling is resumed martensite forms only after an appreciable drop in temperature. Such thermal stabilization has been attributed by some workers to an accumulation of carbon atoms on those dislocations important to martensite formation. This may be regarded as a direct analogue of the yield phenomenon. The temperature interval before transformation is resumed increases with holding time and is analogous to the increase in yield drop accompanying carbon build-up on strain-ageing. Furthermore, when transformation in a stabilized steel does resume, it often starts with a 'burst', which in this case is analogous to the lower yield elongation.

8.3.4 Austenite–bainite transformation

The bainite reaction has many features common to both the pearlite and martensite reactions. The pearlite transformation involves the redistribution of carbon followed by a structure change, the martensite transformation involves the structure change alone, and, in contrast, the bainite transformation involves a structure change followed by the redistribution of carbon, which precipitates as a carbide. Consequently, the austenite–bainite decomposition may be regarded as a martensite transformation involving the diffusion of carbon atoms, so that, in this case, the rate of coherent growth is necessarily slow compared with that of martensite. Lower bainite is hardly distinguishable from martensite tempered at the same temperature, while upper bainite exhibits an acicular structure. The metallographic appearance of the transformed steel is found to alter continuously between these two extremes, the actual structure exhibited being governed by the diffusion rate of the carbon, which in turn depends on the temperature of the transformation. The hardness of the reaction product also increases continuously with decreasing temperature, lower bainite being harder than upper bainite, which is harder than most fine pearlite.

The ferrite in bainite has a martensite-like appearance and is, in most cases, clearly distinguished from both ferrite and pro-eutectoid ferrite formed in the pearlite range. The bainitic ferrite exhibits the same surface relief effects as martensite, while pro-eutectoid ferrite and pearlite do not. Such surface tilting is further evidence for a shear-like transformation, but the orientation relationship between austenite and bainite is not necessarily the same as that between austenite and martensite. In fact, the bainitic ferrite has the same orientation with respect to the parent austenite as does pro-eutectoid ferrite, which suggests that ferrite may nucleate bainite.

8.3.5 Tempering of martensite

The presence of martensite in a quenched steel, while greatly increasing its hardness and TS, causes the material to be brittle. Such behaviour is hardly surprising, since the formation of martensite is accompanied by severe matrix distortions. The hardness and strength of martensite increase sharply with increase in C content. Contributions to the strength arise from the carbon in solution, carbides precipitated during the quench, dislocations introduced during the transformation and the grain size.

Although the martensite structure is thermodynamically unstable, the steel will remain in this condition more or less indefinitely at room temperature because for a change to take place bulk diffusion of carbon, with an activation energy Q of approximately 83 kJ/mol atom is necessary. However, because there is an exponential variation of the reaction rate with temperature, the steel will be able slowly to approach the equilibrium structure at a slightly elevated temperature, i.e. rate of reaction $= A \exp[-Q/kT]$. Thus, by a carefully controlled tempering treatment, the quenching stresses can be relieved and some of the carbon can precipitate from the supersaturated solid solution to form a finely dispersed carbide phase. In this way, the toughness of the steel can be vastly improved with very little detriment to its hardness and tensile properties.

The structural changes which occur on tempering may be considered to take place in three stages. In the primary stage, fine particles of a cph carbide phase (ε-carbide) of composition about $Fe_{2.4}C$, precipitates, with the corresponding formation of low-carbon martensite. This low-carbon martensite grows at the expense of the high-carbon martensite until at the end of this stage the structure consists of retained austenite, ε-carbide and martensite of reduced tetragonality. During the second stage any retained austenite in the steel begins to transform isothermally to bainite, while the third stage is marked by the formation of cementite platelets. The precipitation of cementite is accompanied by a dissolution of the ε-carbide phase so that the martensite loses its remaining tetragonality and becomes bcc ferrite. The degree to which these three stages overlap will depend on the temperature of the anneal and the carbon content. In consequence, the final structure produced will be governed by the initial choice of steel and the properties, and hence thermal treatment, required. Alloying elements, with the exception of Cr, affect the tempering of martensite. Plain carbon steels soften above 100°C owing to the early formation of ε-carbide, whereas in Si-bearing steels the softening is delayed to above 250°C, since

Si stabilizes ε-carbide and delays its transformation to cementite. Alloying additions (see Table 8.2) thus enable the improvement in ductility to be achieved at higher tempering temperatures.

When a steel specimen is quenched prior to tempering, quenching cracks often occur. These are caused by the stresses which arise from both the transformation and the differential expansion produced when different parts of the specimen cool at different rates. To minimize such cracking, the desired properties of toughness and strength are often produced in the steel by alternative heat-treatment schedules; examples of these schedules are summarized in Figure 8.28, from which it will become evident that advantage is taken of the shape of the TTT curve to economize on the time the specimen is in the furnace, and also to minimize quenching stresses. During conventional annealing, for example, the steel is heated above the upper critical temperature and allowed to cool slowly in the furnace. In isothermal annealing the steel is allowed to transform in the furnace, but when it has completely transformed, the specimen is removed from the furnace and allowed to air-cool, thereby saving furnace time. In martempering, the knee of the TTT curve is avoided by rapid cooling, but the quench is interrupted above M_s and the steel allowed to cool relatively slowly through the martensite range. With this treatment the thermal stresses set up by very rapid cooling are reduced. Such a procedure is possible because at the holding temperature there is ample time for the temperature to become equal throughout the sample before the transformation begins, and as a result the transformation occurs much more uniformly. After the transformation is complete, tempering is carried out in the usual way. In austempering, quenching is again arrested above M_s and a bainite product, having similar properties to tempered martensite, is allowed to form.

Alloying elements also lower the M_s temperatures and, consequently, greater stresses and distortion are introduced during quenching. This can be minimized by austempering and martempering as discussed above, but such treatments are expensive. Alloying

elements should therefore be chosen to produce the maximum retardation of tempering for minimum depression of M_s; Table 8.2 shows that (1) C should be as low as possible, (2) Si and Co are particularly effective, and (3) Mo is the preferred element of the Mo, W, V group since it is easier to take into solution than V and is cheaper than W.

Some elements, particularly Mo and V, produce quite high tempering temperatures. In quantities above about 1% for Mo and $\frac{1}{2}$% for V, a precipitation reaction is also introduced which has its maximum hardening effect at 550°C. This phenomenon of increased hardness by precipitation at higher temperatures is known as secondary hardening and may be classified as a fourth stage of tempering. $2-2\frac{1}{2}$ Mo addition produces adequate temper resistance and changes the precipitate to Mo_2C which is more resistant to overageing than Cr_7C_3 which is present in most alloy steels. High V additions lead to undissolved V_4C_3 at the quenching temperature, but 0.5 V in conjunction with 2Mo does not form a separate carbide during tempering but dissolves in the Mo_2C. Cr also dissolves in Mo_2C but lowers the lattice parameters of the carbide and hence lowers the temper resistance by decreasing the matrix/carbide mismatch. However, 1Cr may be tolerated without serious reduction in temper resistance and reduces the tendency to quench crack. Si decreases the lattice parameter of matrix ferrite and hence increases temper resistance. A typical secondary hardening steel usually contains 0.4C, 2Mo, 0.5V, 0.5Si and 1.5Cr, with 1.8 GN/m^2 TS and 15% elongation.

8.3.6 Thermo-mechanical treatments

To produce steels with an improved strength/ductility ratio the heat-treatment may be modified to include a deformation operation in the process. The combined use of mechanical working and heat-treatment is generally called thermo-mechanical treatment (THT). Three types of treatment have proved successful with martensitic and bainitic steels. These may be classified as follows:

1. Deformation in the stable austenite range above A_3 before transformation, i.e. (HTHT).
2. Deformation below A_1 before transformation; this (LTHT) low-temperature thermo-mechanical treatment is called ausforming.
3. Deformation during isothermal transformation to pearlite, i.e. below A_3, known as isoforming.

The main advantage of HTHT is in grain refinement, and steels such as silicon-steels that recrystallize slowly are particularly suitable. It can, however, be applied to low-alloy high-carbon tool steels which are not suitable for ausforming, with significant increases in strength and toughness. The fatigue limit is also improved in many steels provided the deformation is limited to 25–30%. In ausforming, the deformation is

Table 8.2 Influence of alloying additions on tempering

Element	Retardation in tempering per 1% addition	Ratio of retardation of tempering to depression of M_s
C	−40	negative
Co	8	>8
Cr	0	0
Mn	8	0.24
Mo	17	0.8
Ni	8	0.5
Si	20	1.8
V	30	>1.0
W	10	0.9

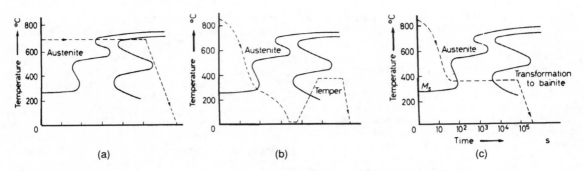

Figure 8.28 *Diagrams showing the heat-treatment procedure during (a) isothermal annealing, (b) martempering and (c) austempering.*

usually carried out in the range 450–550°C and hence the steel must have a large bay in the TTT diagram to enable the deformation to be carried out. A suitable steel is Fe–0.35C–0.5Mn–1.5Ni–1.25Cr–0.25Mo for which the strength increases by about 4.6–7.7 MN/m² for each per cent of deformation. The properties are improved as the deformation temperature is lowered, provided it is not below M_s, and with high deformation treatments (>70%) strengths up to about E/70 with good ductility have been achieved. A very fine carbide dispersion is produced in the austenite during deformation together with a high density of dislocations. The removal of carbon from solution in the austenite means that during transformation the martensite formed is less supersaturated in C and thus has lower tetragonality and is more ductile. The carbides also pin the dislocations in the austenite, helping to retain some of them together with those formed during the transformation. The martensite formed is therefore heavily dislocated with relatively stable dislocations (compared to those which would be formed by deforming martensite at room temperature), and has superior strength and toughness. Such steels are, of course, somewhat difficult to machine.

Isoforming has potential in improving the toughness of low-alloy steels. During isoforming to pearlite the normal ferrite/pearlite structure is modified, by the polygonization of sub-grains in the ferrite and the spheroidizing of cementite particles. Isoforming to bainite is also possible.

8.4 Fracture and toughness

8.4.1 Griffith micro-crack criterion

Most materials break at a stress well below the theoretical fracture stress, which is that stress, σ_t, required to pull apart two adjoining layers of atoms. This stress varies with the distance between the atom planes and, as shown in Figure 8.29, may be approximated to a sine curve of wavelength such that

$$\sigma = \sigma_t \sin\left(\frac{2\pi u}{\lambda}\right) \simeq \sigma_t \left(\frac{2\pi u}{\lambda}\right)$$

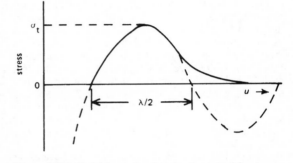

Figure 8.29 *Model for estimating the theoretical fracture strength σ_t.*

where u is the displacement from the equilibrium spacing b. From Hooke's law $\sigma = (Eu/b)$ and hence $\sigma_t = \lambda E/2\pi b$. Now in pulling apart the two atomic planes it is necessary to supply the surface energy γ and hence

$$\gamma = \int_0^{\lambda/2} \sigma_t \sin\left(\frac{2\pi u}{\lambda}\right) du = \frac{\lambda \sigma_t}{2\pi}$$

so that the theoretical tensile strength is given by

$$\sigma_t = \sqrt{(E\gamma/b)} \tag{8.22}$$

Glass fibres and both metallic and non-metallic whiskers have strengths approaching this ideal value of about E/10, but bulk metals even when tested under favourable conditions (e.g. at 4 K) can rarely withstand stresses above E/100. Griffith, in 1920, was the first to suggest that this discrepancy was due to the presence of small cracks which propagate through the crystal and cause fracture. Griffith's theory deals with elastic cracks, where at the tip of the crack atomic bonds exist in every stage of elongation and fracture. As the crack grows, each of the bonds in its path take up the strain, and the work done in stretching and breaking these bonds becomes the surface energy γ of the fractured faces. If separation of the specimen between two atomic layers occurs in this way, the theoretical strength only needs to be exceeded at one point at a time, and the applied stress will be much lower than the theoretical fracture stress. The work done in breaking the bonds has to be supplied by the applied force, or by the elastic energy of the system. Assuming for a crack of length $2c$ that an approximately circular region of radius c is relieved of stress σ and hence strain energy $(\sigma^2/2E)\pi c^2$ by the presence of the crack, the condition when the elastic strain of energy balances the increase of surface energy is given by

$$\frac{\mathrm{d}}{\mathrm{d}c}\left(\frac{\pi c^2 \sigma^2}{2E}\right) = \frac{\mathrm{d}}{\mathrm{d}c}(2c\gamma)$$

and leads to the well-known Griffith formula

$$\sigma = \sqrt{\left(\frac{2\gamma E}{\pi c}\right)} \simeq \sqrt{\left(\frac{\gamma E}{2c}\right)} \qquad (8.23)$$

for the smallest tensile stress σ able to propagate a crack of length $2c$. The Griffith criterion therefore depends on the assumption that the crack will spread if the decrease in elastic strain energy resulting from an increase in $2c$ is greater than the increase in surface energy due to the increase in the surface area of the crack.

Griffith's theory has been verified by experiments on glasses and polymers at low temperatures, where a simple process of fracture by the propagation of elastic cracks occurs. In such 'weak' brittle fractures there is little or no plastic deformation and γ is mainly the surface energy ($\approx 1-10$ J m^{-2}) and the fracture strength $\sigma_f \approx 10^{-5}E$. In crystalline solids, however, the cracks are not of the elastic type and a plastic zone exists around the crack tip as shown in Figure 8.30. In such specimens, fracture cannot occur unless the applied tensile stress σ is magnified to the theoretical strength σ_t. For an atomically sharp crack (where the radius of the root of the crack r is of the order of b) of length $2c$ it can be shown that the magnified stress σ_m will be given by $\sigma_m = \sigma\sqrt{(c/r)}$ which, if the crack is to propagate, must be equal to the theoretical fracture stress of the material at the end of the crack. It follows that substituting this value of σ_t in equation (8.22) leads to the Griffith formula of equation (8.23).

Figure 8.30 shows the way the magnified stress drops off with distance from the tip of the crack. Clearly, at some distance r_y the stress reaches the yield stress and plastic flow occurs. There is thus a zone of plastic flow around the tip of radius r_y. The larger the plastic zone, as in ductile metals, the more energy is absorbed in fracture. In ceramics this zone is usually small.

In 'strong' fractures γ is greatly increased by the contribution of the plastic work around the crack tip which increases the work required for crack propagation. The term γ must now be replaced by $(\gamma + \gamma_p)$ where γ_p is the plastic work term; generally $(\gamma + \gamma_p)$ is replaced by G, the strain energy release rate,

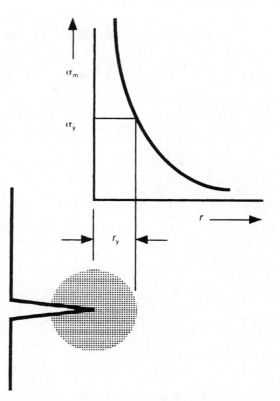

Figure 8.30 *Variation of stress from the tip of a crack and the extent of the plastic zone, radius r_y.*

so that equation (8.23) becomes the Orowan–Irwin relationship

$$\sigma = \sqrt{(EG/\pi c)} \qquad (8.24)$$

Here, G might be $\sim 10^4$ J m^{-2} and $\sigma_f \approx 10^{-2}-10^{-3}E$.

8.4.2 Fracture toughness

In engineering structures, particularly heat-treated steels, cracks are likely to arise from weld defects, inclusions, surface damage, etc. and it is necessary to design structures with the knowledge that cracks are already present and capable of propagation at stresses below the macroscopic yield stress as measured in a tensile test. Since different materials show different crack propagation characteristics (e.g. hard steel and glass) it is necessary for the design engineers to find the limiting design stress in terms of some property or parameter of the material. For this reason, a fracture toughness parameter is now being employed to measure the tendency of cracks of given dimensions to propagate under particular stress conditions.

In Section 8.4.1 it was shown that $\sigma\sqrt{(\pi c)} = \sqrt{(EG)}$, which indicates that fast fracture will occur when a material is stressed to a value σ and the crack reaches some critical size, or alternatively when a material containing a crack is subjected to some critical stress σ, i.e. the critical combination of stress and crack length for fast fracture is a constant, $\sqrt{(EG)}$ for the material, where E is Young's modulus and G is the strain energy release rate. The term $\sigma\sqrt{(\pi c)}$ is given the symbol K and is called the stress intensity factor with units MN m$^{-3/2}$. Fast fracture will then occur when $K = K_c$, where $K_c [= \sqrt{(EG_c)}]$ is the critical stress intensity factor, or more commonly the fracture toughness parameter.

The fracture toughness of a material can alter markedly depending on whether the elastic-plastic field ahead of the crack approximates to plane strain or plane stress conditions, much larger values being obtained under plane stress conditions as operate in thin sheets. The important and critical factor is the size of the plastic zone in relation to the thickness of the section containing the crack. When this is small, as in thick plates and forgings, plane strain conditions prevail and the hydrostatic tension ahead of the crack results in a semi-brittle 'flat' fracture. When the value is large as in thin sheets of ductile metals plane stress conditions operate and the tension at the crack front is smaller, giving rise to a more ductile mode of failure. At intermediate values a mixed fracture, with a flat centre bordered by shear lips, is obtained. Thus without changing the structure or properties of the materials in any way it is possible to produce a large difference in fracture toughness by changing the section thickness. For thick sections, when a state of complete constraint is more nearly approached, the values of K_c and G_c approach minimum limiting values. These values are denoted by K_{Ic} and G_{Ic} and are considered to be material constants; the subscript I denotes the first mode of crack extension, i.e. the opening mode (see Figure 8.31).

The general procedure in measuring the fracture toughness parameter is to introduce a crack of suitable size into a specimen of suitable dimension and geometry. The specimen is then loaded at a slow rate and the crack extension measured up to the critical condition. The measurement of K_{Ic} will be valid if the plastic zone size is small (by a factor 10) in relation to the cross-section of the specimen. The zone size r_y may be obtained by equating the stress field of the crack at $r = r_y$ to the strength σ_y of the material and is given by

$$r_y = K_{Ic}^2 / 2\pi\sigma_y^2$$

An Ashby property chart of fracture toughness versus strength, given in Figure 8.32, shows that the size of the zone (broken lines) varies from atomic dimensions for brittle ceramics to tens of centimetres for ductile metals.

In designing safe structures for a given load, the structure is required to yield before it breaks. For a minimum detectable crack size of $2c$ this condition is given by $(K_{Ic}/\sigma_y) \geq \sqrt{\pi c}$. The safest material is the one with the greatest value of K_{Ic}/σ_y. Clearly, a high

yield stress σ_y is also required. The chart shows that steel satisfies both these requirements and indicates why it is still the best material for highly stressed structures where weight is not important.

Fracture toughness requirements are now written into the general specification of high-technology alloys and hence it is necessary to determine the effect of heat-treatment and alloying additions on fracture toughness parameters. Processes such as ausforming and controlled rolling improve the fracture toughness of certain steels. Carbon has a considerable effect and there are advantages in reducing the C-level below 0.1% where possible. High-strength low-alloy (HSLA) steels have C≤0.1% and the Nb, V and Ti additions form fine carbides which together with the small grain sizes enable good strength levels and acceptable fracture toughness values to be achieved. Maraging steels with high alloy and low carbon (<0.04%) give very high strength combined with high toughness (see Chapter 9).

The brittleness of ceramics is directly linked to their high notch-sensitivity. The fracture toughness of most ceramics is low: expressed in quantitative terms, it is commonly less than 8 MN m$^{-3/2}$. Flaws, often very minute, are almost invariably present in ceramics and act as stress-concentrating notches. It is extremely difficult to prevent these flaws from forming during manufacture and service. In terms of engineering practice, brittle ceramic components are intolerant of misalignment and poor fits within assemblies. The presence of flaws is also responsible for the variability or 'spread' of results from mechanical tests and introduces uncertainty into the design process. (The 'spread' is much less for metallic materials.) Design procedures have moved well beyond the principle that ceramics are only safe when compressive stresses are dominant. Probabilistic assessments of mechanical test results from ceramics now tackle the difficult task of allowing for randomness in the size, shape and distribution of flaws (see Section 10.6).

Despite these underlying problems, progress has been made in producing tough 'ductile ceramics'.

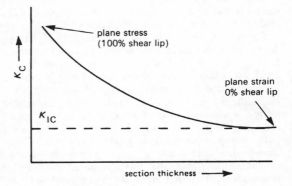

Figure 8.31 *Variation in the fracture toughness parameter with section thickness.*

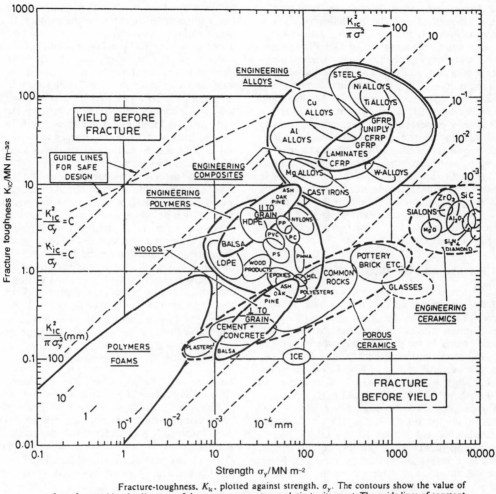

Figure 8.32 *Ashby property chart of fracture toughness versus strength (Ashby, 1989, pp. 1273–93; with permission of Elsevier Science Ltd).*

Fracture toughness values above 20 MN m$^{-3/2}$ have been achieved. Typical measures include elimination of flaws (microcracks, pores), incorporation of crack-retarding phases (in zirconia) and reduction of average grain size (below 1 mm).

At ambient temperature, fracture surfaces of ceramics are conchoidal (glassy), intergranular or cleavage in character, depending on the material. The strain at fracture is very small, being about 0.001. At elevated temperatures, under creep conditions, fracture strain is greater. Conventional ceramics have little or no capacity for slip and crack tips therefore tend to remain sharp. As the crack propagates, the load-supporting cross-section gets smaller and the general level of stress increases so that failure can be sudden.

These remarks refer to tensile stresses since compressive stresses will, of course, tend to close the crack.

In some polycrystalline ceramics, such as magnesia, the von Mises criterion for maintenance of cohesions is not satisfied. Slip is limited and cracks are not effectively blunted. However, raising the temperature often enables the necessary minimum of five independent deformation modes to operate, leading to ductility. Any treatments which eliminate surface flaws will naturally enhance this ductility. (Against this background, one can readily appreciate why fabrication methods for polycrystalline ceramic components, in contrast to those for metals, are not based upon bulk deformation.) At these higher temperatures the environment becomes increasingly more likely to react with surfaces of the ceramic: it may even penetrate an open texture and cause crack-blunting. The mechanisms of flow in polycrystalline ceramics at elevated temperatures are similar to those encountered in metallic systems (e.g. glide and climb of dislocations, vacancy diffusion, grain boundary sliding). The similarity between deformation processes in polycrystalline ceramics and metals is evident if one compares the layout of the corresponding deformation-mechanism maps. A minor phase is usually present at grain boundary surfaces in ceramics, functioning as a ceramic bond, and if it softens with rise in temperature, then the grains may be able to slide past each other. These regions tend to liquefy before the actual grains so that it is advisable not to exceed the solidus temperature. Grain boundary surfaces are particularly susceptible to the nucleation of cracks. As in the final accelerating stage of metallic creep, these cracks form in planes normal to the direction of applied stress.

8.4.3 Cleavage and the ductile–brittle transition

The fracture toughness versus strength chart, shown in Figure 8.32, indicates the wide spread of values for the different classes of material. Metals dissipate much energy in the plastic zone, which accounts for the difference between the fracture energy G_{Ic} and the true surface energy γ. The larger the zone, the more energy is absorbed; G_{Ic} is high and so is K_{Ic} ($\approx$100 MN m$^{-3/2}$). Ceramics and glasses fracture without much plastic flow to blunt the cracks by simply breaking atomic bonds, leading to cleavage; for these materials K_{Ic} is less than 10 MN m$^{-3/2}$.

At low temperatures some metals, notably steels, also become brittle and fracture by cleavage. Since they are ductile at room temperature this transition

to brittle cleavage behaviour is quite spectacular and has led to several engineering catastrophes. In general, brittle cleavage can occur in metals with bcc and cph under the appropriate conditions while in fcc materials it does not. The most important factor linking these three different structures is the Peierls stress and the way the yield stress varies with temperature. In steel, for example, the yield stress increases rapidly with lowering of temperature below room temperature such that plastic deformation at the crack tip is minimized and the fracture mechanism changes from plastic tearing to cleavage. Even in these materials some plastic deformation does occur.

Several models have been suggested for the process whereby glide dislocations are converted into micro-cracks. The simplest mechanism, postulated by Stroh, is that involving a pile-up of dislocations against a barrier, such as a grain boundary. The applied stress pushes the dislocations together and a crack forms beneath their coalesced half-planes, as shown in Figure 8.33a. A second mechanism of crack formation, suggested by Cottrell, is that arising from the junction of two intersecting slip planes. A dislocation gliding in the (1 0 1) plane coalesces with one gliding in the (1 0 $\bar{1}$) plane to form a new dislocation which lies in the (0 0 1) plane according to the reaction,

$$a/2[\bar{1}\,\bar{1}\,1] + a/2[1\,1\,1] \rightarrow a[0\,0\,1]$$

The new dislocation, which has a Burgers vector $a[0\,0\,1]$, is a pure edge dislocation and, as shown in Figure 8.33b, may be considered as a wedge, one lattice constant thick, inserted between the faces of the (0 0 1) planes. It is considered that the crack can then grow by means of other dislocations in the (1 0 1) and (1 0 $\bar{1}$) planes running into it. Although the mechanism readily accounts for the observed (1 0 0) cleavage

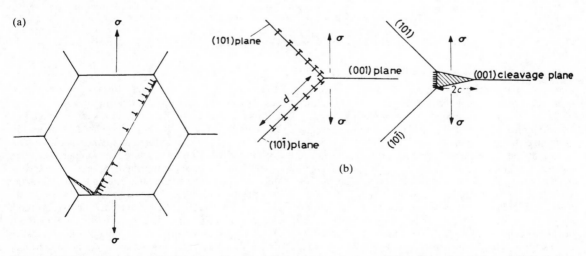

Figure 8.33 *Formation of a crack (a) by the piling up of dislocations against a grain boundary after strain and (b) by the dislocations on (1 0 1) and (10$\bar{1}$) planes running together (after Cottrell, 1958, p. 192).*

plane of bcc metals, examples have not been directly observed.

While these dislocation coalescence mechanisms may operate in single-phase materials, in two-phase alloys it is usually easier to nucleate cracks by piling up dislocations at particles (e.g. grain boundary cementite or cementite lamellae in pearlite). The pile-up stress then leads to cracking of either the particle or the particle/matrix interface. A brittle–ductile transition can then be explained on the basis of the criterion that the material is ductile at any temperature, if the yield stress at that temperature is smaller than the stress necessary for the growth of these micro-cracks, but if it is larger the material is brittle. If cleavage cracks are formed by such a dislocation mechanism, the Griffith formula may be rewritten to take account of the number of dislocations, n, forming the crack. Thus, rearranging Griffith's formula we have

$$\sigma(2c\sigma/E) = \gamma = 2c\sigma^2/E$$

where the product of length $2c$ and the strain σ/E is the maximum displacement between the faces of the crack. This displacement will depend on the number of dislocations forming the cleavage wedge and may be interpreted as a pile-up of n edge dislocations, each of Burgers vector a, so that equation (8.23) becomes

$$\sigma(na) = \gamma \qquad (8.25)$$

and gives a general criterion for fracture. Physically, this means that a number of glide dislocations, n, run together and in doing so cause the applied stress acting on them to do some work, which for fracture to occur must be at least sufficient to supply the energy to create the new cracked faces, i.e. $(\gamma + \gamma_p)$.

Qualitatively, we would expect those factors which influence the yield stress also to have an effect on the ductile–brittle fracture transition. The lattice 'friction' term σ_i, dislocation locking term k_y, and grain size $2d$ should also all be important because any increase in σ_i and k_y, or decrease in the grain size, will raise the yield stress with a corresponding tendency to promote brittle failure.

These conclusions have been put on a quantitative basis by Cottrell, who considered the stress needed to grow a crack at or near the tensile yield stress, σ_y, in specimens of grain diameter, $2d$. Let us consider first the formation of a micro-crack. If τ_y is the actual shear stress operating, the effective shear stress acting on a glide band is only $(\tau_y - \tau_i)$, where, it will be remembered, τ_i is the 'friction' stress resisting the motion of unlocked dislocations arising from the Peierls–Nabarro lattice stress, intersecting dislocations or groups of impurities. The displacement na is then given by

$$na = [(\tau_y - \tau_i)/\mu]d \qquad (8.26)$$

where μ is the shear modulus and d is the length of the slip band containing the dislocations (assumed

here to be half the grain diameter). Once a microcrack is formed, however, the whole applied tensile stress normal to the crack acts on it, so that σ can be written as $(\tau_y \times \text{constant})$, where the constant is included to account for the ratio of normal stress to shear stress. Substituting for na and σ in the Griffith formula (equation (8.25)) then fracture should be able to occur at the yield point when $\sigma = \sigma_y$ and

$$\tau_y(\tau_y - \tau_i)d = C\mu\gamma \qquad (8.27)$$

where C is a constant. The importance of the avalanche of dislocations produced at the yield drop can be seen if we replace τ_y by (constant $\times \sigma_y$), τ_i by (constant $\times \sigma_i$) and use the Petch relationship $\sigma_y = \sigma_i + k_y d^{-1/2}$, when equation (8.27) becomes

$$(\sigma_i d^{1/2} + k_y)k_y = \beta\mu\gamma \qquad (8.28)$$

where β is a constant which depends on the stress system; $\beta \approx 1$ for tensile deformation and $\beta \approx \frac{1}{3}$ for a notched test.

This is a general equation for the propagation of a crack at the lower yield point and shows what factors are likely to influence the fracture process. Alternative models for growth-controlled cleavage fracture have been developed to incorporate the possibility of carbide particles nucleating cracks. Such models emphasize the importance of yield parameters and grain size as well as carbide thickness. Coarse carbides give rise to low fracture stresses, thin carbides with high fracture stresses and ductile behaviour.

8.4.4 Factors affecting brittleness of steels

Many of the effects of alloying, heat treatment, and condition of testing on brittle fracture can be rationalized on the basis of the above 'transition' equation.

Ductile–brittle transition   Under conditions where the value of the left-hand side of equation (8.28) is less than the value of the right-hand side, ductile behaviour should be observed; when the left-hand side exceeds the right-hand side the behaviour should be brittle. Since the right-hand side of equation (8.28) varies only slowly with temperature, it is the way in which changes occur in values of the terms on the left of the equation which are important in determining the ductile–brittle transition. Thus, in a given material brittleness should be favoured by low temperatures and high strain-rate, because these give rise to large values of σ_i and k_y, and by large grain sizes. On the right-hand side, the typical effect of a sharp notch is to raise the transition temperature of structural steel from around 100 K for a normal tensile test into the range of 200–300 K, because the value of β is lowered.

Effects of composition and grain size   At a constant temperature, because the values of σ_i and k_y remain fixed, the transition point will occur at a critical grain size above which the metal is brittle and below which it is ductile.

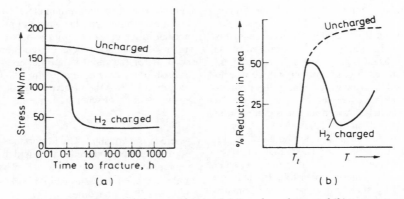

Figure 8.34 *Influence of hydrogen on fracture behaviour showing (a) time-dependence and (b) temperature-dependence.*

The inclusion in equation (8.28) of the grain-size term, $d^{1/2}$, in combination with the σ_i term, enables many previous metallurgical misunderstandings to be cleared up. It shows that there is no simple connection between hardness and brittleness, since hardening produced by refining the grain size reduces the brittleness, whereas hardening due to an increase in σ_i increases the brittleness.

Heat treatment is generally used to control the grain size of the sample and refine the structure. 'Killed' steel has very good notch toughness, because aluminium additions refine the grain size. Manganese reduces the grain size and by combining with carbon also reduces the k_y value so that this addition is especially beneficial in improving low-temperature ductility. It is fairly evident that an improved notch toughness steel, compared with that used for welded ships in World War II, is given by increasing the manganese content and decreasing the carbon content, i.e. a high manganese-to-carbon ratio. Other additions, particularly nickel and chromium, have a similar effect on low-temperature ductility.

The Group 6A metals (Cr, Mo and W) are more susceptible to brittle fracture than the Group 5A metals (V, Nb and Ta). A comparison of these metals in terms of cleavage fracture is difficult, however, since Cr, Mo and W are susceptible to grain boundary fracture because segregation of impurities to such regions reduces the effective surface energy γ. However, even if this effect is eliminated by lowering the impurity level, it appears that Ta, Nb and V are more ductile than Fe, Mo, Cr and W, presumably because they have a lower k_y/μ ratio, and a higher γ value.

Work-hardening and irradiation-hardening Small amounts of plastic deformation at room temperature, which overcomes the yield point and unlocks some of the dislocations, improves the ductility at low temperatures. The room-temperature ductility of chromium is similarly affected by small amounts of plastic deformation at 400°C. In general, however, plastic deformation which leads to work-hardening embrittles the metal because it raises the σ_i contribution, due to the formation of intersecting dislocations, vacancy aggregates and other lattice defects.

The importance of twins in fracture is not clear as there are several mechanisms other than twinning for the formation of a crack which can initiate fracture, and there is good evidence that micro-cracks form in steel in the absence of twins, and that cracks start at inclusions. Nevertheless, twinning and cleavage are generally found under similar conditions of temperature and strain-rate in bcc transition metals, probably because both phenomena occur at high stress levels. The nucleation of a twin requires a higher stress than the propagation of the twin interface.

Irradiation-hardening also embrittles the metal. According to the theory of this type of hardening outlined in Chapter 7, radiation damage can produce an increase in both k_y (migration to dislocations of vacancies or interstitials) and σ_i (formation of dislocation loops and other aggregated defects). However, for steel, radiation-hardening is principally due to an increased σ_i contribution, presumably because the dislocations in mild steel are already too heavily locked with carbon atoms for any change in the structure of the dislocation to make any appreciable difference to k_y. Nevertheless, a neutron dose of 1.9×10^{23} n m^{-2} will render a typical fine-grained, unnotched mild steel, which is normally ductile at $-196°C$, quite brittle. Moreover, experiments on notched fine-grained steel samples (see Figure 7.1c) show that this dose increases the ductile–brittle transition temperature by 65°C.

Microstructure The change in orientation at individual grain boundaries impedes the propagation of the cleavage crack by (1) creating cleavage steps, (2) causing localized deformation, and (3) tearing near the grain boundary. It is the extra work done (γ_p) in such processes which increases the apparent surface energy γ to ($\gamma_s + \gamma_p$). It follows, therefore, that the smaller the distance a crack is able to propagate without being deviated by a change of orientation of the cleavage plane, the greater is the resistance to brittle fracture. In this respect, the coarser high-temperature products of steel, such as pearlite and upper bainite, have inferior fracture characteristics compared with the finer lower bainite and martensite products. The fact that coarse carbides promote cleavage while fine carbides lead to ductile behaviour has already been discussed.

8.4.5 Hydrogen embrittlement of steels

It is well known that ferritic and martensitic steels are severely embrittled by small amounts of hydrogen. The hydrogen may be introduced during melting and retained during the solidification of massive steel castings. Plating operations (e.g. Cd plating of steel for aircraft parts) may also lead to hydrogen embrittlement. Hydrogen can also be introduced during acid pickling or welding, or by exposure to H_2S atmospheres.

The chief characteristics of hydrogen embrittlement are its (1) strain-rate sensitivity, (2) temperature-dependence and (3) susceptibility to produce delayed fracture (see Figure 8.34). Unlike normal brittle fracture, hydrogen embrittlement is enhanced by slow strain-rates and consequently, notched-impact tests have little significance in detecting this type of embrittlement. Moreover, the phenomenon is not more common at low temperatures, but is most severe in some intermediate temperature range around room temperature (i.e. $-100°C$ to $100°C$). These effects have been taken to indicate that hydrogen must be present in the material and must have a high mobility in order to cause embrittlement in polycrystalline aggregates.

A commonly held concept of hydrogen embrittlement is that monatomic hydrogen precipitates at internal voids or cracks as molecular hydrogen, so that as the pressure builds up it produces fracture. Alternatively, it has been proposed that the critical factor is the segregation of hydrogen, under applied stress, to regions of triaxial stress just ahead of the tip of the crack, and when a critical hydrogen concentration is obtained, a small crack grows and links up with the main crack. Hydrogen may also exist in the void or crack but it is considered that this has little effect on the fracture behaviour, and it is only the hydrogen in the stressed region that causes embrittlement. Neither model considers the Griffith criterion, which must be satisfied if cracks are to continue spreading.

An application of the fracture theory may be made to this problem. Thus, if hydrogen collects in microcracks and exerts internal pressure P, the pressure may be directly added to the external stress to produce a total stress $(P + p)$ for propagation. Thus the crack will spread when

$$(P + p)na = (\gamma_s + \gamma_p) \tag{8.29}$$

where the surface energy is made up from a true surface energy γ_s and a plastic work term γ_p. The possibility that hydrogen causes embrittlement by becoming adsorbed on the crack surfaces thereby lowering γ is thought to be small, since the plastic work term γ_p is the major term controlling γ, whereas adsorption would mainly effect γ_s.

Supersaturated hydrogen atoms precipitate as molecular hydrogen gas at a crack nucleus, or the interface between non-metallic inclusions and the matrix. The stresses from the build-up of hydrogen pressure are then relieved by the formation of small cleavage cracks. Clearly, while the crack is propagating, an insignificant amount of hydrogen will diffuse to the crack and, as a consequence, the pressure inside the crack will drop. However, because the length of the crack has increased, if a sufficiently large and constant stress is applied, the Griffith criterion will still be satisfied and completely brittle fracture can, in theory, occur. Thus, in iron single crystals, the presence or absence of hydrogen appears to have little effect during crack propagation because the crack has little difficulty spreading through the crystal. In polycrystalline material, however, the hydrogen must be both present and mobile, since propagation occurs during tensile straining.

When a sufficiently large tensile stress is applied such that $(p + P)$ is greater than that required by the Griffith criterion, the largest and sharpest crack will start to propagate, but will eventually be stopped at a microstructural feature, such as a grain boundary, as previously discussed. The pressure in the crack will then be less than in adjacent cracks which have not been able to propagate. A concentration gradient will then exist between such cracks (since the concentration is proportional to the square root of the pressure of hydrogen) which provides a driving force for diffusion, so that the hydrogen pressure in the enlarged crack begins to increase again. The stress to propagate the crack decreases with increase in length of crack, and since p is increased by straining, a smaller increment ΔP of pressure may be sufficient to get the crack restarted. The process of crack propagation followed by a delay time for pressure build-up continues with straining until the specimen fails when the area between the cracks can no longer support the applied load. In higher strain-rate tests the hydrogen is unable to diffuse from one stopped crack to another to help the larger crack get started before it becomes blunted by plastic deformation at the tip. The decrease in the susceptibility to hydrogen embrittlement in specimens tested at low temperatures results from the lower pressure build-up at these temperature since $PV = 3nRT$, and also because hydrogen has a lower mobility.

8.4.6 Intergranular fracture

Intergranular brittle failures are often regarded as a special class of fracture. In many alloys, however, there is a delicate balance between the stress required to cause a crack to propagate by cleavage and that needed to cause brittle separation along grain boundaries. Although the energy absorbed in crack propagation may be low compared to cleavage fractures, much of the analysis of cleavage is still applicable if it is considered that chemical segregation to grain boundaries or crack faces lowers the surface energy γ of the material. Fractures at low stresses are observed in austenitic chromium–nickel steels, due to the embrittling effect of intergranular carbide precipitation at grain boundaries. High transition temperatures and low fracture stresses are also common in tungsten and molybdenum as a result of the formation of thin second-phase films due to small amounts of oxygen, nitrogen or carbon. Similar behaviour is observed in the embrittlement of copper by antimony and iron by oxygen, although in some cases the second-phase films cannot be detected.

A special intergranular failure, known as temper embrittlement, occurs in some alloy steels when tempered in the range 500–600°C. This phenomenon is associated with the segregation of certain elements or combinations of elements to the grain boundaries. The amount segregated is very small ($\sim$ a monolayer) but the species and amount has been identified by AES on specimens fractured intergranularly within the ultra-high vacuum of the Auger system. Group VIB elements are known to be the most surface-active in iron but, fortunately, they combine readily with Mn and Cr thereby effectively reducing their solubility. Elements in Groups IVB and VB are less surface-active but often co-segregate in the boundaries with Ni and Mn. In Ni–Cr steels, the co-segregation of Ni–P and Ni–Sb occurs, but Mo additions can reduce the tendency for temper embrittlement. Since carbides are often present in the grain boundaries, these can provide the crack nucleus under the stress concentration from dislocation pile-ups either by cracking or by decohesion of the ferrite/carbide interface, particularly if the interfacial energy has been lowered by segregation.

8.4.7 Ductile failure

Ductile failure was introduced in Chapter 4 because of the role played by voids in the failure processes, which occurs by void nucleation, growth and coalescence. The nucleation of voids often takes place at inclusions. The dislocation structure around particle inclusions leads to a local rate of work-hardening higher than the average and the local stress on reaching some critical value σ_c will cause fracture of the inclusion or decohesion of the particle/matrix interface, thereby nucleating a void. The critical nucleation strain ε_n can be estimated and lies between 0.1 and 1.0 depending on the model. For dispersion-hardening materials where dislocation loops are generated the stress on the interface due to the nearest prismatic loop, at distance r, is $\mu b/r$, and this will cause separation of the interface when it reaches the theoretical strength of the interface, of order γ_w/b. The parameter r is given in terms of the applied shear strain ε, the particle diameter d and the length k equal to half the mean particle spacing as $r = 4kb/\varepsilon d$. Hence, void nucleation occurs on a particle of diameter d after a strain ε, given by $\varepsilon = 4k\gamma_w/\mu db$. Any stress concentration effect from other loops will increase with particle size, thus enhancing the particle size dependence of strain to voiding.

Once nucleated, the voids grow until they coalesce to provide an easy fracture path. A spherical-shaped void concentrates stress under tensile conditions and, as a result, elongates initially at about $C(\approx 2)$ times the rate of the specimen, but as it becomes ellipsoidal the growth-rate slows until finally the elongated void grows at about the same rate as the specimen. At some critical strain, the plasticity becomes localized and the voids rapidly coalesce and fracture occurs. The localization of the plasticity is thought to take place when the voids reach a critical distance of approach,

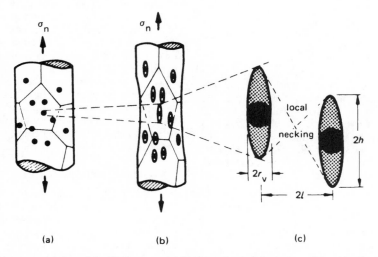

(a)   (b)   (c)

Figure 8.35 *Schematic representation of ductile fracture. (a) Voids nucleate at inclusions, (b) voids elongate as the specimen extends, (c) voids coalesce to cause fracture when their length 2h is about equal to their separation (after Ashby et al., 1979).*

given when the void length $2h$ is approximately equal to the separation, as shown in Figure 8.35. The true strain for coalescence is then

$$\varepsilon = (1/C)\ln[\alpha(2l - 2r_v)/2r_v]$$
$$\simeq (1/C)\ln[\alpha(1/f_v^{1/2}) - 1] \qquad (8.30)$$

where $\alpha \approx 1$ and f_v is the volume fraction of inclusions.

Void growth leading to failure will be much more rapid in the necked portion of a tensile sample following instability than during stable deformation, since the stress system changes in the neck from uniaxial tension to approximately plane strain tension. Thus the overall ductility of a specimen will depend strongly on the macroscopic features of the stress–strain curves which (from Considère's criterion) determines the extent of stable deformation, as well as on the ductile rupture process of void nucleation and growth. Nevertheless, an equation of the form of (8.30) reasonably describes the fracture strain for cup and cone failures.

The work of decohesion influences the progress of voiding and is effective in determining the overall ductility in a simple tension test in two ways. The onset of voiding during uniform deformation depresses the rate of work-hardening which leads to a reduction in the uniform strain, and the void density and size at the onset of necking determines the amount of void growth required to cause ductile rupture. Thus for matrices having similar work-hardening properties, the one with the least tendency to 'wet' the second phase will show both lower uniform strain and lower necking strain. For matrices with different work-hardening potential but similar work of decohesion the matrix having the lower work-hardening rate will show the lower reduction prior to necking but the greater reduction during necking, although two materials will show similar total reductions to failure.

The degree of bonding between particle and matrix may be determined from voids on particles annealed to produce an equilibrium configuration by measuring the contact angle θ of the matrix surface to the particle surface. Resolving surface forces tangential to the particle, then the specific interface energy γ_I is given approximately in terms of the matrix surface energy γ_m and the particle surface energy γ_P as $\gamma_I = \gamma_P - \gamma_m\cos\theta$. The work of separation of the interface γ_w is then given by

$$\gamma_w = \gamma_P + \gamma_m - \gamma_I = \gamma_m(1 + \cos\theta) \qquad (8.31)$$

Measurements show that the interfacial energy of TD nickel is low and hence exhibits excellent ductility at room temperature. Specific additions (e.g. Zr to TD nickel, and Co to Ni–Al$_2$O$_3$ alloys) are also effective in lowering the interfacial energy, thereby causing the matrix to 'wet' the particle and increase the ductility. Because of their low γ_I, dispersion-hardened materials have superior mechanical properties at high temperatures compared with conventional hardened alloys.

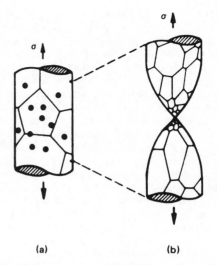

Figure 8.36 *Schematic representation of rupture with dynamic recrystallization (after Ashby et al., 1979).*

8.4.8 Rupture

If the ductile failure mechanisms outlined above are inhibited then ductile rupture occurs (see Figure 8.36). Specimens deformed in tension ultimately reach a stage of mechanical instability when the deformation is localized either in a neck or in a shear band. With continued straining the cross-section reduces to zero and the specimen ruptures, the strain-to-rupture depending on the amount of strain before and after localization. These strains are influenced by the work-hardening behaviour and strain-rate sensitivity. Clearly, rupture is favoured when void nucleation and/or growth is inhibited. This will occur if (1) second-phase particles are removed by zone-refining or dissolution at high temperatures, (2) the matrix/particle interface is strong and ε_n is high, (3) the stress state minimizes plastic constraint and plane strain conditions (e.g. single crystals and thin sheets), (4) the work-hardening rate and strain-rate sensitivity is high as for superplastic materials (in some superplastic materials voids do not form but in many others they do and it is the growth and coalescence processes which are suppressed), and (5) there is stress relief at particles by recovery or dynamic recrystallization. Rupture is observed in most fcc materials, usually associated with dynamic recrystallization.

8.4.9 Voiding and fracture at elevated temperatures

Creep usually takes place above $0.3T_m$ with a rate given by $\dot\varepsilon = B\sigma^n$, where B and n are material parameters, as discussed in Chapter 7. Under such conditions ductile failure of a transgranular nature, similar to the ductile failure found commonly at low temperatures, may occur, when voids nucleated at inclusions within the grains grow during creep deformation and coalesce to produce fracture. However, because these three processes are occurring at $T \approx 0.3T_m$, local recovery is taking place and this delays both the onset of void nucleation and void coalescence. More commonly at lower stresses and longer times-to-fracture, intergranular rather than transgranular fracture is observed. In this situation, grain boundary sliding leads to the formation of either wedge cracks or voids on those boundaries normal to the tensile axis, as shown schematically in Figure 8.37b. This arises because grain boundary sliding produces a higher local strain-rate on an inclusion in the boundary than in the body of the grain, i.e. $\dot{\varepsilon}_{local} \simeq \dot{\varepsilon}(fd/2r)$ where $f \approx 0.3$ is the fraction of the overall strain due to sliding. The local strain therefore reaches the critical nucleation strain ε_n much earlier than inside the grain.

The time-to-fracture t_f is observed to be $\propto (1/\dot{\varepsilon}_{ss})$, which confirms that fracture is controlled by power-law creep even though the rounded-shape of grain boundary voids indicates that local diffusion must contribute to the growth of the voids. One possibility is that the void nucleation, even in the boundary, occupies a major fraction of the lifetime t_f, but a more likely general explanation is that the nucleated voids or cracks grow by local diffusion controlled by creep in the surrounding grains. Figure 8.37c shows the voids growing by diffusion, but between the voids the material is deforming by power-law creep, since the diffusion fields of neighbouring voids do not overlap. Void growth therefore depends on coupled diffusion and power-law creep, with the creep deformation controlling the rate of cavity growth. It is now believed that most intergranular creep fractures are governed by this type of mechanism.

At very low stresses and high temperatures where diffusion is rapid and power-law creep negligible, the diffusion fields of the growing voids overlap. Under these conditions, the grain boundary voids are able to grow entirely by boundary diffusion; void coalescence then leads to fracture by a process of creep cavitation (Figure 8.38). In uniaxial tension the driving force arises from the process of taking atoms from the void surface and depositing them on the face of the grain that is almost perpendicular to the tensile axis, so that the specimen elongates in the direction of the stress and work is done. The vacancy concentration near the tensile boundary is $c_0 \exp(\sigma\Omega/\mathbf{k}T)$ and near the void of radius r is $c_0 \exp(2\gamma\Omega/r\mathbf{k}T)$, as discussed previously in Chapter 7, where Ω is the atomic volume and γ the surface energy per unit area of the void. Thus vacancies flow usually by grain boundary diffusion from the boundaries to the voids when $\sigma \geq 2\gamma/r$, i.e. when the chemical potential difference $(\sigma\Omega - 2\gamma\Omega/r)$ between the two sites is negative. For a void $r \simeq 10^{-6}$ m and $\gamma \approx 1$ J/m^2 the minimum stress for hole growth is ≈ 2 MN/m^2. In spite of being pure diffusional controlled growth, the voids may not always maintain their equilibrium near-spherical shape. Rapid surface diffusion is required to keep the balance between growth rate and surface redistribution and with increasing stress the voids become somewhat flattened.

8.4.10 Fracture mechanism maps

The fracture behaviour of a metal or alloy in different stress and temperature regimes can be summarized conveniently by displaying the dominant mechanisms

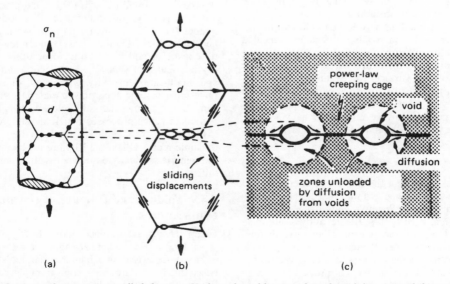

Figure 8.37 *Intergranular, creep-controlled, fracture. Voids nucleated by grain boundary sliding (a) and (b) growth by diffusion in (c) (after Ashby et al., 1979).*

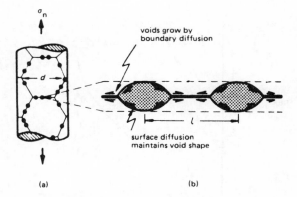

Figure 8.38 *Voids lying on 'tensile' grain boundaries (a) grow by grain boundary diffusion (b) (after Ashby et al., 1979).*

on a fracture mechanism map. Seven mechanisms have been identified, three for brittle behaviour including cleavage and intergranular brittle fracture, and four ductile processes. Figure 8.39 shows schematic maps for fcc and bcc materials, respectively. Not all the fracture regimes are exhibited by fcc materials, and even some of the ductile processes can be inhibited by altering the metallurgical variables. For example, intergranular creep fracture is absent in high-purity aluminium but occurs in commercial-purity material, and because the dispersoid suppresses dynamic recrystallization in TD nickel, rupture does not take place whereas it does in *Nimonic* alloys at temperatures where the γ' and carbides dissolve.

In the bcc metals, brittle behaviour is separated into three fields; a brittle failure from a pre-existing crack, well below general yield, is called either cleavage 1 or brittle intergranular fracture BIF1, depending on the fracture path. An almost totally brittle failure from a crack nucleated by slip or twinning, below general yield, is called either cleavage 2 or BIF2, and

a cleavage or brittle boundary failure after general yield and with measurable strain-to-failure is called either cleavage 3 or BIF3. In many cases, mixed transgranular and intergranular fractures are observed, as a result of small changes in impurity content, texture or temperature which cause the crack to deviate from one path to another, no distinction is then made in the regime between cleavage and BIF. While maps for only two structures are shown in Figure 8.39 it is evident that as the bonding changes from metallic to ionic and covalent the fracture-mechanism fields will move from left to right: refractory oxides and silicates, for example, exhibit only the three brittle regimes and intergranular creep fracture.

8.4.11 Crack growth under fatigue conditions

Engineering structures such as bridges, pressure vessels and oil rigs all contain cracks and it is necessary to assess the safe life of the structure, i.e. the number of stress cycles the structure will sustain before a crack grows to a critical length and propagates catastrophically. The most effective approach to this problem is by the use of fracture mechanics. Under static stress conditions, the state of stress near a crack tip is described by K, the stress intensity factor, but in cyclic loading K varies over a range $\Delta K (= K_{max} - K_{min})$. The cyclic stress intensity ΔK increases with time at constant load, as shown in Figure 8.40a, because the crack grows. Moreover, for a crack of length a the rate of crack growth (da/dN) in μm per cycle varies with ΔK according to the Paris–Erdogan equation

$$da/dN = C(\Delta K)^m \tag{8.32}$$

where C and m are constants, with m between 2 and 4. A typical crack growth rate curve is shown in Figure 8.40b and exhibits the expected linear relationship over part of the range. The upper limit corresponds to K_{Ic}, the fracture toughness of the material, and the lower limit of ΔK is called the threshold for

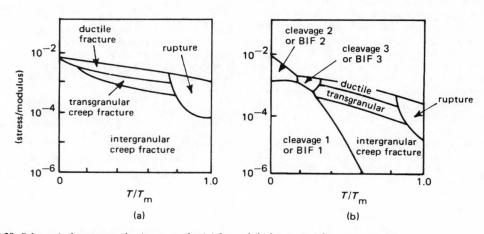

Figure 8.39 *Schematic fracture mechanism maps for (a) fcc and (b) bcc materials.*

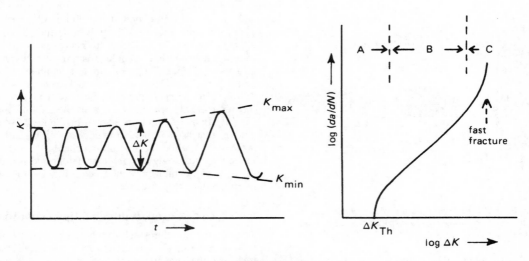

Figure 8.40 *(a) Increase in stress intensity ΔK during fatigue; (b) variation of crack growth rate with increasing ΔK.*

crack growth (ΔK_{th}). Clearly, when the stress intensity factor is less than ΔK_{th} the crack will not propagate at that particular stress and temperature, and hence ΔK_{th} is of significance in design criteria. If the initial crack length is a_0 and the critical length a_c, then the number of cycles to catastrophic failure will be given by

$$N_f = \int_{a0}^{a_c} \mathrm{d}a/C(\Delta K)^m$$
$$= \int_{a0}^{a_c} \mathrm{d}a/C[\Delta\sigma\sqrt{(\pi a)}]^m \qquad (8.33)$$

The mean stress level is known to affect the fatigue life and therefore da/dN. If the mean stress level is increased for a constant value of ΔK, K_{max} will increase and thus as K_{max} approaches K_{Ic} the value of da/dN increases rapidly in practice, despite the constant value of ΔK.

A survey of fatigue fractures indicates there are four general crack growth mechanisms: (1) striation formation, (2) cleavage, (3) void coalescence and (4) intergranular separation; some of these mechanisms have been discussed in Chapter 7. The crack growth behaviour shown in Figure 8.40b can be divided into three regimes which exhibit different fracture mechanisms. In regime A, there is a considerable influence of microstructure, mean stress and environment on the crack growth rate. In regime B, failure generally occurs, particularly in steels, by a transgranular ductile striation mechanism and there is often little influence of microstructure, mean stress or mild environments on crack growth. The degree of plastic constraint which varies with specimen thickness also appears to have little effect. At higher growth rate exhibited in regime C, the growth rates become extremely sensitive to both microstructure and mean stress, with a change from striation formation to fracture modes normally associated with noncyclic deformation, including cleavage and intergranular fracture.

Further reading

Ashby, M. F. and Jones, D. R. H. (1980). *Engineering Materials—An Introduction to their properties and applications*. Pergamon.

Bilby, B. A. and Christian, J. W. (1956) *The Mechanism of Phase Transformations in Metals*, Institute of Metals.

Bowles and Barrett. *Progress in Metal Physics, 3*, 195. Pergamon Press.

Charles, J. A., Greenwood, G. W. and Smith, G. C. (1992). *Future Developments of Metals and Ceramics*. Institute of Materials, London.

Honeycombe, R. W. K. (1981). *Steels, microstructure and properties*. Edward Arnold, London.

Kelly, A. and MacMillan, N. H. (1986). *Strong Solids*. Oxford Science Publications, Oxford.

Kelly, A. and Nicholson, R. B. (eds). (1971). *Strengthening Methods in Crystals*. Elsevier, New York.

Knott, J. (1973). *Fundamentals of Fracture Mechanics*. Butterworths, London.

Knott, J. F. and Withey, P. (1993). *Fracture mechanics, Worked examples*. Institute of Materials, London.

Pickering, F. B. (1978). *Physical Metallurgy and the Design of Steels*. Applied Science Publishers, London.

Porter, D. A. and Easterling, K. E. (1992). *Phase Transformations in Metals and Alloys*, 2nd edn. Chapman and Hall, London.

Chapter 9

Modern alloy developments

9.1 Introduction

In this chapter we will outline some of the developments and properties of modern metallic alloys. Crucial to these materials have been the significant developments that have taken place in manufacturing, made possible by a more detailed understanding of the manufacturing process itself and of the behaviour of the material during both processing and in-service performance. Casting techniques in particular have advanced much over the past decade and now provide reliable clean material with precision. Process modelling is developing to the extent that the process designer is able to take the microstructural specification for a given composition, which controls the properties of the material, and define an optimum manufacturing route to provide the desired material and performance. Modern alloys therefore depend on the proper integration of alloy composition and structure with processing to produce the desired properties and performance.

9.2 Commercial steels

9.2.1 Plain carbon steels

Carbon is an effective, cheap, hardening element for iron and hence a large tonnage of commercial steels contains very little alloying element. They may be divided conveniently into low-carbon (<0.3% C), medium-carbon (0.3–0.7% C) and high-carbon (0.7–1.7% C). Figure 9.1 shows the effect of carbon on the strength and ductility. The low-carbon steels combine moderate strength with excellent ductility and are used extensively for their fabrication properties in the annealed or normalized condition for structural purposes, i.e. bridges, buildings, cars and ships. Even above about 0.2% C, however, the ductility is limiting for deep-drawing operations, and brittle fracture becomes a problem, particularly for

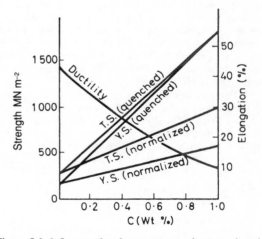

Figure 9.1 *Influence of carbon content on the strength and ductility of steel.*

welded thick sections. Improved low-carbon steels (<0.2% C) are produced by deoxidizing or 'killing' the steel with Al or Si, or by adding Mn to refine the grain size. It is now more common, however, to add small amounts (<0.1%) of Nb which reduces the carbon content by forming NbC particles. These particles not only restrict grain growth but also give rise to strengthening by precipitation-hardening within the ferrite grains. Other carbide formers, such as Ti, may be used but because Nb does not deoxidize, it is possible to produce a semi-killed steel ingot which, because of its reduced ingot pipe, gives increased tonnage yield per ingot cast.

Medium-carbon steels are capable of being quenched to form martensite and tempered to develop toughness with good strength. Tempering in higher-temperature regions (i.e. 350–550°C) produces a spheroidized carbide which toughens the steel sufficiently for use as axles, shafts, gears and rails. The process of ausforming can be applied to steels with this carbon content to produce even higher strengths without significantly reducing the ductility. The high-carbon steels are usually quench hardened and lightly tempered at 250°C to develop considerable strength with sufficient ductility for springs, dies and cutting tools. Their limitations stem from their poor hardenability and their rapid softening properties at moderate tempering temperatures.

9.2.2 Alloy steels

In low/medium alloy steels, with total alloying content up to about 5%, the alloy content is governed largely by the hardenability and tempering requirements, although solid solution hardening and carbide formation may also be important. Some of these aspects have already been discussed, the main conclusions being that Mn and Cr increase hardenability and generally retard softening and tempering; Ni strengthens the ferrite and improves hardenability and toughness; copper behaves similarly but also retards tempering; Co strengthens ferrite and retards softening on tempering; Si retards and reduces the volume change to martensite, and both Mo and V retard tempering and provide secondary hardening.

In larger amounts, alloying elements either open up the austenite phase field, as shown in Figure 9.2a, or close the γ-field (Figure 9.2b). 'Full' metals with atoms like hard spheres (e.g. Mn, Co, Ni) favour close-packed structures and open the γ-field, whereas the

stable bcc transition metals (e.g. Ti, V, Cr, Mo) close the field and form what is called a γ-loop. The development of austenitic steels, an important class of ferrous alloys, is dependent on the opening of the γ-phase field. The most common element added to iron to achieve this effect is Ni, as shown in Figure 9.2a. From this diagram the equilibrium phases at lower temperatures for alloys containing 4–40% Ni are ferrite and austenite. In practice, it turns out that it is unnecessary to add the quantity of Ni to reach the γ-phase boundary at room temperature, since small additions of other elements tend to depress the γ/α transformation temperature range so making the γ metastable at room temperature. Interstitial C and N, which most ferrous alloys contain, also expand the γ-field because there are larger interstices in the fcc than the bcc structure. The other common element which expands the γ-field is Mn. Small amounts (<1%) are usually present in most commercial steels to reduce the harmful effect of FeS. Up to 2% Mn may be added to replace the more expensive Ni, but additions in excess of this concentration have little commercial significance until 12% Mn is reached. Hadfield's steel contains 12–14% Mn, 1% C, is noted for its toughness and its used in railway points, drilling machines and rock-crushers. The steel is water-quenched to produce austenite. The fcc structure has good fracture resistance and, having a low stacking-fault energy, work-hardens very rapidly. During the abrasion and work-hardening the hardening is further intensified by a partial strain transformation of the

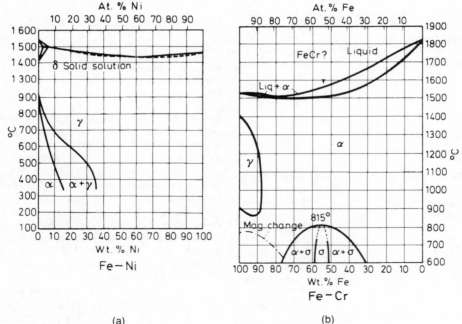

(a)   (b)

Figure 9.2 *Effect of (a) Ni and (b) Cr on γ-field (from Smithells, 1967).*

austenite to martensite; this principle is used also in the sheet-forming of stainless steels (see below).

To make the austenitic steels resistant to oxidation and corrosion (see Chapter 12) the element Cr is usually added in concentrations greater than 12%. Chromium closes the γ-field, however, and with very low carbon contents single-phase austenite cannot be produced with the stainless ($>12\%$) composition. These alloys form the stainless (ferritic) irons and are easily fabricated for use as furnace components. Increasing the carbon content expands the γ-loop and in the medium-carbon range Cr contents with good stainless qualities ($\approx 15-18\%$) can be quench-hardened for cutlery purposes where martensite is required to give a hard, sharp cutting edge. The combination of both Cr and Ni (i.e. 18/8) produces the metastable austenitic stainless steel which is used in chemical plant construction, kitchenware and surgical instruments because of its ductility, toughness and cold-working properties. Metastable austenitic steels have good press-forming properties because the strain-induced transformation to martensite provides an additional strengthening mechanism to work-hardening, and moreover counteracts any drawing instability by forming martensite in the locally-thinned, heavily-deformed regions.

High-strength transformable stainless steels with good weldability to allow fabrication of aircraft and engine components have been developed from the 0.05–0.1% C, 12% Cr, stainless steels by secondary hardening addition (1.5–2% Mo; 0.3–0.5% V). Small additions of Ni or Mn (2%) are also added to counteract the ferrite-forming elements Mo and V to make the steel fully austenitic at the high temperatures. Air quenching to give α followed by tempering at 650°C to precipitate Mo_2C produces a steel with high yield strength (0.75 GN/m^2), high TS (1.03 GN/m^2) and good elongation and impact properties. Even higher strengths can be achieved with stainless (12–16% Cr; 0.05% C) steels which although austenitic at room temperature (5% Ni, 2% Mn) transform on cooling to -78°C. The steel is easily fabricated at room temperature, cooled to control the transformation and finally tempered at 650–700°C to precipitate Mo_2C.

9.2.3 Maraging steels

A serious limitation in producing high-strength steels is the associated reduction in fracture toughness. Carbon is one of the elements which mostly affects the toughness and hence in alloy steels it is reduced to as low a level as possible, consistent with good strength. Developments in the technology of high-alloy steels have produced high strengths in steels with very low carbon contents ($<0.03\%$) by a combination of martensite and age-hardening, called maraging. The maraging steels are based on an Fe–Ni containing between 18% and 25% Ni to produce massive martensite on air cooling to room temperature. Additional

hardening of the martensite is achieved by precipitation of various intermetallic compounds, principally Ni_3Mo or $Ni_3(Mo, Ti)$ brought about by the addition of roughly 5% Mo, 8% Co as well as small amounts of Ti and Al; the alloys are solution heat-treated at 815°C and aged at about 485°C. Many substitutional elements can produce age-hardening in Fe–Ni martensites, some strong (Ti, Be), some moderate (Al, Nb, Mn, Mo, Si, Ta, V) and other weak (Co, Cu, Zr) hardeners. There can, however, be rather strong interactions between elements such as Co and Mo, in that the hardening produced when these two elements are present together is much greater than if added individually. It is found that A_3B-type compounds are favoured at high Ni or (Ni + Co) contents and A_2B Laves phases at lower contents.

In the unaged condition maraging steels have a yield strength of about 0.7 GN/m^2. On ageing this increases up to 2.0 GN/m^2 and the precipitation-strengthening is due to an Orowan mechanism according to the relation $\sigma = \sigma_0 + (\alpha\mu b/L)$ where σ_0 is the matrix strength, α a constant and L the interprecipitate spacing. The primary precipitation-strengthening effect arises from the (Co + Mo) combination, but Ti plays a double role as a supplementary hardener and a refining agent to tie up residual carbon. The alloys generally have good weldability, resistance to hydrogen embrittlement and stress-corrosion but are used mainly (particularly the 18% Ni alloy) for their excellent combination of high strength and toughness.

9.2.4 High-strength low-alloy (HSLA) steels

The requirement for structural steels to be welded satisfactorily has led to steels with lower C ($<0.1\%$) content. Unfortunately, lowering the C content reduces the strength and this has to be compensated for by refining the grain size. This is difficult to achieve with plain C-steels rolled in the austenite range but the addition of small amounts of strong carbide-forming elements (e.g. $<0.1\%$ Nb) causes the austenite boundaries to be pinned by second-phase particles and fine-grain sizes ($<10\ \mu m$) to be produced by controlled rolling. Nitrides and carbonitrides as well as carbides, predominantly fcc and mutually soluble in each other, may feature as suitable grain refiners in HSLA steels; examples include AlN, Nb(CN), V(CN), (NbV)CN, TiC and Ti(CN). The solubility of these particles in the austenite decreases in the order VC, TiC, NbC while the nitrides, with generally lower solubility, decrease in solubility in the order VN, AlN, TiN and NbN. Because of the low solubility of NbC, Nb is perhaps the most effective grain size controller. However, Al, V and Ti are effective in high-nitrogen steels, Al because it forms only a nitride, V and Ti by forming V(CN) and Ti(CN) which are less soluble in austenite than either VC or TiC.

The major strengthening mechanism in HSLA steels is grain refinement but the required strength level is obtained usually by additional precipitation strengthening in the ferrite. VC, for example, is more soluble in austenite than NbC, so if V and Nb are used in combination, then on transformation of the austenite to ferrite, NbC provides the grain refinement and VC precipitation strengthening; Figure 9.3 shows a stress–strain curve from a typical HSLA steel.

Solid-solution strengthening of the ferrite is also possible. Phosphorus is normally regarded as deleterious due to grain boundary segregation, but it is a powerful strengthener, second only to carbon. In car construction where the design pressure is for lighter bodies and energy saving, HSLA steels, rephosphorized and bake-hardened to increase the strength further, have allowed sheet gauges to be reduced by 10–15% while maintaining dent resistance. The bake-hardening arises from the locking of dislocations with interstitials, as discussed in Chapter 7, during the time at the temperature of the paint-baking stage of manufacture.

9.2.5 Dual-phase (DP) steels

Much research into the deformation behaviour of speciality steels has been aimed at producing improved strength while maintaining good ductility. The conventional means of strengthening by grain refinement, solid-solution additions (Si, P, Mn) and precipitation-hardening by V, Nb or Ti carbides (or carbonitrides) have been extensively explored and a conventionally treated HSLA steel would have a lower yield stress of 550 MN m^{-2}, a TS of 620 MN m^{-2} and a total elongation of about 18%. In recent years an improved strength–ductility relationship has been found for low-carbon, low-alloy steels rapidly cooled from an annealing temperature at which the steel consisted of a mixture of ferrite and austenite. Such steels have a microstructure containing principally low-carbon, fine-grained ferrite intermixed with islands of fine martensite and are known as dualphase steels. Typical properties of this group of steels would be a TS of 620 MN m^{-2}, a 0.2% offset flow stress of 380 MN m^{-2} and a 3% offset flow stress of 480 MN m^{-2} with a total elongation ≈28%.

The implications of the improvement in mechanical properties are evident from an examination of the nominal stress–strain curves, shown in Figure 9.3. The dual-phase steel exhibits no yield discontinuity but work-hardens rapidly so as to be just as strong as the conventional HSLA steel when both have been deformed by about 5%. In contrast to ferrite–pearlite steels, the work-hardening rate of dual-phase steel increases as the strength increases. The absence of discontinuous yielding in dual-phase steels is an advantage during cold-pressing operations and this feature combined with the way in which they sustain work-hardening to high strains makes them attractive materials for sheet-forming operations. The flow stress and tensile strength of dual-phase steels increase as the

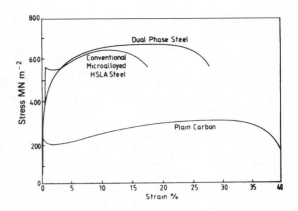

Figure 9.3 *Stress–strain curves for plain carbon, HSLA and dual-phase steels.*

volume fraction of hard phase increases with a corresponding decrease in ductility; about 20% volume fraction of martensite produces the optimum properties.

The dual phase is produced by annealing in the (α + γ) region followed by cooling at a rate which ensures that the γ-phase transforms to martensite, although some retained austenite is also usually present leading to a mixed martensite–austenite (M–A) constituent. To allow air-cooling after annealing, microalloying elements are added to low-carbon–manganese–silicon steel, particularly vanadium or molybdenum and chromium. Vanadium in solid solution in the austenite increases the hardenability but the enhanced hardenability is due mainly to the presence of fine carbonitride precipitates which are unlikely to dissolve in either the austenite or the ferrite at the temperatures employed and thus inhibit the movement of the austenite/ferrite interface during the post-anneal cooling.

The martensite structure found in dual-phase steels is characteristic of plate martensite having internal microtwins. The retained austenite can transform to martensite during straining thereby contributing to the increased strength and work-hardening. Interruption of the cooling, following intercritical annealing, can lead to stabilization of the austenite with an increased strength on subsequent deformation. The ferrite grains (≈5 μm) adjacent to the martensite islands are generally observed to have a high dislocation density resulting from the volume and shape change associated with the austenite to martensite transformation. Dislocations are also usually evident around retained austenitic islands due to differential contraction of the ferrite and austenite during cooling.

Some deformation models of DP steels assume both phases are ductile and obey the Ludwig relationship, with equal strain in both phases. Measurements by several workers have, however, clearly shown a partitioning of strain between the martensite and ferrite, with the mixed (M–A) constituent exhibiting no strain until deformations well in excess of the maximum uniform strain. Models based on the partitioning of strain predict a linear relationship between yield stress, TS and volume fraction of martensite but a linear relationship is not sensitive to the model. An alternative approach is to consider the microstructure as approximating to that of a dispersion-strengthened alloy. This would be appropriate when the martensite does not deform and still be a good approximation when the strain difference between the two phases is large. Such a model affords an explanation of the high work-hardening rate, as outlined in Chapter 7, arising from the interaction of the primary dislocations with the dense 'tangle' of dislocations generated in the matrix around the hard phase islands.

Several workers have examined DP steels to determine the effect of size and volume fraction of the hard phase. Figure 9.4 shows the results at two different strain values and confirms the linear relationship between work-hardening rate $(d\sigma/d\varepsilon)$ and $(f/d)^{1/2}$ predicted by the dispersion-hardening theory (see Chapter 7). Increasing the hard phase volume fraction while keeping the island diameter constant increases the work-hardening rate, increases the TS but decreases the elongation. At constant volume fraction of hard phase, decreasing the mean island diameter produces no effect on the tensile strength but increases the work-hardening rate and the maximum uniform elongation (Figure 9.5). Thus the strength is improved by increasing the volume fraction of hard phase while the work-hardening and ductility are improved by reducing the hard phase island size. Although dual-phase steels contain a complex microstructure it appears from their mechanical behaviour that they can be considered as agglomerates of non-deformable hard particles, made up of martensite and/or bainite and/or retained austenite, in a ductile matrix of ferrite. Consistent with the dispersion-strengthened model, the Bauschinger effect, where the flow stress in compression is less than that in tension, is rather large in dual-phase steels, as shown in Figure 9.6 and increases with increase in martensite content up to about 25%. The Bauschinger effect arises from the long-range back-stress exerted by the martensite islands, which add to the applied stress in reversed straining.

The ferrite grain size can give significant strengthening at small strains, but an increasing proportion of the strength arises from work-hardening and this is independent of grain-size changes from about 3 to 30 μm. Solid solution strengthening of the ferrite (e.g. by silicon) enhances the work-hardening rate; P, Mn and V are also beneficial. The absence of a sharp yield point must imply that the dual-phase steel contains a high density of mobile dislocations. The microstructure exhibits such a dislocation density around the martensite islands but why these remain unpinned at ambient temperature is still in doubt, particularly as strain-ageing is significant on ageing between 423 and 573 K. Intercritical annealing allows a partitioning of the carbon to produce very low carbon ferrite, while aluminium- or silicon- killed steels have limited nitrogen remaining in solution. However, it is doubtful whether the concentration of interstitials is sufficiently low to prevent strain-ageing at low temperature; hence it is considered more likely that continuous yielding is due to the residual stress fields surrounding second-phase islands. Two possibilities then arise: (1) yielding can start in several regions at the same time rather than in one local region which initiates a general yield process catastrophically, and (2) any local region is prevented from yielding catastrophically because the glide band has to overcome a high back stress from the M−A islands. Discontinuous yielding on ageing at higher temperatures is then interpreted in terms of the relaxation of these residual stresses, followed by classical strain-ageing.

In dual-phase steels the n value ≈ 0.2 gives the high and sustained work-hardening rate required when stretch formability is the limiting factor in fabrication. However, when fracture *per se* is limiting, dual-phase steels probably perform no better than other steels with controlled inclusion content. Tensile failure of dual-phase steels is initiated either by decohesion of the martensite−ferrite interface or by cracking of the martensite islands. Improved fracture behaviour is obtained when the martensite islands are unconnected, when the martensite−ferrite interface is free from precipitates to act as stress raisers, and when the hard phase is relatively tough. The optimum martensite content is considered to be 20%, because above this level void formation at hard islands increases markedly.

9.2.6 Mechanically alloyed (MA) steels

For strengthening at high temperatures, dispersion strengthening with oxide, nitride or carbide particles is an attractive possibility. Such dispersion-strengthened materials are usually produced by powder processing,

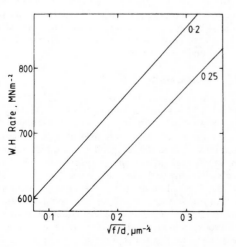

Figure 9.4 *Dependence of work-hardening rate on (volume fraction f/particle size)$^{1/2}$ for a dual-phase steel at strain values of 0.2 and 0.25 (after Balliger and Gladman, 1981).*

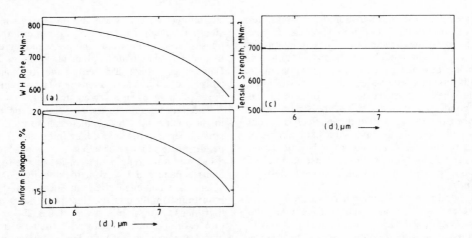

Figure 9.5 *Effect of second phase particles size d at constant volume fraction f on (a) work-hardening rate, (b) elongation and (c) tensile strength (after Balliger and Gladman, 1981).*

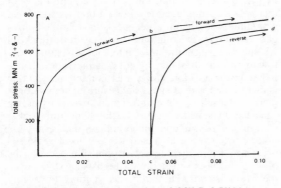

Figure 9.6 *Bauschinger tests for a 0.06%C, 1.5%Mn, 0.85%Si dual-phase steel (courtesy of D. V. Wilson).*

a special form of which is known as mechanical alloying (MA).

Mechanical alloying is a dry powder, high-energy ball-milling process in which the particles of elemental or pre-alloyed powder are continuously welded together and broken apart until a homogeneous mixture of the matrix material and dispersoid is produced. Mechanical alloying is not simply mixing on a fine scale but one in which true alloying occurs. The final product is then consolidated by a combination of high temperature and pressure (i.e. extrusion of canned powder) or hot isostatic pressing (i.e. HIPing). Further processing is by thermo-mechanical processing (TMP) to produce either (1) fine equiaxed grains for good room-temperature strength and good fatigue strength or (2) coarser, elongated grains to give good high-temperature stress–rupture strength and thermal-fatigue resistance.

Various types of ferrous alloy have been made by mechanical alloying, including 17%Cr, 7%Ni, 1.2%Al precipitation-hardened austenitic martensitic

steel and Fe–25Cr–6Al–2Y. However, the most highly developed material is the 20%Cr, 4.5%Al ferritic stainless steel, dispersion-strengthened with 0.5% Y_2O_3 (*MA 956*). *MA 956* which has been made into various fabricated forms has extremely good high-temperature strength (0.2% proof strength is 200 MN m^{-2} at 600°C, 100 MN m^{-2} at 1000°C and 75 MN m^{-2} at 1200°C).

The high-strength capability is combined with exceptional high-temperature oxidation and corrosion resistance, associated with the formation of an aluminium oxide scale which is an excellent barrier to carbon. No carburization occurs in hydrogen–methane mixtures at 1000°C. Sulphidation resistance is also good.

MA 956 was originally developed for use in sheet form in gas-turbine combustors but, with its combination of high strength up to 1300°C, corrosion resistance and formability, the alloy has found many other applications in power stations, including oil and coal burners and swirlers, and fabricated tube assemblies for fluid-bed combustion.

9.2.7 Designation of steels

The original system for labelling wrought steels was devised in 1941 and used En numbers. This system was replaced in 1976 by the British Standard (BS) designation of steels which uses a six-unit system. Essentially, it enables the code to express composition, steel type and supply requirements. The latter is shown by three letters: M means supply to specified mechanical properties, H supply to hardenability requirements and A supply to chemical analysis requirements. For convenience, steels are divided into types; namely, carbon and carbon–manganese steels, free-cutting steels, high-alloy steels and alloy steels. For example, carbon and carbon–manganese steels are designated by mean of Mn/letter/mean of C. Thus 080H41 signifies 0.6–1.0 Mn/hardenability requirement/0.38–0.45 C. Free-cutting steels are designated by 200–240/letter/mean of C. Thus 225M44 signifies free-cutting 0.2–0.3 S/mechanical properties requirement/0.4–0.48 C with 1.3–1.7 Mn. High-alloy steels include stainless and valve steels. The designation is similar to the AISI system and is given by 300–499/letters/variants 11–19. Thus 304S15 (previously known as Type 304 as used by the AISI) signifies 0.06 max. C, 8–11 Ni, 17.5–19 Cr. Alloy steels are designated by 500–999/letter/mean of C. Thus 500–519 are Ni steels, 520–539 Cr steels, 630–659 Ni–Cr steels, 700–729 Cr–Mo steels and 800–839 Ni–Cr–Mo steels. Typically 530M40 signifies 0.36–0.44 C, 0.9–1.2 Cr, supplied to mechanical properties.

Tables 9.1 and 9.2 give the compositions of typical carbon, alloy and stainless steels.

9.3 Cast irons

In the iron-carbon system (Chapter 3) carbon is thermodynamically more stable as graphite than cementite.

At the low carbon contents of typical steels, graphite is not formed, however, because of the sluggishness of the reaction to graphite. But when the carbon content is increased to that typical of cast irons (2–4% C) either graphite or cementite may separate depending on the cooling rate, chemical (alloy) composition and heat treatment (see Figure 9.7). When the carbon exists as cementite, the cast irons are referred to as white because of the bright fracture produced by this brittle constituent. In grey cast irons the carbon exists as flakes of graphite embedded in the ferrite–pearlite matrix and these impart a dull grey appearance to the fracture. When both cementite and graphite are present a 'mottled' iron is produced.

High cooling rates, which tend to stabilize the cementite, and the presence of carbide-formers give rise to white irons. The addition of graphite-forming elements (Si, Ni) produces grey irons, even when rapidly cooled if the Si is above 3%. These elements, particularly Si, alter the eutectic composition which may be taken into account by using the carbon equivalent of the cast iron, given by [total %C + (%Si +

Table 9.1 Compositions of some carbon and alloy steels

BS designation	AISI-SAE number*	% C	% Si	% Mn	% Ni	% Cr	% others
040A20	1020	0.18–0.23		0.30–0.50			
080A62	1060	0.60–0.65		0.70–0.90			
070A78	1080	0.75–0.82		0.60–0.80			
150M36	1340	0.32–0.40		1.30–1.70			
212M44	1140	0.40–0.48	0.25 max	1.00–1.40			0.12–0.2 S
527A19	5120	0.17–0.22		0.70–0.90		0.70–0.90	
665H20	4620	0.17–0.23	0.10–0.35	0.35–0.75	1.50–2.00		0.20–0.30 Mo
708M40	4140	0.36–0.44	0.10–0.35	0.70–1.00		0.90–1.20	0.15–0.25 Mo

*Approximately equivalent composition.

Table 9.2 Compositions and properties of some stainless steels

Steel	BS designation	% C	% Cr	% Ni	Others	Tensile strength (MN m^{-2})	Yield strength (MN m^{-2})	% elongation	Condition
Austenitic									
304	304S15	0.08	18–20	8.0–10.5		517	207	30	Annealed
						1276	966	9	Cold-worked
316	316S16	0.08	16–18	10–14	2–3% Mo	517	207	30	Annealed
321	321S12	0.08	17–19	9–12	Ti (5 × % C)	586	241	55	Annealed
347	347S17	0.08	17–19	9–13	Nb (10 × % C)	621	241	50	Annealed
Ferritic									
430	430S15	0.12	16–18			448	207	22	Annealed
Martensitic									
416	416S21	0.15	12–14		0.60% Mo	1241	966	18	Quenched
431	431S29	0.20	15–17	1.25–2.50		1379	1035	16	and tempered
Precipitation hardening									
17–4		0.07	16–18	3–5	0.15–0.45% Nb	1310	1172	10	Age-
17–7		0.09	16–18	6.5–7.8	0.75–1.25% Al	1655	1586	6	hardened

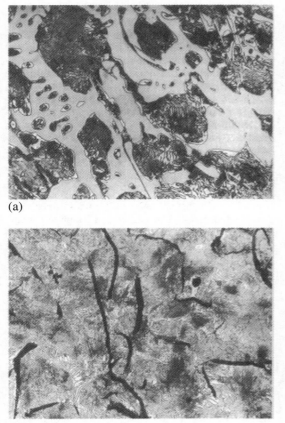

(a)

(b)

Figure 9.7 *Microstructure of cast irons: (a) white iron and (b) grey iron (400 ×). (a) shows cementite (white) and pearlite; (b) shows graphite flakes, some ferrite (white) and a matrix of pearlite.*

%P)/3], rather than the true carbon content. Phosphorus is present in most cast irons as a low melting point phosphide eutectic which improves the fluidity of the iron by lengthening the solidification period; this favours the decomposition of cementite. Grey cast iron is used for a wide variety of applications because of its good strength/cost ratio. It is easily cast into intricate shapes and has good machinability, since the chips break off easily at the graphite flakes. It also has a high damping capacity and hence is used for lathe and other machine frames where vibrations need to be damped out. The limited strength and ductility of grey cast iron may be improved by small additions of the carbide formers (Cr, Mo) which reduce the flake size and refine the pearlite. The main use of white irons is as a starting material for malleable cast iron, in which the cementite in the casting is decomposed by annealing. Such irons contain sufficient Si (<1.3%) to promote the decomposition process during the heat-treatment

but not enough to produce graphite flakes during casting. White-heart malleable iron is made by heating the casting in an oxidizing environment (e.g. hematite iron ore at 900°C for 3–5 days). In thin sections the carbon is oxidized to ferrite, and in thick sections, ferrite at the outside gradually changes to graphite clusters in a ferrite–pearlite matrix near the inside. Black-heart malleable iron is made by annealing the white iron in a neutral packing (i.e. iron silicate slag) when the cementite is changed to rosette-shaped graphite nodules in a ferrite matrix. The deleterious cracking effect of the graphite flakes is removed by this process and a cast iron which combines the casting and machinability of grey iron with good strength and ductility, i.e. TS 350 MN m^{-2} and 5–15% elongation is produced. It is therefore used widely in engineering and agriculture where intricate shaped articles with good strength are required.

Even better mechanical properties (550 MN m^{-2}) can be achieved in cast irons, without destroying the excellent casting and machining properties, by the production of a spherulitic graphite. The spherulitic nodules are roughly spherical in shape and are composed of a number of graphite crystals, which grow radially from a common nucleus with their basal planes normal to the radial growth axis. This form of growth habit is promoted in an as-cast grey iron by the addition of small amounts of Mg or Ce to the molten metal in the ladle which changes the interfacial energy between the graphite and the liquid. Good strength, toughness and ductility can thus be obtained in castings that are too thick in section for malleabilizing and can replace steel castings and forgings in certain applications.

Heat-treating the ductile cast iron produces austempered ductile iron (ADI) with an excellent combination of strength, fracture toughness and wear resistance for a wide variety of applications in automotive, rail and heavy engineering industries. A typical composition is 3.5–4.0% C, 2–2.5% Si, 0.03–0.06% Mg, 0.015% maximum S and 0.06% maximum P. Alloying elements such as Cu and Ni may be added to enhance the heat-treatability. Heat-treatment of the cast ductile iron (graphite nodules in a ferrite matrix) consists of austenization at 950°C for 1–3 hours during which the matrix becomes fully austenitic, saturated with carbon as the nodules dissolve. The fully austenized casting is then quenched to around 350°C and austempered at this temperature for 1–3 hours. The austempering temperature is the most important parameter in determining the mechanical properties of ADI; high austempering temperatures (i.e. 350–400°C) result in high ductility and toughness and lower yield and tensile strengths, whereas lower austempering temperatures (250–300°C) result in high yield and tensile strengths, high wear resistance and lower ductility and toughness. After austempering the casting is cooled to room temperature.

The desired microstructure of ADI is acicular ferrite plus stable, high-carbon austenite, where the presence

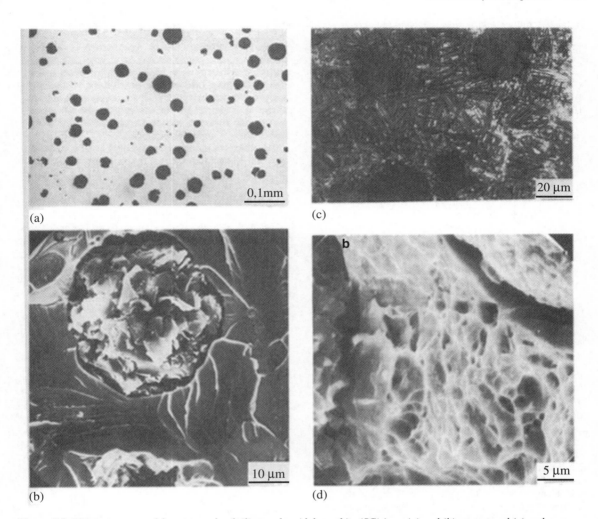

(a)

(c)   20 μm

0,1mm

(b)   10 μm

b

(d)   5 μm

Figure 9.8 *Microstructure and fracture mode of silicon spheroidal graphite (SG) iron, (a) and (b) as-cast and (c) and (d) austempered at 350°C for 1 h (L. Sidjanin and R. E. Smallman, 1992; courtesy of Institute of Metals).*

of Si strongly retards the precipitation of carbides. When the casting is austempered for longer times than that to produce the desired structure, carbides are precipitated in the ferrite to produce bainite. Low austempering temperatures (~250°C) lead to cementite precipitation, but at the higher austempering temperatures (300–400°C) transition carbides are formed, ε carbides at the lower temperatures and η carbides at the higher. With long austempering times the high-carbon austenite precipitates χ-carbide at the ferrite–austenite boundaries. The formation of bainite does not result in any catastrophic change in properties but produces a gradual deterioration with increasing time of austempering. Typically, ADI will have a tensile strength of 1200–1500 MN m^{-2}, an elongation of 6–10% and $K_{Ic} \approx 80$ MN m$^{-3/2}$. With longer austempering the elongation drops to a few per cent and the K_{Ic} reduces

to 40 MN m$^{-3/2}$. The formation of χ-carbide at the ferrite–austenite boundaries must be avoided since this leads to more brittle fracture. Generally, the strength is related to the volume fraction of austenite and the ferrite spacing. Figure 9.8 shows the microstructure of Si spheroidal graphite (SG) iron and the corresponding fracture mode.

9.4 Superalloys

9.4.1 Basic alloying features

These alloys have been developed for high-temperature service and include iron, cobalt and nickel-based materials, although nowadays they are principally nickel-based. The production of these alloys over several decades (see Figure 9.9) illustrates the transition

Table 9.3 Influence of various alloying additions in superalloys

Influence	Cr	Al	Ti	Co	Mo	W	B	Zr	C	Nb	Hf	Ta
Matrix strengthening	√			√	√	√						
γ'-formers		√	√									
Carbide-formers	√		√		√	√				√	√	√
Grain boundary strengthening							√	√	√	√	√	√
Oxide scale formers	√	√										

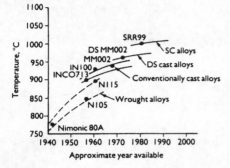

Figure 9.9 *Increases in temperature capability for turbine blade alloys, based on creep rupture in 1000 h at 150 MN m^{-2} (from Driver, 1985, pp. 345–54).*

in the development of engineering materials from basic alloy composition achievements to a more process-dominated control.

In these alloys γ' (Ni$_3$Al) and γ^* (Ni$_3$Nb) are the principal strengtheners by chemical and coherency strain hardening. The ordered γ'-Ni$_3$Al phase is an equilibrium second phase in both the binary Ni–Al and Ni–Cr–Al systems and a metastable phase in the Ni–Ti and Ni–Cr–Ti systems, with close matching of the γ' and the fcc matrix. The two phases have very similar lattice parameters ($\lesssim 0.25\%$, depending on composition) and the coherency (interfacial energy $\gamma_1 \approx 10$–20 mJ m^2) confers a very low coarsening rate on the precipitate so that the alloy overages extremely slowly even at $0.7T_m$. In alloys containing Nb, a metastable Ni$_3$Nb phase occurs but although ordered and coherent, it is less stable than γ' at high temperatures.

Another source of strengthening is due to solid-solution hardening; Cr is a major element, Co may be added up to 20% and Mo, W and Ta up to a total of 15%. These elements also dissolve in γ' so that the hardening effect may be twofold. Additions of Cr, like Co, also increase the γ' solvus and lower the stacking-fault energy.

In high-temperature service, the properties of the grain boundaries are as important as the strengthening by γ' within the grains. Grain boundary strengthening is produced mainly by precipitation of chromium and refractory metal carbides; small additions of Zr and B improve the morphology and stability of these

carbides. Optimum properties are developed by multi-stage heat-treatment; the intermediate stages produce the desired grain boundary microstructure of carbide particles enveloped in a film of γ' and the other stages produce two size ranges of γ' for the best combination of strength at both intermediate and high temperatures. Table 9.3 indicates the effect of the different alloying elements.

Some of the nickel-based alloys have a tendency to form an embrittling σ-phase (based on the composition FeCr) after long-term in-service applications, when composition changes occur removing σ-resisting elements such as Ni and enhancing σ-promoting elements such as Cr, Mo or W. This tendency is predicted in alloy design by a technique known as Phacomp (phase computation) based on Pauling's model of hybridization of 3d-electrons in transition metals. While a fraction of the 3d orbitals hybridize with p and s orbitals to create the metallic bond, the remainder forms non-bonding orbitals which partly fill the electron holes in the d-shell, increasing through the transition series to give electron hole numbers N_v for Cr(4.66), Mn(3.66), Fe(3.66), Co(1.71) and Ni(0.66). Computation shows that the γ/σ phase relation depends on the average hole number $\overline{N}_v$ given by

$$\overline{N}_v = \sum_{i=1}^{n} m_i (N_v)_i$$

where m_i is the atomic fraction of the ith element of electron hole number N_v and n is the number of elements in the alloy. The limit of γ-phase stability is reached at $\overline{\overline{N}}_v \approx 2.5$.

9.4.2 Nickel-based superalloy development

A major application of superalloys is in turbine materials, jet engines, both disc and blades. Initial disc alloys were *Inco 718* and *Inco 901* produced by conventional casting ingot, forged billet and forged disc route. These alloys were developed from austenitic steels, which are still used in industrial turbines, but were later replaced by *Waspaloy* and *Astroloy* as stress and temperature requirements increased. These alloys were turbine blade alloys with a suitably modified heat-treatment for discs. However, blade material is designed for creep, whereas disc material requires tensile strength coupled with low cycle fatigue life to cope with the stress changes in the flight cycle. To meet these requirements

Waspaloy was thermomechanically processed (TMP) to give a fine-grain size and a 40% increase in tensile strength over the corresponding blade material, but at the expense of creep life. Similar improvements for discs have been produced in *Inco 901* by TMP. More highly-alloyed nickel-based discs suffer from excessive ingot segregation which makes grain size control difficult. Further development led to alloys produced by powder processing by gas atomization of a molten stream of metal in an inert argon atmosphere and consolidating the resultant powder by HIPing to near-net shape. Such products are limited in stress application because of inclusions in the powder and, hence, to realize the maximum advantage of this process it is necessary to produce 'superclean' material by electron beam or plasma melting.

Improvements in turbine materials were initially developed by increasing the volume fraction of γ' in changing *Nimonic 80A* up to *Nimonic 115*. Unfortunately, increasing the (Ti + Al) content lowers the melting point, thereby narrowing the forging range which makes processing more difficult. Improved high-temperature oxidation and hot corrosion performance has led to the introduction of aluminide and overlay coatings and subsequently the development of *IN 738* and *IN 939* with much improved hot-corrosion resistance.

Further improvements in superalloys have depended on alternative manufacturing routes, particularly using modern casting technology. Vacuum casting was first used to retain high Ti and Al contents without oxidation loss. With 9–11% (Ti + Al), a 70% volume fraction of γ' has been produced in *IN 100* (*Nimocast PK 24*) which does not require supplementary solid solution strengtheners and therefore gives a saving in density.[2]

Additions of high melting point elements such as W extend the high-temperature capabilities at the expense of density. *M200* contains 12% W and 1% Nb but has limited ductility around 760°C which can be improved by additions of hafnium. The significant improvement in ductility and reduced porosity produced by Hf has led to its addition to other alloys (e.g. *Mar 001* (*IN 100* + Hf) and Mar 004 (*IN 713* + Hf) and *M002* which contains 10% W, 2.5% Ta and 1.5% Hf).

Creep failures are known to initiate at transverse grain boundaries and, hence, it is reasonable to aim to eliminate them in the turbine blade to gain further improvement in performance. Technologically this was achieved by directional solidified (DS) castings with columnar grains aligned along the growth direction with no grain boundaries normal to that direction. By incorporating a geometric constriction into the mould or by the use of a seed crystal it has been possible to eliminate grain boundaries entirely and grow the blade as one single crystal (see Chapter 3).

The elimination of grain boundaries immediately removes the necessity for adding grain-boundary strengthening elements. such as C, B, Zr and Hf. to the superalloy. The removal of such elements raises the melting point and allows a higher solution heat-treatment temperature with consequent improvement in chemical homogeneity and more uniform distribution of γ' precipitates. Particularly important, however, is the control of the growth direction along the [100] direction. The [100] alignment along the axis of the blade gives rise to an intrinsic high creep which enables thermal stresses caused by temperature gradients across the blade to be minimized.

Single-crystal blades have now been used successfully for both civil and military engines, *SSR99* replacing *Mar M002* but with improved tensile, creep and fatigue properties and a lower-density alloy *RR2000* replacing *IN100*.

9.4.3 Dispersion-hardened superalloys

All γ'-hardened alloys experience a reduction in strength at elevated temperatures because of the solution of γ' precipitate. To produce improved high-temperature strength, alloys hardened by oxides (ODS or oxide dispersion-strengthened), particularly thoria, have been developed. TD nickel (i.e. thoria dispersion-strengthened nickel or nickel with 2% ThO_2) and *TD-Nichrome* (i.e. nickel–20% Cr–2% ThO_2) is produced by mixing thoria sols with nickel-containing solutions to ensure a good dispersion. Drying the oxide powder mixture, followed by a hydrogen reduction process, produces a fine composite nickel–thoria powder. Compaction and controlled TMP during extrusion and rolling develops the structure and strength. These materials have excellent high-temperature creep resistance, but poor low-temperature properties which precludes aero-engine applications, although some other applications have been found. It is not possible to add the γ'-forming elements via the above process to produce lower-temperature strengthening because aluminium and titanium oxide are not reduced by hydrogen.

Mechanical alloying (MA) is a dry powder process and overcomes this problem. MA, a high-energy ball-milling process, produces a homogeneous mixture of the matrix material and dispersoid. The final product

[1]Composition

(wt %)	Ni	Cr	Co	Ti	Al	Mo	C	Zr	B	Others
Waspaloy	Balance	19.5	13.5	3.0	1.3	—	0.08	0.06	0.006	
Astroloy	Balance	15.0	17.0	3.5	4.0	5.25	0.06	—	0.030	
Inco 718	Balance	19.0	3.0	0.9	0.6	3.0	0.04	—	—	Fe 20 Nb 5.2
Inco 901	42	13.0	—	3.0	0.3	5.7	0.04	—	—	Bal. Fe

[2]A Pratt and Whitney version of *IN 100 (B1900)* replaced Ti with Ta to improve the castability.

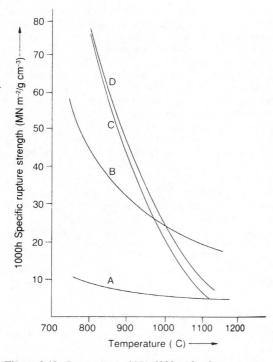

Figure 9.10 *Comparison of MA 6000 with other high-strength nickel alloys: (A) TD-Nickel, (B) MA 6000, (C) directionally-solidified Mar-M200 and (D) single-crystal PWA 454.*

is then produced by TMP with the grain structure on recrystallization elongated in the direction of working.

These materials are generally anisotropic in their properties but a range of MA materials is emerging with more acceptable dispersoids. *MA754* is a *Nichrome* matrix with Y_2O_3 dispersoids and *MA6000* is a more complex nickel superalloy-based-yttria material (Ni–12Cr–2Mo–4W–2Ta–4.5Al–2.5Ti and $1.1Y_2O_3$) suitable for turbine blades, having the creep characteristics of the γ' strengthened materials at low temperatures combined with the advantages of dispersion-strengthening at high temperature (see Figure 9.10). TMP limits the amount of γ' strengthening that can be introduced and while *MA6000* can be run at higher temperatures than conventional nickel-based superalloys, application is then limited to areas where the loadings can be kept low.

9.5 Titanium alloys

9.5.1 Basic alloying and heat-treatment features

Since the emergence of titanium as a 'wonder metal' in the 1950s the titanium industry has developed a wide range of alloys with different compositions (see Figure 9.11).[1] These alloys rely on the high strength/weight ratio, good resistance to corrosion, combined low thermal conductivity and thermal expansion of titanium, properties which make it attractive for aerospace applications in both engine and airframe components.

Titanium exists in the cph α form up to 882°C and then as bcc β to its melting point. Alloying additions change the temperature at which the α to β transition takes place, solutes that raise the transus are termed α-stabilizers and those that lower the β-transus temperature are termed β-stabilizers (Figure 9.12). The predominant α-stabilizer is aluminium. It is also an effective α-strengthening element at ambient and elevated temperatures up to 550°C and thus a major constituent of most commercial alloys. The low density of aluminium is an important additional advantage. α-phase strengthening is also achieved by additions of tin and zirconium. These metals exhibit extensive solubility in both α and β titanium but have little influence on the β-transus and are thus regarded as neutral additions. β-stabilizers may be either β-isomorphous (i.e. have the bcc structure like β–Ti) or β-eutectoid elements. β-isomorphous elements have a limited α-solubility and are completely soluble in β-titanium, typical additions being molybdenum, vanadium and niobium. In contrast, β-eutectoid elements have a restricted solubility in β-titanium and form intermetallic compounds by eutectoid decomposition of the β-phase. In some alloy systems containing β-eutectoid elements, such as silicon or copper, the compound formation (i.e., respectively, Ti_5Si_3 and Ti_2Cu) leads to an improvement in mechanical properties. Titanium will also take interstitial solutes in solid solution, hydrogen being a β-stabilizer while carbon, nitrogen and oxygen are strong α-stabilizers. To minimize gas in Ti leads to a high cost of manufacture and heat-treatment requires vacuum or inert gas conditions and freedom from refractory contact.

In describing titanium alloys it is conventional to classify them in terms of the microstructural phase (α-alloys, β-alloys, ($\alpha + \beta$)-alloys or near-α-alloys, i.e. predominantly α-phase but with a small volume of β-phase). Commercial alloys are usually heat-treated to optimize the mechanical properties by controlling the transformation of the β- to α-phase, the extent of which is governed by the alloy composition and the cooling rate. The α-alloys can transform completely from the β- to α-phase no matter what the cooling rate. Such treatments have a negligible effect on properties and α-alloys tend to be used in the annealed state.

---

[1]The Larson–Miller parameter ϕ is given by $\phi = T(A + \log_{10} t)$, where T is the temperature in degrees Kelvin, t the time in hours and A a constant, and defines the conditions to produce a given amount of plastic strain (e.g. 0.2%).

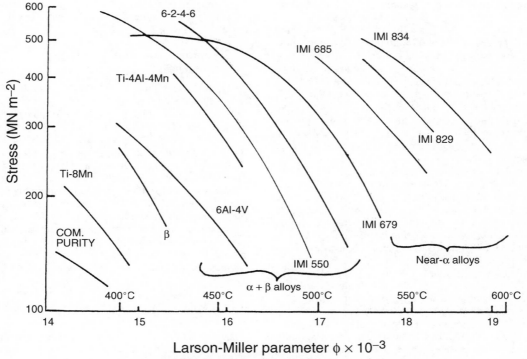

Figure 9.11 *Plot of stress versus Larson–Miller parameter φ for a range of titanium alloys.*

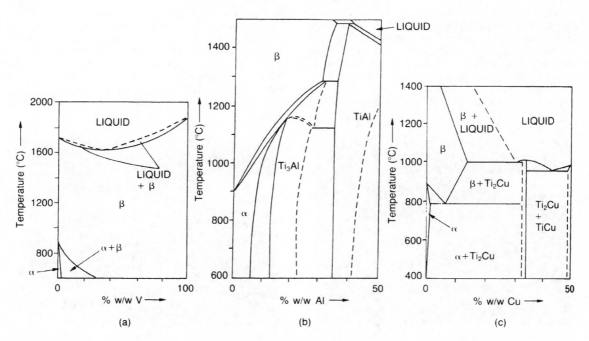

Figure 9.12 *Representative phase diagrams for Ti-alloys. (a) Ti–V. (b) Ti–Al and (c) Ti–Cu (after Smithells, 1992).*

Rapidly-cooled alloys containing β-stabilizers form martensitic α from the β-phase, whereas slower cooling rates favour α formation by a nucleation and growth process. Several morphologies of α can be produced by controlling the nucleation and growth mechanism; slow cooling, for example, tends to produce similarly aligned α platelets in colonies, combined with primary α at the grain boundaries. Faster cooling and higher α-stabilizer contents result in a basket-weave microstructure. Metastable-β, when aged, precipitates fine α, giving increased strength. The less stable the β, the more α can be precipitated and hence the higher the strength attained.

β-stabilizing elements improve strength by strengthening the β-phase. The microstructure consists of primary α combined with the β-phase, which can be strengthened by an ageing treatment to precipitate acicular-α. Further strengthening is achieved by the limited solubility of the β-stabilizing element. Generally, these alloys have poor ductility properties.

The most important alloys contain both α- and β-stabilizers which, after working and annealing, give good strength and fabrication properties. For good creep strength an α-titanium base, strengthened as much as possible by solute elements, is required. To meet this requirement the near-α alloys have been developed. These alloys combine the high α stability with sufficient β-stabilizer to give adequate strength. By β heat-treating, the $(\alpha + \beta)$ microstructure changes to a totally transformed β structure containing basket-weave α. These alloys have good creep resistance and reasonable room-temperature properties. The basket-weave morphology is effective in inhibiting crack growth, and near-α alloys exhibit lower crack propagation rates than the $\alpha + \beta$ microstructures. Most of the commercial alloys which have been recently developed are of this type. The major factor influencing the post-forging microstructure is the cooling rate; an oil quench results in a basket-weave structure (see Figure 9.13a) and an air cool gives a typically aligned

microstructure (see Figure 9.13b). The alloy is usually stress-relieved by annealing for 2 hours at 625°C or above.

9.5.2 Commercial titanium alloys

α *alloys* transform entirely to α on cooling from the β-phase, regardless of cooling rate. Commercially pure titanium with a nominal oxygen content of 500, 1000, 1550 and 2700 ppm, respectively, gives tensile properties which range from 450 to 640 MN m^{-2} (see Figure 9.14). These are used mainly in sheet form. Solid-solution strengthening by aluminium, tin or zirconium increases tensile and creep strength and *IMI 317* (Ti–5Al–2.5Sn) is typical.

Increasing the α-stabilizing composition increases creep strength but makes fabrication more difficult and can lead to embrittlement during prolonged exposure at temperature in service due to the formation of the coherent ordered phase α_2 (Ti$_3$ Al). To avoid this it was empirically established that the aluminium equivalent Al* must be no greater than 9 where Al* (in wt %) is given by:

$$Al^* = Al + \frac{Sn}{3} + \frac{Zr}{6} + (10 \times O_2)$$

IMI 317 is difficult to fabricate and is often replaced by an *IMI 230,* which is an α-phase alloy containing the precipitation-hardening phase Ti$_2$Cu; it can be fabricated and welded and has good strength and ductility up to 400°C.

β *alloys* contain enough β-stabilizing elements to maintain the bcc β-phase to room temperature. Unfortunately, bcc β–Ti alloys are prone to embrittlement. The binary alloys, titanium with Fe, Cr, Mn, Nb, Mo, Cr or V, all precipitate the embrittling ω-phase. More complex β–Ti alloys containing Cr also suffer embrittlement from TiCr$_2$. More stable alloys have been developed (e.g. (Ti–11.5Mo–6Zr–4.5Sn)) but are little used.

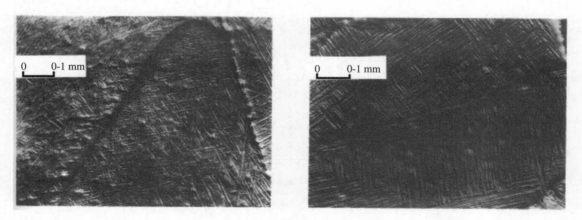

Figure 9.13 *Microstructure of near-α titanium alloy (IMI 829) initially β-heat-treated at 1050° C for 1 h. (a) Oil quenched and (b) air cooled (after Woodfield, Loretto and Smallman, 1988).*

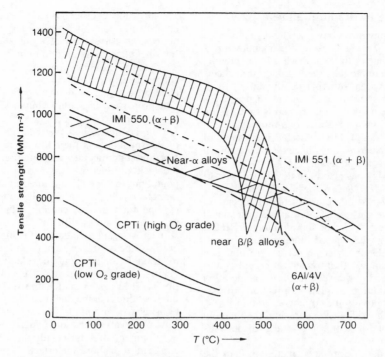

Figure 9.14 *Variation of tensile strength with temperature for a range of commercial titanium alloys.*

However, β-phase alloys, such as Ti–10V–2Fe–3Al, have potential as airframe construction materials offering high strength (1250 MN m^{-2}) and toughness (45 MN m$^{-3/2}$) in relatively thick cross-sections (90 mm) and superior hot-working characteristics which are attractive in expensive forging operations.

$(\alpha + \beta)$ *alloys* are probably the most widely used titanium alloys and contain alloying additions which strengthen both phases. These alloys are thermo-mechanical processed to control the size, shape and distribution of both α and β. The most versatile $(\alpha + \beta)$ alloy is *IMI 318*, which contains 6% Al and 4% V; it can be used at temperatures up to 350°C and has good forging and fabrication properties. It initially replaced steel as a disc material in jet engines leading to 20% weight saving. Another important $(\alpha + \beta)$ alloy is *IMI 550* (Ti–6Al–2Sn–4Mo–0.5Si) which has higher strength and good creep resistance up to 400°C. $(\alpha + \beta)$ alloys remain the principal materials for fan discs and blades and for low- and intermediate-pressure compressor discs and blades of current gas turbine engines. $(\alpha + \beta)$ alloys with extra low interstitial (ELI) content are attractive as 'damage-tolerant' materials for critical airframe components. Ti–6Al–4V with low oxygen has a tensile strength 8% lower than the standard alloy but, more importantly, the minimum fracture toughness is 60 MN m$^{-3/2}$.

Near-α-alloys have increased the strength and the volume of the more creep-resistant α-phase at the

expense of the bcc β-phase, which imports good low-temperature strength and forgeability. *IMI 685* was the first titanium alloy to operate above 500°C. It contains Ti–6Al–5Zr–0.5Mo–0.2Si with Al and Zr instead of Sn as α-stabilizers, reduced Mo, the β-stabilizer, to minimize β at the α-needles and Si to improve creep resistance. These alloys are worked and heat-treated in the β-range and have a tensile strength of about 1000 MN m^{-2} and give less than 0.1% creep strain in 100 h under a stress of 310 MN m^{-2} at 520°C. *IMI 829* (Ti–5.6Al–3.5Sn–3Zr–1Nb–0.25Mo–0.3Si) has been derived from *IMI 685* by replacing some of the Zr with the more potent strengthener Sn. It is β-heat-treated and has sufficient higher-temperature capability to be used in the hotter regions of engines. *IMI 834* (Ti–5.8Al–4Sn–3.5Zr–0.7Nb–0.5Mo–0.35Si–0.06C) has been developed for use up to 600°C and combines the high fracture toughness and crack propagation resistance of a transformed β-structure with the typical equiaxed structure of the $\alpha + \beta$ alloys, providing good fatigue resistance and ductility. The small addition of carbon allows a controlled high α/β heat-treatment. Hot-working is carried out in the $\alpha + \beta$ field and heat-treatment involves solution treatment for 2 hours at 1025°C, consistent with about 15% primary α, followed by oil quenching prior to ageing for 2 hours at 700°C, then air cooling. With such good high-temperature properties the alloy is being specified for engine compressor applications.

9.5.3 Processing of titanium alloys

Some of the titanium alloys have excellent superplastic forming characteristics and fabricating manufacturers have taken advantage of these properties in developing new processing technologies. Fine-grained thin sheets of *IMI 318* have been superplastically deformed at 900°C under slow strain-rate conditions to produce a variety of complex parts. In conjunction with diffusion bonding, weight-saving of about one-third has been achieved in, for example, the wing access panels of the A320 Airbus. Near-α-alloys have also been shown to exhibit superplastic behaviour.

Casting technology has also been developed. Investment casting of *IMI 318* is widely used but there is now increasing interest in the use of high-temperature alloy castings as weight-saving alternatives to steel and nickel alloys above 500°C. The introduction of HIPing for titanium castings has widened their potential application. For *IMI 829,* the best properties are obtained in the α/β HIP and solution-treated and aged condition, generally meeting the specification for the alloy in the wrought condition. Rapid solidification processing may also offer possibilities by incorporating Er_2O_3 and other rare earth oxide dispersions into the titanium matrix. At present, the control of these dispersoids remains a problem.

To develop higher-performance materials than the near-α alloys into the 700–800°C range, attention is being given to strengthening high-temperature alloy materials, including titanium aluminides (see Section 9.6) with ceramic fibres such as SiC and B_4C.

9.6 Structural intermetallic compounds

9.6.1 General properties of intermetallic compounds

In terms of their properties, intermetallic compounds are generally regarded as a class of materials between metals and ceramics which arises from the bonding being a mixture of metallic and covalent. Intermetallics are intrinsically strong (and in the $L1_2$-ordered fcc compounds increases with temperature up to about 600°C) with high elastic modulus. The strong bonding and ordered structure also gives rise to lower self-diffusion coefficients and hence greater stability of diffusion-controlled properties. Some of the compounds of current interest are shown in Table 9.4. Intermetallics containing aluminium or silicon exhibit a resistance to oxidation and corrosion because of their adherent surface oxides. Those based on light elements have attractive low density giving rise to high specific properties particularly important in weight-saving applications.

Like ceramics, however, the greatest disadvantage of intermetallics is their low ductility, particularly at low and intermediate temperatures. The reasons for the lack of ductility vary from compound to compound but include (1) a limited number of easy deformation modes to satisfy the von Mises criterion, (2) operation of dislocations with large slip vectors, (3) restricted cross-slip, (4) difficulty of transmitting slip across grain boundaries, (5) intrinsic grain boundary weakness, (6) segregation of deleterious solutes to grain boundaries, (7) covalent bonding and high Peierls-Nabarro stress and (8) environmental susceptibility. It has been demonstrated, however, that some intermetallics can be ductilized by small alloying additions: Ni_3Al with boron, TiAl with Mn, Ti_3Al with Nb. This observation has encouraged recent research and development of intermetallics and the possibility of application of those materials.

9.6.2 Nickel aluminides

Ni_3Al (*nickel aluminide*) is the ordered fcc γ' phase and is a major strengthening component in superalloys. Ni_3Al single crystals are reasonably ductile but in polycrystalline form are quite brittle and fail by intergranular fracture at ambient temperatures. The basic slip system is $\{111\}\langle110\rangle$ and has more than five independent slip modes but still exhibits grain boundary brittleness. Remarkably, small additions of ~0.1 at.% boron produce elongations up to 50%. General explanations for this effect are that B segregates to grain boundaries and (1) increases the cohesive strength of the boundary and (2) disorders the grain boundary region so that dislocation pile-up stresses can be relieved by slip across the boundary rather than by cracking. This general explanation is no doubt of significance but additionally, there are distinct microstructural changes within the grains which

Table 9.4 Comparison of physical properties of some intermetallic compounds

Compound	Crystal structure	Melting temp. (°C)	Density (kg m⁻³)	Young's modulus/density
Ni_3Al	$L1_2$ (ordered fcc)	1400	7500	45
Ni_3Si	$L1_2$ (ordered fcc)	1140	7300	
NiAl	B2 (ordered bcc	1640	5860	35
Ti_3Si	$D0_{19}$ (ordered cph)	1600	4200	50
TiAl	$L1_0$ (ordered tetragonal)	1460	3910	24
FeAl	B2 (ordered bcc)	1300	5560	47

must lead to a reduced friction stress and ease the operation of polyslip. For example, the addition of B reduces the occurrence of stacking-fault defects. Addition of solutes, such as B, are not expected to raise the stacking-fault energy and hence this effect possibly arises from the segregation of B to dislocations, preventing the superdislocation dissociation reactions (see Chapter 4, Section 4.6.5).

Microhardness measurements inside grains and away from grain boundaries indeed show that boron softens the grains. The ductilization effect is limited to nickel-rich aluminides and cannot be produced by carbon or other elements, although some substitutional solutes such as Pd, which substitutes for Ni, and Cu produce a small improvement in elongation. Small additions of Fe, Mn and Hf have also been claimed to improve fabricability. Grain size has been shown to influence the yield stress according to the Hall–Petch equation and B appears to lower the slope k_y and facilitate slip across grain boundaries. These alloys are also known to be environmentally sensitive. Hf, for example, which does not segregate to grain boundaries but still improves ductility, has a large misfit (11%) and possibly traps H from environmental reactions, such as $Al + H_2O \rightarrow Al_2O_3 + H$. Ti, which has a small misfit, does not improve the ductility.

The most striking property of Ni_3Al is the increasing yield stress with increasing temperature up to the peak temperature of 600°C (see Figure 9.15). This behaviour is also observed in other $L1_2$ intermetallics, particularly Ni_3Si and Zr_3Al. This effect results from the thermally-activated cross-slip of screw dislocations from the {111} planes to the {100} cube

Table 9.5 Anti-phase boundary energies in Ni₃Al

Alloy	γ_{111} (mJ/m^2)	γ_{100} (mJ/m^2)	$\gamma_{111}/\gamma_{100}$
Ni–23.5Al	183 ± 12	157 ± 8	1.17
Ni–24.5Al	179 ± 15	143 ± 7	1.25
Ni–25.5Al	175 ± 13	134 ± 8	1.31
Ni–26.5Al	175 ± 12	113 ± 10	1.51
Ni–23.5Al + 0.25B	170 ± 13	124 ± 8	1.37

planes where the apb-energy is somewhat lower. The glide of super-dislocations is made more difficult by the formation of Kear–Wilsdorf (K–W) locks (see Chapter 7) and their frequency increases with temperature. Electron-microscopy measurements of apb energies given in Table 9.5 shows that the apb energy on {100} decreases with aluminium content and this influences the composition-dependence of the strength, shown in Table 9.15. The cross-slip of screw dislocations from the {111} planes to cube planes also gives rise to a high work-hardening rate.

Although the study of creep in γ'-based materials is limited, it does appear to be inferior to that of superalloys. Above $0.6T_m$ creep displays the characteristic primary and secondary stages, with steady-state creep having a stress exponent of approximately 4 and an activation energy of around 400 KJ mol^{-1}, consistent with climb being the rate-controlling process. At intermediate temperatures (i.e. around the 600°C peak in the yield stress curve) the creep behaviour does not display the three typical stages. Instead, after primary creep, the rate continuously increases with creep strain, a feature known as inverse creep. In primary creep, planar dissociation leads to an initial high creep rate which slows as the screws dissociate on {100} planes to form K–W locks. However, it is the mobile edge dislocations which contribute most to the primary creep strain and their immobilization by climb dissociation which brings about the exhaustion of primary creep. The inverse creep regime is still not fully researched but could well be caused by glide on the {100} planes of the cross-slipped screw components.

The fatigue life in high-cycle fatigue is related to the influence of temperature on the yield stress and is invariant with temperature up to about 800°C, but falls off for higher temperatures with cracks propagating along slip planes. With boron doping the fatigue resistance is very sensitive to aluminium content and decreases substantially as Al increases from 24 to 26 at.%. Nevertheless, crack growth rates of $Ni_3Al + B$ are lower than for commercial alloys.

Hyperstoichiometric Ni_3Al with boron can be prepared by either vacuum melting and casting or from powders by HIPing. Fabrication into sheets is possible with intermediate anneals at 1000°C. At present, however, the application of Ni_3Al is not significant; Ni_3Al powders are used as bond coats to improve adherence of thermal spray coatings. Nevertheless, Ni_3Al alloys have been tested as heating elements, diesel

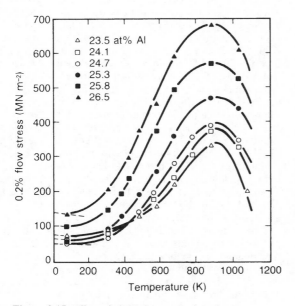

Figure 9.15 *Effect of aluminium content on the temperature-dependence of the flow stress in Ni₃Al (after Noguchi, Oya and Suzuki, 1981).*

Table 9.6 Comparison of super α_2 and γ alloys with conventional titanium alloys

Property	Titanium alloys	$(\alpha_2 + \beta)$	$(\gamma + \alpha_2)$
Density (g cm^{-3})	4.54	4.84	4.04
E, stiffness (GN m^{-2})	110	145	176
RT tensile strength (MN m^{-2})		1100	620
HT (760°C) tensile strength (MN m^{-2})		620	550
Max. creep temp. (°C)	540	730	900
RT ductility (%)	20	4–6	3
Service temp. ductility (%)	high	5–12	5–12

engine components, glass-making moulds and hot-forging dies, slurry-feed pumps in coal-fired boilers, hot-cutting wires and rubber extruders in the chemical industry. Ni$_3$Al-based alloys as matrix materials for composites are also being investigated.

Nickel aluminide (NiAl) has a caesium chloride or ordered β-brass structure and exists over a very wide range of composition either side of the stoichiometric 50/50 composition. It has a high melting point of 1600°C and exhibits a good resistance to oxidation. Even with such favourable properties it has not been commercially exploited because of its unfavourable mechanical properties. Because it is strongly ordered, low-temperature deformation occurs by an $a\langle100\rangle$ dislocation vector and not by $a/2\langle111\rangle$ super-dislocations. $\{110\}\langle100\rangle$ slip therefore leads to insufficient slip modes to satisfy the general plasticity criterion and in the polycrystalline condition β-NiAl is extremely brittle. The ductility does improve with increasing temperature but above 500°C the strength drops off considerably as a result of extensive glide and climb. Improvements in properties are potentially possible by refinement of the grain size and by using alloying additions to promote $\langle111\rangle$ slip, as in FeAl, which has the same structure. In this respect, additions of Fe, Cr or Mn appear to be of interest. For high-temperature applications, ternary additions of Nb and Ta have been shown to improve creep strength through the precipitation of second phases and mechanical alloying with yttria or alumina is also beneficial.

A further commercial problem of this material is that conventional production by casting and fabrication is difficult, but production through a powder route followed by either HIPing or hot extrusion is more promising.

9.6.3 Titanium aluminides

Because of the limited scope for improvements in the properties of conventional titanium alloys above 650°C, either by alloy development or by TMP, increased attention is being given to the titanium intermetallics, Ti$_3$Al (α_2-phase) and TiAl (γ-phase). With low density, high modulus and good creep and oxidation resistance up to 900°C they have considerable potential if the poor ductility at ambient temperatures could be improved. A comparison of Ti$_3$Al- and TiAl-based materials with conventional Ti-alloys is given in Table 9.6.

Electron microscopy studies of Ti_3Al or α_2 have shown that deformation by slip occurs at room temperature by coupled pairs of dislocations with $b = 1/6\langle11\bar{2}0\rangle$ which glide only on $\{10\bar{1}0\}$ planes and by very limited glide on $\{11\bar{2}1\}$ with pairs of dislocation with $b = 1/6\langle11\bar{2}6\rangle$. The ductility increases at higher temperatures due to climb of the $\langle11\bar{2}0\rangle$ dislocations and to the increased glide mobility of $1/6\langle11\bar{2}0\rangle$ and $1/6\langle11\bar{2}6\rangle$ dislocations through thermal activation. Only limited activity of the $\{0001\}\langle11\bar{2}0\rangle$ slip systems is observed, even at high temperatures.

The most successful improvements in the ductility of Ti$_3$Al have been produced by the addition of β-stabilizing elements, particularly niobium, to produce α_2-alloys. An addition of 4 at.% Nb produces significant slip on $\{10\bar{1}0\}\langle11\bar{2}0\rangle$, $\{0001\}\langle11\bar{2}0\rangle$ and $\{11\bar{2}1\}\langle11\bar{2}6\rangle$ as well as some slip on $\{1011\}\langle1120\rangle$. This improvement is attributed to the decrease in covalency as Nb substitutes for Ti with a consequent reduction in the Peierls–Nabarro friction stress. Alloys based on α_2 are Ti–(23–25) Al–(8–18)Nb, of which Ti–24Al–11Nb has excellent spalling resistance. Most Ti$_3$Al + Nb alloys, such as super α_2, also contain other β-stabilizers including Mo and V, i.e. Ti–25Al–10Nb–3V–1Mo, which exhibits about 7% room temperature elongation. Alloying Ti$_3$Al with β-stabilizing elements to produce two-phase alloys significantly increases the fracture strength. These Ti$_3$Al-based alloys can be plasma-melted and cast followed by TMP in the $(\alpha_2 + \beta)$ or β-range. The improved ductility of Ti$_3$Al alloys has led to aerospace applications in after-burners in jet engines where it compares favourably in performance with superalloys and gives a 40% weight-saving.

Developments are taking place in rapid solidification processing to include a second phase (e.g. rare-earth precipitates) and to provide powders, which may be consolidated by HIPing, to produce fully dense components with properties comparable to wrought products. There are also developments in intermetallic matrix composites by the addition of SiC or Al$_2$O$_3$ fibres ($\sim$10 μm). These have some attractive properties, but the fibre/intermetallic interface is still a problem.

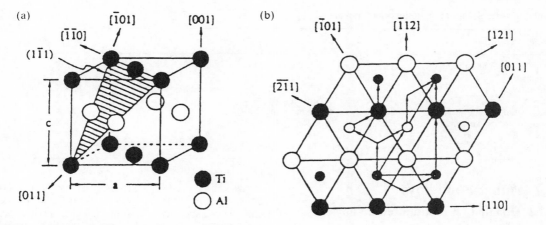

Figure 9.16 *Structure of (a) TiAl (L1$_0$) and (b) (111) plane showing slip vectors for possible dissociation reactions, e.g. ordinary dislocations 1/2[110], super dislocations [011] and 1/2[112], and twin dislocations 1/6[112] (after Kim and Froes, 1990).*

The *γ-phase Ti–(50–56)Al* has an ordered fc tetragonal (L1$_0$) structure up to the m.p. 1460°C. with $c/a =$ 1.02 (Figure 9.16). Deformation by slip occurs on {111} planes and, because of the tetragonality, there are two types of dislocations, namely ordinary dislocations 1/2⟨110⟩ and superdislocations ⟨011⟩ = 1/2⟨011⟩ + 1/2⟨011⟩. Another superdislocation 1/2⟨112⟩ has also been reported.

At room temperature, deformation occurs by both ordinary and super dislocations. However, [011] and [101] super dislocations are largely immobile because segments of the trailing superpartials 1/6⟨112⟩-type form faulted dipoles. The dissociated 1/2⟨110⟩ dislocations bounding complex stacking faults are largely sessile because of the Peierls–Nabarro stress. Some limited twinning also occurs. The flow stress increases with increasing temperature up to 600°C as the superpartials become mobile and cross-slip from {111} to {100} to form K–W-type locks, the 1/2⟨110⟩ slip activity increases and twinning is promoted.

The two-phase $(\gamma + \alpha_2)$ *Ti-Al alloys* have better ductility than single-phase γ with a maximum at 48 at.% Al. This improvement has been attributed to the reduced c/a with decreased Al, further promotion of twinning and the scavenging of O_2 and N_2 interstitials by α_2. Additions of V, Cr or Mn, which reduce the unit cell, also improve ductility. However, the two-phase microstructure is extremely important in optimizing the tensile and creep properties and the fracture toughness. Both the volume fraction and morphology of the lamellar α_2-structure is important and about 30% lamellar α_2 appears to be the optimum.

9.6.4 Other intermetallic compounds

A number of intermetallic compounds are already used in areas which do not rely on stringent mechanical properties. Fe_3Al, for example, is used in fossil-fuel plants where resistance to both sulphur attack and oxidation is important. Ni_3Si is used where resistance to hot sulphuric acid is required. There are several compounds with rare-earth elements used in magnet technology (see Chapter 6). PdIn is gold-coloured and a possible dental material. Zr_3Al has a low neutron capture cross-section and is a possible reactor material.

The β-compound NiTi (*Nitinol*) is an important shape-memory alloy. The shape-memory effect (SME) manifests itself when the alloy is deformed into a shape while in a low-temperature martensitic condition but regains its original shape when the stress is removed and it is heated above the martensitic regime. Strains of the order of 8% can be completely recovered by the reverse transformation of the deformed martensitic phase to the higher-temperature parent phase. The martensite transformation in these alloys is a thermo-elastic martensitic transformation in which the martensite plates form and grow continuously as the temperature is lowered and are removed reversibly as the temperature is raised. NiTi was one of the original SME alloys, but there are many copper-based alloys which undergo a martensitic transformation, e.g. Cu–17Zr–7Al. Application of SME alloys relies on the characteristic that they can change shape repeatedly as a result of heating and cooling and exert a force as the shape changes. By composition control (increasing the Ni content or substitution of Cu lowers the M_s temperature of TiNi), the shape-memory can be triggered by normal body temperature or any other convenient temperature to operate a device. Several biomedical applications have been developed in orthopaedic devices (e.g. pulling fractures together), in orthodontics, in intrauterine contraceptives and in artificial hearts. Industrial applications include pipe couplings for ships which shrink during heating, electrical connectors, servo-mechanisms for driving recording pens, switches, actuators and thermostats.

Table 9.7 Aluminium alloy designation systems

Wrought alloys	Designation	Casting alloys	Designation
99.00% (min.) aluminium	1XXX	99.00% (min.) aluminium	1XX.X
Copper	2XXX	Copper	2XX.X
Manganese	3XXX	Silicon with added copper and/or magnesium	3XX.X
Silicon	4XXX	Silicon	4XX.X
Magnesium	5XXX	Magnesium	5XX.X
Magnesium and silicon	6XXX	Zinc	6XX.X
Zinc	7XXX	Tin	7XX.X
Others	8XXX	Others	8XX.X

9.7 Aluminium alloys

9.7.1 Designation of aluminium alloys

Aluminium has an attractive combination of properties (i.e. low density, strong, easy to fabricate) which can be developed and modified by alloying and processing. The basic physical metallurgy has been outlined in Chapter 8 and hence in this section some of the alloys developed for particular industries such as the transportation, construction, electrical and packaging industries will be considered.

Aluminium alloys are identified by a four-digit system[1] based on the main alloying element. This is summarized in Table 9.7. For wrought alloys the first digit identifies the alloy group and the second digit any modification to the original alloy which is identified by the last two digits. The system is slightly different for casting alloys. The first digit again identifies the group, the second two digits identify the alloy and the last digit, preceded by a decimal point, indicates the product form (i.e. 0 for casting and 1 for ingot).

9.7.2 Applications of aluminium alloys

With the need for fuel economy and weight-saving, aluminium alloys are increasingly used in cars, and its two most important properties are density and thermal conductivity. Over the past 15 years the aluminium content of cars has increased from around 5% to 13% by both volume and weight. In engines they are used for pistons, cylinder heads and sumps. Al–Si casting alloys of the 3XX.X series are being used for engine blocks and Al–Si pistons with cast iron cylinder liners for wear resistance. The superior thermal conductivity reduces the volume of coolant in the system. Aluminium wheels, vacuum-cast or forged, are replacing conventional steel wheels in sports cars. Heat-treatable[2] 2XXX and 6XXX as well as 5XXX

series can be used for body sheet. However, the modulus of aluminium is only one-third that of steel and hence significant design changes are necessary to maintain rigidity and stiffness. A straightforward increase in gauge thickness would lead to a doubling cost which limits the replacement to 'quality' cars. For bulk market cars, gauge for gauge substitution for steel is a future objective with structural reinforcement to enhance body torsional characteristics. This is possible with the use of adhesives in a weld-bonding approach which can reduce the weight by half and fuel consumption and CO_2 emission by almost 15% Aluminium-structured vehicle technology (ASVT) is likely to be essential when emission control to reduce global warming effects is tightened.

The Honda NSX all-aluminium car[3] is manufactured with conventional design and assembly and has three different 6000 series alloys for external panels and a structural subframe of 5182 alloy.

In aircraft construction use is made of the high strength-density ratio of the Al–Cu (2000 series) and Al–Zn–Mg (7000 series) alloys in extruded form for wing spars, fuselage and landing gear and for the skin in plate or sheet form; typically 7075 (5.6Zn–2.5Mg–1.6Cu) is used in the T6 condition and 2024 (4.4Cu–1.5Mg–0.6Mn) in the T3 or T8 conditions. The alloys of the 7000 series have higher strength than the 2000 alloys but lower resistance to fracture. However, higher-purity levels (e.g. 2124 alloy), give enhanced toughness. Alloys of both series lose strength above 100°C and are thus not suitable for supersonic aircraft. The *RR58* alloy used for the Concorde at temperatures up to 175°C was originally an early engine material, for compressor blades and impellers.

Aluminium is used extensively in the construction industry because of its light weight, resistance to atmospheric attack and surface finish. For decorative applications, dyed anodic films produce a permanent durable finish. Generally, the Al–Mg–Si 6000 series

---

[1] The International Alloy Designation System (IADS) was first introduced by the Aluminium Association of the USA and is now standard for wrought alloys.

[2] Alloy treatment is usually described by a suffix letter and digit system (e.g. F—as fabricated, O—annealed, H—work-hardening and T—heat-treated. Digits following

H specify the work-hardened condition, and that following T the type of ageing treatment (e.g. T6 is solution heat-treated and artificially aged, T4 solution heat-treated and naturally aged, T3 solution heat-treated and cold-worked).

[3] Launched in the USA in 1992.

is used, i.e. 6063 medium-strength and 6082 higher-strength alloys in the T6 condition for extrusions or T4 where forming is required during fabrication. The Al–Cu (2014A) alloy is also used for heavily-loaded primary structures.

The packaging industry also provides a large market for aluminium alloys. The main requirement is for low-cost, simple alloys capable of being formed and the Al–Mg (5000) series is often used. Impurity control is essential and liquid metal filtering is necessary in the production of thin sheet.

In the electrical industry, electrical conductivity grade aluminium and higher-strength 5000 and 6000 series alloys are used for transmission lines, replacing the more expensive copper. Dispersion-strengthened alloys containing a fine dislocation substructure stabilized by small precipitates are used for electrical wiring.

9.7.3 Aluminium-lithium alloys

The advantages of aluminium-lithium alloys have been known for a long time but lower density and increased elastic stiffness were offset by poor ductility and fracture performance. Basic Al–Li alloys precipitate the (Al_3Li) δ', a spherical ordered precipitate. Precipitation-hardening leads, however, to localized deformation with limited cross-slip and poor fracture behaviour. Additions of copper to the alloy so that the Li/Cu ratio is high leads to the formation of both δ' and a T_1 phase (Al_2CuLi). This gives some improvement in fracture toughness by independent control of the two precipitates. In the quaternary system Al–Li–Cu–Mg the S-phase precipitates in addition to the δ' and T_1. The S-phase is better at dispersing slip than T_1 and with adjustment of composition can be made to dominate the structure. Both S and T_1 are nucleated heterogeneously on dislocations and the best results are obtained by cold-working the alloy after solution-treatment.

Commercial alloys based on this background are *Lital A, B* and *C* which have been developed to match the (1) conventional medium-strength 2014-T6, (2) high-strength 7075-T6 and (3) damage-tolerant 2024-T3 alloys, with a 10% reduction in density and 10% improvement in stiffness (see Table 9.8).

Lital A in T6 sheet form typically has 365 MN m^{-2} 0.2% proof stress, 465 MN m^{-2} TS, 6% elongation, 66 MN m$^{-3/2}$ fracture toughness, an elastic modulus of 80 GN m^{-2} and density of 2550 kg m^{-3}. *Lital B* has roughly 10% improvement in strength. *Lital C* is a

variant of the 8090 alloy and is heat-treated to increase toughness (~76 MN m$^{-3/2}$) at the expense of strength (TS ~440 MN m^{-2}).

Lithium additions are also being made to conventional aluminium alloys. The addition of lithium has a major influence since Li possesses a significant vacancy binding energy of about 0.25 eV. Lithium atoms therefore trap vacancies and form Li-V aggregates. This decreases the concentration of mobile vacancies available for the transport of zone-forming atoms, and therefore inhibits the diffusion of Zn and Mg in 7075, and Si and Mg in 6061, into zones. Second, the Li-V aggregates, very probably present during quenching and immediately after ageing, act as heterogeneous sites for subsequent clustering of zone-forming atoms during ageing.

Additions of Li into either Al–2Mg–0.6Si–0.3Cu–0.3Cr (6061) or Al–5.9Zn–2.4Mg–1.7Cu (7075) modify the precipitation scheme and age-hardening behaviour of the original alloys. The precipitates which form in the base alloys are inhibited or even suppressed. For the 6061 the addition of 0.7% Li retards the precipitation of needle-shaped GP zones and produces a ternary compound AlLiSi, whereas the addition of 2.0% Li results in the dominant precipitation of δ' and extremely delayed and limited formation of needle-shaped GP zones and AlLiSi. For 7075 the addition of 0.7% Li alters the conventional precipitation scheme from solute-rich GP zone $\rightarrow$ $\eta' \rightarrow \eta MgZn_2$ into vacancy-rich GP zone $\rightarrow$ T' $\rightarrow$ $T(AlZn)_{49}Mg_{32}$, whereas the addition of 2.0% Li produces the dominant δ' precipitate and limited and delayed formation of T-phase. As a result, the age-hardening response relating to these major hardening phases in both base alloys is delayed or decreased. Such additions can produce narrower PFZ's and give improved fracture properties.

A further commercial alloy is *UL40*, which is essentially a binary alloy containing 4% Li. The alloy is cast using a spray-deposition process resulting in a fine-grained microstructure, with uniform distribution of second phase, free from oxide. The high Li content alloy has a very low density (2400 kg m^{-3}) and is almost a third lighter than conventional aluminium and magnesium alloys. It extrudes well and can be welded with Al–Mg–Zr filler, producing components for aircraft and helicopters, such as pump housings and valves, and for yachting with good corrosion resistance.

9.7.4 Processing developments

9.7.4.1 Superplastic aluminium alloys

Superplastic forming is a cost-effective manufacturing process for producing both simple and complex shapes from aluminium alloy sheet, because of its low-cost tooling and short lead times for production. A range of alloys is available including 2004 (Al–6Cu–0.4Zr) or *Supral*, 5083 SPF, 7475 SPF and *Lital* 8090 SPF (Al–Li–Mg–Cu). *Supral* and *Lital*

Table 9.8 Composition of commercial aluminium-lithium alloys.

Alloy	Li	Cu	Mg	Fe	Si	Zr
(8090) *Lital A*	2.5	1.3	0.7	≤ 0.2	≤ 0.1	0.12
(9091) *Lital B*	2.6	1.9	0.9	≤ 0.2	≤ 0.1	0.12

dynamically recrystallize to a fine grain size during the early stages of deformation ($\sim$500°C) which is stabilized with $ZrAl_3$ particles. The grain size in 7475 is stabilized by TMP and submicron Cr particles and, in 5083, by a Mn-dispersoid. Components are formed by clamping the alloy sheet in a pressure chamber and then applying gas or air pressure to force the sheet slowly into contact with a tool surface; both the tool and sheet are maintained at the forming temperature throughout the process. During normal superplastic forming, the alloys tend to develop voiding. This void formation is minimized by forming in hydrostatic conditions by introducing a gas pressure on both the front and back surfaces of the sheet being formed. The sheet is forced against the tool surface with a small pressure differential.

9.7.4.2 Rapid solidification processing of aluminium alloys

Rapid solidification processing (RSP) has been applied to aluminium alloys to produce a fine grain size and extend solid solubility, particularly for the transition metals, iron, molybdenum, chromium and zirconium, which usually have low solid solubility and low diffusion rates in aluminium. Interestingly RSP alloys containing Fe and Cr, on annealing, precipitate metastable spherical quasi-crystals of icosahedral phase with five-fold symmetry. These are extremely stable and hardly coarsen after extensive heat-treatment which indicates a potential for alloy development.

A series of commercially-available high-temperature Al–Fe–V–Si alloys has been developed and consist of very fine, spherical $Al_{13}(FeV)_3Si$ silicides uniformly dispersed throughout the matrix which display much slower coarsening rates than other dispersoids. A typical alloy with 27 vol.% silicides is 8009 (Al–8.5Fe–1.3V–1.7Si) and, without any needle or platelet precipitates in the microstructure, has a K_{1c} $\sim$29 MN m$^{-3/2}$. The tensile properties as a function of temperature are shown in Figure 9.17 in comparison with a conventional 7075-T6 alloy. At all temperatures up to 480°C, the 8009 alloy has a higher specific stiffness than a Ti–6Al–4V alloy. The fatigue and creep rupture properties are better than conventional aluminium alloys with excellent corrosion resistance. These RSP silicide alloys can be readily fabricated into sheet, extruded or forged and the combination of attractive properties makes them serious candidates for aerospace applications. Other alloys developed include Al–Cr–Zn–Mn, Al–8Fe–2Mo and Al–Li.

9.7.4.3 Mechanical alloys of aluminium

Mechanical alloys of aluminium contain dispersions of carbides or oxides, which not only produce dispersion strengthening but also stabilize a fine-grain structure. An advantage of these alloys arises because the strength is derived from the dispersoids and thus the composition of the alloy matrix can be designed principally for corrosion resistance and toughness rather than strength. Thus the alloying elements which are usually added to conventional aluminium alloys for precipitation-strengthening and grain size control may be unnecessary.

Mechanical alloying is carried out with elemental powders and an organic process control agent, such as

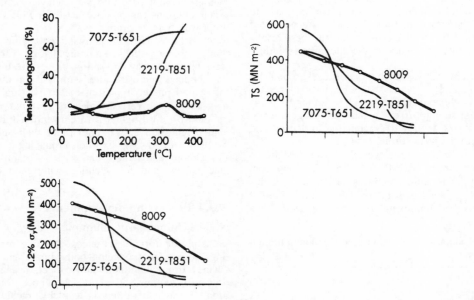

Figure 9.17 *Tensile properties of RSP Al alloy 8009 as a function of temperature compared with conventional aluminium alloys (from Gilman, 1990, p. 505).*

stearic acid, to balance the cold-welding and powder fracture processes. No dispersoid is added because the oxide on the surface of the powders and process control agent are consolidated during mechanical alloying as hydrated oxides and carbonates. The process produces a fine dispersion of ~20 nm particles in a dynamically-recrystallized structure with grains as fine as 0.05 μm. Subsequent vacuum degassing at elevated temperature removes the H_2 and N_2 liberated, improves the homogenization of the matrix and reduces carbonates to Al_4C_3 which forms most of the dispersoid. The final grain size is around 0.1 μm. The powder is then compacted by HIPing, or vacuum hot-pressing and conventionally extruded to produce a material with a stable grain size of 0.3 μm, with grain boundaries pinned by the dispersoid.

Mechanical alloys have been developed corresponding to the 2000, 5000 and 7000 aluminium series alloy. *IN-9021* is heat-treatable by solution treatment and natural or elevated-temperature ageing to give 500–560 MN m^{-2} proof stress, 570–600 MN m^{-2} TS, 12% elongation and 40 MN m$^{-3/2}$ K_{1c}. *IN-9052* is the equivalent of a 5000 series alloy, requiring no heat-treatment and offering good strength in thick sections; 390 MN m^{-2} proof stress, 470 MN m^{-2} TS, 13% elongation and 46 MN m$^{-3/2}$ K_{1c}. Mechanically alloyed Al–Mg–Li offers inherent high strength in thick section 430 MN m^{-2} proof-stress, 500 MN m^{-2} TS, 10% elongation and 30 MN m$^{-3/2}$ K_{1c}.

Further reading

Baker, C. (ed.) (1986). *Proc. of 3rd International Aluminium-Lithium Conference.* Institute of Metals, London.

Bhadeshia, H. K. D. H. (1992). *Bainite in Steels.* Institute of Materials, London.

Honeycombe, R. W. K. (1981). *Steels, Micro-structure and Properties.* Edward Arnold, London.

Janowak, J. R. *et al.* (1984). A review of austempered ductile iron metallurgy. *First International Conference on ADI.*

Materials in defence (1988). *Metals and Materials*, **4**, No. 7. Institute of Materials, London.

Meetham, G. W. (1981). *The Development of Gas Turbine Materials.* Applied Science, London.

Peters, H. (ed.) (1991). *Proc. of 6th International Aluminium-Lithium Conference.* Deutsche Gesellschaft für Materialkunde.

Polmear, I. J. (1989), *Light Alloys*, Edward Arnold, London.

Sims, C. T. and Hagel, W. C. (eds) (1972). *Superalloys.* John Wiley, Chichester.

Sims, C. T., Hagel, W. C. and Stoloff, N. S. (1987). *Superalloys II*, John Wiley, Chichester.

Stoloff, N. S. (ed.) (1984). Ordered alloys. *International Metal Reviews*, **29**, No. 3.

Stoloff, N. S. (1989). Physical and mechanical metallurgy of Ni_3Al and its alloys. *International Metal Reviews*, **34**, No. 4.

Yoo, M. H. *et al.* (1993). Deformation and fracture of intermetallics. Overview No. 15, *Acta Metall. and Mater.*, No. 4.

Young, Won Kim and Froes, F. H. (1989). Physical metallurgy of titanium aluminides. TMS/ASM Symposium on High Temperature Aluminides and Intermetallics.

Chapter 10

Ceramics and glasses

10.1 Classification of ceramics

The term ceramic, in its modern context, covers an extremely broad range of inorganic materials; they contain non-metallic and metallic elements and are produced by a wide variety of manufacturing techniques. Traditionally, ceramics are moulded from silicate minerals, such as clays, dried and fired at temperatures of 1200–1800°C to give a hard finish. Thus we can readily see that the original Greek word *keramos*, meaning 'burned stuff' or 'kiln-fired material', has long been directly appropriate. Modern ceramics, however, are often made by processes that do not involve a kiln-firing step (e.g. hot-pressing, reaction-sintering, glass-devitrification, etc.). Although ceramics are sometimes said to be non-metallic in character, this simple distinction from metals and alloys has become increasingly inadequate and arbitrary as new ceramics with unusual properties are developed and come into use.

Ceramics may be generally classified, according to type or function, in various ways. In industrial terms, they may be listed as pottery, heavy clay products (bricks, earthenware pipes, etc.), refractories (firebricks, silica, alumina, basic, neutral), cement and concrete, glasses and vitreous enamels, and engineering (technical, fine) ceramics. Members of the final group are capable of very high strength and hardness, exceptional chemical stability and can be manufactured to very close dimensional tolerances. These will be our prime concern. Their introduction as engineering components in recent years has been based upon considerable scientific effort and has revolutionized engineering design practice. In general, the development of engineering ceramics has been stimulated by the drive towards higher, more energy-efficient, process temperatures and foreseeable shortages of strategic minerals. In contrast to traditional ceramics, which use naturally-occurring and, inevitably, rather variable minerals, the new generation of engineering ceramics

depends upon the availability of purified and synthesized materials and upon close microstructural control during processing. Ceramics are subject to variability in their properties and statistical concepts often need to be incorporated into design procedures for stressed components. Design must recognize the inherent brittleness, or low resistance to crack propagation, and modify, if necessary, the mode of failure. Ceramics, because of their unique properties, show great promise as engineering materials but, in practice, their production on a commercial scale in specified forms with repeatable properties is often beset with many problems.

Using chemical composition as a basis, it is possible to classify ceramics into five main categories:

1. *Oxides* — alumina, Al_2O_3 (spark plug insulators, grinding wheel grits), magnesia, MgO (refractory linings of furnaces, crucibles), zirconia, ZrO_2 (piston caps, refractory lining of glass tank furnaces), zirconia/alumina (grinding media), spinels, $M^{2+}O.M_2^{3+}O_3$ (ferrites, magnets, transistors, recording tape), 'fused' silica glass (laboratory ware),
2. *Carbides* — silicon carbide, SiC (chemical plant, crucibles, ceramic armour), silicon nitride, Si_3N_4 (spouts for molten aluminium, high-temperature bearings), boron nitride, BN (crucibles, grinding wheels for high-strength steels).
3. *Silicates* — porcelain (electrical components), steatites (insulators), mullite (refractories).
4. *Sialons* — based on Si–Al–O–N and M–Si–Al–O–N where M = Li, Be, Mg, Ca, Sc, Y, rare earths (tool inserts for high-speed cutting, extrusion dies, turbine blades).
5. *Glass-ceramics* — *Pyroceram, Cercor, Pyrosil* (recuperator discs for heat exchangers).

The preceding two methods of classifying ceramics, industrial and chemical, are of very little use to the materials scientist and technologist, who is primarily concerned with structure/property relations. One can predict that a ceramic structure with a fine grain (crystal) size and low porosity is likely to offer advantages of mechanical strength and impermeability to contacting fluids. It is therefore scientifically appropriate to classify ceramic materials in microstructural terms, in the following manner:

1. Single crystals of appreciable size (e.g. ruby laser crystal)
2. Glass (non-crystalline) of appreciable size (e.g. sheets of 'float' glass)
3. Crystalline or glassy filaments (e.g. E-glass for glass-reinforced polymers, single-crystal 'whiskers', silica glass in Space Shuttle tiles)
4. Polycrystalline aggregates bonded by a glassy matrix (e.g. porcelain pottery, silica refractories, hot-pressed silicon nitride)
5. Glass-free polycrystalline aggregates (e.g. ultra-pure, fine-grained, 'zero-porosity' forms of alumina, magnesia and beryllia)
6. Polycrystalline aggregates produced by heat-treating glasses of special composition (e.g. glass-ceramics)
7. Composites (e.g. silicon carbide or carbon filaments in a matrix of glass or glass-ceramic, magnesia-graphite refractories, concrete).

This approach to classifying ceramics places the necessary emphasis upon the crystalline and non-crystalline (glassy) attributes of the ceramic body, the significance of introducing grain boundary surfaces and the scope for deliberately mixing two phases with very different properties.

10.2 General properties of ceramics

The constituent atoms in a ceramic are held together by very strong bonding forces which may be ionic, covalent or a mixture of the two. As a direct consequence, their melting points are often very high, making them eminently suited for use in energy-intensive systems such as industrial furnaces and gas turbines. For instance, alumina primarily owes its importance as a furnace refractory material to its melting point of 2050°C. The type of inter-atomic bonding is responsible for the relatively low electrical conductivity of ceramics. For general applications they are usually regarded as excellent electrical insulators, having no free electrons. However, ion mobility becomes significant at temperatures above 500–600°C and they then become progressively more conductive. This property can prove a problem in electric furnaces.

The strength of ceramics under compressive stressing is excellent; accordingly, designers of ceramic artefacts as different as arches in buildings and metal-cutting tool tips ensure that the forces during service are essentially compressive. In contrast, the tensile strength of ceramics is not exceptional, sometimes poor, largely because of the weakening effect of surface flaws. Thus, in some cases, glazing with a thin vitreous layer can seal surface cracks and improve the tensile strength. The strength of ceramics is commonly expressed as a modulus of rupture (MoR) value, obtained from three-point bend tests, because in the more conventional type of test with uniaxial loading, as used for metals, is difficult to apply with perfect uniaxiality; a slight misalignment of the machine grips will induce unwanted bending stresses. Ceramics are generally regarded as brittle, non-ductile materials, with little or no plastic deformation of the microstructure either before or at fracture. For this reason, which rules out the types of production processes involving deformation that are so readily applied to metals and polymers, ceramic production frequently centres on the manipulation and ultimate bonding together of fine powders. The inherent lack of ductility implies that ceramics are likely to have a better resistance to slow plastic deformation at very high temperatures (creep) than metals.

The modulus of elasticity of ceramics can be exceptionally high (Table 10.1). This modulus expresses stiffness, or the amount of stress necessary to produce unit elastic strain, and, like strength, is a primary design consideration. However, it is the combination of low density with this stiffness that makes ceramics particularly attractive for structures in which weight reduction is a prime consideration.

In aircraft gas turbines, ceramic blades have long been an interesting proposition because, apart from reducing the total mass that has to be levitated, they are

Table 10.1 Specific moduli of various materials

	Modulus of elasticity (E/GN m^{-2})	Bulk density (ρ/kg m^{-3})	Specific modulus (E/ρ)
Alumina	345	3800	0.091
Glass (crown)	71	2600	0.027
Aluminium	71	2710	0.026
Steel (mild)	210	7860	0.027
Oak (with grain)	12	650	0.019
Concrete	14	2400	0.006
Perspex	3	1190	0.003

subject to lower centrifugal forces than metallic versions. It is therefore common practice to appraise competitor materials for aircraft in terms of their specific moduli, in which the modulus of elasticity is divided by density. Ceramics consist largely of elements of low atomic mass, hence their bulk density is usually low, typically about 2000–4000 kg m^{-3}. Ceramics such as dense alumina accordingly tend to become pre-eminent in listings of specific moduli (Table 10.1).

The strong interatomic bonding means that ceramics are hard as well as strong. That is, they resist penetration by scratching or indentation and are potentially suited for use as wear-resistant bearings and as abrasive particles for metal removal. Generally, impact conditions should be avoided. Interestingly, shape can influence performance; thus, the curved edges of dinner plates are carefully designed to maximize resistance to chipping. Although the strength and hardness of materials are often related in a relatively simple manner, it is unwise to assume that a hard material, whether metallic or ceramic, will necessarily prove to be wear-resistant. Grinding of ceramics is possible, albeit costly. Strength can be enhanced in this way but great care is necessary as there is a risk that the machining operation will damage, rather than improve, the critical surface texture.

During the consolidation and densification of a 'green' powder compact in a typical firing operation, sintering of the particles gradually reduces the amount of pore space between contiguous grains. The final porosity, by volume, of the fired material ranges from 30% to nearly zero. Together with grain size, it has a direct influence upon the modulus of rupture; thus, bone china, because of its finer texture, is twice as strong as fired earthenware. Pore spaces, particularly if interconnected, also lower the resistance of a ceramic structure to penetration by a pervasive fluid such as molten slag. On the other hand, deliberate encouragement of porosity, say 25–30% by volume, is used to lower the thermal conductivity of insulating refractories.

Ceramics are often already in their highest state of oxidation. Not surprisingly, they often exhibit low chemical reactivity when exposed to hot oxidizing environments. Their refractoriness, or resistance to degradation and collapse during service at high temperatures, stems from the strong interatomic bonding. However, operational temperatures are subject to sudden excursions and the resulting steep gradients of temperature within the ceramic body can give rise to stress imbalances. As the ceramic is essentially non-ductile, stresses are not relieved by plastic deformation and cracking may occur in planes roughly perpendicular to the temperature gradient, with portions of ceramic becoming detached from the hottest face. The severity of this disintegration, known as spalling, depends mainly upon thermal expansivity (α) and conductivity (k). Silica has a poor resistance to spalling whereas silicon nitride can withstand being heated to a temperature of 1000°C and then quenched in cold water.

The ability of certain ceramic oxides to exist in either crystalline or non-crystalline forms has been commented upon previously. Silica and boric oxide possess this ability. In glass-ceramics, a metastable glass of special composition is shaped while in the viscous condition, then heat-treated in order to induce nucleation and growth of a fine, completely crystalline structure. (This manipulation and exploitation of the crystalline and glassy states is also practised with metals and polymers.) This glass-forming potential is an important aspect of ceramic science. The property of transparency to light is normally associated with glasses, notably with the varieties based upon silica. However, transparency is not confined to glasses and single crystals. It is possible to produce some oxides, normally regarded as opaque, in transparent, polycrystalline forms (e.g. hot-pressed magnesia).

So far as sources in the earth's crust are concerned, mineral reserves for ceramic production are relatively plentiful. While one might observe that important constituent elements such as silicon, oxygen and nitrogen are outstandingly abundant, it must also be recognized that the processes for producing the new ceramics can be very costly, demanding resort to highly specialized equipment and exacting process control.

10.3 Production of ceramic powders

The wide-ranging properties and versatility of modern engineering ceramics owe much to the ways in which they are manufactured. A fine powder is usually the starting material, or precursor; advanced ceramics are mainly produced from powders with a size range of 1–10 μm. Electrical properties are extremely structure-sensitive and there is a strong demand from the electronics industry for even finer particles (in the nanometre range). The basic purpose of the manufacturing process is to bring particle surfaces together and to develop strong interparticle bonds. It follows that specific surface area, expressed per unit mass, is of particular significance. Characterization of the powder in terms of its physical and chemical properties, such as size distribution, shape, surface topography, purity and reactivity, is an essential preliminary to the actual manufacturing process. Tolerances and limits are becoming more and more exacting.

The three principal routes for producing high-grade powders are based upon solid-state reactions, solution and vaporization. The solid-state reaction route, long exemplified by the Acheson process for silicon carbide (Section 10.4.5.2), involves high temperatures. It is used in more refined forms for the production of other carbides (TiC, WC), super-conductive oxides and silicon nitride. An aggregate is produced and the necessary size reduction (comminution) introduces the risk of contamination. Furthermore, as has long been known in mineral-dressing industries, fine grinding is energy-intensive and costly.

The Bayer process for converting bauxite into alumina is a solution-treatment method. In this important process, which will be examined in detail later (Section 10.4.1.2), aluminium hydroxide is precipitated from a caustic solution and then converted to alumina by heating. Unfortunately, this calcination has a sintering effect and fine grinding of the resultant agglomerate is necessary. In the more recent spray-drying and spray-roasting techniques, which are widely used to produce oxide powders, sprayed droplets of concentrated solutions of appropriate salts are rapidly heated by a stream of hot gas. Again, there is a risk of agglomeration and grinding is often necessary.

These difficulties, which stem from the inherent physical problem of removing all traces of solvent in a satisfactory manner, have encouraged development of methods based upon a 'solution-to-gelation' (sol–gel) approach. The three key stages of a typical sol–gel process are:

1. Production of a colloidal suspension or solution (sol) (e.g. concentrated solution of metallic salt in dilute acid)
2. Adjustment of pH, addition of a gelling agent, evaporation of liquid to produce a gel
3. Carefully controlled calcination to produce fine particles of ceramic.

Sol–gel methods are applicable to both ceramics and glasses and are capable of producing filaments as well as powders. One variant involves hydrolysis of distillation-purified alkoxides (formed by reacting metal oxides with alcohol). The hydroxide particles precipitated from the sol are spherical, uniform in shape and sub-micron sized. Sintering does not drastically change these desirable characteristics. Although costs tend to be high and processing times are lengthy, sol–gel methods offer an attractive way to produce oxide powders, such as alumina, zirconia and titania, that will flow, form and sinter readily and give a product with superior properties. Currently, there is great interest in vapour phase methods that enable powders with a particle size as small as 10–20 nm to be produced (e.g. oxides, carbides, nitrides, silicides, borides). The high-energy input required for vaporization is provided by electric arcs, plasma jets or laser beams. The powder is condensed within a carrier gas and then separated from the gas stream by impingement filters or electrostatic precipitators. Sometimes, in a chemical vapour deposition process (CVD), a thin film is condensed directly upon a substrate.

The manufacture of an advanced ceramic usually involves a number of steps, or unit operations. Each operation is subject to a number of interacting variables (time, temperature, pressure, etc.) and, by having a very specific effect upon the developing structure (macro- and micro-), makes its individual contribution to the final quality of the product. When ductile metals are shaped by plastic deformation, each operation stresses the material and is likely to reveal flaws. (For instance, the ability of an austenitic stainless steel to be cold-drawn to the dimensions of a fine hypodermic needle tube is strong evidence of structural integrity.) Individual ceramic particles are commonly brittle and non-deformable; consequently, manufacturing routes usually avoid plastic deformation and there is a greater inherent risk that flaws will survive processing without becoming visible or causing actual disintegration. The final properties of an advanced ceramic are extremely sensitive to any form of structural heterogeneity. The development of special ceramics and highly-innovative production techniques has encouraged greater use of non-destructive evaluation (NDE) techniques at key points in the manufacturing programme. At the design stage, guidelines of the following type are advisedly applied to the overall plan of production:

1. Precursor materials, particularly ultra-fine powders, should be scientifically characterized.
2. Each and every unit operation should be closely studied and controlled.
3. NDE techniques should be carefully integrated within the overall scheme of operations.

10.4 Selected engineering ceramics

10.4.1 Alumina

10.4.1.1 General properties and applications of alumina

Alumina is the most widely used of the twenty or so oxide ceramics and is often regarded as the historic forerunner of modern engineering ceramics. The actual content of alumina, reported as Al_2O_3, ranges from 85% to 99.9%, depending upon the demands of the application.

Alumina-based refractories of coarse grain size are used in relatively massive forms such as slabs, shapes and bricks for furnace construction. Alumina has a high melting point (2050°C) and its heat resistance, or refractoriness, has long been appreciated by furnace designers. In fact, there has been a trend for aluminosilicate refractories (based upon clays) to be replaced by more costly high-alumina materials and high-purity alumina. Interatomic bonding forces, partly ionic and partly covalent, are extremely strong and the crystal structure of alumina is physically stable up to temperatures of 1500–1700°C. It is used for protective sheaths for temperature-measuring thermocouples which have to withstand hot and aggressive environments and for filters which remove foreign particles and oxide dross from fast-moving streams of molten aluminium prior to casting. Large refractory blocks cast from fused alumina are used to line furnaces for melting glass. However, although alumina is a heat-resisting material with useful chemical stability, it is more sensitive to thermal shock than silicon carbide and silicon nitride. A contributory factor is its relatively high linear coefficient of thermal expansion (α).

The respective α-values/ $\times 10^{-6}$ K^{-1} for silicon carbide, silicon nitride and alumina are 8, 4.5 and 3.5.

When intended for use as engineering components at lower temperatures, alumina ceramics usually have a fine grain size (0.5–20 μm) and virtually zero porosity. Development of alumina to meet increasingly stringent demands has taken place continuously over many years and has focused mainly upon control of chemical composition and grain structure. The chemical inertness of alumina and its biocompatibility with human tissue have led to its use for hip prostheses. An oft-quoted example of the capabilities of alumina is the insulating body of the spark-ignition plug for petrol-fuelled engines (Figure 10.1). Its design and fabrication methods have been steadily evolving since the early 1900s. In modern engines, trouble-free functioning of a plug depends primarily upon the insulating capability of its isostatically-pressed alumina body. Each plug is expected to withstand temperatures up to 1000°C, sudden mechanical pressures, corrosive exhaust gases and a potential difference of about 30 kV while 'firing' precisely 50–100 times per second over long periods of time. Plugs are provided with a smooth glazed (glassy) surface so that any electrically-conductive film of contamination can be easily removed.

The exceptional insulating properties and range of alumina ceramics have long been recognized in the electrical and electronics industries (e.g. substrates for circuitry, sealed packaging for semiconductor microcircuits). Unlike metals, there are no 'free' electrons available in the structure to form a flow of current. The dielectric strength, which is a measure of the ability of a material to withstand a gradient of electric potential without breakdown or discharge, is very high. Even at temperatures approaching 1000°C, when the atoms tend to become mobile and transport some electrical charge, the resistivity is still significantly high. Electrical properties usually benefit when the purity of alumina is improved.

Many mass-produced engineering components take advantage of the excellent compressive strength, hardness and wear resistance of alumina (e.g. rotating seals in washing machines and in water pumps for automobile engines, machine jigs and cutting tools, soil-penetrating coulters on agricultural equipment, shaft bearings in watches and tape-recording machines, guides for fast-moving fibres and yarns, grinding abrasives). (Emery, the well-known abrasive, is an impure anhydrous form of alumina which contains as much as 20% $SiO_2 + Fe_2O_3$; pretreatment is often unnecessary.) The constituent atoms in alumina, aluminium and oxygen, are of relatively low mass and the correspondingly low density (3800 kg m^{-3}) is often advantageous. However, like most ceramics, alumina is brittle and should not be subjected to either impact blows or excessive tensile stresses during service.

Alumina components are frequently quite small but their functioning can vitally affect the performance and overall efficiency of a much larger engineering system.

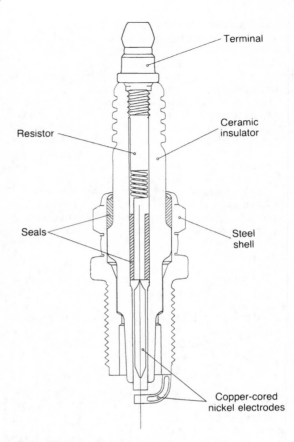

Figure 10.1 *Spark plug for petrol engine (with acknowledgements to Champion Spark Plug Division of Cooper GB Ltd).*

Spark-plug insulators[1] and water-pump sealing rings in internal combustion engines are striking examples of this principle at work.

10.4.1.2 Preparation and shaping of alumina powders

Examination of the general form of the production route for alumina ceramics from ore to finished shape provides an insight into some of the important factors and working principles which guide the ceramics technologist and an indication of the specialized shaping methods that are available for ceramics. As mentioned earlier, each stage of the production sequence makes its own individual and vital contribution to the final quality of the product and must be carefully controlled.

The principal raw material for alumina production is bauxite $Al_2O(OH)_4$, an abundant hydrated rock occurring as large deposits in various parts of the

[1] Over the period 1902–1977 Robert Bosch Ltd developed more than 20 000 different types of spark plug.

world.[2] In the Bayer process, prepared bauxitic ore is digested under pressure in a hot aqueous solution of sodium hydroxide and then 'seeded' to induce precipitation of $Al(OH)_3$ crystals, usually referred to by the mineral term 'gibbsite'. (The conditions of time, temperature, agitation, etc. during this stage greatly influence the quality of the Bayer product.) Gibbsite is chemically decomposed by heating (calcined) at a temperature of 1200°C. Bayer calcine, which consists of α-alumina (>99% Al_2O_3), is graded according to the nature and amount of impurities. Sodium oxide, Na_2O, ranges up to 0.6% and is of special significance because it affects sintering behaviour and electrical resistance. The calcine consists of agglomerates of α-alumina crystallites which can be varied in average size from 0.5 to 100 μm by careful selection of calcining conditions.

Bayer calcine is commonly used by manufacturers to produce high-purity alumina components as well as numerous varieties of lower-grade components containing 85–95% Al_2O_3. For the latter group, the composition of the calcine is debased by additions of oxides such as SiO_2, CaO and MgO which act as fluxes, forming a fluid glassy phase between the grains of α-alumina during sintering.

The chosen grade of alumina, together with any necessary additives, is ground in wet ball-mills to a specified size range. Water is removed by spraying the aqueous suspension into a flow of hot gas (spray-drying) and separating the alumina in a cyclone unit. The free-flowing powder can be shaped by a variety of methods (e.g. dry, isostatic- or hot-pressing, slip- or tape-casting, roll-forming, extrusion, injection-moulding). Extremely high production rates are often possible; for instance, a machine using air pressure to compress dry powder isostatically in flexible rubber moulds ('bags') can produce 300–400 spark plug bodies per hour. In some processes, binders are incorporated with the powder; for instance, a thermoplastic can be hot-mixed with alumina powder to facilitate injection-moulding and later burned off. In tape-casting, which produces thin substrates for micro-electronic circuits, alumina powder is suspended in an organic liquid.

10.4.1.3 Densification by sintering

The fragile and porous 'green' shapes are finally fired in kilns (continuous or intermittent). Firing is a costly process and, wherever possible, there has been a natural tendency to reduce the length of the time cycle for small components. Faster rates of cooling after 'soaking' at the maximum temperature have been found to give a finer, more desirable grain structure.

[2]Long-distance transportation costs have prompted investigation of alternative sources. For instance, roasted kaolinite can be leached in concentrated hydrochloric or sulphuric acid, then precipitated as an aluminium salt which is calcined to form alumina.

It has been mentioned that fluxing oxides are added to lower-grade aluminas in order to form an intergranular phase(s). Although this fluid inter-granular material facilitates densification during firing, its presence in the final product can have a detrimental effect upon strength and resistance to chemical attack. As a consequence, powders of high alumina content are chosen for demanding applications. In general, an increase in alumina content from 88% to 99.8% requires a corresponding increase in firing temperature from 1450°C to 1750°C. 'Harder' firing incurs heavier energy costs and has led to the development of reactive alumina which has an extremely small particle size (1 μm) and a large specific surface. 'Softer' firing temperatures became possible with this grade of alumina and the need to debase the alumina with relatively large amounts of additives was challenged.

Shrinkage is the most apparent physical change to take place when a 'green' ceramic compact is fired. The linear shrinkage of alumina is about 20% and dimensions may vary by up to ±1%. Diamond machining is used when greater precision is needed but requires care as it may damage the surface and introduce weakening flaws.

10.4.2 From silicon nitride to sialons

10.4.2.1 Reaction-bonded silicon nitride (RBSN)

Silicon nitride, which can be produced in several ways, has found application under a variety of difficult conditions (e.g. cutting tools, bearings, heat engines, foundry equipment, furnace parts, welding jigs, metal-working dies, etc.). Its original development was largely stimulated by the search for improved materials for gas turbines. Prior to its development in the 1950s, the choice of fabrication techniques for ceramics was restricted and it was difficult to produce complex ceramic shapes to close dimensional tolerances. The properties available from existing materials were variable and specific service requirements, such as good resistance to thermal shock and attack by molten metal and/or slag, could not be met. The development of silicon nitride minimized these problems; it has also had a profound effect upon engineering thought and practice.

Silicon nitride exists in two crystalline forms (α, β): both belong to the hexagonal system. Bonding is predominantly covalent. Silicon nitride was first produced by an innovative form of pressureless sintering. First, a fragile pre-form of silicon powder (mainly α-Si_3N_4) is prepared, using one of a wide variety of forming methods (e.g die-pressing, isostatic-pressing, slip-casting, flame-spraying, polymer-assisted injection-moulding, extrusion). In the first stage of a reaction-bonding process, this pre-form is heated in a nitrogen atmosphere and the following chemical reaction takes place:

$$3Si + 2N_2 = Si_3N_4$$

A reticular network of reaction product forms throughout the mass, bonding the particles together without liquefaction. Single crystal 'whiskers' of α-silicon nitride also nucleate and grow into pore space. Reaction is strongly exothermic and close temperature control is necessary in order to prevent degradation of the silicon. The resultant nitrided compact is strong enough to withstand conventional machining. In the second and final stage of nitridation, the component is heated in nitrogen at a temperature of 1400°C, forming more silicon nitride *in situ* and producing a slight additional change in dimensions of less than 1%. (Alumina articles can change by nearly 10% during firing.) The final microstructure consists of α-Si$_3$N$_4$ (60–90%), β-Si$_3$N$_4$ (10–40%), unreacted silicon and porosity (15–30%). As with most ceramics, firing is the most costly stage of production.

The final product, reaction-bonded silicon nitride (RBSN), has a bulk density of 2400–2600 kg m^{-3}. It is strong, hard and has excellent resistance to wear, thermal shock and attack by many destructive fluids (molten salts, slags, aluminium, lead, tin, zinc, etc.). Its modulus of elasticity is high.

10.4.2.2 Hot-pressed forms of silicon nitride (HPSN, HIPSN)

In the early 1960s, a greater degree of densification was achieved with the successful production of hot-pressed silicon nitride (HPSN) by G. G. Deeley and co-workers at the Plessey Co. UK. Silicon nitride powder, which cannot be consolidated by solid-state sintering alone, is mixed with one or more fluxing oxides (magnesia, yttria, alumina) and compressed at a pressure of 23 MN m^{-2} within radio-frequency induction-heated graphite dies at temperatures up to 1850°C for about 1 h. The thin film of silica that is usually present on silicon nitride particles combines with the additive(s) and forms a molten phase. Densification and mass transport then take place at the high temperature in a typical 'liquid-phase' sintering process. As this intergranular phase cools, it forms a siliceous glass which can be encouraged to crystallize (devitrify) by slow cooling or by separate heat-treatment. This HP route deliberately produces a limited amount of second phase (up to 3% v/v) as a means of bonding the refractory particles; however, this bonding phase has different properties to silicon nitride and can have a weakening effect, particularly if service temperatures are high. Thus, with 3–5% added magnesia, at temperatures below the softening point of the residual glassy phase, say 1000°C, silicon nitride behaves as a brittle and stiff material; at higher temperatures, there is a fairly abrupt loss in strength, as expressed by modulus of rupture (MoR) values, and slow deformation under stress (creep) becomes evident. For these reasons, controlled modification of the structure of the inter-granular residual phase is of particular scientific concern.

Yttria has been used as an alternative densifier to magnesia. Its general effect is to raise the softening point of intergranular phase significantly. More specifically, it yields crystalline oxynitrides (e.g. Y$_2$Si$_3$O$_3$N$_4$) which dissolve impurities (e.g. CaO) and form refractory solid solutions ('mixed crystals'). Unfortunately, at high temperatures, yttria-containing silicon nitride has a tendency to oxidize in a catastrophic and disruptive manner.

Although the use of dies places a restriction upon component shape, hot-pressing increases the bulk density and improves strength and corrosion resistance. The combination of strength and a low coefficient of thermal expansion (approximately 3.2×10^{-6} °C^{-1} over the range 25–1000°C) in hot-pressed silicon nitride confer excellent resistance to thermal shock. Small samples of HPSN are capable of surviving 100 thermal cycles in which immersion in molten steel (1600°C) alternates with quenching into water.

In a later phase of development, other researchers used hot isostatic-pressing (HIPing) to increase density further and to produce much more consistent properties. Silicon nitride powder, again used as the starting material, together with a relatively small amount of the oxide additive(s) that promote liquid-phase sintering, is formed into a compact. This compact is encapsulated in glass (silica or borosilicate). The capsule is evacuated at a high temperature, sealed and then HIPed, with gas as the pressurizing medium, at pressures up to 300 MN m^{-2} for a period of 1 h. Finally, the glass envelope is removed from the isotropic HIPSN component by sand-blasting. Like HPSN, its microstructure consists of β-Si$_3$N$_4$ (>90%) and a small amount of intergranular residue (mainly siliceous glass).

Production routes involving deformation at very high temperatures and pressures, as used for HPSN and HIPSN, bring about a desirable closure of pores but inevitably cause a very substantial amount of shrinkage (20–30%). (In contrast to HPSN and HIPSN, RBSN undergoes negligible shrinkage during sintering at the lower process temperature of 1400°C and accordingly contains much weakening porosity, say 15–30% v/v.) By the early 1970s, considerable progress had been made in producing silicon nitride by reaction-bonding, hot-pressing and other routes. However, by then it had become evident that further significant improvements in the quality and capabilities of silicon nitride were unlikely. At this juncture, attention shifted to the sialons.

10.4.2.3 Scientific basis of sialons

Although silicon nitride possesses extremely useful properties, its engineering exploitation has been hampered by the difficulty of producing it in a fully dense form to precise dimensional tolerances. Hot-pressing offers one way to surmount the problem but it is a costly process and necessarily limited to simple shapes. The development of sialons provided an attractive and feasible solution to these problems.

Sialons are derivatives of silicon nitride and are accordingly also classified as nitrogen ceramics. The acronym 'sialon' signifies that the material is based upon the Si–Al–O–N system. In 1968, on the basis of structural analyses of silicon nitrides, it was predicted[1] that replacement of nitrogen (N^{3-}) by oxygen (O^{2-}) was a promising possibility if silicon (Si^{4+}) in the tetrahedral network could be replaced by aluminium (Al^{3+}), or by some other substituent of valency lower than silicon. Furthermore, it was also predicted that systematic replacement of silicon by aluminium would allow other types of metallic cation to be accommodated in the structure. Such replacement within the SiN_4 structural units of silicon nitride would make it possible to simulate the highly versatile manner in which SiO_4 and AlO_4 tetrahedra arrange themselves in aluminosilicates. A similarly wide range of structures and properties was anticipated for this new family of ceramic 'alloys'. About two years after the vital prediction, British and Japanese groups, acting independently, produced β'-silicon nitride, the solid solution which was to be the prototype of the sialon family.

In β-silicon nitride, the precursor, SiN_4 tetrahedra form a network structure. Each tetrahedron has a central Si^{4+} which is surrounded by four equidistant N^{3-} (Figure 10.2). Each of these corner N^{3-} is common to

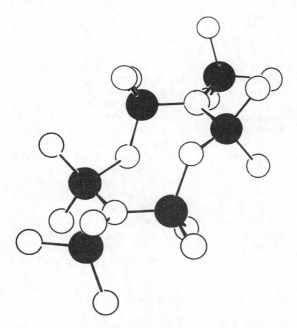

Figure 10.2 *The crystal structure of β-Si_3N_4 and β'-$(Si,Al)_3(O,N)_4$* ● *metal atom,* ○ *non-metal atom (from Jack, 1987, pp. 259–88; reprinted by permission of the American Ceramic Society).*

[1]By K. H. Jack and colleagues at the University of Newcastle-upon-Tyne; separate British and Japanese groups filed patents for producing sialons in the early 1970s. The writings of K. H. Jack on silicon nitride and sialons provide an insight into the complexities of developing a new engineering material.

three tetrahedra. In the unit cell, six Si^{4+} ions balance the electrical charge of eight N^{3-}, giving a starting formula Si_6N_8. Replacement of Si^{4+} and N^{3-} by Al^{3+} and O^{2-}, respectively, forms a β'-sialon structure which is customarily represented by the chemical formula $Si_{6-z}Al_zO_zN_{8-z}$, where $z =$ number of nitrogen atoms replaced by oxygen atoms. The term z ranges in value from 0 to 4. Although considerable solid solution in silicon nitride is possible, the degree of replacement sought in practice is often quite small. With replacement, the formula for the tetrahedral unit changes from SiN_4 to $(Si, Al)(O, N)_4$ and the dimensions of the unit cell increase.

Although replacement causes the chemical composition to shift towards that of alumina, the structural coordination in the solid solution is fourfold (AlO_4) whereas in alumina it is sixfold (AlO_6). The strength of the Al–O bond in a sialon is therefore about 50% stronger than its counterpart in alumina; this concentration of bonding forces between aluminium and oxygen ions makes a sialon intrinsically stronger than alumina.

The problem of representing complex phase relationships in a convenient form was solved by adopting the 'double reciprocal' diagram, a type of phase diagram originally developed for inorganic salt systems by German physical chemists many years ago. Figure 10.3 shows how a tetrahedron for the four elements Si, Al, O and N provides a symmetrical frame of reference for four compounds. By using linear scales calibrated in equivalent % (rather than the usual weight, or atomic %), each compound appears midway on a tetrahedral edge and the resulting section is square. An isothermal version of this type of diagram

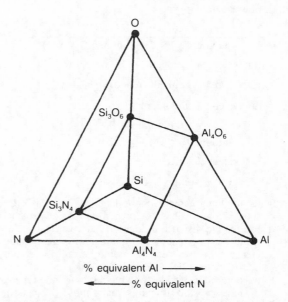
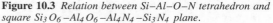

Figure 10.3 *Relation between Si–Al–O–N tetrahedron and square Si_3O_6–Al_4O_6–Al_4N_4–Si_3N_4 plane.*

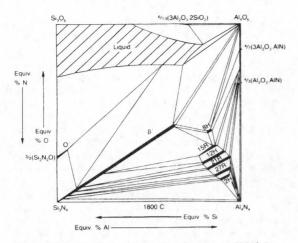

Figure 10.4 *Si–Al–O–N behaviour diagram at 1800°C (from Jack, 1987, pp. 259–88; reprinted by permission of the American Ceramic Society).*

is shown in Figure 10.4. The 'double reciprocal' characteristic refers to the equivalent interplay of N/O and Si/Al along the vertical and horizontal axes, respectively. It is necessarily assumed that the valency of the four elements is fixed (i.e. Si^{4+}, Al^{3+}, O^{2-} and N^{3-}). As the formula for the component Si_3N_4 contains 12 cations and 12 anions, the formulae for the other three components and for the various intermediate phases along the axes are expressed in the forms which give a similar charge balance (e.g. Si_3O_6 rather than SiO_2). The equivalent % of a given element in these formulae can be derived from the following equations:

Equivalent % oxygen

$$= \frac{100(\text{atomic } \%O \times 2)}{(\text{atomic } \%O \times 2) + (\text{atomic } \%N \times 3)}$$

Equivalent % nitrogen

$$= 100\% - \text{equivalent } \% \text{ oxygen}$$

Equivalent % aluminium

$$= \frac{100(\text{atomic } \%Al \times 3)}{(\text{atomic } \%Al \times 3) + (\text{atomic } \%Si \times 4)}$$

Equivalent % silicon

$$= 100\% - \text{equivalent } \%Al$$

Thus the intermediate phase labelled 3/2 (Si_2N_2O) contains 25 equivalent % oxygen and is located one quarter of the distance up the left-hand vertical scale.

An interesting feature of the diagram is the parallel sequence of phases near the aluminium nitride corner (i.e. 27R, 21R, 12H, 15R and 8H). They are referred to as aluminium nitride 'polytypoids', or 'polytypes'. They have crystal structures that follow the pattern of wurtzite (hexagonal ZnS) and are generally stable, refractory and oxidation-resistant.

10.4.2.4 Production of sialons

The start point for sialon production from silicon nitride powder at a temperature of 1800°C will lie in the vicinity of the bottom left-hand corner of Figure 10.4. Simultaneous replacement of N with O and Si with Al produces the desired β'-phase which is represented by the narrow diagonal zone projecting towards the Al_4O_6 corner. Such 'alloying' of the ceramic structure produces progressive and subtle changes in the structure of silicon nitride by altering the balance between covalent and ionic bonding forces. The resultant properties can be exceptional. Importantly, the oxidation resistance and strength of sialons at temperatures above 1000°C are greatly superior to those of conventional silicon nitride. Relatively simple fabrication procedures, similar to those used for oxide ceramics, can be adopted. Pressureless-sintering enables dense complex shapes of moderate size to be produced.

β-Si_3N_4 powder is the principal constituent of the starting mixture for 'alloying'. (As mentioned previously, these particles usually carry a thin layer of silica.) Although fine aluminium nitride would appear to be an appropriate source of replacement aluminium, it readily hydrolyses, making it impracticable to use fabrication routes which involve aqueous solutions or binders. One patented method for producing a β'-sialon ($z = 1$) solves this problem by reacting the silicon nitride (and its associated silica) with a specially-prepared 'polytypoid'. The phase relations for this method are shown in Figure 10.4.

An addition of yttrium oxide to the mixture causes an intergranular liquid phase to form during pressureless-sintering and encourage densification. By controlling conditions, it is possible to induce this phase either to form a glass or to crystallize (devitrify). In sialons, as in many other ceramics, the final character of the intergranular phase has a great influence upon high-temperature strength. A structure of β' grains + glass is strong and resists thermal shock at temperatures approaching 1000°C. However, at higher temperatures the glassy phase deforms in a viscous manner and strength suffers. Improved stability and strength can be achieved by a closely-controlled heat-treatment which transforms the glassy phase into crystals of yttrium-aluminium-garnet (YAG), as represented in the following equation:

$$\underset{\substack{\beta'\text{-sialon} \\ (z = 1)}}{Si_5AlON_7} + \underset{\substack{\text{Oxynitride} \\ \text{glass}}}{Y-Si-Al-O-N}$$

$$\underset{\substack{\text{Modified} \\ \beta'\text{-sialon}}}{Si_{5+x}Al_{1-x}O_{1-x}N_{7+x}} + \underset{\text{YAG}}{Y_3Al_5O_{12}}$$

The two-phase structure of β' grains + YAG is extremely stable. It does not degrade in the presence of molten metals and maintains strength and creep resistance up to a temperature of 1400°C.

More recent work has led to the production of sialons from precursors other than β-silicon nitride (e.g. α'-sialons from α-silicon nitride and O′-sialons from oxynitrides). K. H. Jack proposed that α-silicon nitride, unlike the β-form, is not a binary compound and should be regarded as an oxynitride, a defect structure showing limited replacement of nitrogen by oxygen. The formula for its structural unit approximates to $SiN_{3.9}O_{0.1}$. Dual-phase or composite structures have also been developed in which paired combinations of β'-, α'- and O′- phases provide enhancement of engineering properties. Sometimes, as in α'/β' composites, there is no glassy or crystalline intergranular phase. The sialon principle can be extended to some unusual natural waste materials. For instance, two siliceous materials, volcanic ash and burnt rice husks, have each been used in sinter mixes to produce sialons. Although such products are low grade, it has been proposed that they could find use as melt-resistant refractories.

10.4.2.5 Engineering applications of sialons

The relative ease with which sialons can be shaped is one of their outstanding characteristics. Viable shaping techniques include pressing (uniaxial, isostatic), extrusion, slip-casting and injection-moulding; their variety has been a great stimulus to the search for novel engineering applications. Similarly, their ability to densify fully during sintering at temperatures in the order of 1800°C, without need of pressure application, favours the production of complex shapes. However, due allowance must be made for the large amount of linear shrinkage (20–25%) which occurs as a result of liquid phase formation during sintering. Although final machining with diamond grit, ultrasonic energy or laser beam energy is possible, the very high hardness of sialons encourages adoption of a near-net-shape approach to design. As with many other engineering ceramics, sialon components are extremely sensitive to shape and it is generally appreciated that a change in curvature or section can frequently improve service performance. The structure of a sialon is, of course, the main determinant of its properties. Fortunately, sialons are very responsive to 'alloying' and combinations of attributes such as strength, stability at high temperatures, resistance to thermal shock, mechanical wear and molten metals can be developed in order to withstand onerous working conditions.

During metal-machining, tool tips are subjected to highly destructive and complex conditions which include high local temperatures and thermal shock, high stresses and impact loading, and degradation by wear. At a test temperature of 1000°C, the indentation hardness of β'-sialon (+ glass) is much greater than that of either alumina or cobalt-bonded tungsten carbide (Figure 10.5). The introduction of tool tips made from this sialon was a notable success. They were found to have a longer edge life than conventional tungsten carbide inserts, could remove metal at high speed with large depths of cut and could tolerate the shocks, mechanical and thermal, of interrupted cutting.

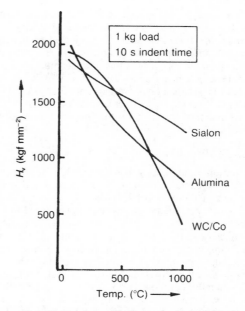

Figure 10.5 *Hot hardness of sialon, alumina and WC/Co cutting tool tips (from Jack, 1987, pp. 259–88; reprinted by permission of the American Ceramic Society).*

The strength and wear resistance of sialons led to their use in the metal-working operations of extrusion (hot- and cold-) and tube-drawing. In each process, the relative movement of the metal stock through the die aperture should be fast with low friction and minimal die wear, producing closely dimensioned bar/tube with a smooth and sound surface texture. Sialon die inserts have been successfully used for both ferrous and non-ferrous metals and alloys, challenging the long-established use of tungsten carbide inserts. Sialons have also been used for the plugs (captive or floating) which control bore size during certain tube-drawing operations. It appears that the absence of metallic microconstituents in sialons obviates the risk of momentary adhesion or 'pick-up' between dies and/or plugs and the metal being shaped. Sialon tools have made it possible to reduce the problems normally associated with the drawing of difficult alloys such as stainless steels.

The endurance of sialons at high temperatures and in the presence of invasive molten metal or slag has led to their use as furnace and crucible refractories. On a smaller scale, sialons have been used for components in electrical machines for welding (e.g. gas shrouds, locating pins for the workpiece). These applications can demand resistance to thermal shock and wear, electrical insulation, great strength as well as immunity to attack by molten metal spatter. Sialons have proved superior to previous materials (alumina, hardened steel) and have greatly extended the service life of these small but vital machine components.

The search for greater efficiency in automotive engines, petrol and diesel, has focused attention on regions of the engine that are subjected to the most severe conditions of heat and wear. Sialons have been adopted for pre-combustion chambers in indirect diesel engines. Replacement of metal with ceramic also improves the power/weight ratio. It is still possible that the original goal of researches on silicon nitride and sialons, the ceramic gas turbine, will eventually be achieved.

10.4.3 Zirconia

Zirconium oxide (ZrO_2) has a very high melting point (2680°C), chemical durability and is hard and strong; because of these properties, it has long been used for refractory containers and as an abrasive medium. At temperatures above 1200°C, it becomes electrically conductive and is used for heating elements in furnaces operating with oxidizing atmospheres. Zirconia-based materials have similar thermal expansion characteristics to metallic alloys and can be usefully integrated with metallic components in heat engines. In addition to these established applications, it has been found practicable to harness the structural transitions of zirconia, thereby reducing notch-sensitivity and raising fracture toughness values into the $15-20$ MN m$^{-3/2}$ band, thus providing a new class of toughened ceramics. This approach is an alternative to increasing the toughness of a ceramic by either (1) adding filaments or (2) introducing microcracks that will blunt the tip of a propagating crack.

Zirconia is polymorphic, existing in three crystalline forms; their interrelation, in order of decreasing temperature, is as follows:

$$\text{Melt} \underset{2680°C}{\overset{}{\rightleftharpoons}} \underset{c}{\text{Cubic}} \underset{2370°C}{\overset{}{\rightleftharpoons}} \underset{t}{\text{Tetragonal}}$$

$$\underset{1150°C}{\overset{950°C}{\rightleftharpoons}} \underset{m}{\text{Monoclinic}}$$

The technique of transformation-toughening hinges upon stabilizing the high-temperature tetragonal (t) form so that it is metastable at room temperature. Stabilization, partial or whole, is achieved by adding certain oxides (Y_2O_3, MgO, CaO) to zirconia. In the metastable condition, the surrounding structure opposes the expansive transition from t- to m-forms. In the event of a propagating crack passing into or near metastable regions, the concentrated stress field at the crack tip enables t-crystals of zirconia-rich solid solution to transform into stable, but less dense, m-ZrO_2 (Figure 10.6). The transformation is martensitic in character. The associated volumetric expansion ($3-5\%$ v/v) tends to close the crack and relieve stresses at its tip. This transformation mechanism is primarily responsible for the beneficial toughening effect of a metastable phase within the microstructure.

The relative stability of zirconia-rich solid solutions can be conveniently expressed in terms of the phase

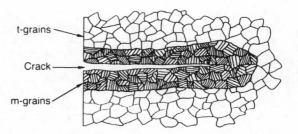

Figure 10.6 *Crack propagating into grains of t-zirconia, causing them to transform into m-zirconia.*

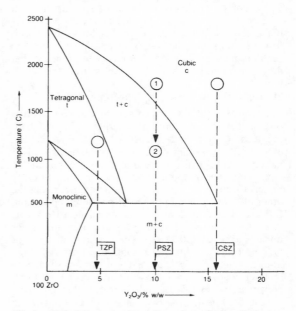

Figure 10.7 *Schematic phase diagram for $ZrO_2-Y_2O_3$ system: all phases depicted are solid solutions. TZP = tetragonal zirconia polycrystal, PSZ = partially-stabilized zirconia, CSZ = cubic-stabilized zirconia.*

diagram for the zirconia-rich end of the $ZrO_2-Y_2O_3$ system (Figure 10.7). The same principles apply in a very general sense to the other two binary systems, ZrO_2-MgO and ZrO_2-CaO. Yttria is particularly effective as a stabilizer. Three zirconia-based types of ceramic have been superimposed upon the diagram; CSZ, TZP and PSZ. The term CSZ refers to material with a fully-stabilized cubic (not tetragonal) crystal structure which cannot take advantage of the toughening transformation. It is used for furnace refractories and crucibles. The version known as tetragonal zirconia polycrystal (TZP) contains the least amount of oxide additive (e.g. $2-4$ mol% Y_2O_3) and is produced in a fine-grained form by sintering and densifying ultra-fine powder in the temperature range $1350-1500°C$; such temperatures are well within the phase field for the tetragonal solid solution (Figure 10.7). After cooling to room temperature, the structure is essentially single-phase, consisting of very fine grains ($\sim$0.2–1 μm) of t-ZrO$_2$ which make this material several times stronger than other types of zirconia-toughened ceramics. A typical TZP microstructure, as revealed by electron microscopy, is shown in Figure 10.8. Added oxide(s) and silicate impurities form an intergranular phase which can promote liquid-phase sintering during consolidation. (A similar effect is utilized in the production of silicon nitride.)

In partially-stabilized zirconia (PSZ), small t-crystals are dispersed as a precipitate throughout a matrix of coarser cubic grains. Zirconia is mixed with 8–10 mol% additive (MgO, CaO or Y$_2$O$_3$) and heat-treated in two stages (Figure 10.7). Sintering in the temperature range 1650–1850°C produces a parent solid solution with a cubic structure which is then modified by heating in the range 1100–1450°C. This second treatment induces a precipitation of coherent t-crystals ($\sim$200 nm in size) within the c-grains. The morphology of the precipitate depends upon the nature of the added solute (e.g. ZrO$_2$–MgO, ZrO$_2$–CaO and ZrO$_2$–Y$_2$O$_3$ solid solutions produce lenticular, cuboid and platey crystals, respectively). The average size of precipitate crystals is determined by the conditions of temperature and time adopted during heat-treatment in the crucial 't + c' field of the phase diagram.

In the third example of transformation-toughening, t-zirconia grains are dispersed in a dissimilar ceramic matrix; for example, in ZT(A) or ZT(Al$_2$O$_3$) they are dispersed among alumina grains (Figure 10.9). An intergranular distribution of the metastable phase results when conventional processing methods are used but it has also been found possible to produce an intragranular distribution. As with PSZ materials, the size of metastable particles and matrix grains must be carefully controlled and balanced.

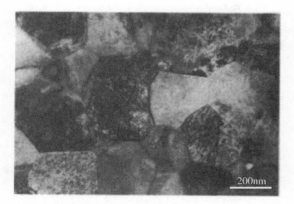

Figure 10.8 *Electron micrograph of tetragonal-zirconia polycrystal stabilized with 3 mol.% yttria (with acknowledgement to M. G. Cain, Centre for Advanced Materials Technology, University of Warwick, UK).*

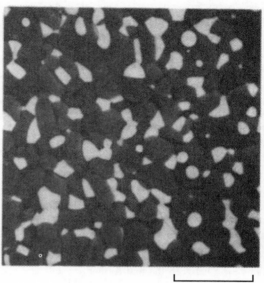

10 μm

Figure 10.9 *Duplex structure of ZT (Al$_2$O$_3$) consisting of alumina and t-zirconia grains (back-scattered electron image). (from Green, 1984, p. 84; by permission of Marcel Dekker Inc.).*

So far, we have concentrated upon mechanical behaviour at or below ambient temperature. If the temperature of a zirconia-toughened material is raised to 900–1000°C, which is close to the t-m transition temperature, the toughening mechanism tends to become ineffective. In addition, thermal cycling in service tends to induce the t-m transition at temperatures in the range 800–900°C and the toughening property is gradually lost. This tendency for fracture toughness to fall as the service temperature increases has naturally led to the investigation of alternative forms of stabilization in systems which have much higher transformation temperatures (e.g. ZT(HfO$_2$)). Intergranular residues (e.g. in TZP), despite their beneficial effect during sintering, become easier to deform as the temperature rises and the material then suffers loss of strength and resistance to creep.

10.4.4 Glass-ceramics

10.4.4.1 Controlled devitrification of a glass

It has long been appreciated that crystallization can take place in conventional glassy structures, particularly when they are heated. However, such crystallization is initiated at relatively few sites and there is a tendency for crystals to grow perpendicular to the free surface of the glass in a preferred manner. The resulting structure, being coarsely crystalline and strongly oriented, is mechanically weak and finds no practical application.

The basic principle of glass-ceramic production is that certain compositions of glass respond to controlled heat-treatment and can be converted, without distortion and with little dimensional change, from a readily-shaped glass into a fine-grained crystalline ceramic possessing useful engineering properties. The key to this structural transformation, which takes place throughout the bulk of the glass (volume crystallization), is the presence of a nucleating agent, or catalyst, in the original formulation.

Controlled devitrification of a special glass involves two or more stages of heat-treatment (Figure 10.10). In the first stage, which can begin while the glass is cooling from the forming and shaping operation, holding at a specific temperature for a definite time period causes the catalyst to initiate the precipitation of large numbers of nuclei throughout the glassy matrix. When these seed regions reach a certain size, different species of crystals may begin to grow upon them. Electron microscopy has demonstrated that epitaxial relationships exist between the first-formed crystals and succeeding generations of crystals (Figure 10.11). Finally, in the second stage of heat-treatment (Figure 10.10), the structure is heated to a different temperature in order to induce further crystallization, crystal growth, crystal transitions and a gradual, almost complete, disappearance of the glassy matrix.

Control of time and temperature is essential during the production by heat-treatment of a glass-ceramic. Figure 10.12 provides a general guide to the temperature-dependence of nucleation and growth processes for any melt, irrespective of whether it forms a glassy or crystalline solid. Curve N represents the rate of homogeneous nucleation; that is, the number of nuclei forming per second in each unit volume of glass. Curve G represents the rate of crystal growth (micron/second). Each curve has a peak value; for viscous glass-forming melts, these maxima are not very pronounced. With regard to the formation of a

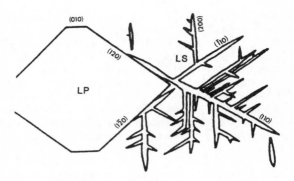

Figure 10.11 *Epitaxial growth of lithium metasilicate* ($LS = Li_2SiO_3$) *on a lithium orthophosphate seed crystal* ($LP = LP_3PO_4$) *in* $SiO_2-Li_2O-Al_2O_3$ *glass containing* P_2O_5 *catalyst (from Headley and Loehmann, 1984, pp. 620–25; reprinted by permission of the American Ceramic Society).*

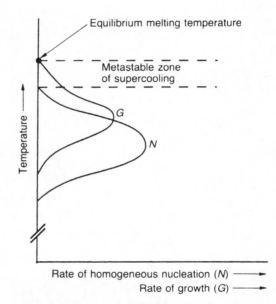

Figure 10.12 *Temperature-dependence of nucleation and growth processes (after Rawson, 1980).*

glass-ceramic, it is of prime importance to select a temperature which is close to the peak of the nucleation curve and then, in the second stage of heat-treatment, a temperature which does not encourage excessive grain growth. Usually the temperature chosen for the second stage is higher than that used for the first. A careful balance of conditions during heat-treatment will favour the production of the desired ultra-fine grain structure. Thus the rate of nucleation should be high, nuclei should be uniformly dispersed and the rate of crystal growth should not be excessive. Crystals are one micron or less in size; interlocking of these crystals will enhance the mechanical strength.

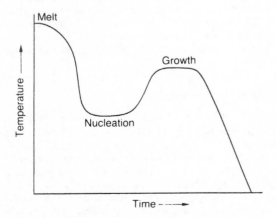

Figure 10.10 *Temperature/time schedule for producing a glass-ceramic.*

Fundamental studies are complicated by the fact that the detailed mechanisms by which nuclei form in a homogeneous glassy matrix, and then develop into crystals, appear to be specific to each type of glass. However, although spinodal decomposition is sometimes possible, it is usually regarded as a nucleation and growth process. Although the exact nature of the early stages of nucleation is highly debatable, the nuclei, once formed, enable hetero-geneous nucleation of the major crystalline phases to take place. Metastable phases may form during heat-treatment; such phases do not feature in phase (equilibrium) diagrams. Well-known crystalline phases may appear at unexpected temperatures. Nominally metastable phases may prove to be quite stable under service conditions. Furthermore, successful composi-tions are usually multi-component in character and modifying oxides, particularly the catalyst, can have a significant effect upon the types of crystal produced and upon their transformation processes. For a given catalyst, a change in the heat-treatment process can result in a change in the major crystalline phase(s).

Nucleation in glasses which use an oxide as catalyst often appears to be preceded by a process of separation into two solid glassy phases (metastable immiscibility). These microphases differ in chemical composition and are therefore believed to have a desirable effect upon the ultimate grain size by favouring a high nucleation density and reducing the growth rate of crystals. Phase separation may occur during either cooling of the melt or reheating and it is logical to presume that it is more likely to take place when two network-formers are present in the glass formulation. The production of *Vycor*, a $SiO_2-B_2O_3-Na_2O$ glass, also takes advantage of phase separation.

10.4.4.2 Development of glass-ceramics

Glass-ceramics date from the late 1950s and are an offshoot of research by S. D. Stookey at the Corn-ing Glass Works, USA, on photosensitive glasses. In the normal procedure for these glasses, after prior irradiation with ultraviolet light, their structure can be altered by heat-treatment. Metals, such as cop-per, silver or gold, act as nucleating agents for lim-ited crystallization. It was fortuitously discovered that higher heat-treatment temperatures could induce com-plete crystallization. Subsequent research at Corn-ing established that certain oxides could also act as effective nucleating agents and led to development of the first glass-ceramics, which were based on the $SiO_2-Al_2O_3-LiO_2$ system. Conveniently, these mate-rials do not require prior irradiation. Oxides which promote crystallization include titania, zirconia and phosphorus pentoxide.

Practical considerations of production greatly influence the development and exploitation of glass-ceramics. As a general rule, the temperatures of melting and refining should not exceed 1400–1500°C. Fortunately, some nucleating agents, such as titania,

have a strong fluxing action on silica. In addition, melt viscosity is important; alumina increases viscosity and will tend to slow down melting and refining operations. As composition control is a vital feature of glass-ceramics, it is essential to maintain melt composition reproducibly from batch to batch. Volatilization and interaction between the melt and the refractory lining of the melting furnace can make this difficult. For instance, high proportions of lithium oxide in the melt will increase attack on the lining. The melt leaving the melting furnace has a temperature-dependent viscosity of 10^{13} poise and the cooling mass can be worked and shaped until its viscosity falls to about 10^8 poise. Although conventional glasses have a useful 'long' range of temperature over which they can be worked, the composition of potential glass-ceramics restricts ('shortens') this range, particularly when aluminium oxide is present. As a consequence, the choice of shaping process may be restricted to gravity or centrifugal casting.

As metastability is an essential feature of a glass-ceramic, it is not surprising to find that certain compositions tend to devitrify prematurely during working. Oxides of aluminium, phosphorus and the alkali metals sodium and potassium inhibit devitrifica-tion in SiO_2-LiO_2 glasses. On the other hand, the glass should crystallise neither too quickly (during cooling of the melt) nor too slowly (during heat-treatment). These tendencies can be eliminated by adding oxides which have a specific effect upon the strength of the glass net-work structure. For instance, lithium oxide introduces non-bridging oxygen ions into a network of SiO_4 tetra-hedra and, by weakening it, favours crystallization.

10.4.4.3 Typical applications of glass-ceramics

The versatility and potential for development of glass-ceramics quickly led to their adoption in heat engines, chemical plant, electronic circuits, seals, cladding for buildings, aerospace equipment, nuclear engineering, etc. They can offer a remarkable combination of prop-erties. For instance, glass-ceramic hob plates for elec-tric cookers are strong, smooth and easy to clean, stable over long periods of heating, transparent to infrared radiation from the tungsten halogen lamp, relatively opaque to visible light (reducing glare) and resistant to thermal shock. The last property originates primarily from the low thermal coefficient of thermal expan-sion (α) of this particular glass-ceramic. In the ver-satile $SiO_2-Al_2O_3-LiO_2$ class of glass-ceramics, the α-value can be 'tailored' from zero to $12 \times 10^{-6}°C^{-1}$ by controlling structure. Thus, in types containing about 10% alumina, crystals of lithium disilicate and quartz (or cristobalite) form to give a relatively high expansion coefficient. Increasing the alumina to about 20% favours the formation of two types of lithium aluminium silicate crystals, β-spodumene and β-eucryptite. Over the temperature range 20–1000°C, the α-values for these two compounds are $0.9 \times 10^{-6}°C^{-1}$ and $-6.4 \times 10^{-6}°C^{-1}$, respectively. Careful balancing of their relative amounts against the residual glass content can reduce the overall expansion coefficient towards zero. This capability is ideal for the large mirror blanks used for telescopes where dimensional stability is essential. In heat engines such as lorry gas turbines, a low expansion coefficient is required for the regenerative heat exchanger which is alternately heated and cooled by exhaust gases and combustion air. In glass/glass and metal/glass seals, precise matching of expansion characteristics is possible. In the mass production of colour television tubes, devitrifiable solder glass-ceramics based upon the $PbO-ZnO-B_2O_3$ system have a fairly high α value and have been used to seal the glass cone to the glass face plate at a relatively low temperature without risk of distortion. The seal is subsequently heated to form the glass-ceramic.

A fine-grained structure of interlocking crystals favours mechanical strength; modulus of rupture values are comparable to those for dense alumina. In a machinable variety of glass-ceramic, interlocking crystals of platey mica deflect or blunt forming cracks and make it possible for complex machinable components to be designed. Chemical stability, as well as wear-resistance, is essential when service involves contact with fluids (e.g. valves, pumps, vessel linings). It is well-known that the fluxing oxides of sodium and potassium lower the chemical resistance of silica glass to aqueous solutions; in glass-ceramics, this susceptibility is countered by stabilizing any residual glass phase with boric oxide. Attempts are being made to extend their use to much higher temperatures (e.g. 1300–1700°C) by exploiting refractory systems, such as $SiO_2-Al_2O_3-BaO$, which can provide a liquidus temperature above 1750°C but such glasses are very difficult to melt.

Conversion from glass to crystals causes mobile ions to disappear from the structure and become 'bound' to crystals, consequently the electrical resistivity and dielectric breakdown strength of glass-ceramics are high, even at temperatures up to 500–700°C. The dielectric loss is low.

Utilization of plentiful and cheap waste byproducts from industry for the bulk production of glass-ceramics has stimulated much interest. Although the chemical complexity of materials such as metallurgical slags makes it difficult to control and complete crystallization, strong and wear-resistant products for architecture and road surfacing have been produced (e.g. *Slagceram, Slagsitall*).

10.4.5 Silicon carbide

10.4.5.1 Structure and properties of silicon carbide

The two principal structural forms of this synthetic compound are α-SiC (non-cubic; hexagonal, rhombohedral) and β-SiC (cubic). The cubic β-form begins to transform to α-SiC when the temperature is raised above 2100°C. Conditions of manufacture determine the exact crystal structure and a number of variants

have been identified by X-ray diffraction analysis. For example, many different stacking sequences are possible in hexagonal α-SiC (e.g. 4H, 6H). The nomenclature for such variants (polytypes) indicates the number of atomic layers in the stacking sequence and the crystal system (e.g. 15R, 3C). Tetrahedral grouping of carbon atoms around a central silicon atom is a common and basic feature of these various structures. This covalent bonding of two tetravalent elements gives exceptional strength and hardness and a very high melting point (>2700°C).

Despite its carbon content, silicon carbide offers useful resistance to oxidation. At elevated temperatures, a thin impervious layer of silica (cristobalite) forms on the grains of carbide. On cooling, the α/β transition occurs in cristobalite and it can crack, allowing ingress of oxygen. Above a temperature of 1500°C, the silica layer is no longer protective and the carbide degrades, forming SiO and CO.

For service applications where resistance to thermal shock is important, it is customary to compare candidate ceramic materials in terms of a parameter which allows for the effect of relevant properties. Many versions of this parameter have been proposed. A typical parameter (R) for sudden thermal shock is $R = \sigma(1 - \nu)/E\alpha$, where σ is the modulus of rupture (N m^{-2}), ν is the Poisson ratio (0.24 for *REFEL* SiC), E is the modulus of elasticity (N m^{-2}), and α is the linear coefficient of thermal expansion (K^{-1}). In the case of silicon carbide, the product in the denominator tends to be high, giving a low index. Silicon nitride gives a higher index and it is understandable that Si_3N_4-bonded silicon carbide is preferred when service involves thermal cycling. The severity of shock can affect the rating of different materials. Thus, for less rapid shock, a thermal conductivity term (k) is included as a multiplier in the numerator of the above parameter.

10.4.5.2 Production of silicon carbide powder and products

Silicon carbide is a relatively costly material because its production is energy-intensive. The Acheson carbothermic process, which is the principal source of commercial-quality silicon carbide, requires 6–12 kW h per kg of silicon carbide. Locations served by hydroelectric power, such as Norway and the Niagara Falls in Canada, are therefore favoured for synthesis plant. In this unique process, a charge of pure silica sand (quartz), petroleum coke (or anthracite coal), sodium chloride and sawdust is packed around a 15 m long graphite conductor. A heavy electric current is passed through the conductor and develops a temperature in excess of 2600°C. The salt converts impurities into volatile chlorides and the sawdust provides connected porosity within the charge, allowing gases/vapours to escape. The essential reaction is:

$$SiO_2 + 3C \longrightarrow \alpha\text{-SiC} + 2CO$$

The reaction product from around the electrode is ground and graded according to size and purity. The colour can range from green (99.8% SiC) to grey (90% SiC).

Numerous chemical conversion and gas-phase synthesis processes have been investigated. The temperatures involved are generally much lower than those developed in the Acheson process; consequently, they yield the cubic β-form of silicon carbide. Chemical vapour deposition (CVD) has been used to produce filaments and ultra-fine powders of β-SiC.

Silicon carbide shapes for general refractory applications are produced by firing a mixture of SiC grains and clay at a temperature of 1500°C. The resultant bond forms mullite and a glassy phase, absorbing the thin layer of silica which encases the grains. Other bonding media include ethyl silicate, silica and silicon nitride. The latter is developed *in situ* by firing a SiC/Si compact in a nitrogen atmosphere. This particular bond is strong at high temperatures and helps to improve thermal shock resistance.

Specialized and often costly processing methods are used to produce fine dense ceramics for demanding engineering applications. The methods available for forming silicon carbide powders include dry-pressing, HIPing, slip-casting, extrusion and injection-moulding. The last process requires an expendable polymeric binder and is particularly attractive in cases where long production runs of complex shapes are envisaged (e.g. automotive applications). With regard to firing, the main methods are akin to those developed for silicon nitride; each, in its own way, is intended to maximize the quality of interparticle bonding. They include hot-pressing (HP SiC), pressureless-sintering (S SiC) and reaction-sintering (Si SiC).

The hot-pressing method for producing α-SiC blanks of high density was originally developed by the Norton Co., USA. A small amount of additive (boron carbide (B_4C) or a mixture of alumina and aluminium) plays a key role while the carbide grains are being heated (>2000°C) and compressed in induction-heated graphite dies. It has been suggested in the case of HP SiC (and S SiC) that the boron encourages grain boundary/surface diffusion and that the carbon breaks down the silica layers which contaminate grain surfaces. These additives leave an intergranular residue which determines the high-temperature service ceiling. The hot-pressed blanks usually require mechanical finishing (e.g. diamond machining). Production of complex shapes by hot-pressing is therefore expensive.

The pressureless-sintering route (for S SiC) uses extremely fine silicon carbide powders of low oxygen content. Again, an additive is necessary (B_4C or aluminium + carbon) in order to promote densification. The mixture is cold-pressed and then fired at approximately 2000°C in an inert atmosphere.

The *REFEL* process for producing siliconized silicon carbide (Si SiC) was developed by the UKAEA and is an example of reaction-sintering (reaction-bonding). A mixture of α-SiC, graphite and a plasticizing binder is compacted and shaped by extrusion, pressing, etc. This 'green' compact can be machined, after which the binder is removed by heating in an oven. The pre-form is then immersed in molten silicon under vacuum at a temperature of 1700°C. Graphitic carbon and silicon react to form a strong intergranular bond of β-SiC. A substantial amount of 'excess' silicon (say, 8–12%) remains in the structure; the maximum operating temperature is thus set by the melting point of silicon (i.e. 1400°C. Beyond this temperature there is a rapid fall in strength. Ideally, neither unreacted graphite nor unfilled voids should be present in the final structure. The dimensional changes associated with the HP process are small and close tolerances can be achieved; the shrinkage of 1–2% is largely due to the bake-out of the binder.

10.4.5.3 Applications of silicon carbide

Silicon carbide has been the subject of continuous development since it was first produced by the then-remarkable Acheson process in 1891.[1] Now it is available in a wide variety of forms which range from monolithics to single-crystal filaments (whiskers). It is used for metal-machining, refractories and heating elements in furnaces, chemical plant, heat exchangers, heat engines, etc.

The extreme hardness (2500–2800 kgf mm^{-2}; Knoop indenter) of its particles and their ability to retain their cutting edges at high contact temperatures (circa 1000°C) quickly established silicon carbide grits as important grinding media. The comparatively high cost of silicon carbide refractories is generally justified by their outstanding high-temperature strength, chemical inertness, abrasion resistance and high thermal conductivity. In the New Jersey process for producing high-purity zinc, SiC is used for components such as distillation retorts, trays and the rotating condensation impellers which have to withstand the action of molten zinc and zinc vapour. In iron-making, silicon carbide has been used to line the water-cooled bosh and stack zones of iron-smelting blast furnaces, where its high thermal conductivity and abrasion-resistance are very relevant. However, it can be attacked by certain molten slags, particularly those rich in iron oxides. This characteristic is illustrated by experience with the skid rails which support steel billets in reheating furnaces. This type of furnace operates at a temperature of 1250°C. Water-cooled steel rails have traditionally been used but warp, wear rapidly and tend to form 'cold spots' where

---

[1]Carborundum Co. Ltd: The Americans E. G. Acheson and W. A. McCallister gave the name Carborundum to their new material, assuming that it was a combination of carbon and corundum (Al_2O_3). Even after its true chemical identity was established, Acheson retained the name, regarding it as 'phonetic, of pleasing effect in print, even though a trifle lengthy'.

they contact the billets. Replacement with uncooled silicon carbide rails solved these problems but it was found that iron oxide scale from the billets could melt and attack silicon carbide. Some improvement was achieved by flush-mounting the rails in the furnace floor.

Silicon carbide also has a key role in recent designs of radiant tube heaters in gas-fired furnaces. The combination of a 60 kW recuperative burner and a radiant tube (1.4 m long × 170 mm diameter) made from Si_3N_4-bonded silicon carbide is shown in Figure 10.13. This British Gas design allows outgoing combustion products to preheat incoming air, giving high thermal efficiency, and also keeps these gases separate from the atmosphere within the furnace chamber. The maximum surface temperature for the radiant tube is 1350°C.

Silicon carbide is electrically conductive and care has to be taken when it is used as a refractory in the structure of electrometallurgical plant. However, the combination of electrical conductivity and refractoriness offers special advantages. For example, silicon carbide resistor elements have been used since about 1930 in indirect resistance-heated furnaces throughout industry (e.g. *Globars*). These elements act as energy-conversion devices, heating the furnace charge by radiation and convection. They can operate in air and inert gas atmospheres at temperatures up to 1650°C but certain conditions can shorten their life (e.g. carbon pick-up from hydrocarbon gases, oxidation by water vapour). Service life also tends to decrease as the operating temperature is increased. Provided that service conditions are not too severe, a life of at least 10 000 h can be anticipated. A double-helical heating section is available as an alternative to the standard cylindrical shape and allows both electrical connections to be

made at the same end. Elements should be of reasonable diameter/length and be neither too fragile nor too massive.

The presence of impurity atoms in silicon carbide enables electron flow to take place; it is accordingly classed as an extrinsic semiconductor. The electrical resistance of silicon carbide is very temperature-dependent, decreasing from room temperature to 650°C and then slowly increasing with further rise in temperature. Because of this characteristic and the great sensitivity of cold-resistance to traces of impurities, a typical production procedure is to check the resistance of each element (in air) with an electrical load per cm^2 of radiating surface which is equivalent to a typical operating surface temperature (e.g. 15.5 W cm^{-2} and 1070°C). This nominal resistance is then used to calculate the number and size of elements required. As the element ages, its resistance slowly increases. A constant rate of energy input to the furnace is maintained by increasing the voltage applied across the elements (e.g. by multi-tap transformer).

Specialized forms of silicon carbide now find widespread use in engineering. At ambient temperatures, they serve in machine components subjected to abrasive wear (e.g. mechanical seals, bearings, slurry pump impellers, wire dies, fibre spinnerets). In high-temperature engineering, silicon carbide is now regarded, together with silicon nitride and the sialons, as a leading candidate material for service in heat engine designs which involve operation at temperatures in excess of 1000°C (e.g. glow plugs, turbocharger rotors, turbine blades and vanes, rocket nozzles). Glow plugs minimize the hazards of 'flame-out' in the gas turbine engines of aircraft. Their function is to reignite the fuel/air mixture. They must withstand considerable thermal shock; for instance,

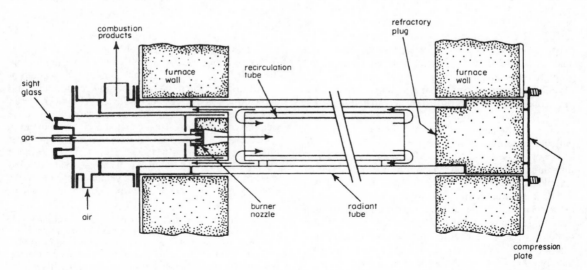

Figure 10.13 *Operating principle of the ceramic radiant tube (from Wedge, Jan 1987, pp. 36–8; by courtesy of the Institute of Materials).*

on engine start-up the temperature rises from ambient to 1600°C in 20 s and falls to 900°C in less than 1 s if flame-out occurs. Si_3N_4-bonded silicon carbide performed well in this application. Large-scale utilization in gas-turbine and diesel engines has been greatly inhibited by the inherent brittleness of silicon carbide. Addition of a second phase to the structure, the composite approach, is regarded as a likely way to solve this problem of low fracture toughness. In a more general sense, it is accepted that the methodology and practice of non-destructive evaluation (NDE) and proof-testing for ceramic components require further refinement.

10.4.6 Carbon

10.4.6.1 The versatility of carbon

Lying in the central zone of the Periodic Table between metallic and non-metallic elements, carbon has been aptly described as the chameleon among materials. Over the years it has become available in numerous forms yet still retains its capacity to surprise. The recent discovery and production of an exciting and completely new structural form, buckminsterfullerene, illustrates this point. Carbon is customarily classed as a ceramic because of its long-established use as a refractory: present-day use of clay/graphite crucibles as containers for molten metals echoes its use for the same purpose in the sixteenth century. Its high-temperature capabilities are exceptional. Although carbon is the traditional fuel for combustion and, ideally, should only be used as a heat-resistant engineering material in atmospheres of low oxygen content, the rate at which it is wasted by oxidation is very often acceptable to industry and compares favourably with that of many metals.

Materials science tends to emphasize the two contrasting crystalline forms (allotropes) of carbon, diamond and graphite, particularly as they represent extremes of hardness and softness. Both forms occur naturally and can be synthesized. Industry has long recognized their unique properties and has successfully developed an extremely wide and potentially confusing range of specialized carbon products. Within this range of structurally-different products, the degree of crystallinity may range from highly-developed to minimal. The following categories of carbon product will now be examined in order to illustrate the general significance of structure and the ways in which it can be manipulated: natural and synthetic diamonds, baked and graphitized carbons, pyrolytic graphite, vitreous carbon, intercalation compounds, buckminsterfullerene.

10.4.6.2 Natural diamond

The cubic form of carbon is renowned for its hardness, strength and beauty. The carbon–carbon bond lengths are all the same (0.1555 nm) so that any sections taken through the tetrahedrally coordinated, highly symmetrical structure cut a large number of these strong bonds; accordingly, there are no planes of easy cleavage. In commercial terms, there are three main qualities of mined diamonds: gemstones, industrial stones and boart (bort, bortz). Rough stones of gemstone quality are an exclusively natural product and their appearance is enhanced by a highly-demanding cutting procedure which involves cleavage on meticulously-selected planes, such as the {111}, and sawing/grinding/polishing with a mixture of diamond paste and olive oil. During cutting at least half of the mass is lost.[1] The symmetrical faceted shapes produced by the lapidary comply with standard mathematically-substantiated patterns (cuts) which optimise the optical effects of internal reflection ('life') and refraction ('fire'). Three cuts currently favoured are brilliant, emerald or baguette.

Single industrial stones of near-gem quality are used as tools for trueing and dressing grinding wheels, engraving, rock-drilling and as dies for drawing wires of copper, steel, tungsten, etc. In the important process of trueing and dressing, a tool holding a single diamond is held against the working face of the grinding wheel as it rotates. The wheel profile is corrected and grit particles are left projecting slightly above the bonding matrix. Natural single crystals are used for drawing the smaller sizes of wire (e.g. <0.2 mm). These dies have a crystallographically-oriented aperture which is typically capable of passing 10 000 km of wire before dimensional tolerances are exceeded.

Diamond indenters are used in macrohardness and microhardness testing machines (Section 5.2.2.4). Surface textures of machined metals are commonly characterized by profilometers which traverse a sampling length with a fine diamond stylus; interpretation of the resultant high-resolution trace should be tempered by the observation that the moving stylus tends to plough a furrow in the test surface.

Boart is generally <1.5 mm in size, badly flawed and imperfectly crystallized. For many years, boart was virtually unsaleable. Then, in the 1930s, its use as an abrasive grit or powder for impregnating the working faces of grinding wheels was fostered. Thereafter it found increasing application on a large scale for sawing, drilling and machining operations and became an accepted abrasive medium for non-ferrous metals and alloys, hard carbides, concrete, rock, glass, polymers, etc. The working face of a bonded grinding wheel that is being propelled into the workpiece is a composite structure, consisting of abrasive particles of grit (diamond, cobalt-bonded tungsten carbide, alumina or silicon carbide) set in a bonding matrix. The matrix can be sintered metal, electroplate, vitreous or resinoid. Selection of the best combination of grit and bond can draw from a large pool of practical experience and broadly depends upon the material being machined and upon the machining conditions. It is reasonable

---

[1]For diamonds, the carat (ct) is a unit of mass, with 1 ct = 0.2 g: for alloyed gold, the carat is one twenty-fourth part by mass of gold.

to regard each particle of diamond grit as a single cutting tool. The tendency of some qualities of diamond to fracture and regenerate new cutting edges, rather than be torn out of the matrix, can be advantageous in certain machining operations. The high thermal conductivity of diamond (which is greater than that of copper) and low coefficient of friction minimize the generation and dissipation of heat from each of a myriad of such tools. Characteristics such as these have encouraged the trend toward higher rates of feed and greater peripheral wheel speeds, to the benefit of productivity, dimensional accuracy and surface texture. In simple terms, a higher cutting speed means that each particle is subjected to stress and heat effects for shorter periods of time. (The availability of synthetic grits has aided this particular development in grinding practice.)

Lapping and polishing operations require diamond powders; under these fine-scale cutting operations, heat generation is not usually a problem. In the metallographic polishing of metallic and ceramic specimens, a progression from coarse to sub-micron powders smoothes the surface as well as gradually reducing the depth of unwanted surface distortion.

10.4.6.3 Synthetic diamond

The quest for a method to synthesize diamonds from carbonaceous material dates back to the nineteenth century. After World War II, research was stimulated by political uncertainties in Africa that threatened to jeopardize supplies of boart. The first announcement of the successful synthesis of diamonds from graphite was made by the General Electric Company, USA, in 1955. (Later, it transpired that the ASEA, Stockholm, Sweden had achieved a comparable result two years earlier but had not publicized it.) These remarkable methods simultaneously subjected graphite to extremely high static pressures and high temperatures.[1]

The physical conditions necessary for synthesis are customarily discussed in terms of the type of phase diagram shown in Figure 10.14. The key feature is the ascending Berman-Simon line[2] representing equilibrium between diamond and graphite. Diamond is stable above this line and graphite is stable below it. Diamond is able, of course, to exist below the line in a metastable condition at ordinary pressures; however, it was found experimentally that heating above a temperature of 1800 K caused rearrangement of its extremely strong C–C bonds and transformation into graphite. From the diagram it may be deduced that the reverse transition, from graphite to diamond, will take place above the line at similar temperatures if a

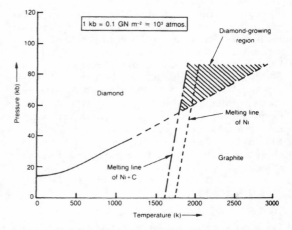

Figure 10.14 *Pressure versus temperature diagram for carbon (after Bovenkerk et al., 1959, pp. 1094–8).*

pressure of at least 60 kb is applied simultaneously. The GEC method achieved these conditions. It also relies upon an addition of a metal (e.g. Ni, Cr, Mn, Fe, Co) which acts as a molten solvent for carbon and a catalyst for diamond crystallization and growth. The liquidus line for the eutectic mixture of carbon and nickel is superimposed on the diagram in order to define the diamond-growing region (shaded). Early synthetic crystals were usually grown under conditions of temperature and pressure well above the Berman-Simon line and were consequently weak and friable, containing stacking faults and metallic inclusions. As improved methods for measuring process temperature and pressure became available, experience showed that synthesis in the shaded region just above the Berman-Simon line gave slower and more controllable growth.

Later GEC experiments at pressures up to 200 kb transformed well-crystallized graphite into small diamond crystallites, about 0.1 mm in size, with a wurtzite-type structure showing hexagonal symmetry. Subsequently, natural 'hexagonal' diamonds[3] (in association with cubic diamonds) were identified in meteorites; presumably they formed on impact with the earth. Small 'hexagonal' diamonds have also been synthesized by explosive shock-loading techniques that are capable of developing pressures as high as 500–1000 kb. These conditions only apply over a period of a few microseconds and thus tend to restrict a physical transition which is time-dependent.

Synthesis is now practised worldwide and is capable of producing diamonds which meet precise physical and chemical requirements. Mined diamonds are inevitably more variable in quality. The maximum size of synthetic diamonds is in the order of 1 mm. Demand for industrial diamonds far exceeds that for

---

[1] Much was owed to the pioneering work of the Nobel Prize winner P. W. Bridgman (1882–1961) at Harvard University, USA, on methods for developing ultra-high pressures.

[2] So named in recognition of the theoretical contribution of two Oxford physicists, R. Berman and F. Simon.

[3] Called lonsdaleite in honour of the eminent crystallographer, Professor Dame Kathleen Lonsdale (1903–1971).

gemstones; approximately 85–90% of industrial diamonds are synthesized and are mainly consumed in abrasion processes. Synthetic and natural diamond abrasives compete with silicon carbide and alumina. As mentioned elsewhere, diamond machining is a final operation in the production of many of the new engineering ceramics.

10.4.6.4 Scientific classification of diamonds

The structural imperfections to be found in diamonds include crystalline inclusions, cracks, impurity atoms and vacant sites. Broadly, inclusions and cracks act as stress-concentrators and mainly affect the mechanical properties, whereas fine-scale defects, notably impurity atoms of nitrogen and boron, influence physical properties such as optical absorption, electrical conductivity, etc. Inclusions in natural diamonds are particles of mineral matter. Inclusions in synthetic diamonds derive from the metals used as catalysts (i.e. Ni, Co, Fe).

The generally-accepted classification of diamonds, which is shown in Table 10.2, recognizes four main categories. It is based upon absorption characteristics determined over the ultraviolet, visible and infrared regions of the electromagnetic spectrum. The choice of this approach is perhaps not surprising when one considers the visible response of cut diamonds to white light. Absorption spectra are highly structure-sensitive and have made it possible to classify different qualities of natural and synthetic diamond in terms of their content of fine defects, such as impurity atoms and vacant sites. Each type of defect provides so-called 'optical centres' which decide the specific manner in which components of incident radiation are absorbed and/or transmitted by the crystal structure. A diamond frequently contains more than one type of defect, hence interpretation of absorption spectra can sometimes be difficult and rather arbitrary.

The decrease in intensity for a particular wavelength, as a result of absorption, is expressed by the classic exponential relation $I = I_0 e^{-\alpha \chi}$, where I is the intensity of transmitted radiation, I_0 is the intensity of incident monochromatic radiation and α is the coefficient of absorption. χ is the path length. Long path lengths within the diamond increase the amount of optical interaction, producing a size effect which explains why larger diamonds tend to be more colourful.

It will be seen that the classification first distinguishes between Type I diamonds which contain nitrogen (say, up to 0.1–0.2%) and Type II diamonds which have an extremely low nitrogen content. Type I diamonds are further sub-divided according to the spatial distribution of nitrogen atoms. In natural diamonds, geologic periods of time at high temperatures and pressure have permitted nitrogen atoms to cluster, sometimes, apparently, into platelets. In most synthetic diamonds, the nitrogen atoms are more dispersed. Nitrogen, singly or in groups, can occupy substitutional or interstitial sites and are responsible for at least five types of optical centre (e.g. A, B, N, N3, platelet). (Thus, the designation Type IaB signifies that the diamond contains B centres wherein a few nitrogen atoms have clustered and replaced carbon atoms.)

From the distinction between Type IIa and Type IIb diamonds, it will be seen that the ability of diamond to act as either an insulator or a semiconductor is governed by its impurity content. In Type II diamonds, the concentration of impurity atoms is smaller than in Type I diamonds, usually being expressed in parts per million. In terms of the electron theory of conduction, pure diamond has a filled valence band that is separated from a partly filled conduction band by a substantial energy gap of 5.5 eV. The mean available thermal energy is approximately 0.025 eV and is therefore insufficient to transport electrons through the gap. Pure diamond is consequently an electrical insulator at room temperature. In a similar sense, photons of white light have associated wavelengths ranging from 400 nm (= 3.1 eV) to 730 nm (= 1.7 eV) and pass through the crystal without bringing about electronic transitions which bridge the gap. Absorption of energy from these incident photons is small. However, photons associated with shorter wavelengths in the ultraviolet range can exceed the energy requirement of 5.5 eV and the gap is bridged.

Moving on from the special case of pure diamond to Type I diamonds, the introduction of nitrogen atoms ($Z = 7$) into the carbon structure ($Z = 6$) allows tetrahedral bonding to be maintained but also adds extra electrons. Optical absorption now becomes likely because electrons can move to higher levels within the band gap. However, the presence of nitrogen atoms

Table 10.2 Classification of diamonds

Type I (nitrogen present)		Type II (negligible nitrogen content)	
Type Ia	Type Ib	Type IIa	Type IIb
Clustering of N atoms	N atoms dispersed and substituting for C atoms		Contain boron
		Non-conducting	Semiconductivity possible in doped synthetic diamonds
Most natural stones	Most synthetic diamonds; rarely natural stones	Rarely found in nature	

does not confer conductivity. In Type IIb diamonds, the principal impurity is boron ($Z = 5$). Measured atomic concentrations of boron are often less than 0.5×10^{-6}. Each boron atom can contribute only three electrons to the surrounding four carbon atoms and therefore provides an acceptor site within the band gap, allowing 'holes' to form in the valence band. Semiconductivity then becomes possible. The two impurity elements nitrogen and boron thus have contrasting effects upon the diamond structure, providing donor and acceptor sites, respectively. When both elements are present, some degree of compensation occurs. For instance, a surfeit of nitrogen can neutralize the effect of boron atoms and the semiconducting characteristic is absent. If a diamond is colourless and electrically conductive one might reasonably infer that it contains traces of nitrogen and uncompensated boron and is classifiable as Type IIb.

Specific types of diamond have proved successful in highly specialized applications. For instance, Type Ia and Type IIa diamonds have very high thermal conductivities (1000 and 2000 W m^{-1}, K^{-1}, respectively, at room temperature) and have been used as heat sinks in electronic devices that release much thermal energy within a confined space. Detector 'windows' of Type II diamond are virtually transparent to infrared radiation and have been used in conditions as disparate as those of diamond synthesis anvils and space probes. Type IIb diamonds have interesting semiconductive prospects, possibly being capable of operating at higher temperatures than silicon devices, but developments in this area have been restricted by the difficulty of developing n-type devices to complement the p-type devices described above.

10.4.6.5 Polycrystalline diamond (PCD)

Polycrystalline aggregates of diamonds rarely occur in nature. It is now possible to synthesise a fine-grained aggregate by sintering diamond crystallites (2–25 mm in size) at a high temperature and pressure in the presence of a metal such as cobalt. The resultant blanks of PCD are typically about 50 mm diameter and can be electric discharge machined (EDM) to the required shape and size, thus overcoming the problems associated with producing large single crystals. During sintering, the randomly-oriented grains deform plastically and interlock. Molten cobalt acts as a solvent/catalyst and also fills the interstices between the grains. This structure has a better wear resistance than either natural diamond or tungsten carbide. It is also appreciably tougher than single crystals of diamond because the random orientation of grains forces any propagating crack to take an erratic, transgranular path which consumes strain energy.

PCD is used to make metal-working dies for drawing the larger sizes of metallic wire (e.g. >0.2 mm). (With smaller sizes of wire, a natural single crystal of diamond gives a better surface texture.) When used for cutting tools, a thin layer of PCD is supported by a thicker layer of tungsten carbide. The latter is brazed onto the tool body. The standard quality of PCD is subject to a temperature ceiling of 700°C that is set by two factors. First, cobalt expands more than the diamond phase and higher temperatures cause stress cracks to form. Second, above 700°C, cobalt promotes the conversion of diamond into graphite. These problems were solved by either removing the cobalt phase by acid leaching or by using silicon instead of cobalt. This variety of PCD is thermally stable at temperatures up to 1200°C and has been successfully used for drilling bits in hard-rock mining.

10.4.6.6 Baked carbons and graphitized carbons

Crucibles made from natural graphite bonded with fire-clay have been used in metal-melting operations for centuries. Although their strength is not exceptional, they can withstand temperatures of up to 1200°C. Other applications of natural graphite include lubricants, brake linings, bearings, foundry facings and pencil 'leads'. Crystallinity is well developed and they are relatively unreactive.

Most of the great demand for manufactured carbons is met by synthesis. Applications are extremely diverse: brushes for electric motors and generators, rocket nozzles and nose cones for space vehicles, refractory linings for furnaces, moulds for metal-founding and hot-pressing, moderators and reflectors for nuclear reactors, electrodes for arc-welding, etc. There are two main types of bulk carbon products for industrial applications. Neither melting nor sintering are involved in their manufacture.

Baked carbons, the first category, are produced by extruding/moulding a mix of carbonaceous filler or grist (e.g. petroleum coke, anthracite, pitch coke, etc.) and binder (e.g. coal-tar pitch, thermosetting resin). The shaping method determines the general anisotropy of the final product because the coke particles usually have one axis longer than the other, consequently extrusion orients the particles parallel to the extrusion direction whereas moulding orients particles perpendicular to the compression forces. (Accordingly, the final product usually has a so-called 'grain'.) The 'green' shape is heated (baked) in a reducing atmosphere at a temperature of 1000–1200°C. During this critical operation, the pitch decomposes or 'cracks' and provides carbon linkages between the particles of filler. Copious amounts of volatiles are evolved, producing about 20–30% porosity. Porosity weakens the structure, influences electrical conductivity and, because of its connected nature, makes the structure permeable to gases and liquids. It may be reduced by impregnating the baked carbon with more pitch and baking again. Further heat treatment at substantially higher temperatures of 2500–3000°C causes the structure to graphitize. An electric resistance furnace is necessary for this purpose, hence the products are sometimes called electrographites. (Most designs of electric furnaces for larger sizes of product are modifications of the original

Acheson design of 1895, a furnace which has a special place in the history of the carbon industry.)

Final shapes, whether baked or graphitized, can be readily and accurately machined. The final properties of the carbon product depend upon the interplay of variables such as initial grain size, type of filler or binder, purity, furnace temperature, etc. Thus, the higher processing temperatures that are used for electrographites eliminate strain, heal imperfections and promote graphitization within the structure. As a result, electrographites have much higher electrical and thermal conductivities than baked carbons, making it possible to halve the cross-sectional area of an electrode and gain valuable savings in mass and size. Accordingly, electrographites are favoured for large steel-melting furnaces. However, both baked carbons and the more costly electrographites find general application in a wide variety of electric furnaces. Their availability and continuous development have made a vital contribution to the electric are processes used to mass-produce steel, aluminium, carbides of silicon and calcium, alkali metals and their hydroxides, phosphorus, magnesium, chlorine, etc. In addition to being refractory, electrodes must withstand chemical attack, mechanical damage during charging of the furnace, severe gradients in temperature and thermal shock. In the last respect, carbon electrodes possess a valuable combination of properties; that is, low thermal expansivity, low modulus of elasticity, good high-temperature strength and high thermal conductivity. The low electrical resistivity minimizes resistive heating and thus helps to restrict the temperature of the electrode.

The chemical industry uses a variety of impregnation techniques to combat the problem of intrinsic porosity so that corrosive liquids such as caustic solutions and sulphite liquors can be handled. Thermoset resins are used to render the carbon impervious to fluids but, being organic, tend to decompose if the service temperature rises above 170°C. Surfaces can also be sealed with more stable compounds such as carbides, silicides, borides and nitrides; for example, impregnation with molten silicon at a temperature of 2000°C forms silicon carbide *in situ* and makes the surface impervious.

10.4.6.7 Pyrolytic graphite and vitreous carbon

Most of the graphitic carbons produced for industry are polycrystalline and, as indicated previously, often exhibit bulk anisotropy, usually as a result of the shaping process. Anisotropy is the outstanding characteristic of individual graphite crystals; in fact, graphite can exhibit more anisotropy in certain properties than most other types of crystal. In polycrystalline products, it has often been necessary to cancel out dangerous anisotropic effects by deliberately randomizing the orientation of the grains (e.g. moderator graphite for nuclear reactors). A converse approach has led to the successful production of graphite with enhanced anisotropy (i.e. pyrolytic graphite (PG)). Compared to conventional graphite, PG is less chemically active and mechanically stronger. It can also be machined to close dimensional tolerances. In the 1960s, for the first time, it became possible to produce pieces of near-perfect graphite that could be used to explore the physics and chemistry of anisotropy in greater detail.[1]

In the basic method of production, pyrolytic graphite is deposited by bringing a mixture of hydrocarbon gas (methane, CH_4, or propane, C_3H_8) and an inert carrier gas into contact with a substrate material which is heated to a temperature of approximately 2000°C. Molecules of the hydrocarbon are thermally 'cracked' within the boundary layer of flow at the substrate surface. (e.g. $CH_4 = C_{gr} + 2H_2$. As indicated previously, the physical character of the deposit is determined by the process conditions, such as temperature, partial pressure of hydrocarbon gas, substrate material and area. These conditions, which are extremely critical, decide whether the deposit-forming carbon reaches the substrate surface as individual atoms, planar fragments or as three-dimensional clusters. Thus, planar fragments form laminar deposits which are highly anisotropic whereas larger clusters form isotropic deposits with little or no overall anisotropy. If the substrate takes the form of a resistance-heated carbon mandrel, a tubular deposit can be produced.

Pyrolytic graphite, as-deposited, consists of crystallites which exhibit some degree of disorder as well as stratification. In order to explain the nature of its structure, it is helpful to compare it with that of graphite in its most stable form. The perfect crystal has a regular **ABABAB**... stacking of hexagonal {0002} layers in which the distance of closest approach between layers is approximately 0.335 nm. This exact registry of hexagonal planes, with its **ABABAB**... mode of stacking, is not found in pyrolytic graphite. Instead, synthesis by chemical vapour deposition (CVD) produces a 'turbostratic' arrangement of layers in which the hexagonal networks, although still parallel, do not register precisely with each other (Figure 10.15). This disregistry has been likened to a pack of playing cards in which the long and short edges are each misaligned. Such structures are both disordered and stratified, hence the term 'turbostratic'. Each crystallite comprises a number of hexagonal layers and the interlayer distance is greater than the ideal of 0.335 nm. It is useful to assign a *c*-axis to the stack of layers forming each crystallite and to determine average values for this dimension (L_c) by X-ray diffraction analysis. In general terms, a turbostratic structure in graphite is envisaged as a conglomerate of crystallites separated by fairly stable transition zones of carbon atoms. If annealed at a very high temperature, say 3200°C,

---

[1]Professor A. R. J. P. Ubbelohde, FRS (1907–1988) and co-workers at Imperial College, London, produced valuable prototypes and made many important contributions to graphite science (e.g. stress-recrystallization, intercalation compounds).

(a)   (b)

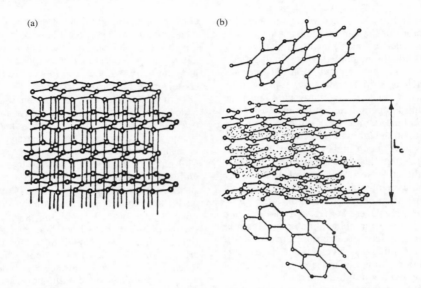

Figure 10.15 *Structures of (a) crystalline graphite and (b) turbostratic carbon (from Cahn and Harris, 11 Jan, 1969, pp. 132–41; by permission of Macmillan Magazines Ltd).*

a turbostratic structure will 'heal' and approach the stable **ABABAB**... configuration.

Various other methods are available for producing pyrolytic graphite. Composite structures for aerospace applications have been produced by depositing pyrolytic graphite on mats of carbon fibre. In the field of nuclear reactors, fluidized bed techniques have been used to coat fuel particles of uranium dioxide with separate PG layers of different anisotropy. In a fluidized bed, a rising flow of hydrocarbon gas and inert gas levitates the charge of particles; the bed temperature is very uniform and close temperature control is possible. Fluidized beds operating at a relatively low temperature of 1500°C have also been used to coat prosthetic devices such as dental implants and disc occluders for heart valves with a 1 mm layer of carbon. It is necessary to polish these coatings before use in order to remove any weakening flaws. Their wear resistance and strength have been enhanced by including silane in the bed atmosphere. Carbon is one of the dominant chemical elements in the human body and it is perhaps not surprising that it is a valuable prosthetic material, being compatible with living tissue and with blood and other body fluids.

Structural disorder, in its various aspects, is largely determined by the conditions of synthesis; thus the relative orientation of the crystallites to each other, as developed by synthesis, may be parallel (laminar), random (isotropic), etc. Both laminar and isotropic deposits of pyrolytic graphite are hard and strong because L_c is small. (Typically, L_c values for PG are about 3–10 nm.) Although highly anisotropic deposits are possible, further enhancement can be achieved by 'stress-recrystallization'. For instance, near-ideal graphite with anisotropy ratios of thermal conductivity

as high as 200 can be produced by encasing a PG deposit in an envelope of commercial polycrystalline graphite and hot-pressing at temperatures in the order of 3000°C. The envelope deforms plastically and the compression forces align the crystallites. In laminar forms of pyrolytic graphite, the anisotropy is such that the material acts as an excellent conductor in directions parallel to the hexagonal networks and as an equally efficient insulator in the c-direction. Thus, the temperature gradient in the direction perpendicular to the hexagonal networks can be very steep (say, 300°C mm^{-1}). One practical consequence of this gradient occurs during the growth of a PG deposit on a heated mandrel, when the temperature at the gas/deposit interface gradually decreases, causing the structure of the deposited carbon to change as well. (This is one reason PG deposits are often very thin: the thickest deposits are usually about 30–40 mm.) On the other hand, engineering advantage can be taken of the anisotropy of laminar PG. For example, pyrolytic graphite was advocated by Ubbelohde for small chambers or crucibles in induction furnaces operating at temperatures of 2000–3000°C. Chamber walls with their internal surfaces parallel to the a-layers favour lateral heat transfer and give temperature uniformity within the working space, a feature usually sought by furnace designers.[1] At the same time, the external wall surfaces operate at much lower temperatures; radiation losses in accordance with Stefan's law (T^4) are minimized, to the benefit of the thermal efficiency of the furnace unit.

---

[1] A tobacco pipe with a PG-lined bowl has been marketed; one would expect the high thermal conductance of its interior to sustain slow combustion.

The glassy, low-porosity material known as vitreous carbon is derived from the pyrolysis of a thermosetting polymer, rather than from a hydrocarbon gas (e.g. phenol formaldehyde resin). In a typical production procedure, a specially prepared resin is shaped by moulding, carbonized in an inert atmosphere at a temperature of 900°C and finally fired in a vacuum at a temperature of 1800°C. Very large amounts of gas are evolved; because of their disruptive effect, the thickness of the product is necessarily restricted to about 5–8 mm. Shrinkage figures in the order of 20–50% have been quoted. These features demand close control of firing conditions.

The structure of vitreous carbon inherits certain bonding characteristics from the complex organic precursor. Thus graphitization and ordering are inhibited. The crystallites are much smaller than those developed in pyrolytic graphite. They are only about 5 nm across. The crystallites have a larger-than-ideal interlayer distance of about 0.35 nm. It appears that cross-links persist between the layers and that these bonds have a stabilizing and strengthening effect. Furthermore, because the crystallites are randomly oriented, vitreous carbon is isotropic. Like silica glass, vitreous carbon is hard, brittle and fractures in a conchoidal manner.

Vitreous carbon is valued for its ability to withstand penetration and attack by aggressive liquids such as concentrated hydrofluoric acid, caustic alkalis and molten metals (e.g. Al, Cu, Pb, Sn. Zn). Its very low porosity and high purity contribute to this inertness. Generally, it has good resistance to attack by metals which do not form carbides. However, like graphite, it is attacked by alkali metals. In the chemical laboratory, vitreous carbon is used for crucibles, furnace boats, filters, non-stick burette taps, etc. in chemical analysis, crystal growing and zone-refining activities, often proving superior to traditional materials such as platinum, nickel, glass and quartz. Serving the electronics industry, vitreous carbon ware has been used in the manufacture of gallium phosphide and arsenide for optical/microwave devices, germanium for transistors and special glasses for fibre optics. In fact, it was the urgent need for an inert container for gallium phosphide which stimulated the development of vitreous carbon. In the chemical industry, it is used for rotating seals in pumps handling difficult liquids. In addition to being harder and stronger than all other forms of carbon, with the exception of diamond, it is conductive and refractory at temperatures up to 3000°C and can be used to make resistance- or induction-heated reaction vessels. Ultrasonic machining of vitreous carbon is feasible, producing sound and clean-cut edges. (Graphite has a more layered structure and machining tends to round or break up edges.)

10.4.6.8 Chemical reactivity and intercalation of graphite

Study of the manner in which foreign atoms interact with and penetrate the layered structure of graphite has proved fruitful. As the interlayer distance in graphite is quite large, with relatively weak forces acting between layers, it is chemically convenient to regard each layer as a very large macroaromatic molecule. (This assumption is reasonable because the covalent C–C bond length of 0.1415 nm in the layers is similar to that found in the aromatic organic molecule of benzene, C_6H_6.) Within individual layers of most forms of graphite, there are usually clusters of vacant sites which form 'holes'. Although chemical reactions with foreign atoms which penetrate between layers can take place at these 'holes', they are more likely to occur at the layer edges where free valency bonds project. For instance, attack by oxygen is favoured when the edges of layers impinge on free surfaces, voids or cracks. If reaction with foreign atoms does occur within the layers, newly-formed bonds can cause the layers to buckle.

In contrast to these rather random forms of attack and reaction, it is possible to interpose (intercalate) substantial numbers of foreign atoms or molecules between the parallel macromolecules to form stoichiometric graphite compounds, as indicated in Figure 10.16. (Ubbelohde coined the term 'synthetic metals' for this new class of material.) These intercalation compounds are produced by exposing graphite to a vapour phase or chemical solution or by electrolysis. By using potassium as the intercalate, a family of potassium graphites has been produced (e.g. $C_{60}K$, $C_{48}K$, $C_{36}K$, $C_{24}K$, C_8K). This regular sequence

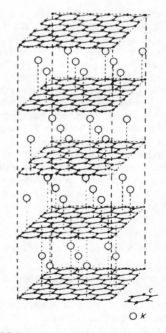

Figure 10.16 *Potassium graphite, C_8K. Hexagon layers of graphite separate and slip into the unstaggered sequence to permit intercalation of potassium atoms (from Ubbelohde, 1964; by permission of BCURA Ltd., Coal Res. Establ., Stoke Orchard, Cheltenham).*

of stoichiometric ratios shows that as intercalation proceeds, the distance between the potassium-filled regions decreases stepwise in a systematic manner. The intercalates force the macromolecules apart and sometimes cause them to shear past each other, changing the **ABABAB**... stacking sequence to **AAAA**... Intercalation species include alkali metals, halogens, caesium and rubidium, nitrate, bisulphate and phosphate radicals.

Intercalation produces interesting electrical effects involving two forms of charge transfer. Graphite itself has moderate electrical conductivity at low temperatures with about one charge carrier per 10^4 atoms of carbon. (In copper, the ratio is 1:1.) Intercalation with potassium to form potassium graphite (C_8K) raises the number of charge carriers to one in ten (−ve), with the electron-donating alkali metal injecting electrons into the 'empty' band of energy levels in graphite. Conversely, when bromine intercalates to form the compound C_8Br, the electrical conductivity again increases significantly as a result of the halogen atoms acting as electron acceptors and removing electrons from the 'full' electron band of graphite. In this case, the number of carriers rises to about one in ten (+ve). The electron transfer associated with intercalation can produce striking colour changes: the potassium graphites $C_{24}K$ (intermediate) and C_8K (saturated) are blue and golden bronze, respectively. These unique processes of charge transfer have provided valuable insights into the chemistry of bonding.

As a general rule, near-ideal graphite intercalates more readily than less perfect forms; that is, the intercalate can penetrate more easily between the macromolecules without hindrance from interlayer bonds. Certain substances have the ability to 'unpin' the layers, acting as intercalation catalysts (e.g. iodine monochloride). These catalysts have been used in large-scale chemical processing. Intercalation with halogens (bromine or chlorine) has been used to strengthen the vital bond between filaments and the resin matrix in carbon fibre 'reinforced' polymers. Intercalation forms the basis of rapid proving tests that have been used by the nuclear industry to simulate the action of neutron bombardment upon prototype graphites. The intercalating bromine atoms force the layers of graphite to open up in the *c*-direction.

10.4.6.9 From buckminsterfullerene to fullerite

Circa 1985, examination of vaporized carbon led to the exciting discovery of a new and unexpected molecular form.[1] Mass spectrometry revealed that stable clusters of carbon atoms, relative molecular mass 720, were present in significant quantity. It was postulated that these molecules were spherical and cage-shaped, with 60 carbon atoms forming the vertices

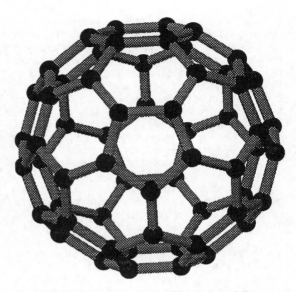

Figure 10.17 *Spherical molecule of buckminsterfullerene (courtesy of Professor H. W. Kroto, University of Sussex, UK).*

of a truncated icosahedron to give a C_{60} shell structure that is reminiscent of geodesic buildings and of a soccer ball (Figure 10.17). The C_{60} molecule was named buckminsterfullerene.[2] Larger fullerene molecules have since been produced which have more than 200 carbon atoms per shell.

The discovery of this third form of carbon, with symmetrical characteristics completely different to those of diamond and graphite, prompted attempts to evolve synthetic routes capable of producing solid buckminsterfullerene in macroscopical quantities large enough for more detailed scientific study and for commercial exploitation. Powder and thin-film samples of a solid crystalline form, fullerite, were subsequently produced in gram quantities.[3] In the basic method, a sooty mixture of graphitic carbon, C_{60} and C_{70} is produced by striking an arc between carbon electrodes in an atmosphere of helium. The other fullerene, C_{70}, which forms in smaller quantities than C_{60}, apparently has elongated molecules. The soot contains about 5–10% of the two fullerenes. After extraction with benzene, micrometre-sized crystals of C_{60} molecules form. Some disordering of stacked hexagonal arrays of

---

[1]Discovered by Professor Harold W. Kroto and colleagues at Sussex University.

[2]Named after the American architect R. Buckminster Fuller (1895–1983) who pioneered the geodesic building, a revolutionary design form which encloses the maximum of space with the minimum of materials. Many spherical viruses also have icosahedral symmetry.

[3]Solid C_{60} was produced by collaboration between D. R. Huffman and W. Kratschmer, and their co-workers, at the University of Arizona, USA, and the Max-Planck-Institut für Kernphysik, Heidelberg, Germany, respectively.

C_{60} molecules is evident. STEM micrographs of mono-layers have been produced which clearly show hexag-onally close-packed arrays of the spherical molecules (i.e. six-fold symmetry)

In parallel with its impact on organic and poly-mer chemistry, C_{60} stimulated many proposals in the materials field. Proposed 'buckyball' products have included lubricants, semiconductors and fila-mentary reinforcement for composites. It has been suggested that the cage-like structure of individual molecules makes them suitable for encapsulating metal or radioisotope atoms, a form of molecular-scale pack-aging (e.g. La^{3+}). There is also interest in bonding ions to their exterior surfaces.

10.5 Aspects of glass technology

10.5.1 Viscous deformation of glass

Viscosity[4] is a prime property of glass. Although the following account deals mainly with the working and annealing of glass, it is broadly relevant to glazes and to the glassy intergranular phase found in many ceram-ics. Two mathematical expressions help to describe the nature of viscous flow in glasses. First, there is the formula for Newtonian flow in an ideal fluid:

$$\delta\gamma/\delta t = \tau \cdot \phi = \tau \cdot 1/\eta$$

where $\delta\gamma/\delta t$ is the shear strain rate, τ is the applied shear stress, ϕ is the fluidity coefficient and η is the viscosity coefficient. In its melting range, a typical SiO_2-Na_2O-CaO glass has a viscosity of 5–50 N s m^{-2}. (The viscosity of liquid metals is roughly 1 mN s m^{-2}.) As glass is hot-drawn, its cross-sectional area decreases at a rate which depends solely upon the drawing force and viscosity, not upon area. For this reason, the glass extends uniformly and does not 'neck'.

The second expression, which is of the Arrhenius-type, concerns the temperature-dependency of visco-sity, as follows:

$$\eta = A \exp(Q/kT)$$

where A is a constant, Q is the activation energy, **k** is the Boltzmann constant and T is the absolute temperature. The activation energy for viscous flow is accordingly obtained by plotting $\log \eta$ against $1/T$ and

measuring the slope of the line. Although the above two expressions are valid for many types of glass, for certain glasses there are conditions of temperature and stress which result in non-Newtonian and/or non-Arrhenian behaviour.

As indicated in Section 2.6, the viscosity of oxide glasses depends upon composition and temperature. Generally, an increase in the concentration of modi-fying cations and/or a rise in temperature will cause the viscosity to fall. Figure 10.18 shows the rela-tion between viscosity and temperature for a typical SiO_2-Na_2O-CaO glass. Two key values of viscosity are marked; they correspond to two practical temper-atures which are known as (1) the Littleton softening point (viscosity = $10^{6.6}$ N s m^{-2}) and (2) the annealing point (viscosity = $10^{12.4}$ N s m^{-2}).

The softening point is determined by a standard-ized procedure and gives the maximum temperature at which the glass can be handled without serious change in dimensions. For ordinary silica glass, it is about 1000 K. At the annealing temperature, the ions are suf-ficiently mobile to allow residual stresses to be relieved in about 15 min. The point in the curve at which the slope is a maximum corresponds to the inflection (fic-tive or glass transition temperature) in the specific vol-ume/temperature curve for the particular glass-forming system. The working range for commercial silica glass corresponds to a viscosity range of 10^3-10^7 N s m^{-2}. The curve for this glass is quite steep, indicating that temperature control to within, say, $\pm10°C$ is necessary during working (i.e. drawing, blowing, rolling, etc.). Figure 10.19 provides a comparison of the viscosity curves for different types of glass. The difficulties of working pure silica glass ('fused quartz') are immedi-ately apparent. Even at a temperature of 1300°C this glass has a viscosity of about 10^{12} N s m^{-2}, which is still too high for working. Glasses with special chemi-cal and physical properties often have a steep viscosity curve and tend to devitrify, presenting difficulties dur-ing drawing at traditional working temperatures. One

[4]Dynamic (absolute) viscosity is the force required to move 1 m^2 of plane surface at a velocity of 1 m s^{-1} to a second plane surface which is parallel to the first and separated 1 m from it by a layer of the fluid phase. Kinematic viscosity = absolute viscosity/density. Unit of viscosity = 1 N s m^{-2} = 10 P (poise). Fluidity is the reciprocal of viscosity.

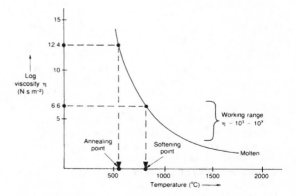

Figure 10.18 *Viscosity versus temperature curve for a typical SiO_2-Na_2O-CaO glass.*

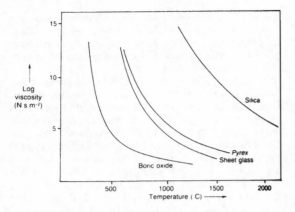

Figure 10.19 *Viscosity curves for typical glasses.*

solution has been to hot-extrude such glasses through graphite dies into rod or tube form.[1] By using very high pressures (up to 10 GN m^{-2}), it is possible to overcome a high viscosity of 10^7 N s m^{-2} and to use a lower extrusion temperature (e.g. 950°C). Such temperatures, which can be as much as 200°C lower than the temperature required for traditional methods, prevent devitrification and, by suppressing surface tension effects, help to form rods and tubes with sharp and accurate profiles.

10.5.2 Some special glasses

Most glasses produced are based upon silica and fluxed with alkali metal oxides, with their chemical formulation adjusted to suit the application (e.g. bottles, plate glass, lamps, etc.). In its purest form, vitreous silica has a very low linear coefficient of thermal expansion ($\alpha = 0.55 \times 10^{-6}$°C^{-1}) and is used in high-temperature applications. 'Fused quartz' is made by melting sand; high-purity 'fused silica' is made by vapour deposition after reacting $SiCl_4$ with oxygen at temperatures above 1500°C. (These two definitions are tentative.)

Window ('soda-lime') glass, typically $72SiO_2$–$14Na_2O$–$10CaO$–$2MgO$–$1Al_2O_3$, owes its convenient range of working temperature (700–1000°C) to the two network-modifiers, Na_2O and CaO. The effectiveness of their fluxing action can be gauged by the fact that the softening point of this glass (700°C) is at least 800°C lower than that of high-purity silica glass. The softening point, as defined in the ASTM test method, is the temperature at which a glass fibre of uniform diameter elongates at a specific rate under its own weight. The chemical formulation given uses the 'system' nomenclature recommended by F. V. Tooley (i.e. oxide components are stated in order of decreasing % w/w.)

---

[1]Philips Research Laboratories, Aachen, Germany.

Pyrex glass, $81SiO_2$–$13B_2O_3$–$4Na_2O$–$2Al_2O_3$, combines two network-formers and is well known for its chemical durability. Because of its relatively low expansivity ($\alpha = 3.3 \times 10^{-6}$°C^{-1}; 0–300°C), it is tolerant to thermal shock and is widely used for chemical glassware in laboratories and industrial processing plant. Sometimes it is referred to as a 'borosilicate' glass, unfortunately implying crystallinity.

In certain glasses, being an intermediate, PbO can actually form a network. This oxide is a flux and, unlike Na_2O, does not reduce the electrical resistivity; glasses in the SiO_2–PbO–Na_2O–Al_2O_3 system are widely used for electronic and electrical components. Their high lead content favours absorption of X-rays and γ-rays, making them suitable for radio-logical shielding (65% PbO) and colour television tubes (23% PbO). Lead also increases the refractive index and dispersion of SiO_2-based glasses. These effects are used to advantage in optical lenses and in the decorative glass known confusingly as 'lead crystal'. In the latter case, 20–30% PbO provides craftsmen with a 'long' glass, that is, one with a large temperature interval for working/manipulation, within which viscosity changes very slightly.

Alumina (Al_2O_3), an intermediate, frequently features in the composition of glasses. In small amounts, say 1–2% w/w, it improves the resistance of SiO_2-based glasses to attack by water. However, its presence increases melt viscosity and makes melting and fining more difficult; for instance, gas bubbles separate less readily. The dependence of viscosity upon composition is always an important matter in glass technology. Al^{3+} cations enter the tetrahedral holes in the network, replacing tetravalent silicon cations. As a result, the number of non-bridging O^{2-} anions is reduced and the strength and rigidity of the network is enhanced. By substantially raising the alumina content to 15–25% w/w, it is possible to develop valuable stability at high temperatures. For example, $62SiO_2$–$17Al_2O_3$–$8CaO$–$7MgO$–$5B_2O_3$–$1Na_2O$ glass has a softening point in excess of 900°C, which is about 100°C higher than that of *Pyrex* glass. This 'aluminosilicate' glass is used for combustion tubes.

10.5.3 Toughened and laminated glasses

Glasses, in general, are weak when placed in tension during bending and do not withstand impact blows. However, in compression, glasses can exhibit exceptional strength. Accordingly, sheet glass is often strengthened and toughened by developing high residual compressive stresses in the surface layers. They oppose any applied tensile stresses (say, on the convex face of a bent sheet of glass), and must be overcome before this vulnerable surface begins to develop dangerous tensile stresses. Various physical and chemical processes, applicable on a mass-production scale, are available to improve the load-bearing capacity and behaviour under impact conditions.

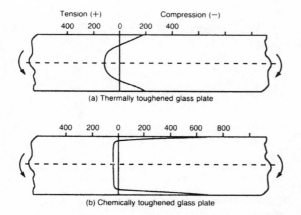

Figure 10.20 *Distribution of residual stress in toughened glass plate (stress values in MN m^{-2}).*

The term 'toughening' generally refers to both heat-treatment and chemical methods. In thermal tempering, a toughening process used extensively for side and rear glazing in cars, glass sheets are cut and shaped, heated uniformly to a temperature above T_g (say, 700°C), soaked for 2–4 min and then quenched with jets of air. Glass has a low thermal conductivity and the eventual contraction of the core regions is restrained by the rigid outer layers, producing the balance of compressive (−ve) and tensile (+ve) stresses shown in Figure 10.20a. The glass sheet must be free from flaws in order to withstand the severe quench stresses and must be at least 2–3 mm thick so that the desired stress gradients develop. Surface stresses up to 200 MN m^{-2} are produced by this method. Heating toughened glass to a temperature of 200°C will relieve these stresses. Small blunt-edged fragments are produced when toughened glass fractures under impact. Toughened products cannot be cut or shaped; notching, drilling, etc. must be done before treatment.

Chemical strengthening/toughening is also used to develop protective compressive stresses. These treatments are more versatile, being suitable for thin and thick sections of glass and for unusual shapes (e.g. *Chemcor*.) However, because ion exchange and diffusion are involved, they require longer process times. Their purpose is to alter the surface composition and reduce its coefficient of thermal expansion. On cooling from a high temperature, the greater contraction of the hot core is restrained and very high compressive stresses develop in the modified surface layers (Figure 10.20b). In one version of chemical strengthening, sodium-containing glass is heated in an atmosphere of water vapour and sulphur dioxide and an exchange between H$^+$ ions and Na$^+$ ions takes place in the glass surface (i.e. 'H-for-Na'). In another high-temperature version, a 'Li-for-Na' exchange occurs when the glass is immersed in a melt of lithium salt. In low-temperature chemical strengthening, large ions from molten salt displace smaller ions from the surfaces of immersed glass, causing them to expand (e.g. 'K-for-Na', 'Na-for-Li'). On cooling from a relatively low temperature, the core is rigid and the presence of absorbed, larger ions in the surface layers develops compressive stresses. Chemical strengthening methods often use glasses of special composition.

The bottle-manufacturing industry uses chemical surface treatments to strengthen returnable bottles (improving trippage) and to provide a useful lubricity as they pass through bottle-filling plant at high speed. (Rubbing two glass bottles together reveals whether a lubricating coating is present.)

The alternative to the above toughening methods, based upon a composite approach, is to interleave two matched sheets of 'float' glass with an interlayer of plastic, nearly 1 mm thick. Polyvinyl butyrate (PVB) is commonly used for the interlayer. In the mid-1960s, this more costly method gained favour for car windscreens as a result of accident statistics which indicated that a laminated windscreen was less likely to cause fatal injury than a toughened one.[1] Cracking is confined to the area of impact. The PVB interlayer deforms elastically, absorbs impact energy and holds glass fragments in place, reducing the risk of facial lacerations. Visibility is retained after impact. De-icing windscreens have fine heating filaments of tungsten embedded between the glass layers (*Triplex Hotscreen*). Laminated glass with a thicker plastic interlayer is used for aircraft windscreens; these must withstand extreme thermal conditions, ranging from high altitudes (−80°C) to kinetic heating (150°C), and the possible high-speed 'strikes' of birds weighing about 2 kg.

The manufacture of laminated glass demands cleanliness so that the PVB will bond strongly to the glass. Curved windscreens are made by separately bending matched pairs of glass sheet at a temperature just above 600°C. After cooling, PVB is applied to one sheet of each pair in an air-conditioned room and the matching sheet added. Air can be removed from the PVB/glass interfaces of the curved assembly by applying a vacuum to its edges. The glazing is finally autoclaved at a pressure of 1.2–1.4 MN m^{-2} and a temperature of 135–145°C in order to complete the bonding.

The oil crisis of the mid-1970s underlined the need for weight saving and fuel efficiency in car design. As a direct consequence, the thickness of laminated windscreens decreased from 3.0 mm/3.0 mm to 2.1 mm/2.1 mm. (Toughened glass decreased in thickness from 5–6 mm to 3–4 mm.) Advances in glass technology have made it possible for modern car designs to increase the area of glazing, use flush,

[1] Strong sunlight striking a quench-toughened window glass is polarized by reflection; the stress-birefringent patterns produced by the quench jets are often apparent, particularly when viewed through *Polaroid* sunglasses. A pastime for traffic jams is to note the choice made by car manufacturers between quench-toughened and laminated glass.

polyurethane adhesive-mounted windows (of low drag coefficient) and achieve rounded, aerodynamic profiles.[1]

10.6 The time-dependency of strength in ceramics and glasses

Time and strain-rate have a special relevance to the strength of ceramics. For example, the measured strength of individuals taken from a set of ceramic specimens can show an appreciable decrease over a period of time. With regard to the mechanism of such time-dependent failure, it is first necessary to distinguish between fast and slow crack growth.

Fast crack growth, as produced in a routine bend test, takes place when the crack reaches a critical size. Slow or delayed crack growth takes place in a stressed ceramic over a relatively long period of time and involves slow propagation of a crack of sub-critical dimensions. Fast fracture occurs when this crack reaches criticality. This phenomenon of progressive weakening and degradation, which is termed static fatigue or delayed fracture, has long been known in glasses. It is now recognized in a wide variety of ceramics (e.g. alumina, silcon nitride, vitreous carbon). Its occurrence in polycrystalline oxide ceramics is often related to the presence of a glassy bonding phase. The stresses capable of causing static fatigue can be continuous or fluctuating: even at quite low stress levels, there is a finite probability of failure by this insidious process. Sometimes chemical agencies have an overriding effect. For example, when moisture is able to diffuse to crack tips in silica glasses and alumina, the velocity of crack growth increases. This type of behaviour is understandably regarded as stress-corrosion. Slow crack growth is a thermally-activated process and an increase in the ambient temperature can have a very significant effect.

The nature of delayed crack growth is customarily expressed by plotting crack velocity (V) against the stress intensity factor (K_I), as shown in Figure 10.21. This plot is idealised but is typical of ceramics which are capable of slow crack growth at ambient temperature, such as glasses and oxide ceramics. Three regions of crack growth lie between the 'static fatigue limit' (K_{Io}), below which crack growth is not of concern, and the critical value of the stress-intensity factor (K_{Ic}) at which sudden fracture occurs. In region I, the velocity is an exponential function of the stress-intensity factor and can be expressed in the form $V = A(K_I)^n$. In region II, velocity is independent of K_I. Regions I and II are environment-sensitive and correspond to conditions of stress-corrosion. In region III, velocity is

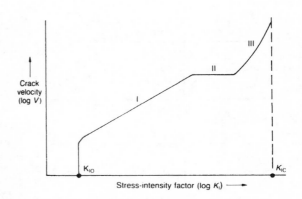

Figure 10.21 *Idealized plot of relation between crack velocity and stress-intensification.*

again dependent upon K_I but is independent of environment.

Time-to-failure depends primarily upon the profile of region I. Its slope (n), sometimes known as the stress-corrosion constant of the particular material, ranges in value from about 10 to 100. Low values, such as those for glasses (10–40), indicate a susceptibility to slow crack growth. Figure 10.22 shows typical curves for a variety of ceramic materials. Changes in environment (temperature, pH, etc.) can displace a curve and/or alter its slope. For instance, at elevated temperatures, the slopes of curves for carbides and nitrides become less steep.

Study of the statistical nature and time-dependency of strength in ceramics led to the development of strength-probability-time (SPT) diagrams[2] of the type shown in Figure 10.23. These diagrams refer to specific conditions of testing and are plotted on Weibull probability paper. As the n-value (for region I crack growth) decreases, indicating increased sensitivity to environment, the sloping lines close up. Such diagrams clearly show the interplay between stress, the probability of survival and the lifetime before fracture.

Recognition of the fact that there is apparently no absolutely safe stress for a ceramic which is subject to long-term service stimulated interest in proof-testing. By eliminating weaker members, such tests provide survivors with a better (narrower) distribution of strength than that of the original population. Statistical predictions of service life benefit accordingly. It must be borne in mind that the proportion of survivors reflects the conditions of testing. Obviously, the period of testing is important; the longer the time, the smaller the number of survivors. Ideally, proof-tests should be carried out under conditions that do not cause subcritical crack growth.

---

[1] The Jaguar XJ220 sports car has compound-curvature front, top and rear screens which blend smoothly with the body contours.

[2] Devised by R. W. Davidge and co-workers at the Atomic Energy Research Establishment, Harwell.

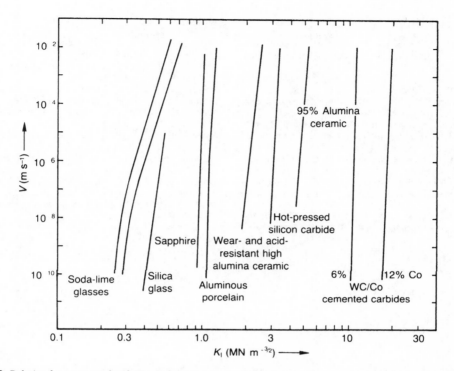

Figure 10.22 *Relation between crack velocity and stress intensity factor for various ceramics in neutral water at 25°C (after Creyke et al., 1982; by courtesy of Elsevier/Chapman and Hall).*

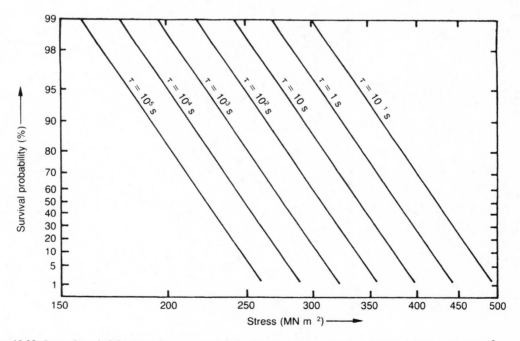

Figure 10.23 *Strength-probability-time diagram for alumina derived from bend tests at 293 K: the lifetime τ of 10^5 s is regarded as a tentative stress-corrosion limit (after Davidge, 1986, p. 145; by courtesy of Cambridge University Press).*

Further reading

Binner, J. (1992). Processing of advanced ceramic powders. *Metals and Materials*, October, 534–537, Institute of Materials.

Burkin, A. R. (ed.) (1987). *Production of Aluminium and Alumina*. Wiley, Chichester.

Cahn, R. W. and Harris, B. (1969). Newer forms of carbon and their uses. *Nature*, 11 January, **221**, 132–141.

Creyke, W. E. C., Sainsbury, I. E. J. and Morrell, R. (1982). *Design with Non-Ductile Materials*, Elsevier/Chapman and Hall, London.

Davidge, R. W. (1986). *Mechanical Behaviour of Ceramics*. Cambridge University Press, Cambridge.

Jack, K. H. (1987). Silicon nitride, sialons and related ceramics. In *Ceramics and Civilisation*, **3**. American Ceramic Society Inc., New York.

Kingery, W. D., Bowen, H. K. and Uhlmann, D. R. (1976). *Introduction to Ceramics*, 2nd edn. Wiley-Interscience, New York

Laminated Glass Information Centre (1993). *Laminated Glass*. LGIC, London.

McColm, I. J. (1983). *Ceramic Science for Materials Technologists*. Leonard Hill, Glasgow.

Mantell, C. L. (1968). *Carbon and Graphite Handbook*. Wiley-Interscience, New York.

Parke, S. (1989). Glass—a versatile liquid. *Metals and Materials*, January, 26–32, Institute of Materials.

Rawson, H. (1980). *Properties and Applications of Glass*. Elsevier Science, Oxford.

Riley, F. L. (ed.) (1977). *Nitrogen Ceramics*. Noordhoff, Leiden.

Shinroku Saito (ed.) (1985). *Fine Ceramics*. Elsevier, New York.

Ubbelohde, A. R. J. P. and Lewis, F. A. (1960). *Graphite and its Crystal Compounds*. Clarendon Press, Oxford.

Wachtman, J. B. (ed.) (c1989). *Structural Ceramics*, **29**, Academic Press, New York.

Chapter 11

Plastics and composites

11.1 Utilization of polymeric materials

11.1.1 Introduction

In Chapter 2 the basic chemistry and structural arrangements of long-chain molecules were described and it was shown how polymers can be broadly classified as thermoplastics, elastomers and thermosets. Certain aspects of their practical utilization will now be examined, with special attention to processing; this stage plays a decisive role in deciding if a particular polymeric material can be produced as a marketable commodity. The final section (11.3) will concern composites, extending from the well-known glass-reinforced polymers to those based upon ceramic and metallic matrices.

11.1.2 Mechanical aspects of T_g

As indicated in Chapter 2, it is customary to quote a glass-transition temperature T_g for a polymer because it separates two very different régimes of mechanical behaviour. (The value of T_g is nominal, being subject to the physical method and procedure used in its determination). Below T_g, the mass of entangled molecules is rigid. Above T_g, viscoelastic effects come into play and it is therefore the lower temperature limit for processing thermoplastics. The structural effect of raising the temperature of a glassy polymer is to provide an input of thermal energy and to increase the vibrations of constituent atoms and molecules. Molecular mobility increases significantly as T_g is approached: rotation about C–C bonds in the chain molecules begins, the free volume of the structure increases and intermolecular forces weaken. It becomes easier for applied forces to deform the structure and elastic moduli to fall.

The mechanical properties of polymers are highly dependent upon time and temperature, the response to stress being partly viscous and partly elastic. For instance, 'natural' time periods are associated with the

various molecular relaxation processes associated with the glass transition. In linear viscoelastic behaviour, total strain comprises a linear elastic (Hookean) component and a linear viscous (Newtonian) component. The stress–strain ratios depend upon time alone. In the more complex non-linear case, which usually applies to polymers, strain is a function of time and stress because molecular movements are involved.

The phenomenon of stress relaxation can be used to chart the way in which the behaviour of a given polymer changes from glassy to rubbery. Figure 11.1 shows the non-linear response of a polymer that is subjected to constant strain ε_0. Stress σ relaxes with time t. The relaxation modulus E_r at time t is given by the expression:

$$E_r = \sigma_t / \varepsilon_0 \qquad (11.1)$$

Thus E_r, which is represented by the slope of dotted join lines, decreases with time. This variation is shown more precisely by a plot of log E_r versus log time t, as in Figure 11.2a. The thermoplastic

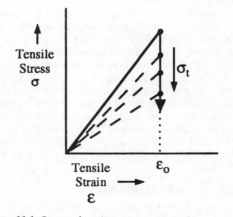

Figure 11.1 *Stress relaxation at constant strain.*

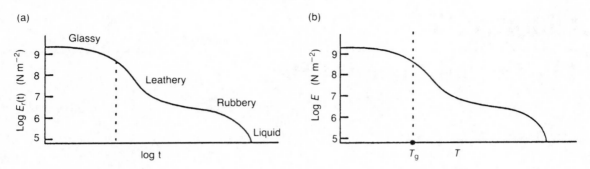

Figure 11.2 *Time–temperature dependence of elastic modulus in thermoplastic polymeric solid: (a) change in relaxation modulus $E_r(t)$ as function of time; (b) change in tensile modulus as function of temperature (from Hertzberg, 1989; by permission of John Wiley and Sons).*

polymer changes in character from a glassy solid, where the relaxation modulus is a maximum, to a rubbery solid. In the complementary Figure 11.2b, data from standard tensile tests on the same polymer at different temperatures are used to provide values of elastic moduli (E). The similarity of profiles in Figures 11.2a and 11.2b illustrates the equivalence of time and temperature. (Theoretically, the modulus for a short time and a high temperature may be taken to equate to that for a combination of a long time and a low temperature; this concept is used in the preparation of relaxation modulus versus time graphs.) The glass transition temperature T_g has been superimposed upon Figures 11.2a and 11.2b.

Although single values of T_g are usually quoted, the process of molecular rearrangement is complex and minor transitions are sometimes detectable. Thus, for PVC, the main glass transition occurs at temperatures above 80°C but there is a minor transition at −40°C. Consequently, at room temperature, PVC exhibits some rigidity yet can elongate slightly before fracture. Addition of a plasticizer liquid, which has a very low T_g, lowers the T_g value of a polymer. Similarly, T_g for a copolymer lies between the T_g values of the original monomers; its value will depend upon monomer proportions.

In elastomeric structures, as the temperature is increased, the relatively few crosslinks begin to vibrate vigorously at T_g and the elastomer becomes increasingly rubbery. As one would anticipate, T_g values for rubbers lie well below room temperature. Increasing the degree of crosslinking in a given polymer has the effect of raising the entire level of the lower 'rubbery' plateau of the modulus versus temperature plot upwards as the polymer becomes more glassy in nature. The thermosets PMMA (*Perspex, Lucite*) and PS have T_g values of 105°C and 81°C, respectively, and are accordingly hard and brittle at room temperature.

11.1.3 The role of additives

Industrially, the term 'plastic' is applied to a polymer to which one or more property-modifying agents have been added. Numerous types of additive are used by manufacturers and fabricators; in fact, virgin polymers are rarely used. An additive has a specific function. Typical functions are to provide (1) protection from the service environment (anti-oxidants, anti-ozonants, anti-static agents, flame-retardants, ultraviolet radiation absorbers), (2) identification (dyes, pigments), (3) easier processability (plasticizers), (4) toughness, and (5) filler.

In many instances the required amount of additive ranges from 0.1% to a few per cent. Although ultraviolet (UV) components of sunlight can structurally alter and degrade polymers, the effect is particularly marked in electric light fittings (e.g. yellowing). Stabilizing additives are advisable as some artificial light sources emit considerable amounts of UV radiation with wavelengths in the range 280–400 nm. For any polymer, there is a critical wavelength which will have the most damaging effect. For instance, a wavelength of 318.5 nm will degrade PS, which is a common choice of material for diffusers and refractors, by either causing cross-linking or by producing free radicals that react with oxygen.

The action of a plasticizer (3) is to weaken intermolecular bonding by increasing the separation of the chain molecules. The plasticizer may take the form of a liquid phase that is added after polymerization and before processing. Additions of a particulate toughener (4) such as rubber may approach 50% and the material is then normally regarded as a composite.

A wide variety of fillers (5) is used for polymers. In the case of thermosets, substances such as mica, glass fibre and fine sawdust are used to improve engineering properties and to reduce the cost of moulded products. PTFE has been used as a filler (15%) to improve the wear resistance of nylon components. Although usually electrically non-conductive, polymers can be made conductive by loading them with an appropriate filler (e.g. electromagnetic shielding, specimen mounts in SEM analysis). Fillers and other additives play an important role in the production of vulcanized rubbers. Inert fillers facilitate handling of the material before vulcanization (e.g. clay, barium sulphate). Reinforcing

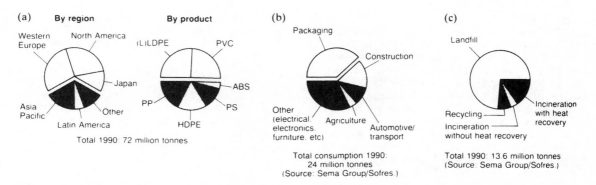

Figure 11.3 *(a) World consumption of plastics, (b) plastics consumption by market sector, Western Europe, and (c) destination of post-consumer plastic waste, Western Europe (courtesy of Shell Briefing Services, London).*

fillers restrict the movement of segments between the branching points shown in Figure 2.24. For instance, carbon black has long been used as a filler for car tyres, giving substantial improvements in shear modulus, tear strength, hardness and resistance to abrasion by road surfaces. Tyres are subject to fluctuating stresses during their working life and it has been found useful to express their stress–strain behaviour in terms of a dynamic shear modulus (determined under cyclic stressing conditions at a specified frequency). Time is required for molecular rearrangements to take place in an elastomer; for this reason the dynamic modulus increases with the frequency used in the modulus test. The dynamic modulus at a given frequency is significantly enhanced by the crosslinking action of vulcanization and by the presence of carbon black. These carbon particles are extremely small, typically 20–50 nm diameter. Small amounts of anti-oxidants[1] and anti-ozonants are often beneficial. The principal agents of degradation under service conditions are extreme temperatures, oxygen and ozone, various liquids. In the last case a particular liquid may penetrate between the chains and cause swelling.

11.1.4 Some applications of important plastics

Figures 11.3a and 11.3b summarize 1990 data on world and West European consumption of plastics. The survey included high-volume, low-price commodity plastics as well as engineering and advanced plastics. Thermoplastics dominate the market (i.e. PE, PVC and PP). Development of an entirely new type of plastic is extremely expensive and research in this direction is

limited. Research is mainly concerned with improving and reducing the cost of established materials (e.g. improved polymerization catalysts, composites, thermoplastic rubbers, waste recycling).

The low- and high-density forms of polyethylene, LDPE and HDPE, were developed in the 1940s and 1950s, respectively. Extruded LDPE is widely used as thin films and coatings (e.g. packaging). HDPE is used for blow-moulded containers, injection-moulded crates and extruded pipes. An intermediate form of adjustable density known as linear low-density polyethylene, LLDPE, became available in the 1980s. Although more difficult to process than LDPE and HDPE, film-extruded LLDPE is now used widely in agriculture, horticulture and the construction industry (e.g. heavy-duty sacks, silage sheets, tunnel houses, cloches, damp-proof membranes, reservoir linings). Its tear strength and toughness have enabled the gauge of PE film to be reduced. A further variant of PE is ultra-high molecular weight polyethylene, UHMPE, which is virtually devoid of residual traces of catalyst from polymerization. Because of its high molecular mass, it needs to processed by sintering. UHMPE provides the wear resistance and toughness required in artificial joints of surgical prostheses.

Polyvinyl chloride (PVC) is the dominant plastic in the building and construction industries and has effectively replaced many traditional materials such as steel, cast iron, copper, lead and ceramics. For example, the unplasticized version (UPVC) is used for window frames and external cladding panels because of its stiffness, hardness, low thermal conductivity and weather resistance. PVC is the standard material for piping in underground distribution systems for potable water (blue) and natural gas (yellow), being corrosion-resistant and offering small resistance to fluid flow. Although sizes of PVC pipes tend to be restricted, PVC linings are used to protect the bore of large-diameter pipes (e.g. concrete). The relatively low softening temperature of PVC has stimulated interest in alternative piping materials for underfloor heating systems. Poly-butylene (PB) has been used for this application; being

---

[1]In the 1930s, concern with oxidation led the Continental Rubber Works, Hannover, to experiment with nitrogen-inflation of tyres intended for use on Mercedes 'Silver Arrows', Grand Prix racing cars capable of 300 km/h. This practice was not adopted by Mercedes-Benz for track events.

supported, it can operate continuously at a temperature of 80°C and can tolerate occasional excursions to 110°C. However, hot water with a high chlorine content can cause failure. Other important building plastics are PP, ABS and polycarbonates. Transparent roofing sheets of twin-walled polycarbonate or PVC provide thermal insulation and diffuse illumination: both materials need to be stabilized with additives to prevent UV degradation.

The thermoplastic polypropylene, PP (*Propathene*) became available in the early 1960s. Its stiffness, toughness at low temperatures and resistance to chemicals, heat and creep ($T_m = 165-170°C$) are exceptional. PP has been of particular interest to car designers in their quest for weight-saving and fuel efficiency. In a typical modern saloon car, at least 10% of the total weight is plastic (i.e. approximately 100 kg). PP, the lowest-density thermoplastic (approximately 900 kg m^{-3}), is increasingly used for interior and exterior automotive components (e.g. heating and ventilation ducts, radiator fans, body panels, bumpers (fenders)). It is amenable to filament-reinforcement, electroplating, blending as a copolymer (with ethylene or nylon), and can be produced as mouldings (injection- or blow-), films and filaments.

Polystyrene (PS) is intrinsically brittle. Engineering polymers such as PS, PP, nylon and polycarbonates are toughened by dispersing small rubber spheroids throughout the polymeric matrix; these particles concentrate applied stresses and act as energy-absorbing sites of crazing. The toughened high-impact form of polystyrene is referred to as HIPS. PS, like PP, is naturally transparent but easily coloured. When supplied as expandable beads charged with a blowing agent, such as pentane, PS can be produced as a rigid heat-insulating foam.

Some of the elastomers introduced in Chapter 2 will now be considered. Polyisoprene is natural rubber; unfortunately, because of reactive C–C links in the chains it does not have a high resistance to chemical attack and is prone to surface cracking and degradation ('perishing'). Styrene-butadiene rubber (SBR), introduced in 1930, is still one of the principal synthetic rubbers (e.g. car tyres).[1] In this copolymer, repeat units of butadiene[2] are combined randomly with those of styrene. Polychloroprene (*Neoprene*), introduced in 1932, is noted for its resistance to oil and heat and is used for automotive components (e.g. seals, water circuit pipes).

As indicated previously, additives play a vital role in rubber technology. The availability of a large family of

rubbers has encouraged innovative engineering design (e.g. motorway bridge bearings, mounts for oil-rigs and earthquake-proof buildings, vehicle suspension systems).

Silicone rubbers may be regarded as being intermediate in character to polymers and ceramics. From Table 2.7 it can be seen that the long chains consist of alternating silicon and oxygen atoms. Although weaker than organic polymers based upon carbon chains, they retain important engineering properties, such as resilience, chemical stability and electrical insulation, over the very useful temperature range of $-100°C$ to $300°C$. These outstanding characteristics, combined with their cost-effectiveness, have led to the adoption of silicone rubbers by virtually every industry (e.g. medical implants, gaskets, seals, coatings).

11.1.5 Management of waste plastics

Concern for the world's environment and future energy supplies has focused attention on the fate of waste plastics, particularly those from the packaging, car and electrical/electronics sectors. Although recovery of values from metallic wastes has long been practised, the diversity and often complex chemical nature of plastics raise some difficult problems. Nevertheless, despite the difficulties of re-use and recycling, it must be recognized that plastics offer remarkable properties and are frequently more cost- and energy-effective than traditional alternatives such as metals, ceramics and glasses. Worldwide, production of plastics accounts for about 4% of the demand for oil: transport accounts for about 54%. Enlightened designers now consider the whole life-cycle and environmental impact of a polymeric product, from manufacture to disposal, and endeavour to economize on mass (e.g. thinner thicknesses for PE film and PET containers ('lightweighting')). Resort to plastics that ultimately decompose in sunlight (photodegradation) or by microbial action (biodegradation) represents a loss of material resource as they cannot be recycled; accordingly, their use tends to be restricted to specialized markets (e.g. agriculture, medicine).

Figure 11.3c portrays the general pattern of plastics disposal in Western Europe. Landfilling is the main method but sites are being rapidly exhausted in some countries. The principal routes of waste management are material recycling, energy recycling and chemical recycling. The first opportunity for material recycling occurs during manufacture, when uncontaminated waste may be re-used. However, as in the case of recycled paper, there is a limit to the number of times that this is possible. Recycling of post-consumer waste is costly, involving problems of contamination, collection, identification and separation.[3]. Co-extruded

---

[1] Synthetic (methyl) rubber was first produced in Germany during World War I as a result of the materials blockade; when used for tyres, vehicles had to be jacked-up overnight to prevent flat areas developing where they contacted the ground.

[2] Butadiene, or but-2-ene, is an unsaturated derivative of butane C_4H_{10}; the central digit indicates that the original butene monomer C_4H_8 contains two double bonds.

[3] German legislation requires that, by 1995, 80% of all packaging (including plastics) must be collected separately from other waste and 64% of total waste recycled as material.

blow-moulded containers are being produced with a three-layer wall in which recycled material is sandwiched between layers of virgin polymer. In the German car industry efforts are being made to recycle flexible polyurethane foam, ABS and polyamides. New ABS radiator grilles can incorporate 30% from old recycled grilles.

Plastics have a high content of carbon and hydrogen and can be regarded as fuels of useful calorific value. Incinerating furnaces act as energy-recycling devices, converting the chemical energy of plastics into thermal/electrical energy and recovering part of the energy originally expended in manufacture. Noxious fumes and vapours can be evolved (e.g. halogens); control and cleaning of flue gases are essential.

Chemical recycling is of special interest because direct material recycling is not possible with some wastes. Furthermore, according to some estimates, only 20–30% of plastic waste can be re-used after material recycling. Chemical treatment, which is an indirect material recycling route, recovers monomers and polymer-based products that can be passed on as feedstocks to chemical and petrochemical industries. Hydrogenation of waste shows promise and is used to produce synthetic oil.

11.2 Behaviour of plastics during processing

11.2.1 Cold-drawing and crazing

A polymeric structure is often envisaged as an entangled mass of chain molecules. As the T_g values for many commercial polymers are fairly low, one assumes that thermal agitation causes molecules to wriggle at ambient temperatures. Raising the temperature increases the violence of molecular agitation and, under the action of stress, molecules become more likely to slide past each other, uncoil as they rotate about their carbon–carbon bonds, and extend in length. We will first concentrate upon mechanistic aspects of two important modes of deformation; namely, the development of highly-preferred molecular orientations in semi-crystalline polymers by cold-drawing and the occurrence of crazing in glassy polymers.

Cold-drawing can be observed when a semi-crystalline structure containing spherulites is subjected to a tensile test at room temperature. A neck appears in the central portion of the test-piece. As the test-piece extends, this neck remains constant in cross-section but increases considerably in length. This process forms a necked length that is stronger and stiffer than material beyond the neck. At first, the effect of applied tensile stress is to produce relative movement between the crystalline lamellae and the interlamellar regions of disordered molecules. Lamellae that are normal to the direction of principal stress rotate in a manner reminiscent of slip-plane rotation in metallic single crystals, and break down into smaller blocks.

These blocks are then drawn into tandem sequences known as microfibrils (Figure 11.4). The individual blocks retain their chain-folding conformation and are linked together by the numerous tie molecules which form as the original lamellae unfold. A bundle of these highly-oriented microfibrils forms a fibril (small fibre). The microfibrils in a bundle are separated by amorphous material and are joined by surviving interlamellar tie molecules. The pronounced molecular orientation of this type of fibrous structure maximizes the contribution of strong covalent bonds to strength and stiffness while minimizing the effect of weak intermolecular forces.

Industrially, cold-drawing techniques which take advantage of the anisotropic nature of polymer crystallites are widely used in the production of synthetic fibres and filaments (e.g. *Terylene*). (Similarly, biaxial stretching is used to induce exceptional strength in film and sheet and bottles e.g. *Melinex*). Crystallization in certain polymers can be very protracted. For instance, because *nylon* 6,6 has a T_g value slightly below ambient temperature, it can continue to crystallize and densify over a long period of time during normal service, causing undesirable after-shrinkage. This metastability is obviated by 'annealing' nylon briefly at a temperature of 120°C, which is below T_m: less perfect crystals melt while the more stable crystals grow. Stretching *nylon* 6,6 at room temperature during the actual freezing process also encourages crystallization and develops a strengthening preferred orientation of crystallites.

Let us now turn from the bulk effect of cold-drawing to a form of localized inhomogeneous deformation, or yielding. In crazing, thin bands of expanded material form in the polymer at a stress much lower than the bulk yield stress for the polymer. Crazes are usually associated with glassy polymers (PMMA and PS) but may occur in semi-crystalline polymers (PP). They are

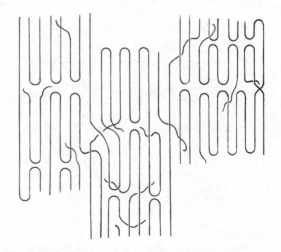

Figure 11.4 *Persistence of crystalline block structure in three microfibrils during deformation.*

several microns wide and fairly constant in width: they can scatter incident light and are visible to the unaided eye (e.g. transparent glassy polymers). As in the stress-corrosion of metals, crazing of regions in tension may be induced by a chemical agent (e.g. ethanol on PMMA). The plane of a craze is always at right angles to the principal tensile stress. Structurally, each craze consists of interconnected microvoids, 10–20 nm in size, and is bridged by large numbers of molecule-orientated fibrils, 10–40 nm in diameter. The voidage is about 40–50%. As a craze widens, bridging fibrils extend by drawing in molecules from the side walls. Unlike the type of craze found in glazes on pottery, it is not a true crack, being capable of sustaining some load. Nevertheless, it is a zone of weakness and can initiate brittle fracture. Each craze has a stress-intensifying tip which can propagate through the bulk polymer. Crazing can take a variety of forms and may even be beneficial. For instance, when impact causes crazes to form around rubber globules in ABS polymers, the myriad newly-created surfaces absorb energy and toughen the material. Various theoretical models of craze formation have been proposed. One suggestion is that triaxial stresses effectively lower T_g and, when tensile strain has exceeded a critical value, induce a glass-to-rubber transition in the vicinity of a flaw, or similar heterogeneity. Hydrostatic stresses then cause microcavities to nucleate within this rubbery zone.

As Figure 11.5 shows, it is possible to portray the strength/temperature relations for a polymeric material on a deformation map. This diagram refers to PMMA and shows the fields for cold-drawing, crazing, viscous flow and brittle fracture, together with superimposed contours of strain rate over a range of 10^{-6} to $1\ s^{-1}$.

11.2.2 Processing methods for thermoplastics

Processing technology has a special place in the remarkable history of the polymer industry: polymer

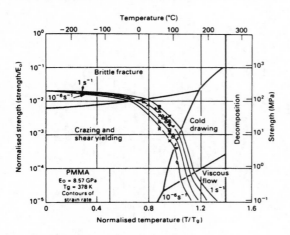

Figure 11.5 *Deformation map for PMMA showing deformation regions as a function of normalized stress versus normalized temperature (from Ashby and Jones, 1986; permission of Elsevier Science Ltd, UK).*

chemistry decides the character of individual molecules but it is the processing stage which enables them to be arranged to maximum advantage. Despite the variety of methods available for converting feedstock powders and granules of thermoplastics into useful shapes, these methods usually share up to four common stages of production; that is, (1) mixing, melting and homogenization, (2) transport and shaping of a melt, (3) drawing or blowing, and (4) finishing.

Processing brings about physical, and often chemical, changes. In comparison with energy requirements for processing other materials, those for polymers are relatively low. Temperature control is vital because it decides melt fluidity. There is also a risk of thermal degradation because, in addition to having limited thermal stability, polymers have a low thermal conductivity and readily overheat. Processing is usually rapid, involving high rates of shear. The main methods that will be used to illustrate technological aspects of processing thermoplastics are depicted in Figure 11.6.

Injection-moulding of thermoplastics, such as PE and PS, is broadly similar in principle to the pressure die-casting of light metals, being capable of producing mouldings of engineering components rapidly with repeatable precision (Figure 11.6a). In each cycle, a metered amount (shot) of polymer melt is forcibly injected by the reciprocating screw into a 'cold' cavity (cooled by oil or water channels). When solidification is complete, the two-part mould opens and the moulded shape is ejected. Cooling rates are faster than with parison moulds in blow-moulding because heat is removed from two surfaces. The capital outlay for injection-moulding tends to be high because of the high pressures involved and machining costs for multi-impression moulding dies. In die design, special attention is given to the location of weld lines, where different flows coalesce, and of feeding gates. Computer modelling can be used to simulate the melt flow and distributions of temperature and pressure within the mould cavity. This prior simulation helps to lessen dependence upon traditional moulding trials, which are costly. Microprocessors are used to monitor and control pressure and feed rates continuously during the moulding process; for example, the flow rates into a complex cavity can be rapid initially and then reduced to ensure that flow-dividing obstructions do not produce weakening weld lines.

Extrusion is widely used to shape thermoplastics into continuous lengths of sheet, tube, bar, filament, etc. with a constant and exact cross-sectional profile (Figure 11.6b). A long Archimedean screw (auger) rotates and conveys feedstock through carefully proportioned feed, compression and metering sections. The polymer is electrically heated in each of the three barrel sections and frictionally heated as it is 'shear-thinned' by the screw. Finally, it is forced through a die orifice. Microprocessor control systems are available to measure pressure at the die inlet and to keep it constant by 'trimming' the rotational speed of the screw. Dimensional control of the product benefits

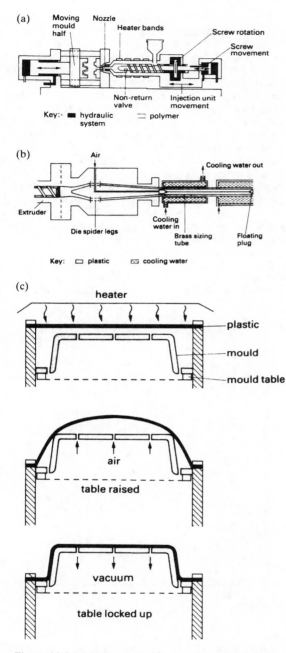

Figure 11.6 *(a) Injection-moulding machine, (b) production of plastic pipe by extrusion and (c) thermoforming of plastic sheet (from Mills, 1986; by permission of Edward Arnold).*

from this device. On leaving the die, the continuously-formed extrudate enters cooling and haul-off sections. Frequently, the extrudate provides the preform for a second operation. For example, in a continuous melt-inflation technique, tubular sheet of LDPE or HDPE

from an annular die is drawn upwards and inflated with air to form thin film: stretching and thinning cease when crystallization is complete at about 120°C. Similarly, in the blow-moulding of bottles and air-ducting, etc., tubular extrudate (parison) moves vertically downwards into an open split-mould. As the mould closes, the parison is inflated with air at a pressure of about 5 atmospheres and assumes the shape of the cooled mould surfaces. Relatively inexpensive aluminium moulds can be used because stresses are low.

Thermoforming (Figure 11.6c) is another secondary method for processing extruded thermoplastic sheet, being particularly suitable for large thin-walled hollow shapes such as baths, boat hulls and car bodies (e.g. ABS, PS, PVC, PMMA). In the basic version of the thermoforming, a frame-held sheet is located above the mould, heated by infrared radiation until rubbery and then drawn by vacuum into close contact with the mould surface. The hot sheet is deformed and thinned by biaxial stresses. In a high-pressure version of thermoforming, air at a pressure of several atmospheres acts on the opposite side of the sheet to the vacuum and improves the ability of the sheet to register fine mould detail. The draw ratio, which is the ratio of mould depth to mould width, is a useful control parameter. For a given polymer, it is possible to construct a plot of draw ratio versus temperature which can be used as a 'map' to show various regions where there are risks of incomplete corner filling, bursts and pin-holes. Unfortunately, thinning is most pronounced at vulnerable corners. Thermoforming offers an economical alternative to moulding but cycle times are rather long and the final shape needs trimming.

11.2.3 Production of thermosets

Development of methods for shaping thermosetting materials is restricted by the need to accommodate a curing reaction and the absence of a stable viscoelastic state. Until fairly recently, these restrictions tended to limit the size of thermoset products. Compression moulding of a thermosetting P-F resin (*Bakelite*) within a simple cylindrical steel mould is a well-known laboratory method for mounting metallurgical samples. Resin granules, sometimes mixed with hardening or electrically-conducting additives, are loaded into the mould, then heated and compressed until crosslinking reactions are complete. In transfer moulding, which can produce more intricate shapes, resin is melted in a primary chamber and then transferred to a vented moulding chamber for final curing. In the car industry, body panels with good bending stiffness are produced from thermosetting sheet-moulding compounds (SMC). A composite sheet is prepared by laying down layers of randomly-oriented, chopped glass fibres, calcium carbonate powder and polyester resin. The sheet is placed in a moulding press and subjected to heat and pressure. Energy requirements are attractively low.

Greater exploitation of thermosets for large car parts has been made possible by reaction injection-moulding

(RIM). In this process, polymerization takes place during forming. Two or more streams of very fluid chemical reactants are pumped at high velocity into a mixing chamber. The mixture bottom-feeds a closed chamber where polymerization is completed and a solid forms. Mouldings intended for high-temperature service are stabilized, or post-cured, by heating at a temperature of 100°C for about 30 min. The reactive system in RIM can be polyurethyane-, nylon- or polyurea-forming. The basic chemical criterion is that polymerization in the mould should be virtually complete after about 30 s. Foaming agents can be used to form components with a dense skin and a cellular core. When glass fibres are added to one of the reactants, the process is called reinforced reaction injection-moulding (RRIM). RIM now competes with the injection-moulding of thermoplastics. Capital costs, energy requirements and moulding pressures are lower and, unlike injection-mouldings, thick sections are not subject to shrinkage problems ('sinks' and voids). Cycle times for RIM-thermosets are becoming comparable with those for injection-moulded thermoplastics and mouldings of SMC. Stringent control is necessary during the RIM process. Temperature, composition and viscosity are rapidly changing in the fluid stream and there is a challenging need to develop appropriate dynamic models of mass transport and reaction kinetics.

11.2.4 Viscous aspects of melt behaviour

Melts of thermoplastic polymers behave in a highly viscous manner when subjected to stress during processing. Flow through die orifices and mould channels is streamline (laminar), rather than turbulent, with shear conditions usually predominating. Let us now adopt a fluid mechanics approach and consider the effects of shear stress, temperature and hydrostatic pressure on melt behaviour. Typical rates of strain (shear rates) range from $10–10^3$ s^{-1} (extrusion) to $10^3–10^5$ s^{-1} (injection-moulding). When a melt is being forced through a die, the shear rate at the die wall is calculable as a function of the volumetric flow rate and the geometry of the orifice. At the necessarily high levels of stress required, the classic Newtonian relation between shear stress and shear (strain) rate is not obeyed: an increase in shear stress produces a disproportionately large increase in shear rate. In other words, the shear stress/shear rate ratio, which is now referred to as the 'apparent shear viscosity', falls. Terms such as 'pseudo-plastic' and 'shear-thinning' are applied to this non-Newtonian flow behaviour.[1] The general working range of apparent shear viscosity for extrusion, injection-moulding, etc. is $10–10^4$ Ns m^{-2}. (Shear viscosities at low and high stress levels are measured by cone-and-plate and capillary extrusion techniques, respectively.)

[1]In thixotropic behaviour, viscosity decreases with increase in the *duration* of shear (rather than the shear rate).

Figure 11.7 shows the typical fall in apparent shear viscosity which occurs as the shear stress is increased. If Newtonian flow prevailed, the plotted line would be horizontal. This type of diagram is plotted for fixed values of temperature and hydrostatic pressure. A change in either of these two conditions will displace the flow curve significantly. Thus, either raising the temperature or decreasing the hydrostatic (bulk) pressure will lower the apparent shear viscosity. The latter increases with average molecular mass. For instance, fluidity at a low stress, as determined by the standard melt flow index (MFI) test,[1] is inversely proportional to molecular mass. At low stress and for a given molecular mass, a polymer with a broad distribution of molecular mass tends to become more pseudo-plastic than one with a narrow distribution. However, at high stress, a reverse tendency is possible and the version with a broader distribution may be less pseudo-plastic.

Figure 11.8 provides a comparison of the flow behaviour of five different thermoplastics and is useful for comparing the suitability of different processes. It indicates that acrylics are relatively difficult to extrude

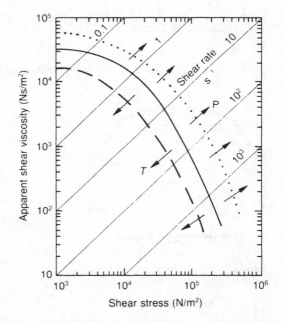

Figure 11.7 *Typical plot of apparent shear viscosity versus shear stress for LDPE at 210°C and atmospheric pressure: effects of increasing temperature T and hydrostatic pressure P shown (after Powell, 1974; courtesy of Plastics Division, Imperial Chemical Industries Plc).*

[1]This important test, which originated in ICI laboratories during the development of PE, is used for most thermoplastics by polymer manufacturers and processors. The MFI is the mass of melt extruded through a standard cylindrical die in a prescribed period under conditions of constant temperature and compression load.

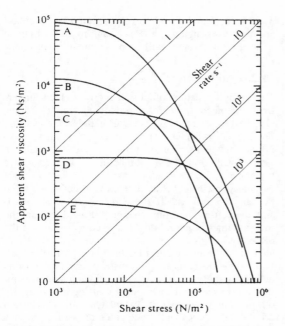

Figure 11.8 *Typical curves of apparent shear viscosity versus shear stress for five thermoplastics at atmospheric pressure. A Extrusion-grade LDPE at 170°C; B extrusion-grade PP at 230°C; C moulding-grade acrylic at 230°C; D moulding-grade acetal copolymer at 200°C; E moulding-grade nylon at 285°C (after Powell, 1974; courtesy of Plastics Division, Imperial Chemical Industries Plc.).*

and that PP is suited to the much faster deformation process of injection-moulding. In all cases, Newtonian flow is evident at relatively low levels of shear stress.

The following type of power law equation has been found to provide a reasonable fit with practical data and has enabled pseudo-plastic behaviour to be quantified in a convenient form:

$$\tau = C\gamma^n \tag{11.2}$$

where C and n are constants. Now $\tau = \eta\gamma$, hence the viscosity $\eta = C\gamma^{n-1}$. The characteristic term $(n-1)$ can be derived from the line gradient of a graphical plot of log viscosity versus log shear rate. In practice, the power law index n ranges from unity (Newtonian flow) to <0.2, depending upon the polymer. This index decreases in magnitude as the shear rate increases and the thermoplastic melt behaves in an increasingly pseudo-plastic manner.

So far, attention has been concentrated on the viscous aspects of melt behaviour during extrusion and injection-moulding, with emphasis on shear processes. In forming operations such as blow-moulding and filament-drawing, extensional flow predominates and tensile stresses become crucial; for these conditions, it is appropriate to define tensile viscosity, the counterpart of shear viscosity, as the ratio of tensile stress

to tensile strain rate. At low stresses, tensile viscosity is independent of tensile stress. As the level of tensile stress rises, tensile viscosity either remains constant (*nylon* 6,6), rises (LDPE) or falls (PP, HDPE). This characteristic is relevant to the stability of dimensions and form. For example, during blow-moulding, thinning walls should have a tolerance for local weak spots or stress concentrations. PP and HDPE lack this tolerance and there is a risk that 'tension-thinning' will lead to rupture. On the other hand, the tensile viscosity of LDPE rises with tensile stress and failure during wall-thinning is less likely.

11.2.5 Elastic aspects of melt behaviour

While being deformed and forced through an extrusion die, the melt stores elastic strain energy. As extrudate emerges from the die, stresses are released, some elastic recovery takes place and the extrudate swells. Dimensionally, the degree of swell is typically expressed by the ratio of extrudate diameter to die diameter; the elastic implications of the shear process are expressed by the following modulus:

$$\mu = \tau/\gamma_R \tag{11.3}$$

where μ is the elastic shear modulus, τ is the shear stress at die wall, and γ_R is the recoverable shear strain. The magnitude of modulus μ depends upon the polymer, molecular mass distribution and the level of shear stress. (Unlike viscosity, dependency of elasticity upon temperature, hydrostatic pressure and average molecular mass is slight.) If the molecular mass distribution is wide, the elastic shear modulus is low and elastic recovery is appreciable but slow. For a narrow distribution, with its greater similarities in molecular lengths, recovery is less but faster. With regard to stress level, the modulus remains constant at low shear stresses but usually increases at the high stresses used commercially, giving appreciable recovery.

The balance between elastic to viscous behaviour during deformation can be gauged by comparing the deformation time with the relaxation time or 'natural time' (λ) of the polymer. λ is the ratio of apparent viscosity to elastic shear modulus (η/μ), and derives from the Maxwell model of deformation. The term viscoelasticity originated from the development of such models (e.g. Maxwell, Voigt, standard linear solid (SLS)). The Maxwell model is a mechanical analogue that provides a useful, albeit imperfect, simulation of viscoelasticity and stress relaxation in linear polymers above T_g (Figure 11.9). It is based upon conditions of constant strain. A viscously damped 'Newtonian' dashpot, representing the viscous component of deformation, and a spring, representing the elastic component, are combined in series. At time t, the stress σ is exponentially related to the initial stress σ_0, as follows:

$$\sigma = \sigma_0 \exp(-t/\lambda) \tag{11.4}$$

where λ is the relaxation time. When $t \gg \lambda$, there is sufficient time for viscous movement of chain

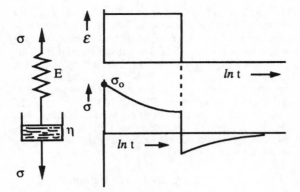

Figure 11.9 *Representation of stress relaxation under constant strain conditions (Maxwell model).*

molecules to take place and stress will fall rapidly. When $t \ll \lambda$, elastic behaviour predominates. The magnitude of λ ranges from infinity, at the start of rubbery behaviour, to zero at the start of viscous behaviour. A real polymer contains different lengths of molecules and therefore features a spectrum of relaxation times. Nevertheless, although best suited to polymers of low molecular mass, the Maxwell model offers a reasonable first approximation for melts.

Let us now apply the relaxation time concept to an injection-moulding process in which a thermoplastic acrylic at a temperature of 230°C is sheared rapidly at a rate of 10^5 s^{-1}. Assume that the injection (deformation) time is 2 s. For the shear rate given, Figure 11.8 indicates that the apparent shear viscosity is 9 Ns m^{-2} and the corresponding maximum shear stress is 0.9 MN m^{-2}. At this shear stress, the elastic shear modulus for acrylic is 0.21 MN m^{-2}. The value of $\lambda (= \eta/\mu)$ is 43 μs, which is very small compared to the injection time of 2 s, hence viscous behaviour will predominate. A similar procedure can be applied to deformation by extrusion. For instance, PP with a relaxation time of 0.5 s might pass though the extrusion die in 20 s. The time difference is smaller than the previous example of injection-moulding, indicating that although deformation is mainly viscous, elasticity will play a greater part than in injection-moulding. The previously-mentioned phenomenon of die swell then becomes understandable. (Swelling is equivalent to the spring action in the Maxwell model.) Although the degree of elastic behaviour may be relatively small during injection-moulding and extrusion, it can, nevertheless, sometimes cause serious flow defects.

Relaxation times for extensional flow, as employed in blow-moulding, can be derived from the ratio of apparent tensile viscosity to elastic tensile modulus ($E = \sigma/\varepsilon_R$). Suppose that a PP parison at a temperature of 230°C hangs for 5 s before inflation with air. If the tensile viscosity and tensile modulus are

36 kN s m^{-2} and 4.6 kN m^{-2}, respectively, the relaxation time is roughly 8 s. Hence sagging of the parison under its own weight will be predominantly elastic.

11.2.6 Flow defects

The complex nature of possible flow defects underlines the need for careful product design (sections, shapes, tools) and close control of raw materials and operational variables (temperatures, shear rates, cooling arrangements). The quality of processing makes a vital contribution to the engineering performance of a polymer.

Ideally, melt flow should be streamlined throughout the shaping process. If the entry angle of an extrusion die causes an abrupt change in flow direction, the melt assumes a natural angle as it converges upon the die entry and a relatively stagnant 'dead zone' is created at the back of the die. In this region, the melt will have a different thermal history. In addition to its dominant shear component, the convergent flow contains an extensional component that increases rapidly during convergence. If the extensional stress reaches a critical value, localized 'melt fracture' will occur at a frequency depending upon conditions. The fragments produced recover some of the extensional strain. The effect upon the emerging extrudate can range from a matt finish to gross helical distortions. The associated flow condition is often termed 'non-laminar' despite the fact that the calculated value of the dimensionless Reynolds number is very low. The choice of entry angle for the die is crucial and depends partly upon the polymer.

As a melt passes through the die, velocity gradients develop, with the melt near the die surface moving slower than the central melt. Upon leaving the die, the outer layers of extrudate accelerate, eliminating the velocity gradient. Above a critical velocity, the resultant stresses rupture the surface to give a 'sharkskin' effect which can range in severity from a matt finish to regular ridging perpendicular to the extrusion direction. 'Sharkskin' is most likely when the polymer has a high average molecular mass (i.e. highly viscous) and a narrow molecular mass distribution (i.e. highly elastic); these factors cause surface stress to build up rapidly and to relax slowly. Fast extrusion at a low temperature favours this defect. Heating of the tip of the die lowers viscosity and reduces its likelihood.

An inhomogeneous melt will produce a non-uniform recovery of elastic strain at the cooling surface and influence its final texture. Thorough mixing before shaping is essential. However, inhomogeneity may exist on a molecular scale. For instance, in both injection-moulding and extrusion, a broad distribution of molecular mass gives a more matt finish than a narrow distribution. Thus, extrusion of a polymer with a narrow mass distribution at a rate slow enough to prevent the development of 'sharkskin' will favour a high-gloss finish.

Volatile constituents tend to vaporize, or 'boil off', from the melt if processing temperatures are high. For instance, water vapour may derive from hygroscopic raw materials. Sometimes, a polymer degrades and releases a volatile monomer. Hydrostatic pressure usually keeps the volatiles in solution; as the hot polymer leaves the die, this pressure is released and the escaping volatiles form internal bubbles and may pit the surface.

When a polymer is heated to the processing temperature, the weak intermolecular forces are readily overcome by thermal vibrations. Its density may decrease by as much as 25%. The subsequent cooling can produce shrinkage defects, particularly in crystalline polymers, which assume more closely packed conformations than amorphous polymers, and in thick sections. Polymers have a low thermal conductivity and the hot core can contract to produce depressions or 'sink marks' in the surface. If the surface layer cools rapidly, its rigidity can encourage internal voids to form. Careful product design can minimize shrinkage problems. Cooling under pressure after injection-moulding is beneficial.

Orientation effects, which are common in shaped polymers, have a special significance when the polymer is 'reinforced' with short lengths of glass fibre. (It has been estimated that roughly half of the engineering thermoplastics are fibre-filled.) In extruded pipes, fibres will tend to be aligned parallel to the extrusion direction and improve longitudinal strength. However, this orientation weakens the pipe transversely and the burst strength will suffer. In addition, the tubular form necessitates use of a 'spider' to support a central die, or mandrel, which causes the melt to split and coalesce before entering the die. The resultant weld lines introduce weakening interfaces. A die system is now available in which one or both dies are rotated. Their shearing action has the beneficial effect of inclining fibres at an angle to the extrusion direction. Weld lines are also reoriented so that they are placed in shear rather than tension when in service.

11.3 Fibre-reinforced composite materials

11.3.1 Introduction to basic structural principles

11.3.1.1 The functions of filaments and matrix

In engineering practice, where considerations of mechanical strength and stiffness are usually paramount, the term 'composite' is generally understood to refer to a material which combines a matrix phase with admixed filaments of a reinforcing (strengthening) phase. A composite derives from the essentially simple and practical idea of bonding together two or more homogeneous materials which have very different properties. Thus, in the glass-reinforced polymer (GRP) known generally as *Fibreglass*, large numbers of short, strong and stiff fibres of glass are randomly dispersed in a weaker but tougher matrix of thermoset resin. In general, the diameter of reinforcing filaments for composites is in the order of 10 μm. They may be continuous, extending the full length of a component, or short (discontinuous) and may share a common orientation, be randomly oriented or even woven into cloth. There is a statistical distribution of strength values among filaments made from brittle materials such as glass, boron and carbon.

The term 'composite' is sometimes taken to include materials in which the second phase is in the form of particles or laminae. In such cases the composite structure may offer special advantages, other than strength, such as economy and corrosion-resistance (e.g. filler in plastics, plastic-coated steel sheet). The recorded history of technology contains numerous accounts of innovative, often remarkable, ideas for composites.[1] The present account primarily concerns the reinforcement of polymeric, metallic and ceramic matrices with fibres and 'whisker' crystals. Finally, on an even finer scale, we will consider application of the composite principle to microstructural constituents (i.e. nanocomposites). Relatively simple mechanical models will be used because they provide sufficient introduction to the ground rules of composite design and behaviour.

Although fibres are a striking feature of a composite, it is initially helpful to examine the functions of the matrix. Ideally, it should be able to (1) infiltrate the fibres and solidify rapidly at a reasonable temperature and pressure, (2) form a coherent bond, usually chemical in nature, at all matrix/fibre interfaces, (3) envelop the fibres, which are usually very notch-sensitive, protecting them from mutual damage by abrasion and from the environment (chemical attack, moisture), (4) transfer working stresses to the fibres, (5) separate fibres so that failure of individual fibres remains localized and does not jeopardize the integrity of the whole component, (6) debond from individual fibres, with absorption of significant amounts of strain energy, whenever a propagating crack in the matrix chances to impinge upon them, and (7) remain physically and chemically stable after manufacture.

11.3.1.2 Continuous-fibre composites

In mechanical terms, the prime function of the matrix is to transfer stresses to the fibres (item (4) above) because these are stronger and of higher elastic modulus than the matrix. The response of a composite to applied stress depends upon the respective properties of the fibre and matrix phases, their relative volume fractions, fibre length and the orientation of fibres relative to the direction of applied stress.

---

[1] During the Battle of the Atlantic in World War II, Geoffrey Pyke proposed the construction of ocean-going aircraft carriers from paper pulp/frozen sea water (*Pykecrete*). In 1985, ice/wood fibres (*Icecrete*) was proposed for wharf and off-shore oil platform construction in Norwegian waters.

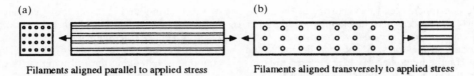

(a)   (b)

Filaments aligned parallel to applied stress   Filaments aligned transversely to applied stress

Figure 11.10 *'Parallel' (a) and 'series' (b) models of unidirectional filament alignment in composites.*

Some basic principles of the elastic response to stress can be obtained from mechanical models in which continuous fibres are set unidirectionally in an isotropic, void-free matrix (Figures 11.10a and 11.10b). It will be assumed that the Poisson ratio for the fibre material is similar to that of the matrix. Using subscript letters c, f, m, l and t we can signify where property values refer respectively to composite, fibre, matrix, longitudinal and transverse directions. Thus V_f/V_m is the ratio of volume fractions of fibre and matrix, where $(1 - V_f) = V_m$. Certain longitudinal properties for a composite can be obtained by using the 'parallel' model shown in Figure 11.10a and applying the Rule of Mixtures. For this condition of isostrain, stresses are additive and the equations for stress (strength) and elastic modulus are:

$$\sigma_{cl} = \sigma_{fl} V_f + \sigma_m V_m \tag{11.5}$$

$$E_{cl} = E_{fl} V_f + E_m V_m \tag{11.6}$$

It is now possible to derive the following relation:

$$(\sigma_{fl}/\sigma_m) = (V_f/V_m)(E_{fl}/E_m) \tag{11.7}$$

Figure 11.11 illustrates this relation, showing that as the modulus ratio and/or the volume fraction of fibres increase, more and more stress is transferred to the fibres. If the modulus ratio is unity, the composite must contain at least 50% v/v fibres if the fibres are to carry the same load as the matrix. Three typical composites A, B and C with 50% v/v reinforcement are superimposed on the graph to show the extent to which two increases in modulus ratio raise the stress ratio.

An alternative arrangement of fibres relative to applied tensile stress is shown in Figure 11.10b. The transverse elastic modulus for the composite is given by the equation:

$$(1/E_{ct}) = (V_f/E_{ft}) + (V_m/E_m) \tag{11.8}$$

This 'series' version of the Rule assumes a condition of isostress and is derived by adding strains: it is less accurate than the 'parallel' version. Both versions can be used to calculate shear moduli and conductivities (thermal, electrical). More refined mathematical treatments are available: they are particularly helpful for transverse properties. Sometimes fibre properties are highly anisotropic and this feature influences the corresponding value for the composite; for instance, $E_{ft} \ll E_{fl}$ for aramid (*Kevlar, Nomex*) and carbon fibres. In general, $\sigma_{fl} \gg \sigma_m$, $E_{fl} \gg E_m$ and $E_{ft} > E_m$,

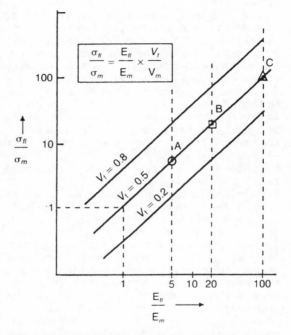

Figure 11.11 *Relation between modulus ratio and stress ratio (continuous parallel fibres).*

so that the equations for the 'parallel' and 'series' models express, in mathematical form, the dominant effect of fibres on longitudinal properties and the dominant effect of the matrix on transverse properties.

If typical tensile stress versus strain curves for fibre and matrix materials (Figure 11.12a) are compared, it can be seen that the critical strain, beyond which the composite loses its effectiveness, is determined by the strain at fracture of the fibres, ε_f. At this strain value, when the matrix has usually begun to deform plastically and to strain-harden, the corresponding stress on the matrix is σ'_m. Thus, in the related Figure 11.12b, it follows that the strength of the composite lies between the limits σ'_m and σ_f, depending upon the volume fraction of fibres. When a few widely-spaced fibres are present, the matrix carries more load than the fibres. Furthermore, in accordance with the Rule of Mixtures, the strength of the composite falls as the volume fraction of fibres deceases. Construction lines representing these two effects meet at a minimum point, V_{min}. Obviously, V_f must exceed V_{crit} if the tensile strength of the matrix is to benefit from the

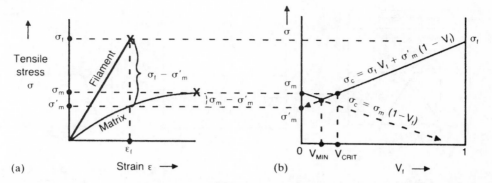

Figure 11.12 *(a) Stress–strain curves for filament and matrix and (b) dependence of composite strength on volume fraction of continuous filaments.*

presence of fibres. In practical terms, the upper limit for V_f is about 0.7–0.8. At higher values, fibres are likely to damage each other. The Rule is only applicable when $V_f > V_{min}$.

At the critical volume for fibres, $\sigma_m = \sigma_c$ and $V_f = V_{crit}$. From the Rule equation we derive:

$$V_{crit} = (\sigma_m - \sigma'_m)/(\sigma_f - \sigma'_m) \qquad (11.9)$$

In general, a low V_{crit} is sought in order to minimize problems of dispersal and to economize on the amount of reinforcement. Very strong fibres will maximize the denominator and are clearly helpful. Strain-hardening of the matrix (Figure 11.12a) is represented approximately by the numerator of the above ratio. Thus, a matrix with a strong tendency to strain-harden will require a relatively large volume fraction of fibres, a feature that is likely to be very significant for metallic matrices. For example, an fcc matrix of austenitic stainless steel (Fe-18Cr-8Ni) will tend to raise V_{crit} more than a cph matrix of zinc.

11.3.1.3 Short-fibre composites

So far, attention has been focused on the behaviour of continuous fibres under stress. Fabrication of these composites by processes such as filament-winding is exacting and costly. On the other hand, composites made from short (discontinuous) fibres enable designers to use cheaper, faster and more versatile methods (e.g. injection-moulding, transfer-moulding). Furthermore, some reinforcements are only available as short fibres. At this point, it is appropriate to consider an isolated short fibre under axial tensile stress and to introduce the idea of an aspect ratio (length/diameter $= l/d$). It is usually in the order of 10 to 10^3 for short fibres: many types of fibre and whisker crystal with aspect ratios greater than 500 are available. For a given diameter of fibre, an increase in length will increase the extent of bonding at the fibre/matrix interface and favour the desired transfer of working stresses. As will be shown, it is necessary for the length of short fibres to exceed a certain critical

length[1] if efficient transfer of stress is to take place. With respect to diameter, fibre strength increases as the diameter of a brittle fibre is reduced. This effect occurs because a smaller surface area makes it less likely that weakening flaws will be present.

For the model, in which a matrix containing a short fibre is subjected to a tensile stress (Figure 11.13a), it is assumed that the strain to failure of the matrix is greater than the strain to failure of the fibre. The differences in displacement between matrix and fibre thus cause shear stress, τ, to develop at the cylindrical interfaces toward each fibre end. A corresponding tensile stress, σ, builds up within the fibre. At each end of the fibre, these stresses change over a distance known as the stress transfer length $(l/2)$; that is, the tensile stress increases as the interfacial shear stress decreases. In Figure 11.13a, for simplicity, we assume that the gradient of tensile stress is linear. If the length of the fibre is increased, the peak tensile stress coincides with fracture stress for the filament (Figure 11.13b). The total length of fibre now has a critical value of l_c and the transfer length becomes $l_c/2$. If the length is sub-critical, fibre failure cannot occur. At the critical condition, the average tensile stress on the fibre is only $\sigma_f/2$. With any further increase in fibre length, a plateau develops in the stress profile (Figure 11.13c). The average tensile stress on the fibre, which is stated beneath Figure 11.13c, approaches the fracture strength σ_f as the fibre length increases beyond its critical value. In effect, the load-carrying efficiency of the fibre is approaching that of its direct-loaded continuous counterpart. Provided that the shear stresses do not cause 'pull-out' of the fibre, fracture will eventually occur in the mid-region of the fibre.

The condition of critical fibre length can be quantified. Suppose that an increment of tensile force, $\delta\sigma$, is applied to an element of fibre, δl. The balance

[1] A. Kelly introduced the concept of 'critical fibre length'.

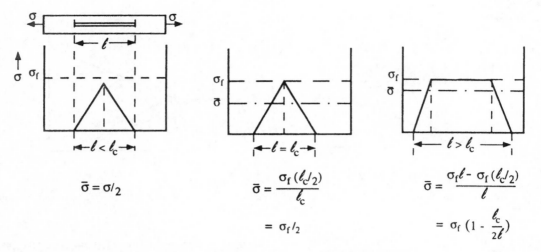

Figure 11.13 *Distribution of tensile stress in a short fibre.* $\sigma_f = $ *fracture stress of fibre in tension,* $\overline{\sigma} = $ *mean tensile stress.*

between tensile force and interfacial shear force is:

$$\delta\sigma(\pi d^2/4) = \tau(\pi d.\delta l) \tag{11.10}$$

Hence the gradient $\delta\sigma/\delta l$ for the build-up of tensile stress is $4\tau/d$. For the critical condition (Figure 11.13b), the gradient is σ_f divided by the critical transfer length $l_c/2$. The critical length l_c is therefore $\sigma_f d/2\tau$. Expressed in terms of the critical aspect ratio, the criterion for efficient stress transfer takes the form:

$$l_c/d \geq \sigma_f/2\tau \tag{11.11}$$

This relation provides an insight into the capabilities of short-fibre composites. For instance, for a given diameter of fibre, if the interfacial shear strength of the fibres is lowered, then longer fibres are needed in order to grip the matrix and receive stress. When the operating temperature is raised and the shear strength decreases faster than the fracture strength of the fibres, the critical aspect ratio increases. The presence of tensile and shear terms in the relation highlights the indirect nature of short-fibre loading: matrix strength and interfacial shear strength are much more crucial factors than in continuous-fibre composites where loading is direct. Interfacial shear strength depends upon the quality of bonding and can have an important effect upon the overall impact resistance of the composite. Ideally, the bond strength should be such that it can absorb energy by debonding, thus helping to inhibit crack propagation. Interfacial adhesion is particularly good in glass fibre/polyester resin and carbon fibre/epoxy resin systems. Coupling agents are used to promote chemical bonding at the interfaces. For example, glass fibres are coated with silane (size) which reacts with the enveloping resin. Sometimes these treatments also improve resistance to aqueous environments. However, if interfacial bonding is extremely strong, there is an attendant risk that an impinging crack will pass into and through fibres with little hindrance.

Provided that $V_f > V_{crit}$, the tensile strength at fracture of a short-fibre reinforced composite can be calculated from the previous Rule of Mixtures equation by substituting the mean tensile stress term for the fracture stress, as follows:

$$\sigma_c = \sigma(1 - l_c/2l)V_f + \sigma'_m V_m \tag{11.12}$$

It is assumed that the shear strength τ remains constant and that fibres are perfectly aligned. First, the equation shows that short fibres strengthen less than continuous ones. For example, if all fibres are ten times the critical length, they carry 95% of the stress carried by continuous fibres. On the other hand, if their length falls below the critical value, the strength suffers seriously. For instance, if the working load on the composite should cause some of the weaker fibres to fail, so that the load is then carried by a larger number of fibres, the effective aspect ratio and the strength of the composite tend to fall. The same type of equation can also be applied to the elastic modulus of a short-fibre composite; again the value will be less than that obtainable with its continuous counterpart. The equation also shows that matrix properties become more prominent as fibres are shortened.

11.3.1.4 Effect of fibre orientation on strength

Let us reconsider composites in which continuous fibres are aligned in the same direction as the direction of applied stress. The strength of this highly anisotropic type of composite varies with the volume fraction of fibres in a linear manner, as shown in Figure 11.12b. If the fibres are now oriented at an angle ϕ to the direction of applied tensile stress, the general effect is to reduce the gradient of the strength curve for values of V_f greater than V_{min}, as shown in Figure 11.14. This weakening effect is represented by

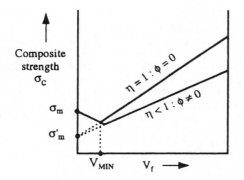

Figure 11.14 *Effect of fibre orientation on strength of unidirectional composite (continuous fibres..*

inserting an orientation factor η in the basic strength equation to give:

$$\sigma_c = \eta\sigma_f V_f + \sigma'_m V_m \tag{11.13}$$

As ϕ increases from zero, η falls below unity.

In order to provide a more detailed analysis of the variation of composite strength with fibre orientation, it is customary to apply a 'maximum stress' theory based on the premise that there are three possible modes of composite failure. Apart from the angle of fibre orientation ϕ, three properties of the composite are invoked: the strength parallel to the fibres σ_{fl}, the shear strength of the matrix parallel to the fibres τ_m, and the strength normal to the fibres σ_{ft}. Each mode of failure is represented by an equation which relates the composite strength σ_{cl} to a resolved stress.

For the first mode of failure, which is controlled by tensile fracture of the fibres, the equation is

$$\sigma_{cl} = \sigma_{fl} \sec^2 \phi \tag{11.14}$$

The equation for failure controlled by shear on a plane parallel to the fibres is:

$$\sigma_{cl} = 2\tau_m \operatorname{cosec} 2\phi \tag{11.15}$$

As the temperature is raised, this mode of failure becomes more likely in off-axis composites because the shear strength τ_m falls more rapidly than σ_{fl}.

In the third mode of failure, transverse rupture occurs, either in the matrix or at the fibre/matrix interface (debonding). The relevant equation is:

$$\sigma_{cl} = \sigma_{ft} \operatorname{cosec}^2 \phi \tag{11.16}$$

Figure 11.15 shows the characteristic form of the relation between composite strength and fibre orientation. While illustrating the highly anisotropic character of the unidirectional continuous reinforcement, it also shows the merit of achieving low values of ϕ. Predictions, using the maximum stress theory, and experimental results are in good agreement and confirm the general validity of this curve. (Measured values of σ_{fl}, τ_m and σ_{ft} are required for these calculations.) The

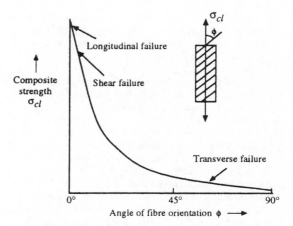

Figure 11.15 *Relation between failure mode, strength and fibre orientation (schematic diagram for unidirectional continuous-fibre composite).*

mode of failure is decided by the equation which gives the lowest value of composite strength σ_{cl}. Thus, transverse rupture becomes dominant when ϕ is large. At relatively low values of ϕ, there is a rapid fall in composite strength which is associated with the transition from tensile failure in fibres to failure by shear. Combination of the first two of the three equations eliminates σ_{cl} and gives the critical angle for this transition:

$$\phi_{crit} = \tan^{-1}(\tau_m/\sigma_{fl}) \tag{11.17}$$

If the longitudinal strength is about ten times the shear strength of the matrix, then the angle is about 6°.

When an application involves applied stresses that are not confined to one direction, the problem of anisotropy can often be effectively solved or minimized by using continuous fibres in the form of woven cloth or laminates. Although these forms are more isotropic than unidirectional composites, there is inevitably a slight but usually tolerable loss in strength and stiffness. Glass, carbon and aramid fibres are used; sometimes two or more different types of fibre are used in combination (hybrid composites). Fibre cloth in a variety of weave patterns is available. In a nominally two-dimensional sheet of woven cloth there is a certain amount of fibre oriented in the third dimension. A more truly three-dimensional reinforcement, with improved through-thickness properties, can be obtained by placing woven cloths on top of each other and stitching them together with continuous fibre.

Laminates based on carbon and aramid fibres are commonly used for high-performance applications that involve complex stress systems (e.g. twisting, bending). The unit of construction is a thin ply of unidirectional composite, 50–130 µm thick. Plies are carefully stacked and oriented with respect to orthogonal reference axes (0° and 90°). The simplest lay-up sequence is (0/90/90/0). Other more isotropic sequences are $(0/+45/-45/-45/+45/0)$ and

$(0/+60/-60/-60/+60/0)$. The stacking of plies is symmetrical about the mid-plane of the laminate in order to prevent distortion and to ensure a uniform response to working stresses.

Randomly-oriented short fibres of glass are commonly used in sheets and in three-dimensional mouldings. In fact, fibre misorientation is quite common in composites, frequently being an unavoidable result of fabrication. For instance, when resins loaded with short fibres are injection-moulded, the mixture follows complex flow paths. If the finished moulding is sectioned, fibres will clearly indicate the patterns of flow: these patterns are determined by melt viscosity, mould contours and processing conditions. Fortunately, flow patterns are repeatable from moulding to moulding. Close to the surface of the moulding, the short fibres tend to follow streamline paths; in the central core, where flow is more turbulent, fibres are likely to assume transverse orientations.

11.3.2 Types of fibre-reinforced composite

11.3.2.1 Polymer-matrix composites

Glass-reinforced polymers (GRP) date from the early 1940s[1] and were the forerunners of present-day polymer-, metal- and ceramic-matrix composites. A typical fabrication procedure for a GRP is to add a mixture of polyester resin, curing agent and catalyst to fibres of low-alkali E-glass (typically $53SiO_2 - 18CaO - 14Al_2O_3 - 10B_2O_3 - 5MgO$). The thermosetting reactions of curing are then allowed to take place at a temperature below 150°C. The relatively low cost, stiffness and ease of fabrication of GRP led to their widespread adoption in engineering, even for large structures (e.g. storage tanks and silos, mine counter-measure vessels).

Glass fibres for composites are made by allowing molten glass to pass through the nozzles of an electrically heated bushing made from Pt-10Rh alloy. (The number of nozzle holes in the base of the bushing is 204, or a multiple of 204.) The filaments emerge at a velocity of $50-100$ m s^{-1} and are rapidly cooled with a water mist to prevent crystallization, hauled over a 'size' applicator and finally collected by a rotating cylinder (collet). 'Sizing' applies a coating which loosely bonds the fibres, protects their fragile glass surfaces from damage and introduces a surface-modifying 'coupling' agent to promote eventual fibre/matrix bonding. The primary bundle of continuous, untwisted fibres is the unit of collection from the bushing and is known as a strand. That is, it contains 204, 408, 816 or more fibres. (The equivalent unit for carbon fibres is called a tow.) Strands can be combined to form a larger bundle (roving). Strands or rovings are used for unidirectional composites and as yarn for weaving; alternatively, they can be chopped

into short lengths of 25–50 mm and oriented randomly in a plane. (e.g. chopped strand mat for press moulding). Interstage 'sizing' is often practised. The range of fibre diameters is 5–20 μm. Although originally developed for electrical applications, E-glass is the principal material for the production of continuous glass fibres. Other compositions are available, giving a higher modulus of elasticity (H-modulus glass), greater tensile strength (S-glass), or better alkali resistance (AR-glass), etc.

The principal methods for fabricating polymer-matrix composites are (1) hand lay-up or spray-up, (2) press-moulding with heated matching dies, (3) vacuum moulding, (4) autoclave moulding, (5) resin transfer moulding, (6) reinforced reaction injection moulding (RRIM), (7) pultrusion, and (8) filament winding. Broadly, these listed methods either bring fibres and 'wet' resin together during processing or use pre-impregnated shapes (pre-pregs) in which fibres and thermoset resin are pre-combined. Pre-pregs are made by infiltrating rovings, mats or woven fabrics with resin and then heating in order to initiate partial curing. This B-stage of polymerization is preserved until processing by storing at a low temperature. Pre-pregs facilitate control of infiltration, orientation and fibre content, making it possible to produce polymer-matrix composite (PMC) of repeatable quality on an automated mass-production basis. In each of the above methods there is need to prevent air or vapours becoming trapped in the composite and forming weakening voids. Voids form preferentially at fibre/matrix interfaces and between the plies of layered composites. Low-viscosity resins, outgassing and high pressures are some of the means used to minimise this porosity.

Polyester-based matrices remain the principal choice for polymer-matrix composites. In the 1970s, polyesters became available in the form of moulding compounds that are particularly suitable for hot press-moulding. Dough moulding compounds (DMC) and sheet moulding compounds (SMC) contain roughly equal volume fractions of polyester resin, inert filler particles and chopped glass fibres. When heated, these compounds rapidly become fluid, reproduce the contours and details of the moulding dies accurately and then cure. This technique is used for domestic articles, panels and doors of vehicles, cabinets for office and electronic equipment, etc.

As part of the search for cheaper fabrication methods, much effort has been devoted to the development of PMCs with thermoplastic matrices (e.g. *nylon 66*, PP, PTFE, PET, polyether sulphone (PES), etc). For example, water boxes of car radiators and shell housings for street lamps have been made from a composite of 33% glass fibres in a nylon 66 matrix (*Maranyl*). In the early 1980s, pre-pregs made of polyether ether ketone (PEEK) reinforced with PAN-derived carbon fibres (*APC2*) became available: however, they are costly and use tends to be restricted to highly-specialized applications (e.g. aircraft components). The general advantages of a thermoplastic

---

[1] First used in quantity for aircraft radomes which required strength, low electrical and thermal conductivity and 'transparency' to radar waves.

matrix are its toughness, indefinite shelf life and, in the absence of curing, a shorter time cycle for fabrication. However, during the necessary heating, the viscosity of the matrix is higher than that of a thermoset resin, making infiltration between fibres more difficult. Furthermore, the pre-pregs are stiff and lack the drapability of thermoset pre-pregs which enables them to bend easily into shape.

For the exacting requirements of aerospace and high-performance aircraft, the principal PMCs have epoxide matrices reinforced with continuous fibres of carbon or aramids (*Kevlar*). The basic advantage of epoxides is that they can be used at higher service temperatures than polyester matrices. Although the T_g value of a polymer provides an indication of its temperature ceiling, it is substantially higher than the maximum temperature for safe service under load. For instance, the maximum temperature for a load-bearing epoxide matrix is about 160°C, whereas the corresponding T_g values lie in the range 200–240°C (depending upon the method of determination). The search for matrices with a higher temperature capability has led to the development of bismaleimides (BMI) and polyimides (PI). These and other new polymers raised the ceiling temperature closer to 200°C but have sometimes introduced a brittleness problem and they can be difficult to process.

Carbon fibre reinforced polymers (CFRPs) are firmly established as construction materials for specialized and demanding applications (e.g. helicopter rotors, monocoque chassis of racing cars, aircraft floor panels, spacecraft components, sports goods, high-speed loom components). Laminates of continuous carbon fibres (*Grafil*) are widely used. Carbon fibres are also used in metal-matrix and ceramic-matrix composites. Frequently, they are combined with other types of fibre to form hybrid composites (e.g. glass and carbon, aramid (*Kevlar*) and carbon). Carbon fibres, 5–10 μm diameter, are available in untwisted tows containing 1000, 3000, 6000, 12000 or 120000 filaments and as pre-pregs with resin. In the UK and the USA they are mostly produced from the textile polyacrylonitrile (PAN) and its copolymers. Many types of PAN-derived carbon fibre are produced (e.g. commodity, high-modulus, high-extension, etc.).

The three-stage process[1] for manufacturing carbon fibre is based on the controlled degradation or pyrolysis of spun fibres of PAN. Hot-stretching is a central feature of processing: it counteracts the tendency of the fibres to shrink and induces a high degree of molecular orientation. The tow is first oxidized under tension at a temperature of 200°C and forms a stable, crosslinked 'ladder' structure. In the second stage, heating in an inert atmosphere at temperatures between 800°C and 1600°C carbonizes the structure, releasing vapours and gases (hydrogen, nitrogen) and reducing the original mass by 40–50%. Finally, the oriented carbon fibres

form graphite crystallites during heat-treatment at temperatures in the range 1300°C to more than 2000°C. Raising this treatment temperature encourages graphitization and improves the elastic modulus but lowers the strain to failure. Finally, the fibres are surface-treated (e.g. electrolytic oxidation) to improve subsequent bonding to the matrix and 'sized' to facilitate handling.

Each carbon fibre produced is very pure and consists essentially of interwoven 'ribbons' of turbostratic graphite (Figure 11.16a) and some amorphous carbon. The ribbons are aligned parallel to the fibre axis. Being an imperfect structure, the amount of porosity is appreciable. In general terms, a-axes of the planar crystallites are parallel to the fibre axis, the other a-axis is radial or circumferential, and the c-axes are perpendicular to the fibre axis. As the structure becomes more truly graphitic, the ribbon orientation approaches that of the fibre axis and the axial modulus increases. The fibre structure is highly anisotropic: the moduli of elasticity along the fibres and perpendicular to the fibres are 200–800 GN m^{-2} and 10–20 GN m^{-2}, respectively. (The modulus for E-glass is about 73 GN m^{-2}.)

Another textile, rayon, is used as a precursor for carbon fibres, but to a lesser extent than PAN. Alternatively, melt-spun filaments of high-purity, mesophase pitch can be oxidized and pyrolized in a process broadly similar to the PAN process to yield carbon fibres with a very high modulus approaching 1000 GN m^{-2}. Pitch-derived carbon fibre is expensive, more difficult to handle than PAN-derived fibres and its use is confined to specialized applications.

Aramid fibres (*Kevlar 29* and *49, Twaron*) based on aromatic polyamides are important reinforcements for polymers. Their linear molecular structure (Figure 11.16b) is produced from spun fibre by a process of drawing and heating under tension at a temperature of approximately 550°C. This linear structure, which contrasts with the more planar structure of carbon fibres, gives aramid fibres a fibrillar character and they can absorb considerable amounts of impact energy. When struck with a projectile, aramid fibres split into numerous microfibrils, giving exceptional 'stopping power'. This property has led to the use of aramid fibres and aramid/resin laminates for ballistic applications (e.g. armour). The elastic modulus is 50–130 GN m^{-2} and it is stable at temperatures approaching 400°C (depending upon the environment). Their mechanical properties are degraded by ultraviolet radiation; nevertheless, aramid fibres are widely used, particularly in hybrid composites.

The rapid deceleration of racing cars and landing aircraft develops very high frictional forces and temperatures in braking systems: this challenge has been met by composites in which a carbonaceous matrix is reinforced with carbon fibres. These carbon–carbon composites combine the refractory potential of carbon with the high specific strength/stiffness of carbon fibres. PAN or pitch are used as precursors when

---

[1]Originally developed at the Royal Aircraft Establishment (RAE), Farnborough, in the 1960s over a five-year period.

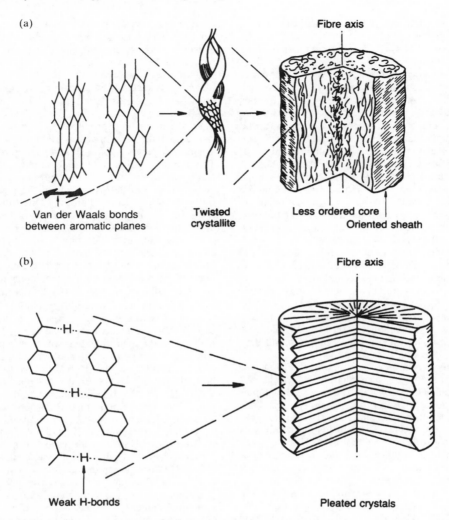

Figure 11.16 *Structure of (a) carbon fibre and (b) aramid fibre (from Hughes, June 1986, pp. 365–8; by permission of the Institute of Materials).*

high-modulus fibres are required. Chemical vapour deposition (CVD) can be used to produce the matrix phase: hydrocarbon gas infiltrates the carbon fibres and is thermally 'cracked' to form a matrix of pyrolytic graphite. C–C composites maintain their strength at high temperatures and have high thermal conductivity and excellent friction/wear characteristics. They are used for aircraft and racing car brakes, rocket nozzles and the heat shield of the Space Shuttle. In the presence of oxygen, C–C composites begin to oxidize and sublime at relatively low temperatures, say 400°C. Efforts are being made to develop long-life multilayer coatings for this type of composite that will inhibit oxygen diffusion, 'self-heal' and enable service temperatures in oxidising environments to be raised to 1400–1500°C.

11.3.2.2 Metal-matrix composites

Large-scale research and development studies of metal-matrix composites (MMCs) date back to the 1960s, being stimulated at that time by the new availability of carbon and boron fibres, whisker crystals and, more indirectly, by successes achieved with reinforced polymer–matrix composites. The aerospace and defence industries were attracted by the prospects of a new type of constructional material possessing high specific strength/stiffness. The fall in strength produced by rising temperature is more gradual in MMCs than with unreinforced matrix material, promising higher service temperatures. In addition, the MMC concept offers prospects of unique wear resistance and thermal expansion characteristics.

For example, parts of space structures are required to maintain dimensional stability while being cycled through an extremely wide range of temperature. One criterion of performance is called the thermal deformation resistance, which is the ratio of thermal conductivity to the coefficient of thermal expansion. The expansion coefficient of graphite fibres with a very high elastic modulus is negative, which makes it feasible to design continuous-fibre MMCs with a zero coefficient of thermal expansion.

Typical early versions of MMC were boron fibres in titanium and carbon fibres in nickel. Over the years, interest has ranged from continuous and discontinuous fibres to whisker crystals and particles. The nominal ranges of diameter for continuous fibres, short fibres and whiskers, and particles are $3–140$ μm, $0–20$ μm and $0.5–100$ μm. Continuous unidirectional fibre-reinforcement gives the greatest improvement in properties over those of unreinforced matrix material. Interest in whisker reinforcement, once very great, has tended to decline because of the carcinogenic risks associated with handling small whiskers during manufacture and composite fabrication. At present, particle-reinforced MMCs find the largest industrial application, being essentially isotropic and easier to process. Matrices based upon low-density elements have gradually come into prominence. The principal matrix materials are aluminium and its alloys, titanium and its alloys, and magnesium. Most MMCs are based on aluminium and its alloys. In recent years, cheaper particles and fibres of silicon carbide have become available, making them the commonest choice of reinforcement material. Alumina (*Saffil*) reinforcement is also used in many MMCs.

Figure 11.17 illustrates the considerable improvements in specific tensile strength (longitudinal) and specific modulus that result when aluminium alloys are reinforced with fibres, whiskers or particles. Property changes of this magnitude are unlikely to be achieved by more orthodox routes of alloy development. At the same time, as the diagram shows, these changes are accompanied by rising costs and bring anisotropy into prominence. The ratio of longitudinal strength to transverse strength can be 15:1 or more for MMCs. Although continuous-fibre reinforcement can confer maximum unidirectional strength, service conditions frequently involve multi-axial stresses.

A wide and growing variety of methods is available for producing MMCs, either as components or as feedstock for further processing (e.g. billets for extrusion, rolling, forging). Many of these methods are still on a laboratory or development scale. In general terms, they usually involve either melting of the matrix metal, powder blending or vapour/electrodeposition. Particle-reinforced aluminium matrix composites can be made by (1) pressing and sintering blends of pre-alloyed powder and reinforcement particles (powder blending), (2) mechanical alloying (MA), (3) mixing particles with molten metal (melt-stirring), (4) compocasting (rheocasting) and (5) spray co-deposition.

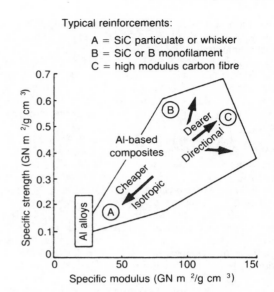

Typical reinforcements:

A = SiC particulate or whisker
B = SiC or B monofilament
C = high modulus carbon fibre

Figure 11.17 *Range of longitudinal tensile specific strengths and moduli achieved in aluminium-based composites (from Feest, May 1988, pp. 273–8; by permission of the Institute of Materials).*

Powder blending techniques (1) are favoured, rather than melting, when the metallic matrix has a high melting point, thus minimizing the fibre/matrix interaction problem. After the critical blending operation, the MMC powder is canned, vacuum degassed and consolidated by hot-pressing or HIPing. Finally, the MMC is worked and shaped (e.g. extrusion, forging). Particle volume fractions of $0.25–0.50$ are typical. Mechanical alloying is broadly similar to powder blending; the essential difference is that alloying of pure metal powders is achieved in high-speed ball mills. In the melt-stirring technique (3), the presence of particles raises the melt viscosity. Possible difficulties include non-wetting of particles, agglomeration and/or gravity-settling of particles and unwanted particle/metal interactions. In the related compocasting process, the stirred melt is maintained in a two-phase 'mushy' state at temperatures between solidus and liquidus. The method is not applicable to metallic systems with a narrow range of solidification.

Production of particle-reinforced MMCs by spray co-deposition (5) is based upon a versatile process[1] originally developed for building deposits of steels that are difficult to cast, nickel-based superalloys and copper. A stream of induction-melted metal or alloy is broken into fine droplets by relatively cold inert gas (nitrogen). Droplets begin to freeze before striking a movable collector placed in the line of flight. In the MMC variant, reinforcement particles are

[1] Patented by Osprey Metals, adopted under licence and further developed by Alcan International Ltd, Banbury.

injected into this stream. Fine-grained MMC deposits of SiC/aluminium alloy and Al_2O_3/steel can be built as plate, tube, billets for hot-working, cladding, etc. The pathogenic risks associated with certain types of fine particles, fibres and whiskers have necessarily been taken into consideration. Filaments of $0.1–3$ μm diameter and with lengths greater than 5 μm are hazardous. Accordingly, SiC particle sizes exceed a certain threshold (e.g. 5 μm).

The fabrication of continuous-fibre reinforced MMCs is difficult and technically demanding. Fibre preforms with a high volume fraction of fibres are difficult to 'wet' and infiltrate; on the other hand, if V_f is low, the preform will lack the 'green' strength required for handling. A high melt temperature will lower viscosity and assist infiltration of the fibres but increases the risk of fibre/metal reactions. (SiC, Al_2O_3 and carbon fibres react with aluminium alloy matrices at a temperature of $500°C$.) There is also a risk that secondary working operations will damage the fibres. In general, the methods for continuous fibres are costly, give low production rates and limit the size and shape of the MMC product. Typical methods for producing continuous-fibre MMCs are (1) diffusion bonding (DB), (2) squeeze-casting, and (3) liquid pressure forming (LPF).

Diffusion bonding (1) can be used to consolidate metal with continuous fibres of SiC, Al_2O_3, boron and carbon. Fibres are pre-coated with matrix material or carbon. The process conditions must achieve a delicate balance between promotion of solid-state diffusion and limitation of fibre/matrix reactions. DB has been used to produce boron/aluminium struts for the NASA space shuttle and SiC/titanium alloy composites.

In principle, squeeze-casting (2) is a single-shot process combining gravity die-casting with closed-die forging. It is mainly used for discontinuous-fibre MMCs. A metered charge of melt containing short fibres is fed into the lowermost of a pair of dies and then compressed at high pressure ($35–70$ MN m^{-2}) by the descending upper die. The pressure is maintained while the charge solidifies. Interpretation of the relevant phase diagram for the matrix has to allow for the pressure variable. Boundaries in the diagram are shifted and, for alloys that contract on freezing, the liquidus temperature is raised. With aluminium alloys, this shift is about $10–25°C$. Thus, the applied pressure has an undercooling effect which, together with the loss of heat through the dies, favours rapid solidification. The high pressure also discourages the nucleation of gas bubbles. The final matrix structure is accordingly fine-grained and dense. Moderate-sized components can be produced at high rates.

In liquid pressure forming (3), a preheated fibre preform is placed within split dies. The dies are closed and air pressure in the cavity reduced below 1 mb. Molten aluminium casting alloy under pressure is then infiltrated upwards through the preform and allowed to solidify. LPF uses a lower pressure (<8 MN m^{-2}) than squeeze-casting and there is less limitation on size and shape of product. Particle-reinforced MMCs can be produced. Using a single preform and one injection shot, it is possible to vary the volume fraction and type of fibre (continuous, discontinuous) within a component.

With regard to applications, the emphasis in the aerospace industry is the development of MMCs based upon titanium or intermetallic compounds such as TiAl and Ti_3Al. For example, coated SiC fibres are used to reinforce Ti-6Al-4V alloy. Proposed applications for MMCs include compressor discs and blades in aero-engines, engine pylons and stabilizers. Channel extrusions made from discontinuous-fibre reinforced MMC are in use for electrical racking in aircraft. In general manufacturing, there is interest in using SiC fibre-reinforced aluminium for critical components operating at very high speeds where high specific stiffness would be mechanically advantageous (e.g. carpet-making, food packaging, textile production). Co-sprayed steel matrix deposits have been used for components of plant handling highly-abrasive materials (e.g. coke, minerals, wood, fibreboard).

In the car industry MMCs are now accepted as candidate materials for valve rocker arms, connecting rods, gear selector forks, pulleys, propshafts, etc. They have also made it possible to replace cast iron engine cylinder blocks with selectively-reinforced blocks of aluminium alloy (*Honda*). The hybrid composite, which contains 12% v/v alumina particles (for strength) and 9% v/v carbon fibres (for lubricity), saves weight, improves the power rating and dispenses with cast iron cylinder liners.

In diesel engines[2] the achievement of higher combustion efficiencies, with better fuel economy and reduction in the exhaust emissions of undesirable gases and particulate matter, has resulted in higher temperatures and peak pressures in the combustion chambers. The standard alloy for diesel pistons is Al–12Si–1Cu–1Mg (LM 13 in BS1490: *Lo-Ex*). Although this eutectic alloy is satisfactory for a working temperature range of $250–300°C$, advances in engineering design have raised crown temperatures to $300–350°C$. For instance, in one relatively new design feature, now established for direct-injection engines, a deep combustion bowl is located in the crown of each piston. This turbulence-inducing cavity usually has re-entrant angles and sharp edges. The higher mechanical and thermal loading on the bowl lip can cause cracking. This problem has been solved by squeeze-casting the piston body and, at the same time, incorporating a fibre-reinforced bowl. A controlled quantity of aluminium alloy melt is fed into an open die, which contains a preformed compact of alumina fibres, and then compressed with a pressure of $120–150$ MN m^{-2} by a hydraulically-actuated plug die, so that the fibres are infiltrated, 'wetted' and

---

[2]The first engine patent of Rudolf Diesel (1858–1913) was officially authenticated in 1893.

bonded to the alloy. (The fibres are 2–4 μm diameter, 200–500 μm long, and occupy at least 20% v/v of the composite) Squeeze-casting prevents shrinkage and dissolved gases from causing microporosity and gives a fine-grained structure with better high-temperature fatigue properties than gravity-casting. The MMC structure provides the bowl lips with the necessary hot strength and resistance to cracking.

Figure 11.18 shows a typical aluminium alloy piston design, with local reinforcement of the combustion bowl and the top ring groove. Accepted practice for the large pistons of heavy-duty and turbocharged diesel engines is to reinforce this groove, and possibly the second groove, with a wear-resistant insert of nickel-rich, and hence austenitic, cast iron (*Ni-Resist*) which is bonded to the aluminium alloy body by the *Alfin* process. In small high-speed indirect-injection diesel engines, which have a pre-combustion chamber, lighter and cheaper MMC inserts have been used successfully to reinforce the groove. The performance of this insert is crucial because modern designs are tending to locate the top groove much closer to the crown, even in the 'headland' position, thus reducing the volume of the annular 'dead space' above the ring and giving fuel economy and reduced emissions.

When considered in terms of the costly effort that has been put into development of MMCs, commercial exploitation has been disappointingly limited. MMC structures and components face strong competition from alternative, more conventional engineering materials. Exploitation has been mainly hindered by the high cost of reinforcement materials, particularly continuous fibres, and fundamental features of MMC

structures that tend to make fabrication difficult and service performance questionable.

During fabrication, formation of the matrix often necessitates a high processing temperature (e.g. melt infiltration). This temperature can be high enough to promote unwanted chemical reactions at the interfaces between matrix and reinforcement. These conditions may arise during service at elevated temperatures, of course. SiC-reinforced titanium is an example of a composite in which brittle interfacial products tend to form (i.e. silicides and carbides). Ideally, the aim is to develop a strong interfacial bond without degrading the fibre or forming weakening phases at the interfaces. The coating of fibres is generally regarded as the most promising means to control chemical interaction at reinforcement/matrix surfaces. However, it is difficult to develop a coating with long-term stability.

As might be anticipated, the deliberate introduction of short stress-concentrating fibres into a metal or alloy tends to reduce the ductility and toughness below that of the unreinforced matrix. When a short-fibre MMC is deformed by applied stress, the amount of strain in fibre and matrix may differ substantially, leading to rupture of the interfacial bonding at the end regions of fibres. This debonding results in the nucleation and rapid growth of voids. Void formation is the dominant mode of tensile failure in SiC/aluminium composites. A difference in the coefficients of thermal expansion of fibre and matrix can produce the same effect, possibly during the cooling stages of composite fabrication or during thermal excursions and cycling in service. Thus, because $\alpha_f < \alpha_m$, fibres and particles develop residual compressive stress and the matrix is left in a state of tension. This disparity in expansion characteristics between the reinforcement phase and the matrix also helps to generate a high density of dislocations in the matrix. Plastic deformation of the matrix involves dislocation movement across slip planes. If a slip plane is threaded by fibres, a glissile dislocation will bow between them and recombine beyond the fibres, leaving an Orowan loop around each fibre. These loops reduce the effective gap between fibres and successive dislocations are forced to assume smaller radii of curvature as they bypass the fibres. In combination with a high dislocation density and a fine matrix grain size, this mechanism can produce a high rate of initial strain-hardening in the matrix.

The presence of extremely hard reinforcement particles or fibres can benefit wear-resistance in service but leads to difficulties during the finish-machining operations that are usually needed for MMC products. These particles act as chip-breakers during machining of materials, such as SiC-reinforced aluminium alloys, and reduce cutting forces: unfortunately, they also shorten tool life significantly, even when tools are tipped with polycrystalline diamond (PCD). Machining tends to generate sub-surface damage, a matter of special concern in the preparation of mechanical test-pieces.

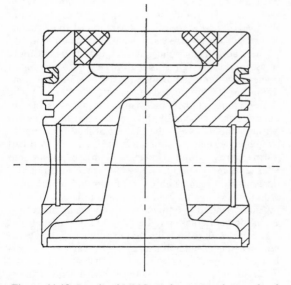

Figure 11.18 *Localized MMC reinforcement of piston head (courtesy of R. Munro and AE Piston Products, T & N Group).*

Despite these formidable problems, the basic idea of reinforcing a strong deformable matrix with elastic fibres retains its appeal to the aerospace, defence and automotive industries and active research on MMCs continues worldwide. There is a need to expand and consolidate the database for properties of MMCs (e.g. fracture toughness, fatigue resistance, corrosion resistance, etc.).[1]

11.3.2.3 Ceramic-matrix composites

Reinforcement of cements and concretes with short filaments of glass, steel or carbon is applied in the building and construction industries. Costs are very closely scrutinized in these industries and the increase in cost associated with fibre reinforcement is often regarded as unacceptable unless there is promise of exceptional properties. In one example, a 2% v/v dispersion of steel wires (low-carbon or austenitic stainless) in a concrete matrix (*Wirand Concrete*[2]) confers improved resistance to crack propagation and greater flexural and compressive strength. These wires are typically 0.25–0.5 μm diameter, with an aspect ratio of 100:1. The resistance of this composite to salt water and to pebble abrasion have led to its use for 40 t twin-fluked shapes (dolosse) for shore defences against heavy seas.

Since the 1980s there has been a sustained interest in the development of ceramic-matrix composites (CMCs) for high-temperature applications that demand exceptional strength and chemical stability (e.g. advanced engine designs, cutting tools). Ceramics in general are inherently notch-sensitive and brittle: the introduction of fibres is primarily intended to improve the toughness of the matrix and reliability in service. (In polymer-matrix and metal-matrix composites, reinforcement is used to improve strength, stiffness and, if possible, toughness.) Currently, the principal matrices are inorganic glasses and glass-ceramics but it is also possible to use engineering ceramics, such as alumina and silicon nitride. Reinforcement materials, which include silicon carbide (*Nicalon, Tyranno*), alumina and carbon, have taken the form of continuous fibres, short fibres, whisker crystals, woven fibre cloth and laminates.

Failure in CMC structures is complex in character and a variety of mechanisms has been identified. It is not possible to apply directly the theory of fracture mechanics that was developed for monolithic ceramic bodies. Researches on CMC systems using continuous fibres primarily aim to achieve substantial improvements in toughness and reliability. To this end, there is particular interest in structures that have a capacity for accumulating damage and that eventually fail

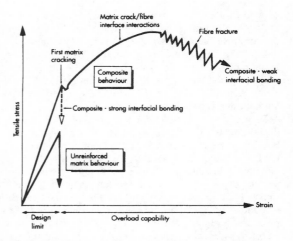

Figure 11.19 *Stages of failure in CMC reinforced with continuous fibres (from King, Dec 1989, pp. 720-6; by permission of the Institute of Materials).*

in a gradual ('graceful') manner, rather than suddenly and catastrophically. This overload capability is illustrated schematically in Figure 11.19, which gives the tensile stress versus strain curve for a unidirectional composite. It will be noted that the fracture strain of the unreinforced matrix is considerably less than that of the composite. Superimposed on the curve for the composite are the principal micromechanisms of failure, namely (1) microcracking of the matrix, (2) interfacial failure as fibre pull-out takes place and (3) fibre fracture.

In general terms, events taking place in the wake of each crack tip become more significant than flaws in the matrix. Thus, unbroken fibres bridge the cracks in the matrix, oppose opening separation of these cracks and reduce the effective stress intensity factor at the tip of the crack. As the matrix cracks, load is increasingly transferred to the fibres. A high E_f/E_m ratio and a high content of fibres favour this transfer. Volume fractions up to 0.5–0.6 are quoted in reports.

Debonding at the fibre/matrix interfaces is clearly an important aspect of the overall fracture process. Strong bonding assists the transfer of load from matrix to fibres and increases the amount of energy absorbed during fibre pull-out. However, if this bond strength is exceptionally strong, cracks may travel in an unhindered, brittle manner through matrix and fibres alike, with the CMC behaving like a monolithic ceramic. On the other hand, if bonding is relatively weak, the course of an incident crack may be deflected into the interfacial zone, with useful absorption of energy. For these reasons, special attention is given to ways of controlling and/or modifying the surface character of fibres.

The degree to which the linear coefficients of thermal expansion (α) of fibre and matrix materials match is an important aspect of CMC design. For instance, a mismatch between α-values can favour the type of

---

[1]In the early 1980s, MMCs were classed as strategic materials by the US government: samples and detailed information are consequently unavailable.

[2]Patent based on researches of Professor James P. Romualdi of the Carnegie-Mellon University and the Battelle Development Corporation, USA, in the 1960s).

crack-deflection process just described. Again, if short fibres are used and there is a substantial mismatch in α-values, cracks will tend to be initiated at the ends of the fibres. Thermal cycling will aggravate this type of damage. A mismatch in the expansion characterisics of well-bonded fibres and matrix can also introduce a 'pre-stressing' effect that inhibits crack growth. Thus, if $\alpha_f > \alpha_m$, then cooling from a high temperature will induce residual tensile stress in the fibres and residual compressive stress in the matrix. This imbalance has a toughening effect because microcracks advancing in the matrix have to overcome compressive stress.

Scanning electron microscopy provides valuable information on deformation and failure mechanisms in CMC structures. A sample of composite can be imaged *in situ* with high resolution while being subjected to tensile, bending, compressive or cyclic stress. Figure 11.20a shows a SiC fibre in a matrix of calcium aluminosilicate during an early stage of fatigue failure, with the fibre bridging the microcrack in the matrix. Eventually, as Figure 11.20b shows, the fibre itself fractures after a certain amount of pull-out from the matrix.

Fabrication of CMCs is expensive, demanding very close control. During fabrication of a continuous fibre CMC, it is necessary to infiltrate the fibres, which may be in the form of woven cloth, with a fluid form of the matrix-building material. The most common method uses a slurry of ceramic particles, a carrier liquid (water or alcohol), an organic binder and, sometimes, a surface-wetting agent. After drying, the shape is hot-pressed. A range of alternative production routes is now available (e.g. sol–gel reaction, melt infiltration, chemical vapour deposition (CVD), reaction-sintering, transfer-moulding). Unique fabrication problems can arise. For instance, during high-temperature firing, chemical reaction at the fibre/matrix interfaces may cause weakening intermediate phases to form. With regard to industrial utilization, there is still a widely-accepted need to improve toughness and reliability further. Although some CMCs can be used in air at temperatures approaching 1000°C, oxidation is a potential problem when carbon and SiC fibres are used; fibre oxidation may occur at temperatures as low as 350–400°C. Coatings have proved effective in preventing this wasting attack and in raising the ceiling temperature for the composite. Alternatively, an inert gaseous environment may be used but is obviously not always feasible. Glassy matrices are favoured because processing is generally simpler; unfortunately, it is possible for certain glasses to begin to soften in the temperature range 500–800°C.

As the foregoing remarks indicate, a number of problems are associated with the production and utilization of ceramic-matrix composites. Commercial exploitation of CMCs to date has consequently been rather restricted. Combinations that have found applications in engineering are SiC whiskers/polycrystalline alumina (cutting-tool tips), graphite fibres/borosilicate

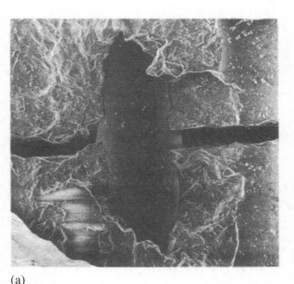

(a)

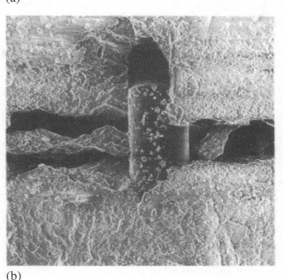

(b)

Figure 11.20 *Field-emission scanning electron micrographs of fatigue cracking in SiC fibre/Ca aluminosilicate composite (fibre diameter 15 μm) (courtesy of M. D. Halliday, Interdisciplinary Research Centre, University of Birmingham, UK).*

glass (substrate of laser mirrors), and graphite/glass-ceramic (bearings, seals and brakes). The SiC fibre/lithium aluminosilicate (LAS) glass composite is a candidate material for new types of heat-engine.

11.3.2.4 *In-situ* composites and nanocomposites

Alloy microstructures, such as eutectics and eutectoids, containing a dispersion of fibres or lamellae within a matrix phase are well-known products of phase transformation. They form an intimate combination of two or more quite different phases and, understandably, there has been strong interest in developing *in-situ* (self-assembled) composites which offer the advantages of the orthodox composites produced by adding a matrix phase to preformed fibres. This new class of materials, with its micron-scale structures, offers the prospect of unique physical (electrical, thermal, magnetic), mechanical and chemical properties. Provided that the alloy system is carefully chosen and manipulated, it is possible to produce unidirectionally solidified structures with a uniform distribution of aligned fibres over extended distances. The problem of fibre/matrix interaction so often associated with conventional fabrication routes is eliminated. Eutectic systems are immediately attractive because the solid phases produced simultaneously by this type of reaction are essentially in thermodynamic equilibrium and are held together by strong interfacial bonds. Fibres are preferred to lamellae because the latter are more likely to be weakened by the presence of transverse cracks.

In-situ composites are commonly produced by the classical laboratory technique in which molten alloy is withdrawn vertically downwards from a furnace held at a constant temperature. The rate of withdrawal is necessarily very low for some alloys (<10 mm h^{-1}), increasing the risk that the materials of containment will contaminate the melt. Because of the low production rates, costs tend to be high. As the eutectic reaction proceeds, solid phases nucleate and grow perpendicular to the horizontal melt/eutectic interface in a cooperative manner. Interdiffusion of alloying elements takes place in the melt adjacent to this advancing interface. If both matrix and reinforcement phases have a low entropy of fusion, the eutectic structure grows in a uniform manner and the interface is essentially planar. Fortunately, most alloy systems of industrial interest tend to freeze with this morphology when solidification is closely controlled, even in cases where the reinforcing phase (e.g. carbide) has a known tendency to form crystallographically oriented facets because of its high entropy of fusion. When both phases have a high entropy of fusion, each phase has a strong tendency to form facets and the final structure is likely to be very irregular with much branching. This last combination is intrinsically unsuitable for producing *in-situ* composites.

Severe limitations on phase composition and phase ratio may apply to these melt-grown composites. For instance, binary alloy systems can often only develop a relatively small volume fraction of a strong fibrous phase. In the Ni–Al system, which has been extensively studied, the volume fraction of strong Al$_3$Ni filaments within the aluminium-based matrix is 0.1. (Beyond a certain volume fraction in a given system, laminae have a lower interfacial energy and tend to form in preference to fibres.) Complex ternary alloy systems are more promising: a monovariant eutectic reaction at a 'valley' line in the liquidus surface of the Co–Cr–C system can produce 30% v/v carbide fibres in a solid solution matrix of cobalt and chromium. For high-temperature applications, such as turbine blades in aero engines, *in-situ* composites can compete with directionally-solidified (DS) nickel-based superalloys and single-crystal superalloys (e.g. stress-rupture properties in directions parallel to the reinforcement). Aligned two-phase structures can be grown from the alloy melt; phase compositions can be varied quite widely without preventing such structures from forming. There is also valuable scope for alloying additions (Cr, Al, W, Re) which enhance properties such as resistance to oxidation, corrosion, creep and thermal fatigue. Examples are the NITAC and COTAC series of eutectic superalloys, which are based upon nickel and cobalt, respectively. In typical NITAC alloy, the ductile matrix consists of γ-Ni and a fine precipitate of Ni$_3$Al particles (gamma-prime phase, γ'), and is reinforced with a carbide phase (e.g. TaC, Cr$_3$C$_2$, NbC).

In service, the integrity of *in-situ* composites can be jeopardized by oxidative attack, temperature gradients and/or thermal cycling. The last condition is potentially disruptive because a mismatch in the thermal expansion characteristics of the two phases will cause stiff fibres to fracture and the ductile matrix phase to fail by fatigue.

In the late 1980s, researchers began to extend the composite principle beyond the micron scale and into the nanometre range (1–50 nm). The 'jumps' in various properties achieved with these new nanocomposite microstructures were considerable. An early prototype was a nominally single-phase ceramic structure based upon a conventional polycrystalline matrix of micron-sized alumina grains. A uniform dispersion of 5% silicon carbide inclusions, about 10 nm in size, throughout this matrix gives a fourfold increase in strength. The processing route for this ceramic is based upon established sintering technology: the main stages are the

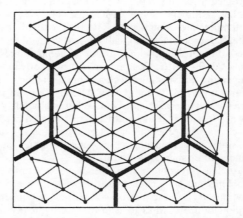

Figure 11.21 *SiC inclusion/alumina nanocomposite (from Brook and MacKenzie, Jan 1993, pp. 27–30; by permission of the Institute of Materials).*

grinding and mixing of fine β-SiC and α-Al_2O_3 powders, hot-pressing (1750°C) and annealing (1300°C). The structural character of the ceramic is shown schematically in Figure 11.21. It has been proposed that the large number of dislocations present form a sub-structure within each grain, with SiC particles at the nodal points. In general terms, this combination of micron-sized grains and much finer nanometre-scale sub-structures gives exceptional strength and, importantly in the case of ceramics, good reproducibility.

Study of nanoscale structures has been extended to a wide variety of metallic systems. Again, fabrication is based on pure, ultrafine powders. These powders are mainly produced by either condensation from a vapour, which minimizes the risk of contamination, or high-energy ball-milling. Two or more metal powders are compacted at high pressure ($1-5$ GN m^{-2}) to give a unique nanometre-scale grain structure. An extremely large proportion of atoms are sited at or near grain boundary surfaces and properties dependent upon grain boundary characteristics come into prominence. For instance, these multiphase metallic nanostructures are stable and highly resistant to grain growth. The grains are randomly-oriented and fairly uniform in size (say, $5-10$ nm). Contiguous grains can differ in chemical composition.

Ceramic and metallic nanocomposites can display unusual properties and behaviour that are not explicable in terms of conventional scientific theory. Thus current explanations of dislocation behaviour, diffusion, solid solubility and fracture in nanoscale structures are often tentative.

Further reading

Ashbee, K. H. G. (1993). *Fundamental Principles of Fiber Reinforced Composites*, 2nd edn. Technomic Publ. Co. Inc., Lancaster, USA.

Ashby, M. F. and Jones, D. R. H. (1986). *Engineering Materials*, **2**, Elsevier Science, Oxford.

Brook, R. J. and MacKenzie, R. A. D. (1993). Nanocomposite materials. *Materials World*, January, 27–30, Institute of Materials.

Feest, E. A. (1988). Exploitation of the metal-matrix composites concept. *Metals and Materials*, May, 273–278, Institute of Materials.

Harris, B. (1986). *Engineering Composite Materials*. Institute of Metals, London.

Hertzberg, R. W. (1989). *Deformation and Fracture Mechanics of Engineering Materials*, 3rd edn. Wiley, Chichester.

Hughes, J. D. H. (1986). *Metals and Materials*, June, pp. 365–368, Institute of Materials.

Hull, D. (1981). *An Introduction to Composite Materials*. Cambridge University Press, Cambridge.

Imperial Chemical Industries, Plastics Division (1974). *Thermoplastics: Properties and Design*, (ed. R. M. Ogorkiewicz), Chap. 11 by P. C. Powell, Wiley, Chichester.

Kelly, A. (1986) *Strong Solids*. Clarendon Press, Oxford.

King, J. E. (1989). *Metals and Materials*, 720–6. Institute of Materials.

Lemkey, F. D. (1984). Advanced in situ composites. In Chap. 14, *Industrial Materials Science and Engineering* (ed. L. E. Murr). Marcel Dekker.

Mascia, L. (1989). *Thermoplastics: Materials Engineering*. 2nd edn, Elsevier Applied Science, London.

McLean, M. (1983). *Directionally-Solidified Materials for High-Temperature Service*. Metals Society, London.

Metals and Materials. (1986). Set of articles on composites, Institute of Materials:

Hancox, N. L. Pt I Principles of fibre reinforcement (May).

Hughes, J. D. H. Pt II Fibres (June).

Hancox, N. L. Pt III Matrices (July).

Bowen, D. H. Pt IV Manufacturing methods (September).

Davidson, R. Pt V Performance (October)

Bowen, D. H. Pt VI Applications of Polymer-Matrix Composites (December)

Mills, N. J. (1986). *Plastics: Microstructure, Properties and Applications*. Edward Arnold, London.

Morley, J. G. (1987). *High-Performance Fibre Composites*. Academic Press, New York.

Morton-Jones, D. H. (1989). *Polymer Processing*. Chapman and Hall, London.

Powell, P. C. (1974). *Thermoplastics: properties and design*. Chapter 11. ed. by R. M. Ogorkiewicz, Wiley-Interscience.

Shah, Vishu, (c1984) *Handbook of Plastics Testing Technology*. Wiley, Chichester.

Wood, A. K. (1989). Advances in engineering plastics manufacture. *Metals and Materials*, May, 281–284, Institute of Materials.

Chapter 12

Corrosion and surface engineering

12.1 The engineering importance of surfaces

The general truth of the engineering maxim that 'most problems are surface problems' is immediately apparent when one considers the nature of metallic corrosion and wear, the fatigue-cracking of metals and the effect of catalysts on chemical reactions. For instance, with regard to corrosion, metal surfaces commonly oxidize in air at ambient temperatures to form a very thin oxide film (tarnish). This 'dry' corrosion is limited, destroys little of the metallic substrate and is not normally a serious problem. However, at elevated temperatures, nearly all metals and alloys react with their environment at an appreciable rate to form a thick non-protective oxide layer (scale). Molten phases may form in the scale layer, being particularly dangerous because they allow rapid two-way diffusion of reacting species between the gas phase and the metallic substrate. 'Wet' or aqueous corrosion, in which electrochemical attack proceeds in the presence of water, can also destroy metallic surfaces and is responsible for a wide variety of difficult problems throughout all branches of industry. The principles and some examples of 'dry' and 'wet' corrosion will be discussed in Section 12.2.

Conventionally, the surface properties of steels are improved by machining to produce a smooth surface texture (superfinishing), mechanically working (shot-peening), introducing small atoms of carbon and/or nitrogen by thermochemical means (carburizing, nitriding, carbo-nitriding), applying protective coatings (galvanizing, electroplating), chemically converting (anodizing), etc. Many of these traditional methods employ a liquid phase (melt, electrolyte). In contrast, many of the latest generation of advanced methods for either coating or modifying material surfaces use vapours or high-energy beams of atoms/ions as the active media. Their successful application on a commercial scale has revealed the merits of developing a new philosophy of surface design and engineering. In

Section 12.3, we examine some typical modern methods for improving surface behaviour.

12.2 Metallic corrosion

12.2.1 Oxidation at high temperatures

12.2.1.1 Thermodynamics of oxidation

The tendency for a metal to oxidize, like any other spontaneous reaction, is indicated by the free energy change ΔG accompanying the formation of the oxide. Most metals readily oxidize because ΔG is negative for oxide formation. The free energy released by the combination of a fixed amount (1 mol) of the oxidizing agent with the metal is given by $\Delta G°$ and is usually termed the standard free energy of the reaction. $\Delta G°$ is, of course, related to $\Delta H°$, the standard heat of reaction and $\Delta S°$ the standard change in entropy, by the Gibbs equation. The variation of the standard free energy change with absolute temperature for a number of metal oxides is shown in Figure 12.1. The noble metals which are easily reduced occur at the top of the diagram and the more reactive metals at the bottom. Some of these metals at the bottom (Al, Ti, Zr), however, resist oxidation at room temperature owing to the impermeability of the thin coherent oxide film which first forms.

The numerical value of $\Delta G°$ for oxidation reactions decreases with increase in temperature, i.e. the oxides become less stable. This arises from the decreased entropy accompanying the reaction, solid (metal) + gas (oxygen) $\rightarrow$ solid (oxide). The metal and oxide, being solids, have roughly the same entropy values and $d(\Delta G)/dT$ is thus almost equivalent to the entropy of the oxygen, i.e. 209.3 J deg^{-1} mol^{-1}. Most of the ΔG versus T lines therefore slope upwards at about this value, and any change in slope is due to a change in state. As expected, melting has a small effect on ΔS and hence ΔG, but transitions through the boiling point (e.g. ZnO at 970°C) and sublimation

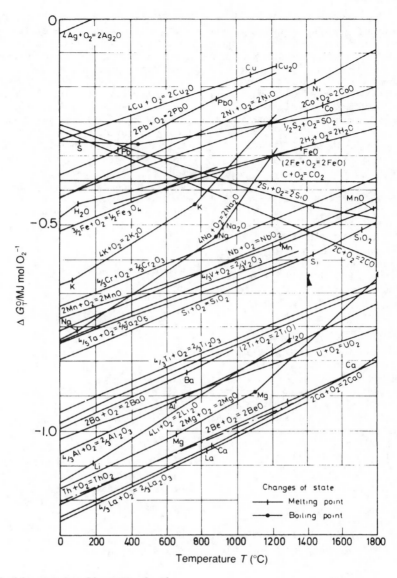

Figure 12.1 *Standard free energies of formation of oxides.*

(e.g. Li_2O at 1330°C) have large effects. Exceptions to the positive slope of the ΔG versus T line occur for carbon oxidation to CO or CO_2. In both cases the oxide product is gaseous and thus also has a high free energy. In the reaction $2C + O_2 \rightarrow 2CO$, two moles of gas are produced from one of oxygen so that

$$\Delta S = (S_{oxide} - S_{carbon} - S_{oxygen})$$

$$= (2S°_{oxide} - S°_{carbon} - S°_{oxygen}) \simeq S°_{oxide} \quad (12.1)$$

The CO free energy versus temperature line has a downward slope of approximately this value. For the $C + O_2 \rightarrow CO_2$ reaction, one mole of CO_2 is produced

from one mole of oxygen and hence $\Delta S \simeq 0$; the CO_2 free energy line is thus almost horizontal. The carbon monoxide reaction is favoured at high temperatures and consequently carbon is a very effective reducing agent, having a greater affinity for oxygen than most oxides.

Because of the positive slope to the $\Delta G°$ versus T line for most oxides in Figure 12.1, $\Delta G°$ tends to zero at some elevated temperature. This is known as the standard dissociation temperature when the oxide is in equilibrium with the pure element and oxygen at 1 atm pressure. In the case of gold, the oxide is not stable at room temperature, for silver

Ag_2O dissociates when gently heated to about 200°C, and the oxides of the Pt group of metals around 1000°C. The other oxides dissociate at much higher temperatures. However, the temperature is affected by pressure since the free energy per mole of any gaseous phase varies with pressure P(atm) according to $G(P) = G° + \mathbf{R}T \ln P$, whereas that for the solid phase is relatively unaffected. Thus, for the metal + oxygen → metal oxide reaction under standard conditions, $\Delta G° = G°_{oxide} - G°_{metal} - G°_{oxygen}$, and at P atm oxygen, $\Delta G = \Delta G° - \mathbf{R}T \ln P_{O_2}$. The reaction is in equilibrium when $\Delta G = 0$ and hence

$$P_{O_2} = \exp[\Delta G°/\mathbf{R}T] \qquad (12.2)$$

is the equilibrium dissociation pressure of the oxide at the temperature T. If the pressure is lowered below this value the oxide will dissociate, if raised above, the oxide is stable. The common metal oxides have very low dissociation pressures $\approx 10^{-10} N/m^2$ ($\sim 10^{-15}$ atm) at ordinary annealing temperatures and thus readily oxidize in the absence of reducing atmospheres.

The standard free energy change $\Delta G°$ is also related to the equilibrium constant K of the reaction. For the reaction discussed above, i.e.

$$Me + O_2 \rightarrow MeO_2$$

the equilibrium constant $K = [MeO_2]/[Me][O_2]$ derived from the law of mass action. The active masses of the solid metal and oxide are taken equal to unity and that of the oxygen as its partial pressure under equilibrium conditions. The equilibrium constant at constant pressure, measured in atmospheres is thus $K_P = 1/P_{O_2}$. It then follows that $\Delta G° = -\mathbf{R}T \ln K_P$. To illustrate the use of these concepts, let us consider the reduction of an oxide to metal with the aid of a reducing agent (e.g. Cu_2O by steam). For the oxidation reaction.

$$4Cu + O_2 \rightarrow 2Cu_2O$$

and from Figure 12.1 at 1000 K, $\Delta G° = -0.195$ MJ/mol $= 1/P_{O_2}$, giving $P_{O_2} = 6.078 \times 10^{-6}$ N/m². At 1000 K the equilibrium constant for the steam reaction $2H_2O \rightarrow 2H_2 + O_2$ is

$$K = P_{O_2}P_{H_2}^2/P_{H_2O}^2 = 1.013 \times 10^{-15} \text{ N/m}^2$$

Thus to reduce Cu_2O the term $P_{O_2} < 6 \times 10^{-11}$ in the steam reaction gives $P_{H_2}/P_{H_2O} > 10^{-5}$, so that an atmosphere of steam containing 1 in 10^5 parts of hydrogen is adequate to bright-anneal copper.

In any chemical reaction, the masses of the reactants and products are decreasing and increasing respectively during the reaction. The term chemical potential $\mu(= dG/dn)$ is used to denote the change of free energy of a substance in a reaction with change in the number of moles n, while the temperature, pressure and the number of moles of the other substances are kept constant. Thus,

$$\mu_i = \mu_i^0 + \mathbf{R}T \ln P_i \qquad (12.3)$$

and the free energy change of any reaction is equal to the arithmetical difference of the chemical potentials of all the phases present. So far, however, we have been dealing with ideal gaseous components and pure metals in our reaction. Generally, oxidation of alloys is of interest and we are then dealing with the solution of solute atoms in solvent metals. These are usually non-ideal solutions which behave as if they contain either more or less solute than they actually do. It is then convenient to use the activity of that component, a_i, rather than the partial pressure, P_i, or concentration, c_i. For an ideal solution $P_i = P_i^0 c_i$, whereas for non-ideal solutions $P_i = P_i^0 a_i$, such that a_i is an effective concentration equal to the ratio of the partial, or vapour, pressure of the ith component above the solution to its pressure in the standard state. The chemical potential may then be rewritten

$$\mu_i = \mu_i^0 + \mathbf{R}T \ln a_i \qquad (12.4)$$

where for an ideal gas mixture $a_i = P_i/P_i^0$ and by definition $P_i^0 = 1$. For the copper oxide reaction, the law of mass action becomes

$$K = \frac{a_{Cu_2O}^2}{a_{Cu}^4 a_{O_2}} = \frac{1}{P_{O_2}} = \exp[-\Delta G°/\mathbf{R}T] \qquad (12.5)$$

where a_i^n is replaced by unity for any component present in equilibrium as a pure solid or liquid. Some solutions do behave ideally (e.g. Mn in Fe) obeying Raoult's law with $a_i = c_i$. Others tend to in dilute solution (e.g. Fe in Cu) but others deviate widely with a_i approximately proportional to c_i (Henry's law).

12.2.1.2 Kinetics of oxidation

Free energy changes indicate the probable stable reaction product but make no prediction of the rate at which this product is formed. During oxidation the first oxygen molecules to be absorbed on the metal surface dissociate into their component atoms before bonding chemically to the surface atoms of the metal. This process, involving dissociation and ionization, is known as chemisorption. After the build-up of a few adsorbed layers the oxide is nucleated epitaxially on the grains of the base metal at favourable sites, such as dislocations and impurity atoms. Each nucleated region grows, impinging on one another until the oxide film forms over the whole surface. Oxides are therefore usually composed of an aggregate of individual grains or crystals, and exhibit phenomena such as recrystallization, grain growth, creep involving lattice defects, just as in a metal.

If the oxide film initially produced is porous the oxygen is able to pass through and continue to react at the oxide–metal interface. Usually, however, the film is not porous and continued oxidation involves diffusion through the oxide layer. If oxidation takes place at the oxygen–oxide surface, then metal ions and electrons have to diffuse through from the underlying metal. When the oxidation reaction occurs at the metal–oxide interface, oxygen ions have to diffuse through the oxide and electrons migrate in the opposite direction to complete the reaction.

The growth of the oxide film may be followed by means of a thermobalance in conjunction with metallographic techniques. With the thermobalance it is possible to measure to a sensitivity of 10^{-7} g in an accurately controlled atmosphere and temperature. The most common metallographic technique is ellipsometry, which depends on the change in the plane of polarization of a beam of polarized light on reflection from an oxide surface; the angle of rotation depends on the thickness of the oxide. Interferometry is also used, but more use is being made of replicas and thin films in the transmission electron microscope and the scanning electron microscope.

The rate at which the oxide film thickens depends on the temperature and the material, as shown in Figure 12.2. During the initial stages of growth at low temperatures, because the oxygen atoms acquire electrons from the surface metal atoms, a strong electric field is set up across the thin oxide film pulling the metal atoms through the oxide. In this low-temperature range (e.g. Fe below 200°C) the thickness increases logarithmically with time ($x \propto \ln t$), the rate of oxidation falling off as the field strength diminishes.

At intermediate temperatures (e.g. 250–1000°C in Fe) the oxidation develops with time according to a parabolic law ($x^2 \propto t$) in nearly all metals. In this region the growth is a thermally-activated process and ions pass through the oxide film by thermal movement, their speed of migration depending on the nature of the defect structure of the oxide lattice. Large stresses, either compressive or tensile, may often build up in oxide films and lead to breakaway effects when the protective oxide film cracks and spalls. Repeated breakaway on a fine scale can prevent the development of extensive parabolic growth and the oxidation assumes an approximately linear rate or even faster. The stresses in oxide film are related to the Pilling–Bedworth (P–B) ratio, defined as the ratio of the molecular volume of the oxide to the atomic volume of the metal from which the oxide is formed.

If the ratio is less than unity as for Mg, Na, K, the oxide formed may be unable to give adequate protection against further oxidation right from the initial stages and under these conditions, commonly found in alkali metals, linear oxidation ($x \propto t$) is obeyed. If, however, the P–B ratio is very much greater than unity, as for many of the transition metals, the oxide is too bulky and may also tend to spall.

At high temperatures oxide films thicken according to the parabolic rate law, $x^2 \propto t$ and the mechanism by which thickening proceeds has been explained by Wagner. Point defects (see Chapter 4) diffuse through the oxide under the influence of a constant concentration gradient. The defects are annihilated at one of the interfaces causing a new lattice site to be formed. Specifically, zinc oxide thickens by the diffusion of zinc interstitials created at the metal/oxide interface through the oxide to the oxide/oxygen interface where they are removed by the reaction

$$2Zn_i^{++} + 4e + O_2 \rightarrow 2ZnO$$

The concentration of zinc interstitials at the metal/oxide interface is maintained by the reaction

$$Zn_{(metal)} \rightarrow Zn_i^{++} + 2e$$

with the creation of vacancies in the zinc lattice. The migration of charged interstitial defects is accompanied by the migration of electrons and for thick oxide films it is reasonable to assume that the concentrations of the two migrating species are constant at the two surfaces of the oxide, i.e. oxide/gas and oxide/metal, governed by local thermodynamic equilibria. There is thus a constant concentration difference Δc across the oxide and the rate of transport through unit area will be $D\Delta c/x$, where D is a diffusion coefficient and x the film thickness. The rate of growth is then

$$dx/dt \propto D(\Delta c/x)$$

and the film thickens parabolically according to the relation

$$x^2 = \mathbf{k}t \tag{12.6}$$

where $\mathbf{k}$ is a constant involving several structural parameters. Wagner has shown that the oxidation process can be equated to an ionic plus an electronic current, and obtained a rate equation for oxidation in chemical equivalents cm^{-2} s^{-1} involving the transport numbers of anions and electrons, respectively, the conductivity of the oxide, the chemical potentials of the diffusing ions at the interfaces and the thickness of the oxide film. Many oxides thicken according to a parabolic law over some particular temperature range. It is a thermally-activated process and the rate constant $k = k_0 \times \exp[-Q/\mathbf{R}T]$ with Q equal to the activation energy for the rate-controlling diffusion process.

At low temperatures and for thin oxide films, a logarithmic rate law is observed. To account for this the Wagner mechanism was modified by Cabrera

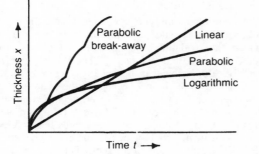

Figure 12.2 *Different forms of oxidation behaviour in metals.*

and Mott. The Wagner mechanism is only applicable when the concentrations of point defects and electrons are equal throughout the film; for thin oxide films this is not so, a charged layer is established at the oxide/oxygen interface. Here the oxygen atoms on the outer surface become negative ions by extracting electrons from the metal underneath the film and so exert an electrostatic attraction on the positive ions in the metal. When the oxide thickness is $\leqslant 10$ nm this layer results in an extremely large electric field being set up which pulls the diffusing ions through the film and accelerates the oxidation process. As the film thickens the field strength decreases as the distance between positive and negative ions increases and the oxidation rate approximates to that predicted by the Wagner theory.

As the scale thickens, according to a parabolic law, the resultant stress at the interface increases and eventually the oxide layer can fail either by fracture parallel to the interface or by a shear or tensile fracture through the layer. In these regions the oxidation rate is then increased until the build-up in stress is again relieved by local fracture of the oxide scale. Unless the scale fracture process occurs at the same time over the whole surface of the specimen then the repeated parabolic nature of the oxidation rate will be smoothed out and an approximately linear law observed. This breakaway parabolic law is sometimes called paralinear, and is common in oxidation of titanium when the oxide reaches a critical thickness. In some metals, however, such as U, W and Ce, the linear rate process is associated with an interface reaction converting a thin protective inner oxide layer to a non-protective porous oxide.

12.2.1.3 Parameters affecting oxidation rates

The Wagner theory of oxidation and its dependence on the nature of the defect structure has been successful in explaining the behaviour of oxides under various conditions, notably the effects of small alloying additions and oxygen pressure variations. The observed effects can be explained by reference to typical n- and p-type semiconducting oxides. For oxidation of Zn to ZnO the zinc atoms enter the oxide interstitially at the oxide/metal interface ($Zn \rightarrow Zn_i{}^{++} + 2e$), and diffuse to the oxide/oxygen interface. The oxide/oxygen interface reaction ($2Zn_i{}^{++} + 4e + O_2 \rightarrow 2ZnO$) is assumed to a rapid (equilibrium) process and consequently, the concentration of defects at this interface is very small, and independent of oxygen pressure. This is found to be the case experimentally for oxide thicknesses in the Wagner region. By considering the oxide as a semiconductor with a relatively small defect concentration, the law of mass action can be applied to the defect species. For the oxide/oxygen interface reaction this means that

$$[Zn^{++}]^2[e]^4 = \text{constant}$$

The effect of small alloying additions can be explained (Wagner–Hauffe rule) by considering this equation.

Suppose an alloying element is added to the metal that enters the oxide on the cation lattice. Since there are associated with each cation site only two electron sites available in the valence band of the oxide, if the element is trivalent the excess electrons enter the conduction band, increasing the concentration of electrons. For a dilute solution the equilibrium constant remains unaffected and hence, from the above equation, the net effect of adding the element will be to decrease the concentration of zinc interstitials and thus the rate of oxidation. Conversely, addition of a monovalent element will increase the oxidation rate. Experimentally it is found that Al decreases and Li increases the oxidation rate.

For Cu_2O, a p-type semiconductor, the oxide formation and cation vacancy ($Cu^+\square$) creation take place at the oxide/oxygen interface, according to

$$O_2 + 4Cu = 2Cu_2O + 4(Cu^+\square) + 4(e\square)$$

The defect diffuses across the oxide and is eliminated at the oxide-metal interface; the equilibrium concentration of defects is at the metal–oxide interface and the excess at the oxide–oxygen interface. It follows therefore that the reaction rate is pressure-dependent. Applying the law of mass action to the oxidation reaction gives

$$[Cu^+\square]^4[e\square]^4 = \text{const } P_{O_2}$$

and, since electrical neutrality requires $[Cu^+\square] = [e\square]$, then

$$[Cu^+\square] = \text{const } P_{O_2}^{1/8} \qquad (12.7)$$

and the reaction rate should be proportional to the 1/8th power of the oxygen pressure. In practice, it varies as $P_{O_2}^{1/7}$, and the discrepancy is thought to be due to the defect concentration not being sufficiently low to neglect any interaction effects. The addition of lower valency cations (e.g. transition metals) would contribute fewer electrons and thereby increase the concentration of holes, decreasing the vacancy concentration and hence the rate. Conversely, higher valency cations increase the rate of oxidation.

12.2.1.4 Oxidation resistance

The addition of alloying elements according to the Wagner–Hauffe rule just considered is one way in which the oxidation rate may be changed to give increased oxidation resistance. The alloying element may be added, however, because it is a strong oxide-former and forms its own oxide on the metal surface in preference to that of the solvent metal. Chromium, for example, is an excellent additive, forming a protective Cr_2O_3 layer on a number of metals (e.g. Fe, Ni) but is detrimental to Ti which forms an n-type anion-defective oxide. Aluminium additions to copper similarly improve the oxidation behaviour by preferential oxidation to Al_2O_3. In some cases, the oxide formed is a compound oxide of both the solute and solvent metals. The best-known examples are the spinels with cubic structure (e.g. $NiO.Cr_2O_3$ and $FeO.Cr_2O_3$). It is probable that the spinel formation is temperature-dependent, with Cr_2O_3 forming at low temperatures and the spinel at higher ones.

Stainless steels (ferritic, austenitic or martensitic) are among the best oxidation-resistant alloys and are based on Fe–Cr. When iron is heated above about 570°C the oxide scale which forms consists of wüstite, FeO (a p-type semiconductor) next to the metal, magnetite Fe_3O_4 (a p-type semiconductor) next and haematite Fe_2O_3 (an n-type semiconductor) on the outside. When Cr is added at low concentrations the Cr forms a spinel $FeO.CrO_3$ with the wüstite and later with the other two oxides. However, a minimum Cr addition of 12% is required before the inner layer is replaced by Cr_2O_3 below a thin outer layer of Fe_2O_3. Heat-resistant steels for service at temperatures above 1000°C usually contain 18% Cr or more, and austenitic stainless steels 18% Cr, 8% Ni. The growth of Cr_2O_3 on austenitic stainless steels containing up to 20% Cr appears to be rate-controlled by chromium diffusion. Kinetic factors determine whether Cr_2O_3 or a duplex spinel oxide form, the nucleation of Cr_2O_3 is favoured by higher Cr levels, higher temperatures and by surface treatments (e.g. deformation), which increase the diffusivity. Surface treatments which deplete the surface of Cr promote the formation of spinel oxide. Once Cr_2O_3 is formed, if this film is removed or disrupted, then spinel oxidation is favoured because of the local lowering of Cr.

When chromium-bearing alloys, such as austenitic stainless steels, are exposed to the hot combustion products of fossil fuels, the outer layer of chromium oxide which forms is often associated with an underlying sulphide phase (Figure 12.3a). This duplex structure can be explained by using phase (stability) diagrams and the concept of 'reaction paths'. Previously, in Section 3.2.8.5, it was indicated that a two-dimensional section could be taken through the full three-dimensional diagram for a metal–sulphur–oxygen system (Figure 3.23). Accordingly, in a similar way, we can extract an isothermal section from the full phase diagram for the Cr–S–O system, as shown in Figure 12.3b. The chemical activities of sulphur and oxygen in the gas phase are functions of their partial pressures (concentration). If the partial pressure of sulphur is relatively low, the composition of the gas phase will lie within the chromium oxide field and the alloy will oxidize (Figure 12.3b). Sulphur and oxygen diffuse through the growing layer of oxide scale but S_2 diffuses faster than O_2, accordingly, the composition of the gas phase in contact with the alloy follows a 'reaction path', as depicted by the dashed line. Figure 12.3c shows the reaction path for gases with a higher initial partial pressure of sulphur. Its slope is such that first chromium oxide forms, and then chromium sulphide (i.e. $Cr + S = CrS$). Sometimes the oxide scale may crack or form voids. In such cases, the activity of S_2

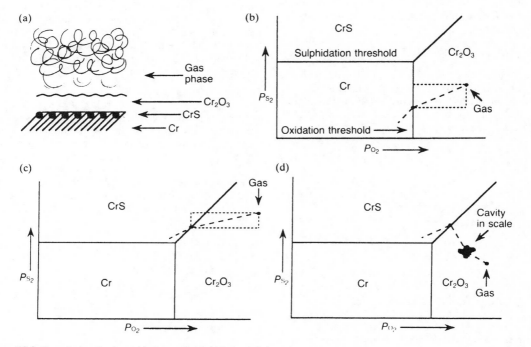

Figure 12.3 *Reaction paths for oxidation and sulphidation of chromium.*

may rise locally within the scale and far exceed that of the main gas phase. Sulphidation of the chromium then becomes likely, despite a low concentration of sulphur in the main gas stream (Figure 12.3d).

Relative tendencies of different metallic elements to oxidize and/or sulphidize at a given temperature may be gauged by superimposing their isothermal $p_{S_2} - p_{O_2}$ diagrams, as in Figure 12.4. For example, with the heat-resistant 80Ni–20Cr alloy (*Nichrome*), it can be reasoned that (1) Cr_2O_3 scale and CrS subscale are both stable in the presence of nickel, and (2) Cr_2O_3 forms in preference to NiO; that is, at much lower partial pressures of oxygen. The physical state of a condensed phase is extremely important because liquid phases favour rapid diffusion and thus promote corrosive reactions. Although nickel has a higher sulphidation threshold than chromium, the Ni–NiS eutectic reaction is of particular concern with Ni-containing alloys because it takes place at the relatively low temperature of 645°C.

12.2.2 Aqueous corrosion

12.2.2.1 Electrochemistry of corrosion

Metals corrode in aqueous environments by an electrochemical mechanism involving the dissolution of the metal as ions (e.g. $Fe \rightarrow Fe^{2+} + 2e$). The excess electrons generated in the electrolyte either reduce hydrogen ions (particularly in acid solutions) according to

$$2H^+ + 2e \rightarrow H_2$$

so that gas is evolved from the metal, or create hydroxyl ions by the reduction of dissolved oxygen according to

$$O_2 + 4e + 2H_2O \rightarrow 4OH^-$$

The corrosion rate is therefore associated with the flow of electrons or an electrical current. The two reactions involving oxidation (in which the metal ionizes) and reduction occur at anodic and cathodic sites, respectively, on the metal surface. Generally, the metal surface consists of both anodic and cathodic sites, depending on segregation, microstructure, stress, etc., but if the metal is partially-immersed there is often a distinct separation of the anodic and cathodic areas with the latter near the waterline where oxygen is readily dissolved (differential aeration). Figure 12.5 illustrates the formation of such a differential aeration cell; Fe^{2+} ions pass into solution from the anode and OH^- ions from the cathode, and where they meet they form ferrous hydroxide $Fe(OH)_2$. However, depending on the aeration, this may oxidize to $Fe(OH)_3$, red-rust $Fe_2O_3.H_2O$, or black magnetite Fe_3O_4. Such a process is important when water, particularly seawater, collects in crevices formed by service, manufacture or design. In this form of corrosion the rate-controlling process is usually the supply of oxygen to the cathodic areas and, if the cathodic area is large, can often lead to intense local attack of small anode areas, such as pits, scratches, crevices, etc.

In the absence of differential aeration, the formation of anodic and cathodic areas depends on the ability to ionize. Some metals ionize easily, others with difficulty and consequently anodic and cathodic areas may be produced, for example, by segregation, or the joining of dissimilar metals. When any metal is immersed in an aqueous solution of its own ions, positive ions go into solution until the resulting electromotive force (emf) is sufficient to prevent any further solution; this emf is the electrode potential or half-cell potential. To measure this emf it is necessary to use a second reference electrode in the solution, usually a standard hydrogen electrode. With no current flowing, the applied potential cancels out the extra potential developed by the spontaneous ionization at the metal electrode over and above that at the standard hydrogen electrode. With different metal electrodes a table of potentials (E_0) can be produced for the half-cell reactions

$$M \rightarrow M^{n+} + ne \qquad (12.8)$$

where E_0 is positive. The usual convention is to write the half-cell reaction in the reverse direction so that the sign of E_0 is also reversed, i.e. E_0 is negative; E_0 is referred to as the standard electrode potential.

It is common practice to express the tendency of a metal to ionize in terms of this voltage, or potential, E_0, rather than free energy, where $\Delta G = -nFE_0$ for the half-cell reaction with nF coulomb of electrical

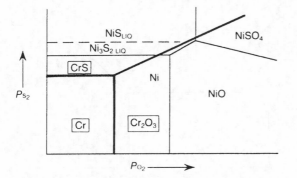

Figure 12.4 *Superimposition of isothermal sections from Cr–S–O and Ni–S–O systems.*

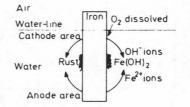

Figure 12.5 *Corrosion of iron by differential aeration.*

Table 12.1 Electrochemical Series

Electrode reaction		Standard electrode potential E_0 (V)
$Cs = Cs^+ + e$		−3.02
$Li = Li^+ + e$		−3.02
$K = K^+ + e$		−2.92
$Na = Na^+ + e$	Reactive metals	−2.71
$Ca = Ca^{2+} + 2e$		−2.50
$Mg = Mg^{2+} + 2e$		−2.34
$Al = Al^{3+} + 3e$		−1.07
$Ti = Ti^{2+} + 2e$		−1.67
$Zn = Zn^{2+} + 2e$		−0.76
$Cr = Cr^{3+} + 3e$		−0.50
$Fe = Fe^{2+} + 2e$		−0.44
$Cd = Cd^{2+} + 2e$		−0.40
$Ni = Ni^{2+} + 2e$		−0.25
$Sn = Sn^{2+} + 2e$		−0.136
$Pb = Pb^{2+} + 2e$		−0.126
$H = 2H^+ + 2e$		0.00
$Cu = Cu^{2+} + 2e$		+0.34
$Hg = Hg^{2+} + 2e$		+0.80
$Ag = Ag^+ + e$	Noble metals	+0.80
$Pt = Pt^{2+} + 2e$		+1.20
$Au = Au^+ + e$		+1.68

charge transported per mole. The half-cell potentials are given in Table 12.1 for various metals, and refer to the potential developed in a standard ion concentration of one mole of ions per litre (i.e. unit activity), relative to a standard hydrogen electrode at 25°C which is assigned a zero voltage. The voltage developed in any galvanic couple (i.e. two half cells) is given by the difference of the electrode potentials. If the activity of the solution is increased then the potential increases according to the Nernst equation $E = E_0 + (RT/nF) \ln a$.

The easily ionizable 'reactive' metals have large negative potentials and dissolve even in concentrated solutions of their own ions, whereas the noble metals have positive potentials and are deposited from solution. These differences show that the valency electrons are strongly bound to the positive core in the noble metals because of the short distance of interaction, i.e. $d_{atomic} \simeq d_{ionic}$. A metal will therefore displace from solution the ions of a metal more noble than itself in the Series. When two dissimilar metals are connected in neutral solution to form a cell, the more metallic metal becomes the anode and the metal with the lower tendency to ionize becomes the cathode. The Electrochemical Series indicates which metal will corrode in the cell but gives no information on the rate of reactions. When an anode M corrodes, its ions enter into the solution initially low in M^+ ions, but as current flows the concentration of ions increases. This leads to a change in electrode potential known as polarization, as shown in Figure 12.6a, and corresponds to a reduced tendency to ionize. The current density in the cell is a maximum when the anode and cathode potential curves intersect. Such a condition would exist if the two metals were joined together or anode and cathode regions existed on the same metal, i.e. differential aeration. This potential is referred to as the corrosion potential and the current, the corrosion current.

In many reactions, particularly in acid solutions, hydrogen gas is given off at the cathode rather than the anode metal deposited. In practice, the evolution of hydrogen gas at the cathode requires a smaller additional overvoltage, the magnitude of which varies considerably from one cathode metal to another, and is high for Pb, Sn and Zn and low for Ag, Cu, Fe and Ni; this overvoltage is clearly of importance in electrodeposition of metals. In corrosion, the overvoltage arising from the activation energy opposing the electrode reaction decreases the potential of the cell,

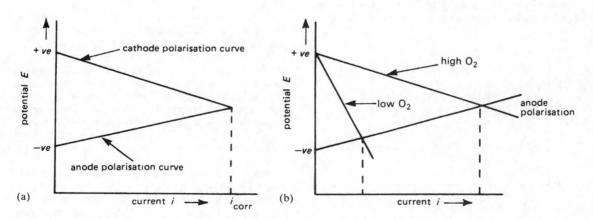

Figure 12.6 Schematic representation of (a) cathode and anode polarization curves and (b) influence of oxygen concentration on cathode polarization.

i.e. hydrogen atoms effectively shield or polarize the cathode. The degree of polarization is a function of current density and the potential E to drive the reaction decreases because of the increased rate of H_2 evolution, as shown in Figure 12.7 for the corrosion of zinc and iron in acid solutions. Corrosion can develop up to a rate given by the current when the potential difference required to drive the reaction is zero; for zinc this is i_{Zn} and for iron i_{Fe}. Because of its large overvoltage zinc is corroded more slowly than iron, even though there is a larger difference between zinc and hydrogen than iron and hydrogen in the Electrochemical Series. The presence of Pt in the acid solution, because of its low overvoltage, increases the corrosion rate as it plates out on the cathode metal surface. In neutral or alkaline solutions, depolarization is brought about by supplying oxygen to the cathode area which reacts with the hydrogen ions as shown in Figure 12.6b. In the absence of oxygen both anodic and cathodic reactions experience polarization and corrosion finally stops; it is well-known that iron does not rust in oxygen-free water.

It is apparent that the cell potential depends on the electrode material, the ion concentration of the electrolyte, passivity and polarization effects. Thus it is not always possible to predict the precise electrochemical behaviour merely from the Electrochemical Series (i.e. which metal will be anode or cathode) and the magnitude of the cell voltage. Therefore it is necessary to determine the specific behaviour of different metals in solutions of different acidity. The results are displayed usually in Pourbaix diagrams as shown in Figure 12.8. With stainless steel, for example, the anodic polarization curve is not straightforward as discussed previously, but takes the form shown in Figure 12.9, where the low-current region corresponds to the condition of passivity. The corrosion rate depends on the position at which the cathode polarization curve for hydrogen evolution crosses this anode curve, and can be quite high if it crosses outside the passive region. Pourbaix diagrams map out the regions of passivity for solutions of different acidity. Figure 12.8 shows that the passive region is restricted to certain conditions of

pH; for Ti this is quite extensive but Ni is passive only in very acid solutions and Al in neutral solutions. Interestingly, these diagrams indicate that for Ti and Ni in contact with each other in corrosive conditions then Ni would corrode, and that passivity has changed their order in the Electrochemical Series. In general, passivity is maintained by conditions of high oxygen concentration but is destroyed by the presence of certain ions such as chlorides.

The corrosion behaviour of metals and alloys can therefore be predicted with certainty only by obtaining experimental data under simulated service conditions. For practical purposes, the cell potentials of many materials have been obtained in a single environment, the most common being sea water. Such data in tabular form are called a Galvanic Series, as illustrated in Table 12.2. If a pair of metals from this Series were connected together in sea water, the metal which is higher in the Series would be the anode and corrode, and the further they are apart, the greater the corrosion tendency. Similar data exist for other environments.

12.2.2.2 Protection against corrosion

The principles of corrosion outlined above indicate several possible methods of controlling corrosion. Since current must pass for corrosion to proceed, any factor, such as cathodic polarization which reduces the current, will reduce the corrosion rate. Metals having a high overvoltage should be utilized where possible. In neutral and alkaline solution de-aeration of the electrolyte to remove oxygen is beneficial in reducing corrosion (e.g. heating the solution or holding under a reduced pressure preferably of an inert gas). It is sometimes possible to reduce both cathode and anode reactions by 'artificial' polarization (for example, by adding inhibitors which stifle the electrode reaction). Calcium bicarbonate, naturally present in hard water, deposits calcium carbonate on metal cathodes and stifles the reaction. Soluble salts of magnesium and zinc act similarly by precipitating hydroxide in neutral solutions.

Anodic inhibitors for ferrous materials include potassium chromate and sodium phosphate, which convert the Fe^{2+} ions to insoluble precipitates stifling the anodic reaction. This form of protection has no effect on the cathodic reaction and hence if the inhibitor fails to seal off the anode completely, intensive local attack occurs, leading to pitting. Moreover, the small current density at the cathode leads to a low rate of polarization and the attack is maintained. Sodium benzoate is often used as an anodic inhibitor in water radiators because of its good sealing qualities, with little tendency for pitting.

Some metals are naturally protected by their adherent oxide films; metal oxides are poor electrical conductors and so insulate the metal from solution. For the reaction to proceed, metal atoms have to diffuse through the oxide to the metal–liquid interface and electrons back through the high-resistance oxide. The

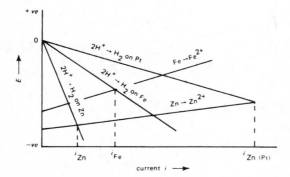

Figure 12.7 *Corrosion of zinc and iron and the effect of polarization.*

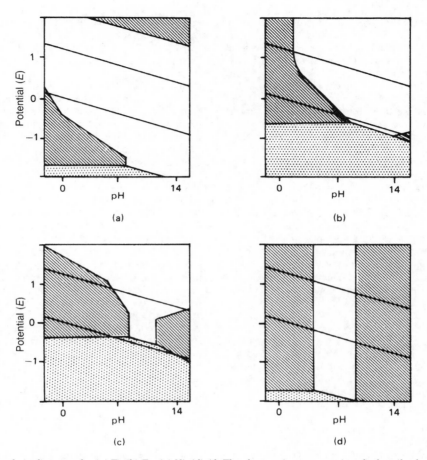

Figure 12.8 *Pourbaix diagrams for (a) Ti, (b) Fe, (c) Ni, (d) Al. The clear regions are passive, the heavily-shaded regions corroding and the lightly-shaded regions immune. The sloping lines represent the upper and lower boundary conditions in service.*

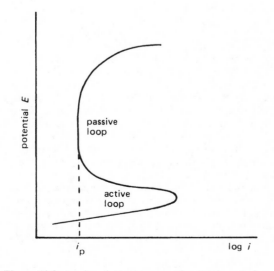

Figure 12.9 *Anode polarization curve for stainless steel.*

Table 12.2 Galvanic Series in sea water

	Anodic or most reactive		
Increasing reactivity	Mg and its alloys	Decreasing reactivity	Cu
	Zn		Ni (active)
	Galvanized steel		*Inconel* (active)
	Al		Ag
	Mild steel		Ni (passive)
	Cast iron		*Inconel* (passive)
	Stainless steel (active)		*Monel*
	Pb		Ti
	Sn		Stainless steel (passive)
	Brass		Cathodic or most noble

corrosion current is very much reduced by the formation of such protective or passive oxide films. Al is cathodic to zinc in sea water even though the Electrochemical Series shows it to be more active. Materials which are passivated in this way are chromium, stainless steels, *Inconel* and nickel in oxidizing conditions. Reducing environments (e.g. stainless steels in HCl) destroy the passive film and render the materials active to corrosion attack. Certain materials may be artificially passivated by painting. The main pigments used are red lead, zinc oxide and chromate, usually suspended in linseed oil and thinned with white spirit. Slightly soluble chromates in the paint passivate the underlying metal when water is present. Red lead reacts with the linseed oil to form lead salts of various fatty acids which are good anodic inhibitors.

Sacrificial or cathodic protection is widely used. A typical example is galvanized steel sheet when the steel is protected by sacrificial corrosion of the zinc coating. Any regions of steel exposed by small flaws in the coating polarize rapidly since they are cathodic and small in area; corrosion products also tend to plug the holes in the Zn layer. Cathodic protection is also used for ships and steel pipelines buried underground. Auxiliary sacrificial anodes are placed at frequent intervals in the corrosive medium in contact with the ship's hull or pipe. Protection may also be achieved by impressing a d.c. voltage to make it a cathode, with the negative terminal of the d.c. source connected to a sacrificial anode.

12.2.2.3 Corrosion failures

In service, there are many types of corrosive attack which lead to rapid failure of components. A familiar example is intergranular corrosion and is associated with the tendency for grain boundaries to undergo localized anodic attack. Some materials are, however, particularly sensitive. The common example of this sensitization occurs in 18Cr–8Ni stainless steel, which is normally protected by a passivating Cr_2O_3 film after heating to 500–800°C and slowly cooling. During cooling, chromium near the grain boundaries precipitates as chromium carbide. As a consequence, these regions are depleted in Cr to levels below 12% and are no longer protected by the passive oxide film. They become anodic relative to the interior of the grain and, being narrow, are strongly attacked by the corrosion current generated by the cathode reactions elsewhere. Sensitization may be avoided by rapid cooling, but in large structures that is not possible, particularly after welding, when the phenomenon (called weld decay) is common. The effect is then overcome by stabilizing the stainless steel by the addition of a small amount (0.5%) of a strong carbide-former such as Nb or Ti which associates with the carbon in preference to the Cr. Other forms of corrosion failure require the component to be stressed, either directly or by residual stress. Common examples include stress-corrosion cracking (SCC) and corrosion-fatigue. Hydrogen embrittlement is sometimes included in this category but this type of failure has somewhat different characteristics and has been considered previously. These failures have certain features in common. SCC occurs in chemically active environments; susceptible alloys develop deep fissures along active slip planes, particularly alloys with low stacking-fault energy with wide dislocations and planar stacking faults, or along grain boundaries. For such

selective chemical action the free energy of reaction can provide almost all the surface energy for fracture, which may then spread under extremely low stresses.

Stress corrosion cracking was first observed in α-brass cartridge cases stored in ammoniacal environments. The phenomenon, called season-cracking since it occurred more frequently during the monsoon season in the tropics, was prevented by giving the cold-worked brass cases a mild annealing treatment to relieve the residual stresses of cold forming. The phenomenon has since extended to many alloys in different environments (e.g. Al–Cu, Al–Mg, Ti–Al), magnesium alloys, stainless steels in the presence of chloride ions, mild steels with hydroxyl ions (caustic embrittlement) and copper alloys with ammonia ions.

Stress corrosion cracking can be either transgranular or intergranular. There appears to be no unique mechanism of transgranular stress corrosion cracking, since no single factor is common to all susceptible alloys. In general, however, all susceptible alloys are unstable in the environment concerned but are largely protected by a surface film that is locally destroyed in some way. The variations on the basic mechanism arise from the different ways in which local activity is generated. Breakdown in passivity may occur as a result of the emergence of dislocation pile-ups, stacking faults, micro-cracks, precipitates (such as hydrides in Ti alloys) at the surface of the specimen, so that highly localized anodic attack then takes place. The gradual opening of the resultant crack occurs by plastic yielding at the tip and as the liquid is sucked in also prevents any tendency to polarize.

Many alloys exhibit coarse slip and have similar dislocation substructures (e.g. co-planar arrays of dislocations or wide planar stacking faults) but are not equally susceptible to stress-corrosion. The observation has been attributed to the time necessary to repassivate an active area. Additions of Cr and Si to susceptible austenitic steels, for example, do not significantly alter the dislocation distribution but are found to decrease the susceptibility to cracking, probably by lowering the repassivation time.

The susceptibility to transgranular stress corrosion of austenitic steels, α-brasses, titanium alloys, etc. which exhibit co-planar arrays of dislocations and stacking faults may be reduced by raising the stacking-fault energy by altering the alloy composition. Cross-slip is then made easier and deformation gives rise to fine slip, so that the narrower, fresh surfaces created have a less severe effect. The addition of elements to promote passivation or, more importantly, the speed of repassivation should also prove beneficial.

Intergranular cracking appears to be associated with a narrow soft zone near the grain boundaries. In α-brass this zone may be produced by local dezincification. In high-strength Al-alloys there is no doubt that it is associated with the grain boundary precipitate-free zones (i.e. PFZs). In such areas the strain-rate may be so rapid, because the strain is localized, that repassivation cannot occur. Cracking then proceeds even though the slip steps developed are narrow, the crack dissolving anodically as discussed for sensitized stainless steel. In practice there are many examples of intergranular cracking, including cases (1) that depend strongly on stress (e.g. Al-alloys), (2) where stress has a comparatively minor role (e.g. steel cracking in nitrate solutions) and (3) which occur in the absence of stress (e.g. sensitized 18Cr–8Ni steels); the last case is the extreme example of failure to repassivate for purely electrochemical reasons. In some materials the crack propagates, as in ductile failure, by internal necking between inclusions which occurs by a combination of stress and dissolution processes. The stress sensitivity depends on the particle distribution and is quite high for fine-scale and low for coarse-scale distributions. The change in precipitate distribution in grain boundaries produced, for example, by duplex ageing can thus change the stress-dependence of intergranular failure.

In conditions where the environment plays a role, the crack growth rate varies with stress intensity K in the manner shown in Figure 12.10. In region I the crack velocity shows a marked dependence on stress, in region II the velocity is independent of the stress intensity and in region III the rate becomes very fast as K_{IC} is approached. K_{ISC} is extensively quoted as the threshold stress intensity below which the crack growth rate is negligible (e.g. $\lesssim 10^{-10} \mathrm{m\ s^{-1}}$) but, like the endurance limit in fatigue, does not exist for all materials. In region I the rate of crack growth is controlled by the rate at which the metal dissolves and the time for which the metal surface is exposed. While anodic dissolution takes place on the exposed metal at the crack tip, cathodic reactions occur at the oxide film on the crack sides leading to the evolution of hydrogen which diffuses to the region of triaxial tensile stress and hydrogen-induced cracking. At higher stress intensities (region II) the strain-rate is higher, and then other processes become rate-controlling, such as diffusion of new reactants into the crack tip region. In hydrogen embrittlement this is probably the rate of hydrogen diffusion.

The influence of a corrosive environment, even mildly oxidizing, in reducing the fatigue life has been briefly mentioned in Chapter 7. The $S-N$ curve shows no tendency to level out, but falls to low S-values. The damage ratio (i.e. corrosion fatigue strength divided by the normal fatigue strength) in salt water environments is only about 0.5 for stainless steels and 0.2 for mild steel. The formation of intrusions and extrusions gives rise to fresh surface steps which form very active anodic sites in aqueous environments, analogous to the situation at the tip of a stress corrosion crack. This form of fatigue is influenced by those factors affecting normal fatigue but, in addition, involves electro-chemical factors. It is normally reduced by plating, cladding and painting but difficulties may arise in localizing the attack to a small number of sites, since the surface is continually being deformed. Anodic inhibitors may also reduce the corrosion fatigue but their use is more limited than in the absence of fatigue because of the probability of incomplete inhibition leading to increased corrosion.

Fretting corrosion, caused by two surfaces rubbing together, is associated with fatigue failure. The oxidation and corrosion product is continually removed, so that the problem must be tackled by improving the mechanical linkage of moving parts and by the effective use of lubricants.

With corrosion fatigue, the fracture mechanics threshold ΔK_{th} is reduced and the rate of crack propagation is usually increased by a factor of two or so. Much larger increases in crack growth rate are produced, however, in low-frequency cycling when stress-corrosion fatigue effects become important.

12.3 Surface engineering

12.3.1 The coating and modification of surfaces

The action of the new methods for coating or modifying material surfaces, such as vapour deposition and beam bombardment, can be highly specific and energy-efficient. They allow great flexibility in controlling the chemical composition and physical structure of surfaces and many materials which resisted conventional treatments can now be processed. Grain size and the degree of crystalline perfection can be varied over a wide range and beneficial changes in properties produced. The new techniques often eliminate the need for the random diffusion of atoms so that temperatures can be relatively low and processing times short. Scientifically, they are intriguing because their nature makes it possible to bypass thermodynamic restrictions on alloying and to form unorthodox solid solutions and new types of metastable phase.

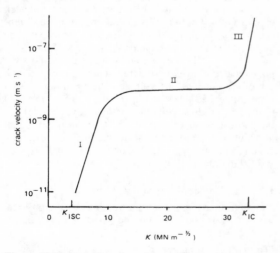

Figure 12.10 *Variation of crack growth rate with stress intensity during corrosion.*

Table 12.3 Methods of coating and modifying surfaces (after R. F. Bunshah, 1984; by permission of Marcel Dekker)

Atomistic deposition	Particulate deposition	Bulk coatings	Surface modification
Electrolytic environment	Thermal spraying	Wetting processes	Chemical conversion
Electroplating	Plasma-spraying	Painting	Electrolytic
Electroless plating	Detonation-gun	Dip coating	Anodizing (oxide)
Fused salt electrolysis	Flame-spraying	Electrostatic spraying	Fused salts
Chemical displacement	Fusion coatings	Printing	Chemical-liquid
Vacuum environment	Thick film ink	Spin coating	Chemical-vapour
Vacuum evaporation	Enamelling	Cladding	Thermal
Ion beam deposition	Electrophoretic	Explosive	Plasma
Molecular beam epitaxy	Impact plating	Roll bonding	Leaching
Plasma environment		Overlaying	Mechanical
Sputter deposition		Weld coating	Shot-peening
Activated reactive evaporation		Liquid phase epitaxy	Thermal
Plasma polymerization			Surface enrichment
Ion plating			Diffusion from bulk
Chemical vapour environment			Sputtering
Chemical vapour deposition			Ion implantation
Reduction			Laser processing
Decomposition			
Plasma enhanced			
Spray pyrolysis			

The number and diversity of methods for coating or modifying surfaces makes general classification difficult. For instance, the energies required by the various processes extend over some five orders of magnitude. Illustrating this point, sputtered atoms have a low thermal energy (<1 eV) whereas the energy of an ion beam can be >100 keV. A useful introductory classification of methods for coating and modifying material surfaces appears in Table 12.3, which takes some account of the different forms of mass transfer. The first column refers to coatings formed from atoms and ions (e.g. vapour deposition). The second column refers to coatings formed from liquid droplets or small particles. A third category refers to the direct application of coating material in quantity (e.g. paint). Finally, there are methods for the near-surface modification of materials by chemical, mechanical and thermal means and by bombardment (e.g. ion implantation, laser processing).

Some of the methods that utilize deposition from a vapour phase or direct bombardment with particles, ions or radiation will be outlined: it will be apparent that each of the processes discussed has three stages: (1) a source provides the coating or modifying specie, (2) this specie is transported from source to substrate and (3) the specie penetrates and modifies the substrate or forms an overlay. Each stage is, to a great extent, independent of the other two stages, tending to give each process an individual versatility.

12.3.2 Surface coating by vapour deposition

12.3.2.1 Chemical vapour deposition

In the chemical vapour deposition (CVD) process a coating of metal, alloy or refractory compound is produced by chemical reaction between vapour and a carrier gas at or near the heated surface of a substrate (Figures 12.11a and 12.11b). CVD is not a 'line-of-sight' process and can coat complex shapes uniformly, having good 'throwing power'.[1] Typical CVD reactions for depositing boron nitride and titanium carbide, respectively, are:

$$BCl_{3(g)} + NH_{3(g)} \xrightarrow{500-1500°C} BN_{(s)} + 3HCl_{(g)}$$

$$TiCl_{4(g)} + CH_{4(g)} \xrightarrow{800-1000°C} TiC_{(s)} + 4HCl_{(g)}$$

It will be noted that the substrate temperatures, which control the rate of deposition, are relatively high. Accordingly, although CVD is suitable for coating a refractory compound, like cobalt-bonded tungsten carbide, it will soften a hardened and tempered high-speed tool steel, making it necessary to repeat the high-temperature heat-treatment. In one variant of the process (PACVD) deposition is plasma-assisted by a plate located above the substrate which is charged with a radio-frequency bias voltage. The resulting plasma zone influences the structure of the coating. PACVD is used to produce ceramic coatings (SiC, Si_3N_4) but the substrate temperature of 650°C (minimum) is still too high for heat-treated alloy steels. The maximum coating thickness produced economically by CVD and PACVD is about 100 μm.

12.3.2.2 Physical vapour deposition

Although there are numerous versions of the physical vapour deposition (PVD) process, their basic design is

---

[1]The term 'throwing power' conventionally refers to the ability of an electroplating solution to deposit metal uniformly on a cathode of irregular shape.

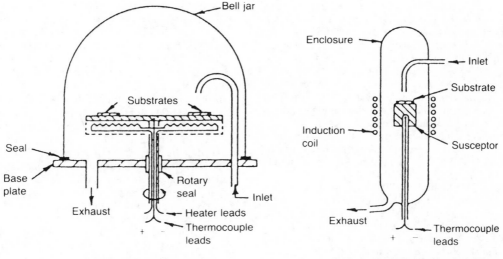

(a) Resistively heated plate   (b) Inductively heated pedestal

Figure 12.11 *Experimental CVD reactors (from Bunshah, 1984; by permission of Marcel Dekker).*

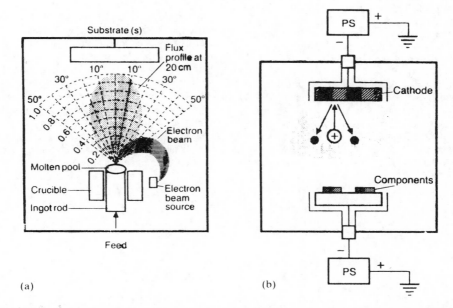

(a)   (b)

Figure 12.12 *(a) Evaporation-dependent and (b) sputter-dependent PVD (from Barrell and Rickerby, Aug 1989, pp. 468–73; by permission of the Institute of Materials).*

either evaporation- or sputter-dependent. In the former case, the source material is heated by high-energy beam (electron, ion, laser), resistance, induction, etc. in a vacuum chamber (Figure 12.12a). The rate of evaporation depends upon the vapour pressure of the source and the chamber pressure. Metals vaporize at a reasonable rate if their vapour pressure exceeds 1 N m^{-2} and the chamber pressure is below 10^{-3} N m^{-2}. The evaporant atoms travel towards the substrate (component), essentially following lines-of-sight.

When sputtering is used in PVD (Figure 12.12b), a cathode source operates under an applied potential of up to 5 kV (direct-current or radio-frequency) in an atmosphere of inert gas (Ar). The vacuum is 'softer', with a chamber pressure of $1-10^{-2} \text{ N m}^{-2}$. As positive argon ions bombard the target, momentum is transferred and the ejected target atoms form a coating on the substrate. The 'throwing power' of sputter-dependent PVD is good and coating thicknesses are uniform. The process benefits from the fact that the sputtering yield (Y) values for metals are fairly similar. (Y is the average number of target atoms ejected from the surface per incident ion, as determined experimentally.) In contrast, with an evaporation source, for a given temperature, the rates of vaporization can differ by several orders of magnitude.

As in CVD, the temperature of the substrate is of special significance. In PVD, this temperature can be as low as 200–400°C, making it possible to apply the method to cutting and metal-forming tools of hardened steel. A titanium nitride (TiN) coating, <5 μm thick, can enhance tool life considerably (e.g. twist drills). TiN is extremely hard (2400 HV), has a low coefficient of friction and a very smooth surface texture. TiN coatings can also be applied to non-ferrous alloys and cobalt-bonded tungsten carbide. Experience with the design of a TiN-coated steel has demonstrated that the coating/substrate system must be considered as a working whole. A sound overlay of wear-resistant material on a tough material may fail prematurely if working stresses cause plastic deformation of the supporting substrate. For this reason, and in accordance with the newly-emerging principles of surface engineering, it has been recommended that steel surfaces should be strengthened by nitriding before a TiN coating is applied by PVD.

Two important modifications of the PVD process are plasma-assisted physical vapour deposition (PAPVD) and magnetron sputtering. In PAPVD, also known as 'ion plating', deposition in a 'soft' vacuum is assisted by bombardment with ions. This effect is produced by applying a negative potential of 2–5 kV to the substrate. PAPVD is a hybrid of the evaporation-and sputter-dependent forms of PVD. Strong bonding of the PAPVD coating to the substrate requires the latter to be free from contamination. Accordingly, in a critical preliminary stage, the substrate is cleansed by bombardment with positive ions. The source is then energized and metal vapour is allowed into the chamber.

In the basic magnetron-assisted version of sputter-dependent PVD, a magnetic field is used to form a dense plasma close to the target. The magnetron, an array of permanent magnets or electromagnets, is attached to the rear of the target (water-cooled) with its north and south poles arranged to produce a magnetic field at right angles to the electric field between the target and substrate (Figure 12.13a). This magnetic field confines electrons close to the target surface, increases the rate of ionization and produces a much denser plasma. The improved ionization efficiency allows a lower chamber pressure to be used; sputtered target atoms then become less likely to be scattered by gas molecules. The net effect is to improve the rate of deposition at the substrate. Normally the region of dense plasma only extends up to about 6 cm from the target surface. Development of unbalanced magnetron systems (Figure 12.13b) has enabled the depth of the dense plasma zone to be extended so that the substrate itself is subjected to ion bombardment. These energetic ions modify the chemical and physical properties of the deposit. (In one of the various unbalanced magnetron configurations, a ring of strong rare-earth magnetic poles surrounds a weak central magnetic pole.) This larger plasma zone can accommodate large complex workpieces and rapidly forms dense, non-columnar coatings of metals or alloys. Target/substrate separation distances up to 20 cm have been achieved with unbalanced magnetron systems.

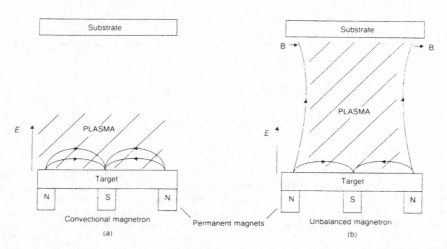

Figure 12.13 *Comparison of plasma confinement in conventional and unbalanced magnetrons (PVD) (from Kelly, Arnell and Ahmed, March 1993, pp. 161–5; by permission of the Institute of Materials).*

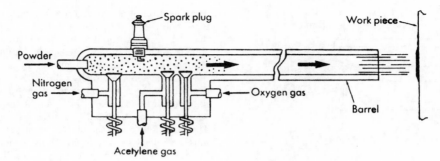

Figure 12.14 *Coating by detonation-gun (from Weatherill and Gill, 1988; by permission of the Institute of Materials).*

12.3.3 Surface coating by particle bombardment

Since the first practical realization of gas turbine engines in the 1940s, the pace of engineering development has largely been prescribed by the availability of suitable high-temperature materials. Components in the most critical sections of the engine are exposed to hot products of combustion moving at high velocity. In addition, there are destructive agents passing through the engine, such as sea salt and sand. In this hostile environment, it is extremely difficult, if not impossible, to develop an alloy that combines the necessary high-temperature strength with corrosion resistance. Much effort has therefore been devoted to the search for alloy systems that will develop a thin self-healing 'protective' oxide scale. In practice, this outer layer does not prevent diffusing atoms from reaching and reacting with the alloy substrate and it may also be subject to thinning by erosion. The difference in thermal expansion between the oxide (ceramic) scale and the metallic substrate can lead to rupture and spalling of the scale if the scale lacks plasticity or is weakly bonded to the alloy. Refractory coatings which resist wear and corrosion provide one possible answer to these problems.

The two established thermal spray methods[1] of coating selected here for brief description are used for gas turbine components. In thermal spraying, powders are injected into very hot gases and projected at very high velocities onto the component (substrate) surface. On impact, the particles plastically deform and adhere strongly to the substrate and each other. The structure, which often has a characteristic lenticular appearance in cross-section, typically comprises refractory constituents, such as carbide, oxide and/or aluminide, and a binding alloy phase. Many types of thermally sprayed coatings can operate at temperatures >1000°C. They range in thickness from microns to millimetres, as required.

In the detonation-gun method (Figure 12.14) a mixture of metered quantities of oxygen and acetylene

(C_2H_2) is spark-ignited and detonated. Powder of average diameter 45 mm is injected, heated by the hot gases and projected from the 1 m long barrel of the gun onto the cooled workpiece at a velocity of roughly 750 m s^{-1}. Between detonations, which occur four to eight times per second, the barrel is purged with nitrogen. Typical applications, and coating compositions, for wear-resistant *D-Gun* coatings are bearing sealing surfaces (WC-9Co), compressor blades (WC-13Co) and turbine blade shroud interlocks (Cr$_3$C$_2$/80Ni–20Cr).

In the plasma-spray technique, powder is heated by an argon-fed d.c. arc (Figure 12.15) and then projected on to the workpiece at velocities of 125–600 m s^{-1}. A shielding envelope of inert gas (Ar) is used to prevent oxidation of the depositing material. The process is used to apply MCrAlY-type coatings to turbine components requiring corrosion resistance at high temperatures (e.g. blades, vanes), where M signifies the high-m.p. metals Fe, Ni and/or Co. These coatings can accommodate much more of the scale-forming elements chromium and aluminium than superalloys (e.g. 39Co–32Ni–21Cr–7.5Al–0.5Y). They provide a reservoir of oxidizable elements and allow the 'protective' scale layer to regenerate itself. The small amount of yttrium improves scale adhesion. This particular composition of coating is used for hot gas path seals in locations where a small clearance between the rotating blades and the interior walls of the engine gives greater fuel efficiency. These coatings will withstand occasional rubbing contact.

12.3.4 Surface modification with high-energy beams

12.3.4.1 Ion implantation

The chemical composition and physical structure at the surface of a material can be changed by bombarding it, *in vacuo*, with a high-velocity stream of ions. The beam energy is typically about 100 keV; efforts are being made to increase the beam current above 5 mA so that process times can be shortened. Currently, implantation requires several hours. The ions

---

[1]The Union Carbide Corporation has been granted patent rights for the *D-Gun* and plasma-spraying methods.

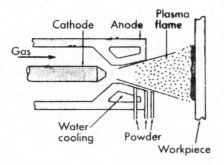

Figure 12.15 *Coating by plasma-spray torch (from Weatherill and Gill, 1988; by permission of the Institute of Materials).*

may be derived from any element in the Periodic Table: they may be light (most frequently nitrogen) or heavy, even radioactive. Ion implantation[1] is a line-of-sight process; typically, a bombardment dose for each square centimetre of target surface is in the order of 10^{17}–10^{19} ions. These ions penetrate to a depth of 100–200 nm and their concentration profile in a plane normal to the surface is Gaussian. Beyond this modified region, the properties of the substrate are unaffected.

The beam usually has a sputtering effect which ejects atoms from the surface and skews the concentration profile. This effect is most marked when heavy ions or heavy doses are used. It is possible for a steady state to be achieved, with the rate of sputter erosion equal to the rate of implantation. Thus, depending upon the target, the type and energy of ion and the substrate material, sputter erosion is capable of limiting the amount of implantation possible. As a general guide, the maximum concentration of implanted ion is given, roughly, by the reciprocal of the sputtering yield (Y). As one would expect, Y increases in value with increases in ion energy. However, Y values for pure metals are broadly similar, being about 1 or 2 for typical argon ion energies and not differing from each other by more than an order of magnitude. Thus, because of sputter, the maximum concentration of implanted ions possible is in the order of 40–50 at.%. In cases where it is difficult to attain this concentration, a thin layer of the material to be implanted is first deposited and then driven into the substrate by bombardment with inert gas ions (argon, krypton, xenon). This indirect method is called 'ion beam mixing'.

During ion bombardment each atom in the near-surface region is displaced many times. Various forms of structural damage are produced by the cascades of collisions (e.g. displacement spikes, vacancy/interstitial (Frenkel) pairs, dislocation tangles and loops, etc.). Damage cascades are most concentrated when heavy ions bombard target atoms of high atomic number (Z).

[1]Pioneered by the UKAEA, Harwell, in the 1960s.

The injection of atoms and the formation of vacancies tend to increase the volume of the target material so that the restraint imposed by the substrate produces a state of residual compressive stress. Fatigue resistance is therefore likely to be enhanced.

As indicated previously, the ions penetrate to a depth of about 300–500 atoms. Penetration is greater in crystalline materials than in glasses, particularly when the ions 'channel' between low-index planes. The collision 'cross-section' of target atoms for light ions is relatively small and ions penetrate deeply. Ion implantation can be closely controlled, the main process variables being beam energy, ion species, ion dose, temperature and substrate material.

Ion implantation is used in the doping of semiconductors, as discussed in Chapter 6, and to improve engineering properties such as resistance to wear, fatigue and corrosion. The process temperature is less than 150°C; accordingly, heat-treated alloy steels can be implanted without risk of tempering effects. Nitrogen-implantation is applied to steel and tungsten carbide tools, and, in the plastics industry, has greatly improved the wear resistance of feed screws, extrusion dies, nozzles, etc. The process has also been used to simulate neutron damage effects in low-swelling alloys being screened for use in atomic fission and fusion reactors. A few hours' test exposure to an ion beam can represent a year in a reactor because the ions have a larger 'cross-section' of interaction with the atoms in the target material than neutrons. However, ions cannot simulate neutron behaviour completely; unlike neutrons, ions are electrically charged and travel smaller distances (see Chapter 6).

12.3.4.2 Laser processing

Like ion implantation, the laser[2] process is under active development. A laser beam heats the target material locally to a very high temperature; its effects extend to a depth of 10–100 μm, which is about a thousand times greater than that for an ion beam. Depending on its energy, it can heat, melt, vaporize or form a plasma. The duration of the energy pulse can be 1 ns or less. Subsequent cooling may allow a metallic target zone to recrystallize, possibly with a refined substructure, or undergo an austenite/martensite transformation (e.g. automotive components). There is usually an epitaxial relation between the altered near-surface region and the substrate. Cooling may even be rapid enough to form a glassy structure (laser glazing). Surface alloying can be achieved by pre-depositing an alloy on the substrate, heating this deposit with a laser beam to form a miscible melt and allowing to cool. In this way, an integral layer of austenitic corrosion-resistant steel can be built on a ferritic steel substrate. In addition to

[2]Light Amplification by Stimulated Emission of Radiation (LASER) devices provide photons of electromagnetic radiation that are in-phase (coherent) and monochromatic (see Chapter 6).

its use in alloying and heat-treatment, laser processing is used to enhance etching and electroplating. (e.g. semiconductors).

The principal variables in laser processing are the energy input and the pulse duration. For established techniques like cutting, drilling and welding metals, the rate of energy transfer per unit area ('power density') is in the order of $1 \, MW \, cm^{-2}$ and pulses are of relatively long duration (say, 1 ms). For more specialized functions, such as metal hardening by shock wave generation, the corresponding values are approximately $100 \, MW \, cm^{-2}$ and 1 ns. Short pulses can produce rapid quenching effects and metastable phases.

Further reading

Bell, T. (1992). Surface engineering: its current and future impact on tribology. *J. Phys D.: Appl. Phys.* **25**, A297–306.

Bunshah, R. F. (1984). Overview of deposition technologies with emphasis on vapour deposition techniques. *Industrial Materials Science and Engineering*, Chapter 12 (L.E. Murr, (ed.)). Marcel Dekker, New York.

Picraux, S. T. (1984). Surface modification of materials — ions, lasers and electron beams. *Industrial Materials Science and Engineering*, Chapter 11 (L.E. Murr, (ed.)). Marcel Dekker, New York.

Shreir, L. L. (1976). *Corrosion*, Vol. 1 and 2, 2nd edn. Newnes-Butterworth, London.

Trethewey, K. R. and Chamberlain, J. (1988). *Corrosion for students of science and engineering*. Longman, Harlow.

Chapter 13

Biomaterials

13.1 Introduction

Biomaterials are materials used in medicine and dentistry that are intended to come in contact with living tissue. The familiar tooth filling is where most humans first encounter biomaterials but increasingly many people now rely on more critical implants such as joint replacements, particularly hips, and cardiovascular repairs. Undoubtedly, these biomaterial implants improve the quality of life for an increasing number of people each year, not just for an ageing population with greater life expectancy, but for younger people with heart problems, injuries or inherited diseases.

Biomaterials have now been successfully developed and used for more than a generation. First-generation biomaterials largely depended on being inert, or relatively inert, with minimal tissue response. For these materials a minimal fibrous layer forms between the biomaterials and the body when the material is not totally accepted by the body. The success of this type of implant depends largely on the selection of materials for their manufacture. Thus the now standard hip replacement (almost a million worldwide each year) initially used a multi-component assembly made with austenitic stainless steel for the stem, PMMA for fixation and polyethylene for the acetabular cup (see Figure 13.1). All the materials proved relatively bio-inert and gave an average life time of 10 years or more.

Nowadays, while continuing with improved bio-inert materials, development has focused on bioactive materials which influence the biological response in a positive way, e.g. encourage bonding to surrounding tissue with stimulation of new bone growth. With this bio-active approach the interface between the body cells and the implant is critical and the materials science of the biomaterials surface extremely important.

In this chapter, various applications of biomaterials will be examined, from dental materials to drug delivery systems. All types of materials are used in these applications and the criteria governing their selection

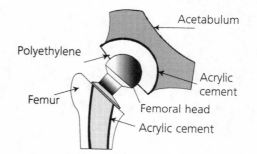

Figure 13.1 *Schematic diagram of a replacement hip joint.*

will be considered together with future development in the biomaterials field.

13.2 Requirements for biomaterials

The requirements for a biomaterial are extremely demanding. Replacement or repair of a body feature, tissue, organ or function often necessitates the material used to have specialised mechanical, physical and chemical properties. However, the very first requirement is biocompatibility with the human body, i.e. the ability of the material to perform with an appropriate host response. Unfortunately, no material is universally biocompatible, since a material may be biocompatible in one application but not with another. Biocompatibility is therefore application specific.

For the successful use of the biomaterial, consideration has to be given to the appropriate material selection, engineering design and manufacturing process. While proper design and manufacture is essential, it is particularly important to select the correct material to provide the appropriate properties as well as being biocompatible, recognizing that the combined influence of mechanical and chemical factors can be quite serious, e.g. causing fatigue, corrosion fatigue, stress corrosion, wear, fracture. It is also important to recognize that the biological environment is not constant and that oxygen levels, availability of free radicals and cellular activity will vary. Corrosion and degradation can lead to loss of integrity of the implant and, of course, release ions into the body, often setting up an allergic reaction.

Biomaterial applications make use of all classes of material, metals, ceramics, polymers and composites, divided roughly into three usertypes. These are (i) inert or relatively inert with minimal host response, (ii) bioactive which actually stimulates bonding to the surrounding tissue and (iii) biodegradable which resorb in the body over a period of time. Metals are generally chosen for their inert qualities whereas ceramics and polymers may offer bioactivity or resorption.

The most common metallic materials used are austenitic stainless steels, cobalt–chromium alloys or titanium; typical compositions are shown in Table 13.1. Recently, titanium alloys, particularly Ti–6Al–4V, have been introduced because of their corrosion resistance, strength and elastic modulus (see Table 13.2) but poor tribology can still be a problem. It is also favoured for its superior biocompatibility and, unlike Co–Cr or stainless steel, does not cause hypersensitivity. Of the ceramics, aluminium oxide, calcium phosphate, apatite, carbon/graphite and bioglass are in use mainly for their inertness, good wear characteristics, high compressive strength and in some cases bioactivity. Their poor tensile properties and fracture toughness are design limitations. Polymers are widely used, both alone and in combination with ceramics or metals. These include; polymethyl methacrylate (PMMA) for cement and lenses; polyethylene for orthopaedics; polyurethane as blood contact material, e.g. vascular tubing, cardiovascular devices, catheters; polysiloxanes in plastic surgery, maxillofacial and cardiovascular surgery; polyesters and polyamides in wound closure management. Composites such as ultra-high-molecular-weight polyethylene reinforced with either carbon fibres or the ceramic hydroxyapatite are increasingly being considered for applications involving high contact stress and wear resistance.

13.3 Dental materials

13.3.1 Cavity fillers

Dentistry has always been very dependent on the use of biomaterials and particularly receptive to the application of new developments in metals, ceramics, polymers and composites.

Dental amalgams have been used in cavity fillings for more than 100 years, initially with silver–mercury amalgams, later modified by tin additions to control the amount of expansion. These amalgams produced the weak, corrodible intermetallic γ_2 phase, Sn_7Hg, and hence the modern dental amalgam now also contains copper ($>12\%$) in order to suppress this phase. The amalgam is made by mixing silver, tin, copper alloy powder with mercury and this mixture is packed into the cavity where it hardens to produce a strong, corrosion-resistant, biocompatible filling. There is some evidence that even this filling may be susceptible to corrosion as a result of the Cu_6Sn_5 (η') phase and the addition of Pd has been advocated. Attempts to replace the Hg amalgam by gallium, indium, silver, tin, copper pastes have not yet been completely successful.

Alternative resin-based composite filling materials have been continuously developed since first introduced in the 1960s. These composite fillings have a

Table 13.1 Composition of orthopaedic implant alloys (wt%); from Bonfield, 1997

| Element | Cobalt-base alloys | | | Stainless steel | | Titanium alloys | |
	ASTM F75 cast	ASTM F90 wrought	ASTM F563 isostatically pressed	ASTM F138/A	ASTM F138/9B	Commercial purity titanium	Ti–6Al–4V
Co	Balance	Balance	Balance	—	—	—	—
Cr	27–30	19–12	18–22	17–20	17–20	—	—
Fe	0.75 max	3.0 max	4–6	Balance	Balance	0.3–0.5	0.25 max
Mo	5–7	—	3–4	2–4	2–4	—	—
Ni	2.5 max	9–11	15–25	10–14	10–14	—	—
Ti	—	—	0.5–3.5	—	—	Balance	Balance
Al	—	—	—	—	—	—	5.5–6.5
V	—	—	—	—	—	—	3.5–4.5
C	0.35 max	0.05–0.15	0.05 max	0.03 max	0.08 max	0.01 max	0.08 max
Mn	1.0 max	2.0 max	1.0 max	2.0 max	2.0 max	—	—
P	—	—	—	0.03 max	0.025 max	—	—
S	—	—	0.01 max	0.03 max	0.01 max	—	—
Si	1.0 max	1.0 max	0.5 max	0.75 max	0.75 max	—	—
O	—	—	—	—	—	0.18–0.40	0.13 max
H	—	—	—	—	—	0.01–0.015	0.012 max
N	—	—	—	—	—	0.03–0.05	0.05 max

Table 13.2 The mechanical properties of some natural and biomaterials

Material	Elastic modulus (GN m^{-2})	Tensile strength (MN m^{-2})	Elongation (%)	Fracture toughness (MN $m^{-3/2}$)	Fatigue strength (MN m^{-2})
Austenitic stainless steel	200	200–1100	40	100	200–250
Cobalt–Chromium	230	450–1000	10–30	100	600
Ti–6Al–4V	105–110	750–1050	12	80	350–650
Alumina	365	—	< 1		400
Hydroxyapatite	85	40–100	—	—	
Glass fibre	70	2000	2	1–4	
PMMA	2.8	55	8	—	20–30
Bone cement	3–2.3	1.5	1–2	400	
Polyethylene	1	20–30		1–4	16
Nylon 66	4.4	700	25		
Silicone rubber	6×10^{-3}	1.4			
Polycarbonate	2	60			
Bone (cortical)	7–25	50–150	—	2–12	
Bone (cancellous)	0.1–1.0	50–150		2–12	
Tooth enamel	13	240	—	—	
Tooth dentine	—	135		—	
Collagen, tendon, wet	2	100	10	—	

strength similar to amalgams but poorer wear properties. The paste is created by mixing a dimethacrylate monomer with resin and adding a filler of micron-sized silane-coated ceramic particles. The paste is activated by strong light when the resin polymerizes. Bonding of the composite resin to the tooth structure employs a phosphoric acid etch of the tooth enamel. This produces mini-chasms into which the resin material flows and locks to form a strong mechanical bond. This technique is not successful, however, for bonding to the dentine in the tooth cavity (Figure 13.2) and so, in the absence of enamel, dentine bonding agents have to be used. These are primers containing bifunctional compounds with (i) hydrophilic molecules which form links with the wet dentine in the tooth cavity and (ii) hydrophobic molecules which form links with the resin in the composite.

Cavities in front teeth are usually filled with glass cements to match the colour and translucency of the enamel. Silicate cements are formed when phosphoric acid displaces metal ions from an alumina–silica glass, containing metal oxides and fluorides. The cement sets when aluminium phosphate is precipitated between the glass particles. Developments based on this basic chemistry employ polymeric acids with carboxylate groups. In this case, the metal ions displaced from the glass cross-link with the polymeric acid chain, causing the cement to set. In addition, the acids undergo an ion-exchange reaction with the calcium phosphate in the apatite of the dental material. These glass ionomer cements therefore form direct chemical bonds to the tooth material. Resin-modified versions are also available which have improved durability; these contain carboxylate groups to give a good bond to the tooth and dimethacrylate, as in the composite resin.

13.3.2 Bridges, crowns and dentures

Missing teeth may be replaced by artificial teeth in a number of different ways. For a group of missing teeth, removal partial dentures (RPDs) may be the answer which consist of a cast metal framework of Co–Cr or Ni–Cr alloy carrying the artificial teeth and having end clasps to retain it to good natural teeth nearby. Fixed partial dentures (FPDs or bridges) may be used for a few missing teeth. Sometimes the supporting teeth are cut down to accommodate a close-fitting artificial tooth casting which is cemented into place. In other cases, the alloy framework carrying the artificial teeth is bonded to acid-etched teeth to avoid cutting down good teeth. The teeth are acid etched and the metal framework electrolytically etched to produce structural grooves and chasms which allow strong mechanical bonds to be formed with resin-based composite cements. In some situations, etching can be avoided when the oxides on the metal framework can be treated with bifunctional primers to form chemical links to the cement.

Over the last 20 years or so the quality of bonded restorations, i.e. porcelain-covered metal castings, has

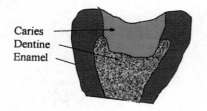

Caries
Dentine
Enamel

Figure 13.2 *Schematic diagram of a tooth.*

been refined to combine the impact strength of a metal substructure with the appearance of dental porcelain. These porcelains have good mechanical properties and their thermal expansion characteristics are matched to the metal in order to avoid interfacial stresses and cracking.

Dental porcelains are basically vitrified feldspar with metallic oxide pigments to simulate natural tooth enamel. They are usually supplied to the dental laboratories in the reacted and ground form for final fabrication by the technician; this involves mixing the powder with distilled water to form a paste which is used to make the crown, then drying and firing in order to sinter and densify the crown material. Generally, firing is carried out in stages starting with the innermost structure of the crown, followed by the body and finally the outer glaze and surface staining. Developments have included the strengthening of the inner core material with alumina to prevent cracking; also the addition of magnesia to form a magnesia–alumina spinel which has a low shrinkage on firing. Glass-ceramics have also been used either to fabricate the crown by casting followed by heat treatment to produce crystallization, or by machining from a pre-fired block of glass-ceramic under CAM/CAD conditions.

For complete replacement dentures, the basic material, which has existed for many decades, is methyl methacrylate. Substitute materials have been limited and most improvements and developments have occurred in the processing technology. While the mechanical properties of denture base resins are not particularly good (modulus of elasticity 3×10^9 N m^{-2}, tensile strength $\sim$100 kN m^{-2}, elongation $\sim$3%) they do have suitable surface and abrasion properties and are chemically inert, non-toxic and cheap. Improvements have been forthcoming in elastomers used for taking impressions; these now include vinyl addition silicone and polyether elastomers.

13.3.3 Dental implants

Dental implants have been far less developed than those associated with body implants (see hip joints, etc.). Probably the simplest forms are posts, Co–Cr, stainless steel, titanium alloy or gold alloys cemented, or even screwed, into the tooth canal after the tooth has been root treated to remove the nerve. Dental porcelain caps may then be cemented onto the root post. Ti implants have been screwed into the bone beneath extracted teeth. After some time the passive surface layer of the titanium implants becomes osseo-integrated with the bone and can be used as a strong base onto which a titanium mini structure can be fitted, complete with tooth assembly. Osseo-integration is improved by using a coating on the titanium implant such as hydroxyapatite or bioglass (Figure 13.3). Ceramic and carbon implants set into the bone have been used with sapphire single crystals and pyrolitic graphite as favoured materials.

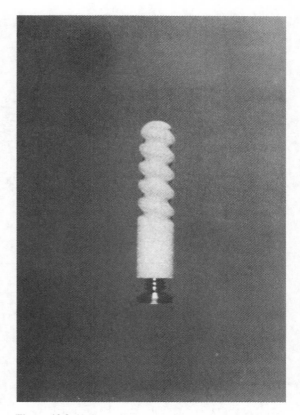

Figure 13.3 *Hydroxyapatite-coated titanium root implant courtesy P. Marquis, Dental School, Birmingham.*

13.4 The structure of bone and bone fractures

In repairing bone fractures it is necessary to hold the bone together while the natural healing process takes place, usually for a few months. Metallic alloys have been universally used for plates, pins, screws, etc. They are considered to have the required strength and rigidity together with sufficient corrosion resistance. Generally, these fixtures are considered temporary until the bone has joined after which they are removed. The main reason for this is that the very rigidity with which the bone is held to allow healing will eventually lead to progressive weakening of the bone structure.

The biological structure of bone is reproduced in Figure 13.4. In material science terms, the apparent complexity of bone can be described as a composite made up of a matrix of collagen (polymer) reinforced with approximately 50% volume fraction of hydroxyapatite (ceramic) nanometre-scale crystals. Most bones are made of a porous cellular structure (cancellous bone) covered with a denser compact shell. The porosity and density of the cancellous bone varies with location in the body depending on the stress state in that region. Low density regions have a relatively open

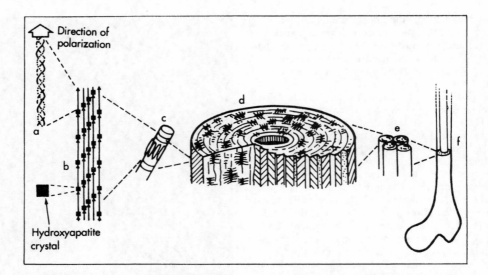

Figure 13.4 *The hierarchy of structure in bone. At the molecular level **a**, polarized triple helix of tropocollagen molecules assembles into a microfibril **b**, with small gaps between the ends of the molecules into which small (5 by 30 nm) crystals of hydroxyapatite later form. These microfibrils assemble into larger fibrils **c**, which then form the layers in the osteon (**d**—part cut away to show the alternating orientation of fibres in the annular layers). Osteons form in association with each other **e**, forming bone **f**. The cells which are responsible for most of this process, the osteocytes, are shown sitting between the layers of the osteon **d**. (after Vincent, 1990; courtesy Institute of Materials).*

cell structure and high density regions more closed. The density determines the strength and stiffness of the bone which grows and develops to support the stress imposed on it. This may be uniaxial, when the cell walls will be oriented and thicker along this direction, or more uniform when the cells are roughly equiaxed.

Bone is not a static entity but dynamic in nature, continually undergoing remodelling where 'old' bone is resorbed and replaced by new bone. Various factors control the process but extremely important in stimulating the bone-producing cells (osteoblasts) is the application of stress. In bone, the mineral phosphate material is slowly dissolved and resorbed by the body and when subjected to normal stress is replenished by new bone material synthesized by the osteoblasts. This recycling process ensures a healthy, strong bone structure with ageing. Bones for which the major stress is carried by metal implants do not show the same tendency for replenishment so that the bone surrounding the implant is resorbed without replenishment leading to a loosening of the implant. A rigid metal plate attached to a bone that has healed will nevertheless carry the majority of the load. To avoid this 'stress shielding' leading to bone weakness it is recommended that the surgeon removes the holding plates soon after the fracture has healed.

With the aim of trying to avoid the removal operation, alternative approaches are being considered for bone plates, screws, etc., including the use of biodegradable, or resorbable, materials such as polymers or composites which could be resorbed into the

surrounding tissue or dissolve completely over a period of time after the fracture has healed. However, the combination of strength, ductility, toughness, rigidity and corrosion resistance of metals is hard to match with the non-metallics.

The large difference in elastic modulus between competing biomaterials and bone (see Table 13.2) is evident. Of the various metals, titanium and its alloys are clearly the most suitable and are being increasingly used. Titanium does, however, have a tribological weakness but the application of coatings and surface engineering is being increasingly adopted to overcome this problem.

13.5 Replacement joints

13.5.1 Hip joints

The basic design of an artificial hip joint includes an alloy femoral stem, with a metallic or ceramic femoral head moving in an ultra-high-molecular-weight polyethylene (UHMWPE) acetabular cup, as shown in Figures 13.1 and 13.5. The average life of the joint is better than 10 years but implants tend to loosen as a result of bone resorption due to modulus mismatch. Friction and wear also cause wear debris between the cancellous, i.e. more porous, bone and the cup and also between the femoral head and the softer UHMWPE. Overall the failure rate is now about 1–2% per year, as shown in Figure 13.6.

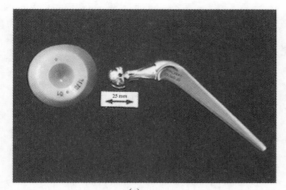

(a)

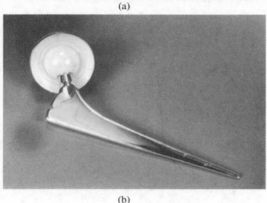

(b)

Figure 13.5 *(a) Conventional hip joint made of stainless steel and UHMWPE acetabular cup (courtesy R. Grimer, Royal Orthopaedic Hospital, Birmingham). (b) Modular artifical hip made of titanium stem, alumina head and UHMWPE acetabular cup (courtesy P. Marquis, Dental School, Birmingham).*

Friction and wear problems are being improved by using ultra-hard materials as the bearing material. A good resistance to frictional wear together with biocompatibility makes alumina well suited for the femoral head. Unfortunately, Al_2O_3 has a low impact strength and is liable to failure when subjected to stresses introduced by extra activity, e.g. jumping. Research has also shown that ion implantation of the UHMWPE in a nitrogen atmosphere at 10^{-5} mbar at 80 keV to a dose of 1×10^{17} ions cm^{-2} reduces the wear behaviour to virtually zero. The enchanced surface properties together with higher hardness and elastic moduli would very much improve the interaction problems. A compromise solution is to have a titanium stem with stainless steel head coated with a thin layer of ceramic.

The problem of loosening of the joint due to bone resorption is being tackled by the second generation of biomaterials and implants which mimic the body's tissues. One such material developed by Professor Bonfield and his colleagues at Queen Mary and Westfield College is a composite of hydroxyapatite (HA) in high-density polyethylene. The HA provides the strengthening and stiffness reinforcement to the tough polyethylene matrix. *HAPEX*, as it is called, is biocompatible and bioactive, encouraging bone growth onto the implant material which, of course, contains HA. The biological response has been demonstrated by cell culture studies with human oesteoblasts in which the cells grew and spread over the composite initiated at the HA particles. *In vivo* testing has shown a stable interface between the implant and bone.

13.5.2 Shoulder joints

These joints, like hip joints, are increasingly being replaced by metal prostheses (Figure 13.7). Initially, similar principles were applied to shoulder joints as to hip joints, using a stainless steel ball and socket with polyethylene cup. However, because it is not a load-bearing joint the trend in shoulder replacement is simply to resurface the worn part on the humeris side with a stainless steel cap rather than replace both sides of the joint.

13.5.3 Knee joints

Total knee replacements (TKR) are increasingly being made to give pain-free improved leg function to many thousands of patients suffering from osteoarthritis, or rheumatoid arthritis. As with hip replacements, TKR involves removing the articulating surfaces of the affected knee joint and replacing them with artificial components made from biomaterials. Such operations have a successful history and a failure rate of less than 2% per year (Figure 13.6). A typical replacement joint consists of a tibial base plate or tray, usually made of stainless steel, Co–Cr, or titanium alloy, with a tibial insert (UHMWPE) that acts as the bearing surface (see Figure 13.8). The femoral component largely takes the shape of a natural femoral condyle made from the above alloys, and articulates with the bearing surface, together with the knee cap or patella. The patella may be all polyethylene or metal backed. The components are either fixed by cement or uncemented. Cemented components use acrylic cement (polymethyl methacrylate). Uncemented components rely on bone ingrowth into the implant. Titanium used on its own shows evidence of bone ingrowth, but more recently this has been improved by use of hydroxyapatite (HA) coating. Problems include wear debris from the PTFE and possible fracture of the component.

One of the continuing challenges is improved design against failure, currently about 2%, as a result of overloading, or increased physical activity, and possibly bone resorption.

13.5.4 Finger joints and hand surgery

Replacement finger and hand joints are far more complex than other joints because of the degree of flexibility required through large angles while maintaining overall stability. Early designs were based on

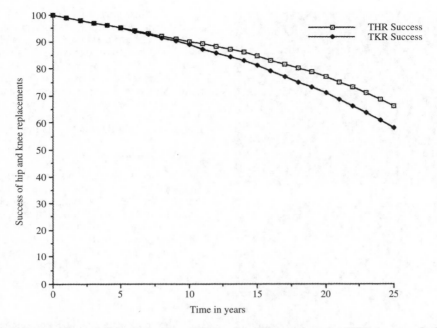

Figure 13.6 *The failure rate for a total hip replacement (THR) and total knee replacement (TKR) (courtesy R. Grimer, Royal Orthopaedic Hospital, Birmingham).*

the movement of a metal 'hinge' but had a problem of fatigue and/or corrosion. Nowadays silicone rubber is more commonly used, showing little tendency to fatigue failure while being biocompatible with an ability to absorb lipids and fatty acids.

Polysiloxane is used in hand surgery to form a 'tunnel' to allow transplanted tendons to slide back and forward. It is also used to replace carpal bones. It is, however, not suitable for longer joints or bones unless reinforced.

13.6 Reconstructive surgery

13.6.1 Plastic surgery

The many different biomaterials used in plastic surgery include collagen, silicone (polydimethylsiloxane), *Teflon*, polyethylene, *Dacron* and polyglocolic acid. Polysiloxanes are widely used in reconstructive plastic surgery because of their lack of any tissue reactivity, mechanical properties and structure. Polysiloxane may be either heat vulcanized or vulcanized at room temperature for more delicate structures and is ideal for soft tissue replacement where repeated flexure occurs. In applications where strength is also required, steel wire reinforcement is necessary. The material is used not only for implants but also for explants where it is moulded to fit a specific shape, coloured to give a skin match and fixed with tissue adhesive. Common examples are ear prosthesis or facial reconstruction after loss due to injury or cancer surgery. Silicone mammary implants are now widely used for cosmetic

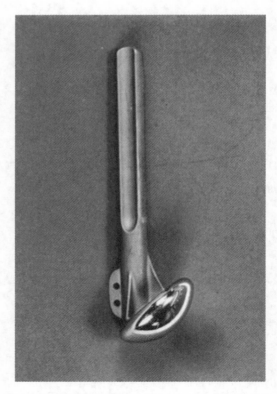

Figure 13.7 *Shoulder joint prosthesis (courtesy P. Marquis, Dental School, Birmingham).*

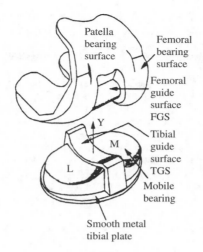

(a)

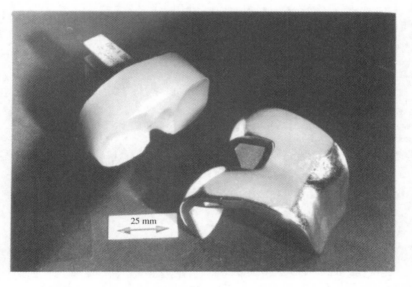

(b)

Figure 13.8 *A total knee replacement joint (a) schematic diagram (after Walker and Sathasiwan, 1999 and (b) photograph of a stainless steel prosthesis (courtesy R. Grimer, Royal Orthopaedic Hospital, Birmingham).*

reasons or after mastectomy for breast cancer. These have an outer shell with the appropriate shape and resilience covering an inner volume of silicone gel. In these reconstructions *Dacron* may be used as a lining for the ear and as a backing in a breast implant to provide a better fastening of the implant to the surrounding tissue.

13.6.2 Maxillofacial surgery

Disease or injury to the oral and facial area may be repaired by implants from a range of biomaterials.

Often the effective prostheses can improve the confidence and well-being of the patient beyond the immediate functional repair. The jaw, and jaw bone area, can utilize many of the metal implants already discussed. Commercially pure titanium in perforated sheet form has been used because of its biocompatibility and ease of manipulation and fixing.

Polymers, particularly silicones and polyurethanes, may be used to replace flexible tissues of the nose, cheek and ear regions of the face. Polysiloxanes have been used for onlays in the area of the molar bone in the lateral side of the mandible or over the forehead to smooth it. Reinforced with stainless steel wires they can also replace the mandible. A porous composite of polytetrafluoroethylene (PTFE) strengthened with carbon fibres may be used to replace damaged bone structures. Both PTFE and carbon are biocompatible and fibrous tissue growth into the pores ensures bonding of the artificial and natural bone. Composites of hydroxyapatite in a polyethylene matrix *HAPEX* have been used for patients who had either fractured the orbital floor supporting the eye or had lost an eye. A great advantage of such material implants is that they can be shaped during the operation and inserted on the base of the eye socket and bonded firmly to the supporting bone.

13.6.3 Ear implants

HAPEX has been also used successfully for other clinical replacements, mainly middle-ear implants. These transmit sound from the outer to inner ear where the vibrations are translated into electrical signals to be processed by the brain. Middle-ear malfunction can lead to deafness (conduction deafness) which may be cured by implant surgery. Otosclerosis (middle-ear deafness) may result from fibrosis of the middle ear caused by repeated infections or from an hereditary disease. A stapedectomy removes a small amount of bone and immobilized tissue and replaces it with biomaterial. The implant has a hydroxyapatite head on a *HAPEX* shaft which can be trimmed and shaped in the operating theatre using an ordinary scalpel to fit the individual patient.

13.7 Biomaterials for heart repair

13.7.1 Heart valves

Heart disease is one of the major killers in the developed world. Many of the serious conditions arise from the strain imposed on the heart by obstructions to the flow of blood either in the main passages of the circulatory system or as a result of valvular disease. The heart acts as a blood pump with four valves which open in response to a unidirectional flow of blood. Two of the valves allow the blood into the heart and the other two control the blood leaving the heart. Problems can arise with the heart valves if they become structurally damaged by disease affecting the opening and closing mechanisms.

The early prostheses developed in the 1960s for valve replacement were based on a stainless steel ball-in-cage or polysiloxane ball in a Co–Cr alloy cage (see Figure 13.9). These restrict the blood flow, even in the open position, and were superseded by a tilting disc device which opens and closes to the beat of the heart to allow the required blood flow. The major problem with valves arose from the tendency to initiate a blood clot and thus it was found necessary

to give valve recipients anticoagulant drugs. Nowadays, advanced valves are made with Co–Cr or titanium bodies with metal or graphite discs, or occluders, coated with pyrolytic carbon. These coatings are made by heating a hydrocarbon, such as methane, to about 1500°C depositing the carbon vapour on the graphite surface to a thickness of about 1 mm. Small amounts of silane mixed with the CH_4 adds Si to the deposit, increasing its strength. Pyrolytic carbon is strong and wear resistant but, more importantly, resists the formation of blood clots on its surface. The discs are attached by a coated Ti metal arm to a fabric ring made of polymer (PTFE) which is sewn to the tissue of the heart valve opening. *Dacron* cloth has also been used and encourages tissue growth with better anchorage and thromboembolic resistance.

The prosthesis has moving parts and thus catastrophic failure by fracture is a finite possibility. As an alternative, the construction of artificial valves from biological tissue has been developed. Collageneous tissue from the heart wall of cows and heart valves from pigs have been used to make these 'bioprosthetic' valves. These valves are naturally biocompatible with

(a)

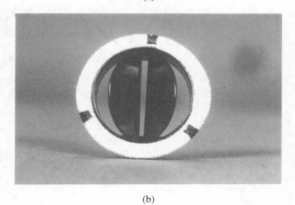

(b)

Figure 13.9 *Photographs of replacement heart valves: (a) ball in a cage type (courtesy Institute of Materials, London), and (b) pyrolitic carbon disc type (courtesy P. Marquis, Dental School, Birmingham).*

a reduced risk of thrombosis. Unfortunately two problems have hindered this development. The first is that the collagen valve material suffers from slow calcification whereby hard deposits form on the surface of the valves causing them to stick and tear. The use of anticalcification drugs are a possibility but the second problem has thrown the whole area of implant surgery using animal tissue in doubt. This problem is the emergence of BSE (mad cow disease) in cattle, which has led to restrictions and worries in the use of animal tissue for reconstructive implant surgery because of the fear of transmission of viral illness from animal tissue to humans.

13.7.2 Pacemakers

Other developments in improving heart performance include the use of cardiac pacemakers which produces a 5V electrical impulse for 1/500 second at regular heartbeat rate. These devices have been available for some time but have improved significantly over recent years. The basic requirement is to provide electrical signals to the heart at the appropriate level to stimulate the patient's own electrical activity to produce the proper physiological change, normally linking the pacemaker into the cardiac system so that it works when needed. The biomaterial aspects of the pacemaker are, however, also important, not only in overcoming the problems introduced by any device/body environmental interaction, but also in designing the proper electrical supply and insulation. Power supplies have advanced considerably in the last few years and lithium cells are now exclusively used. Titanium is again the most common biomaterial to encase the device, manufactured and electron beam welded to seal it hermetically. Polymers have been used for encapsulation, e.g. epoxy resin or silicone rubber but these materials do not completely prevent moisture from entering the pacemaker and shortening the lifetime of the device. These have now been superseded by titanium alloys because of their better strength and environmental properties. These are sutured into the aorta with a *Dacron* sleeve.

Another problem area is that provided by the electrodes which have to flex with every heartbeat and hence are liable to fatigue failure (see Chapter 7). Good design and choice of electrode materials can minimize this problem. The electrical supply passes through the titanium casing via a ceramic insulator and the leads to the heart are insulated with a polymer (polyetherurethane). Degradation with time is still a possibility and has to be considered in an effective design.

13.7.3 Artificial arteries

In branches of surgery, particularly cardiovascular surgery, there is often a need to replace arteries blocked by atherosclerosis. Sometimes this can be achieved by using tissue grafts from the patients, thereby avoiding any immune response. In other cases it is necessary to use artificial arteries made from polymers; such arteries must be tough and flexible enough to avoid kinking, with the added requirement of avoiding the formation of blood clots. In modern surgery, the blood clotting tendency can be removed by anticoagulants, such as heparin, but the ultimate goal is to provide artificial arteries with natural clot resistance. Several different polymers have been used to make blood vessels but none is entirely satisfactory.

The polyester *Dacron*, can make small tubes but these have porous walls which have to be sealed. This is achieved by treating with the protein albumin and heating it to form a coagulated coating. In the body, the albumin degrades and is replaced by the natural protein collagen forming a smooth lining (pseudointima). This process leads to some initial inflammation which is one disadvantage of this biomaterial. Woven *Dacron* is quite rigid and unsuitable for small arteries and is difficult to suture; it is mostly used for resected aortic aneurysms. Knitted *Dacron* is easier to suture. It may be coated with polyurethane, tetrafluoroethylene, or heparin to reduce the thrombogenic tendency. The use of poly(hydroxyethyl methacrylate) coating establishes an endothelial-like cell layer in a few weeks.

Other arterial polymers include PTFE and polyurethanes. PTFE is used as an expanded foam to form the porous tubes. These rapidly develop a smooth neointima layer and thus acquire blood compatibility. Polyurethanes have a natural compatibility with blood and are tough and flexible in tube form but unfortunately slowly degrade in the body producing toxic products. PTFE coatings on polyester and polyurethane vessels have also been tried. Silicone-lined tubing has been used for extra-corporeal circulation during open-heart surgery.

To produce artificial arteries with built-in clot resistance, heparin molecules have been attached to their surface either directly by chemical bonds or by cross-linking to form a polymerized heparin film. To mimic total thrombosis resistance, however, requires not only anticoagulation but also avoidance of platelet deposition normally achieved by the endothelial cells lining the blood vessel releasing the protein prostacyclin. Ideal artificial arteries should have both of these anti-clotting agents attached at their surface.

13.8 Tissue repair and growth

Tissues include skin, tendons, ligaments and cartilages. Skin has the dual property of keeping the body fluids in while allowing the outward movement of moisture through a porous membrane, which is important in cooling and maintaining the body temperature. Skin also protects against infections, such as bacteria but is not, of course, particularly strong. It is made up of layers including an outer epidermis and an inner dermis, a dense network of nerve and blood vessels. It is therefore virtually impossible to make an artificial skin from biomaterials to match this complexity. Nevertheless skin replacements have been made from polymers with an open structure which provides a basic framework onto which real skin is able to grow. Moreover, with a biodegradable polymer the framework degrades as the new tissue regrows. The porous film can be coated with silicone rubber to provide infection protection and retain fluids while the skin grows. When sutured in place, tissue-forming cells (fibroblasts) migrate into the porous polymer framework to generate new skin layers. For severe burns, artificial skin can be made by growing epidermal skin cells within a biodegradable collagen mesh in a culture medium. The synthetic skin can then be grafted onto the patient. Other biodegradable products include the copolymers lactic acid–glycolic acid and lysine–lactic acid. The adhesion of the polymer framework can be improved by incorporating an adhesive protein fibronectin.

Other tissues such as ligaments and cartilages are largely elastic filaments of fibrous proteins. Synthetic substitutes have included *Dacron* polyesters, PTFE fibres and pyrolyzed carbon fibres, with mixed success. The fibres may be coated with polylactic acid polymer which breaks down in the body to be replaced by collagen. At this stage such techniques are relatively new but it does suggest that in future the growth of cells in a culture vessel may possibly supply complex biomaterials for various implanted functions. This approach is termed tissue engineering and together with biometrics, i.e. the mimicking of the working of biological systems, offers a way of producing materials which totally synergize with the human body.

13.9 Other surgical applications

Polymers are used in a wide variety of surgical applications. Polyurethane has good tissue and blood contact properties and is used in both short-term applications, e.g. catheters, endotracteal tubes, vascular tubing, haemodialysis parts, and long-term applications, e.g. heart-assist devices. *Dacron* is used in composite form with a poly(2-hyroxyethyl methacrylate) matrix for orthopaedic tendon reconstruction. Varying the composite mix can alter the properties to match the requirements. Reinforced, *Dacron* fabric is used for reconstructing the trachea and, in woven *Dacron* form, for small bowel repair or replacement. It is also used in the genitourinary systems, in mesh form in repairing hernias and abdominal wall defects. Polysiloxanes are used in neurological surgery, e.g. in valves to drain fluids produced intercranially and also as tubes to drain other canals such as the middle ear. Not having a porous structure, these tubes resist bacterial contamination.

Shape–memory–effect (SME) alloys (see Chapter 9, section 9.6.4), particularly Ni–Ti, have been used in several biomedical applications because of their unique behaviour and for their biocompatibility. In orthopaedics, for example, pre-stressed fracture bone plates can be made to shrink on heating to provide a rigid, compressive load fixing. By contrast, Ni–Ti rods can be programmed to provide traction on local heating. In other applications Ni–Ti has been used in artificial heart muscles, teeth-straightening devices, intrauterine contraceptive devices and as a filter in the vena cava (inserted cold the filter opens its mesh at the temperature of the deoxygenated blood flowing back to the heart).

In spinal surgery, commercially pure titanium cables and screws have been used for the correction of scoliosis by gradual tightening of the cable to straighten the spine. A big advantage of Ti for these devices is its resistance to crevice corrosion.

13.10 Ophthalmics

The main usage of biomaterials in ophthalmics is for hard contact lenses, soft contact lenses, disposable contact lenses and artificial interocular lenses. Apart from being easier to wear soft contact lenses have several advantages over hard lenses. Soft lenses ride on a thinner tear film fitting closer to the shape of the cornea generally giving better corneal health, less irritation from dust under the lens and less likelihood of being dislodged.

For satisfactory lens performance the biomaterial must be inert, dimensionally stable to retain its optical properties, have a low density ($\sim$1g cm^{-3}), a refractive index close to that of the cornea (1.37) and good oxygen permeability to maintain a healthy cornea. Hard contact lenses first became available using PMMA because it was light (density 1.19g cm^{-3}), easy to shape, has a refractive index of 1.49 and is reasonably biocompatible. They are, however, difficult to wear for long periods because PMMA is (i) hydrophobic and not easily wetted by eye fluids and (ii) has a very low oxygen permeability. These have now been superseded by soft lenses.

A wide variety of materials have been used for soft lenses including silicones, hydroxyethyl methacrylate (HEMA) and copolymers with HEMA as the major component and vinyl pyrrolidine to increase the water content. Most soft lenses are made from hydrogels; poly(2-hydroxyethyl methacrylate) is still the most popular, containing a small amount of ethylene glycol dimethacrylate to act as a cross-linking agent. These hydrogel materials have excellent compatibility and other properties but even with good oxygen permeability (100 times better than PMMA) additional oxygen transplant via tear exchange is necessary. To increase the permeability the water content of the lens is raised but too high a water content can result in reduced strength and poor handleability.

After cataract surgery and removal, polysiloxane coated with wetting agents such as polyvinyl pyrrolidine (PVP) has been used. For intraocular problems, silicone injectable has been used.

13.11 Drug delivery systems

Oral administration of tablets is a familiar method for taking medicine. Less familiar is regular injection. Both methods have many disadvantages not least of which is that the drug is being used as a general body medicine when, ideally, it is required only at some specific site in the body. A second disadvantage is the variation in drug level with time, as it is metabolized from a high to an insignificant rate rather than a steady, more moderate rate. Nowadays, these disadvantages are being addressed by developing controlled drug delivery; in some cases to specified tissues or organs.

One form of controlled drug delivery and targeting system uses polymeric biomaterial to contain the drug which then escapes by diffusion. The diffusion pathway through the polymer is provided by the gaps in the chain-like structure which may be varied to control the release rate. One such release system uses a copolymer of lactic and glycolic acids to target peptide-based drugs to reduce prostate cancer. Another approach uses a biodegradable material chemically bonded to the drug. In the body the drug is gradually released as the material degrades, giving a continued release over a known degradation period at a specific site within the body.

A more ambitious system delivers a drug in response to blood chemistry levels, such as the release of insulin in response to glucose level and has been used with some success. The implanted micro-infusion system consists of a titanium reservoir of insulin together with a micro-pump which delivers the insulin via a fine catheter. In principle, a glucose sensor could provide a feedback control to the reservoir to complete the creation of an artificial pancreas.

Further reading

Ball, P. (1998). Spare parts—biomedical materials, in Chapter 5, *Measure for Measure*, Princeton University Press, Princeton, NJ, USA.

Bhumbra, R. S. *et al.*, (1998). Enhanced bone regeneration and formation around implants. *J. Biomed. Mater. Res. (Appl. Biomater.)*, **43**, 162–167.

Bonfield, W. (1992). Can materials stimulate advances in orthopaedics? p. 168, *Science of New Materials*, Blackwell, Oxford, UK.

Bonfield, W. (1997). Biomaterials—a new generation. *Materials World*, January, p. 18, Institute of Materials.

Brown, D. (1994). Polymers in dentistry. *Progress in Rubber and Plastics Technology*, **10**, 185.

Brown, D. (1996). Filling the gap. *Materials World*, May, p. 259, Institute of Materials.

Helson, J. E. F. A. and Jürgan Brime, H. (eds) (1998). *Metals as Biomaterials*. John Wiley and Sons Ltd.

Ratner, B. (ed.) (1996). *Biomaterials Science*. Academic Press, New York, USA.

Vincent, J. (1990). Materials technology from nature. *Metals and Materials*, June, p. 395, Institute of Materials.

Williams, D. F. (1991). Materials for surgical implants. *Metals and Materials*, January, p. 24, Institute of Materials.

Chapter 14

Materials for sports

14.1 The revolution in sports products

Within a relatively short period of time, say one generation, sport has come into remarkable worldwide prominence; in the wake of this phenomenon, innovative and highly specialized industries have emerged. Athletic sports, irrespective of the particular form of competition, involve the application and transmission of force and the expenditure of energy and, as a natural consequence, one of the attendant trends is to take advantage of the latest developments in materials science and engineering and to use the latest generation of new materials for sporting artefacts such as rackets, golf clubs, skis, vaulting-poles, etc. Fortuitously, parallel long-term activities in laboratories of the aerospace industries, among others, have helped to provide and sustain a basic armoury of versatile new materials: it is now generally appreciated by the viewing public that these materials have sport-transforming capabilities. (Indeed, their introduction has sometimes led to urgent reappraisal of the rule book.) With their aid, a human can kick, throw or strike a ball farther, cleave air or water at greater speed and, in the event of the occasional mishap, survive with less risk of personal injury.

In some sports, traditional materials hold sway and there are few material changes. In others, new materials and designs appear with bewildering rapidity, being driven by the competitive spirit of the sports equipment industry and competitors alike. With this dynamic background in mind, we have emphasized principles wherever possible. Also, rather than attempting to encompass sports such as automobile and yacht racing, which involve extremely large financial outlays, we have focused upon the more individual sports; that is, upon sports which engage both professional and amateur and which are essentially personal and physical. This restriction still leaves an extremely broad arena of sporting activity, so we have concentrated on certain items of sports equipment which provide an insight into the processes of engineering design and manufacture; topics include a revolutionary polymer-moulding technique (tennis racket), bending stiffness and flexure (golf club), energy storage and transfer (archery bow), joining problems (bicycle frames), duplex steel composites (fencing foils), durability at low temperatures (snowboard bindings) and shock absorption (safety helmets). The vital interplay between structure and properties will be self-evident. Time and time again, ultimate commercial success has depended upon a precise identification of crucial material properties during the design and development stage.

As a preliminary, we have taken this opportunity to describe the structure and properties of wood. Wood provides a benchmark for alternative materials in sport and many mimic its cellular structure. In the more general context of engineering materials, wood remains extremely important in tonnage and particularly volumetric terms. Wood has always been, and remains, the most widely used material for construction; arguably, it should appear in any preliminary list of candidate materials for any engineering design project.

14.2 The tradition of using wood

Wood is a biopolymeric composite. Large long-chain molecules of cellulose comprise about 50–60% of the structure and form hollow, elongated cells (tracheids). The principal orientation of tracheids is parallel to the axis of the trunk, giving wood its well-known grain. Typically, these fluid-conveying cells are 0.5–5 mm long and have an aspect ratio of 100:1. In temperate zones of climate, new layers of cells grow beneath the bark during spring (large cells, thin walls) and summer (small cells, thick walls). Other, more elongated cells, known as ray cells, form at right angles to the tracheids and carry fluids transversely across the trunk. Growth ceases during winter, giving rise to the visible 'growth rings' in the trunk's cross-section. Sap diffuses up the outer annulus of sapwood while the heartwood provides a supportive core. According to botanical classification, broad-leaved trees provide hardwoods (ash, oak, elm, maple, balsa, etc.) and coniferous trees with needle-shaped leaves provide softwoods (pine, spruce, fir, etc.).

Differences in the wall structure, shape and size of the cells are responsible for the many different types of wood. Within the strong and elastic wall of each tracheid, different types of layers of cellulose microfibrils have been identified (Figure 14.1). In the primary (P) layer, microfibrils are randomly orientated. In the three secondary (S) layers, microfibril orientation varies from layer to layer. Thus, in the thick, middle S2 layer, the alignment of microfibrils at a slight angle to the major axis of the cell is mainly responsible for the high tensile strength along the grain. The cells are cemented together, between their primary walls, by a non-crystalline phase which is rich in lignin. This middle lamina (ML) is weaker than the cellulose layers. The overall density of wood depends upon cell structure, the ratio of dense lignin to cellulose and the amount of water absorbed. Moisture modifies the layer structure of the cell walls and thereby drastically reduces strength and elasticity.

Because of the fibrous, highly anisotropic structure of wood, control of dimensional stability is a potential problem. Typically, coefficients of thermal expansion for tangential, radial and longitudinal directions in a tree trunk are in the order of 70×10^{-6}, 50×10^{-6} and $3 \times 10^{-6}\,°C^{-1}$, respectively. When felled, timber can contain more than its own weight of water, i.e. $>100\%$ moisture. As moisture is lost from within cells and from their walls during either air drying or kiln drying, shrinkage occurs. Dimensional stability in service is achieved by aiming for an equilibrium between the moisture content of the wood and atmospheric humidity. Thus, 10% moisture content (kiln dried) might be specified for wood used in a centrally heated building whereas 20% moisture (air dried) might be more appropriate for outdoor service.

The mechanical strength of wood increases roughly in proportion to density. Longitudinal tensile strength with the grain is usually several times greater than tensile strength across the grain: specific tensile strength is generally excellent and comparable to that of mild steel. Compressive stress along the grain can cause the fibrous structure to buckle: compressive strength is therefore not exceptional. In the composite known as plywood, thin layers of wood are glued together with adjacent grains crossed, giving uniformity of strength and minimizing shrinkage.

Continual stressing can cause structural creep of wood; accordingly, archery bows are unstrung when not in use. Wood has good fracture toughness, being able to withstand the localized damage produced by screws and nails. Although wood has often been displaced from its dominant role, certain sports still take advantage of its unique characteristics. The excellent damping capacity of certain woods, together with their elasticity, has long favoured their use for bats and clubs, e.g. golf club heads, hockey sticks, ash hurleys, table tennis bats. A classic example was the choice of supple blue willow (*Salix alba caerulea*), with its thick-walled cell structure, for the English cricket bat.

14.3 Tennis rackets

14.3.1 Frames for tennis rackets

Rackets utilizing the latest engineering materials have increased the capacity of the professional player to make strokes over a remarkable range of pace and spin. At an amateur level, their introduction has made the sport more accessible by providing a greater tolerance for error and by easing beginners' problems. For many years, hardwoods, such as ash, maple and okume, were used in the laminated construction of racket frames. The standard area was 70 in^2. Although damping capacity was excellent, their temperature- and humidity-dependent behaviour could be awkward: wear and warping problems were commonly accepted. The introduction of tubular-section aluminium alloy and steel frames in the 1960s enabled designers to break through the barriers imposed by the use of wood, and tennis boomed. In the 1970s, 'large-head' rackets of different shape with a larger area of 105 in^2 came into vogue. The greater stiffness and strength enabled string tensions to be raised and the effective playing area, the so-called 'sweet spot', was considerably increased. The 'sweet spot' is the central portion

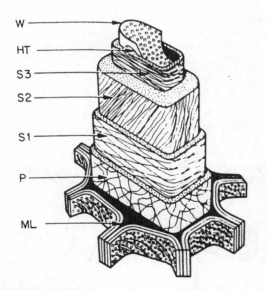

Figure 14.1 *Schematic representation of cell wall structure in wood, showing various layers and their differing orientations of cellulose microfibrils. M.L. = middle layer, P = primary wall, S1 = outer layer of secondary wall, S2 = middle layer of secondary wall, S3 = innermost layer of secondary wall, HT = helical thickening, W = warty layer (from Butterfield and Meylan, 1980: by permission of Kluwer Academic Publishers, Norwell, MA, USA).*

of the stringed area where the vibrational nodes are minimal, or zero, when the ball is struck. A mishit of the ball outside this area generates undesirable vibrations and, quite apart from spoiling the accuracy and power of the subsequent shot, can cause the player to develop the painful and debilitating muscular condition known as 'tennis elbow'. Despite their merits, alloy frames had insufficient damping capacity and failed to displace wooden frames.

In the next stage of evolution, during the 1970s, composite frames were developed which used polyester, epoxy or phenolic resin matrices in a variety of combinations with continuous fibres of E-glass, carbon and aramid (*Kevlar*); although more costly than alloy frames, they had better damping capacity. 'Mid-size' composite rackets with a strung area of 80–100 in^2 came into favour. The generally accepted manufacturing method was to lay-up the continuous fibres in a frame mould, surround them with thermosetting resin, compression mould, cure, surface finish and drill string holes.

These 'conventional' composite frames were costly to produce and used material that was twice as dense as a laminated wood frame. In 1980, the world market in tennis rackets was 55% wood, 30% composite and 15% metal. In the same year, injection moulding of hollow 'mid-size' frames (Figure 14.2) was pioneered (Haines *et al.*, 1983). This method directly challenged the 'conventional' manufacturing route for composites. In the original patented process, a composite mix of 30% v/v short carbon fibres (PAN) and polyamide thermoplastic (nylon 66) was injected into a complex frame mould carrying an accurately located central core. (Rheological studies have shown that, at high shear rates, the viscosity of the fibre-laden mixture

is not greatly different from that of thermoplastic alone.) The fusible core was made of Bi–Sn alloy, m.p. 138.5°C, some 130°C lower than the melting point of the polymeric matrix phase. (Being thermally conductive, the core acted as a heatsink and remained intact.) The core was slowly melted out in an oil bath (150°C) and later reused, leaving a central cavity in the frame which could be filled with vibration-damping, low-density polyurethane foam. (The melt-out process also had the beneficial effect of relieving stresses in the frame.) The mould/core design allowed the flowing composite to form strong tubular pillars in the frame through which the strings could be threaded. The wall thickness of the frame was 2.5 mm. The cycle time for core preparation, injection moulding and core melt-out was only 3 minutes per frame. The balanced handle was given a leather grip and filled with medium-density polyurethane foam. Injection-moulded frames of this type, using a short-fibre composite, were much stronger than laminated wood frames and as strong as, often stronger than, 'conventional' continuous-fibre composite frames.[1]

14.3.2 Strings for tennis rackets

Significant string properties include retention of elasticity and tension, impact efficiency, directional control of the ball, resistance to ultraviolet radiation, gamma radiation, abrasion, moisture, creep, chemical agents, etc. String tension is a subject of prime importance to

[1]For the racket and manufacturing process, Dunlop Sports Co. Ltd, UK, won the 1981 Design Council Award. This racket type was used by champions Steffi Graf and John McEnroe.

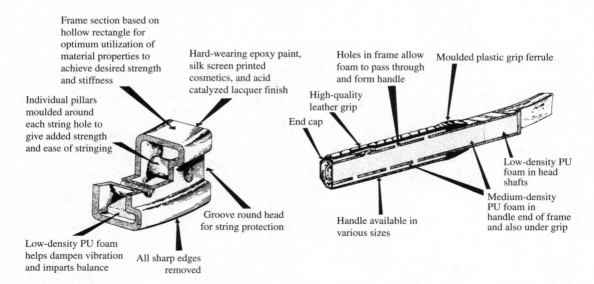

Figure 14.2 *Injection-moulded frame and handle of tennis racket (from Haines et al., 1983: by permission of the Council of the Institution of Mechanical Engineers).*

tennis players and, because of its intrinsic time dependence, much study and debate. It is contended that some ready-strung rackets, as purchased, are already slack.

Ox-gut, the traditional string material, is very effective. However, a greater rebound velocity can be imparted to the ball by using synthetic materials of higher elastic modulus, such as nylon. Each nylon string, typically 1.3 mm diameter, is a braid composed of plaited and intertwined bundles of strong fibres. After immersion in an elastomeric solution, the string is given a polymeric coating to protect it from moisture and wear. Rebound velocity is, of course, also directly influenced by string tension, which, depending upon racket type and personal preference, can range from a force of 200 N to more than 300 N. Another promising material for strings, initially supplied in the form of yarn (*Zyex, Victrex*), is polyether-etherketone (PEEK).

14.4 Golf clubs

14.4.1 Kinetic aspects of a golf stroke

Consider the violent but eventful history of an effective drive with a golf club (Horwood, 1994). Figure 14.3 shows how the bending moment of the shaft varies with time during the 2 second period of the stroke. (A bending moment of 1 N m is equivalent to a deflection of about 13 mm at the end of the shaft.) In this particular test, toward the end of its downward swing and prior to impact, the shaft deflected backwards and then forwards, the club head reaching a swing speed at impact of 42.5 m s^{-1} (95 miles h^{-1}). The period of contact between head and ball was approximately 0.5 ms. After a well-executed stroke, the driven golf ball, aided aerodynamically by its surface dimples, is capable of travelling a distance of some 220–240 m. The shaft of a club bends and twists elastically during the complete swing and, after impact, vibrations travel at the speed of sound along the shaft toward the grip. The much-quoted 'feel' of a club tends to be a highly subjective judgement; for instance, in addition to transmitted vibrations, it usually takes into especial account the sound heard after the instant of impact. A mishit occurs if the striking face of the head is imperfectly aligned, horizontally and vertically, with the ball; a fair proportion of drives fall into this category. The desired 'sweet spot' of impact lies at the point where a line projected from the centre of gravity of the head meets the striking face perpendicularly. Off-centre impact (outside the 'sweet spot') rotates the large head of a 'wood' about its centre of gravity, causing the ball to either hook sidespin (toed shot) or slice sidespin (heeled shot) in accordance with the well-known 'gear effect'.

The intrinsic difficulties of achieving distance, accuracy and consistency have challenged golf players for centuries and continue to do so. One beneficial effect of the new materials recently adopted for club heads and shafts has been to make this demanding sport accessible to a wider range of player ability e.g. senior citizens. Despite golf being frequently described as a triumph of art over science, concerted efforts are being made to explain and rationalize its unique physical aspects in engineering and scientific terms (Cochran, 1994). Such research activities, which are commercially stimulated, have led to the realization that many phenomena attending the impact of club upon ball are still imperfectly understood. One consequence is that, at a practical level, equipment makers apply a variety of testing methods that are often unco-ordinated and potentially confusing: there appears to be a need for

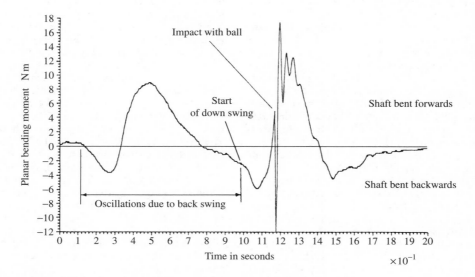

Figure 14.3 *Shaft bending in typical swing of golf club (after Horwood, 1994).*

standard testing procedures for shafts, club heads, etc. to be established internationally.

14.4.2 Golf club shafts

The principal design parameters for a shaft are weight, bending stiffness, bend point and torsional stiffness. For a given head weight, lightening the shaft reputedly makes the swing faster and hence gives some increase in ball speed and distance. Figure 14.3 has already indicated the violent forward and backward bending that occurs in the plane of swing during a drive. Professional golfers and players with a rapid, compact swing favour very stiff shafts. The bend point (also called the flex point or kick point) of a shaft is generally taken as the region where the minimum radius of curvature occurs during bending. A butt-flexible shaft has a high bend point (near the grip) and suits long hitters. Conversely, a tip-flexible shaft has a low bend point (near the head) and suits weaker hitters. A low bend point also increases the dynamic loft of the struck ball. During a drive shot, the shaft tends to twist a few degrees (as well as bend) because the centre of gravity of the club head and the long axis of the shaft are offset. Torsional stiffness (wrongly called 'torque') helps the striking face to remain 'square' during impact. Long hitters tend to favour torsionally stiff ('low torque') shafts.

A typical driver, total weight 350 g, comprises a grip (50 g), steel shaft (100 g) and head (200 g). A variety of materials, including alloy steel, aluminium alloy (7075), Ti–6Al–4V alloy and composites, have been used for the hollow shafts. Carbon, *Kevlar*, glass, boron and silicon carbide fibres have been used as 'reinforcement' in shaft composites. Both epoxy and alloy matrices have been used e.g. 7075 aluminium alloy + 17%v/v short SiC fibres. The main advantages of hollow CFRP shafts, loosely termed 'graphite' shafts, is their relative lightness (about 60 g) and high damping capacity. They are made by (i) wrapping prepreg sheets of carbon fibre around tapered mandrels, (ii) filament winding around a mandrel or (iii) resin transfer moulding (RTM) in which resin is forced around a carbon fibre sleeve. Reproducing exactly the same bending characteristics from shaft to shaft is a difficult, labour-intensive task and demands considerable care and skill. Quality was not always assured with the early CFRP shafts. With CFRP, torsional stiffness and the location of the bend point can be manipulated by varying the textural form and lay-up of the graphite filaments. In this respect, steel shafts, which represent the principal competition to the more recently introduced CFRP shafts, are more restricted. However, they can match CFRP for specific stiffness and strength. Paradoxically, irons with 'heavy graphite' shafts are now available: addition of expensive boron fibres to the CFRP near the shaft tip allows a lighter club head to be used. The net effect is to allow the balance point of the club to be shifted 40–50 mm higher up the shaft and away from the club head. This device seeks to retain the desired damping capacity of CFRP while giving an overall balance similar to that of a steel shaft.

The main categories of golf club are (i) wood-type clubs ('woods'), comprising wooden 'woods' and metal 'woods', (ii) iron-type clubs ('irons') and (iii) putters. Materials used for club heads range from hardwoods to alloys (stainless steel, titanium alloy, aluminium alloy, copper alloy) and composites (CFRP). There has been a tendency for club heads to get larger and for shafts to get longer.

14.4.3 Wood-type club heads

'Woods' have the bulkiest shape and the largest front to back dimension. The traditional wooden 'wood' club heads, which are still highly regarded, are made from either persimmon (date-plum tree) or maple (laminated). These two hardwoods have similar cell structure, density and hardness. Their capacity for damping vibrations is excellent. Manufacture involves 120–200 manual operations and is skill demanding. In a preliminary curing process, the wood structure is impregnated with linseed oil in order to make it waterproof and hard. A striking face and a sole plate, both metallic, are inserted and a central cavity filled with cork. The head is protectively coated with polyurethane and finally weighs about 200 g. As with all club heads, the principal design variables are size, shape, location of the centre of gravity and mass distribution. For instance, the cork insert helps to displace mass to the outside of the head to give some degree of 'peripheral weighting', a feature which reduces the twisting action of off-centre shots i.e. outside the 'sweet spot'.

In the 1970s, metallic 'woods' made from alloys became available as drivers. Materials ranged from stainless steel to the light alloys of aluminium and titanium. Nowadays 17Cr–4Ni stainless steel is a popular choice. Compared to traditional wooden 'woods', alloy heads are easier to make and repeatable quality is more easily achieved, largely because of the introduction of investment casting. In this modern version of the ancient 'lost wax' process, which is eminently suitable for high m.p. alloys that cannot be diecast, a thin refractory shell is formed around a wax pattern. The wax is melted to leave a cavity which serves as a detailed mould for molten steel. The precision castings are welded to form a thin-walled, hollow shell. The central cavity is usually filled with low-density foam. Compared with wooden 'woods', the moment of inertia is greater and the centre of gravity is located closer to the striking face. The shell construction gives peripheral weight distribution away from the centre of gravity of the head. In a recent design the centre of gravity has been lowered by locating dense Cu/W inserts in the sole, e.g. *Trimetal* clubs. Overall, such innovative features confer a greater tolerance for mishits. For instance, a greater moment of inertia reduces twisting of the head when impact is off-centre and also reduces spin.

Composite 'woods' have been made from CFRP. These heads, which emphasize lightness and strength, are similar in shape and style to classic wooden heads and have a similar placing of the centre of gravity. Typically, they have wear-resistant alloy sole plates and a foam-filled core. Compression moulding or injection moulding are used in their manufacture.

14.4.4 Iron-type club heads

The relatively narrow heads for 'irons' are usually made from steels and copper alloys which are shaped by either hot forging or investment casting. Stainless materials include 17Cr–4Ni and AISI Types 431 and 304. As an alternative to the traditional blade-type head, 'cavity-back irons' provide peripheral weighting. In a recent innovation, a non-crystalline zirconium-based alloy[2] containing Cu, Ti, Ni and Be has been used for the heads of irons (and putters). This alloy has high specific strength and good damping capacity and can be successfully vacuum cast in a glassy state without the need for ultra-fast freezing rates.

14.4.5 Putting heads

Although a set of golf clubs may only contain one putter, it is typically used for 40–50% of the strokes in a game. The dynamic demands are less than those for 'woods' and 'irons', consequently putter designs have tended to be less innovative. Design parameters include 'sweet spot', weight distribution, bending stiffness, etc., as previously. Alloys used include stainless steel (17Cr–4Ni), manganese 'bronze' (Cu–Zn–Mn) and beryllium–copper (Cu–2Be).

14.5 Archery bows and arrows

14.5.1 The longbow

For centuries, archery bows have combined design skill with knowledge of material properties. From the evidence of many well-preserved yew longbows retrieved in 1982 from the wreck of the Tudor warship '*Mary Rose*', which sank in Portsmouth harbour (1545), we know that the original seasoned stave was shaped in such a way as to locate sapwood on the outer convex 'back' of the bow and darker heartwood at the concave 'belly' surface. When the bow was braced and drawn, this natural composite arrangement gave the greatest resistance to the corresponding tensile and compressive stresses. Considerable force, estimated to be in the order of 36–72 kgf (80–160 lbf), was needed

to draw a heavy longbow.[3] The mystique of the longbow and its near-optimum design have intrigued engineers and scientists; their studies have greatly helped in providing a theoretical basis for modern designs of bows and arrows (Blyth & Pratt, 1976).

14.5.2 Bow design

A bow and its arrows should be matched to the strength and length of the archer's arm. A well-designed bow acts as a powerful spring and transfers stored strain energy smoothly and efficiently to the arrow. As the archer applies force and draws the bowstring from the braced condition (which already stores energy) through a draw distance of about 35 cm, additional energy is stored in the two limbs (arms) of the bow. In general, increasing the length of the bow reduces stress and increases the potential for energy storage. Upon release of the bowstring, stored energy accelerates the arrow as well as the string and the two limbs of the bow. The efficiency (η) of the bow at the moment of loose may be taken simply as the kinetic energy of the arrow divided by work expended in drawing the bow. Alternatively, allowance can be made for the energy-absorbing movement of the two limbs, as in the Klopsteg formula:

$$\eta = m/k + m \tag{14.1}$$

where m is the weight of the arrow and k is the 'virtual weight' of the particular bow. The constant term k treats energy losses in the bow as an extra burden on the driven arrow, travelling with the same velocity. For a given bow and draw force, η increases with arrow mass. Thus, for a certain yew bow ($k = 23.5$ g), increasing the arrow weight from 23.5 g to 70.5 g increased the efficiency of the bow from 50% to 75%.

Bow materials are often compared in terms of specific modulus of rupture and specific modulus of elasticity. Thus, for wooden bows, timbers which combine a high MoR/ρ with a comparatively low E/ρ are generally preferred as they provide lightness, the necessary resistance to bending stresses and a capacity to store energy. The fine-grained hardwoods ash and wych elm meet these criteria. Although nominally a softwood, yew was favoured for longbows, its very fine grain giving remarkable bending strength. Strain energy per unit volume can be derived from the stress v. strain diagram and expressed as $0.5\varepsilon^2 E$. It follows that maximizing strain ε (below the elastic limit) is an effective way of maximizing stored energy.

There are two main types of modern bow, the standard recurve (Olympic) bow and the more complex compound bow. In contrast to the D-section of the

---

[2]*Vitreloy* or *Liquidmetal*, developed at Caltech, now produced by Howmet Corp. Greenwich, CT-06830, USA.

[3]Skeletal remains of an archer, taken from the same shipwreck, indicate that a lifetime of drawing the longbow produced permanent physical deformation. This powerful weapon was developed in conflicts in the Welsh Marches; its ability to penetrate plate armour had both military and social significance.

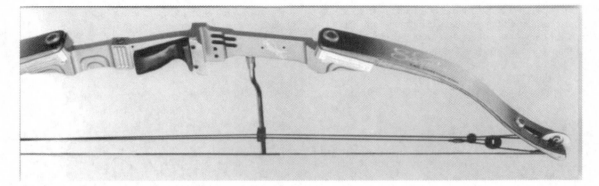

Figure 14.4 *Modern competition bow, compound-type: laminated upper limb (wood, gfrp, cfrp) and CNC-machined central riser (Al–Mg–Si alloy 6082) (courtesy of Merlin Bows, Loughborough, U.K.).*

Tudor longbow, limb sections of a recurve bow are wide, flat and thin, giving resistance to twisting. The energy efficiency of a compound bow (Figure 14.4) is twice that of the longbow and, as a consequence, can propel an arrow much faster, at velocities of 90 m s^{-1} or more. Pulley cams at the ends of the two limbs sustain the load at full draw during the sighting period of 10 s or more. The most powerful compound bows use light alloys for the mid-section ('riser' or grip) e.g. forged Al alloy, diecast Mg alloy. Laminated wood is used for some bow grips, e.g. maple plus rosewood. The two limbs are very often laminated in construction and are much less susceptible to temperature change and humidity than wood alone. Many different material combinations are used for laminae, e.g. CFRP, wood, GRP, foam, etc. For instance, in one type of composite bow limb, facing and backing strips of GRP are joined to each side of a thin core strip of maple with epoxy adhesive. Tubular alloy steel limbs have been superseded as they were prone to internal aqueous corrosion: sudden fracture of a drawn bow (or its string) can be extremely dangerous.

14.5.3 Arrow design

Successful discharge of an arrow from a bow involves a careful balancing of three arrow characteristics; namely, length, mass and stiffness ('spine'). Subsequent flight depends on the aerodynamic qualities of the design of head, shaft and fletching. The length of the arrow is determined by the geometry of the human body; typically, lengths range from 71 to 76 cm. The mass chosen depends initially upon the type of archery, e.g. maximum range, target shooting, etc. The product of efficiency (η) and stored energy (E_s) gives the kinetic energy of the arrow, hence:

$$\eta E_s = 0.5 \, mv_o^2 \qquad (14.2)$$

where m = mass of the arrow and v_o = its initial velocity. Thus velocity increases with bow efficiency and decreasing arrow mass.

Finally, an arrow must possess an optimum, rather than maximum, stiffness ('spine') which must be matched to the bow. Bending stiffness of an arrowshaft is measured in a three-point bend test (Figure 7.6). Central to the design of an arrow is the phenomenon known as the Archer's Paradox (Figure 14.5). At the loose, with the arrow pointing slightly away from the target, the arrow is subjected to a sudden compressive force along its length which, together with the deflecting action of the archer's fingers, generates lateral vibrations in the moving arrow. Correct matching of the dimensions, stiffness and vibration characteristics (frequency, amplitude) of the arrow enables the arrow to clear the bow cleanly. In addition to dependence on the arrow's dimensions, the frequency of flexural vibration is proportional to the square root of the specific stiffness (E/ρ). Frequencies are in the order of 60 Hz. Reasonable agreement has been obtained between theory and high-speed cinéphotographic studies.[4]

Of the 15 kinds of wood used as arrows for longbows in medieval times, ash was generally regarded as the best. Nowadays, arrowshafts are tubular and made from (i) drawn and anodized aluminium alloy (7075-T9, 7178-T9), (ii) similar alloys bonded to a smooth outer wrap of unidirectional CFRP and (iii) pultruded CFRP. (Early CFRP arrows were unpopular because they tended to develop splintering damage.) Most arrowshafts are constant in diameter along their length but have the disadvantage that their bending moment varies, increasing from zero at the ends to a maximum at the centre. Tapering ('barrelling') the tubular shaft from the middle to the ends reduces this undesirable flexing characteristic: barrelled arrows are used by top professionals.

Feathers are the traditional fletching material and still used but are fragile and suffer from the weather. They rotate the arrow and give stability during flight

---

[4]At the Royal Armaments Research & Development Establishment (RARDE), UK.

Time / ms

0

5

10

15

20

25

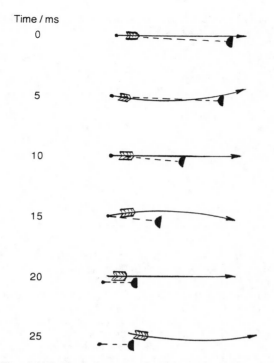

Figure 14.5 *The Archer's Paradox (after Pratt, 1976).*

but consume kinetic energy. Smooth polymeric vanes made from polyethylene terephthalate (*Mylar*) are strong, weather resistant and, because of their lower aerodynamic drag, give greater range; the same polymer is also commonly used in stranded form (*Dacron*) for bowstrings.

14.6 Bicycles for sport

14.6.1 Frame design

The modern bicycle is a remarkable device for converting human energy into propulsion. The familiar diamond frame, with its head, top, seat and down tubes, evolved in the late nineteenth century. When in use, it distorts elastically; this compliance provides rider comfort. Compliance absorbs energy and frame stiffness is accordingly maximized in racing machines. The stress distribution in a working frame is complex, being in-plane as well as out-of-plane. Sudden impact stresses must be withstood. Poor design, workmanship and/or maintenance can lead to component failure which, because of the fluctuating nature of stressing, often has fatigue characteristics.

McMahon & Graham (1992) have provided a detailed comparison of typical tube materials for frames. Beam theory is used to identify the key design parameters. The basic linking formula which expresses the stresses and strains at points along a beam deflecting under load is:

$$M/I = \sigma/y = E/r \qquad (14.3)$$

where M = bending moment, I = moment of inertia of beam section, σ = stress at a point, y = distance of point from neutral axis of beam, E = modulus of elasticity and r = radius of curvature of loaded beam. In particular, we are concerned with (i) the maximum tensile stress σ_{max} in the convex surface of a tube subjected to a nominal bending moment and (ii) the corresponding radius r of bending. Obviously, σ_{max} should bear a good relation to the yield strength of the material and r should be maximized.

A tubular cross-section offers special advantages. Within a bent beam, it locates as much material as possible in the highly stressed regions which lie distant from the neutral axis: this axis lies in the plane marking the transition from tension to compression. Being symmetrical in section, a tube can be loaded transversely in any direction and can withstand torsion. Its moment of inertia is $\pi(D^4 - d^4)/64$, where D and d are the outside and inside diameters of the tube, respectively. From relation (14.3) it can be seen that, for a given bending moment, increasing the moment of inertia reduces stress and increases the radius of curvature. In similar fashion, it can be reasoned that reducing the tube wall thickness increases the surface stress (σ_{max}). Sometimes it is beneficial to raise the moment of inertia by changing from a circular cross-section to a more expensive elliptical cross-section. Thus, in front wheel forks, which are subjected to severe bending stresses, an increase in the major diameter of the ellipse reduces stress in the crucial plane.

Table 14.1 compares the bending characteristics of tubes made from four typical materials used for cycle frames; that is, from plain carbon steel, 0.3C–Cr–Mo alloy steel (AlSl 4130), 6061 (T6) aluminium alloy and Ti–3Al –2.5V alloy. Calculated values for bend curvature and maximum stress, which are the criteria of stiffness and permissible loading, are compared. The frames of mountain bicycles must sustain sudden impact shocks; accordingly, larger-diameter (D) and/or thicker-walled tubing is used for certain frame members in order to reduce stress levels. The specific elastic moduli (E/ρ) for the four materials are similar. Aluminium alloy offers weight saving but, because of its relatively low E value (70 GN m^{-2}), at the expense of greater flexure of the frame. Titanium alloy allows reductions in tube diameter and wall thickness; its specific yield strength (σ_y/ρ) is about two and a half times greater than that of Cr–Mo steel.

Although not included in Table 14.1, cold-drawn seamless Mn–Mo tube steels have a special place in the history of competitive cycling. e.g. *Reynolds 531*. Their nominal composition is 0.25C–1.4Mn –0.2Mo. Introduced in the 1930s, they have been used for the frames of many Tour de France winners and are still

Table 14.1 Comparison of weight and bending characteristics of four metallic frame materials: r and σ_{max} calculated for tubes subjected to a bending moment of 100 N m (from McMahon & Graham, 1992)

Material	D (mm)	d (mm)	Moment of inertia I (cm^4)	Mass per unit length (g m^{-1})	Radius of curvature r (mm)	Maximum stress σ_{max} (MN m^{-2})	σ_{max}/σ_y
Racing cycles							
C steel	28.70	26.16	1.041	860	48	138	0.58
Cr–Mo steel	28.78	26.93	0.750	601	67	190	0.39
Al alloy (6061-T6)	28.80	25.91	1.124	332	127	128	0.50
Mountain cycles (Top tube)							
Al alloy (6061-T6)	38.10	35.56	2.50	397	57	76	0.30
Ditto	34.93	30.81	2.87	575	49	61	0.24
Ti–3Al–2.5V	31.75	29.47	1.41	549	64	45	0.18

widely used for racing cycles.[5]

14.6.2 Joining techniques for metallic frames

The above guidelines provide a general perspective but do not allow for the potentially weakening effect of the thermal processes used for joining the ends of individual frame tubes. Such joints often coincide with the highest bending moments. In mass production, the cold-drawn low-carbon steel tubes of standard bicycle frames are joined by brazing. Shaped reinforcing sockets (lugs) of low–carbon steel, together with thin inserts of solid brazing alloy, are placed around the tube ends, suitably supported, and heated. A 60Cu–40Zn alloy such as CZ7A (British Standard 1845) freezes over the approximate temperature range of 900–870°C as the frame cools and forms a strong, sufficiently ductile mixture of α and β phases (Figure 3.20). Butted tubes are commonly used to counteract softening of the steel in the heat-affected zones (HAZ); they have a smaller inside diameter (d) toward the tube ends. For limited production runs of specialized racing frames made from butted alloy tubes, fillet brazing with an oxy-acetylene torch at a lower temperature is more appropriate, using a silver brazing alloy selected from the AG series of British Standard 1845, such as 50Ag–15Cu–16Zn –19Cd (melting range 620–640°C). Cadmium-free alloys are advocated if efficient fume-extraction facilities are not available because CdO fumes are dangerous to health.

Tungsten-inert gas (TIG) welding[6] is widely used and has tended to replace brazing, e.g. lugless Cr–Mo steel frames for mountain bicycles. Unlike oxyacetylene flames, heating is intense and very localized. The hardenability of Cr–Mo steels is such that a strong

mixture of dispersed alloy carbides, pearlite and possibly bainite forms in the weld fillet as they air cool from temperatures above 850°C. The latest type of low-alloy steel for frames, available in either cold-drawn or heat-treated condition (*Reynolds 631* and *853*), is air hardening. Although inherently very hard (400 VPN), TIG-welding increases its hardness in the HAZ. It possesses better fatigue resistance than other alloy steels and its strength/weight ratio makes it competitive with Ti–3Al–2.5V alloy and composites.

Aluminium alloy tubes, which are solution treated and artificially aged (T6 condition), present a problem because heating during joining overages and softens the structure, e.g. 6061, 7005. The high thermal conductivity of aluminium worsens the problem. Titanium alloys, such as the frame alloy Ti–3Al –2.5V, absorb gases and become embrittled when heated in air, e.g. oxygen, nitrogen, hydrogen. Again, it is essential to prevent this absorption by shrouding the weld pool with a flowing atmosphere of inert gas (argon).

14.6.3 Frame assembly using epoxy adhesives

These joining problems encouraged a move toward the use of epoxy adhesives with sleeved tube joints.[7] As well as helping to eliminate the HAZ problem, adhesives make it possible to construct hybrid frames from various combinations of dissimilar materials (adherends), including composites. Brake assemblies can be glued to CFRP forks. Adhesive bonds also damp vibrations, save weight, reduce assembly costs and are durable. Extremes of humidity and temperature can cause problems and care is essential during adhesive selection. Adhesives technology meets the

---

[5] *'531'* tubes were used for the chassis of the jet-powered Thrust 2 vehicle in which Richard Noble broke the one-mile land speed record (1983), achieving a speed of 1019 km h^{-1}

[6] Patented in the 1930s in the USA., where argon and helium were available, this fluxless arc process is widely used for stainless steels and alloys of Al, Ti, Mg, Ni and Zr.

[7] Adhesive-bonded racing cycles, sponsored by Raleigh Cycles of America, were highly successful in the 1984 Olympic Games. Subsequently, Raleigh made mountain cycles from aluminium alloy tubes (6061-T8) bonded with Permabond single-part *ESP-311* epoxy adhesive.

stringent demands of modern aircraft manufacturers[8] and makes a vital contribution throughout the world of sport.

The structural adhesives most widely used in general engineering are the epoxy resins; their thermosetting character has been described previously (Section 2.7.3). Normally they are water resistant. They form strong bonds but, being in a glassy state, are brittle. Accordingly, thermoplastic and/or elastomeric constituents are sometimes included with the thermosetting component. When using the two-part version of a thermosetting adhesive, it is important to control the proportions of basic resinous binder and catalytic agent (hardener) exactly, to mix thoroughly and to allow adequate time for curing. In single-part epoxy adhesives the resin and hardener are pre-mixed: rapid curing is initiated by raising the temperature above 100°C. Thermoplastic adhesives, used alone, are weaker, more heat sensitive and less creep resistant. Elastomeric adhesives, based on synthetic rubbers, are inherently weak. Meticulous preparation of the adherend surfaces is essential for all types of adhesive.

14.6.4 Composite frames

Epoxy resins are also used to provide the matrix phase in the hollow, composite frames of high-performance bicycles. Carbon fibre reinforced polymers (Section 11.3.2.1) combine high strength and stiffness; their introduction facilitated the construction of monocoque (single shell) frames and led to the appearance of a remarkable generation of record-breaking machines.[9] Typically, they feature a daring cantilevered seat, a disc rear wheel and three-spoke open front wheels, all of which are made from CFRP. An example is depicted in Figure 14.6.

14.6.5 Bicycle wheels

The familiar array of wire spokes between axle and rim normally uses hard-drawn wire of either plain 0.4% carbon steel (AISI 1040) or austenitic 18Cr–8Ni stainless steel (McMahon & Graham, 1992). Each spoke is tangential to the axle, thus preventing 'wind-up' displacement between axle and rim, and is elastically pretensioned (e.g. 440 MN m^{-2}) so that it is always in tension during service. During each wheel revolution, the stress on a given spoke is mostly above the pretension stress, falling once below it. Under these cyclic conditions, carbon steel has a greater nominal fatigue endurance than the corrosion-resistant 18/8 steel but

Figure 14.6 *High-performance Zipp bicycle with monocoque frame (courtesy of Julian Ormandy, School of Metallurgy and Materials, University of Birmingham, UK).*

has a smaller resistance to corrosion fatigue. The latter property is boosted by plating the carbon steel with a sacrificial layer of zinc or cadmium. Both types of steel respond well to the strain-hardening action of wire drawing through tungsten carbide dies. Wheel rims should be strong, stiff, light and corrosion resistant. They are often formed by bending strips of extruded, precipitation-hardenable aluminium alloy to shape and joining e.g. 6061-T6.

Conventional multi-spoked wheels generate energy-absorbing turbulence during rotation. The distinctive CFRP front and rear wheels of highly specialized time-trial machines, made by such firms as Lotus, Zipp, RMIT-AIS and Ultimate Bike, have a much lower aerodynamic coefficient of drag. They are the products of extensive computer-aided design programmes, wind tunnel simulations and instrumented performance testing.

14.7 Fencing foils

A typical steel foil is about 0.9 m long and tapers to a rectangular cross-section of 4 mm × 3 mm. This design gives a low resistance to buckling under the large axial stress produced when an opponent is struck directly, an action which can bend the foil forcibly into a radius as small as 20 cm. Traditionally, swordmakers use medium-carbon alloy steels of the type employed in engineering for springs. The extensive range of elastic behaviour that is associated with a high yield strength is obviously desirable. The foil is formed by hot working 10 mm square bar stock and then oil quenching and tempering to develop a martensitic structure with a yield strength in the order of 1500–1700 MN m^{-2}. On occasion, during a fencing bout, the applied stress exceeds the yield strength and the foil deforms plastically: provided that the foil is defect free, the fencer can restore straightness by careful reverse bending. In practice, however, used foils are not defect free. During bouts, repeated blows from the opposing blade produce small nicks in the surface

---

[8]Urea-formaldehyde resins (*Beetle* cements) revolutionized aircraft building in the 1940s when they were used for bonding and gap-filling functions with birchwood/balsa composites and spruce airframes e.g. De Havilland Mosquito, Airspeed Horsa gliders.

[9]The prototype was the Lotus bicycle on which Chris Boardman won the 4000 m individual pursuit in the 1992 Olympic Games at Barcelona.

of a foil. In time, it is possible for one of these stress-raising notches to reach a critical size and to initiate fatigue cracking within the tempered martensite. Final failure occurs without warning and the buttoned foil instantly becomes a deadly weapon.

One research response to this problem was to concentrate upon improving fracture toughness and resistance to fatigue failure, thus eliminating instantaneity of failure.[10] In this alternative material, a steel–steel composite, lightly tempered fibres of martensite are aligned within a continuous matrix phase of tough austenite. 10 mm square feedstock of duplex steel for the blade-forging machine is produced by diffusion annealing packs of nickel-electroplated bars of spring steel at a temperature of 1000°C, extruding and hot working. While the bars are at elevated temperatures, nickel interdiffuses with the underlying steel. Nickel is a notable austenite (γ)-forming element, as indicated previously in Figure 9.2. The optimum volume fraction of tough austenite is about 5%. This duplex material has the same specific stiffness as the conventional steel and has greater fracture toughness. In the event of a surface nick initiating a crack in a longitudinal filament of brittle martensite, the crack passes rapidly across the filament and, upon encountering the tough interfilamentary austenite, abruptly changes direction and spreads parallel to the foil axis, absorbing energy as the austenite deforms plastically and new surfaces are formed. In practical terms, if the fencer should fail to notice marked changes in the handling characteristics of a deteriorating foil, the prolonged nature of final fracture is less likely to be dangerous and life threatening. Although safer, the duplex foil involves increased material-processing costs and has a yield strength about 5–10% lower than that of heat-treated spring steel.

Highly alloyed maraging steels (Section 9.2.3) are used nowadays for top-level competition fencing. By combining solid solution strengthening with fine precipitation in low-carbon martensite, they provide the desired high yield strength and fracture toughness.

A typical composition is 0.03C (max)–18 Ni–9 Co –5 Mo –0.7 Ti–0.1 Al.

14.8 Materials for snow sports

14.8.1 General requirements

The previously quoted examples of equipment frequently share common material requirements and properties, such as bending stiffness, yield strength, toughness, fatigue resistance, density and comfort. However, each sport makes its own unique demands on materials. In snowboarding and skiing equipment, for instance, additional requirements include toughness at sub zero temperatures (say down to −30°C), low frictional drag

and resistance to prolonged contact with snow and moisture. More specifically, in cross-country (Nordic) skiing, lightness is very important as it makes striding less tiring. From a commercial aspect, it is desirable that materials for individual items of equipment should be able to display vivid, durable colours and designer logos.

14.8.2 Snowboarding equipment

The bindings which secure a snowboarder's boots to the top surface of the board are highly stressed during a downhill run. Good binding design provides a sensitive interaction between the board and the snowboarder's feet, facilitating jumps and turns. Modern designs are complex and usually employ a variety of polymers. Thus, the recent snowboard design shown in Figure 14.7 includes components made from an acetal homopolar (*Delrin*), a nylon-based polymer (*Zytel*) and a thermoplastic polyester elastomer (*Hytrel*).[11] Highly crystalline *Delrin* is tough, having a low glass-transition temperature (T_g), and strong and fatigue resistant. It is also suitably UV resistant and moisture resistant. *Zytel* is tough at low temperatures, can be moulded into complex shapes and can be stiffened by glass-fibre reinforcement. The third polymer, *Hytrel*, has properties intermediate to those of thermoplastics and elastomers, combining flexibility, strength and fatigue resistance. Both *Hytrel* and the nylon *Zytel* can be fibre reinforced. Thus, some snowboard blades are made from *Zytel* reinforced with fibres of either glass or aramid (*Kevlar*). Colourants mixed with the resins give attractive moulded-in colours.

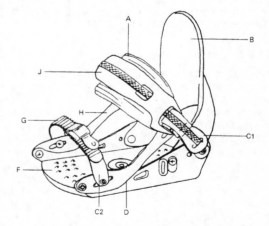

Figure 14.7 *Snowboard binding utilizing: thermoplastic elastomer (Hytrel)—ankle strap A, spoiler B, ratchet strap G, nylon (Zytel)—side frames D and H, base and disc F, top frame J; acetal homopolar (Delrin)—strap buckles C1 and C2 (courtesy of Fritschi Swiss Bindings AG and Du Pont UK Ltd).*

[10]Materials research conducted at Imperial College, London, on behalf of the fencing sword manufacturers, Paul Leon Equipment Co. Ltd, London (Baker, 1989).

[11]*Delrin, Hytrel* and *Zytel* are registered trademarks of DuPont.

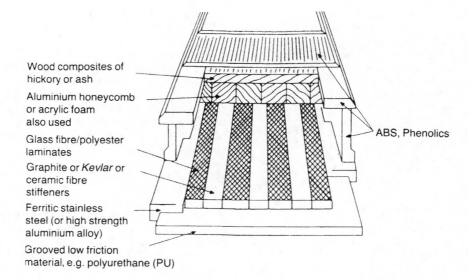

Wood composites of hickory or ash

Aluminium honeycomb or acrylic foam also used

Glass fibre/polyester laminates

Graphite or *Kevlar* or ceramic fibre stiffeners

Ferritic stainless steel (or high strength aluminium alloy)

Grooved low friction material, e.g. polyurethane (PU)

ABS, Phenolics

Figure 14.8 *Transverse section showing multi-component structure of a downhill ski (from Easterling, 1990 by permission of the Institute of Materials).*

14.8.3 Skiing equipment

In the older sport of skiing, the principal items of equipment are the boots, bindings, skis and poles. External *Hytrel–Kevlar* components have been used to enhance the stiffness of ski boots; this feature gives the boots a firmer grip on the skier's ankles and leads to better control of the skis. Polymers feature prominently in many design of ski bindings. For instance, in the Fritschi *Diamir* touring binding,[12] acetal polymer *(Delrin)* is used for the locking bar, heel release lever, heel block and front swivel plate while glass-reinforced nylon *(Zytel)* is used for the front block and the two base plates.

Modern ski designs aim at solving the conflicting requirements of (i) longitudinal and torsional stiffness that will distribute the skier's weight correctly and (ii) flexibility that will enable the ski to conform to irregularities in the snow contour (Easterling, 1993). Originally, each ski runner was made from a single piece of wood, e.g. hickory. Laminated wood skis appeared in the 1930s. The adoption of polymers for ski components in the 1950s, combining lightness and resistance to degradation, was followed by the introduction of metal frames for downhill skis, e.g. alloy steel, aluminium alloy. By the 1960s, GRP and CFRP were coming into prominence. The internal structure of a ski is determined by the type of skiing and, as Figure 14.8 shows, often uses a surprising number of different materials. Skis usually have a shock-absorbing, cellular core that is natural (ash, hickory) and/or synthetic (aramid, aluminium, titanium

[12]Used by Hans Kammerlander in his 1996 ski descent of Mount Everest.

or paper honeycomb). Polyethylene and polyurethane have been used for the soles of skis.

Modern ski poles are tubular and designed to give high specific stiffness and good resistance to impacts. Nowadays, CFRP–GRP hybrids are favoured. The pointed tips are sometimes made of wear-resistant carbide.

14.9 Safety helmets

14.9.1 Function and form of safety helmets

Most sports entail an element of personal risk. The main function of a safety helmet is to protect the human skull and its fragile contents by absorbing as much as possible of the kinetic energy that is violently transferred during a collision. The three principal damaging consequences of sudden impact are fracture of the skull, linear acceleration of the brain relative to the skull, and rotational acceleration of the brain. Although linear and rotational acceleration may occur at the same time, many mechanical testing procedures for helmets concentrate upon linear acceleration and use it as a criterion of protection in specifications.

A typical helmet consists of an outer shell and a foam liner. The shell is usually made from a strong, durable and rigid material that is capable of spreading and redistributing the impacting forces without suffering brittle fracture. This reduction in pressure lessens the risk of skull fracture. The foam liner has a cellular structure that absorbs energy when crushed by impact. Specialized designs of helmets are used in cycling, horse riding, canoeing, mountaineering, skiing, skate boarding, ice hockey, etc. Some designs are quite rudimentary and offer minimal protection.

In general, the wearer expects the helmet to be comfortable to wear, lightweight, not restrict peripheral vision unduly and be reasonably compact and/or aerodynamic. Production costs should be low. Increasing the liner thickness is beneficial but, if the use of helmets is to be promoted, there are size constraints. Thus, for a cricket helmet, acceptable shell and liner thicknesses are about 2–3 mm and 15 mm, respectively (Knowles *et al.*, 1998).

Strong and tough helmet shells have been produced from ABS and GRP. The great majority of shock-absorbent foam linings are made from polystyrene (Figure 14.9): polypropylene and polyurethane are also used.

14.9.2 Mechanical behaviour of foams

Polymeric foams provide an extremely useful class of engineering materials (Gibson & Ashby, 1988; Dyson, 1990). They can be readily produced in many different structural forms by a wide variety of methods using either physical or chemical blowing agents. Most thermoplastic and thermosetting resins can be foamed. The properties of a foam are a function of (i) the solid polymer's characteristics, (ii) the relative density of the foam; that is, the ratio of the foam density to the density of the solid polymer forming the cell walls (ρ/ρ_s), and (iii) the shape and size of the cells. Relative density is particularly important; a wide range is achievable (typically 0.05–0.2). Polymer foams are often anisotropic. Broadly speaking, two main types of structure are available: open-cell foams and closed-cell foams. Between these two structural extremes lies a host of intermediate forms.

In the case of safety helmets, the ability of a liner foam to mitigate shock loading depends essentially upon its compression behaviour. Initially, under compressive stress, polymer foams deform in a linear–elastic manner as cell walls bend and/or stretch.

With further increase in stress, cell walls buckle and collapse like overloaded struts; in this second stage, energy absorption is much more pronounced and deformation can be elastic or plastic, depending upon the particular polymer. If the cells are of the closed type, compression of the contained air makes an additional and significant contribution to energy absorption. Eventually the cell walls touch and stress rises sharply as the foam densifies. This condition occurs when a liner of inadequate thickness 'bottoms out' against the helmet shell. The design of a helmet liner should provide the desired energy absorption without 'bottoming out' and at the same time keep peak stresses below a prescribed limit.

Some polymeric structures can recover their original form viscoelastically and withstand a number of heavy impacts; with others, a single impact can cause permanent damage to the cell structure, e.g. expanded PS. After serious impact, helmets with this type of liner should be destroyed. Although this requirement is impracticable in some sporting activities, there are cases where single-impact PS liners are considered to be adequate.

14.9.3 Mechanical testing of safety helmets

Various British Standards apply to protective helmets and caps for sports such as climbing (BS 4423), horse- and pony riding (BS EN 1384) and pedal cycling, skateboarding and rollerskating (BS EN 1078). These activities involve different hazards and accordingly the testing procedures and requirements for shock absorption and resistance to penetration differ. In one typical form of test, a headform (simulating the mass and shape of the human head) is encased in a helmet and allowed to fall freely through a certain distance against a rigid anvil. Specified headform materials (BS EN 960) depend on the nature of the impact test and extend from laminated hardwood (beech) to alloys with a low resonance frequency (Mg–0.5Zr). A triaxial accelerometer is affixed to the headform/helmet assembly in the zone of impact. The area beneath the curve of a continuous graphical record of striking force v. local deformation taken during the test provides a useful measure of the kinetic energy absorbed as the helmet structure is crushed. Specified values for permissible peak acceleration at impact which appear in test procedures vary but generally extend up to about 300 g, where g (acceleration due to gravity) = 9.81 m s^{-2}. Drop heights range from about 1 to 2.5 m, depending upon the striking force required. Test specifications often include requirements for helmets to be mechanically tested after exposure to extremes of temperature, ultraviolet radiation and water. On occasions, unsafe and/or inadequate helmets are marketed: naturally, closer international collaboration and regulation is being sought.

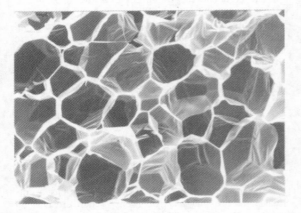

Figure 14.9 *Cell structure of polystyrene foam, as used for shock-absorbent packaging: average cell diameter 100 μm (courtesy of Chris Hardy, School of Metallurgy and Materials, University of Birmingham, UK).*

Further reading

Baker, T. J. (1989). Fencing blades—a materials challenge. *Metals and Materials*, **Dec.**, 715–718, Institute of Materials.

Blyth, P. H. and Pratt, P. L. (1992). The design and materials of the bow/the arrow, Appendices to *Longbow: A Social and Military History*, 3rd edn. by Robert Hardy. Patrick Stephens Ltd, Cambridge.

Cochran, A. (ed.) (1994). *Golf: the Scientific Way*. Aston Publ. Group, Hemel Hempstead, Herts. UK.

Easterling, K. E. (1993). *Advanced Materials for Sports Equipment*. Chapman & Hall Ltd, London.

Gibson, L. J. and Ashby, M. J. (1988). *Cellular Solids–Structure and Properties*. Pergamon Press.

Knowles, S., Fletcher, G., Brooks, R. and Mather, J. S. B. (1998). Development of a superior performance cricket helmet, in *The Engineering of Sport* (ed. S. J. Haake). Blackwell Science, Oxford.

Lees, A. W. (ed.) (1989). *Adhesives and the Engineer*. Mechanical Engineering Publications Ltd, London.

McMahon, C. J. and Graham, C. D. (1992). *Introduction to Materials: the Bicycle and the Walkman*. Merion Books, Philadelphia.

Pearson, R. G. (1990). *Engineering Polymers* (ed. R. W. Dyson), Chapter 4 on foams, pp. 76–100, Blackie & Son Ltd, Glasgow and London.

Shields, J. (1984). *Adhesives Handbook*. 3rd edn. Butterworths, Oxford.

Appendix 1

SI units

The Système Internationale d'Unités (SI) was introduced in the UK in the late 1960s. Historically, the SI can be traced from the metric enthusiasms of Napoleonic times, through a centimetre–gram (c.g.) system, a centimetre–gram–second (c.g.s.) system, a metre–kilogram–second (MKS) system in 1900 and a metre–kilogram–second–ampere (MKSA Giorgi) system in 1950. Table A1 lists the seven basic units and Table A2 lists the prefixes.

The SI is 'rational, comprehensive and coherent'. Coherency means that the product or quotient of *basic* units gives an appropriate *derived* unit of the resultant quantity. A coherent system facilitates manipulation of units, checking the dimensions of equations and, most importantly, the correlation of different disciplines. Some of the more frequently-used derived units are given in Table A3.

The force unit, the newton, is the cornerstone of the SI. Appropriately, the gravitational attraction for an apple is roughly one newton. The SI unit of stress is $N\ m^{-2}$: the pascal (Pa) is an orphan, being non-SI and non-coherent. Energy is defined in mechanical terms, being the work done when the point of application of a force of 1 N is displaced through a distance of 1 m in the direction of the force.

Table A1

Quantity	Unit	Symbol
Length	metre	m
Mass	kilogram	kg
Time	second	s
Electric current	ampere	A
Temperature	degree Kelvin	K
Luminous intensity	candela	cd
Amount of substance	mole	mol

Table A2

Factor	Prefix	Symbol
10^{12}	tera	T
10^{9}	giga	G
10^{6}	mega	M
10^{3}	kilo	k
10^{2a}	hecto[a]	h[a]
10^{1a}	deca[a]	da[a]
10^{-1a}	deci[a]	d[a]
10^{-2a}	centi[a]	c[a]
10^{-3}	milli	m
10^{-6}	micro	μ
10^{-9}	nano	n
10^{-12}	pico	p
10^{-15}	femto	f
10^{-18}	atto	a

[a]Discouraged

Table A3

Physical quantity	SI unit	Definition of unit
Volume	cubic metre	m^{3}
Force	newton (N)	$kg\ m\ s^{-2}$
Pressure, stress	newton per square metre	$N\ m^{-2}$
Energy	joule (J)	$N\ m$
Power	watt (W)	$J\ s^{-1}$
Electric charge	coulomb (C)	$A\ s$
Electric potential	volt (V)	$W\ A^{-1}$
Electric resistance	ohm (Ω)	$V\ A^{-1}$
Electric capacitance	farad (F)	$A\ s\ V^{-1}$
Frequency	hertz (Hz)	s^{-1}

The surprising frequency with which SI units are misused in textbooks, learned papers, reports, theses and even examination papers, justifies a reminder of some rules:

1. Try to locate basic units after the solidus: 1 MN/m² preferred to 1 N/mm².
2. A space is significant: ms is not the same as m s.
3. Prior to calculations, convert to basic SI units: 1 mm becomes 1×10^{-3} m.
4. Where possible, work in steps of $10^{\pm 3}$.
5. If possible, group digits in threes and avoid commas: 37 532 rather than 37,532 because a comma means decimal point in some countries.
6. When selecting a prefix, arrange for the preceding number to lie between 0.1 and 1000: use 10 mm rather than 0.01 m. In a comparison, one may break this rule, e.g. 'increase from 900 kN to 12 000 kN'.
7. Do not use double prefixes: pF, not $\mu\mu$F.
8. Avoid multiples of the solidus: acceleration written as m/s/s self-destructs.

Appendix 2

Conversion factors, constants and physical data

Quantity	Symbol	Traditional units	SI units
1 atmosphere (pressure)	atm		101.325 kN m^{-2}
Avogadro constant	N_A		0.602×10^{24} mol^{-1}
1 Angstrom	Å	10^{-8} cm	10^{-10} m
1 barn	b	10^{-24} cm^2	10^{-28} m^2
1 bar	bar or b		10^5 N m^{-2}
Boltzmann constant	**k**		1.380×10^{-23} J K^{-1}
1 calorie	cal		4.1868 J
1 dyne	dyn	$0.224\,809 \times 10^{-5}$ lbf	10^{-5} N
1 day		$86\,400$ s	86.4 ks
1 degree (plane angle)		$0.017\,45$ rad	17 mrad
Electron rest mass	m_e	$9.109\,56 \times 10^{-28}$ g	$9.109\,56 \times 10^{-31}$ kg
1 erg (dyn cm)		6.242×10^{11} eV	10^{-7} J
		2.39×10^{-8} cal	
1 erg/cm^2		6.242×10^{11} eV cm^{-2}	10^{-3} J m^{-2}
Gas constant	R	$8.314\,3 \times 10^7$ erg K^{-1} mol^{-1}	$8.314\,3$ J K^{-1} mol^{-1}
		1.987 cal K^{-1} mol^{-1}	
Density $\begin{cases} \text{Al} \\ \text{Fe} \\ \text{Cu} \\ \text{Ni} \end{cases}$	ρ	2.71 g cm^{-3}	2710 kg cm^{-3}
		7.87 g cm^{-3}	7870 kg cm^{-3}
		8.93 g cm^{-3}	8930 kg cm^{-3}
		8.90 g cm^{-3}	8900 kg cm^{-3}
Electronic charge	e	1.602×10^{-20} emu	0.1602 aC
1 electron volt	eV	3.83×10^{-20} cal	0.1602 aJ
		1.602×10^{-12} erg	
Faraday	$F = N_A e$		9.6487×10^4 C mol^{-1}
1 inch	in	2.54 cm	25.4 mm
1 kilogram	kg	$2.204\,62$ lb	1 kg
1 kilogram-force/cm^2	kgf/cm^2	14.22 lbf/in^2	
1 litre	l	0.220 gal	1 dm^3
1 micron	μm	10^4 Angstrom	10^{-6} m
		10^{-4} cm	

Quantity	Symbol	Traditional units	SI units
1 minute (angle)		2.908×10^{-4} radian	2.908×10^{-4} rad
Modulus of elasticity (average) $\begin{cases} Al \\ Fe \\ Ni \\ Cu \\ Au \end{cases}$	E		70 GN m^{-2} 210 GN m^{-2} 209 GN m^{-2} 127 GN m^{-2} 79 GN m^{-2}
Planck's constant	**h**	6.6262×10^{-27} erg s	6.6262×10^{-34} J s
Poisson ratio $\begin{cases} Al \\ Au \\ Cu \\ Mg \\ Pb \\ Ti \\ Zn \end{cases}$	ν		0.34 0.44 0.35 0.29 0.44 0.36 0.25
1 pound	lb	453.59 g	0.453 kg
1 pound (force)	lbf		4.448 22 N
1 psi	lbf/in^2	7.03×10^{-2} kgf/cm^2	6 894.76 N m^{-2}
1 radian	rad	57.296 degrees	1 rad
Shear modulus (average) $\begin{cases} Al \\ Fe \\ Ni \\ Cu \\ Au \end{cases}$	μ	2.7×10^{11} dyn cm^{-2} 8.3×10^{11} dyn cm^{-2} 7.4×10^{11} dyn cm^{-2} 4.5×10^{11} dyn cm^{-2} 3.0×10^{11} dyn cm^{-2}	27 GN m^{-2} 83 GN m^{-2} 74 GN m^{-2} 45 GN m^{-2} 30 GN m^{-2}
1 ton (force)	1 tonf		9.964 02 kN
1 tsi	1 tonf/in^2	1.574 9 kgf/mm^2	15.444 3 MN m^{-2}
1 tonne	t	1000 kg	10^3 kg
1 torr	torr	1 mm Hg	133.322 N m^{-2}
Velocity of light (*in vacuo*)	c	$2.997\,925 \times 10^{10}$ cm/s	$2.997\,925 \times 10^8$ m s^{-1}

Figure references

Chapter 1

Rice, R. W. (1983). *Chemtech.* 230

Chapter 2

Askeland, D. R. (1990). *The Science and Engineering of Materials*, 2nd (SI) edn. Chapman and Hall, London.

Chapter 3

Brandes, E. A. and Brook, G. B. (1992). *Smithells Metals Reference Book.* Butterworth-Heinemann, Oxford.

Copper Development Association (1993). *CDA publication 94.* Copper Development Association.

Keith, M. L. and Schairer, J. F. (1952). *J. Geology*, **60**, 182, University of Chicago Press

Raynor, G. V. *Annot. Equilib. Diag. No. 3*, Institute of Metals, London.

Williamson, G. K. and Smallman, W. (1953). *Acta Cryst.*, **6**, 361.

Chapter 4

Barnes, R. and Mazey, D. (1960). *Phil. Mag.* **5**, 124. Taylor and Francis.

Berghezan, A., Fourdeux, A. and Amelinckx, S. (1961). *Acta Met.*, **9**, 464, Pergamon Press.

Bradshaw, F. J. and Pearson, S. (1957). *Phil. Mag.* **2**, 570.

Cottrell, A. H. (1959). Forty-sixth Thomas Hawksley Memorial Lecture, *Proc. Institution of Mechanical Engineers.* **14**, Institution of Mechanical Engineers.

Diehl, J. Chapter 5 in *Moderne Probleme d. Metallphysik* (1965). ed. by Seeger, A., Bd.l, Berlin, Heidelberg, New York, Springer.

Dobson, P., Goodhew, P. and Smallman, R.E. (1968). *Phil. Mag.* **16**, 9. Taylor and Francis.

Edington, J. W. and Smallman, R. E. (1965). *Phil. Mag.* **11**, 1089. Taylor and Francis.

Hales, R., Smallman, R. E. and Dobson, P. (1968). *Proc. Roy. Soc.,* A307, 71.

Hameed, M. Z., Loretto, M. H. and Smallman, R. E. (1982). *Phil. Mag.* **46**, 707. Taylor and Francis.

Johnston, I., Dobson, P. and Smallman, R. E. (1970). *Proc. Roy. Soc.* **A315**, 231, London.

Mazey, D. and Barnes, R. (1968). *Phil. Mag.,* **17**, 387. Taylor and Francis.

Mitchell, T., Foxall, R. A. and Hirsch, P.B. (1963). *Phil. Mag.* **8**, 1895, Taylor and Francis.

Nelson, R. S. and Hudson, J. A. (1976). *Vacancies '76.* 126. The Metals Society.

Panseri, C. and Federighi, T. (1958). *Phil. Mag.,* **3**, 1223.

Partridge, P. (1967). *Met. Reviews*, **118**, 169, American Society for Metals.

Weertman, J. (1964). *Elementary Dislocation Theory*, Collier-Macmillan International.

Westmacott, K. H., Smallman, R. E. and Dobson, P. (1968). *Metal Sci. J.,* **2**, 117, Institute of Metals.

Chapter 5

Askeland, D. R. (1990). *The Science and Engineering of Materials*, 2nd edn. p. 732. Chapman and Hall, London.

Barnes, P. (1990). *Metals and Materials.* Nov, 708–715, Institute of Materials.

Cahn, R. W. (1949). *J. Inst. Metals*, **77**, 121.

Dash, J. (1957). *Dislocations and Mechanical Properties of Crystals*, John Wiley and Sons.

Gilman, J. (Aug. 1956). *Metals*, 1000.

Hirsch, P. B., Howie, A. and Whelan, M. (1960). *Phil. Trans.,* A252, 499, Royal Society.

Hirsch, P. B. and Howie, A. *et al.* (1965). *Electron Microscopy of Thin Crystals.* Butterworths, London.

Howie, A. and Valdre, R. (1963). *Phil. Mag., 8*, 1981, Taylor and Francis.
Vale, R. and Smallman, R. E. (1977). *Phil. Mag., 36*, 209, Zeiss, C. (Dec 1967). *Optical Systems for the Microscope*, 15. Carl Zeiss, Germany.

Chapter 6

Ashby, M. F. (1989). *Acta Met. 37*, 5, Elsevier Science, Oxford, pp. 1273–93.
Barrett, C. S. (1952). *Structure of Metals*, 2nd edn. McGraw-Hill.
Mathias, B. T. (1959) . *Progress in Low-Temperature Physics*, ed. By Gorter, C. J., North Holland Publishing Co.
Morris, D., Besag, F. and Smallman, F. (1974). *Phil. Mag. 29*, 43 Taylor and Francis, London.
Pashley, D. and Presland, D. (1958–9). *J. Inst. Metals. 87*, 419. Institute of Metals.
Raynor, G. V. (1958). *Structure of Metals*, Inst. of Metallurgists, 21, Iliffe and Sons, London.
Rose, R. M., Shepard, L. A. and Wulff, J. (1966). *Structure and Properties of Materials*. John Wiley and Sons.
Shull, C. G. and Smart, R. (1949). *Phys. Rev., 76*, 1256.
Slater, J. C. *Quantum Theory of Matter.*
Wert, C. and Zener, C. (1949). *Phys. Rev., 76*, 1169 American Institute of Physics.

Chapter 7

Adams, M. A. and Higgins, P. (1959). *Phil. Mag, 4*, 777.
Adams, M. A., Roberts, A. C. and Smallman, R. E. (1960). *Acta Metall., 8*, 328.
Broom, T. and Ham, R. (1959). *Proc. Roy. Soc., A251*, 186.
Buergers, *Handbuch der Metallphysik*. Akademic-Verlags-gesellschaft.
Burke and Turnbull (1952). *Progress in metal Physics 3*, Pergamon Press.
Cahn, J. (1949). *Inst. Metals, 77*, 121.
Churchman, T., Mogford, I. and Cottrell, A. H. (1957). *Phil. Mag., 2* 1273.
Clareborough, L. M. Hargreaves, M. and West (1955). *Proc. Roy. Soc., A232*, 252.
Cottrell, A. H. (1957). Conference on Properties of Materials at High Rates of Strain. Institution of Mechanical Engineers.
Cottrell, A. H. *Fracture*. John Wiley & Sons.
Dillamore, I. L., Smallman, R. E. and Wilson, D. (1969). *Commonwealth Mining and Metallurgy Congress*, London, Institute of Mining and Metallurgy.
Hahn (1962). *Acta Met., 10*, 727, Pergamon Press, Oxford.
Hancock, J., Dillamore, I. L. and Smallman, R. E. (1972). *Metal Sci. J., 6*, 152.

Hirsch, P. B. and Mitchell, T. (1967). *Can. J. Phys., 45*, 663, National Research Council of Canada.
Hull, D. (1960). *Acta Metall., 8*, 11.
Hull, D. and Mogford, I. (1958). *Phil. Mag., 3*, 1213.
Johnston, W. G. and Gilman, J. J. (1959). *J. Appl. Phys., 30*, 129, American Institute of Physics.
Lücke, K. and Lange, H. (1950). *Z. Metallk, 41*, 65.
Morris, D. and Smallman, R. E. (1975). *Acta Met., 23*, 573.
Maddin, R. and Cottrell, A. H. (1955). *Phil. Mag., 66*, 735.
Puttick, K. E. and King, R. (1952). *J. Inst. Metals, 81*, 537.
Steeds, J. (1963). *Conference on Relation between Structure and Strength in Metals and Alloys*, HMSO.
Stein, J. and Low, J. R. (1960). *J. Appl. Physics, 30*, 392, American Institute of Physics.
Wilson, D. (1966). *J. Inst. Metals, 94*, 84, Institute of Metals.

Chapter 8

Ashby, M. F. *et al.* (1979). *Acta Met., 27*, 669.
Ashby, M. F. (1989). *Acta Met., 1273–93*: Elsevier Science Ltd.
Brookes, J. W., Loretto, M. H. and Smallman, R. E. (1979). *Acta Met. 27*, 1829.
Cottrell, A. H. (1958). Brittle Fracture in Steel and Other Materials. *Trans. Amer. Inst. Mech. Engrs., April*, p. 192.
Fine, M., Bryne, J. G. and Kelly, A. (1961). *Phil. Mag., 6*, 1119.
Greenwood, G. W. (1968). *Institute of Metals Conference on Phase Transformation*, Institute of Metals.
Guinier, A. and Fournet, G. (1955). *Small-angle Scattering of X-rays*. John Wiley & Sons.
Guinier, A. and Walker, R. (1953). *Acta Metall., 1*, 570.
Kelly, P. and Nutting, J. (1960). *Proc. Roy. Soc., A259*, 45, Royal Society.
Kurdjumov, G. (1948). *J. Tech. Phys. SSSR, 18*, 999.
Metals Handbook, American Society for Metals.
Mehl, R. F. and Hagel, K. (1956). *Progress in Metal Physics, 6*, Pergamon Press.
Nicholson, R. B., Thomas, G. and Nutting, J. (1958–9). *J. Inst. Metals, 87*, 431.
Silcock, J., Heal, T. J. and Hardy, H. K. (1953–4). *J. Inst. Metals, 82*, 239.

Chapter 9

Balliger, N. K. and Gladman, T. (1981). *Metal Science, March*, 95.
Driver, D. (1985). *Metals and Materials, June*, 345–54, Institute of Materials, London.
Gilman, P. (1990). *Metals and Materials, Aug*, 505, Institute of Materials, London.

Kim, Y-W. and Froes, F. H. (1990). *High-Temperature Aluminides and Intermetallics*, TMS Symposium, ed. by Whang, S. H., Lin, C. T. and Pope D.

Noguchi, O., Oya. Y. and Suzuki, T. (1981). *Metall. Trans.* **12A**, 1647.

Sidjanin, L. and Smallman, R. E. (1992). *Mat. Science and Technology*, **8**, 105.

Smithells, C. J., *Smithells Metals Reference Book*, 7th edn. Butterworth-Heinemann.

Woodfield, A. P., Postans, P. J., Loretto, M. H. and Smallman, R. E. (1988). *Acta Metall.*, **36**, 507.

Chapter 10

Bovenkerk, H. P. *et al.* (1959). *Nature*, **184**, 1094–1098.

Green, D. J. (1984). *Industrial Materials Science and Engineering*, ed. by L. E. Murr, Chapter 3, Marcel Dekker.

Headley, T. J. and Loehmann, R. E. (1984). *J. Amer. Ceram. Soc.* **Sept, 67**, 9, 620–625.

Ubbelohde, A. R. J. P. (1964). *BCURA Gazette*, **51**, BCURA Ltd, Coal Research Establishment. Stoke Orchard, Cheltenham, UK.

Wedge, P. J. (1987). *Metals and Materials*, **Jan**, 36–8, Institute of Materials.

Chapter 11

Polymeric Materials (1975). copyright © American Society for Matals Park, OH.

Chapter 12

Barrell, R. and Rickerby, D. S. (1989). Engineering coatings by physical vapour deposition. *Metals and materials*, August, 468–473, Institute of Materials.

Kelly, P. J., Arnell, R. D. and Ahmed, W. *Materials World*. (March 1993), pp. 161–5. Institute of Materials.

Weatherill, A. E. and Gill, B. J. (1988). Surface engineering for high-temperature environments (thermal spray methods). *Metals and Materials*, September, 551–555, Institute of Materials.

Chapter 13

Bonfield, W. (1997). *Materials World*, **Jan**, 18, Institute of Materials.

Vincent, J. (1990) *Metals and Materials*, **June**, 395, Institute of Materials.

Walker, P. S. and Sathasiwan, S. (1999), *J. Biomat.* **32**, 28.

Chapter 14

Butterfield, B. G. and Meylan, B. A. (1980). *Three-Dimensional Structure of Wood: an Ultrastructural Approach*, 2nd Edn. Chapman and Hall, London.

Easterling, K. E. (1990). *Tomorrow's Materials*. Institute of Metals, London.

Haines, R. C., Curtis, M. E., Mullaney, F. M. and Ramsden, G. (1983). The design, development and manufacture of a new and unique tennis racket. *Proc. Instn. Mech. Engrs.* **197B**, May, 71–79.

Horwood, G. P. (1994). Flexes, bend points and torques. In *Golf: the Scientific Way* (ed. A. Cochran) Aston Publ. Group, Hemel Hempstead, Herts, UK. pp. 103–108.

Lüthi, J. (1998). Just step in, push down, and go. *Engineering Design*, 98-3, Du Pont de Nemours Internat. SA, P.O. Box CH-1218, Le Grande-Saconnex, Switzerland. ('Crocodile' snowboard binding, Fritschi Swiss Bindings AG, CH-3714, Frutigen).

McMahon, C. J. and Graham, C. D. (1992). *Introduction to Materials: the Bicycle and the Walkman*. Merion Books, Philadelphia.

Pratt, P. L. (1992). The arrow, Appendix to *Longbow: A Social and Military History*, 3rd edn by Robert Hardy, Patrick Stephens, Cambridge.

Index